Oxidative Stress in Applied Basic Research and Clinical Practice

Editor-in-Chief
Donald Armstrong

More information about this series at http://www.springer.com/series/8145

Note from the Editor-in-Chief

All books in this series illustrate point-of-care testing and critically evaluate the potential of antioxidant supplementation in various medical disorders associated with oxidative stress. Future volumes will be updated as warranted by emerging new technology, or from studies reporting clinical trials.

Donald Armstrong
Editor-in-Chief

Emanuele Albano • Maurizio Parola
Editors

Studies on Hepatic Disorders

Humana Press

Editors
Emanuele Albano
Department of Health Sciences
University of Eastern Piedmont
Amedeo Avogadro
Novara, Italy

Maurizio Parola
Department of Clinical and Biological Sciences
Unit of Experimental Medicine and Clinical Pathology
School of Medicine, University of Torino
Torino, Italy

ISSN 2197-7224 ISSN 2197-7232 (electronic)
Oxidative Stress in Applied Basic Research and Clinical Practice
ISBN 978-3-319-15538-8 ISBN 978-3-319-15539-5 (eBook)
DOI 10.1007/978-3-319-15539-5

Library of Congress Control Number: 2015936281

Springer Cham Heidelberg New York Dordrecht London

Printed on acid-free paper

Humana Press is a brand of Springer
Springer International Publishing AG Switzerland is part of Springer Science+Business Media (www.springer.com)

Preface

The role of oxidative stress in causing liver diseases emerged in the early 1960s from the laboratories of Trevor Slater, Mario Umberto Dianzani and others. These pioneering works demonstrated that free radical intermediates and lipid peroxidation reactions were involved in causing tissue injury by hepatotoxic agents. With time, two additional concepts became increasingly clear, placing liver redox mechanisms at the forefront of biomedical research. First, several laboratories showed that the liver, like other tissues, has developed evolutionarily conserved mechanisms and strategies to carefully control intra- and extracellular generation of ROS as well as of other oxidative stress-related radical or non-radical reactive intermediates in order to maintain redox homeostasis. These concepts led to the definition by Helmut Sies of oxidative stress as "a disturbance in the pro-oxidant–antioxidant balance in favour of the former". A second aspect, potentially even more relevant for its clinical impact, was the characterization of the physiological function of redox mechanisms in modulating signal transduction, gene expression and cellular responses in various tissues including the liver. These notions were fundamental to understanding that a derangement in redox homeostasis, resulting in sustained levels of oxidative stress, plays a significant role in the pathogenesis of a variety of liver diseases and significantly contributes to wound healing and hepatic fibrogenesis.

Studies in Hepatic Disorders, included in a multi-volume series by Springer Science entitled *Oxidative Stress in Applied Basic Research and Clinical Practice*, has been designed to offer to clinicians, researchers and Ph.D. students dealing with liver diseases a state-of-the-art overview of the impact of oxidative stress and redox reactions in liver and biliary tree disorders. Basic scientists as well as pathologists and clinicians devoted to experimental and clinical liver-targeted research have provided a collection of selected contributions on specific topics. These contributions have been arranged first to cover the basic principles of oxidative stress and redox signalling as well as address the significance of oxidative stress and redox changes in liver pathophysiology with a focus on hepatocellular injury, liver inflammation and liver fibrogenesis. The role of oxidative stress has been then analysed in specific liver disease conditions like acute and chronic liver failure as well as in the most clinically relevant conditions of chronic liver disease, including chronic viral infections, alcohol abuse, autoimmune injury, chronic biliary tree damage and metabolic alterations. Finally, the last three chapters offer an up-to-date classification of liver cancer and then an overview of the role of oxidative stress in experimental and human liver carcinogenesis.

In addition to thanking all authors who have been involved in the preparation of this volume and the publisher we would like to dedicate this volume to the memory of Prof. Mario Umberto Dianzani who died in June 2014.

Novara, Italy — Emanuele Albano
Torino, Italy — Maurizio Parola

Contents

Part I Oxidative Stress: Basic Principles in the Liver Scenario

1 **Principles of Redox Signaling** ... 3
Paola Chiarugi, Maria Letizia Taddei, and Elisa Giannoni
2 **Oxidative Stress, Antioxidant Defenses, and the Liver** ... 41
Carlo Vascotto and Claudio Tiribelli
3 **Oxidative Mechanisms in Liver Senescence and Regeneration** ... 65
Fabio Marongiu, Maria Paola Serra, and Ezio Laconi

Part II Oxidative Stress and Liver Pathophysiology: From Cell Damage and Death to Inflammatory and Fibrogenic Responses

4 **Oxidative Stress and Hepatocellular Injury** ... 99
Marjolein H. Tiebosch, Golnar Karimian, and Han Moshage
5 **Oxidative Stress and the Unfolded Protein Response in the Liver** ... 113
William M. Hudson and Michael J. Pagliassotti
6 **Oxidative Stress and Liver Inflammation** ... 123
Stephen N. Greenhalgh, Alexandra I. Thompson, Neil C. Henderson, and John P. Iredale
7 **Oxidative Stress and Liver Ischemia–Reperfusion Injury** ... 149
Carmen García-Ruíz, Albert Morales, and José C. Fernández-Checa
8 **Oxidative Stress and Liver Fibrogenesis** ... 171
Erica Novo, Fabio Marra, and Maurizio Parola

Part III Oxidative Stress in Specific Liver Disease Conditions

9 **Oxidative Stress in Acute Liver Failure** 199
Mitchell R. McGill and Hartmut Jaeschke
10 **Oxidative Stress in Alcoholic Liver Disease** 215
Emanuele Albano
11 **Oxidative Stress in Chronic Viral Hepatitis** 241
Nicole L.B. Corder, Bhargav Koduru, Seung Bum Park, Yiyan Wang, and Jinah Choi
12 **Oxidative Stress in Nonalcoholic Fatty Liver Disease** 279
Montserrat Marí, Albert Morales, Anna Colell, Carmen García-Ruiz, and José C. Fernandez-Checa
13 **Oxidative Stress in Nonautoimmune Biliary Diseases** 309
M. Cadamuro, L. Fabris, and M. Strazzabosco
14 **Oxidative Stress in Autoimmune Liver Disease** 325
Verónica Martín-Domínguez, Eduardo Vilar-Gómez, Luisa García-Buey, and Ricardo Moreno-Otero
15 **Oxidative Stress and Hepatic Iron Overload** 345
S. Pelusi, L. Valenti, and Silvia Fargion
16 **Oxidative Stress in the Central Nervous System Complications of Chronic Liver Failure** 357
Chantal Bemeur
17 **Oxidative Stress and Liver Transplantation** 371
Piotr Czubkowski, Piotr Socha, and Joanna Pawłowska
18 **Antioxidant Approach to the Therapy of Chronic Liver Diseases** 389
Le Thi Thanh Thuy, Hoang Hai, and Norifumi Kawada

Part IV Oxidative Stress and Liver Cancer

19 **Liver Cancer Classification** 417
Maria Guido, Samantha Sarcognato, Matteo Fassan, and Marco Pizzi
20 **Oxidative Stress Mechanisms in Hepatocarcinogenesis** 449
Raquel Urtasun, Carmen Berasain, and Matías A. Avila
21 **The Role of Oxidative Stress in Hepatocarcinogenesis** 479
Helmut K. Seitz and Sebastian Mueller

Index 505

Contributors

Emanuele Albano Department of Health Sciences, University of Eastern Piedmont Amedeo Avogadro, Novara, Italy

Matías A. Avila Division of Hepatology, Center for Applied Medical Research, University of Navarra, Pamplona, Spain

Networked Biomedical Research Center for Hepatic and Digestive Diseases, Clinical University of Navarra, Instituto de Salud Carlos III, Pamplona, Spain

Chantal Bemeur, Dt.P., Ph.D. Department of Nutrition, Montréal University Health Center (CHUM), University of Montréal, Montréal, Quebec, Canada

Carmen Berasain Division of Hepatology, Center for Applied Medical Research, University of Navarra, Pamplona, Spain

Networked Biomedical Research Center for Hepatic and Digestive Diseases, Clinical University of Navarra, Instituto de Salud Carlos III, Pamplona, Spain

M. Cadamuro Department of Surgery and Translational Medicine, University of Milan-Bicocca, Milan, Italy

Department of Molecular Medicine, University of Padova School of Medicine, Padova, Italy

Paola Chiarugi Department of Experimental and Clinical Biomedical Sciences, University of Florence, Florence, Italy

Jinah Choi School of Natural Sciences, University of California, Merced, Merced, CA, USA

Anna Colell Department of Cell Death and Proliferation, Institute of Biomedical Research of Barcelona - Spanish National Research Council, The August Pi i Sunyer Biomedical Research Institute, Liver Unit-Hospital Clínic, Networked Biomedical Research Center for Hepatic and Digestive Diseases, Barcelona, Spain

Nicole L. B. Corder School of Natural Sciences, University of California, Merced, Merced, CA, USA

Piotr Czubkowski Department of Gastroenterology, Hepatology, Feeding Disorders and Pediatrics, The Children's Memorial Health Institute, Warsaw, Poland

L. Fabris Department of Molecular Medicine, University of Padova School of Medicine, Padova, Italy

Digestive Disease Section, Yale University School of Medicine, New Haven, CT, USA

Silvia Fargion IRCCS Ca' Granda Foundation Policlinico Hospital Milan and Department of Pathophysiology and Transplantation, University of Milan, Milan, Italy

Matteo Fassan Department of Medicine, General Pathology and Cytopathology Unit, University of Padova, Verona Padova, Italy

ARC-NET Research Centre, University of Verona, Verona, Italy

José C. Fernández-Checa Department of Cell Death and Proliferation, Institute of Biomedical Research of Barcelona (IIBB), Spanish National Research Council and Liver Unit-Hospital Clinic and CIBEREHD, Barcelona, Spain

Southern California Research Center for ALPD and Cirrhosis, Keck School of Medicine of the University of Southern California, Los Angeles, CA, USA

Luisa García-Buey Digestive Diseases Service and Networked Biomedical Research Center for Hepatic and Digestive Diseases, Institute of Health Carlos III, Research Institute at La Princesa University Hospital, Autonomous University of Madrid, Madrid, Spain

Carmen García-Ruíz Department of Cell Death and Proliferation, Institute of Biomedical Research of Barcelona (IIBB), Consejo Superior Investigaciones Científicas (CSIC) and Liver Unit-Hospital Clinic and Networked Biomedical Research Center for Hepatic and Digestive Diseases, Barcelona, Spain

Southern California Research Center for ALPD and Cirrhosis, Keck School of Medicine of the University of Southern California, Los Angeles, CA, USA

Elisa Giannoni Department of Experimental and Clinical Biomedical Sciences, University of Florence, Florence, Italy

Stephen N. Greenhalgh The University of Edinburgh/Medical Research Council Centre for Inflammation Research, The Queen's Medical Research Institute, Edinburgh, United Kingdom

Maria Guido Department of Medicine, General Pathology and Cytopathology Unit, University of Padova, Padova, Italy

Hoang Hai Department of Hepatology, Graduate School of Medicine, Osaka City University, Osaka, Japan

Neil C. Henderson The University of Edinburgh/Medical Research Council Centre for Inflammation Research, The Queen's Medical Research Institute, Edinburgh, United Kingdom

William M. Hudson Department of Food Science and Human Nutrition, Colorado State University, Fort Collins, CO, USA

John P. Iredale The University of Edinburgh/Medical Research Council Centre for Inflammation Research, The Queen's Medical Research Institute, Edinburgh, United Kingdom

Hartmut Jaeschke, Ph.D. Department of Pharmacology, Toxicology and Therapeutics, University of Kansas Medical Center, Kansas City, KS, USA

Golnar Karimian Department of Gastroenterology and Hepatology, University Medical Center Groningen, University of Groningen, Groningen, The Netherlands

Department of Laboratory Medicine, University Medical Center Groningen, University of Groningen, Groningen, The Netherlands

Norifumi Kawada Department of Hepatology, Graduate School of Medicine, Osaka City University, Osaka, Japan

Bhargav Koduru School of Natural Sciences, University of California, Merced, Merced, CA, USA

Ezio Laconi, M.D., Ph.D. Department of Biomedical Sciences—Unit of Experimental Medicine, University of Cagliari—Italy, Cagliari, Italy

Montserrat Marí Department of Cell Death and Proliferation, Institute of Biomedical Research of Barcelona - Spanish National Research Council, IDIBAPS, Liver Unit-Hospital Clinic, Networked Biomedical Research Center for Hepatic and Digestive Diseases, Barcelona, Spain

Centre Esther Koplowitz, Barcelona, Spain

Fabio Marongiu Department of Biomedical Sciences—Unit of Experimental Medicine, University of Cagliari—Italy, Cagliari, Italy

Fabio Marra Department of Experimental and Clinical Medicine, University of Florence, Florence, Italy

Center for Research, Transfer and Higher Education DenoTHE, University of Florence, Florence, Italy

Verónica Martín-Domínguez Digestive Diseases Service and Networked Biomedical Research Center for Hepatic and Digestive Diseases, Institute of Health Carlos III, Research Institute at La Princesa University Hospital, Autonomous University of Madrid, Madrid, Spain

Mitchell R. McGill, Ph.D. Department of Pharmacology, Toxicology and Therapeutics, University of Kansas Medical Center, Kansas City, KS, USA

Albert Morales Department of Cell Death and Proliferation, Institute of Biomedical Research of Barcelona - Spanish National Research Council and Liver Unit-Hospital Clinic and Networked Biomedical Research Center for Hepatic and Digestive Diseases, Barcelona, Spain

Ricardo Moreno-Otero, M.D. Digestive Diseases Service and Networked Biomedical Research Center for Hepatic and Digestive Diseases, Institute of Health Carlos III, Research Institute at La Princesa University Hospital, Autonomous University of Madrid, Madrid, Spain

Han Moshage Department of Gastroenterology and Hepatology, University Medical Center Groningen, University of Groningen, Groningen, The Netherlands

Department of Laboratory Medicine, University Medical Center Groningen, University of Groningen, Groningen, The Netherlands

Sebastian Mueller Department of Medicine, Salem Medical Centre and Alcohol Research Centre, University of Heidelberg, Heidelberg, Germany

Erica Novo Department of Clinical and Biological Sciences, Unit of Experimental Medicine, University of Torino, Torino, Italy

Interuniversity Centre for Liver Pathophysiology, University of Torino, Torino, Italy

Michael J. Pagliassotti Department of Food Science and Human Nutrition, Colorado State University, Fort Collins, CO, USA

Seung Bum Park School of Natural Sciences, University of California, Merced, Merced, CA, USA

Maurizio Parola Department of Clinical and Biological Sciences, Unit of Experimental Medicine, University of Torino, Torino, Italy

Interuniversity Centre for Liver Pathophysiology, University of Torino, Torino, Italy

Joanna Pawłowska Department of Gastroenterology, Hepatology, Feeding Disorders and Pediatrics, The Children's Memorial Health Institute, Warsaw, Poland

S. Pelusi IRCCS Ca' Granda Foundation Policlinico Hospital Milan and Department of Pathophysiology and Transplantation, University of Milan, Milan, Italy

Marco Pizzi Department of Medicine, General Pathology and Cytopathology Unit, University of Padova, Padova, Italy

Samantha Sarcognato Department of Medicine, General Pathology and Cytopathology Unit, University of Padova, Padova, Italy

Helmut K. Seitz, M.D. Department of Medicine, Salem Medical Centre and Alcohol Research Centre, University of Heidelberg, Heidelberg, Germany

Maria Paola Serra Department of Biomedical Sciences, Unit of Experimental Medicine, University of Cagliari, Cagliari, Italy

Piotr Socha Department of Gastroenterology, Hepatology, Feeding Disorders and Pediatrics, The Children's Memorial Health Institute, Warsaw, Poland

M. Strazzabosco Department of Surgery and Translational Medicine, University of Milan-Bicocca, Milan, Italy

Digestive Disease Section, Yale University School of Medicine, New Haven, CT, USA

Maria Letizia Taddei Department of Experimental and Clinical Biomedical Sciences, University of Florence, Florence, Italy

Alexandra I. Thompson The University of Edinburgh/Medical Research Council Centre for Inflammation Research, The Queen's Medical Research Institute, Edinburgh, United Kingdom

Le Thi Thanh Thuy Department of Hepatology, Graduate School of Medicine, Osaka City University, Osaka, Japan

Marjolein H. Tiebosch Department of Gastroenterology and Hepatology, University Medical Center Groningen, University of Groningen, Groningen, The Netherlands

Department of Laboratory Medicine, University Medical Center Groningen, University of Groningen, Groningen, The Netherlands

Claudio Tiribelli Center for Liver Studies, AREA Science Park, Trieste, Italy

Department of Medical Sciences, University of Trieste, Trieste, Italy

Raquel Urtasun Division of Hepatology, Center for Applied Medical Research, University of Navarra, Pamplona, Spain

L. Valenti IRCCS Ca' Granda Foundation Policlinico Hospital Milan and Department of Pathophysiology and Transplantation, University of Milan, Milan, Italy

Carlo Vascotto Department of Medical and Biological Sciences, University of Udine, Udine, Italy

Eduardo Vilar-Gómez Digestive Diseases Service and Networked Biomedical Research Center for Hepatic and Digestive Diseases, Institute of Health Carlos III, Research Institute at La Princesa University Hospital, Autonomous University of Madrid, Madrid, Spain

Yiyan Wang School of Natural Sciences, University of California, Merced, Merced, CA, USA

Abbreviations

ADMA	Asymmetric dimethylarginine
AIF	Apoptosis-inducing factor
AIH	Autoimmune hepatitis
AILD	Autoimmune liver disease
ALD	Alcoholic liver disease
ALF	Acute liver failure
ALP	Alkaline phosphatase
ALT	Alanine aminotransferase
AMAP	3′-Hydroxyacetanilide
AMPK	AMP kinase
ANA	Antinuclear antibodies
ANCA	Anti-neutrophil cytoplasmic antibodies
Anti-LC-1	Anti-liver cytosol type 1 antibody
Anti-LKM-1	Anti-liver kidney microsomal type 1 antibody
AP	Abasic
AP-1	Activator protein 1
APAP	Acetaminophen, paracetamol
APE1	Apurinic/apirimidinic endonuclease 1
AQP	Aquaporin
ARE	Antioxidant response elements
ASK1	Apoptosis signal-regulating kinase 1
ASMA	Anti-smooth muscle antibody
AST	Aspartate aminotransferase
ATF	Activating transcription factor
ATF4	Activating transcription factor 4
ATF6	Activating transcription factor 6
ATM	Ataxia-telangiectasia mutated
ATP	Adenosine triphosphate
BA	Bile acid
BDL	Bile duct ligation

BER	Base excision repair
BHA	Butylated hydroxyanisole
BMP6	Bone morphogenetic protein 6
CAT	Catalase
cccDNA	Covalently closed circular DNA
CCl_4	Carbon tetrachloride
CD81	Cluster of differentiation 81
CDC	Centers for Disease Control
CDD	Choline-deficient diet
CDDO-TFEA	CDDO-trifluoroethylamide
CGD	Chronic granulomatous disease
CHB	Chronic hepatitis B
CHC	Chronic hepatitis C
CINC	Cytokine-induced neutrophil chemoattractant
cLD	Cytoplasmic lipid droplet
CLDN1	Claudin-1
CMD	Choline methionine-deficient diet
COX	Cyclooxygenase
COX-2	Cyclooxygenase-2
cPLA2	Cytosolic phospholipase A2
CPT1	Carnitine palmitoyltransferase 1
CREB	cAMP-response element binding protein
CTLA-4	Cytotoxic T-lymphocyte antigen 4
Cul3	Cullin 3
CuZnSOD, SOD1	CuZn superoxide dismutase
CVH	Chronic viral hepatitis
CYP2E1	Cytochrome P450 2E1
CypD	Cyclophilin D
DAA	Direct acting antiviral
DAMP	Damage-associated molecular pattern
ε-dA	N^6-ethenodeoxyadenosine
ε-dC	N^4-ethenodeoxycytidine
DCIYB	Duodenal cytochrome b
DEN	Diethylnitrosamine
DHA	Docosahexaenoic acid
DILI	Drug-induced liver injury
DKG	2,3-Diketogluconic acid
DMT1	Divalent metal transporter-1
DPI	Diphenyleneiodonium chloride
DUOX	Dual oxidase
E1	Envelope glycoprotein 1
E2	Envelope glycoprotein 2
6-ECDCA	6-Ethyl-chenodeoxycholic
ECM	Extracellular matrix
EDEM	ER degradation-enhancing α-like protein

EGFR	Epidermal growth factor receptor
eIF	Eukaryotic initiation factor
EMCV	Encephalomyocarditis virus
ER	Endoplasmic reticulum
ERAD	ER-associated degradation
ERFE	Erythroferrone
Ero1	ER oxidase 1
ESCRT	Endosomal sorting complex required for transport
ETC	Electron transport chain
F/ARFP	F protein/alternate reading frame protein
FC	Free cholesterol
FFA	Free fatty acid
FoxO	Forkhead box class-O
Fp-1	Ferroportin-1 (protein)
FPN1	Ferroportin-1 (gene)
FXR	Farnesoid X receptor
GADD34	Growth arrest and DNA damage 34
GAG	Glycosaminoglycan
GAPDH	Glyceraldehyde-3-phosphate dehydrogenase
γ-GCL	γ-Glutamylcysteine ligase
GCN2	General control non-derepressible 2 kinase
$GdCl_3$	Gadolinium chloride
GPCRs	G-protein coupled receptors
GPI	Glycosylphosphatidylinisotol
GPX	Glutathione peroxidase
GRP78	Glucose-regulated protein 78/immunoglobulin-heavy-chain-binding protein
Grxs	Glutaredoxins
GSH	Glutathione
GSK3β	Glycogen synthase kinase 3β
GSNO	*S*-nitrosoglutathione adduct
GSS	GSH synthetase
GSSG	Glutathione disulfide
GST	Glutathione *S*-transferase
GTPx	Glutathion peroxidase
HAMP	Hepcidin antimicrobial peptide
HBsAg	Hepatitis B surface antigen
HBV	Hepatitis B virus
HBx	HBV X
HCC	Hepatocellular carcinoma
HCV	Hepatitis C virus
HDV	Hepatitis delta virus
HH	Hereditary hemochromatosis
HIF	Hypoxia-inducible factor
HIF-1	Hypoxia-inducible factor 1

HIV	Human immunodeficiency virus
HJV	Hemojuvelin
HLA	Human leukocyte antigen
HLA-DR	Human leukocyte antigen
HMGB1	High mobility group box 1
4-HNE	4-Hydroxynonenal
HNE	4-Hydroxy-2-nonenal
$HO^{\bullet}$	Hydroxyl radical
HO	Heme oxygenase
HO^-	Hydroxyl radical
HO-1	Heme oxygenase-1
HOCl	Hypochlorous acid
hOGG1	Human OGG1
H_2O	Water
H_2O_2	Hydrogen peroxide
HRI	Heme-regulated inhibitor kinase
HSC	Hepatic stellate cell
HSF-1	Heat shock factor 1
hsp	Heat shock protein
HSPG	Heparan sulfate proteoglycan
Huh-7	Human hepatoma
HVB	Hepatitis B virus
I/R	Ischemia–reperfusion
IBD	Inflammatory bowel disease
ICAM-1	Intercellular adhesion molecule 1
IFN	Interferon
IFN-γ	Interferon-γ
IgG4-SC	IgG4-related sclerosis cholangitis
IL	Interleukin
IL-β	Interleukin-β
iNOS	Inducible nitric oxide synthase
IR	Ischemia–reperfusion
γ-IRE	γ-Interferon response element
IRE1α	Inositol requiring 1α
IRES	Internal ribosomal entry site
IRF3	Interferon regulatory factor 3
IRI	Ischemia–reperfusion injury
IRIDA	Iron-refractory iron-deficiency anemia
IRK	Insulin receptor kinase
IRP1	Iron regulatory protein 1
JAK	Janus kinase
JFH-1	Japanese fulminant hepatitis-1
JNK	c-Jun N-terminal kinase
KC	Kupffer cell

Keap1	Kelch-like ECH-associated protein 1
LCA-CoAs	Long-chain fatty acyl-CoAs
LDLR	Low density lipoprotein receptor
LMP	Lysosomal membrane permeabilization
LMWA	Low molecular weight antioxidants
LMW-PTP	Low molecular weight protein tyrosine phosphatase
LOX	Lipoxygenase
LPO	Lipid peroxidation
LPS	Lipopolysaccharide
LT	Liver transplantation
M_1dG	Malondialdehyde-derived DNA adduct
Maf	Musculo-aponeurotic fibrosarcoma
MAP	Mitogen-activated protein
MAPEG	Membrane associated proteins in eicosanoid and glutathione metabolism
MAPK	Mitogen-activated protein kinase
MAVS	Mitochondrial antiviral signal
MC	Mixed cryoglobulinemia
MCA-CoAs	Medium-chain acyl-CoAs
MCP	Monocyte chemotactic/chemoattractant protein
MDA	Malondialdehyde
MDM2	Murine double minute 2
Met-$tRNA_i$	Methionine initiator tRNA
MGST	Membrane-associated microsomal GST
MHC	Major histocompatibility complex
MIM	Mitochondrial inner membrane
Mite	Mitoquinone
MitoQ	Mitoquinone
MLK3	Mixed lineage kinase 3
6-MMP	6-Methilmercaptopurine
MMPs	Matrix metalloproteases
MnSOD, SOD2	Mn superoxide dismutase
MOM	Mitochondrial outer membrane
MOMP	MOM-permeabilization
MPO	Myeloperoxidase
MPT	Membrane permeability transition
MPTP	Mitochondrial membrane permeability transition pore
MRC	Mitochondrial respiratory chain
MsrA	Methionine sulfoxide reductase
mTOR	Rapamycin
3-NT	3-Nitrotyrosine
NAC	*N*-acetylcysteine
NADPH	Nicotinamide adenine dinucleotide phosphate
NAFLD	Non-alcoholic fatty liver disease
NANBH	Non-A, non-B hepatitis

NAPQI	*N*-acetyl-*p*-benzoquinone imine
NASH	Non-alcoholic steatohepatitis
NFIL-6	Nuclear factor interleukin-6
NF-κB	Nuclear factor-kappa B
NHANES	National Health and Nutrition Examination Survey
NNDSS	National Notifiable Diseases Surveillance System
NO	Nitric oxide
NOS	Nitric oxide synthase
NOX	NADPH oxidase
NPSH	Non-protein sulfhydryl
Nqo1	NAD(P)H:quinone oxidoreductase 1
Nrf2	Nuclear factor (erythroid derived 2)-like 2
Nrf2/ARE	NF-E2-regulated factor 2/antioxidant response element
NS	Nonstructural protein
NTCP	Sodium taurocholate cotransporting polypeptide
NTY	Nitrotyrosine
2OG	2-Oxoglutarate
5-OHC	5-Hydroxycytosine
O_2	Oxygen
$O_2^{\bullet-}$	Superoxide anion
OCA	Obeticholic acid
OCLN	Occludin
Oct1	Octamer-binding protein 1
OGG1	8-Oxoguanine glycosylase
•OH	Hydroxyl radical
8-OHdG	8-Hydroxy-2-deoxy-guanosine
OLT	Orthotopic liver transplantation
$ONOO^-$	Peroxynitrite
ORF	Open reading frame
OS	Oxidative stress
8-oxoG	8-Hydroxyguanine
PBC	Primary biliary cirrhosis
PCB	Pentachlorophenol
PDGF	Platelet-derived growth factor
PDHs	Prolyl hydroxylases
PDTC	Pyrrolidine dithiocarbamate
peg-IFN	Pegylated interferon
p-eIF2α	Phosphorylation of the α-subunit of eukaryotic initiation factor 2
PERK	Double-stranded RNA-dependent protein kinase-like ER kinase
pgRNA	Pregenomic RNA
phox	Phagocytic oxidase
PI3K	Phosphatidylinositol 3-kinase
PK	Pyruvate kinase
PKR	Double-stranded RNA-activated protein kinase
PPARα	Peroxisome proliferator-activated receptor α

PPC	Polyenylphosphatidylcholine
PPP	Penthose phosphate pathway
PRB	Phagocytic respiratory burst
PRXs	Peroxiredoxins
PSC	Primary sclerosing cholangitis
PTEN	Phosphatidyl-inositol (PI) 3-phosphatase and tensin homolog
PTKs	Protein tyrosine kinases
PTPs	Protein tyrosine phosphatases
PUFA	Polyunsaturated fatty acids
RdRp	RNA-dependent RNA polymerase
RHS	Reactive hydrogen species
RIP3	Receptor-interacting protein kinase 3
RNS	Reactive nitrogen species
ROI	Reactive oxygen intermediate
ROS	Reactive oxygen species
RS	Reactive species
RTKs	Receptor tyrosine kinases
RXR	Retinoid x receptor
SAMe	*s*-Adenosyl methionine
SepSecS	Selenocysteinyl-tRNA synthase
SHP-2	Src homology-2 domain-containing phosphatase 2
SLA/LP	Soluble liver antigen/liver pancreas
SOD	Superoxide dismutase
SP	Signal peptidase
SR-B1	Scavenger receptor B type 1
SREBP	Sterol regulatory element-binding protein
SREBP-1	Sterol regulatory element-binding protein-1
SRX	Sulfiredoxin
Srxn1	Sulfiredoxin 1
SVR	Sustained virological response
T2DM	Type 2 diabetes mellitus
TAA	Thioacetamide
TACE	TNF-α-converting enzyme
TBARS	Thiobarbituric acid reactive substances
TF	Transferrin
TFR	Transferrin receptor
TFs	Transcription factors
TG	Triglyceride
TGFβ	Transforming growth factor beta
TGF-β1	Transforming growth factor-β1
TLR	Toll-like receptor
TLR4	Toll-like receptor 4
TMPRSS6	Transmembrane protease, serine 6
TNFα	Tumor necrosis factor-α
TNF-β	Tumor necrosis factor-β

TNF-ζ	Tumor necrosis factor-ζ
T-regs	Regulatory T cells
TRX	Thioredoxin
TrxR	Thioredoxin reductase
TRXs	Thioredoxins
TXS	Thromboxane synthase
UCP2	Uncoupling protein-2
UDCA	Ursodeoxycholic acid
uPA	Urokinase plasminogen activator
UPR	Unfolded protein response
UTR	Untranslated regions
UVDDB	UV-damaged DNA-binding protein
VDR	Vitamin D receptor
VEGFR2	Vascular endothelial growth factor receptor 2
VLDL	Very low-density lipoprotein
WHO	World Health Organization
XBP1	X-box-binding protein 1
XO	Xanthine oxidase
XOR	Xanthine oxidoreductase
6-TGN	6-Thioguanine nucleotides
α-TTP	α-Tocopherol transfer protein

Part I
Oxidative Stress: Basic Principles in the Liver Scenario

Chapter 1
Principles of Redox Signaling

Paola Chiarugi, Maria Letizia Taddei, and Elisa Giannoni

1.1 Reactive Oxygen Species

Reactive oxygen species (ROS) include a number of chemically reactive molecules derived from oxygen. ROS are formed and degraded by all aerobic organisms, leading to either physiological concentrations required for normal cell function, or excessive quantities, the state called oxidative stress.

ROS can be classified into oxygen-centered radicals and oxygen-centered non-radicals. Oxygen-centered radicals are superoxide anion ($^{\bullet}O_2^-$), hydroxyl radical ($^{\bullet}OH$), alkoxyl radical ($RO^{\bullet}$), and peroxyl radical ($ROO^{\bullet}$). Oxygen-centered non-radicals are hydrogen peroxide (H_2O_2) and singlet oxygen ($^1O_2$). Other reactive species are reactive nitrogen species (RNS) such as nitric oxide ($NO^{\bullet}$) and peroxynitrite ($OONO^-$) [181].

Clinical studies reported that ROS are associated with many age-related degenerative diseases, including atherosclerosis, cancers, arthritis, heart attack, hepatitis, and liver injury [38]. Benign functions of free radicals have also been reported, including control of gene expression and cell growth. In addition, ROS are acknowledged defence mechanisms to target tumour cells and microbial infections [181].

1.1.1 Superoxide Anion ($^{\bullet}O_2^-$)

Superoxide anion is a reduced form of molecular oxygen created from molecular oxygen by the addition of an electron. The superoxide anion, in spite of being a free radical, is not highly reactive. It lacks the ability to penetrate lipid membranes and

P. Chiarugi (✉) • M.L. Taddei • E. Giannoni
Department of Experimental and Clinical Biomedical Sciences, University of Florence, Viale Morgagni 50, Florence 50134, Italy
e-mail: paola.chiarugi@unifi.it

E. Albano, M. Parola (eds.), *Studies on Hepatic Disorders*,
Oxidative Stress in Applied Basic Research and Clinical Practice,
DOI 10.1007/978-3-319-15539-5_1

is therefore enclosed in the compartment where it was produced. Superoxide anion derives spontaneously from the mitochondrial electron transport system, but is also produced endogenously by flavoenzymes, i.e., xanthine oxidase (XO), lipoxygenase (LOX), and cyclooxygenase (COX) [136]. Two molecules of superoxide rapidly dismutate to hydrogen peroxide and molecular oxygen ($2^{\bullet}O_2^- + 2H^+ \rightarrow H_2O_2 + O_2$) and this reaction is further accelerated by superoxide dismutase (SOD).

1.1.2 Hydroxyl Radical (•OH)

Hydroxyl radical is the most reactive free radical and can be formed from hydrogen peroxide in a reaction catalyzed by metal ions (Fe^{2+} or Cu^+), often bound in complex with different proteins or other molecules. This is known as the Fenton reaction: $H_2O_2 + Cu^+/Fe^{2+} \rightarrow OH + OH^- + Cu^{2+}/Fe^{3+}$.

Due to its strong reactivity with biomolecules, •OH is probably capable of doing more damage to biological systems than any other ROS [13].

1.1.3 Hydrogen Peroxide (H_2O_2)

Hydrogen peroxide can be generated through a dismutation reaction from superoxide ($2^{\bullet}O_2^- + 2H^+ \rightarrow H_2O_2 + O_2$) catalyzed by SOD or through the activity of enzymes such as amino acid oxidase and XO.

Hydrogen peroxide is the least reactive molecule among ROS and is stable under physiological pH and temperature in the absence of metal ions. However, it plays a radical-forming role as an intermediate in the production of more reactive ROS molecules, among which include HOCl (hypochlorous acid), the hydroxyl radical, and the singlet oxygen [187].

1.1.4 Singlet Oxygen ($^1O_2$)

Singlet oxygen is a non-radical and excited status. Compared with other ROS, singlet oxygen is rather mild and non-toxic for mammalian tissue [187]. However, singlet oxygen has been known to be involved in cholesterol oxidation, accelerated by the co-presence of fatty acid methyl ester [66]. In some cases, singlet oxygen acts as a weapon, showing therapeutic potency against various pathogens such as microbes, viruses, and cancer cells [187].

1.1.5 Peroxyl and Alkoxyl Radicals

Peroxyl radicals (ROO•) are formed by a direct reaction of oxygen with alkyl radicals (R•), for example, the reaction between lipid radicals and oxygen. Decomposition of alkyl peroxides (ROOH) also results in peroxyl (ROO•) and alkoxyl (RO•) radicals.

Peroxyl and alkoxyl radicals are good oxidizing agents [43] and they can abstract hydrogen from other molecules with lower standard reduction potential. This reaction is frequently observed in the propagation stage of lipid peroxidation.

1.1.6 *Nitric Oxide (NO•)*

Nitric oxide is a free radical with a single unpaired electron. However, despite its unpaired electron, it does not readily react with most biomolecules. On the other hand, it easily reacts with other free radicals (e.g., peroxyl and alkyl radicals), generating mainly less reactive molecules, thus acting as a free radical scavenger. Nitric oxide is formed from L-arginine by NO synthase (NOS) [50]. In physiologic concentrations NO functions mainly as an intracellular messenger, relaxing smooth muscle in blood vessels. When produced in larger amounts, NO becomes an important factor in redox control of cellular function. Nitrosylation of proteins is known to regulate, for example, enzymatic activity [186]. Excessive production of NO is counteracted by its conjugation with glutathione that results in the *S*-nitrosoglutathione adduct (GSNO).

1.1.7 *Peroxynitrite (OONO⁻)*

When NO is produced in large amounts in parallel with $^{\bullet}O_2^-$, the two species react with each other to generate peroxynitrite ($^{\bullet}O_2^- + NO^{\bullet} \rightarrow OONO^-$), which is highly cytotoxic and diffusible across cell membranes [4]. Peroxynitrite may react directly with diverse biomolecules, as for example with CO_2 to form highly reactive nitroso peroxocarboxylate ($ONOOCO_2^-$), or protonated as peroxonitrous acid (ONOOH) undergoes homolysis to form either •OH and nitric dioxide ($^{\bullet}NO_2$). Peroxynitrite, directly or via its reaction products, may oxidize low-density lipoprotein and generally cause direct protein oxidation and DNA oxidation. Accordingly, peroxynitrite appears to be an important tissue-damaging species generated at the sites of inflammation and has been shown to be involved in various neurodegenerative disorders [4].

1.2 Mechanism of ROS Generation

ROS are products of normal cellular metabolism and derive from many sources in different cellular compartments. Cellular production of ROS occurs from both enzymatic and non-enzymatic sources. Any electron-transferring protein or enzymatic system can virtually result in the formation of ROS as “by-products” of electron transfer reactions. Key sources of intracellular ROS include mitochondrial respiratory enzymes, membrane oxidases as NADPH oxidase (NOX), and cytosolic enzymatic systems as LOX, COX, XO, and the CytP450 superfamily of monooxygenases.

1.2.1 Mitochondria

The generation of mitochondrial ROS is mainly, but not only, a consequence of oxidative phosphorylation. At several sites along the electron transport chain (ETC), electrons derived from pyridine or flavin-reduced coenzymes can react directly with molecular oxygen and generate superoxide ions. It was demonstrated that complex I, also referred to as NADH-ubiquinone oxidoreductase, is involved in superoxide production [141]. The mitochondrial superoxide produced by complex I is released into the matrix [184]. Similarly, succinate dehydrogenase (complex II enzyme) is responsible for the reduction of CoQ and has also shown to be involved in generating low levels of superoxide anion [216]. On the other hand, complex III (ubiquinol-cytochrome c oxidoreductase) was shown to be responsible for superoxide generation in the intermembrane space, through the Q-cycle [84]. Interestingly, the contribution of each of these enzymes to ROS production is different in different tissues and during disease conditions. For instance, while complex III has been implicated as the major source of superoxide in the heart, complex I seems to be of prime importance in the brain [111]. Additionally, enzymes such as glycerol-3-phosphate dehydrogenase, monoamine oxidase, dihydrolipoamide dehydrogenase, and electron-transferring-flavoprotein dehydrogenase have also been implicated in ROS production [45, 135].

Once formed, $^{\bullet}O_2^-$ is usually converted to H_2O_2 by Mn-containing mitochondrial SOD2, Cu/Zn-containing cytosolic SOD1, or extracellular SOD3 [58]. Although H_2O_2 is more stable than superoxide, it can freely diffuse out of the mitochondria into the cytosol, thereby reducing the harmful effects of these reactive species to the mitochondria. Alternatively, superoxide may be carried to the cytoplasm by voltage-dependent anion channels [58].

1.2.2 NOX Enzymes

Of the many ROS-generating reactions that occur in cells, the sole enzyme that produces ROS as its primary function is NOX. Originally designated as "phagocytic oxidase" activities involved in defence and innate immunity, the NOX family is now formally recognized as a category of transmembrane proteins expressed by a wide variety of cell types [6]. All NOX isoforms have conserved structural properties to allow for the reduction of molecular oxygen to superoxide anion in concert with the transport of electrons across the plasma membrane. To date, seven members of this family have been described: NOX 1–5 and dual oxidase (DUOX) 1 and 2. The catalytic subunit of all NOXes contains a C-terminal flavin adenine nucleotide dehydrogenase domain with a NADPH binding site. NOXes are expressed in many tissues and mediate diverse biological functions.

NOX2, also known as the phagocytic NOX, is the prototypical NOX that is endowed with a clear role in the host defence against bacteria. NOX2 is activated when the regulatory subunit p47phox is phosphorylated and associates with the regulatory subunits p67phox and p40phox, the membrane-bound p22phox, and the RAC GTPase [6]. A pathological increase in NOX2 contributes to oxidative injury and vascular damage, including angiogenesis, endothelial dysfunction, and vascular inflammation [196].

NOX1 and NOX3 are related to the generation of ROS by extracellular stimuli through membrane-bound receptors for hormones, cytokines, and integrins [8]. Notably, peroxiredoxin I synergizes with NOX-driven redox signaling. Indeed, growth factors lead to localized inactivation of peroxiredoxin I, allowing the transient accumulation of hydrogen peroxide around membranes, where signaling components are concentrated [213].

NOX4 is mainly detected in the endoplasmic reticulum, mitochondria, and nucleus [196]. NOX4 activation does not seem to require p47phox, p67phox, p40phox, or RAC. Another distinctive trait of NOX4 is its preferential production of H_2O_2 versus $^{\bullet}O_2^-$, likely due to a highly conserved histidine residue, promoting the rapid dismutation of $^{\bullet}O_2^-$ before it leaves the enzyme [17]. Although the precise mechanisms are unclear, NOX4 has been implicated in hypertension, atherosclerosis, and in cardiovascular and renal complications of diabetes [27].

NOX5 is the most recently identified of the NOX isoforms and its features are highly distinctive. Unlike the other isoforms, NOX5 possesses an amino-terminal calmodulin-like domain and is regulated in a Ca^{2+}-sensitive manner. NOX5 is also unique in that its activation is independent of p22phox or other subunits. The biological significance of vascular NOX5 is unknown, although it has been implicated in endothelial cell proliferation, angiogenesis, and in oxidative damage in atherosclerosis [196].

DUOX1 and DUOX2, highly expressed in the thyroid, show both a NOX domain and a thyroid peroxidase domain. They share with NOX4 the ability to internally dismutase superoxide, as the main ROS detected upon their activation is hydrogen peroxide [119].

1.2.3 Lipoxygenase and Cyclooxygenase

Arachidonic acid, derived from the PLA2-dependent hydrolysis of phospholipids, forms the substrate for COX- and LOX-dependent synthesis of the four major classes of eicosanoids: prostaglandins, prostacyclins, thromboxanes, and leukotrienes. These synthetic pathways involve a series of oxidation steps that involve a number of free radical intermediates.

Arachidonic acid metabolism which leads to leukotriene synthesis, particularly involving the LOX pathway, has been reported to generate ROS [194]. The products of the reaction catalyzed by LOXes are hydroperoxyeicosatetraenoaic acid, thereafter converted into leukotrienes, and superoxide, converted into H_2O_2 again by enzymatic dismutation mediated by cytosolic SOD.

COX is a rate-limiting enzyme in prostaglandin biosynthesis, driving a two-step enzymatic process culminating in ROS generation [95]. Oxygenase activity of

COX first converts arachidonic acid to prostaglandin G2 and then to prostaglandin H_2 by COX peroxidase activity. While COX-1 is constitutive and widespread among different tissues, COX-2, generally undetectable in most tissues, is inducible by growth factors and cytokines and is dramatically known as a conspirator of cancer progression [95].

1.2.4 Xanthine Oxidase

In addition to membrane-associated oxidases, soluble enzymes such as XO, aldehyde oxidase, dihydroorotate dehydrogenase, flavoprotein dehydrogenase, and tryptophan dioxygenase can generate ROS during catalytic cycling. The most extensively studied of these is the $^{\bullet}O_2^-$-generating XO [21]. XO is a molybdoflavin enzyme that catalyzes the terminal two reactions in purine degradation. XO is a homodimer with each subunit consisting of four redox centers: a molybdenum cofactor (Mo-co), one FAD site, and two Fe/S clusters. The Mo-co is the site of purine oxidation, while NAD^+ and O_2 reduction occur at the FAD site. The two Fe/S clusters provide the conduit for electron flux between the Mo-co and the FAD [146]. During inflammatory conditions, post-translational modifications by oxidation or limited proteolysis decrease the affinity of the FAD site for NAD^+ while increasing the affinity for oxygen. This culminates in the enhancement of univalent and divalent electron transfer to O_2, generating $^{\bullet}O_2^-$ and H_2O_2, respectively [87]. This capacity to reduce O_2 led to XO being identified as the first source of biological $^{\bullet}O_2^-$ formation [21].

1.2.5 Cytochrome P450

The endoplasmic reticulum is an intracellular organelle that contains enzymes that catalyze a series of reactions to detoxify lipid-soluble drugs and other harmful metabolic products. The most extensively studied of these enzymes is the cytochrome P450 superfamily, a group of heme-containing proteins with multiple functions including the metabolism of xenobiotics such as drugs, toxins, carcinogens, and endogenous substrates, such as fatty acids and steroids. The cytochrome P450 enzymes catalyze a number of chemical reactions such as peroxidation, de-alkylation, mono-oxygenation, reduction, epoxidation, and de-halogenation [5]. The catalytic activity of the cytochrome P450 enzymes requires molecular oxygen activation, which results in the generation of ROS such as superoxide anion, hydrogen peroxide, and hydroxyl radical.

Oxidative stress is a pathogenic factor for the onset of alcoholic (ALD) and non-alcoholic fatty liver diseases (NAFLD), clinical conditions leading to hepatocellular injury and inflammation resulting from alcohol consumption, high fat diet, obesity, and diabetes [121]. Of note, cytochrome P450 is greatly involved in alcohol metabolism. Among the cytochrome P450 family, CYP2E1 has been identified as the most relevant for ALD as it is highly inducible and it has high catalytic activity for alcohol [122].

CYP2E1 is mainly expressed in the liver, with hepatocytes showing the highest expression. During the catalytic cycle of CYP2E1, significant amounts of ROS are generated, which subsequently cause cellular damage, such as lipid peroxidation, oxidant and nitrosative stress, and mitochondrial damage.

In addition to ethanol, CYP2E1 also metabolizes polyunsaturated fatty acids such as linoleic acid and arachidonic acid to generate hydroxylated fatty acids [5]. The hydroxylated fatty acids are further metabolized to dicarboxylic fatty acids that are cytotoxic at high concentrations. Indeed, increased CYP2E1 protein expression and activity were found in obesity, fatty liver, and non-alcoholic steatohepatitis [5].

1.2.6 *Peroxisomal Enzymes*

Mammalian peroxisomes play a key role in various metabolic pathways, including fatty acid α- and β-oxidation, amino acid catabolism, and polyamine oxidation [56]. Interestingly, many of the enzymes participating in these pathways generate specific ROS as by-products of their normal catalytic function, including acyl-CoA oxidases, urate oxidase, and D-amino acid oxidase [56].

Specific signaling roles have not been ascribed to peroxisome-derived oxidants and only a small fraction of H_2O_2 generated in these intracellular organelles appears to escape peroxisomal catalase. Indeed, peroxisomal catalase utilizes H_2O_2 produced by these oxidases to oxidize a variety of other substrates in "peroxidative" reactions. These types of oxidative reactions are particularly important in liver and kidney cells in which peroxisomes detoxify a variety of toxic molecules (including ethanol) that enter the circulation.

Among the peroxisomal enzymatic sources of oxidants, the inducible form of NOS (NOS2) is the major source of RNS. NOS2 is a homodimeric enzyme that catalyzes the oxidation of L-arginine to $NO^\bullet$ and citrulline in a complex reaction requiring O_2, NADPH, tetrahydrobiopterin, FMN, and FAD. Interestingly, in the absence of adequate substrate or when in its monomeric form, the enzyme can also produce significant amounts of $^\bullet O_2^-$. Several studies have shown that NOS2 displays a dual cytosolic-peroxisomal localization in hepatocytes. Interestingly, it has been shown that while the peroxisomal pool of NOS2 mainly consists of "inactive" monomers, the cytosolic pool is composed of both monomers and "active" homodimers [128]. These findings, in combination with the observation that monomeric NOS2 can generate $^\bullet O_2^-$, led to the hypothesis that NOS2 might be sequestered by peroxisomes to protect the larger cellular environment from monomeric NOS2-generated superoxide [128]. Nevertheless, at the moment, it cannot be rigorously excluded that, under certain circumstances, peroxisomal NOS2 may also actively produce $NO^\bullet$ and function as a source of RNS signaling molecules.

1.3 ROS Scavenging

Both aerobic and anaerobic metabolisms produce ROS, which are balanced by endogenous antioxidant systems. Imbalances between ROS and antioxidants may result in cellular oxidative stress, and therefore, prevention of excessive ROS and repair of cellular damages is essential for cell's life. Halliwell and Gutteridge in 1995 defined antioxidants as "any substance that, when present at low concentrations compared with that of an oxidizable substrate, significantly delays or inhibits oxidation of that substrate" [83], but later, defined them as "any substance that delays, prevents or removes oxidative damage to a target molecule" [82]. Antioxidants can be classified based on their localization into plasma antioxidants (vitamin C, bilirubin, uric acid, transferrin, ceruloplasmin, β carotene), cell membrane antioxidants (α tocopherol/vitamin E), and intracellular antioxidants (SOD, catalase, glutathione peroxidase (GPX)). Antioxidants can be further classified according to their nature in enzymatic antioxidants (SOD, catalase, GPX, glutathione reductase) and non-enzymatic antioxidants, such as nutrient antioxidants (β-carotene, α-tocopherol, ascorbic acid) and metabolic antioxidants (bilirubin, uric acid, ceruloplasmin, ferritin, transferrin, albumin, glutathione). Moreover, oxidants can be sorted on the basis of their mode of action: the different antioxidants are thus divided into (1) antioxidants involved in the prevention of oxidants formation by favoring the generation of less active ROS (i.e., SOD, catalase, GPX); (2) antioxidants which are chelators of transition metal ions, like transferrin, ferritin, ceruloplasmin; and (3) antioxidants that are ROS scavengers (i.e., ascorbic acid, vitamin E, uric acid).

1.3.1 Non-enzymatic Antioxidants

The non-enzymatic antioxidants are a heterogeneous group of molecules, containing among the others: glutathione, proteins chelating transition metals which prevent the production of free radical, uric acid which is the major antioxidant in the human plasma able to scavenge singlet oxygen and hydroxyl radical, and bilirubin which efficiently scavenges peroxyl radical. The vitamin compounds with antioxidant activity include vitamin E (α-tocopherol), an organic chemical molecule soluble in fats localized mainly in cell membranes which has a key role mainly by quenching lipid peroxidation initiated by ROS and by capturing free hydroxyl radicals and superoxide [123]. Also vitamin C (ascorbic acid) is involved in the reduction of radicals from a variety of source, inhibiting lipid peroxidation and participating in recycling radicals produced by oxidation of vitamin E [197]. Carotenoids are a group of fat-soluble organic compounds found mainly in yellow, red, orange, and pink vegetable dyes; they are characterized by an extended system of conjugated double bonds which are responsible for their antioxidant properties.

Carotenoids also protect biological system against singlet oxygen-mediated damage by direct quenching [144]. Finally, a wide class of compounds derived from the secondary metabolism of plants, the polyphenols, are able to scavenge free radicals, up-regulate many metal chelating reactions, and inhibit the lipid peroxidation of LDL.

Glutathione is one of the most important cellular antioxidants [160]. It is a cysteine-containing peptide that participates directly in the neutralization of free radicals and reactive oxygen compounds and maintains exogenous antioxidants such as vitamins C and E in their reduced (active) forms. Glutathione exists in both reduced (GSH) and oxidized (GSSG) states. In the reduced state, the thiol group of cysteine is able to donate a reducing equivalent ($H^+ + e^-$) to other unstable molecules, such as ROS. In donating an electron, glutathione itself becomes reactive and readily binds to another reactive GSH to form glutathione disulfide (GSSG). During oxidative stress, GSH is converted to GSSG, resulting in a decrease in GSH/GSSG ratio, commonly considered an indicator of redox balance. In healthy cells and tissues, more than 90 % of the total glutathione pool is in the reduced form (GSH) and less than 10 % exists in the disulfide form (GSSG). GSH can be regenerated from GSSG by the enzyme glutathione reductase (GSR) as NADPH reduces the FAD moiety of GSR to produce a transient FADH-anion. This anion then quickly breaks a disulfide bond (Cys58–Cys63) and leads to Cys63 nucleophilic attack of the nearest sulfide unit in the GSSG molecule, which creates a mixed disulfide bond (GS-Cys58) and a GS-anion. His467 of GSR then protonates the GS-anion to form the first GSH. Next, Cys63 reacts with the sulfide of Cys58, releasing a GS-anion, which, in turn, picks up a solvent proton and is released from the enzyme, thereby creating the second GSH. So, for every GSSG and NADPH, two reduced GSH molecules are gained, which can again act as antioxidants scavenging ROS in the cell. In addition, the sulfridryl group of GSH, in the presence of O_2, may react with nitric oxide to produce GSNO, which can serve as a physiological scavenger of NO. Indeed, GSNO can react with GSH, yielding GSSG, nitrite, nitrous oxide, and ammonia [183], products that may contribute to protein nitrosylation. The steady state of nitrosylated protein depends on the enzyme GSNO-reductase, playing a central role in controlling protein S-nitrosylation-based signalling. In addition to GSNO-reductase action, denitrosylation of SNO proteins depends also on the thioredoxin (TRX) system, which comprises TRX and TRX reductase (TRXR) [145].

GSH may also react with sulfenic acid of oxidized proteins to form a mixed disulphide in a reaction known as S-glutathionylation or S-thiolation. This reaction prevents the further oxidation to sulfinic acid and sulfonic acid. The S-glutathionylation can be either spontaneous or catalyzed by glutathione *S*-transferase (GST). GSTs are the most important family of enzymes known to have a crucial function in the detoxification of a variety of both endogenous products of oxidative stress and exogenous carcinogens [89], playing an important role in cellular protection against oxidative stress.

1.3.2 *Enzymatic Antioxidants*

1.3.2.1 Superoxide Dismutase, Catalase, and Glutathione Peroxidase

For what the endogenous enzymatic antioxidants is concerned, SOD is one of the primary antioxidant defences against superoxide radicals; indeed, increased enzymatic activity of SOD correlates with enhanced resistance to oxidative stress. The cytoplasmic isoform of SOD is copper/zinc-dependent enzyme, while the mitochondrial isoform is a manganese-dependent enzyme [9]. It is also present as a third form of SOD in extracellular fluids, which contains copper and zinc in its active site. Superoxides, through the reaction catalyzed by SOD, are transformed into the less reactive hydrogen peroxide moiety, then the decomposition of hydrogen peroxide to form water and oxygen is accomplished in the cell by catalases, enzymes dependent on either an iron or manganese cofactor. Catalase is also able to oxidize alcohol in vitro in the presence of an H_2O_2-generating system [105], even if under physiologic liver conditions its role in ethanol oxidation is secondary with respect to CYTP450 and the microsomal ethanol oxidizing system. This antioxidative enzyme is widely distributed in the cell, with the majority of the activity occurring in the mitochondria and peroxisomes.

By the way, the enzyme with the much greater affinity for hydrogen peroxide is GPX, an enzyme containing selenium as cofactor that catalyzes the breakdown of hydrogen peroxide and organic hydroperoxides [191]. During normal function of the antioxidant defence system, GPX utilizes reduced glutathione to detoxify hydrogen peroxide. As already mentioned, GSR is then necessary to convert GSSG into GSH using electrons from the oxidation of NADPH [131]. For this reason, a constant supply of NADPH must be available to continuously detoxify hydrogen peroxide.

1.3.2.2 Peroxiredoxins, Sulfiredoxins, Thioredoxins, and Glutaredoxins

Beside the activity of catalase and GPX, recently peroxiredoxins (PRX) have been identified as a new family of peroxidases able to efficiently decrease growth factors-induced hydrogen peroxide. At least six mammalian PRXes have been identified, subdivided into typical 2-cysteine PRXes, in which the two active site cysteines are conserved, one in the N-terminal region and the other in the C-terminal region; atypical 2-cysteine PRXes, with conservation of the N-terminal cysteine but not of the C-terminal cysteine and 1-cysteine PRXes that contains only the N-terminal-conserved cysteine as redox site [167]. PRXes are highly abundant and are believed to control the hydrogen peroxide intracellular level; indeed, their overexpression lowers the intracellular levels of H_2O_2 and inhibits the signalling induced by platelet-derived growth factor (PDGFR), tumour necrosis factor-α, or ceramide [167]. The redox-active cysteine (the peroxidatic cysteine) in the active site of PRX is oxidized to a sulfenic acid by the peroxide substrate, resulting in a subsequent formation of PRX inter-subunit disulfide bond.

As mentioned, the over-oxidation of the peroxidatic cysteine residue to sulfinic acid in PRXes inactivates these enzymes, but this can be reversed by the action of Sulfiredoxin (SRX), a cysteine sulfinyl reductases by repairing the sulfinic group of PRX, regenerating the active enzyme. This PRX inactivation–reactivation cycle represents a cell signaling loop [99]; indeed, cell cycle arrest is combined with recruitment of hyper-oxidized PRX II to high molecular weight oligomers, and that recovery is associated with regeneration of the reduced enzyme [156]. An alternative way to inhibit PRX through tyrosine phosphorylation following NOX activation has been recently proposed [213]. Originally, PRX were considered only weak scavengers of ROS, but recently it has been shown that they are active peroxidases as catalase or GPX. Moreover, PRXes are characterized by their interaction with numerous regulatory proteins, including c-Abl, c-Myc, c-Jun N-terminal kinase (JNK), and the platelet-derived growth factor [97]. Indeed, PRX II is a negative regulator of PDGF signalling, since PRX II deficiency leads to an increase in PDGFR-mediated H_2O_2 production. It has been demonstrated that PRX II is recruited to PDGFR following its stimulation and then it is able to suppress protein tyrosine phosphatase (PTPs) inactivation removing endogenous H_2O_2, thus terminating the PDGF signalling [35].

Interestingly, PRX I is the most active 2-Cys PRX in the elimination of ROS from the liver. Ethanol consumption induces the accumulation of CYP2E1, a major contributor to ethanol-induced ROS production in the liver. PRX I was found to be closely associated with CYP2E1 on the cytosolic side of the liver endoplasmic reticulum membrane. Thus, the selective role of PRX I in ROS removal in liver is likely due to the proximity of PRX I and CYP2E1 [7]. Also SRX takes part in this process; as mentioned, SRX is an enzyme that can reduce sulfinic acid back to thiol, in an ATP-dependent manner. It has been demonstrated that ethanol consumption markedly increased the expression of SRX in the liver of mice in a nuclear factor erythroid 2-related factor 2 (Nrf2)-dependent manner, and the SRX action, in concert with PRX I, exerts a protective action of the liver in ethanol-fed mice as shown by the severe oxidative damage observed in mice lacking either SRX or PRX [7]. The antioxidant function of SRX depends on a C-terminal cysteine residue that is conserved in all family members [26]. SRX also cause the deglutathionylation of PRX II and other proteins [55]. The biological function of SRX may involve in the regulation of various cell-signaling pathways to promote tumorigenesis and cancer progression. Abnormally high expression of SRX has been demonstrated in many malignant tumours including those of skin, lung, and colon [211].

The disulphide reductase TRX is able to drive reduction of sulphenic acids (SOH) and disulphides (S–S). Glutaredoxin (GRX) can support the reversibility of mixed disulphides with GSH and S-nitrosothiols, while only SRX can reverse the formation of sulphinic acids (SO_2H). Finally, sulphonic acids (SO_3H) are currently believed to be irreversible terminal products of over-oxidation, since no known enzyme is able to catalyze the reduction of sulfonic acids in proteins [97]. TRX and GRX are disulfide reductases which play a key role in controlling redox homeostasis. Both TRX and GRX contain a common dithiol/disulfide active site motif (Cys-X-X-Cys) [11].

TRX catalyzes the reversible reduction of disulfides, causing a reduced pool of dithiol target protein. The resulting disulfides in the active form of TRX are then reduced by TRXR, using NADPH as electron donor. TRX1 is ubiquitously expressed and is upregulated under inflammatory conditions [11]. TRX1/TRXR1 is the main TRX-based system in the cytoplasm, while TRX2/TRXR2 acts mainly within mitochondria. Several proteins are redox regulated by TRX and include Nuclear Factor kappa B (NF-κB), Activator Protein-1, p53, glucocorticoid and estrogen receptors, and hypoxia inducible factor (HIF-1) [97]. TRX1 can also be translocated to the nucleus, following certain stimuli, or to the extracellular compartment following a secretory pathway. Circulating levels of TRX were reported to be increased in patients with several diseases, like rheumatoid arthritis, acute lung injury, or HIV infection. Furthermore, extracellular TRX can act as a unique chemoattractant for neutrophils, monocytes, and T cells [12]. TRX80, a 10-kDa C-terminally truncated variant of TRX, is also secreted. This is a potent mitogenic cytokine that stimulates growth of resting human peripheral blood mononuclear cells [153] and differentiation of human CD14(+) monocytes into a novel phenotype in association with activation of mitogen-activated protein kinases [152].

Mammalian GRXes, also known as thiol transferases, reverse *S*-glutathionylated cysteines, restoring the protein cysteine to the sulfhydryl group [176]. The GRX system consists of GRX enzymes, GSH, GSR, and NADPH. GRX performs the deglutationylation reactions, also known as the reduction of protein-mixed disulfides (PSSG, glutathionylated cysteines). The reduction of PSSG by GRX results in a mixed disulfide between GSH and the N-terminal active site cysteine (glutathionylation of GRX). This is then reduced by a second GSH molecule. GSSG is finally regenerated by GSR using NADPH [176]. Levels of GRX are dynamically regulated and an increase in GRX has been observed both in cancer and cardiovascular diseases [11, 148]. Recently, the group of Janssen-Heininger has shown that multiple stimuli, such as the Toll-like receptor ligands, lipopolysaccharide and CpG DNA, interleukin-4, and transforming growth factor-beta 1 cause a decrease in GRX activity in primary lung epithelial cells. In contrast, increase in GRX activity occurs in cells exposed to Interferon-gamma or in mice with allergic airways disease [164]. Interestingly, both GRX1 and 2 can be *S*-nitrosylated, resulting in a decrease of GRX activity [88]. Also, TRX and protein disulfide isomerase (PDI) are targets of S-nitrosylation. Intriguingly, in addition to GRX itself, TRX and PDI are targets for S-glutathionylation [24, 57]. S-glutathionylation of TRX at Cys72 abolishes the disulfide reductase activity of TRX [24], whereas S-nitrosylation of Cys63 was reported to increase TRX activity [81].

All together, these observations underline the important cross-regulation of the various redox systems, which may have impact on steady state levels of several cysteine oxidation states. Indeed, the aforementioned enzyme systems operate together with classical antioxidant defences consisting of SODs, catalase, or GPX to control steady state levels of $^{\bullet}O_2^-$, H_2O_2, $NO^{\bullet}$, and SNO. It is evident that cells have evolved several different systems, consisting of cysteine/cystine, GSH/GSSG, and TRX-SH/TRX-SS, to keep under control the redox blend in order to allow and finely tune redox signalling.

1.3.3 The Keap1–Nrf2 Signalling

Another mechanism for cytoprotection is the Keap1 (kelch-like ECH-associated protein 1)–Nrf2 pathway which plays a central role in the defence of cells against oxidative and xenobiotic stresses. Nrf2 is a potent transcription activator possessing a well-conserved basic region-leucine zipper (bZip) motif that recognizes a unique DNA sequence known as the antioxidant response element (ARE) or electrophile response element (EpRE; TGA(G/C)NNNGC), through heteromerizing with the small Maf protein [139]. This complex plays a central role in the inducible expression of many cytoprotective genes in response to oxidative and electrophilic stresses. Under non-oxidative conditions, Nrf2 activation is prevented by its binding to Keap1. The Keap1/Nrf2 complex is constantly ubiquitinated in the cytoplasm and degraded in the proteasome. Keap1 is a thiol-rich protein, endowed with multiple highly reactive cysteine residues. Oxidative insults target Keap1 by modifying its Cys273 and Cys288, leading to Keap1 inactivation, dissociation of Keap1 from Nrf2, and finally causing Nrf2 stabilization by preventing its degradation. It has been demonstrated that Keap1 acts as a redox-sensitive adaptor for a Cul3-based E3 ligase [109]. Thus, the complex of Keap1, Nrf2, and Cul3 determines rapid degradation of Nrf2 via the ubiquitin proteasome system. Accumulation of Nrf2 in the nucleus activates transcription of genes that encode detoxifying enzymes and antioxidant proteins, such as NAD(P)H:quinone oxidoreductase 1, GST, heme oxygenase-1, etc. In general, the target genes of Nrf2 (a) provide direct antioxidants, (b) encode enzymes that directly inactivate oxidants, (c) increase levels of glutathione synthesis and regeneration, (d) stimulate NADPH synthesis, (e) enhance toxin export via the multidrug response transporters, (f) inhibit cytokine-mediated inflammation, and (g) enhance the recognition, repair, and removal of damaged proteins [206].

Recently, it has been highlighted as a crucial role of Nrf2 signalling for protecting against liver injury induced by alcohol, as well as other hepatotoxicants [112]. Indeed, Nrf2(−/−) mice displayed aggravation of ethanol-induced liver injury, increased oxidative stress, inflammation and necrosis, and finally a dramatically increased mortality. Indeed, acute ethanol treatment induces the expression of several antioxidant and anti-inflammatory enzymes in the liver tissue in a Nrf2-dependent manner [112], suggesting the existence of an early compensatory or adaptive mechanism of Nrf2 signalling to suppress ethanol-induced oxidative injury and inflammation in liver tissue. In keeping with a central role of Nrf2 in the pathophysiology of liver oxidative stress, it has been found that overexpression of CYP2E1 induces an increase in Nrf2 level and the activation of Nrf2-dependent antioxidants response attenuates CYP2E1-mediated oxidative hepatic injury [69].

1.4 Redox-Sensitive Targets in Proteins

Several oxidative modifications of amino acids have been associated to oxidative damage of proteins and their targeting to degradation. These include formation of dityrosines, carbonylation of lysine, arginine, proline or threonine, or the racemization of lysine, arginine, or aspartate [71]. The recent discovery of de-carbonylation enzymatic activity, occurring via TRXes, supports the idea that carbonylation may also serve as a signalling trick, at least for pulmonary artery smooth muscle cells growth [212].

Hundreds of reports concentrate on sulfur-containing amino acids, such as cysteine and methionine, the main role as signalling devices for protein function regulation. Oxidation of methionines by $^{\bullet}O_2^-$, H_2O_2, peroxynitrite ($ONOO^-$), or $^{\bullet}OH$ leads to methionine-sulfoxide and is rescued by methionine sulfoxide reductase, which use TRX as intermediate [113]. Oxidation of methionine to methionine sulfoxide, likely for substitution with charged aminoacids, results in tertiary structure perturbation, as well as charge-dependent protein–protein interactions. Methionine oxidation occurs due to intracellular oxidative stress and proceeds further to methionine sulfone. To date, the signalling role of methionine oxidation in physiological and pathological contexts is not fully understood and is likely currently underestimated [113].

Growing evidence indicates that cysteine thiol groups act as redox-sensitive switches, thereby providing a common trigger for a wide range of ROS-mediated signalling events. The pK_a of individual cysteines at a physiological pH dictates if the residue is likely to exist as a thiol or more reactive thiolate anion (S^-) [30, 163, 166]. For this reason, low pK_a cysteines are more vulnerable to oxidation than those with a pK_a value closer to the surrounding pH, which remain unreactive. Oxidation of low pK_a cysteines forms reactive sulfenic acid that can form disulfide bonds with other near cysteines or undergo further oxidation (sulfinic or sulfonic acid) [29, 166] (e.g., 5.1–5.6 for cysteine residues of PTPs). Local perturbations in pH can affect the reactivity of cysteine residues, rendering proteins more or less vulnerable to undergo oxidation and conferring a spatially defined oxidation of redox-sensitive proteins within discrete subcellular locations. In case of presence of close nitrogen residues, the sulfenic acid may react to form a sulfenamide. Noteworthy, Sanchez et al. determined an algorithm to identify reversibly oxidized cysteine, using the Cysteine Oxidation Prediction Algorithm, which classifies cysteines on the basis of three parameters critical for prediction of thiol oxidation susceptibility, i.e., the distance to the nearest cysteine sulfur atom, accessibility to the solvent, and of course cysteine pK_a [170].

Glutathionylation of proteins, which occurs through the formation of a mixed disulphide between one cysteine of glutathione and one cysteine of the other protein, constitutes an efficient mechanism to protect proteins from irreversible modifications and also seems to play an important role in cell signaling [67]. Cysteines that are *S*-glutathionylated to prevent hyperoxidation are likely to be crucial to the stability or activity of the redox-sensing protein, while this modification may also lead to transient enzyme inactivation. S-glutathionylation, which has been recently reviewed elsewhere [67], has been reported for several proteins as β-actin, protein kinase C, and cAMP-dependent protein kinase.

All these oxidative modifications of cysteines always result in structural or functional changes of the redox-sensing protein [163], although protein oxidation can be reversed by thiol donors such as glutathione, GRX, PRX, and TRX (see later). Reversible cysteine oxidation plays a pivotal role in redox-signaling cascades, and the main molecular targets are PTPs or lipid phosphatases, proteases, enzymes, signaling adaptors, and transcription factors (TFs). The effect of cysteine oxidation varies among redox-regulated proteins, from enzymatic inactivation, as for PTPs, several enzymes involved in metabolism and some TFs, to enzymatic activation, as for receptors or cytosolic protein tyrosine kinases (PTKs), small GTPases, and a small group of TFs. Cysteine oxidation of proteases may be either inhibitory, as for the caspase family [132], or activatory, as for matrix metalloproteases (MMPs), through the disruption of the inhibitory interaction between the pro-domain and the catalytic site [76]. Reversible cysteine oxidation of TFs disturbs DNA binding and transcriptional activity of several nuclear factors [126, 215].

Last, cysteine oxidation of signaling proteins can also modulate signal transduction by interrupting protein–protein interactions. One example of this kind of regulation is the nuclear translocation of Nrf2 following disruption of its interaction with the adaptor Keap1, containing two reactive cysteines, and allowing the activation of Nrf2 antioxidant response [44, 126].

Cysteine oxidation via attachment of a nitrogen monoxide group, extensively reviewed in [92], contributes to intracellular signal transduction by functionally affecting several proteins. This modification, resulting in S-nitrosylation, has been reported for the activation of Src kinase, the inactivation of epidermal growth factor receptor (EGFR), of Jun-N-terminal kinase-1, and apoptosis signal-regulated kinase-1. All of these proteins are also sensitive to hydrogen peroxide oxidation, often via the same sensitive cysteines. An "acid–base" motif, assumed to specify S-nitrosylation-susceptible sites on proteins, facilitates local decrease of cysteine pK_a through vicinity of charged amino acids. S-nitrosylation of cysteines functions as the "primary effector" and could facilitate further evolution to hydrogen peroxide sensing and sulphenic acid formation, as observed in the ryanodine receptor/Ca^{2+} channel. The role of S-nitrosylation as a priming mechanism for other signalling kinases remains to be elucidated.

Beside cysteine oxidation, ferrous iron-containing proteins may also be targeted by ROS, through oxidation of ferrous to ferric iron and the consequent disruption of protein function. Proteins that sense redox signaling through this mechanism mainly belong to 2-oxoglutarate (2OG) dioxygenase family. These are ubiquitous 2OG-dependent oxygenases that couple substrate oxidation to the conversion of 2OG to succinate and carbon dioxide and their role includes collagen biosynthesis, DNA repair, chromatin modifications, and hypoxic sensing [25, 133]. Among this group of enzymes, prolyl hydroxylases (PHDs) and JmjC histone demethylases have been reported to play important signalling roles [93, 137, 174].

1.4.1 Spatial Regulation of Redox Signals

Redox signalling is at least in part facilitated by the discrete subcellular compartmentalization of ROS production, the restricted availability of NOX-activating/regulatory subunits, control at an expressional level of LOXes and COXes [193].

Scavenging and neutralizing enzymes such as Cu/Zn SODs, GPX, and PRXes, which prevent the escape or accumulation of ROS to toxic levels, may also be compartmentalized in the cytosol or the mitochondria, while the reticular environment is always maintained pro-oxidant [168]. Indeed, proteins regularly undergo oxidative processing as part of their maturation within the endoplasmic reticulum, resulting in disulfide bond formation. Noteworthy, reticular proteins, such as PTP1B, could be regulated by NOX enzymes targeted to the reticulum for signalling purposes [28].

Since several decades scientists assumed hydrogen peroxide to permeate biological membranes by simple diffusion, although recent evidence challenged this notion and disclosed the role of aquaporin water channels (AQP) in mediating hydrogen peroxide transport across plasma/mitochondrial membranes. Decisive studies have shown that specific AQP isoforms are capable of funnelling hydrogen peroxide across the plasma membrane in yeast, plant cells, smooth muscle cells, or in human tumour cells [14]. AQP1 and AQ8 appear to mediate the main role, at least for human cells, as it is able to modulate hydrogen peroxide transport through the plasma membrane, but also to affect downstream redox signalling linked to leukaemia cell proliferation [203].

Local redox regulation is also controlled through co-localization of ROS-producing enzymes and the protein target to distinct separated intracellular compartments. The main role for this aspect is played by mitochondria, showing several ROS internal or external delivery systems, and NOXes. Mitochondria can deliver ROS in the mitosol, thereby affecting mitochondria proteins, through deregulation of complex I and IV of the electron transfer chain. On the contrary, deregulation of complex III Rieske Fe–S protein, due to its membrane facing, causes a delivery of ROS in the intramembrane space, which finally results in cytosolic oxidative signalling [25]. NOXes localization to endosomes, podosomes, lipid rafts, and caveolae has also been reported and correlated to specific targeting of oxidants to affect precise subcellular signalling [198].

Finally, regulation of PRXes via phosphorylation by c-Src kinase in response to GF extracellular signals allows the local accumulation of hydrogen peroxide within the receptors, inactivation of PTPs, and thereby enables its messenger function [168].

1.4.2 Redox-Sensing Proteins

Proteins that are able to sense redox signals includes PTPs and kinases, structural components of cytoskeleton, or proteins regulating cytoskeleton dynamics, TFs, as well as enzymes involved in cell growth, nutrient sensing, and metabolic homeostasis. Here we shortly summarize some of the key molecules exploiting redox-based regulation.

Protein tyrosine phosphatases. PTPs are a huge group of hydrolases, encoded by about 100 different genes. They regulate the tyrosine phosphorylation of several signaling proteins and exert complex functions in cell proliferation, adhesion, survival, and motility. Although members of the PTP family differ in their intrinsic susceptibility to oxidation, they invariably contain a catalytic Cys residue conferring redox sensitivity to various oxidant species, such as H_2O_2, superoxide, or nitric oxide [31]. In vivo the reversibility of PTP inactivation is mediated by formation of a mixed disulfide with glutathione, an intramolecular S–S bridge among cysteines, or a sulphenyl–amide intermediate [31, 166]. PTPs mainly undergo redox-dependent inactivation in cells in response to growth factors signalling, thereby allowing the endurance of receptor tyrosine kinases (RTKs) intracellular signalling due to their transient inability to dephosphorylate and inactivate the RTKs. Hence, PTP redox-mediated inactivation is the basis for RTK tyrosine phosphorylation and the reversibility of PTP inactivation, through their re-reduction to basal state by peroxiredoxins/glutaredoxins, is a switch off mechanism granting for RTK signalling efficient termination. Redox-regulated PTPs include PTP1B, LMW-PTP, PTEN, SHP2, and many others, able to regulate RTKs as EGFR, PDGFR, fibroblast growth factor receptor, hepatocyte growth factor receptor, Ephrin receptor, etc. Several of these PTPs behave as regulators of cytoskeleton dynamics during cell motility, adhesion, and proliferation and their regulation via redox-based mechanisms plays a mandatory role in such context.

The lipid phosphatidyl-inositol (PI) 3-phosphatase and tensin homolog (PTEN), functioning on cell survival, adhesion, and motility through its downstream kinase Akt, is a key redox-sensing protein [117]. PTEN is sensitive to rapid inactivation by hydrogen peroxide or by *S*-nitrosothiols, forming a stable intramolecular disulfide bond between the active site Cys124 and the proximal Cys71 [108]. PTEN redox-mediated inhibition leads to inability to contrast PI3-kinase and hence helps in sustaining Akt activation, acknowledged to favour proliferation, adhesion, and survival to stresses. Inactivation of PTEN is rescued by TRX-interacting protein or PRX I, both acting through reduction of the PTEN disulphide [22]. PTEN undergoes oxidation in response to growth factors stimulation, correlating with ROS-dependent activation of downstream Akt phosphorylation [118]. Although PTEN has been acknowledged as a human tumour suppressor gene, deleted in several cancers, for the moment PTEN oxidation has been implicated in the development of T-cell acute lymphoblastic leukemia, as well as in the multi-organ tumourigenesis in mice lacking peroxirexoxin-1 [22, 180].

Redox control of Src homology-2 domain-containing phosphatase 2 (SHP-2), a key regulator in signalling pathways mediated by several growth factors and cytokines, has also been correlated with adhesion and motility [16, 188]. Upon integrin receptor engagement by matrix proteins, SHP-2 undergoes a reversible oxidization/inactivation to which mitochondrial and 5-LOX-derived ROS contribute differentially. Reversible oxidation/inhibition of SHP-2 prevents the dephosphorylation/inactivation of its substrate p125 focal adhesion kinase, thereby enabling the continued propagation of signals arising by integrin engagement. SPH-2 has also been found oxidized in response to hypoxia, which is a common feature of several aggressive cancers [171, 192].

Low molecular weight PTP (LMW-PTP) undergoes redox regulation during PDGFR signalling, due to oxidation of its catalytic Cys12 and Cys17, thereby maintaining receptor activation through tyrosine phosphorylation and allowing signal transduction [33]. Its oxidation plays a compulsory role during anchorage to matrix and cell motility, due to control of de-phosphorylation of FAK [34] and p190RhoGAP during integrin signaling. By this way, LMW-PTP oxidation is involved in the antagonistic crosstalk between Rac-1 and RhoA small GTPases [31, 32], a crucial event for cellular dynamics, the former promoting membrane protrusion, cell polarity, and spreading, the second cytoplasm contractility and tail retraction [49]. During polarized motility, also called mesenchymal motility, Rac-1-driven ROS, by inactivating LMW-PTP, causes hyper-phosphorylation of the PTP substrate p190Rho-GTPase Activating Protein (GAP), thereby increasing its inhibitory activity on RhoA [32]. On the contrary, during non-polarized/amoeboid motility, there is a reduction of Rac-1 activity owing to attenuated generation of ROS, causing LMW-PTP activation, p190RhoGAP de-phosphorylation, and to an increase of Rho signaling [19]. Hence, redox regulation of this PTP plays a central role in the decision of the most useful motility style to be used for moving cells, driving towards polarized or amoeboid motility.

β-Actin. The movement of cells during processes as embryogenesis and organogenesis, vasculogenesis/angiogenesis, or tumour metastasis is driven by polymerization of β-actin monomers at the leading edge of the moving cells, an event that is again under redox control. A proteomic screen for cysteine glutathionylated proteins in cells exposed to oxidants identified several cytoskeleton components (actin, vimentin, myosin, tropomyosin, cofilin, profilin) as targets of these oxidants [57]. Hydrogen peroxide produced by LOXes upon integrins ligation due to ECM contact has been involved in in vivo β-actin redox regulation. Actin oxidation takes place via the formation of a mixed disulfide between cysteine 374 and reduced glutathione, allowing the formation of stress fibres and consequently cell spreading [52, 53]. β-Actin oxidation through glutathionylation causes the disassembly of the actinomyosin complex, with severe implications for cell contractility and movement, thus exerting a crucial effect on assembly/disassembly dynamics of the contractile machinery [51, 53].

Src family kinases. Src-family kinases are critically involved in the control of cytoskeleton organization and in the generation of integrin-dependent signaling responses, inducing tyrosine phosphorylation of numerous signaling and cytoskeletal proteins. Beside the phosphorylation circuitry involving Tyr416 and Tyr527 of c-Src kinase, cysteine oxidation has been recently reported as a further mechanism of enzyme activation. Beside in vitro evidence on mercuric chloride and NO-releasing agents [1, 138], a redox regulation of Src kinase via oxidation has also been reported during anchorage-dependent growth of tumour cells [63]. Indeed, the kinase undergoes oxidation/activation in response to the formation of an S–S bond between Cys245 and Cys487, respectively located in the SH2 and in the kinase domain of the Src molecule [63]. Mounting evidence describes Src activation via redox regulation as a key outcome in several circumstances, including growth factor and cytokines signaling, integrin-mediated cell adhesion and motility, membrane receptor cross-talk as well as in cell transformation and tumour progression [29, 65]. Src oxidation has also been involved in survival to hypoxia owing to mitochondrial delivery of ROS and regulation of NF-κB transcription factor [127]. Activation via oxidation is mandatory also for the oncoproteins v-Src and SrcY527F, which are oxidized by exogenous oxidative bursts or during anchorage-independent cell growth, greatly affecting solid tumour formation [63]. In keeping with this, v-Src oncogenic potential is reduced by curcumin, a powerful dietetic antioxidant [120].

Src oxidation has also been reported for molecular cross-talk between membrane receptors, as RTKs, integrin receptors, and G-protein coupled receptors [29, 65, 127, 161]. This redox-based activation of Src leads to ligand-independent activation of receptors via intracellular cross-talk, thereby sustaining proliferation, survival, and cell adhesion in several circumstances, including cancer malignancy [62, 64].

Another member of the family kinase, Lyn, has been identified as a redox sensor that neutrophils use to reach the unhealed wound [219]. Lyn activation in neutrophils is dependent on wound-derived oxidants after tissue injury and Cys466 is responsible for a direct oxidation/activation of Lyn, inducing intramolecular conformational changes [117]. By this way, after sensing an initial tissue injury, wound signals can lead to tissue regeneration. The same authors recently included Fynb, another member of the Src family kinase, in redox-dependent wound repair [218]. Hence, ROS generated at wounds play a dual role, allowing firstly the oxidation/activation of Lyn in the recruited neutrophils and thereafter the redox-mediated Fynb activation and calcium signaling in epithelia, cooperating as integrated "wound signals" that rapidly initiate a regenerative response.

Receptor tyrosine kinases. Oxidants can also modulate the activity of protein kinases and, among these, PTKs. In agreement with their opposing role to PTPs in the modulation of protein tyrosine phosphorylation, PTKs are activated by oxidation. However, activation of PTKs appears to be essentially due to two mechanisms. First, similar to what happens for PTPs, cysteine oxidation may occur, leading to direct kinase activity enhancement. Second, as PTKs themselves are frequently tyrosine phosphorylated proteins and their activity is increased due to phosphorylation, most likely the concomitant inhibition of PTPs indirectly leads to sustained activation of PTKs. Herein we will focus on RTKs as prototypical membrane kinases and on Src family kinases as cytosolic kinases (see below).

A direct oxidation of RTK cysteines has been reported for insulin receptor kinase (IRK), epidermal and platelet-derived growth factor receptors (respectively EGFR, PDGFR), and Ret kinase. Insulin sensitivity requires a process of "redox priming" of the β-subunit of IRK, due to oxidation of any of the four cysteine residues 1,056, 1,138, 1,234, and 1,245 into sulphenic acid [172]. This "redox priming" of the IRK facilitates its autophosphorylation in the activation loop, producing activatory structural changes.

The c-RET proto-oncogene encodes a cadherin-like domain receptor-type tyrosine kinase, and its mutations in the germ line are responsible for the inheritance of multiple endocrine neoplasia type 2A and2B (MEN2A/B) [104]. UV-mediated production of ROS of cells expressing c-Ret results in the dimerization of c-Ret, mediated by the formation of a disulphide between the cystein residues of neighbouring monomers [104]. This results in c-Ret strong activation in response to ROS and contributes to the oncogenic potential of c-Ret as UV light irradiation induces superactivation of the constitutively activated Ret-MEN2A and Ret-MEN2B.

Finally, nitration of Tyr residues in PDGFR has been reported in mild oxidant-treated cells, leading to Src kinase-dependent activation of MAPK and Akt [114]. The direct redox sensitivity of PDGF-R has also been suggested by the ability of *N*-ethylmaleimide, which selectively alkylates free thiol groups of cysteine residues, to inhibit PDGFR-β kinase activity. Two reactive cysteines have been identified so far, Cys822, positioned in the catalytic loop, and Cys940, located in the C-terminal kinase subdomain. Mutation of these residues significantly reduces PDGFR autophosphorylation and downstream signalling, thereby confirming a key role of direct oxidation in this RTKs [114].

Besides their function as downstream modulators of RTK signalling, ROS act as upstream mandatory activators in RTK *trans*-activation, leading to a ligand-independent signalling. RTK ligand-independent *trans*-activation is redox-dependent in three circumstances: (1) during exogenous oxidant delivery by lymphocytes, macrophages, or upon treatment with ROS-producing agents as alkylating molecules, UV, and heavy metals; (2) during the integration of cross-talking signals elicited by G-Protein-Coupled Receptors (GPCRs) or integrins and several RTKs, mainly mediated by Src kinase redox regulation; (3) during PTP oxidation/inhibition allowing signal transduction (see above).

Cysteine modification in the PTP catalytic site has been proposed for both UV- and alkylating agents-induced RTK *trans*-activation [72], corroborating the key role for PTP redox regulation in the control of RTK tyrosine phosphorylation, even in the absence of the natural ligand. Furthermore, in both phagocytes and lymphocytes, as well as in several cardiovascular diseases, NOXes generate oxidants involved in host defence or inflammation, resulting in pro-oxidant conditions for bystander cells. These pro-oxidant backgrounds can cause a redox regulation of RTKs (through PTP inhibition) in neighbouring cells during inflammation. If hold true, this redox-based cross-talk would account for signaling during the inflammatory reaction or cardiovascular diseases with important pathophysiological and therapeutic implications. In keeping with this, ROS generated from hyperglycemia promoted ligand-independent phosphorylation of vascular endothelial growth factor receptor 2 (VEGFR2). Hyperglycemia-induced phosphorylation of VEGFR2 did not require intrinsic receptor kinase activity and was instead mediated by Src family kinases [210].

Activation of different membrane receptors, belonging to the G-protein-coupled or tyrosine kinase subclasses, is able to activate in *trans* several RTKs via redox-based mechanisms. Indeed, GPCRs, RTKs, and integrin receptors are all able to increase generation of H_2O_2 [166]. Intracellular redox signalling, either acting via PTP inhibition or via Src activation, leads to ligand-independent RTKs *trans*-activation [31]. In addition, adhesion molecules are able to associate with RTKs and trigger their ligand-independent activation [140]. This complex interaction is believed to be a key step of protection from *anoikis* of anchorage-dependent cells, mainly acting through a Src-mediated ligand-independent (integrin-dependent) EGFR *trans*-phosphorylation [64, 140]. Moreover, hydrogen peroxide signalling has also been implicated as a key molecule leading to ligand-independent RTK activation in "lateral propagation" of EGF signalling waves [165].

PKM2. Pyruvate kinase (PK) is the final enzyme in glycolysis, converting phosphoenolpyruvate to pyruvate with the production of ATP and it is the prototype of metabolic enzymes regulated via redox modifications. Glicolytic flux in cancer cells is greatly determined by expression, in virtually every cancer cell type so far analysed, of the isoform M2 of PK (PKM2) [36]. Contrariwise to its splice variant PKM1, expressed in several adult and/or non-proliferating tissues, PKM2 is susceptible to inhibitory post-translation cues, including tyrosine phosphorylation in response to growth factor signalling, lysine acetylation in response to high glucose environment, or cysteine oxidation in response to oxidative stress/signalling [3, 129, 209]. Cys358 oxidation leads to inhibition of PKM2, thereby allowing proliferating cells to divert glucose from the glycolytic pathway, causing accumulation of glycolytic intermediates which fuel other metabolic pathways, as penthose phosphate pathway (PPP) or serine synthesis [3]. PPP is acknowledged to convert glucose into penthoses or glycolysis intermediates, mandatory molecules for anabolism of aminoacids and nucleotides, and to produce NADPH, the mandatory molecule of cellular reducing activity, driving re-reduction of intracellular glutathione. The concerted inhibition of glycolysis and activation of PPP leads to increased anabolism, allows growing cells to meet the increased biosynthetic demands ofproliferation, and allows cells to defend themselves from oxidative insults detoxifying ROS [200, 201]. Oxidation of PKM2 has also been observed in hypoxic conditions, known to increase Warburg-like glycolytic metabolism and to signal through ROS delivery by mitochondria or cytosolic NADPH oxidase [3]. Noteworthy, treatment with reducing agents restored PKM2 activity, confirming that PKM2 undergoes a reversible redox regulation in proliferating cells due to oxidative stress.

The adaptation of proliferating cells to increase resistance to oxidants, following oxidative block of PKM2, is due to accumulation of its substrate phosphoenolpyruvate, acting as a feedback inhibitor of the glycolytic enzyme triosephosphate isomerase. This latter enzyme is directly responsible for the stimulation of PPP, leading to increased antioxidant metabolism and to a general reconfiguration of central carbon metabolism, helpful in stressing conditions [75]. Indeed, this kind of metabolism, commonly referred as Warburg metabolism, is implicated in resistance of cancer cells to therapeutic drugs [110, 169]. Beside the role of ataxia teleangiectasia mutated protein in the regulation of PPP (see below), PPP can be engaged in cancer cells either following p53-deficiency or hypoxia exposure, due to interaction between PKM2 and CD44, an acknowledged cancer stem cells marker. CD44 ablation resulted in depletion of reduced glutathione, increase in oxidative stress, and enhancement of the effect of chemotherapeutic drugs [190].

GAPDH. Glyceraldehyde-3-phosphate dehydrogenase (GAPDH) oxidoreductases are known for their involvement in glycolysis. The catalytic mechanism involves Cys-152, resulting in the formation of a catalytic hemiacetal, undergoing oxidation in vivo [20]. S-nitrosylation and mixed disulfide formation represents a target for nitrosative stress due to $NO^{\bullet}$ or to peroxinitrite, or for subsequent post-translational modification with NAD^{+}, leading to enzyme inactivation and culminating in non-enzymatic ADP-ribosylation [70, 98]. Oxidative inactivation of GAPDH is able to regulate glucose balance towards glycolysis or PPP. Indeed, hydrogen peroxide exposure inactivates GAPDH, blocking glycolysis and redirecting glucose flux into the oxidant-resistant PPP [74, 110, 162]. Then, glycolysis could be re-established by re-reduction due to NADPH increase. Nevertheless, the balance between glycolysis and PPP is maintained in favour of the latter by further PKM2 inhibition, sensitive to oxidants but also to tyrosine phosphorylation and acetylation [3, 129, 209]. These circumstances are helpful in allowing cells to resist stressful environments.

AMP-kinase. AMP kinase (AMPK) is the main energy-responsive enzyme. It is sensitive to increased AMP/ATP ratio, thereby allowing cells to respond to nutrient deprivation signals. AMPK also senses redox signals, as it has been reported to respond to pro-oxidative or nitrosative conditions, thus enhancing cell survival via autophagy in response to nutrient starvation [158]. In this way, AMPK behaves as a flexible molecular effector for integrating metabolic and oxidative cues.

Several environmental conditions characterized by oxidative stress, as block of mitochondrial ETC, reduced nutrient availability, or hypoxia, and enhanced AMPK activity via redox modifications [85]. Mild hypoxia causes accumulation of ROS, mainly of mitochondrial origin, and leads to a ROS-dependent activation of AMPK, also in the absence of the classical changes in the adenylate pool [46]. Redox regulation of AMPK after exposure to hydrogen peroxide results in the oxidation of Cys299 and Cys304 of the α- and β-subunits of AMPK, leading to enhancement of its kinase activity by inducing an allosteric rearrangement of the AMPK αβγ heterotrimer [221]. In parallel to hydrogen peroxide, NO and peroxynitrite have been implicated in AMPK activation. Although not yet confirmed, it is conceivable that an increase in RNS can promote S-nitrosylation of reactive Cys299 and Cys304 of AMPK [23].

AMPK activation by ROS can promote cell survival by inducing autophagy, mitochondrial biogenesis, and expression of genes involved in antioxidant defence (reviewed in [179]). Data on redox regulation of AMPK during cancer progression are still incomplete, although some recent results support this idea. Indeed, new evidence shows that autophagy activation by AMPK, a process for which its redox sensitivity may play a key role, enables long-term survival for cancer cells during hypoxia or ischemia. Moreover, AMPK-mediated autophagy also enhances the resistance of cancer cells to chemotherapeutic agents or kaempferol, a redox active compound causing metabolic oxidative stress [54, 86]. On the other hand, AMPK inhibition in the liver by ethanol plays a key role in the development of steatosis induced by chronic alcohol consumption, thereby suggesting AMPK as a possible therapeutic target in alcohol-induced liver diseases.

ATM. Ataxia-telangiectasia mutated (ATM) is a serine/threonine protein kinase activated under genotoxic or oxidative stress conditions. It phosphorylates several proteins involved in cell proliferation, cell survival, and DNA repair [94, 177]. ATM responds to ROS through oxidation at specific cysteines leading to ATM dimerization and activation, finally leading to phosphorylation of downstream signaling proteins, such as p53, checkpoint kinase 2, or H2AX [149, 177]. Cys2991 is critical for ATM dimerization through disulfide bond formation, as its mutation results in insensitivity of ATM to oxidative insults [78]. ATM, behaving as a functional sensor of cellular redox homeostasis through regulation of the rate-limiting enzyme glucose-6-phosphate dehydrogenase, stimulates PPP [73, 100], providing NADPH for both detoxification or re-reduction of oxidized molecules, as well as ribose for nucleotide synthesis required for DNA repair [40]. Noteworthy, p53 acts by opposing this function of ATM behaving as a post-transcriptional inhibitor of glucose-6-phosphate dehydrogenase, blocking its dimerization/activation [100].

ATM-driven regulation of NADPH synthesis has also been correlated with Warburg metabolism, leading to re-route central carbon metabolism towards PPP upon oxidative stress. This correlation plays a key role in several diseases linked to oxidative stress, as tumour progression and resistance to immune-, radiation-, and chemo-therapy [74]. Indeed, we should consider that chemotherapeutic drugs often show both the ability to cause oxidative stress and DNA double strands breaks. In keeping with this idea, increased activity of glucose-6-phosphate dehydrogenase is mandatory to some multi-drug-resistant cancer cells to increase NADPH and GSH content, necessary to safely handle oxidative stress and to extrude anticancer drugs from the cell [159].

ASK1. Apoptosis signal-regulating kinase 1 (ASK1) is a member of MAP kinase kinase kinase family. It activates JNK and p38 mitogen-activated protein kinases in response to oxidative stress and endoplasmic reticulum stress. ASK1 undergoes both direct and indirect redox-regulation. The oxidation of TRX-1 causes its dissociation from the kinase, facilitating the direct oxidation of ASK1 and triggering the formation of a multimeric kinase through disulfide bonds [195]. The ASK1 oligomer stabilization through covalent disulfide linkages is required for activation of the JNK pathway and to signal apoptotic cues. Upon removal of oxidative stress, rescued TRX1 directly reduces the oxidized ASK1 and switches off the kinase signal [143].

The MAPK pathway. The mitogen-activated protein kinase (MAPK) pathway has been reported to be redox-sensitive both through the oxidative inactivation of its regulatory phosphatases and direct oxidation of the kinases [115]. Oxidation of p44MAPK occurs at low hydrogen peroxide concentrations (0.1 μm) and drives activation of the kinase and its nuclear translocation [39, 115]. The same cysteine residues of the kinase are apparently insensitive to oxidation in presence of toxic doses of hydrogen peroxide (up to 10 μm). These doses of oxidants are compatible with apoptotic signals and engage JNK/p38 activation and cell-cycle arrest programme. This differential sensitivity to oxidant concentrations for different proteins, even among enzymes of the same family, helps in acknowledging hydrogen peroxide as a versatile messenger.

1.4.3 Redox-Sensing Transcription Factors

The activity of TFs is modulated by redox cascades either in a direct or indirect fashion, i.e., through the regulation of their phosphorylation, acetylation, hydroxylation. Among the overabundance of redox-sensitive TFs, we will mainly focus on NF-κB, Nrf2, forkhead box class-O (FoxO), p53, and HIFs.

NF-κB. NF-κB was the first transcription factor that has been recognized to be sensitive to redox signals [185]. Although NF-κB has been originally involved in the inflammatory response, more recently this TF has been involved in several steps of tumour progression, as the activation of cancer-associated macrophages and their polarization towards M2 phenotype, reactivity of cancer-associated fibroblasts, the recruitment of endothelial precursor cells to instruct de novo angiogenesis and vasculogenesis, as well as in the motogenic epigenetic transcriptional programs of metastatic cancer cells, leading these tumour cells to enhance their motility through epithelial mesenchymal transcription and their stem-cell traits. All these events, for which NF-κB activation is mandatory, concur to successful metastatic dissemination of tumour cells [47, 68, 214]. NF-κB forms homo- or heterodimers composed by Rel or NF-κB subfamily members, which interact with the inhibitory proteins IκBs, preventing its nuclear translocation and DNA binding in the absence of proinflammatory stimuli. Activation of NF-κB leads to phosphorylation of IκBs by IκB-kinase (IKK complex), rapidly leading to proteosomal degradation of IκBs. Free NF-κB is allowed to migrate into the nucleus and activate transcription of target genes.

Redox regulation of NF-κB embraces both cytoplasmic and nuclear steps in NF-κB activation, including degradation of its inhibitor IκBα, regulation of DNA-binding ability, and transcriptional activity, as well as chromatin remodelling [68, 101]. Hydrogen peroxide directly targets NEMO, an essential regulator of the IKK complex, causing formation of NEMO dimers through disulfide bonds between Cys54 and Cys347. Dimerization of NEMO causes inhibition of IKK complex and culminates in NF-κB activation [91]. Moreover, tyrosine phosphorylation of IκB-α is under indirect redox control. Indeed, exposure to hypoxia of cancer cells, a very common feature of cancer microenvironment characterized by mitochondrial delivery of ROS due to incomplete oxygen reduction [25], leads to IκB-α phosphorylation by oxidized/activated c-Src kinase [127]. The activation of NF-κB through c-Src-mediated phosphorylation promotes liver carcinogenesis favouring cell survival and tumour progression. H_2O_2 similarly inhibits the nuclear import of IκB-α, thereby facilitating its proteasome degradation, and concurring to keep NF-κB in the nucleus [68]. NF-κB is also sensitive to oxidation of Cys62 in its p50 subunit, essential for DNA binding [134]. Once in the nucleus, Cys62 of p50 needs to be re-reduced by Ref-1 to enable NF-κB transactivation of the DNA [157].

FoxOs. The FoxOs family of TFs is involved in crucial cellular process including cell cycle regulation, apoptosis, resistance to oxidative stress, cancer progression, and metastatic spread [42, 205]. Oxidative stress-mediated activation of FoxOs was shown to regulate a variety of genes, including genes that promote cell motility as

MMPs, stress resistance proteins as SOD and PRXes, stemness, and self renewal regulators as p27 or p21 cyclin inhibitors [142].

As usual for redox control, indirect redox regulation of FoxOs is mainly mediated by the control of their phosphorylation. Generally, an increase in intracellular ROS facilitates the localization of FoxOs to the nucleus where they are transcriptionally active [48]. Indeed, an increase in intracellular reactive ROS induces the activation of the c-Jun-N-terminal kinase, a key regulator of FoxO4. FoxO3a and FoxO1 are targeted by mammalian sterile 20-like kinase signaling [48, 116]. Of note, redox-dependent phosphorylation of FoxO can also produce opposite effects, as reported for Akt-mediated phosphorylation of the TF. FoxOs are directly phosphorylated by Akt on three conserved residues, leading to its nuclear export, cytoplasmic retention, and inhibition of transcriptional activity by FoxOs [42]. ROS can also act on this pathway through PTEN inactivation (see above), leading to Akt activation, FoxO phosphorylation, and nuclear exclusion of the factor [18]. However, several other post-translational modifications are needed to exert the full spectrum of FoxO-mediated responses to ROS, including acetylation and monoubiquitination [150].

FoxO activity can also be regulated via direct oxidation. Following hydrogen peroxide treatment or mitochondrial ROS production due to glucose deprivation, cysteines within FoxO4 undergo oxidation and disulfide formation with the histone acetylase p300. The covalent binding of p300 with FoxO leads to acetylation of the TF, driving also a change in transcriptional targets [41]. The NAD-dependent deacetylase SIRT-1 is able to rescue classical FoxO transcriptional activity [199].

HIFs. Convincing evidence indicates that also HIFs are subdued to multiple mechanisms of redox regulation. The HIF family, composed of three members HIF-1/2/3, plays a key role in the reprogramming of cancer metabolism, motility, and survival to stress and chemotherapeutic agents. HIFs consist of an O_2-regulated HIF-α subunit and a constitutively expressed HIF-β subunit [174, 175]. In normoxic circumstances, HIF-1α is hydroxylated on proline residue 402 and Pro564 by prolyl hydroxylases (PHD), which use O_2 and 2-OG as substrates [102]. Prolyl-hydroxylation targets HIF-1α for proteasomal degradation. Beside hydroxylation of prolines, Asp803 is hydroxylated by factor inhibiting HIF-1, leading to inhibition of binding of p300 histone acetylase coactivator [175]. O_2 deprivation causes inhibition of both prolyl and asparaginyl hydroxylation, thereby causing HIF-1 stabilization and supporting its transcriptional activity.

S-nitrosylation of HIF-1α at Cys800, promoting HIF binding to p300 histone acetylase through a disulphide with p300 Cys388 or 393, has been shown to increase HIF-1 stability and transcriptional activity [217]. In parallel, hydrogen peroxide indirectly promotes HIF stabilization by oxidizing the catalytic Fe^{2+} of PHD and inhibiting their activity [60]. This mechanism is implicated in physiological stabilization of HIFs under mild (1–3 %), but not deep, hypoxia, a circumstance accompanied by production of ROS both from mitochondria and NOXes [79, 173]. In three different tumorigenesis mouse models, HIF-1 redox sensitivity has been confirmed as a mandatory step for tumour progression [59]. Beside hypoxic circumstances,

HIF-1 redox sensitivity plays an important role in normoxic conditions, as during growth factor supply or treatment with chemotherapeutic drugs [61, 80, 151]. Again PHD inhibition by ROS delivery has been called upon to explain HIF-1 stabilization due to both growth factor signalling and citotoxic drug sensing [207]. ROS also mediate oncogene-driven accumulation of HIF-1 in normoxia, a phenomenon crucial for the establishment of the Warburg metabolism of neoplastic cells, as for H-Ras transformation [207].

p53. p53 is a key redox-sensitive TF, controlling a broad variety cellular functions in response to stress [107]. Upon activation by a multiplicity of stress-related signals (DNA damaging agents, hypoxia, heat shock, etc.), p53 transactivates genes responsible for cell cycle arrest, apoptosis, and DNA repair. ROS may exert their role both upstream or downstream with respect to p53 activation (for a review see [125]).

Oxidation of cysteine residues causes p53 conformational changes and affects its transcriptional activity and biological responses [130]. p53 is glutathionylated on its Cys141 under oxidative stress, with a consequent inhibition of p53 dimerization and association with DNA [202]. Oxidation of p53, in conditions of persistent oxidative stress, results in an intramolecular disulfide, which needs to be re-reduced by TRX to allow DNA-binding activity of p53 and protection from apoptosis [130]. Finally, nitric oxide and peroxinitrite can also modulate the redox status of p53 via nitration of critical tyrosine residues present in its DNA-binding domain, resulting in p53 aggregation into multimers and loss of its DNA-binding ability [37]. Of course these redox modifications shielding of reactive cysteines contributes to a negative regulation for human p53 and may represent an acute defensive response with major consequences for oncogenesis.

Nrf2. In order to prevent oxidative stress, the cell must respond to ROS by mounting an nrf-2-mediated antioxidant response (see above) [96]. This redox-mediated response culminates in activation of transcription of target genes with antioxidants and detoxification properties [204, 220]. In addition to this chief regulatory mechanism, Nrf2 can also be directly targeted by ROS. Indeed, its Cys514 and Cys119 should undergo Ref-1-mediated control, affecting binding to antioxidant responsive elements on DNA and transcriptional activity [15].

To highlight the importance of Nrf2 antioxidant response pathway as a determinant of susceptibility to carcinogenesis, recent studies reported a correlation between Nrf2 disorders and enhanced incidence and tumour burden in several models of cancer, including hepatic carcinoma [155, 189]. Deregulation of expression of Nrf2, as well as missense mutations in Keap1 and Nrf2 genes, has been identified so far, both leading to constitutive activation of Nrf2 and increased antioxidant and detoxification response, with a clear advantage in terms of stress resistance and cell proliferation in normal and cancer cells [90, 155, 189]. Indeed, high expression of Nrf2 target genes grants an advantage to cancer cells for survival against anti-cancer drugs and irradiation [208]. Furthermore, constitutively stabilized Nrf2 promotes cell proliferation, as *NRF2* knockdown inhibits the proliferation of human lung cancer cell lines [182]. Really, the prognoses of patients with Nrf2-positive cancers are significantly poor. It has been reported that the increased activity of Nrf2 in cancers depends on several mechanisms, in particular: (1) somatic mutations in *KEAP1* or *NRF2*, (2) DNA hypermethylation at the promoter region of *KEAP1*, (3) the aberrant accumulation of proteins that disrupt the Keap1–Nrf2 interaction, (4) transcriptional up-regulation of *NRF2* gene through oncogene-dependent signalling, and (5) the modification of Keap1 protein through onco-metabolites [139]. Moreover, recent studies have shown that constitutively high level of Nrf2 promotes not only cancer formation, but also contributes to chemoresistance [106]. Down-regulation of Nrf2 sensitizes cells to chemotherapeutic agents, whereas up-regulation enhances resistance in a variety of cancer cells [208]. In keeping with a role for Nrf2 in chemoresistance, the expression of Nrf2 in cancer cells is increased during acquisition of drug resistance [178].

Chromatin remodelling. Modifications of chromatin remodelling factors like histone acetylases/deacetylases, or directly to DNA, are arising as an additional mechanism of redox-dependent transcriptional regulation. Nitrosylation of histone deacetylase-2 in response to neurotrophin promotes transcription by the cAMP-responsive element-binding factor, through the release of the histone deacetylase from chromatin. This enhances acetylation of histones surrounding neurotrophin-dependent gene promoters and promotes transcription [147]. Furthermore, it has been reported that oxidative stress, due to several inflammatory diseases, causes formation of peroxynitrite, which inactivates histone deacetylase-2 through nitration of critical tyrosine residues [10].

1.4.4 Non-Proteins Substrates

Beside proteins, oxidants may also target DNA or lipids eliciting the formation of molecules endowed with signalling abilities. For example, the formation of 8-oxoguanine upon oxidative injury of DNA has been reported to play a role in transcriptional activation downstream to estrogens receptor and Myc oncogene [2, 154]. Demethylation of H3K4 by lysine-4 demethylase-1 (LSD1) produces hydrogen peroxide, which nearby oxidizes guanine to 8-oxoguanine and induces the recruitment of DNA repair enzymes, as glycosylase-1 and topoisomerase-IIβ, triggering chromatin and DNA conformational changes that are essential for transcription [2, 154]. LSD1 demethylation, DNA oxidation, and chromatin looping have been reported as mandatory mechanisms for transcription driven by estrogens, Myc, and Snail TFs, likely affecting Myc-driven tumorigenesis and tumour invasion associated with Snail-mediated epithelial mesenchymal transition, two hallmarks of cancer aggressiveness [2, 103, 124, 154].

In addition, several lipid peroxides play role as signalling molecules downstream to LOXes or COXes metabolism of arachidonic acid. In addition, non-enzymatic lipid peroxidation alters the non-covalent interactions within the membrane bilayer, contributing to local membrane destabilization [77].

References

1. Akhand AA, Pu M, Senga T, Kato M, Suzuki H, Miyata T, Hamaguchi M, Nakashima I (1999) Nitric oxide controls src kinase activity through a sulfhydryl group modification-mediated Tyr-527-independent and Tyr-416-linked mechanism. J Biol Chem 274:25821–25826
2. Amente S, Bertoni A, Morano A, Lania L, Avvedimento EV, Majello B (2010) LSD1-mediated demethylation of histone H3 lysine 4 triggers Myc-induced transcription. Oncogene 29:3691–3702
3. Anastasiou D, Poulogiannis G, Asara JM, Boxer MB, Jiang JK, Shen M, Bellinger G, Sasaki AT, Locasale JW, Auld DS, Thomas CJ, Vander Heiden MG, Cantley LC (2011) Inhibition of pyruvate kinase M2 by reactive oxygen species contributes to cellular antioxidant responses. Science 334:1278–1283
4. Ascenzi P, di MA, Sciorati C, Clementi E (2010) Peroxynitrite—an ugly biofactor? Biofactors 36:264–273
5. Aubert J, Begriche K, Knockaert L, Robin MA, Fromenty B (2011) Increased expression of cytochrome P450 2E1 in nonalcoholic fatty liver disease: mechanisms and pathophysiological role. Clin Res Hepatol Gastroenterol 35:630–637
6. Babior BM (2002) The leukocyte NADPH oxidase. Isr Med Assoc J 4:1023–1024
7. Bae SH, Sung SH, Cho EJ, Lee SK, Lee HE, Woo HA, Yu DY, Kil IS, Rhee SG (2011) Concerted action of sulfiredoxin and peroxiredoxin I protects against alcohol-induced oxidative injury in mouse liver. Hepatology 53:945–953
8. Banfi B, Malgrange B, Knisz J, Steger K, Dubois-Dauphin M, Krause KH (2004) NOX3, a superoxide-generating NADPH oxidase of the inner ear. J Biol Chem 279:46065–46072
9. Bannister JV, Bannister WH, Rotilio G (1987) Aspects of the structure, function, and applications of superoxide dismutase. CRC Crit Rev Biochem 22:111–180
10. Barnes PJ, Ito K, Adcock IM (2004) Corticosteroid resistance in chronic obstructive pulmonary disease: inactivation of histone deacetylase. Lancet 363:731–733
11. Berndt C, Lillig CH, Holmgren A (2007) Thiol-based mechanisms of the thioredoxin and glutaredoxin systems: implications for diseases in the cardiovascular system. Am J Physiol Heart Circ Physiol 292:H1227–H1236
12. Bertini R, Howard OM, Dong HF, Oppenheim JJ, Bizzarri C, Sergi R, Caselli G, Pagliei S, Romines B, Wilshire JA, Mengozzi M, Nakamura H, Yodoi J, Pekkari K, Gurunath R, Holmgren A, Herzenberg LA, Herzenberg LA, Ghezzi P (1999) Thioredoxin, a redox enzyme released in infection and inflammation, is a unique chemoattractant for neutrophils, monocytes, and T cells. J Exp Med 189:1783–1789
13. Betteridge DJ (2000) What is oxidative stress? Metabolism 49:3–8
14. Bienert GP, Schjoerring JK, Jahn TP (2006) Membrane transport of hydrogen peroxide. Biochim Biophys Acta 1758:994–1003
15. Bloom D, Dhakshinamoorthy S, Jaiswal AK (2002) Site-directed mutagenesis of cysteine to serine in the DNA binding region of Nrf2 decreases its capacity to upregulate antioxidant response element-mediated expression and antioxidant induction of NAD(P)H:quinone oxidoreductase1 gene. Oncogene 21:2191–2200
16. Boivin B, Yang M, Tonks NK (2010) Targeting the reversibly oxidized protein tyrosine phosphatase superfamily. Sci Signal 3:l2
17. Brandes RP, Takac I, Schroder K (2011) No superoxide–no stress?: Nox4, the good NADPH oxidase! Arterioscler Thromb Vasc Biol 31:1255–1257
18. Brunet A, Bonni A, Zigmond MJ, Lin MZ, Juo P, Hu LS, Anderson MJ, Arden KC, Blenis J, Greenberg ME (1999) Akt promotes cell survival by phosphorylating and inhibiting a Forkhead transcription factor. Cell 96:857–868
19. Buricchi F, Giannoni E, Grimaldi G, Parri M, Raugei G, Ramponi G, Chiarugi P (2007) Redox regulation of ephrin/integrin cross-talk. Cell Adh Migr 1:33–42
20. Butterfield DA, Hardas SS, Lange ML (2010) Oxidatively modified glyceraldehyde-3-phosphate dehydrogenase (GAPDH) and Alzheimer's disease: many pathways to neurodegeneration. J Alzheimers Dis 20:369–393

21. Cantu-Medellin N, Kelley EE (2013) Xanthine oxidoreductase-catalyzed reactive species generation: a process in critical need of reevaluation. Redox Biol 1:353–358
22. Cao J, Schulte J, Knight A, Leslie NR, Zagozdzon A, Bronson R, Manevich Y, Beeson C, Neumann CA (2009) Prdx1 inhibits tumorigenesis via regulating PTEN/AKT activity. EMBO J 28:1505–1517
23. Cardaci S, Filomeni G, Ciriolo MR (2012) Redox implications of AMPK-mediated signal transduction beyond energetic clues. J Cell Sci 125:2115–2125
24. Casagrande S, Bonetto V, Fratelli M, Gianazza E, Eberini I, Massignan T, Salmona M, Chang G, Holmgren A, Ghezzi P (2002) Glutathionylation of human thioredoxin: a possible crosstalk between the glutathione and thioredoxin systems. Proc Natl Acad Sci U S A 99:9745–9749
25. Chandel NS (2010) Mitochondrial regulation of oxygen sensing. Adv Exp Med Biol 661:339–354
26. Chang TS, Jeong W, Woo HA, Lee SM, Park S, Rhee SG (2004) Characterization of mammalian sulfiredoxin and its reactivation of hyperoxidized peroxiredoxin through reduction of cysteine sulfinic acid in the active site to cysteine. J Biol Chem 279:50994–51001
27. Chen F, Haigh S, Barman S, Fulton DJ (2012) From form to function: the role of Nox4 in the cardiovascular system. Front Physiol 3:412
28. Chen K, Kirber MT, Xiao H, Yang Y, Keaney JF Jr (2008) Regulation of ROS signal transduction by NADPH oxidase 4 localization. J Cell Biol 181:1129–1139
29. Chiarugi P (2008) From anchorage dependent proliferation to survival: lessons from redox signalling. IUBMB Life 60:301–307
30. Chiarugi P (2009) Survival or death: the redox paradox. Antioxid Redox Signal 11:2651–2654
31. Chiarugi P, Buricchi F (2007) Protein tyrosine phosphorylation and reversible oxidation: two cross-talking posttranslation modifications. Antioxid Redox Signal 9:1–24
32. Chiarugi P, Cirri P, Taddei L, Giannoni E, Camici G, Manao G, Raugei G, Ramponi G (2000) The low M(r) protein-tyrosine phosphatase is involved in Rho-mediated cytoskeleton rearrangement after integrin and platelet-derived growth factor stimulation. J Biol Chem 275:4640–4646
33. Chiarugi P, Fiaschi T, Taddei ML, Talini D, Giannoni E, Raugei G, Ramponi G (2001) Two vicinal cysteines confer a peculiar redox regulation to low molecular weight protein tyrosine phosphatase in response to platelet-derived growth factor receptor stimulation. J Biol Chem 276:33478–33487
34. Chiarugi P, Pani G, Giannoni E, Taddei L, Colavitti R, Raugei G, Symons M, Borrello S, Galeotti T, Ramponi G (2003) Reactive oxygen species as essential mediators of cell adhesion: the oxidative inhibition of a FAK tyrosine phosphatase is required for cell adhesion. J Cell Biol 161:933–944
35. Choi MH, Lee IK, Kim GW, Kim BU, Han YH, Yu DY, Park HS, Kim KY, Lee JS, Choi C, Bae YS, Lee BI, Rhee SG, Kang SW (2005) Regulation of PDGF signalling and vascular remodelling by peroxiredoxin II. Nature 435:347–353
36. Christofk HR, Vander Heiden MG, Harris MH, Ramanathan A, Gerszten RE, Wei R, Fleming MD, Schreiber SL, Cantley LC (2008) The M2 splice isoform of pyruvate kinase is important for cancer metabolism and tumour growth. Nature 452:230–233
37. Cobbs CS, Samanta M, Harkins LE, Gillespie GY, Merrick BA, MacMillan-Crow LA (2001) Evidence for peroxynitrite-mediated modifications to p53 in human gliomas: possible functional consequences. Arch Biochem Biophys 394:167–172
38. Cohen JH, Kristal AR, Stanford JL (2000) Fruit and vegetable intakes and prostate cancer risk. J Natl Cancer Inst 92:61–68
39. Corcoran A, Cotter TG (2013) Redox regulation of protein kinases. FEBS J 280:1944–1965
40. Cosentino C, Grieco D, Costanzo V (2011) ATM activates the pentose phosphate pathway promoting anti-oxidant defence and DNA repair. EMBO J 30:546–555
41. Dansen TB, Smits LM, van Triest MH, de Keizer PL, Van LD, Koerkamp MG, Szypowska A, Meppelink A, Brenkman AB, Yodoi J, Holstege FC, Burgering BM (2009) Redox-sensitive cysteines bridge p300/CBP-mediated acetylation and FoxO4 activity. Nat Chem Biol 5:664–672

42. de Keizer PL, Burgering BM, Dansen TB (2011) Forkhead box o as a sensor, mediator, and regulator of redox signaling. Antioxid Redox Signal 14:1093–1106
43. Decker EA, Livisay SA, Zhou S (2000) A re-evaluation of the antioxidant activity of purified carnosine. Biochemistry (Mosc) 65:766–770
44. Dinkova-Kostova AT, Holtzclaw WD, Cole RN, Itoh K, Wakabayashi N, Katoh Y, Yamamoto M, Talalay P (2002) Direct evidence that sulfhydryl groups of Keap1 are the sensors regulating induction of phase 2 enzymes that protect against carcinogens and oxidants. Proc Natl Acad Sci U S A 99:11908–11913
45. Drahota Z, Chowdhury SK, Floryk D, Mracek T, Wilhelm J, Rauchova H, Lenaz G, Houstek J (2002) Glycerophosphate-dependent hydrogen peroxide production by brown adipose tissue mitochondria and its activation by ferricyanide. J Bioenerg Biomembr 34:105–113
46. Emerling BM, Weinberg F, Snyder C, Burgess Z, Mutlu GM, Viollet B, Budinger GR, Chandel NS (2009) Hypoxic activation of AMPK is dependent on mitochondrial ROS but independent of an increase in AMP/ATP ratio. Free Radic Biol Med 46:1386–1391
47. Erez N, Truitt M, Olson P, Arron ST, Hanahan D (2010) Cancer-associated fibroblasts are activated in incipient neoplasia to orchestrate tumor-promoting inflammation in an NF-kappaB-dependent manner. Cancer Cell 17:135–147
48. Essers MA, Weijzen S, de Vries-Smits AM, Saarloos I, de Ruiter ND, Bos JL, Burgering BM (2004) FOXO transcription factor activation by oxidative stress mediated by the small GTPase Ral and JNK. EMBO J 23:4802–4812
49. Etienne-Manneville S, Hall A (2002) Rho GTPases in cell biology. Nature 420:629–635
50. Fang YZ, Yang S, Wu G (2002) Free radicals, antioxidants, and nutrition. Nutrition 18:872–879
51. Farah ME, Amberg DC (2007) Conserved actin cysteine residues are oxidative stress sensors that can regulate cell death in yeast. Mol Biol Cell 18:1359–1365
52. Fiaschi T, Cozzi G, Chiarugi P (2012) Redox regulation of nonmuscle myosin heavy chain during integrin engagement. J Signal Transduct 2012:754964
53. Fiaschi T, Cozzi G, Raugei G, Formigli L, Ramponi G, Chiarugi P (2006) Redox regulation of beta-actin during integrin-mediated cell adhesion. J Biol Chem 281:22983–22991
54. Filomeni G, Desideri E, Cardaci S, Graziani I, Piccirillo S, Rotilio G, Ciriolo MR (2010) Carcinoma cells activate AMP-activated protein kinase-dependent autophagy as survival response to kaempferol-mediated energetic impairment. Autophagy 6:202–216
55. Findlay VJ, Townsend DM, Morris TE, Fraser JP, He L, Tew KD (2006) A novel role for human sulfiredoxin in the reversal of glutathionylation. Cancer Res 66:6800–6806
56. Fransen M, Nordgren M, Wang B, Apanasets O (2012) Role of peroxisomes in ROS/RNS-metabolism: implications for human disease. Biochim Biophys Acta 1822:1363–1373
57. Fratelli M, Demol H, Puype M, Casagrande S, Eberini I, Salmona M, Bonetto V, Mengozzi M, Duffieux F, Miclet E, Bachi A, Vandekerckhove J, Gianazza E, Ghezzi P (2002) Identification by redox proteomics of glutathionylated proteins in oxidatively stressed human T lymphocytes. Proc Natl Acad Sci U S A 99:3505–3510
58. Fukai T, Ushio-Fukai M (2011) Superoxide dismutases: role in redox signaling, vascular function, and diseases. Antioxid Redox Signal 15:1583–1606
59. Gao P, Zhang H, Dinavahi R, Li F, Xiang Y, Raman V, Bhujwalla ZM, Felsher DW, Cheng L, Pevsner J, Lee LA, Semenza GL, Dang CV (2007) HIF-dependent antitumorigenic effect of antioxidants in vivo. Cancer Cell 12:230–238
60. Gerald D, Berra E, Frapart YM, Chan DA, Giaccia AJ, Mansuy D, Pouyssegur J, Yaniv M, Mechta-Grigoriou F (2004) JunD reduces tumor angiogenesis by protecting cells from oxidative stress. Cell 118:781–794
61. Giannoni E, Bianchini F, Masieri L, Serni S, Torre E, Calorini L, Chiarugi P (2010) Reciprocal activation of prostate cancer cells and cancer-associated fibroblasts stimulates epithelial-mesenchymal transition and cancer stemness. Cancer Res 70:6945–6956
62. Giannoni E, Buricchi F, Grimaldi G, Parri M, Cialdai F, Taddei ML, Raugei G, Ramponi G, Chiarugi P (2008) Redox regulation of anoikis: reactive oxygen species as essential mediators of cell survival. Cell Death Differ 15:867–878

63. Giannoni E, Buricchi F, Raugei G, Ramponi G, Chiarugi P (2005) Intracellular reactive oxygen species activate Src tyrosine kinase during cell adhesion and anchorage-dependent cell growth. Mol Cell Biol 25:6391–6403
64. Giannoni E, Fiaschi T, Ramponi G, Chiarugi P (2009) Redox regulation of anoikis resistance of metastatic prostate cancer cells: key role for Src and EGFR-mediated pro-survival signals. Oncogene 28:2074–2086
65. Giannoni E, Taddei ML, Chiarugi P (2010) Src redox regulation: again in the front line. Free Radic Biol Med 49:516–527
66. Girotti AW, Korytowski W (2000) Cholesterol as a singlet oxygen detector in biological systems. Methods Enzymol 319:85–100
67. Giustarini D, Milzani A, Aldini G, Carini M, Rossi R, Dalle-Donne I (2005) S-nitrosation versus S-glutathionylation of protein sulfhydryl groups by S-nitrosoglutathione. Antioxid Redox Signal 7:930–939
68. Gloire G, Piette J (2009) Redox regulation of nuclear post-translational modifications during NF-kappaB activation. Antioxid Redox Signal 11:2209–2222
69. Gong P, Cederbaum AI (2006) Nrf2 is increased by CYP2E1 in rodent liver and HepG2 cells and protects against oxidative stress caused by CYP2E1. Hepatology 43:144–153
70. Grant CM, Quinn KA, Dawes IW (1999) Differential protein S-thiolation of glyceraldehyde-3-phosphate dehydrogenase isoenzymes influences sensitivity to oxidative stress. Mol Cell Biol 19:2650–2656
71. Grimsrud PA, Xie H, Griffin TJ, Bernlohr DA (2008) Oxidative stress and covalent modification of protein with bioactive aldehydes. J Biol Chem 283:21837–21841
72. Gross S, Knebel A, Tenev T, Neininger A, Gaestel M, Herrlich P, Bohmer FD (1999) Inactivation of protein-tyrosine phosphatases as mechanism of UV-induced signal transduction. J Biol Chem 274:26378–26386
73. Gruning NM, Lehrach H, Ralser M (2010) Regulatory crosstalk of the metabolic network. Trends Biochem Sci 35:220–227
74. Gruning NM, Ralser M (2011) Cancer: sacrifice for survival. Nature 480:190–191
75. Gruning NM, Rinnerthaler M, Bluemlein K, Mulleder M, Wamelink MM, Lehrach H, Jakobs C, Breitenbach M, Ralser M (2011) Pyruvate kinase triggers a metabolic feedback loop that controls redox metabolism in respiring cells. Cell Metab 14:415–427
76. Gu Z, Kaul M, Yan B, Kridel SJ, Cui J, Strongin A, Smith JW, Liddington RC, Lipton SA (2002) S-nitrosylation of matrix metalloproteinases: signaling pathway to neuronal cell death. Science 297:1186–1190
77. Gueraud F, Atalay M, Bresgen N, Cipak A, Eckl PM, Huc L, Jouanin I, Siems W, Uchida K (2010) Chemistry and biochemistry of lipid peroxidation products. Free Radic Res 44:1098–1124
78. Guo Z, Kozlov S, Lavin MF, Person MD, Paull TT (2010) ATM activation by oxidative stress. Science 330:517–521
79. Guzy RD, Hoyos B, Robin E, Chen H, Liu L, Mansfield KD, Simon MC, Hammerling U, Schumacker PT (2005) Mitochondrial complex III is required for hypoxia-induced ROS production and cellular oxygen sensing. Cell Metab 1:401–408
80. Haddad JJ, Land SC (2001) A non-hypoxic, ROS-sensitive pathway mediates TNF-alpha-dependent regulation of HIF-1alpha. FEBS Lett 505:269–274
81. Haendeler J, Hoffmann J, Tischler V, Berk BC, Zeiher AM, Dimmeler S (2002) Redox regulatory and anti-apoptotic functions of thioredoxin depend on S-nitrosylation at cysteine 69. Nat Cell Biol 4:743–749
82. Halliwell B (2007) Biochemistry of oxidative stress. Biochem Soc Trans 35:1147–1150
83. Halliwell B, Gutteridge JM (1995) The definition and measurement of antioxidants in biological systems. Free Radic Biol Med 18:125–126
84. Hamanaka RB, Chandel NS (2010) Mitochondrial reactive oxygen species regulate cellular signaling and dictate biological outcomes. Trends Biochem Sci 35:505–513
85. Hardie DG (2007) Biochemistry. Balancing cellular energy. Science 315:1671–1672

86. Harhaji-Trajkovic L, Vilimanovich U, Kravic-Stevovic T, Bumbasirevic V, Trajkovic V (2009) AMPK-mediated autophagy inhibits apoptosis in cisplatin-treated tumour cells. J Cell Mol Med 13:3644–3654
87. Harris CM, Massey V (1997) The oxidative half-reaction of xanthine dehydrogenase with NAD; reaction kinetics and steady-state mechanism. J Biol Chem 272:28335–28341
88. Hashemy SI, Johansson C, Berndt C, Lillig CH, Holmgren A (2007) Oxidation and S-nitrosylation of cysteines in human cytosolic and mitochondrial glutaredoxins: effects on structure and activity. J Biol Chem 282:14428–14436
89. Hayes JD, Flanagan JU, Jowsey IR (2005) Glutathione transferases. Annu Rev Pharmacol Toxicol 45:51–88
90. Hayes JD, McMahon M (2009) NRF2 and KEAP1 mutations: permanent activation of an adaptive response in cancer. Trends Biochem Sci 34:176–188
91. Herscovitch M, Comb W, Ennis T, Coleman K, Yong S, Armstead B, Kalaitzidis D, Chandani S, Gilmore TD (2008) Intermolecular disulfide bond formation in the NEMO dimer requires Cys54 and Cys347. Biochem Biophys Res Commun 367:103–108
92. Hess DT, Matsumoto A, Kim SO, Marshall HE, Stamler JS (2005) Protein S-nitrosylation: purview and parameters. Nat Rev Mol Cell Biol 6:150–166
93. Hou H, Yu H (2010) Structural insights into histone lysine demethylation. Curr Opin Struct Biol 20:739–748
94. Hurley PJ, Bunz F (2007) ATM and ATR: components of an integrated circuit. Cell Cycle 6:414–417
95. Hussain T, Gupta S, Mukhtar H (2003) Cyclooxygenase-2 and prostate carcinogenesis. Cancer Lett 191:125–135
96. Itoh K, Wakabayashi N, Katoh Y, Ishii T, Igarashi K, Engel JD, Yamamoto M (1999) Keap1 represses nuclear activation of antioxidant responsive elements by Nrf2 through binding to the amino-terminal Neh2 domain. Genes Dev 13:76–86
97. Janssen-Heininger YM, Mossman BT, Heintz NH, Forman HJ, Kalyanaraman B, Finkel T, Stamler JS, Rhee SG, van der Vliet A (2008) Redox-based regulation of signal transduction: principles, pitfalls, and promises. Free Radic Biol Med 45:1–17
98. Jenkins JL, Tanner JJ (2006) High-resolution structure of human D-glyceraldehyde-3-phosphate dehydrogenase. Acta Crystallogr D Biol Crystallogr 62:290–301
99. Jeong W, Park SJ, Chang TS, Lee DY, Rhee SG (2006) Molecular mechanism of the reduction of cysteine sulfinic acid of peroxiredoxin to cysteine by mammalian sulfiredoxin. J Biol Chem 281:14400–14407
100. Jiang P, Du W, Wang X, Mancuso A, Gao X, Wu M, Yang X (2011) p53 regulates biosynthesis through direct inactivation of glucose-6-phosphate dehydrogenase. Nat Cell Biol 13:310–316
101. Kabe Y, Ando K, Hirao S, Yoshida M, Handa H (2005) Redox regulation of NF-kappaB activation: distinct redox regulation between the cytoplasm and the nucleus. Antioxid Redox Signal 7:395–403
102. Kaelin WG Jr, Ratcliffe PJ (2008) Oxygen sensing by metazoans: the central role of the HIF hydroxylase pathway. Mol Cell 30:393–402
103. Kalluri R (2009) EMT: when epithelial cells decide to become mesenchymal-like cells. J Clin Invest 119:1417–1419
104. Kato M, Iwashita T, Takeda K, Akhand AA, Liu W, Yoshihara M, Asai N, Suzuki H, Takahashi M, Nakashima I (2000) Ultraviolet light induces redox reaction-mediated dimerization and superactivation of oncogenic Ret tyrosine kinases. Mol Biol Cell 11:93–101
105. Keilin D, Hartree EF (1945) Properties of catalase. Catalysis of coupled oxidation of alcohols. Biochem J 39:293–301
106. Kensler TW, Wakabayashi N (2010) Nrf2: friend or foe for chemoprevention? Carcinogenesis 31:90–99
107. Kim DH, Kundu JK, Surh YJ (2011) Redox modulation of p53: mechanisms and functional significance. Mol Carcinog 50:222–234

108. Kim I, Han SJ, Kim Y, Ahn Y, Chay KO, Lee SR (2011) Tyr740 and Tyr751 residues of platelet-derived growth factor beta receptor are responsible for the redox regulation of phosphatase and tensin homolog in the cells stimulated with platelet-derived growth factor. Redox Rep 16:181–186
109. Kobayashi A, Kang MI, Okawa H, Ohtsuji M, Zenke Y, Chiba T, Igarashi K, Yamamoto M (2004) Oxidative stress sensor Keap1 functions as an adaptor for Cul3-based E3 ligase to regulate proteasomal degradation of Nrf2. Mol Cell Biol 24:7130–7139
110. Kruger A, Gruning NM, Wamelink MM, Kerick M, Kirpy A, Parkhomchuk D, Bluemlein K, Schweiger MR, Soldatov A, Lehrach H, Jakobs C, Ralser M (2011) The pentose phosphate pathway is a metabolic redox sensor and regulates transcription during the antioxidant response. Antioxid Redox Signal 15:311–324
111. Kushnareva Y, Murphy AN, Andreyev A (2002) Complex I-mediated reactive oxygen species generation: modulation by cytochrome c and NAD(P)+ oxidation–reduction state. Biochem J 368:545–553
112. Lamle J, Marhenke S, Borlak J, von WR, Eriksson CJ, Geffers R, Borlak J, Manns MP, Yamamoto M, Vogel A (2008) Nuclear factor-eythroid 2-related factor 2 prevents alcohol-induced fulminant liver injury. Gastroenterology 134:1159–1168
113. Lee BC, Gladyshev VN (2011) The biological significance of methionine sulfoxide stereochemistry. Free Radic Biol Med 50:221–227
114. Lee JW, Kim JE, Park EJ, Kim JH, Lee CH, Lee SR, Kwon J (2004) Two conserved cysteine residues are critical for the enzymic function of the human platelet-derived growth factor receptor-beta: evidence for different roles of Cys-822 and Cys-940 in the kinase activity. Biochem J 382:631–639
115. Lee YJ, Cho HN, Soh JW, Jhon GJ, Cho CK, Chung HY, Bae S, Lee SJ, Lee YS (2003) Oxidative stress-induced apoptosis is mediated by ERK1/2 phosphorylation. Exp Cell Res 291:251–266
116. Lehtinen MK, Yuan Z, Boag PR, Yang Y, Villen J, Becker EB, DiBacco S, de la Iglesia N, Gygi S, Blackwell TK, Bonni A (2006) A conserved MST-FOXO signaling pathway mediates oxidative-stress responses and extends life span. Cell 125:987–1001
117. Leslie NR, Batty IH, Maccario H, Davidson L, Downes CP (2008) Understanding PTEN regulation: PIP2, polarity and protein stability. Oncogene 27:5464–5476
118. Leslie NR, Bennett D, Lindsay YE, Stewart H, Gray A, Downes CP (2003) Redox regulation of PI 3-kinase signalling via inactivation of PTEN. EMBO J 22:5501–5510
119. Leto TL, Morand S, Hurt D, Ueyama T (2009) Targeting and regulation of reactive oxygen species generation by Nox family NADPH oxidases. Antioxid Redox Signal 11:2607–2619
120. Leu TH, Su SL, Chuang YC, Maa MC (2003) Direct inhibitory effect of curcumin on Src and focal adhesion kinase activity. Biochem Pharmacol 66:2323–2331
121. Leung TM, Nieto N (2013) CYP2E1 and oxidant stress in alcoholic and non-alcoholic fatty liver disease. J Hepatol 58:395–398
122. Lieber CS (1997) Cytochrome P-4502E1: its physiological and pathological role. Physiol Rev 77:517–544
123. Lien EJ, Ren S, Bui HH, Wang R (1999) Quantitative structure–activity relationship analysis of phenolic antioxidants. Free Radic Biol Med 26:285–294
124. Lin T, Ponn A, Hu X, Law BK, Lu J (2010) Requirement of the histone demethylase LSD1 in Snai1-mediated transcriptional repression during epithelial-mesenchymal transition. Oncogene 29:4896–4904
125. Liu B, Chen Y, St Clair DK (2008) ROS and p53: a versatile partnership. Free Radic Biol Med 44:1529–1535
126. Liu H, Colavitti R, Rovira II, Finkel T (2005) Redox-dependent transcriptional regulation. Circ Res 97:967–974
127. Lluis JM, Buricchi F, Chiarugi P, Morales A, Fernandez-Checa JC (2007) Dual role of mitochondrial reactive oxygen species in hypoxia signaling: activation of nuclear factor-{kappa} B via c-SRC and oxidant-dependent cell death. Cancer Res 67:7368–7377

128. Loughran PA, Stolz DB, Vodovotz Y, Watkins SC, Simmons RL, Billiar TR (2005) Monomeric inducible nitric oxide synthase localizes to peroxisomes in hepatocytes. Proc Natl Acad Sci U S A 102:13837–13842
129. Lv L, Li D, Zhao D, Lin R, Chu Y, Zhang H, Zha Z, Liu Y, Li Z, Xu Y, Wang G, Huang Y, Xiong Y, Guan KL, Lei QY (2011) Acetylation targets the M2 isoform of pyruvate kinase for degradation through chaperone-mediated autophagy and promotes tumor growth. Mol Cell 42:719–730
130. Maillet A, Pervaiz S (2012) Redox regulation of p53, redox effectors regulated by p53: a subtle balance. Antioxid Redox Signal 16:1285–1294
131. Mannervik B (1987) The enzymes of glutathione metabolism: an overview. Biochem Soc Trans 15:717–718
132. Mannick JB, Hausladen A, Liu L, Hess DT, Zeng M, Miao QX, Kane LS, Gow AJ, Stamler JS (1999) Fas-induced caspase denitrosylation. Science 284:651–654
133. Masson N, Ratcliffe PJ (2003) HIF prolyl and asparaginyl hydroxylases in the biological response to intracellular O(2) levels. J Cell Sci 116:3041–3049
134. Matthews JR, Kaszubska W, Turcatti G, Wells TN, Hay RT (1993) Role of cysteine62 in DNA recognition by the P50 subunit of NF-kappa B. Nucleic Acids Res 21:1727–1734
135. Maurel A, Hernandez C, Kunduzova O, Bompart G, Cambon C, Parini A, Frances B (2003) Age-dependent increase in hydrogen peroxide production by cardiac monoamine oxidase A in rats. Am J Physiol Heart Circ Physiol 284:H1460–H1467
136. McIntyre M, Bohr DF, Dominiczak AF (1999) Endothelial function in hypertension: the role of superoxide anion. Hypertension 34:539–545
137. Melvin A, Rocha S (2012) Chromatin as an oxygen sensor and active player in the hypoxia response. Cell Signal 24:35–43
138. Minetti M, Mallozzi C, Di Stasi AM (2002) Peroxynitrite activates kinases of the src family and upregulates tyrosine phosphorylation signaling. Free Radic Biol Med 33:744–754
139. Mitsuishi Y, Motohashi H, Yamamoto M (2012) The Keap1-Nrf2 system in cancers: stress response and anabolic metabolism. Front Oncol 2:200
140. Moro L, Dolce L, Cabodi S, Bergatto E, Boeri EE, Smeriglio M, Turco E, Retta SF, Giuffrida MG, Venturino M, Godovac-Zimmermann J, Conti A, Schaefer E, Beguinot L, Tacchetti C, Gaggini P, Silengo L, Tarone G, Defilippi P (2002) Integrin-induced epidermal growth factor (EGF) receptor activation requires c-Src and p130Cas and leads to phosphorylation of specific EGF receptor tyrosines. J Biol Chem 277:9405–9414
141. Murphy MP (2009) How mitochondria produce reactive oxygen species. Biochem J 417:1–13
142. Myatt SS, Brosens JJ, Lam EW (2011) Sense and sensitivity: FOXO and ROS in cancer development and treatment. Antioxid Redox Signal 14:675–687
143. Nadeau PJ, Charette SJ, Toledano MB, Landry J (2007) Disulfide bond-mediated multimerization of Ask1 and its reduction by thioredoxin-1 regulate H(2)O(2)-induced c-Jun NH(2)-terminal kinase activation and apoptosis. Mol Biol Cell 18:3903–3913
144. Ndhlala AR, Moyo M, Van SJ (2010) Natural antioxidants: fascinating or mythical biomolecules? Molecules 15:6905–6930
145. Nikitovic D, Holmgren A (1996) S-nitrosoglutathione is cleaved by the thioredoxin system with liberation of glutathione and redox regulating nitric oxide. J Biol Chem 271:19180–19185
146. Nishino T, Okamoto K (2000) The role of the [2Fe-2s] cluster centers in xanthine oxidoreductase. J Inorg Biochem 82:43–49
147. Nott A, Watson PM, Robinson JD, Crepaldi L, Riccio A (2008) S-Nitrosylation of histone deacetylase 2 induces chromatin remodelling in neurons. Nature 455:411–415
148. Okuda M, Inoue N, Azumi H, Seno T, Sumi Y, Hirata K, Kawashima S, Hayashi Y, Itoh H, Yodoi J, Yokoyama M (2001) Expression of glutaredoxin in human coronary arteries: its potential role in antioxidant protection against atherosclerosis. Arterioscler Thromb Vasc Biol 21:1483–1487
149. Paravicini TM, Touyz RM (2006) Redox signaling in hypertension. Cardiovasc Res 71:247–258

150. Parri M, Chiarugi P (2013) Redox molecular machines involved in tumor progression. Antioxid Redox Signal 19:1828–1845
151. Patten DA, Lafleur VN, Robitaille GA, Chan DA, Giaccia AJ, Richard DE (2010) Hypoxia-inducible factor-1 activation in nonhypoxic conditions: the essential role of mitochondrial-derived reactive oxygen species. Mol Biol Cell 21:3247–3257
152. Pekkari K, Goodarzi MT, Scheynius A, Holmgren A, Avila-Carino J (2005) Truncated thioredoxin (Trx80) induces differentiation of human CD14+ monocytes into a novel cell type (TAMs) via activation of the MAP kinases p38, ERK, and JNK. Blood 105:1598–1605
153. Pekkari K, Holmgren A (2004) Truncated thioredoxin: physiological functions and mechanism. Antioxid Redox Signal 6:53–61
154. Perillo B, Ombra MN, Bertoni A, Cuozzo C, Sacchetti S, Sasso A, Chiariotti L, Malorni A, Abbondanza C, Avvedimento EV (2008) DNA oxidation as triggered by H3K9me2 demethylation drives estrogen-induced gene expression. Science 319:202–206
155. Petrelli A, Perra A, Cora D, Sulas P, Menegon S, Manca C, Migliore C, Kowalik MA, Ledda-Columbano GM, Giordano S, Columbano A (2014) MicroRNA/gene profiling unveils early molecular changes and nuclear factor erythroid related factor 2 (NRF2) activation in a rat model recapitulating human hepatocellular carcinoma (HCC). Hepatology 59:228–241
156. Phalen TJ, Weirather K, Deming PB, Anathy V, Howe AK, van der Vliet A, Jonsson TJ, Poole LB, Heintz NH (2006) Oxidation state governs structural transitions in peroxiredoxin II that correlate with cell cycle arrest and recovery. J Cell Biol 175:779–789
157. Pineda-Molina E, Klatt P, Vazquez J, Marina A, Garcia de Lacoba M, Perez-Sala D, Lamas S (2001) Glutathionylation of the p50 subunit of NF-kappaB: a mechanism for redox-induced inhibition of DNA binding. Biochemistry 40:14134–14142
158. Poels J, Spasic MR, Callaerts P, Norga KK (2009) Expanding roles for AMP-activated protein kinase in neuronal survival and autophagy. Bioessays 31:944–952
159. Polimeni M, Voena C, Kopecka J, Riganti C, Pescarmona G, Bosia A, Ghigo D (2011) Modulation of doxorubicin resistance by the glucose-6-phosphate dehydrogenase activity. Biochem J 439:141–149
160. Pompella A, Visvikis A, Paolicchi A, De Tata T, Casini AF (2003) The changing faces of glutathione, a cellular protagonist. Biochem Pharmacol 66:1499–1503
161. Pyne NJ, Pyne S (2011) Receptor tyrosine kinase-G-protein-coupled receptor signalling platforms: out of the shadow? Trends Pharmacol Sci 32:443–450
162. Ralser M, Wamelink MM, Kowald A, Gerisch B, Heeren G, Struys EA, Klipp E, Jakobs C, Breitenbach M, Lehrach H, Krobitsch S (2007) Dynamic rerouting of the carbohydrate flux is key to counteracting oxidative stress. J Biol 6:10
163. Ray PD, Huang BW, Tsuji Y (2012) Reactive oxygen species (ROS) homeostasis and redox regulation in cellular signaling. Cell Signal 24:981–990
164. Reynaert NL, Wouters EF, Janssen-Heininger YM (2007) Modulation of glutaredoxin-1 expression in a mouse model of allergic airway disease. Am J Respir Cell Mol Biol 36:147–151
165. Reynolds AR, Tischer C, Verveer PJ, Rocks O, Bastiaens PI (2003) EGFR activation coupled to inhibition of tyrosine phosphatases causes lateral signal propagation. Nat Cell Biol 5:447–453
166. Rhee SG, Bae YS, Lee SR, Kwon J (2000) Hydrogen peroxide: a key messenger that modulates protein phosphorylation through cysteine oxidation. Sci STKE 2000:e1
167. Rhee SG, Kang SW, Jeong W, Chang TS, Yang KS, Woo HA (2005) Intracellular messenger function of hydrogen peroxide and its regulation by peroxiredoxins. Curr Opin Cell Biol 17:183–189
168. Rhee SG, Woo HA, Kil IS, Bae SH (2012) Peroxiredoxin functions as a peroxidase and a regulator and sensor of local peroxides. J Biol Chem 287:4403–4410
169. Riganti C, Gazzano E, Polimeni M, Aldieri E, Ghigo D (2012) The pentose phosphate pathway: an antioxidant defense and a crossroad in tumor cell fate. Free Radic Biol Med 53(3):421–436
170. Sanchez R, Riddle M, Woo J, Momand J (2008) Prediction of reversibly oxidized protein cysteine thiols using protein structure properties. Protein Sci 17:473–481

171. Sandin A, Dagnell M, Gonon A, Pernow J, Stangl V, Aspenstrom P, Kappert K, Ostman A (2011) Hypoxia followed by re-oxygenation induces oxidation of tyrosine phosphatases. Cell Signal 23:820–826
172. Schmid E, Hotz-Wagenblatt A, Hacj V, Droge W (1999) Phosphorylation of the insulin receptor kinase by phosphocreatine in combination with hydrogen peroxide: the structural basis of redox priming. FASEB J 13:1491–1500
173. Schroedl C, McClintock DS, Budinger GR, Chandel NS (2002) Hypoxic but not anoxic stabilization of HIF-1alpha requires mitochondrial reactive oxygen species. Am J Physiol Lung Cell Mol Physiol 283:L922–L931
174. Semenza GL (2011) Hypoxia. Cross talk between oxygen sensing and the cell cycle machinery. Am J Physiol Cell Physiol 301:C550–C552
175. Semenza GL (2012) Hypoxia-inducible factors: mediators of cancer progression and targets for cancer therapy. Trends Pharmacol Sci 33:207–214
176. Shelton MD, Chock PB, Mieyal JJ (2005) Glutaredoxin: role in reversible protein s-glutathionylation and regulation of redox signal transduction and protein translocation. Antioxid Redox Signal 7:348–366
177. Shiloh Y (2003) ATM and related protein kinases: safeguarding genome integrity. Nat Rev Cancer 3:155–168
178. Shim GS, Manandhar S, Shin DH, Kim TH, Kwak MK (2009) Acquisition of doxorubicin resistance in ovarian carcinoma cells accompanies activation of the NRF2 pathway. Free Radic Biol Med 47:1619–1631
179. Sid B, Verrax J, Calderon PB (2013) Role of AMPK activation in oxidative cell damage: implications for alcohol-induced liver disease. Biochem Pharmacol 86:200–209
180. Silva A, Yunes JA, Cardoso BA, Martins LR, Jotta PY, Abecasis M, Nowill AE, Leslie NR, Cardoso AA, Barata JT (2008) PTEN posttranslational inactivation and hyperactivation of the PI3K/Akt pathway sustain primary T cell leukemia viability. J Clin Invest 118: 3762–3774
181. Simon HU, Haj-Yehia A, Levi-Schaffer F (2000) Role of reactive oxygen species (ROS) in apoptosis induction. Apoptosis 5:415–418
182. Singh A, Boldin-Adamsky S, Thimmulappa RK, Rath SK, Ashush H, Coulter J, Blackford A, Goodman SN, Bunz F, Watson WH, Gabrielson E, Feinstein E, Biswal S (2008) RNAi-mediated silencing of nuclear factor erythroid-2-related factor 2 gene expression in non-small cell lung cancer inhibits tumor growth and increases efficacy of chemotherapy. Cancer Res 68:7975–7984
183. Singh SP, Wishnok JS, Keshive M, Deen WM, Tannenbaum SR (1996) The chemistry of the S-nitrosoglutathione/glutathione system. Proc Natl Acad Sci U S A 93:14428–14433
184. St-Pierre J, Buckingham JA, Roebuck SJ, Brand MD (2002) Topology of superoxide production from different sites in the mitochondrial electron transport chain. J Biol Chem 277:44784–44790
185. Staal FJ, Roederer M, Herzenberg LA, Herzenberg LA (1990) Intracellular thiols regulate activation of nuclear factor kappa B and transcription of human immunodeficiency virus. Proc Natl Acad Sci U S A 87:9943–9947
186. Stamler JS (1994) Redox signaling: nitrosylation and related target interactions of nitric oxide. Cell 78:931–936
187. Stief TW (2003) The physiology and pharmacology of singlet oxygen. Med Hypotheses 60:567–572
188. Taddei ML, Parri M, Mello T, Catalano A, Levine AD, Raugei G, Ramponi G, Chiarugi P (2007) Integrin-mediated cell adhesion and spreading engage different sources of reactive oxygen species. Antioxid Redox Signal 9:469–481
189. Taguchi K, Motohashi H, Yamamoto M (2011) Molecular mechanisms of the Keap1-Nrf2 pathway in stress response and cancer evolution. Genes Cells 16:123–140
190. Tamada M, Nagano O, Tateyama S, Ohmura M, Yae T, Ishimoto T, Sugihara E, Onishi N, Yamamoto T, Yanagawa H, Suematsu M, Saya H (2012) Modulation of glucose metabolism by CD44 contributes to antioxidant status and drug resistance in cancer cells. Cancer Res 72:1438–1448

191. Tappel ME, Chaudiere J, Tappel AL (1982) Glutathione peroxidase activities of animal tissues. Comp Biochem Physiol B 73:945–949
192. Ten FH, Dagnell M, Leuchs M, Vantler M, Berghausen EM, Caglayan E, Weissmann N, Dahal BK, Schermuly RT, Ostman A, Kappert K, Rosenkranz S (2011) Hypoxia enhances platelet-derived growth factor signaling in the pulmonary vasculature by down-regulation of protein tyrosine phosphatases. Am J Respir Crit Care Med 183:1092–1102
193. Terada LS (2006) Specificity in reactive oxidant signaling: think globally, act locally. J Cell Biol 174:615–623
194. Thannickal VJ, Fanburg BL (2000) Reactive oxygen species in cell signaling. Am J Physiol Lung Cell Mol Physiol 279:L1005–L1028
195. Tobiume K, Saitoh M, Ichijo H (2002) Activation of apoptosis signal-regulating kinase 1 by the stress-induced activating phosphorylation of pre-formed oligomer. J Cell Physiol 191:95–104
196. Touyz RM, Briones AM (2011) Reactive oxygen species and vascular biology: implications in human hypertension. Hypertens Res 34:5–14
197. Urso ML, Clarkson PM (2003) Oxidative stress, exercise, and antioxidant supplementation. Toxicology 189:41–54
198. Ushio-Fukai M (2006) Localizing NADPH oxidase-derived ROS. Sci STKE 2006:re8
199. van der Horst A, Tertoolen LG, de Vries-Smits LM, Frye RA, Medema RH, Burgering BM (2004) FOXO4 is acetylated upon peroxide stress and deacetylated by the longevity protein hSir2(SIRT1). J Biol Chem 279:28873–28879
200. Vander Heiden MG, Cantley LC, Thompson CB (2009) Understanding the Warburg effect: the metabolic requirements of cell proliferation. Science 324:1029–1033
201. Vander Heiden MG, Locasale JW, Swanson KD, Sharfi H, Heffron GJ, Amador-Noguez D, Christofk HR, Wagner G, Rabinowitz JD, Asara JM, Cantley LC (2010) Evidence for an alternative glycolytic pathway in rapidly proliferating cells. Science 329:1492–1499
202. Velu CS, Niture SK, Doneanu CE, Pattabiraman N, Srivenugopal KS (2007) Human p53 is inhibited by glutathionylation of cysteines present in the proximal DNA-binding domain during oxidative stress. Biochemistry 46:7765–7780
203. Vieceli Dalla SF, Zambonin L, Fiorentini D, Rizzo B, Caliceti C, Landi L, Hrelia S, Prata C (2014) Specific aquaporins facilitate Nox-produced hydrogen peroxide transport through plasma membrane in leukaemia cells. Biochim Biophys Acta 1843:806–814
204. Villeneuve NF, Lau A, Zhang DD (2010) Regulation of the Nrf2-Keap1 antioxidant response by the ubiquitin proteasome system: an insight into cullin-ring ubiquitin ligases. Antioxid Redox Signal 13:1699–1712
205. Vurusaner B, Poli G, Basaga H (2012) Tumor suppressor genes and ROS: complex networks of interactions. Free Radic Biol Med 52:7–18
206. Wakabayashi N, Dinkova-Kostova AT, Holtzclaw WD, Kang MI, Kobayashi A, Yamamoto M, Kensler TW, Talalay P (2004) Protection against electrophile and oxidant stress by induction of the phase 2 response: fate of cysteines of the Keap1 sensor modified by inducers. Proc Natl Acad Sci U S A 101:2040–2045
207. Wang FS, Wang CJ, Chen YJ, Chang PR, Huang YT, Sun YC, Huang HC, Yang YJ, Yang KD (2004) Ras induction of superoxide activates ERK-dependent angiogenic transcription factor HIF-1alpha and VEGF-A expression in shock wave-stimulated osteoblasts. J Biol Chem 279:10331–10337
208. Wang XJ, Sun Z, Villeneuve NF, Zhang S, Zhao F, Li Y, Chen W, Yi X, Zheng W, Wondrak GT, Wong PK, Zhang DD (2008) Nrf2 enhances resistance of cancer cells to chemotherapeutic drugs, the dark side of Nrf2. Carcinogenesis 29:1235–1243
209. Ward PS, Thompson CB (2012) Metabolic reprogramming: a cancer hallmark even warburg did not anticipate. Cancer Cell 21:297–308
210. Warren CM, Ziyad S, Briot A, Der A, Iruela-Arispe ML (2014) A ligand-independent VEGFR2 signaling pathway limits angiogenic responses in diabetes. Sci Signal 7:ra1
211. Wei Q, Jiang H, Matthews CP, Colburn NH (2008) Sulfiredoxin is an AP-1 target gene that is required for transformation and shows elevated expression in human skin malignancies. Proc Natl Acad Sci U S A 105:19738–19743

212. Wong CM, Marcocci L, Liu L, Suzuki YJ (2010) Cell signaling by protein carbonylation and decarbonylation. Antioxid Redox Signal 12:393–404
213. Woo HA, Yim SH, Shin DH, Kang D, Yu DY, Rhee SG (2010) Inactivation of peroxiredoxin I by phosphorylation allows localized H(2)O(2) accumulation for cell signaling. Cell 140:517–528
214. Wu Y, Deng J, Rychahou PG, Qiu S, Evers BM, Zhou BP (2009) Stabilization of snail by NF-kappaB is required for inflammation-induced cell migration and invasion. Cancer Cell 15:416–428
215. Xanthoudakis S, Curran T (1992) Identification and characterization of Ref-1, a nuclear protein that facilitates AP-1 DNA-binding activity. EMBO J 11:653–665
216. Yankovskaya V, Horsefield R, Tornroth S, Luna-Chavez C, Miyoshi H, Leger C, Byrne B, Cecchini G, Iwata S (2003) Architecture of succinate dehydrogenase and reactive oxygen species generation. Science 299:700–704
217. Yasinska IM, Sumbayev VV (2003) S-nitrosation of Cys-800 of HIF-1alpha protein activates its interaction with p300 and stimulates its transcriptional activity. FEBS Lett 549:105–109
218. Yoo SK, Freisinger CM, LeBert DC, Huttenlocher A (2012) Early redox, Src family kinase, and calcium signaling integrate wound responses and tissue regeneration in zebrafish. J Cell Biol 199:225–234
219. Yoo SK, Starnes TW, Deng Q, Huttenlocher A (2011) Lyn is a redox sensor that mediates leukocyte wound attraction in vivo. Nature 480:109–112
220. Zhang DD, Hannink M (2003) Distinct cysteine residues in Keap1 are required for Keap1-dependent ubiquitination of Nrf2 and for stabilization of Nrf2 by chemopreventive agents and oxidative stress. Mol Cell Biol 23:8137–8151
221. Zmijewski JW, Banerjee S, Bae H, Friggeri A, Lazarowski ER, Abraham E (2010) Exposure to hydrogen peroxide induces oxidation and activation of AMP-activated protein kinase. J Biol Chem 285:33154–33164

Chapter 2
Oxidative Stress, Antioxidant Defenses, and the Liver

Carlo Vascotto and Claudio Tiribelli

2.1 Introduction

Several million years ago, some unicellular organisms in the primordial seas began dumping a new toxic agent into the environment, and some others found a way to use this "pollutant" to their own advantage. The toxic agent in question was oxygen, and it is now accepted that multicellular animal life evolved partly as a consequence of the many-fold increase in cellular energy that can be produced from nutrients by enzymatic reduction of oxygen [1]. If on the one hand the appearance of oxygen prompted the evolution, on the other oxygen still remains an extremely dangerous agent with which to coexist. This very basic dilemma has come to be known as "The Oxygen Paradox" [2]. Aerobic leaving organisms produce reactive oxygen species (ROS) and reactive nitrogen species (RNS) as a result of normal cellular metabolism. ROS and RNS are critical intermediates in the normal physiology and pathophysiology of many cells including the hepatocyte. ROS of physiological significance are superoxide anion ($O_2^{\bullet -}$), hydroxyl radical ($^{\bullet}OH$), and hydrogen peroxide (H_2O_2), while RNS includes nitric oxide ($NO^{\bullet}$) and peroxynitrite ($ONOO^-$) (Fig. 2.1) [3]. When the levels of oxidation products exceed the capacity of normal antioxidant systems, a net oxidative stress results [4]. In the liver, free radicals triggered by ROS and RNS are created by neutrophils, Kupffer cells, mitochondria, and cytochromes P450 [5]. Oxidative stress acts by affecting major cellular components including lipids, DNA, and proteins [6].

C. Vascotto (✉)
Department of Medical and Biological Sciences, University of Udine, Udine 33100, Italy
e-mail: carlo.vascotto@uniud.it

C. Tiribelli
Center for Liver Studies, AREA Science Park, Trieste 34149, Italy

Department of Medical Sciences, University of Trieste, Trieste 34149, Italy
e-mail: ctliver@csf.units.it

E. Albano, M. Parola (eds.), *Studies on Hepatic Disorders*,
Oxidative Stress in Applied Basic Research and Clinical Practice,
DOI 10.1007/978-3-319-15539-5_2

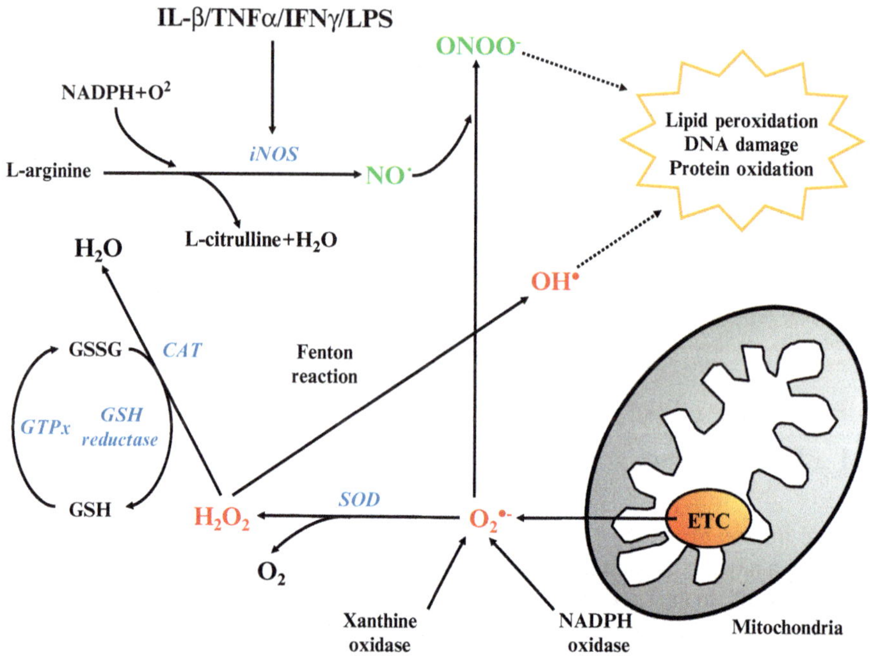

Fig. 2.1 Main pathways for the formation of ROS and RNS. Mitochondria are the major source of cellular ROS during oxidative phosphorylation and ETC. Electron leak leads at the formation of superoxide anion ($O_2^{\bullet-}$) that is generated by the reduction of molecular oxygen. This process may also be mediated by NADPH oxidase and xanthine oxidase. Superoxide dismutase (SOD) catalyzes the dismutation of two superoxide anions into hydrogen peroxide (H_2O_2) and oxygen (O_2). H_2O_2 could be enzymatically converted into water by catalase (CAT) or glutathione peroxidase (GTPx). If not inactivated, H_2O_2 can react with reduced transition metals, via Fenton's reaction, to produce hydroxyl radical (•OH). Nitric oxide synthase (iNOS) induction by IL-β, TNFα, INFγ, and LPS is responsible for the formation of nitric oxide (NO•) from L-arginine in a NADPH-dependent reaction. NO• can react with $O_2^{\bullet-}$ forming peroxynitrite ($ONOO^-$). Due to their high reactivity, H_2O_2, $O_2^{\bullet-}$, and $ONOO^-$ could all damage lipids, DNA, and proteins leading to cell death

To cope with these harmful by-products of oxygen, all living organisms developed cell defenses mechanisms to counterbalance the effects of oxidants. The first defense is the sharp gradient of oxygen tension, seen in all mammals, from the environmental level of 20 % to a tissue concentration of only 3–4 % oxygen. These relative low tissue levels of oxygen prevent the occurrence of most oxidative damage [1]. Nevertheless, our tissues are continuously exposed to oxidative stress but luckily, thanks to enzymatic and non-enzymatic antioxidant defenses, we have defense against oxygen radicals and other activated-oxygen species. Aerobic organisms synthesize several antioxidant enzymes. These defense mechanisms include Superoxide dismutases (SOD), Catalase (CAT), Glutathion peroxidise (GTPx), Thioredoxins (TRXs), Peroxiredoxins (PRXs), Thioredoxin reductase (TrxR), Glutaredoxins (Grxs), and Glutathione transferase (GST) [3]. In addition, our cells utilize a series of antioxidant compounds that directly react with oxidizing agents acting as "scavengers" in an unavoidably suicide mechanism. Vitamin E, vitamin C (ascorbic acid), glutathione (GSH), and carotenoids are some of the most effective antioxidant molecules [3]. Cells can also adapt to oxidative stress by rapidly increasing the production of antioxidant enzymes, the coding genes of which are rapidily turned on in response to oxidation. This adaptive response enables cells to survive under oxidative stress conditions and to restore the net balance between pro- and antioxidants [1].

Chronic oxidative stress is a major pathogenetic event occurring in several liver diseases, ranging from metabolic to proliferative disorders [7]. The relevance of cellular redox state in liver pathologies is outlined by a number of studies in patients affected by viral, alcoholic or non-alcoholic fatty liver diseases, ischemia/reperfusion injury, metabolic disorders, and cancer (for details on oxidative stress in liver pathology refer to Chapters 12–26), pointing to a correlation between organ damage and increase in pro-oxidant cellular markers associated with a concomitant decrease of antioxidants [5, 8–10]. Chemical modification of essential biomolecules by ROS may cause their functional inactivation leading to cell death or an adaptive cellular response. Identification of chronic and acute oxidative stress biomarkers is of the utmost importance to monitor the degree of liver damage, the pharmacological response to therapies, and the development of new therapeutic approaches.

2.2 Cellular Sources of ROS and RNS

ROS are produced from molecular oxygen (O_2) as a result of normal cellular metabolism. ROS can be divided into two groups: free radicals and non-radicals compounds. Radical species are characterized by the presence of one or more unpaired electrons that confer chemical reactivity to the molecule. When two free radicals share their unpaired electrons, a non-radical form is created [3]. $O_2^{\bullet-}$ is formed by the addition of one electron to the oxygen generating a free radical oxygen form [11]. This process is mediated by nicotine amide dinucleotide phosphate (NADPH) oxidase, or xantine oxidase, or mitochondria electron transport chain (ETC). The respiratory chain provides the main portion of cellular ATP, but is also a major source of ROS. During oxidative phosphorylation and transfer of energy along the ETC, approximately 1–3 % of electrons escape from complexes I and III forming superoxide. This process of superoxide generation, termed "electron leak," has been implicated in the pathophysiology of numerous diseases [12]. Superoxide has a very short half-life, cannot pass outward through the mitochondrial inner membrane, and upon formation it is readily converted into H_2O_2 and O_2 by the action of SOD [13]. H_2O_2 is freely diffusible in the cellular environment and may reach the nucleus and interact with DNA-bound transition metal ions, iron and copper in particular, leading to the formation of $^{\bullet}OH$ via the Fenton-type reaction [14]. Hydroxyl radical is the most reactive of ROS and can damage proteins, DNA and start the lipid peroxidation by taking an electron from polyunsaturated fatty acids [3]. Alternatively, H_2O_2 could be converted into water by CAT and GTPx (Fig. 2.1) [15].

$NO^{\bullet}$ is a RNS critical in the redox biology of hepatocytes. It is a hydrophobic, freely diffusible, small molecule created by nitric oxide synthase (NOS), which is present in three forms: neuronal (nNOS), endothelial (eNOS), and inducible (iNOS) [4]. The first two forms have been found to be constitutively active in neuronal tissue and endothelium, respectively. On the contrary, iNOS is an inducible NOS form found in a variety of cells, including hepatocytes [4]. NOS utilizes L-aginine and oxygen, in combination with electrons from NADPH, to create $NO^{\bullet}$ and L-citrulline (Fig. 2.1) [16]. In the liver, iNOS was found to be critical in the development and propagation of inflammation [17] and is expressed in hepatocytes, Kupffer cells, vascular endothelial cells, and stellate cells [18]. iNOS expression is induced by Interleukin-β (IL-β) or IL-β in combination with Tumor necrosis factor-α (TNFα), Interferon-γ (IFNγ), and Lipopolisaccarides (LPS) [19–21]. NF-κB and γ-interferon response element (γ-IRE) have both been found to affect iNOS promoter [22], contributing to mitochondrial dysfunction by interfering with the mitochondrial respiratory chain and by forming $O_2^{\bullet-}$ [23]. Some of the physiological effects of RNS are mediated through the formation of *S*-nitroso-Cys or *S*-nitroso-GSH intermediates [24]. RNS may functionally inactivate proteins of the mitochondrial respiratory chain through nitration of their Tyr residues or the intermediate formation of *S*-nitrosated protein adducts at Cys residues [24]. An indirect effect of promoting mitochondrial dysfunction is the increased production of RNS as a consequence of the induction of iNOS and the formation of $ONOO^-$ [25].

2.3 Damages Generated by ROS

ROS are strong oxidants capable of damaging DNA, proteins, and lipids. The oxidative products derived from each biomolecule are complex and multiple. Reactivity, site, and mechanism of production and products formed vary depending both on the free radical and the molecular target. Some of these oxidative products are markers of oxidative stress.

ROS can modify DNA in several ways determining base degradation, single- or double-stranded breaks, purine, pyrimidine or sugar-bound modification, mutation, deletion or translocation, and cross-linking with proteins [3]. Purines undergo oxidation of the ring atoms, leading to various chemical modifications. The highly mutagenic guanine derivate 8-hydroxyguanine (8-oxoG) is formed in large quantities as consequence of the high oxidation potential of this base (Fig. 2.2a) [26]. The miscoding effect of 8-oxoG lesion is due to DNA polymerase activity which inserts adenine opposite to 8-oxoG, resulting in G:C to A:T transition mutations generating DNA base mutation site with high frequency. Formation of 8-oxoG in transcription factors-binding sites can modify binding activity of transcription factors and thus change the expression of related genes as has been shown for AP-1 and Sp-1 target sequences [27]. The most frequent pyrimidine oxidation is represented by the formation of 5-hydroxycytosine (5-OHC), which leads to the insertion of a

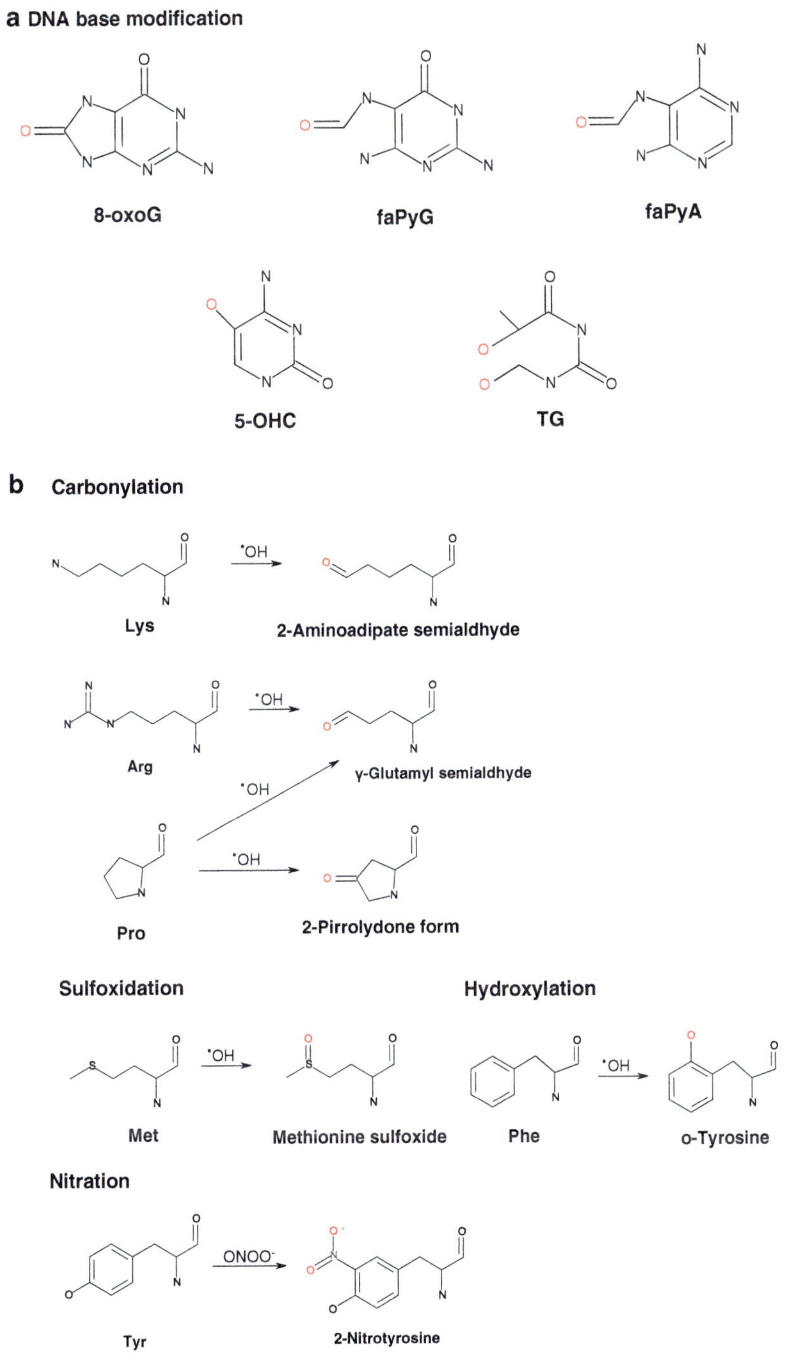

Fig. 2.2 Oxidative and nitrosative modification of biomolecules. Panel **a**: Oxidative modification of DNA bases. Panel **b**: Carbonylation, sulfoxidation, hydroxylation, and nitration of amino acid side chains. Panel **c**: Mechanism of lipid peroxidation. In the initiation step a fatty acid radical is

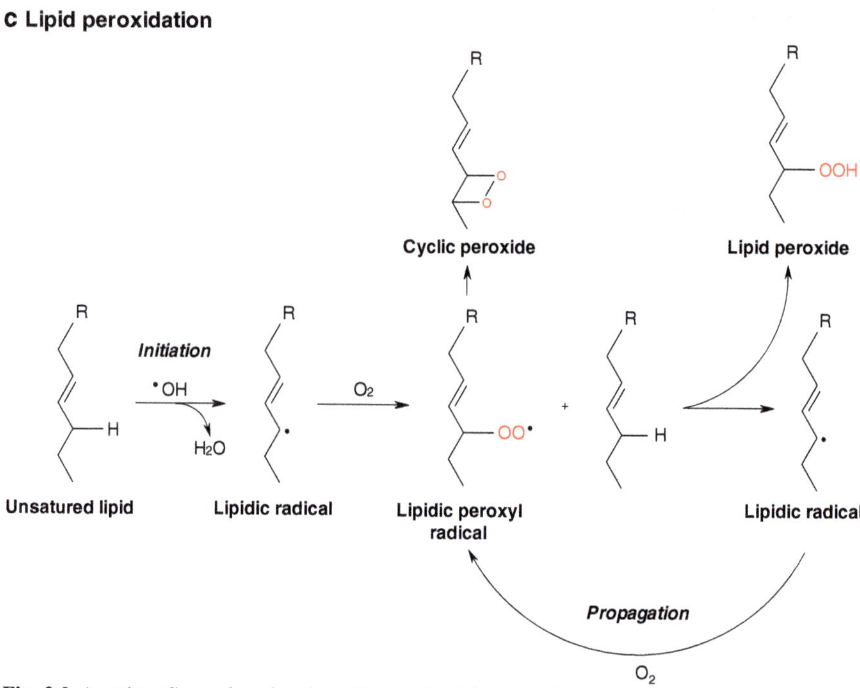

Fig. 2.2 (continued) produced and readily reacts with molecular oxygen creating a peroxyl-fatty acid radical. During the propagation, this reacts with another lipid producing a new lipid radical and a lipid peroxide, or reacts with itself generating a cyclic peroxide. Oxygen and nitrogen atoms transferred during the oxidation/nitrosylation processes are represented in *red* and *blue*, respectively

thymine creating a potential premutagenic lesion [28]. Other important oxidation lesions are the formamidopyrimidine such as faPyA and faPyG and the oxidized thymine glycol (TG) (Fig. 2.2a) [29]. All these mutations are recognized by different DNA glycosylases, which remove the damaged nitrogenous base by cleaving the *N*-glycosylic bond and generating an abasic (AP) site while leaving the sugar-phosphate backbone intact [30]. This reaction represents the initiation step of the base excision repair (BER) pathway and forms the substrate for the AP endonuclease (APE1) enzyme that generates a nick in the phosphodiester backbone of the AP site [31].

ROS-induced protein modification can result in unfolding or alteration of protein structure and function. Reversible protein modification, such as S-glutathionylation, S-nitrosation, and methionine sulfoxidation, may have a dual role of protection from irreversible oxidation and modulation of protein function [32, 33]. Differently, irreversible protein modification can lead to inactivation of various proteins and could have long-lasting detrimental cellular effects [34]. Protein carbonylation is a type of protein oxidation that can be promoted by ROS. Carbonyl groups are introduced into proteins by direct oxidation of the Lys, Arg, Pro, and Thr side chains. Cys residues undergo S-thiolation by forming a mixed disulfide through the conjugation with low molecular weight thiols, such as GSH [35]. In addition, Cys residues could be oxidized forming sulfenic acid that could generate a Cys disulfide bridge or be sequentially oxidized to sulfinic- and sulonic-acid forms [36]. The methyl-thioether group of Met is particularly susceptible to ROS and could generate a sulfoxide form of Met (MetO). Sulfoxidation of Met leads most proteins to conformational alteration and loss of function. An active defense against this modification cells expresses methionine sulfoxide reductase (MsrA), an enzyme that acts by reducing MetO [37]. Also RNS could modify protein residue generating 2-nitrotyrosine as consequence of Tyr residue nitrosation by $ONOO^-$ (Fig. 2.2b) [38].

Lipid peroxidation by ROS can disrupt the membrane lipid bilayer arrangement inactivating membrane receptors and enzymes and increasing cellular permeability [39]. Phospholipid peroxidation generates a large number of diverse products, including breakdown products resulting from cleavage of the oxidized fatty acyl chain. Lipid peroxidation requires an initiation step in which a fatty acid radical is produced as consequence of a radical reaction by $O_2^{\bullet-}$ or $^{\bullet}OH$ [40, 41]. The fatty acid radical is not a very stable molecule, so it readily reacts with molecular oxygen, thereby creating a peroxyl-fatty acid radical. This in turn reacts with another free fatty acid producing a different fatty acid radical and a lipid peroxide, or a cyclic peroxide if it reacts with itself (Fig. 2.2c). This cycle continues, as the new fatty acid radical reacts in the same way. The end products of lipid peroxidation are reactive aldehydes, such as malondialdehyde (MDA) and 4-hydroxynonenal (HNE) [41]. HNE is one of the most studied products and was found to occur in rat liver subjected to oxidative damage by carbon tetrachloride [42].

2.4 Enzymatic Antioxidant Defenses

Cellular defenses coping with ROS generation are multiple and include enzymatic and non-enzymatic antioxidants. SOD, CAT, GTPx, TRXs, PRXs, TrxR, GRXs, and GST are the major enzymatic antioxidant cellular defenses.

SOD is an oxido-reductase that catalyzes the dismutation of $O_2^{\bullet-}$ into oxygen and H_2O_2. Three forms of SOD are present in humans: (1) SOD1 is highly abundant, comprising ~1 % of total protein in the cell [43] and resides mainly in the cytosol; (2) SOD2 is a nuclear-encoded mitochondria enzyme; (3) SOD3 is predominately located outside the cell in the extracellular matrix [44]. All three proteins are codified by different genes located on chromosomes 21, 6, and 4, respectively. Other key differences among the SOD proteins include their quaternary structures and mechanism of superoxide dismutation: SOD1 is a homodimer while SOD2 and SOD3 are homotetrameric proteins. SOD1 and SOD3 catalyze the dismutation of $O_2^{\bullet-}$ through the alternate reduction and reoxidation of Cu^{2+}, whereas SOD2 utilizes manganese as a redox active transition metal [44]. The physiological relevance of SOD1 activity extends beyond oxidative stress protection. Indeed, SOD1 catalytic activity plays a key role in signal transduction. H_2O_2 can reversibly and specifically react with proteins, generally by oxidizing Cys residues and therefore altering biochemical and functional properties of those proteins in a redox-dependent manner [45]. Several signal transduction pathways are modulated by H_2O_2 including gene expression, cell proliferation, differentiation, and death [46, 47]. SOD2 is a nuclear-encoded mitochondrial protein whose expression is highly regulated both at transcriptional and translational levels, and through post-translational modifications [48, 49]. TNFα, IL1-β, IFNγ, LPS, and anti-cancer drugs are able to induce SOD2 expression [49]. NF-κB is considered the most responsive redox-sensitive transcriptional factor that binds the NF-κB consensus motif present in *SOD2* regulatory region [49]. Both the promoter and the enhancer regions of *SOD2* gene present NF-κB-binding sequences at which the protein can bind enhancing transcription [50]. SP1 is another redox-responsive transcription factor that binds to *SOD2* promoter and is necessary for both basal and induced expression of SOD2 [51]. AP1 is a redox-regulated protein consisting of a c-Fos, c-Jun, and Fra family proteins. The oncogenic stimulation of Ap1 expression occurs in response to growth factors and oxidative stress [51–54]. An Ap1 consensus-binding site has been identified in *SOD2* gene promoter suggesting that Ap1 activation by oxidative stress may play a positive role in inducing *SOD2* transcription [55].

H_2O_2 is mainly produced by SOD and could be converted into water and oxygen by the action of catalase [56]. CAT is a tetrameric enzyme of approximately 60 kDa subunits, each has 527 amino acid residues and one heme group with Fe^{3+}. The human catalase gene is localized on chromosome 13 and is composed by 13 exons and 12 introns [57]. CAT is not the only enzyme in the redox cycle responsible for the reduction of H_2O_2. GTPx are a family of tetrameric enzymes with four identical subunits, each of which contains one unique selenocysteine residue within the active site, and use low-molecular-weight thiols (such as reduced glutathione (GSH)) to reduce H_2O_2 and lipid peroxides to their relative alcohols [58]. Four GTPx encoded by different genes have been described: GTPx-1 is ubiquitous and reduces H_2O_2 and fatty acid peroxides, but not esterified peroxyl lipids [59]. Despite this range of peroxide substrates, GTPx-1 is very specific for GSH as reducing substrate. Therefore, GTPx-1 activity is strictly associated with GST, whose enzymatic activity maintains a constant supply of GSH from GSSG [59]. GTPx-2 is localized in gastrointestinal epithelial cells where it serves to reduce dietary peroxides [60]. After the discovery of GTPx-1, it was long believed that GTPx activity in plasma was due to leakage of the enzyme from the liver and other organs. GTPx-3 is a glycoprotein and is the only member of the GTPx family located in the extracellular compartment. It is believed to be one of the most important extracellular antioxidant enzymes in mammals [59, 61]. The partial sequencing of the protein and then of its cDNA confirmed that GTPx-3 is distinct from GTPx-1 with only a 40–50 % of homology between the two forms. The mRNA for GTPx-3 is predominantly found in kidney [62], but also in a number of other cell types as heart, placenta, lung, gastrointestinal cells, and thyroid [63–66]. Esterified lipids are reduced by membrane-bound GTPx-4, which can use several different low-molecular-weight thiols as reducing equivalents [59]. This is a 22 kDa protein that in contrast with the tetrameric structure of the other GTPx proteins exerts its enzymatic activity as a monomer [67]. GTPx-4 can use phospholipids hydroperoxidases, but also H_2O_2 and a wild range of lipid hydroperoxides as substrates [68].

In addition to GTPx enzymes, cellular ROS detoxification is closely associated with thiol-containing enzymes such as TRXs, GRXs, PRXs, and TTRs. Trx-1 is a small multifunctional ubiquitous redox-active protein of 12 kDa [69]. TRXs have two redox-active cysteine residues in their conserved active-site sequence (Cys-Gly-Pro-Cys), which in the reduced state exist in their -SH form. The human form of Trx-1 has three additional Cys residues, which are absent in *E. coli* and in the mammalian mitochondrial thioredoxin (Trx-2) [70]. These Cys residues are converted to the oxidized intramolecular disulfide bond state in a reaction that results in the reduction of a disulfide of the target protein. This oxidized form of TRX is then recycled back to its reduced form by TrxR, which uses NADPH as a cofactor [71].

GRXs and TRXs share a number of common features, but GRXs are more versatile than TRXs for substrate and reaction mechanism [72]. Similarly to TRXs, double Cys active-site sequence is also present in the dithiol GRXs. However, in monothiol GRXs, the C-terminal Cys residue is replaced by a Ser [72]. GRXs use the reducing power of GSH to catalyze the reduction of protein disulfides by a dithiol mechanism, or the reduction of mixed GS-S protein disulfides through a monothiol mechanism (Fig. 2.3a) [72]. Unlike TRXs, GRXs have a stronger affinity toward the mixed disulfides. To reduce a disulfide, both active-site Cys are required in the dithiol mechanism, while for the monothiol mechanism involving GSH-mixed disulfides, only the N-terminal Cys is necessary [73].

PRXs are a family of antioxidant enzymes composed by six members (1–6) that catalyze the reduction of peroxides [74]. Family members share the same basic catalytic mechanism consisting of the presence of a redox-active Cys residue that becomes oxidized by peroxides to sulfenic acid (–SOH). This oxidized form of PRX must then be reduced back to the –SH form by reduced TRX. This mechanism of reconversion differs among the PRX subfamilies. PRXs containing 2 Cys residues in the catalytic site are the predominant members (1–4) [74]. In addition to a catalytic residue, 2-Cys PRXs possess a “resolving Cys” residue which reacts with the oxidized peroxidatic Cys-OH to form a disulfide bridge. In the typical 2-Cys PRXs (1–4), the resolving Cys and catalytic Cys are on different molecules and so the active enzyme exists as functional homodimer (Fig. 2.3b). In contrast, PRX 5 is an atypical 2-Cys Prx that exists as a monomer with both the resolving and catalytic Cys residues on the same molecule. PRX 6 is expressed only in mammalian cells and is characterized by the presence of a single Cys residue in the catalytic site. This 1-Cys PRX do not contain a resolving cysteine and cannot be reduced by TRX, but may instead be reduced by other agents, potentially glutathione or cyclophilin [74]. All of them contain fixed Cys residues on the N-terminal regions of the molecules, and the isoforms 1–4 have additional analogous Cys residues on the C-ends [74].

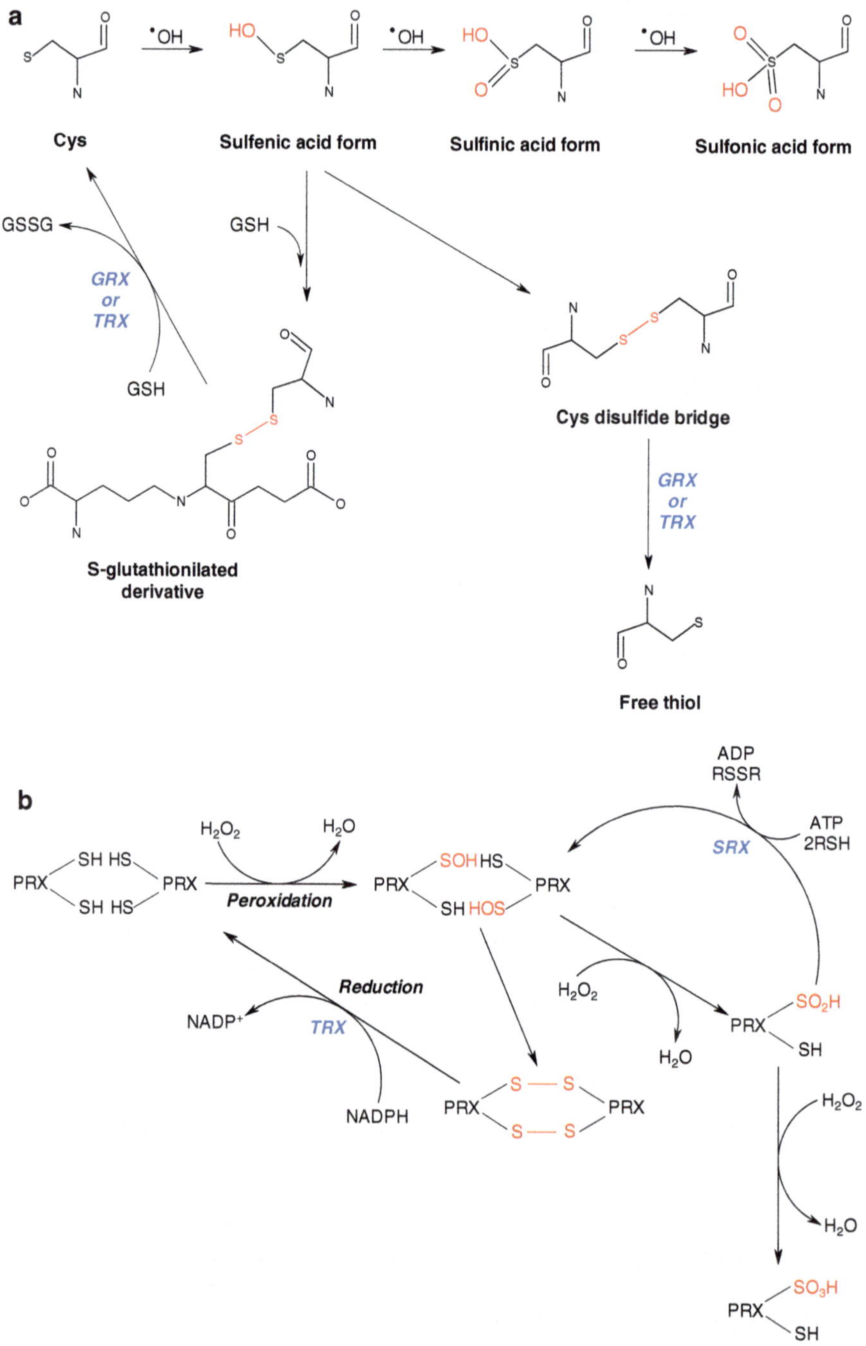

Fig. 2.3 Oxidative thiol modifications. Panel **a**: Exposure of redox-sensitive Cys residues to oxidants leads to the formation of sulfenic (–SOH), sulfinic (–SO_2H), and sulfonic (–SO_3H) acid. Sulfenic acids readily react with nearby thiols of the same protein to form intramolecular disulfide bridge.

Under increased oxidative stress, PRX-SOH can sometimes be further oxidized by peroxide to sulfinic acid (–SO_2H), causing inactivation of peroxidase activity [75]. PRX–SO_2H is not a substrate for the resolving Cys and cannot be reduced by TRX. For this reason, PRX–SO_2H was thought to be an irreversible form. Subsequently, it was demonstrated that PRX–SO_2H can be reduced back to the catalytically active form by the ATP-dependant reductase sulfiredoxin (SRX) [76, 77]. Hyperoxydation to sulfonic acid (–SO_3H) determines an irreversible modification and the inactivation of the protein. The expression of genes encoding different Prx isoforms has cellular, tissue, and organ specificity (Fig. 2.3b). PRX 1 is the most widely represented and highly expressed member of the PRXs family in virtually all organs and tissues of mice and humans, both in normal tissues and malignant tumors [78, 79]. High expression of the *PRX4* gene, the mitochondrial form of PRX proteins, is characteristic of liver, testes, ovaries, and muscles, whereas low expression is observed in small intestine, placenta, lung, kidney, spleen, and thymus [78].

GSTs are another class of antioxidant enzymes involved in cellular detoxification of endogenous toxic metabolites, superoxide radicals, and exogenous toxic chemicals. They act by inactivating secondary metabolites, such as unsaturated aldehydes, epoxides, and hydroperoxides, and regulating the levels of the cellular antioxidant GSH in different cellular compartments. In addition, GSTs also play an important role in the activation of signals by mitogen-activated protein (MAP) kinases and various TFs that regulate apoptosis and cell survival pathways [80, 81]. Three major families of GSTs have been described: soluble cytosolic (cGST), mitochondrial (mGST), and membrane-associated microsomal GST (MGST). The classification was made on the basis of sequence similarity and immunological cross-reactivity. cGST family is composed of nine isoforms (α, μ, π, ω, τ, δ, σ, ζ), and the mGST of four isoforms (α, μ, π, κ). The MGST family contains six members including 1, 2, 3, leukotriene C-4 synthase, 5-lipoxygenase-activating protein, and prostaglandin E synthase, now referred to as membrane-associated proteins in eicosanoid and glutathione metabolism (MAPEG) members [82]. Each member of the family has multiple isoenzymes with overlapping substrate specificity. All GST isoforms are codified by nuclear genes, synthesized in the cytosol, and then distributed in other intracellular organelles [83].

Fig. 2.3 (continued) They can also form a mixed disulphide through the conjunction with GSH forming a *S*-glutathionylated derivative. This oxidative thiol modifications are reduced by members of GRXs or TRXs family. Panel **b**: Reaction cycle of 2-Cys PRXs. Under oxidative stress conditions peroxidatic Cys undergoes oxidation forming sulfenic acid (–SOH). The "resolving Cys" reacts with the oxidized Cys forming a disulfide bridge that could be reduced by Thioredoxin (TRX) in a NADPH-dependant reaction. Under increased oxidative stress, PRXs Cys are further oxidated to sulfinic and sulfonic acid. Sulfinic acid can be reduced back to the catalytically active form by Sulfiredoxin (SRX) in an ATP-dependent mechanism, while the sulfonic acid could not be modified and represent an irreversible modification

2.5 Non-enzymatic Antioxidant Defenses

ROS and RNS are constitutively produced as consequence of the physiological cellular metabolism. The key point is to intercept and inactivate such damaging species, once formed. For radical compounds, the final deactivation consists of the formation of non-radical end products. Due to the nature of the free radicals, there is a tendency towards chain reaction where a compound carrying an unpaired electron will react with another compound to generate an unpaired electron in that compound [84]. In addition to enzymatic antioxidant defenses, our cells utilize a series of antioxidant compounds that directly react with oxidizing agents acting as scavengers (Fig. 2.4).

Vitamin E is a term encompassing a group of potent, lipid-soluble, chain-breaking antioxidants. It is the principal defense against oxidant-induced membrane injury and acts by donating an electron to peroxyl radical produced during lipid peroxidation. Structural analyses have revealed that molecules having vitamin E antioxidant activity include four tocopherols (α, β, γ, δ) and four tocotrienols (α, β, γ, δ). One form, α-tocopherol, is the most abundant form in nature, has the highest biological activity, and reverses vitamin E deficiency in humans [85]. Vitamin E is absorbed in the intestine and enters the circulation via the lymphatic system. It is absorbed together with lipids, packed into chylomicrons, and transported to the liver with the chylomicrons and the remnants derived thereof [86]. Only after passage through the liver does α-tocopherol appears in the plasma. Most of the ingested β-,

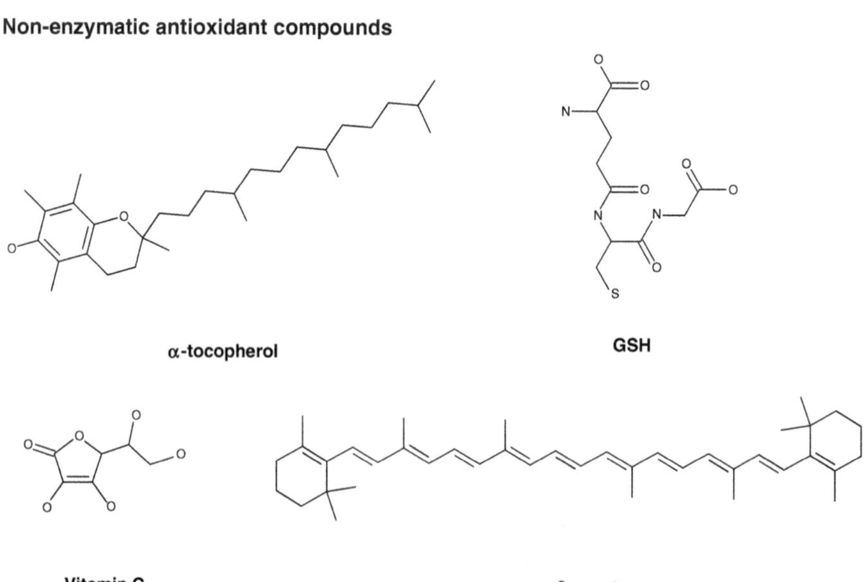

Fig. 2.4 Chemical structure of α-tocopherol, glutathione (GSH), vitamin C, and β-carotene

γ-, and δ-tocopherol are excreted in the feces [87]. In humans, severe vitamin E deficiency leads to neuromuscular abnormalities characterized by spinocerebellar ataxia and myopathies [88, 89]. Vitamin E deficiency anemia occurs, largely in premature infants, as a result of free radical damage [90]. Diminished erythrocyte life span and increased susceptibility to peroxide-induced hemolysis are apparent not only in severe deficiency, but also in modest vitamin E deficiency in hypercholesterolemic subjects [91]. Vitamin E deficiency may also occur as a result of genetic abnormalities in the α-tocopherol transfer protein (α-TTP) and as consequence of various fat malabsorption syndromes [86]. Also chronic liver disease, fat malabsorption, or abetalipoproteinemia can determine symptoms of vitamin E deficiency and could be ameliorated by high doses of vitamin E [92].

Vitamin C is an essential micronutrient required for normal metabolic functioning of the body [93]. As a result of a mutation of the gene coding for L-GULONOLACTONE oxidase, an enzyme required for the biosynthesis of vitamin C, human and other primates have lost the ability to synthesize this vitamin C that must be introduced through the diet [94]. A lack of vitamin C determines the insurgence of scurvy [95]. Vitamin C is a cofactor of several enzymes involved in the biosynthesis of collagen, carnitine, and neurotransmitters [96, 97]. A deficiency of vitamin C results in a weakening of collagenous structures causing characteristic symptoms of scurvy such as joint pains and tooth loss. Carnitine is essential for the transport of activated long-chain fatty acids into the mitochondria. Vitamin C deficiency determines fatigues and lethargy, early symptoms of scurvy [98]. In addition, vitamin C is a cofactor in the enzymatic conversion of dopamine to noradrenalin [96]. Depression, hypochondria, and mood change often occur in scurvy and are related to deficient dopamine hydroxylation [98]. It has been demonstrated that vitamin C is implicated in the metabolism of cholesterol to bile acids, and also the hydroxylation of aromatic drugs and carcinogens by hepatic cytochrome P450 is enhanced by reducing agents such as vitamin C [97]. Beside the role in metabolic pathways, vitamin C is an important water-soluble antioxidant present in biological fluids [99, 100]. It reacts with hydroxyl radicals and can also act as a coantioxidant by regenerating α-tocopherol from the α-tocopheroxyl radical, produced via scavenging of lipid-soluble radicals [101]. Vitamin C has also been shown to regenerate glutathione and β-carotene in vitro from their respective one-electron oxidation products [102]. Vitamin C reacts with a ROS or a RNS forming the ascorbic radical, which readily dismutates forming ascorbate and dehydroascorbic acid, or is reduced back to ascorbate by an NADH-dependant semidehydroascorbate reductase [103]. The dehydroascorbic acid can itself be reduced back to ascorbate by TrxR or Grx [97]. Alternatively, dehydroascorbic acid is rapidly and irreversibly hydrolyzed to 2,3 diketogluconic acid (DKG) [104].

Differently from vitamins, GSH is not an essential nutrient but is synthesized in the body from its precursor amino acids Cys, Glu, and Gly in two consecutive ATP-consuming enzymatic reactions [105]. The first reaction is catalyzed by γ-glutamylcysteine ligase (γ-GCL), the rate-limiting and regulatory enzyme in GSH synthesis, and the second by GSH synthetase (GSS) [106]. While all cells in the human body are capable of synthesizing glutathione, liver glutathione synthesis has been shown to be predominant and essential. Mice with genetically induced hepatic loss of GSH synthesis only die within 1 month of birth. Liver has the highest GSH level among tissues and hepatocytes are the main source of GSH in liver [107]. Almost 85–90 % of cellular GSH is present in the cytosol, 10–15 % in the mitochondria, and a small percentage is in the endoplasmic reticulum (ER) and nucleus [83]. The compartmentalization of GSH constitutes distinct redox pools in terms of balance between oxidized and reduced forms and their turnover rates. Cytosolic GSH (cGSH) has a rapid turnover of 2–3 h, while mitochondrial GSH (mGSH) has relatively longer half-life of 30 h. A shift in this balance is a good indicator of cellular redox stress [108]. Mitochondria are the primary site of oxygen metabolism and their proper function is closely linked to maintenance of the GSH pool. Consequently, it is believed that GSTs play a key role in protecting mitochondrial genetic and metabolic machinery against oxidative insults [109]. GSH reacts rapidly with hydroxyl radical derived from the Fenton reaction and peroxynitrite formed by the reaction of $NO^{\bullet}$ with superoxide [110, 111]. In reactions catalyzed by the several isoforms of GTPx, GSH also participates in the reductive detoxification of H_2O_2 and lipid peroxides [105]. The initial products are chemically stable sulfides of GSH, but further metabolism removes the Glu and Gly residues, forming *S*-substituted Cys. Acetylation of the cysteinyl amino group can then form a mercapturic acid, which is easily excreted in the urine. Such metabolism results in the irreversible loss of the Cys residue of GSH, a consequence of some importance because amino acid is often limiting for GSH synthesis [105]. If not metabolized, the glutathione disulfide (GSSG) is reduced intracellularly to GSH by GSSG reductase in a NADPH-dependent reaction [105]. Cellular GSH levels reflect a steady state balance between synthesis and consumption. Synthesis includes both de novo production and GSH regeneration from GSSG via GSSG reductase. The latter reaction is responsible for the maintenance of the total glutathione pool in a predominantly reduced state and usually does not have a major influence on cellular GSH levels. Moreover, extreme levels of oxidative or nitrosative stress can quickly and substantially diminish GSH levels in favor of GSSG, particularly if GSSG reductase activity is low due to an inherited deficiency [112] or administration of an inhibitor (e.g., cancer chemotherapeutic agent [113]) determining an increased demand for de novo GSH synthesis [114]. De novo GSH synthesis is regulated by at least three factors: (1) the level of γ-GCL present in the cell; (2) the availability of its substrates, particularly Cys; and (3) the feedback inhibition of GSH on γ-GCL. In addition, there is recent evidence that γ-GCL activity can also be modulated by phosphorylation and nitrosation [105]. Knowledge about how these control mechanisms integrate to establish a tissue- or cell-specific rate of GSH synthesis is still scanty. Moreover, our understanding of the inter-relationship between the control mechanisms governing GSH synthesis and the mechanisms governing GSH and GSSG utilization is poor. The steady state level of GSH cannot be predicted from the level of substrates, enzymes, and transporters [105].

Carotenoids are among the most common natural pigments, and more than 600 different compounds have been characterized until now, with β-carotene being the most prominent. They are responsible for many of the red, orange, and yellow hues of plant leaves and fruits [115]. Only plants, bacteria, fungi, and algae can synthesize carotenoids, but many animals incorporate them from their diet [115]. Carotenoids serve as antioxidants in animals as efficient antioxidants scavenging singlet molecular oxygen and peroxyl radicals. Carotenoids are lipophilic molecules that tend to accumulate in lipophilic compartments like membranes or lipoproteins. The lipophilicity of these compounds also influences their absorption, transport, and excretion in the organism [116]. The scavenger activity of carotenoids toward $O_2^{\bullet-}$ largely depends on physical quenching which involves direct energy transfer between the two molecules. The energy of radical oxygen is transferred to the carotenoid molecule leaving a ground state oxygen and a triplet-excited carotene. Instead of further chemical reactions, the carotenoid returns to ground state dissipating its energy by interaction with the surrounding solvent. Since the carotenoids remain intact during physical quenching of $O_2^{\bullet-}$, they can be reused several fold in such quenching cycles [117, 118]. Among the various carotenoids, xanthophylls as well as carotenes proved to be efficient quenchers of singlet oxygen. Among the various radicals which are formed under oxidative conditions in the organism, carotenoids most efficiently react with peroxyl radicals generated in the process of lipid peroxidation. Due to their lipophilicity and specific property to scavenge peroxyl radicals, carotenoids are thought to play an important role in the protection of cellular membranes and lipoproteins against oxidative damage [119].

2.6 Role of Nrf2 Transcriptional Factor in Liver Antioxidant Defense

The liver performs many essential functions related to digestion, metabolism, immunity, and the storage of nutrients within the body. All these physiological activities are associated with ROS and RNS production with whom hepatocytes have to cope to maintain the redox balance. In addition, oxidative stress is a major pathogenetic event occurring in several liver disease and ROS can influence gene expression profile by affecting intracellular signal transduction pathways through the modulation of protein structure/function [120]. ROS and RNS activate signal transduction through receptor protein tyrosine kinase and protein tyrosine phosphatase [121], although the precise mechanisms of ROS-mediated signaling responsible for the transcriptional regulation of antioxidant response elements (ARE) are not fully understood. AP1 and NF-κB families of TFs are considered as the most critical downstream components of the redox-sensitive signal transduction. However, over the last decade the nuclear erythroid 2-related factor 2 (Nrf2) has emerged as a significant transcription factor for the induction of a variety of detoxification enzymes in response to oxidative stress conditions and in metabolic processes of detoxification and elimination of potentially harmful exogenous chemicals and their metabolites [122]. In light of its role in detoxification and its ability to facilitate hepatoprotection, Nrf2 has more recently been considered as a target for new therapeutic approaches aimed to treat liver diseases [123]. Under normal physiological conditions, Nrf2 is sequestered in the cytosol by the actin-binding protein kelch-like ECH associating protein 1 (Keap1) [124], which functions as an adaptor for Cullin 3 (Cul3), an E3-based ligase, which targets Nrf2 for ubiquitination and subsequent proteasomal degradation [125]. This mechanism of proteasomal degradation of Nrf2 is very efficient as the half-life of Nrf2 under homeostatic conditions is approximately 20 min [124, 126]. However, under oxidative stress conditions, the interaction between Nrf2 and Keap1 is disrupted determining a decreased proteasomal degradation of Nrf2 with subsequent accumulation of free Nrf2 in the cytosol, and an increase in Nrf2 translocation into the nucleus [127]. Once in the nucleus, Nrf2 heterodimerizes with a small musculo-aponeurotic fibrosarcoma (Maf) protein and binds to ARE. The Nrf2/Maf complex then recruits CREB-binding protein and p300, which have been implicated in the recruitment of histone acetyltransferases and RNA polymerases [128, 129]. The entire complex then initializes transcription of a large battery of cytoprotective genes [130].

NAD(P)H:quinone oxidoreductase 1 (Nqo1) is a cytosolic flavoprotein catalyzing the two-electron reductive metabolism and detoxification of endogenous and exogenous chemicals [131]. Nqo1 is critical for cytoprotection against many highly reactive and potentially damaging quinones. As would be expected, Nrf2-null mice showed reduced constitutive expression and activity of Nqo1 in liver, stomach, and small intestine [130, 132–134].

Heme oxygenase-1 (Ho-1) catalyzes the first and rate-limiting step in the catabolism of the pro-oxidant heme to carbon monoxide, biliverdin, and free iron. Ho-1 can have both anti-oxidative and anti-inflammatory effects, as biliverdin can be reduced to the antioxidant bilirubin, by biliverdin reductase, and small amounts of carbon monoxide can have anti-inflammatory effects [135]. Both Ho-1 mRNA and protein expression levels are commonly up-regulated following oxidative stress and cellular injury [136] and Nrf2 has been shown to directly regulate *Ho-1* promoter activity through the interaction with the *Ho-1* promoter-binding sites resembling AREs [137]. However, other transcriptional mechanisms can regulate Ho-1 expression [138, 139]. As previously reported, GST and γ-GCL are involved in the maintenance of redox balance and protection against oxidative insult, and their expression is regulated by Nrf2. Both genes contain ARE sequences in their promoters [140] and decreased levels of GSH and reduced expression of GST mRNA and protein were reported in Nrf2-null mice [140, 141].

Sulfiredoxin 1 (Srxn1) catalyzes the reduction of the active site of PRXs, thereby converting this important enzyme from an inactive to an active state. Srxn1 has a functional ARE in its promoter region and can be rapidly induced (20-fold) in vivo in the liver by administration of the Nrf2 activator CDDO-trifluoroethylamide (CDDO-TFEA) [142].

Numerous studies demonstrated that targeted deletion of Nrf2 in Nrf2-null mice leads to enhanced susceptibility to hepatic injury. Acetaminophen is a well-known and often used hepatotoxicant and was the first compound used to verify the effects of a loss of Nrf2. In two separate studies a dose of 300 mg/kg or greater of acetaminophen administrated in Nrf2-null mice caused death in the homozygous knockout mice only. Those that survived showed a greater severity in hepatic damage than the wild-type mice, as demonstrated by increased plasma alanine aminotransferase activity, decreased hepatic non-protein sulfhydryl (NPSH) content, and centrilobular hepatocellular necrosis [143, 144]. More recent studies using arsenic and pentachlorophenol (PCB) demonstrated the deleterious effects of knocking out Nrf2 in the liver [145]. Treatments on Nrf2-null mice determined DNA hypomethylation, oxidative DNA damage, and apoptotic cell death. In a chronic model of liver fibrosis utilizing carbon tetrachloride (CCL_4), Nrf2-null mice had aggravated liver fibrosis and an increased inflammatory response, as quantified by hepatic IL-1α, TNFα, and IFNγ mRNA expression, than when CCL_4 was given to wild-type mice [146]. The role of Nrf2 was also investigated in liver regeneration processes that resulted significantly impaired in partially hepatectomized Nrf2-null mice, as a result of increased oxidative stress and impaired insulin/insulin growth factor-1 signaling [147]. In conclusion, several evidences support the link between protection toward oxidative damage and Nrf2 transcriptional control activity over cytoprotective genes. For this reason, the possibility to use small molecules to activate Nrf2 as a counteraction against oxidative damage for therapeutic applications is currently actively investigated.

2.7 Conclusions

A correct balance between pro- and anti-oxidants is crucial for the correct functioning of any living cell. Oxidative events lead to the production of ROS and are rapidly counteracted by several, different mechanisms. ROS act by damaging biomolecules, but they also control the function of transcriptional factors, therefore leading to profound changes in gene expression. Acute and chronic liver diseases are characterized by an imbalance of the cellular redox state and are a common characteristic in inflammation. Unfortunately, only few and scattered data are available on molecular biomarkers of ROS production and redox alteration. In this scenario the study of redox state is of the utmost importance and represent one of the first examples of the so-called translational medicine. It is expected that the future will teach us more on the applicability of the "bench" discovery to the "bed" scenario.

References

1. Davies KJ (2000) Oxidative stress, antioxidant defenses, and damage removal, repair, and replacement systems. IUBMB Life 50(4–5):279–289
2. Davies KJ (1995) Oxidative stress: the paradox of aerobic life. Biochem Soc Symp 61:1–31
3. Birben E, Sahiner UM, Sackesen C, Erzurum S, Kalayci O (2012) Oxidative stress and antioxidant defense. World Allergy Organ J 5(1):9–19
4. Diesen DL, Kuo PC (2010) Nitric oxide and redox regulation in the liver: Part I. General considerations and redox biology in hepatitis. J Surg Res 162(1):95–109
5. Kang KJ (2002) Mechanism of hepatic ischemia/reperfusion injury and protection against reperfusion injury. Transplant Proc 34(7):2659–2661

6. Poli G, Leonarduzzi G, Biasi F, Chiarpotto E (2004) Oxidative stress and cell signalling. Curr Med Chem 11(9):1163–1182
7. Tell G, Vascotto C, Tiribelli C (2013) Alterations in the redox state and liver damage: hints from the EASL Basic School of Hepatology. J Hepatol 58(2):365–374
8. Loguercio C, Federico A (2003) Oxidative stress in viral and alcoholic hepatitis. Free Radic Biol Med 34:1–10
9. Kurose I, Higuchi H, Miura S, Saito H, Watanabe N, Hokari R et al (1997) Oxidative stress-mediated apoptosis of hepatocytes exposed to acute ethanol intoxication. Hepatology 25:368–378
10. Poli G, Parola M (1997) Oxidative damage and fibrogenesis. Free Radic Biol Med 22:287–305
11. Miller DM, Buettner GR, Aust SD (1990) Transition metals as catalysts of "autoxidation" reactions. Free Radic Biol Med 8(1):95–108
12. Monsalve M, Borniquel S, Valle I, Lamas S (2007) Mitochondrial dysfunction in human pathologies. Front Biosci 12:1131–1153
13. Loschen G, Azzi A, Richter C, Flohé L (1974) Superoxide radicals as precursors of mitochondrial hydrogen peroxide. FEBS Lett 42(1):68–72
14. Wang Y (2008) Bulky DNA lesions induced by reactive oxygen species. Chem Res Toxicol 21(2):276–281
15. Cesaratto L, Vascotto C, Calligaris S, Tell G (2004) The importance of redox state in liver damage. Ann Hepatol 3(3):86–92
16. Stamler JS, Singel DJ, Loscalzo J (1992) Biochemistry of nitric oxide and its redox-activated forms. Science 258(5090):1898–1902
17. Chartrain NA, Geller DA, Koty PP, Sitrin NF, Nussler AK, Hoffman EP, Billiar TR, Hutchinson NI, Mudgett JS (1994) Molecular cloning, structure, and chromosomal localization of the human inducible nitric oxide synthase gene. J Biol Chem 269(9):6765–6772
18. Taylor BS, Alarcon LH, Billiar TR (1998) Inducible nitric oxide synthase in the liver: regulation and function. Biochemistry (Mosc) 63(7):766–781
19. Curran RD, Billiar TR, Stuehr DJ, Hofmann K, Simmons RL (1989) Hepatocytes produce nitrogen oxides from L-arginine in response to inflammatory products of Kupffer cells. J Exp Med 170(5):1769–1774
20. Billiar TR, Curran RD, Stuehr DJ, West MA, Bentz BG, Simmons RL (1989) An L-arginine-dependent mechanism mediates Kupffer cell inhibition of hepatocytes protein synthesis in vitro. J Exp Med 169(4):1467–1472
21. Spitzer JA (1994) Cytokine stimulation of nitric oxide formation and differential regulation in hepatocytes and nonparenchymal cells of endotoxemic rats. Hepatology 19(1):217–228
22. Geller DA, Di Silvio M, Nussler AK, Wang SC, Shapiro RA, Simmons RL, Billiar TR (1993) Nitric oxide synthase expression is induced in hepatocytes in vivo during hepatic inflammation. J Surg Res 55(4):427–432
23. Sánchez-Alcázar JA, Schneider E, Martínez MA, Carmona P, Hernández-Muñoz I, Siles E, De La Torre P, Ruiz-Cabello J, García I, Solis-Herruzo JA (2000) Tumor necrosis factor-alpha increases the steady-state reduction of cytochrome b of the mitochondrial respiratory chain in metabolically inhibited L929 cells. J Biol Chem 275(18):13353–13361
24. Gow AJ, Stamler JS (1998) Reactions between nitric oxide and haemoglobin under physiological conditions. Nature 391(6663):169–173
25. Palmer RM, Rees DD, Ashton DS, Moncada S (1988) L-arginine is the physiological precursor for the formation of nitric oxide in endothelium-dependent relaxation. Biochem Biophys Res Commun 153(3):1251–1256
26. Kasai H, Chung MH, Jones DS, Inoue H, Ishikawa H, Kamiya H, Ohtsuka E, Nishimura S (1991) 8-Hydroxyguanine, a DNA adduct formed by oxygen radicals: its implication on oxygen radical-involved mutagenesis/carcinogenesis. J Toxicol Sci 16(Suppl 1):95–105
27. Ghosh R, Mitchell DL (1999) Effect of oxidative DNA damage in promoter elements on transcription factor binding. Nucleic Acids Res 27(15):3213–3218
28. Lindahl T (1993) Instability and decay of the primary structure of DNA. Nature 362(6422): 709–715

29. Dalhus B, Laerdahl JK, Backe PH, Bjørås M (2009) DNA base repair—recognition and initiation of catalysis. FEMS Microbiol Rev 33(6):1044–1078
30. Lindahl T (1974) An N-glycosidase from Eschcrichia coli that releases free uracil from DNA containing deaminated cytosine residues. Proc Natl Acad Sci U S A 71(9):3649–3653
31. Demple B, Herman T, Chen DS (1991) Cloning and expression of APE, the cDNA encoding the major human apurinic endonuclease: definition of a family of DNA repair enzymes. Proc Natl Acad Sci U S A 88(24):11450–11454
32. Levine RL, Moskovitz J, Stadtman ER (2000) Oxidation of methionine in proteins: roles in antioxidant defense and cellular regulation. IUBMB Life 50(4–5):301–307
33. Dalle-Donne I, Scaloni A, Giustarini D, Cavarra E, Tell G, Lungarella G, Colombo R, Rossi R, Milzani A (2005) Proteins as biomarkers of oxidative/nitro sative stress in diseases: the contribution of redox proteomics. Mass Spectrom Rev 24(1):55–99
34. Dalle-Donne I, Aldini G, Carini M, Colombo R, Rossi R, Milzani A (2006) Protein carbonylation, cellular dysfunction, and disease progression. J Cell Mol Med 10(2):389–406
35. Reddy S, Jones AD, Cross CE, Wong PS, Van Der Vliet A (2000) Inactivation of creatine kinase by S-glutathionylation of the active-site cysteine residue. Biochem J 347(Pt 3):821–827
36. Toda T, Nakamura M, Morisawa H, Hirota M, Nishigaki R, Yoshimi Y (2010) Proteomic approaches to oxidative protein modifications implicated in the mechanism of aging. Geriatr Gerontol Int 10(Suppl 1):S25–S31
37. Sun H, Gao J, Ferrington DA, Biesiada H, Williams TD, Squier TC (1999) Repair of oxidized calmodulin by methionine sulfoxide reductase restores ability to activate the plasma membrane Ca-ATPase. Biochemistry 38(1):105–112
38. Skov K, Hofmann T, Williams GR (1969) The nitration of cytochrome c. Can J Biochem 47(7):750–752
39. Spickett CM (2013) The lipid peroxidation product 4-hydroxy-2-nonenal: Advances in chemistry and analysis. Redox Biol 1(1):145–152
40. Farmer EE, Mueller MJ (2013) ROS-mediated lipid peroxidation and RES-activated signaling. Annu Rev Plant Biol 64:429–450
41. Guéraud F, Atalay M, Bresgen N, Cipak A, Eckl PM, Huc L, Jouanin I, Siems W, Uchida K (2010) Chemistry and biochemistry of lipid peroxidation products. Free Radic Res 44(10):1098–1124
42. Benedetti A, Comporti M, Esterbauer H (1980) Identification of 4-hydroxynonenal as a cytotoxic product originating from the peroxidation of liver microsomal lipids. Biochim Biophys Acta 620(2):281–296
43. Pardo CA, Xu Z, Borchelt DR, Price DL, Sisodia SS, Cleveland DW (1995) Superoxide dismutase is an abundant component in cell bodies, dendrites, and axons of motor neurons and in a subset of other neurons. Proc Natl Acad Sci U S A 92(4):954–958
44. Rotunno MS, Bosco DA (2013) An emerging role for misfolded wild-type SOD1 in sporadic ALS pathogenesis. Front Cell Neurosci 7:253
45. Georgiou G (2002) How to flip the (redox) switch. Cell 111(5):607–610
46. Rhee SG (2006) Cell signaling. H2O2, a necessary evil for cell signaling. Science 312(5782):1882–1883
47. Brown DI, Griendling KK (2009) Nox proteins in signal transduction. Free Radic Biol Med 47(9):1239–1253
48. Dhar SK, St Clair DK (2012) Manganese superoxide dismutase regulation and cancer. Free Radic Biol Med 52(11–12):2209–2222
49. Jones PL, Ping D, Boss JM (1997) Tumor necrosis factor alpha and interleukin-1beta regulate the murine manganese superoxide dismutase gene through a complex intronic enhancer involving C/EBP-beta and NF-kappaB. Mol Cell Biol 17(12):6970–6981
50. Xu Y, Kiningham KK, Devalaraja MN, Yeh CC, Majima H, Kasarskis EJ, St Clair DK (1999) An intronic NF-kappaB element is essential for induction of the human manganese superoxide dismutase gene by tumor necrosis factor-alpha and interleukin-1beta. DNA Cell Biol 18(9):709–722

51. St Clair DK, Porntadavity S, Xu Y, Kiningham K (2002) Transcription regulation of human manganese superoxide dismutase gene. Methods Enzymol 349:306–312
52. Kim SJ, Angel P, Lafyatis R, Hattori K, Kim KY, Sporn MB, Karin M, Roberts AB (1990) Autoinduction of transforming growth factor beta 1 is mediated by the AP-1 complex. Mol Cell Biol 10(4):1492–1497
53. Qadri I, Iwahashi M, Capasso JM, Hopken MW, Flores S, Schaack J, Simon FR (2004) Induced oxidative stress and activated expression of manganese superoxide dismutase during hepatitis C virus replication: role of JNK, p38 MAPK and AP-1. Biochem J 378(Pt 3):919–928
54. Eckert D, Buhl S, Weber S, Jäger R, Schorle H (2005) The AP-2 family of transcription factors. Genome Biol 6(13):246, Epub 2005 Dec 28
55. Zhao Y, St Clair DK (2003) Detection of the content and activity of the transcription factor AP-1 in a multistage skin carcinogenesis model. Methods Mol Biol 218:177–184
56. Díaz A, Loewen PC, Fita I, Carpena X (2012) Thirty years of heme catalases structural biology. Arch Biochem Biophys 525(2):102–110
57. Góth L, Nagy T (2013) Inherited catalase deficiency: is it benign or a factor in various age related disorders? Mutat Res 753(2):147–154
58. Flohé L (1988) Glutathione peroxidase. Basic Life Sci 49:663–668
59. Arthur JR (2000) The glutathione peroxidases. Cell Mol Life Sci 57(13–14):1825–1835
60. Chu FF, Doroshow JH, Esworthy RS (1993) Expression, characterization, and tissue distribution of a new cellular selenium-dependent glutathione peroxidase, GSHPx-GI. J Biol Chem 268(4):2571–2576
61. Takahashi K, Newburger PE, Cohen HJ (1986) Glutathione peroxidase protein. Absence in selenium deficiency states and correlation with enzymatic activity. J Clin Invest 77(4):1402–1404
62. Avissar N, Ornt DB, Yagil Y, Horowitz S, Watkins RH, Kerl EA, Takahashi K, Palmer IS, Cohen HJ (1994) Human kidney proximal tubules are the main source of plasma glutathione peroxidase. Am J Physiol 266(2 Pt 1):C367–C375
63. Avissar N, Eisenmann C, Breen JG, Horowitz S, Miller RK, Cohen HJ (1994) Human placenta makes extracellular glutathione peroxidase and secretes it into maternal circulation. Am J Physiol 267(1 Pt 1):E68–E76
64. Maser RL, Magenheimer BS, Calvet JP (1994) Mouse plasma glutathione peroxidase cDNA sequence analysis and renal proximal tubular expression and secretion. J Biol Chem 269(43):27066–27073
65. Tham DM, Whitin JC, Kim KK, Zhu SX, Cohen HJ (1998) Expression of extracellular glutathione peroxidase in human and mouse gastrointestinal tract. Am J Physiol 275(6 Pt 1): G1463–G1471
66. Howie AF, Walker SW, Akesson B, Arthur JR, Beckett GJ (1995) Thyroidal extracellular glutathione peroxidase: a potential regulator of thyroid-hormone synthesis. Biochem J 308(Pt 3):713–717
67. Schuckelt R, Brigelius-Flohé R, Maiorino M, Roveri A, Reumkens J, Strassburger W, Ursini F, Wolf B, Flohé L (1991) Phospholipid hydroperoxide glutathione peroxidase is a selenoenzyme distinct from the classical glutathione peroxidase as evident from cDNA and amino acid sequencing. Free Radic Res Commun 14(5–6):343–361
68. Maiorino M, Thomas JP, Girotti AW, Ursini F (1991) Reactivity of phospholipid hydroperoxide glutathione peroxidase with membrane and lipoprotein lipid hydroperoxides. Free Radic Res Commun 12–13(Pt 1):131–135
69. Ahsan MK, Lekli I, Ray D, Yodoi J, Das DK (2009) Redox regulation of cell survival by the thioredoxin superfamily: an implication of redox gene therapy in the heart. Antioxid Redox Signal 11(11):2741–2758
70. Holmgren A (1985) Thioredoxin. Annu Rev Biochem 54:237–271
71. Holmgren A (1989) Thioredoxin and glutaredoxin systems. J Biol Chem 264(24): 13963–13966
72. Lillig CH, Berndt C, Holmgren A (2008) Glutaredoxin systems. Biochim Biophys Acta 1780(11):1304–1317

73. Fernandes AP, Holmgren A (2004) Glutaredoxins: glutathione-dependent redox enzymes with functions far beyond a simple thioredoxin backup system. Antioxid Redox Signal 6(1):63–74
74. Wood ZA, Schröder E, Robin Harris J, Poole LB (2003) Structure, mechanism and regulation of peroxiredoxins. Trends Biochem Sci 28(1):32–40
75. Rhee SG, Jeong W, Chang TS, Woo HA (2007) Sulfiredoxin, the cysteine sulfinic acid reductase specific to 2-Cys peroxiredoxin: its discovery, mechanism of action, and biological significance. Kidney Int Suppl 106:S3–S8
76. Biteau B, Labarre J, Toledano MB (2003) ATP-dependent reduction of cysteine-sulphinic acid by S cerevisiae sulphiredoxin. Nature 425(6961):980–984
77. Chang TS, Jeong W, Woo HA, Lee SM, Park S, Rhee SG (2004) Characterization of mammalian sulfiredoxin and its reactivation of hyperoxidized peroxiredoxin through reduction of cysteine sulfinic acid in the active site to cysteine. J Biol Chem 279(49):50994–51001
78. Li B, Ishii T, Tan CP, Soh JW, Goff SP (2002) Pathways of induction of peroxiredoxin I expression in osteoblasts: roles of p38 mitogen-activated protein kinase and protein kinase C. J Biol Chem 277(14):12418–12422
79. Chang JW, Jeon HB, Lee JH, Yoo JS, Chun JS, Kim JH, Yoo YJ (2001) Augmented expression of peroxiredoxin I in lung cancer. Biochem Biophys Res Commun 289(2):507–512
80. Zimniak P (2006) Substrate and reaction mechanism of GSTs. In: Taylor Awasthi YC (ed) Toxicology of glutathione transferases. Francis CRC Press, Boca Raton, pp 71–102
81. Yang Y, Awasthi YC (2006) Glutathione S-transferases as modulators of signal transduction. In: Taylor Awasthi YC (ed) Toxicology of glutathione transferases. Francis CRC Press, Boca Raton, pp 205–230
82. Hayes JD, Flanagan JU, Jowsey IR (2005) Glutathione transferases. Annu Rev Pharmacol Toxicol 45:51–88
83. Lu SC (2009) Regulation of glutathione synthesis. Mol Aspects Med 30(1–2):42–59
84. Brigelius-Flohé R, Traber MG (1999) Vitamin E: function and metabolism. FASEB J 13(10):1145–1155
85. IUPAC-IUB Joint Commission on Biochemical Nomenclature (JCBN) (1982) Nomenclature of tocopherols and related compounds. Recommendations 1981. Eur J Biochem 123(3):473–475
86. Traber MG, Sies H (1996) Vitamin E in humans: demand and delivery. Annu Rev Nutr 16:321–347
87. Drevon CA (1991) Absorption, transport and metabolism of vitamin E. Free Radic Res Commun 14(4):229–246
88. Laplante P, Vanasse M, Michaud J, Geoffroy G, Brochu P (1984) A progressive neurological syndrome associated with an isolated vitamin E deficiency. Can J Neurol Sci 11(4 Suppl):561–564
89. Krendel DA, Gilchrist JM, Johnson AO, Bossen EH (1987) Isolated deficiency of vitamin E with progressive neurologic deterioration. Neurology 37(3):538–540
90. Kayden HJ, Silber R, Kossmann CE (1965) The role of vitamin E deficiency in the abnormal autohemolysis of acanthocytosis. Trans Assoc Am Physicians 78:334–342
91. Simon E, Paul JL, Atger V, Simon A, Moatti N (1998) Erythrocyte antioxidant status in asymptomatic hypercholesterolemic men. Atherosclerosis 138(2):375–381
92. Muller DP, Lloyd JK, Wolff OH (1985) The role of vitamin E in the treatment of the neurological features of abetalipoproteinaemia and other disorders of fat absorption. J Inherit Metab Dis 8(Suppl 1):88–92
93. Jaffe GM (1984) Vitamin C. In: Machlin L (ed) Handbook of vitamins. Marcel Dekker, New York, pp 199–244
94. Woodall AA, Ames BN (1997) Diet and oxidative damage to DNA: the importance of ascorbate as an antioxidant. In: Packer L, Fuchs J (eds) Vitamin C in health and disease. Marcel Dekker, New York, pp 193–203
95. Levine M (1986) New concepts in the biology and biochemistry of ascorbic acid. N Engl J Med 314(14):892–902

96. Burri BJ, Jacob RA (1997) Human metabolism and the requirement for vitamin C. In: Packer L, Fuchs J (eds) Vitamin C in health and disease. Marcel Dekker, New York, pp 341–366
97. Tsao CS (1997) An overview of ascorbic acid chemistry and biochemistry. In: Packer L, Fuchs J (eds) Vitamin C in health and disease. Marcel Dekker, New York, pp 25–58
98. Carr AC, Frei B (1999) Toward a new recommended dietary allowance for vitamin C based on antioxidant and health effects in humans. Am J Clin Nutr 69(6):1086–1107
99. Frei B, England L, Ames BN (1989) Ascorbate is an outstanding antioxidant in human blood plasma. Proc Natl Acad Sci U S A 86(16):6377–6381
100. Frei B, Stocker R, England L, Ames BN (1990) Ascorbate: the most effective antioxidant in human blood plasma. Adv Exp Med Biol 264:155–163
101. Bowry VW, Mohr D, Cleary J, Stocker R (1995) Prevention of tocopherol-mediated peroxidation in ubiquinol-10-free human low density lipoprotein. J Biol Chem 270(11):5756–5763
102. Edge R, Truscott TG (1997) Prooxidant and antioxidant reaction mechanisms of carotene and radical interactions with vitamins E and C. Nutrition 13(11–12):992–994
103. Packer L (1997) Vitamin C and redox cycling antioxidants. In: Packer L, Fuchs J (eds) Vitamin C in health and disease. Marcel Dekker, New York, pp 95–121
104. Halliwell B (1996) Vitamin C: antioxidant or pro-oxidant in vivo? Free Radic Res 25(5):439–454
105. Griffith OW (1999) Biologic and pharmacologic regulation of mammalian glutathione synthesis. Free Radic Biol Med 27(9–10):922–935
106. Meister A (1982) Metabolism and function of glutathione: an overview. Biochem Soc Trans 10(2):78–79
107. Chen Y, Yang Y, Miller ML, Shen D, Shertzer HG, Stringer KF, Wang B, Schneider SN, Nebert DW, Dalton TP (2007) Hepatocyte-specific Gclc deletion leads to rapid onset of steatosis with mitochondrial injury and liver failure. Hepatology 45(5):1118–1128
108. Raza H (2011) Dual localization of glutathione S-transferase in the cytosol and mitochondria: implications in oxidative stress, toxicity and disease. FEBS J 278(22):4243–4251
109. Lash LH (2006) Mitochondrial glutathione transport: physiological, pathological and toxicological implications. Chem Biol Interact 163(1–2):54–67
110. Kalyanaraman B, Karoui H, Singh RJ, Felix CC (1996) Detection of thiyl radical adducts formed during hydroxyl radical- and peroxynitrite-mediated oxidation of thiols—a high resolution ESR spin-trapping study at Q-band (35 GHz). Anal Biochem 241(1):75–81
111. Briviba K, Klotz LO, Sies H (1999) Defenses against peroxynitrite. Methods Enzymol 301:301–311
112. Tanaka KR, Paglia DE (1995) Pyruvate kinase and other enzymopathies of the erythrocyte. In: Scriver CR, Beaudet al, Sly WS, Valle D (eds) The metabolic basis of inherited disease. McGraw-Hill, New York, pp 3485–3512
113. Babson JR, Reed DJ (1978) Inactivation of glutathione reductase by 2-chloroethyl nitrosourea-derived isocyanates. Biochem Biophys Res Commun 83(2):754–762
114. Akerboom T, Sies H (1990) Glutathione transport and its significance in oxidative stress. In: Viña J (ed) Glutathione: metabolism and physiological functions. CRC Press, Boca Raton, pp 45–55
115. Stahl W, Sies H (2003) Antioxidant activity of carotenoids. Mol Aspects Med 24(6):345–351
116. Mayne ST (1996) Beta-carotene, carotenoids, and disease prevention in humans. FASEB J 10(7):690–701
117. Krasnovsky AA Jr, Rodgers MA, Galpern MG, Rihter B, Kenney ME, Lukjanetz EA (1992) Quenching of singlet molecular oxygen by phthalocyanines and naphthalocyanines. Photochem Photobiol 55(5):691–696
118. Conn PF, Schalch W, Truscott TG (1991) The singlet oxygen and carotenoid interaction. J Photochem Photobiol B 11(1):41–47. Erratum in: J Photochem Photobiol B 1993;17(1):89
119. Sies H, Stahl W (1995) Vitamins E and C, beta-carotene, and other carotenoids as antioxidants. Am J Clin Nutr 62(6 Suppl):1315S–1321S
120. Cho HY, Reddy SP, Kleeberger SR (2006) Nrf2 defends the lung from oxidative stress. Antioxid Redox Signal 8(1–2):76–87

121. Denu JM, Tanner KG (1998) Specific and reversible inactivation of protein tyrosine phosphatases by hydrogen peroxide: evidence for a sulfenic acid intermediate and implications for redox regulation. Biochemistry 37(16):5633–5642
122. Bauer AK, Hill T 3rd, Alexander CM (2013) The involvement of NRF2 in lung cancer. Oxid Med Cell Longev 2013:746432
123. Aleksunes LM, Manautou JE (2007) Emerging role of Nrf2 in protecting against hepatic and gastrointestinal disease. Toxicol Pathol 35(4):459–473
124. Itoh K, Wakabayashi N, Katoh Y, Ishii T, Igarashi K, Engel JD, Yamamoto M (1999) Keap1 represses nuclear activation of antioxidant responsive elements by Nrf2 through binding to the amino-terminal Neh2 domain. Genes Dev 13(1):76–86
125. Kobayashi A, Kang MI, Okawa H, Ohtsuji M, Zenke Y, Chiba T, Igarashi K, Yamamoto M (2004) Oxidative stress sensor Keap1 functions as an adaptor for Cul3-based E3 ligase to regulate proteasomal degradation of Nrf2. Mol Cell Biol 24(16):7130–7139
126. McMahon M, Itoh K, Yamamoto M, Hayes JD (2003) Keap1-dependent proteasomal degradation of transcription factor Nrf2 contributes to the negative regulation of antioxidant response element-driven gene expression. J Biol Chem 278(24):21592–21600
127. Li W, Kong AN (2009) Molecular mechanisms of Nrf2-mediated antioxidant response. Mol Carcinog 48(2):91–104
128. Zhu M, Fahl WE (2001) Functional characterization of transcription regulators that interact with the electrophile response element. Biochem Biophys Res Commun 289(1):212–219
129. Vo N, Goodman RH (2001) CREB-binding protein and p300 in transcriptional regulation. J Biol Chem 276(17):13505–13508, Epub 2001 Mar 8
130. Itoh K, Chiba T, Takahashi S, Ishii T, Igarashi K, Katoh Y, Oyake T, Hayashi N, Satoh K, Hatayama I, Yamamoto M, Nabeshima Y (1997) An Nrf2/small Maf heterodimer mediates the induction of phase II detoxifying enzyme genes through antioxidant response elements. Biochem Biophys Res Commun 236(2):313–322
131. Ross D (2004) Quinone reductases multitasking in the metabolic world. Drug Metab Rev 36(3–4):639–654
132. Li Y, Jaiswal AK (1992) Regulation of human NAD(P)H:quinone oxidoreductase gene. Role of AP1 binding site contained within human antioxidant response element. J Biol Chem 267(21):15097–15104. Erratum in: J Biol Chem 1993;268(28):21454
133. McMahon M, Itoh K, Yamamoto M, Chanas SA, Henderson CJ, McLellan LI, Wolf CR, Cavin C, Hayes JD (2001) The Cap'n'Collar basic leucine zipper transcription factor Nrf2 (NF-E2 p45-related factor 2) controls both constitutive and inducible expression of intestinal detoxification and glutathione biosynthetic enzymes. Cancer Res 61(8):3299–3307
134. Ramos-Gomez M, Kwak MK, Dolan PM, Itoh K, Yamamoto M, Talalay P, Kensler TW (2001) Sensitivity to carcinogenesis is increased and chemoprotective efficacy of enzyme inducers is lost in nrf2 transcription factor-deficient mice. Proc Natl Acad Sci U S A 98(6):3410–3415
135. Wunder C, Potter RF (2003) The heme oxygenase system: its role in liver inflammation. Curr Drug Targets Cardiovasc Haematol Disord 3(3):199–208
136. Guo X, Shin VY, Cho CH (2001) Modulation of heme oxygenase in tissue injury and its implication in protection against gastrointestinal diseases. Life Sci 69(25–26):3113–3119
137. Alam J, Stewart D, Touchard C, Boinapally S, Choi AM, Cook JL (1999) Nrf2, a Cap'n'Collar transcription factor, regulates induction of the heme oxygenase-1 gene. J Biol Chem 274(37):26071–26078
138. Reichard JF, Motz GT, Puga A (2007) Heme oxygenase-1 induction by NRF2 requires inactivation of the transcriptional repressor BACH1. Nucleic Acids Res 35(21):7074–7086
139. Moinova HR, Mulcahy RT (1998) An electrophile responsive element (EpRE) regulates beta-naphthoflavone induction of the human gamma-glutamylcysteine synthetase regulatory subunit gene. Constitutive expression is mediated by an adjacent AP-1 site. J Biol Chem 273(24):14683–14689
140. Reisman SA, Yeager RL, Yamamoto M, Klaassen CD (2009) Increased Nrf2 activation in livers from Keap1-knockdown mice increases expression of cytoprotective genes that

detoxify electrophiles more than those that detoxify reactive oxygen species. Toxicol Sci 108(1):35–47
141. Hayes JD, Chanas SA, Henderson CJ, McMahon M, Sun C, Moffat GJ, Wolf CR, Yamamoto M (2000) The Nrf2 transcription factor contributes both to the basal expression of glutathione S-transferases in mouse liver and to their induction by the chemopreventive synthetic antioxidants, butylated hydroxyanisole and ethoxyquin. Biochem Soc Trans 28(2):33–41
142. Soriano FX, Baxter P, Murray LM, Sporn MB, Gillingwater TH, Hardingham GE (2009) Transcriptional regulation of the AP-1 and Nrf2 target gene sulfiredoxin. Mol Cells 27(3):279–282
143. Enomoto A, Itoh K, Nagayoshi E, Haruta J, Kimura T, O'Connor T, Harada T, Yamamoto M (2001) High sensitivity of Nrf2 knockout mice to acetaminophen hepatotoxicity associated with decreased expression of ARE-regulated drug metabolizing enzymes and antioxidant genes. Toxicol Sci 59(1):169–177
144. Chan K, Han XD, Kan YW (2001) An important function of Nrf2 in combating oxidative stress: detoxification of acetaminophen. Proc Natl Acad Sci U S A 98(8):4611–4616
145. Jiang T, Huang Z, Chan JY, Zhang DD (2009) Nrf2 protects against As(III)-induced damage in mouse liver and bladder. Toxicol Appl Pharmacol 240(1):8–14
146. Xu W, Hellerbrand C, Köhler UA, Bugnon P, Kan YW, Werner S, Beyer TA (2008) The Nrf2 transcription factor protects from toxin-induced liver injury and fibrosis. Lab Invest 88(10):1068–1078
147. Beyer TA, Xu W, Teupser D, auf dem Keller U, Bugnon P, Hildt E, Thiery J, Kan YW, Werner S (2008) Impaired liver regeneration in Nrf2 knockout mice: role of ROS-mediated insulin/IGF-1 resistance. EMBO J 27(1):212–223, Epub 2007 Dec 6

Chapter 3
Oxidative Mechanisms in Liver Senescence and Regeneration

Fabio Marongiu, Maria Paola Serra, and Ezio Laconi

3.1 Oxygen and the Evolution of the Biosphere

Life on earth emerged in an anoxic environment. The earliest prokaryotes that inhabited our planet for hundreds of million years relied on H_2, H_2S, and CH_4 as potential electron donors and were perfectly adapted to live in the absence of what is today an essential element of the biosphere, i.e., O_2. Oxygen began to accumulate in oceans and then in the atmosphere about 2.2 billion years ago, following the emergence of oxygenic photosynthetic microorganisms, which were able to split water and release molecular O_2. This event is rightly considered a major step in the evolution of the biosphere, for a series of important reasons. (1) The introduction of O_2 in an anaerobic environment acted as a cataclysm on the preexisting biosphere, given the unprotected vulnerability of life forms to oxygen-derived reactive species. (2) The ensuing strong selection pressure precipitated the development of novel enzymes and new pathways, in order to cope with the bio-toxicity of O_2 derivatives [151]. (3) Evidence indicates that rising concentrations of O_2 also caused the reshuffling of integral biochemical pathways, with many enzymatic reactions that were central to anoxic metabolism being effectively replaced in aerobic organisms [148]. (4) The availability of molecular oxygen also made possible a highly exergonic respiratory chain based on O_2 as a terminal electron acceptor, an event that is considered as crucial for the evolution of eukaryotes and complex multicellular life species [46, 48] and possibly for the colonization of terrestrial environments. Thus, both the quality and the quantity of existing biochemical networks were heavily impacted upon by the appearance of molecular oxygen [54, 149]. Using model systems, it has been

F. Marongiu • M.P. Serra • E. Laconi, M.D., Ph.D. (✉)
Department of Biomedical Sciences, Unit of Experimental Medicine,
University of Cagliari, Via Porcell 4, Cagliari 09124, Italy
e-mail: elaconi@unica.it

E. Albano, M. Parola (eds.), *Studies on Hepatic Disorders*,
Oxidative Stress in Applied Basic Research and Clinical Practice,
DOI 10.1007/978-3-319-15539-5_3

estimated that the introduction of O_2 into a set of 2,162 biochemical reactions involving 1,672 metabolites results in over 1,000 additional reactions producing 2,317 metabolites [149]. This is consistent with the intriguing hypothesis proposed by Stuart Kauffman in his book "Reinventing the Sacred". The author postulates that the biosphere is endowed with inherent creative power, due to the possibility of continuous expansion of the (bio)-chemical landscape into what he refers to as the "adjacent possible", i.e., any new (bio)-chemical path that emerges as thermodynamically accessible. Accordingly, the addition of a new and exquisitely reactive element such as oxygen into the primordial anaerobic biochemistry of our planet must have profoundly altered the biophysical environment and expanded the realm of the "adjacent possible" for the existing biosphere. It is a fact that the evolution of eukaryotes and complex multicellular organisms on earth was preceded by the appearance of O_2. Furthermore, several biochemical pathways (e.g., sterol biosynthesis) evolved to become dependent on oxygen availability. It has been speculated that aerobic metabolism may have facilitated biological evolution through promoting the establishment of nuclear signaling systems, which are essential to the eukaryotic cell. This hypothesis is based on the observation that most nuclear receptor ligands (which include in fact many steroid derivatives) are produced only through aerobic metabolism [79].

The O_2-dependent respiratory pathway generates four times as much energy per molecule of glucose oxidized than any of the ancient anaerobic pathways, providing new opportunities and avenues for the evolutionary process to unfold. On the other hand, it also exposes biological systems to the potential toxicity of newly formed reactive oxygen species (ROS), which are highly destructive on membrane lipids, on proteins, and on nucleic acids [96, 117]. This new challenge prompted two adaptive consequences, which also contributed to evolutionary changes: (1) it selected for both enzymatic and non-enzymatic mechanisms to protect the cell from ROS toxic potential; (2) it resulted in the incorporation of ROS as part of redox mechanisms regulating several biochemical pathways and basic biological processes, including cell cycle control and cellular senescence. The latter aspects will be discussed in the following paragraphs.

3.2 ROS and Cell Senescence

Two assumptions are often made in discussing the role of ROS in cell senescence: (1) senescence represents the cellular counterpart of organismal aging, and it is basically sustained by similar triggering mechanisms; (2) ROS represent the main contributing factor to both cell senescence and organismal aging. However, neither of these assumptions can be considered as proven beyond any reasonable doubt. In fact, it is still a matter of debate whether and to what extent aging of tissues is indeed linked to cellular senescence [10, 141]; nor is a direct causative involvement of ROS in aging universally accepted [69]. Thus, the role of oxidative stress in cell senescence will be discussed in light of these uncertainties.

3.2.1 What Is Cell Senescence?

In Greek mythology, Tithonus received from Zeus the gift of immortality but continued to experience the aging process. So, according to the wisdom behind ancient mythology, aging is an inevitable consequence of age, unless some miracle intervenes.

However, despite the fact that we witness the aging phenomenon in everyday life and in virtually all living organisms, including possibly prokaryotes [90], we still do not have a satisfactory answer as to the intimate nature of this process. Such basic biological question was tentatively laid on scientific grounds about half a century ago, following the description of cell senescence in cells in vitro, when Hayflick and colleagues showed that normal cells had a limited ability to proliferate in culture (Hayflick's limit) [67]. The concept met initially with considerable scepticism; in fact, it had been dogma for 60 years that the eventual death of normal human cells in culture was not due to some inherent property, but caused by ignorance of the proper conditions under which to culture them. And almost three additional decades had to elapse before telomere shortening was identified as the possible molecular counterpart of the Hayflick's limit [2, 65].

Nevertheless, the finding that normal cells could not replicate indefinitely, despite the presence of optimum growth conditions either in vitro or in vivo, carried two main conceptual implications. Given the fact that many cancer cells proliferate indefinitely in culture, the senescence response was interpreted as one mechanism that protected tissues and organisms from cancer. On the other hand, considering that the ability to perform tissue repair deteriorates with age, cellular senescence was seen as the possible basis for the aging process at organ and organismal levels, as exemplified by the loss of regenerative capacity of tissues in vivo.

The idea that a biological process such as cellular senescence can be both beneficial (tumour suppressive) and deleterious (limit tissue renewal) is consistent with a significant evolutionary theory of aging named antagonistic pleiotropy [169].

In agreement with the antagonistic pleiotropy theory of aging, natural selection has favoured genes providing advantages to the organism during the reproductive years, at the expenses of deterioration in the distant future. The "disposable soma" theory expresses this as a life-history strategy in which somatic maintenance is below the level required to prevent aging, hence allowing increased fertility.

However, the interpretation of cell senescence within the scheme of antagonistic pleiotropy is not universally accepted. Recently, it has been proposed that cell senescence is the end result of two opposing signals: one that reads for cell growth and the other imposing a persistent block in the cell cycle [15, 102, 163]. As such, it is perfectly functional to the biology of both young and older tissues and organisms. Furthermore, the discovery and characterization of the senescence-associated secretory phenotype (SASP) [40] has clearly indicated that cell senescence leads to profound changes in the tissue microenvironment, potentially modulating tissue repair and regeneration as well as the progression of chronic disease processes such as neoplasia.

An additional and most intriguing layer of complexity to the issue on the essence and significance of cell senescence has been added by the recent reports describing the presence of senescent cells during embryonic development [121, 171]. These findings add to the contention that the senescence phenotype is not necessarily the end result of accidental damage suffered by the cell, but might also represent the outcome of a programmed developmental process relevant to tissue remodeling and morphogenesis [11, 32, 144].

It is very possible that what we identify as senescent cells under various conditions (e.g., physiological aging, embryonic development, exposure to DNA-damaging agents) might actually represent distinct cellular reactions with specific biological significance, despite the presence of common phenotypic expressions that we use as markers of senescence (see next section). It is very likely that this uncertainty will persist until a comprehensive and satisfactory understanding about the nature of cell senescence is achieved [15, 60].

3.2.2 Biomarkers of Cellular Senescence

As we have seen, the increasing complexity of the senescence phenotype defies any satisfactory functional definition. However, several markers have been described and are being currently used to operationally identify senescent cells, both in cell cultures and in tissues. It should be emphasized that no single marker can be considered as pathognomonic of cellular senescence: it is rather a combination of phenotypic features, possibly with functional significance, that helps identifying senescent cells. The following is a short summary of the most reliable markers currently in use.

1. *Cell enlargement*. Flatness and enlarged cell shape of senescent cells was already reported in the seminal work of Hayflick and Moorhead [68], who first described the phenomenon of replicative senescence of human fibroblasts. An obvious marker for senescent cells is the lack of DNA replication, which is typically detected by the incorporation of 5-bromodeoxyuridine or ^{3}H-thymidine. However, the latter markers do not discriminate between senescent cells and quiescent or differentiated post-mitotic cells.
2. *SA-β-gal activity*. Since it was first reported [47], SA-β-gal activity has been the most widely utilized biomarker in detecting senescent cells because of the simplicity of the assay method and its apparent specificity for senescent cells. This histochemical stain identifies replicative senescence in a variety of cell types in vitro. Moreover, SA-β-gal activity has been detected in organs of old individuals and animals, suggesting that cellular senescence is an aging trait of organisms and that senescent cells accumulate with age in various tissues [142]. A major drawback of this marker is that it has no known functional significance towards the establishment of the senescence phenotype. In fact, SA-β-gal is thought to increase in senescent cells due to the expansion of lysosomes, which contain a specific β-galactosidase enzyme activity that can be measured at pH 6.0 and appears to be selective for senescent cells [47].

3. *DNA damage foci*. The use of DNA damage foci has been suggested as a marker for the detection of senescent cells [182]. Both DNA double-strand breaks (DSB) and telomere uncapping [42] are able to induce a DNA damage response (DDR). This reaction is characterized by activation of the ataxia telangiectasia-mutated (ATM)/ataxia telangiectasia-mutated and Rad3-related (ATR) protein complex, which is recruited to the site of damage, causing the phosphorylation of Ser-139 of histone H2AX molecules (γ-H2AX) next to the site of DNA damage. The phosphorylation of histone H2AX facilitates the focal assembly of checkpoint and DNA repair factors, such as 53BP1, MDC1/NFBD1 and NBS1, and also mediates the activation by phosphorylation of Chk1 and Chk2, which converge the signal on p53/p21 [12].
4. *Expression of inhibitors of cell cycle progression*. The cyclin-dependent kinase inhibitor (CDKI) p16 is an important regulator of senescence and is expressed by many senescent cells [19]. Investigation of the signalling pathways leading to oncogene-induced senescence (OIS, see the following sections) has shown that two main pathways are involved in cell cycle arrest, p16INK4a–RB (retinoblastoma) and ARF–p53. Among the important components of these pathways are CDKIs. CDKIs inhibit critical cell cycle-regulatory phosphorylation events, such as those that inactivate the growth-suppressive activity of pRB [166]. One such CDKI, p21CIP1/WAF1 (CDKN1A), is a direct target of p53 transactivation, generally in response to genomic damage, and is crucial for establishing and maintaining the p53-mediated senescence growth arrest. Another CDKI, p16INK4a (CDKN2A), is a tumour suppressor in its own right, can be induced by stress that does not entail DNA damage, and acts upstream of pRB to establish the pRB-regulated growth arrest [165].

3.2.3 *What Are the Triggers of Cell Senescence?*

Besides the classical replicative senescence described by Hayflick after repeated passages in vitro, a number of other stimuli are also able to elicit the senescence phenotype, both in cultured cells and in animal tissues (Fig. 3.1). Known senescence-inducing stimuli are the following:

1. *Telomere uncapping*. Replicative DNA polymerases lack the capacity to replicate completely the terminal ends of linear DNA molecules, a function that is proprietary of a specialized DNA polymerase known as telomerase. However, most mammalian somatic cells do not express telomerase and this leads to the progressive and cumulative loss of telomere-protective sequences from chromosome ends. The telomere-binding complex of proteins, known as shelterin, together with telomere DNA, functions as a dynamic unit that protects chromosome ends from being recognized as broken DNA [132]. When telomeres are critically eroded, they are recognized by DNA repair machinery as DSB, thereby enforcing a DDR leading to irreversible cell cycle arrest, the so-called replicative senescence, which coincides with Hayflick's limit [68, 127].

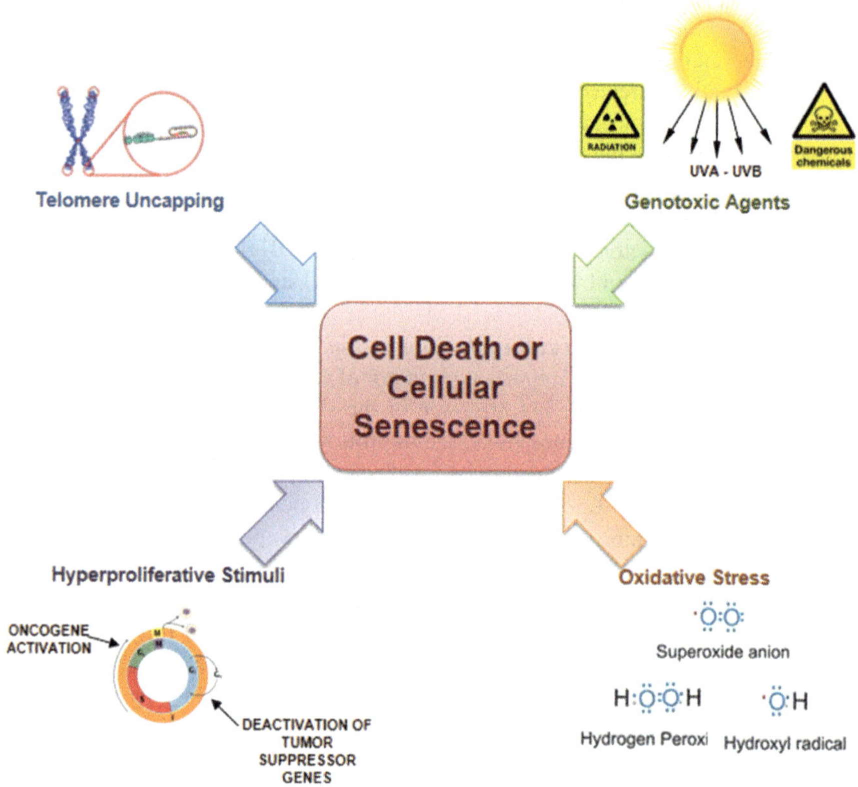

Fig. 3.1 Various stimuli can perturb the integrity or stability of the genetic information and trigger cell death or cell senescence. The extent of damage is critical in determining the type of outcome

2. *DNA damage by genotoxic agents*. Genotoxic agents such as radiation (X or γ-rays, UV) or DNA-interacting drugs (e.g., bleomycin, doxorubicin, mitomycin) cause DNA double-strand breaks (DSBs). DSB is a type of DNA damage which occurs when two nearby complementary strands of the double helix of DNA are damaged simultaneously. DSB is the most dangerous type of DNA damage, because it has been suggested that a single unrepaired DSB is sufficient for the initiation of the cell death program [153].
3. *DNA damage by oxidative stress*. The generation of ROS has been associated to the DDR induced by DSBs and has been implicated in the shortening of telomeres and hence in "replicative senescence" [177, 181]. Telomeres are indeed emerging as perfect oxidative stress sensors as they are particularly sensitive to oxidative DNA damage. The reason is twofold: first, they acquire DNA damage at a faster rate than the rest of genome. This is due to triple-G structure, present in telomeres of all eukaryotes, which are exquisitely sensitive to oxidative damage. Second, the repair of the damage is less efficient at telomeres probably

due to telomere-binding proteins that restrict access to telomeric DNA by repair complexes [139].

In fact, according to a proposed hypothesis, all DNA-damaging agents induce cell senescence via interaction with telomeres, i.e., telomeric DNA damage, however induced, is the ultimate trigger of senescence, due to its inherent irreparable nature [59, 70].

4. *DNA damage by oncogenic or hyper-proliferative signals*. Oncogenes are mutant versions of normal genes that have the potential to alter the cell behaviour towards a transformed phenotype. Normal cells respond to the activation of some oncogenes by eliciting a senescent phenotype76. Because oncogenes that induce senescence stimulate cell division, the senescent response could be aimed at counteracting excessive activation of the cell cycle, which would disrupt normal homeostatic mechanisms for the control of tissue mass [30, 109]. The pathways mediating "oncogene-induced senescence" (OIS) are not completely elucidated, but the proliferative arrest involves activation of both Rb and p53 pathways [18]. OIS is often accompanied by the up-regulation of the CDK inhibitors, p15/INK4B, p16/INK4A, and p21/CIP1. It has been proposed that OIS might be mediated, at least in part, by oxidative mechanisms exerted not directly on the DNA molecule, but on DNA precursor deoxynucleotide pools [145]. This hypothesis is based on the finding that overexpression of the enzyme 8-oxo-dGTPase, human MutT homolog 1 (MTH1), which repairs ROS-damaged dGTP, is able to prevent H-Ras-induced DDR and cell senescence, although levels of RAS-induced ROS were unaffected [146].

3.2.4 The Senescence-Associated Secretory Phenotype

As already mentioned, an additional feature of senescent cells indicating their active participation in the homeostasis of the host tissue is the presence of a distinct secretory phenotype, variously referred to as SASP [40] or senescence messaging secretome (SMS) [92]. The latter includes a striking enhancement in the secretion of growth factors and cytokines by senescent cells, hence exerting a broad range of effects on the local tissue environment [155]).

Proteins that are known to stimulate inflammation, including IL-6, IL-8, IL-1, granulocyte macrophage colony-stimulating factor (GM-CSF), growth-regulated oncogene (GRO)α, monocyte chemotactic protein (MCP)-2, MCP-3, MMP-1, MMP-3, and many of the insulin-like growth factor (IGF)-binding proteins [93, 155] are among the most robustly induced and secreted factors in the SASP response.

The SASP inflammatory cytokines are of particular interest because they can play a role in many age-related pathologies [31]. In addition to being inflammatory mediators, IL-6 and IL-8 were shown to belong to a minor number of secreted factors which also reinforce the senescence phenotype via autocrine and paracrine mechanisms [1, 91].

The mechanisms that initiate and maintain the SASP are unclear. IL-1α and IL-β are minor SASP components which, compared to IL-6 and IL-8, are secreted at low levels. IL-1 (α/β forms) is a multifunctional cytokine that regulates inflammatory and immune responses mainly by originating a signal transduction cascade which lastly induces IL-6 and IL-8 expression. Recombinant IL-1α and IL-1β interact with the same receptor (IL-1R) and mediate similar biological effects. However, IL-1β is active only as a mature secreted form, while IL-1α is rarely secreted at high levels and produces effects either intracellularly or as a cell surface-bound protein. In addition, IL-1α, differently from IL-1β, can function both as an uncleaved precursor protein (pIL-1α) or as a cleaved protein. Once IL-1R binds its receptor, IL-1 initiates the formation of a complex containing IL-1R and its coreceptor (IL-1RAcP) [4, 105, 107, 129]. This complex triggers a series of cytoplasmic events that ultimately activate the transcription factor NF-κB. The latter then transactivates numerous genes, including those encoding IL-6 and IL-8 [123]. As already mentioned, a most prominent cytokine of the SASP is IL-6, a pleiotropic pro-inflammatory cytokine. IL-6 secretion has been shown to increase markedly after DNA damage and OIS of mouse and human keratinocytes, melanocytes, monocytes, fibroblasts, and epithelial cells [91].

The senescence response, and the SASP in particular, may impact health and lifespan by stimulating both the low level inflammation associated with aging [58] as well as the development of specific age-related diseases [123].

3.2.5 *Oxidative Stress, Aging, and Senescence*

Since the first proposal of the "Free Radical Theory of Aging" by Harman almost 60 years ago [66], oxidants have been inextricably linked to aging. Evidence obtained from a variety of experimental systems has supported the existence of an association between increased accumulation of ROS and accelerated aging [16]. Furthermore, experimental manipulations that increase lifespan in invertebrates and rodents, such as caloric restriction, correlate to increased resistance to oxidative stress or reduced oxidative damage [104, 168]. Obviously, a simple correlation does not establish a causal link per se and complex manipulations, as in case of caloric restriction, do exert a plethora of effects that are difficult to interpret on mechanistic grounds.

On the other hand, analysis of several (at least 18) animal models with altered expression of genes involved in ROS generation and detoxification have failed to confirm the presence of a constant and linear inverse association between levels of oxidative stress and longevity [143]. Thus, mice emizygous for the enzyme manganese-superoxide dismutase (MnSOD) or glutathione peroxidase 1 (GPX1) are more sensitive to oxidative stress, have increased levels of DNA damage, and yet their life span is comparable to that of WT animals [180]. Moreover, transgenic mice overexpressing CuZnSOD were more resistant to oxidative stress induced by chemicals and had reduced levels of lipid peroxidation, but their lifespan was not prolonged compared to WT controls [75]. Similar data were obtained in mice over-expressing catalase and, most intriguingly, in animals that were double transgenic for CuZnSOD and catalase [143].

Another notable exception to the ROS-based theory of aging is the longest-living rodent known, the naked mole rat, whose maximum lifespan can be as long as 28 years. These animals, with a body mass slightly larger than a mouse, exhibit persistent higher levels of lipid peroxidation, protein carbonylation, and DNA oxidative damage than do mice [3].

Adding further to the contention, no evidence has been obtained so far that dietary supplementation with anti-oxidants can extend lifespan, with some studies showing in fact a negative effect in this regard [14, 140].

As mentioned earlier, there is also controversy as to the nature of the relationship between aging and cellular senescence. While the two processes have been often associated under the common denominator of ROS-related molecular damage [167], with the ensuing slow accumulation of biological "garbage" underlying both phenomena [173], recent data suggest a more complex interplay between aging, cell senescence, and ROS.

Such a new orientation is best exemplified by the hypothesis proposing the existence of a positive feedback mechanism between ROS production and the emergence/stabilization of the senescence phenotype [138]. This hypothesis builds on early observations describing the critical role of p21 for the establishment of a complete senescence phenotype: late passage, p21-deficient fibroblasts displayed markers of replicative senescence but were still able to complete S-phase when culture in vitro, while p21-expressing fibroblasts were cell cycle arrested [49]. Along the same line, it was later reported that p21-induced cell senescence was critically dependent on the generation of ROS: inhibiting ROS production was able to rescue the p21-dependent senescence phenotype [106]. Subsequent studies have confirmed and extended these concepts: a cell experiencing genotoxic stress activates the DDR and a stable cell cycle arrest through the induction of p21. Prolonged expression of p21 is able to alter mitochondrial stability, via activation of a pathway which includes GADD 45, p38MAPK, and TGFβ, leading to increased production of ROS. The final step of this feedback loop is the reinforcement of the DDR via mitochondrial ROS, causing the emergence of a stable senescence phenotype [138]. Relevant to the point, a critical role of ROS, and more specifically H_2O_2 signalling, was also proposed for Ras-induced replicative senescence in human diploid cells [100].

This type of data clearly implies a profound reconsideration of the role of ROS in cell senescence. The "classical" view is centred on the concept that ROS-induced damage, including DNA damage, is a primary trigger for cell senescence; however, the above data indicate that the sequence of events can be reversed, i.e., increased generation of ROS can be a consequence of DNA damage and can follow the activation of a DDR (Fig. 3.2). In addition, ROS emerge as important signalling molecules to establish and/or reinforce the senescence phenotype, at least in presence of DNA damage. In fact, ROS production is stimulated through a specific signalling pathway and these molecular species appear to act in turn as signalling mechanism to reinforce and stabilize cell senescence. These findings are very intriguing and

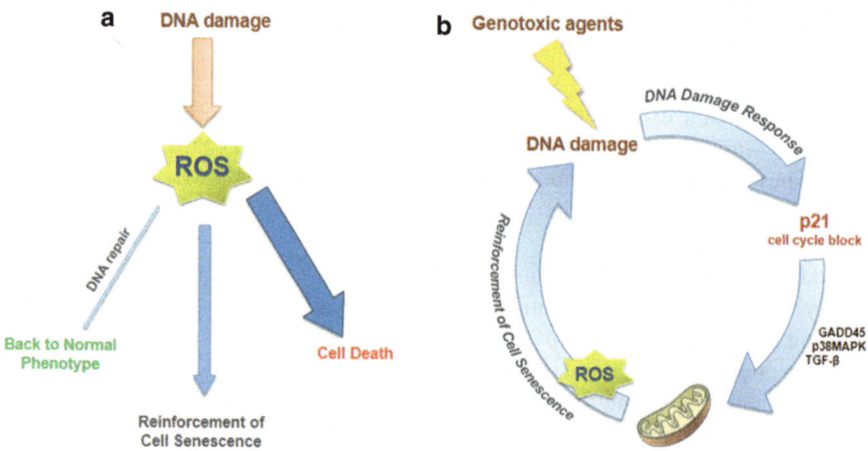

Fig. 3.2 Panel **a** (*left*). Activation of DNA damage response (DDR) includes increased generation of ROS. Depending on the extent of DNA damage and/or the amount of ROS generated, the outcome can result in (1) efficient DNA repair (very low damage and/or low ROS); (2) massive damage and/or massive generation of ROS leading to cell death; or (3) unstable senescence phenotype with low levels of ROS reinforcing senescence. Panel **b** (*right*) highlights the pathway related to the DDR and leading to increased generation of ROS in mitochondris to reinforce the senescence phenotype

suggest that (1) ROS should no longer be regarded as mere accidental by-products of cellular metabolism and they can be actively produced as signalling molecules in the context of cell senescence. Furthermore, (2) the phenotype of senescent cells appears to be susceptible to modulation and is dependent, at least initially, on the generation of mitochondrial ROS. Clearly, these concepts represent a significant depart from the paradigm of a cell as a passive victim of ROS and open new possibilities for active intervention in controlling these signalling pathways.

The concept of ROS as signalling molecules reinforcing the senescence phenotype was further extended in a recent report describing what has been referred to a bystander effect in cell senescence [125]. It was shown that senescent fibroblasts induce a DDR, in neighbouring cells. Furthermore, the effect was dependent on the presence of functional gap junctions and involved generation of ROS. Interestingly, hepatocytes expressing high levels of 4-hydroxy-nonenal (4-HN), a product of lipid peroxidation and a possible signalling molecule [137], clustered together in the liver of mice [125]. While this phenomenon is very intriguing, its functional significance, if any, is difficult to discern at the moment.

3.2.6 Oxidative Stress and Cell Senescence in the Liver

Oxidative stress is invariably associated to the pathogenesis of both acute and chronic liver disease ([108]; see other chapters in this book). For example, the activity of many enzymes involved in lipid metabolism is regulated by the redox status of hepatocytes [164], and it is therefore not surprising that ROS have been implicated in the origin of both alcoholic and non-alcoholic steatohepatitis [154, 189]. On the other hand, lipid metabolism can impact the redox balance of hepatocytes [164], raising the possibility that increased generation of ROS could be a consequence, and not necessarily the cause, of steatohepatitis.

The role of cell senescence as an important third party in between ROS and liver injury is emerging with increasing evidence. The occurrence of both parenchymal (hepatocyte) and non-parenchymal cell senescence has been reported in numerous clinical and experimental studies, particularly in association with chronic liver disease, including neoplastic disease [7, 134, 136, 184].

The picture emerges can be summarized as follows. The presence of cell senescence in hepatocytes is generally associated with progression and/or worst prognosis of the underlying disease. For example, hepatocyte senescence and a decreased regenerative capacity were reported in the liver of c-myc/TGF-α double transgenic mice, and this was associated with a high incidence of hepatocellular carcinoma (HCC) [53]. Similarly, shortened telomeres and other markers of cell senescence were highly expressed in human hepatocytes from livers with chronic viral hepatitis and HCC [21, 134] and were associated to progression of liver cirrhosis [184]. Furthermore, cell senescence, including hepatocyte senescence, was reported in liver of children with end stage liver disease, suggesting that this phenotypic change is linked to the disease process, rather than to aging per se [63]. More recently, a positive correlation was found between expression of senescence markers in hepatocytes and progression of both alcoholic [6] and non-alcoholic [7] fatty liver disease; in fact, the use of these markers was proposed as a prognostic tool in the clinical setting.

In many of the above studies, correlating hepatocyte senescence to increasing gravity of liver disease, a mechanistic link has been implied, the underlying hypothesis being that senescent hepatocytes secrete a variety of cytokines, growth factors, and ECM-degrading proteases that may impact the evolution of the disease process. In line with this interpretation, we have recently reported that extensive hepatocyte senescence is present in the neoplastic-prone microenvironment induced by retrorsine in rat liver (Fig. 3.3) [95, 163]. Retrorsine is a naturally occurring pyrrolizidine alkaloid, known for its ability to exert a long-lasting cell cycle block on hepatocytes in vivo [94]. It is enzymatically activated to toxic metabolites that can bind DNA and induce a persistent DDR [185]; in addition, DNA adducts consistent with the presence of chronic oxidative stress were detected in the liver of retrorsine-exposed rats (Peluso M. and Laconi E., unpublished observation). Using an orthotopic model for hepatocyte transplantation developed in our laboratory, it was observed that the senescent microenvironment induced by retrorsine in rat liver, including activation of SASP, is able to support the growth and progression of transplanted preneoplastic hepatocytes [95]. Moreover, clearance of retrorsine-induced senescent hepatocytes and attenuation of SASP resulted in a delay in the emergence of HCC [110]. While a direct mechanistic link between cell senescence and the neoplastic-prone microenvironment induced by retrorsine remains to be proven, the evidence obtained so far is strongly supportive of this hypothesis.

A radical turn away from this line of thought has been proposed in a recent study describing the clearance of putative pre-malignant senescent hepatocytes by the

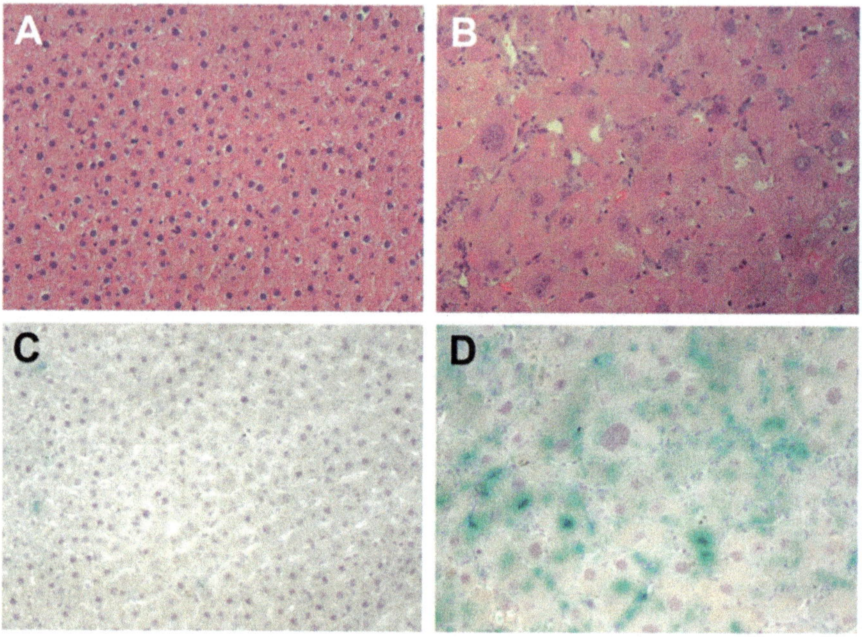

Fig. 3.3 Panels **a** and **b**. Standard histological staining (H&E) showing a normal rat liver (*left*) and a rat liver exposed to retrorsine (*right*). Note the presence of enlarged hepatocytes (megalocytes), one of the hallmarks of cell senescence. In the lower panels, **c** and **d**, histochemical staining for senescence-associated beta-galactosidase (SA-beta-Gal) is shown. Panel **c** is from a normal rat liver. In Panel **d**, from a liver exposed to retrorsine, both hepatocytes and non-parenchymal cell express high levels of this SA-beta-Gal, a marker of cell senescence

immune system. It was reported that N-ras-expressing senescent mouse hepatocytes were deleted by a specific T-cell-mediated immunological response. Furthermore, N-ras-specific Th1 lymphocytes were detected in these animals and inhibition of such process of "senescence surveillance" by the immune system was associated with a high incidence of HCC [83].

While these results are very intriguing, the suggested conclusions are at odds with current concepts regarding the nature of cell senescence. The main contention of this study is in fact the existence of putative "pre-malignant senescent hepatocytes", which are able to progress to cancer if left unchecked [25, 83, 124]. This concept is difficult to reconcile with the current understanding of cell senescence. In fact, phenotypic senescence is thought to represent a fail-safe mechanism against the possibility for malignant transformation in damaged cells [18]; however, the results of the study imply that senescent cells are actually a population at risk for neoplastic progression, unless they are removed by a specific immune response mediated by T lymphocytes. Within this perspective, pre-malignant senescent cells no longer express a stable cell cycle arrest, i.e., they have lost one of the most consistent and almost defining phenotypes of cell senescence [30, 158].

As already mentioned, these and other inconsistencies regarding the role and significance of cell senescence will continue to persist and stimulate much needed additional research, at least until a more clear picture emerges on the intimate nature of this phenotype in various tissues and cell types and under different clinical or experimental conditions [144].

Cellular senescence has also been described in the bile ductular epithelium of patients with primary biliary cirrhosis, indicating its possible involvement in the loss of bile ducts in the evolution of this disease [159].

By contrast, phenotypic senescence in stellate cells has been generally correlated with a better disease outcome, notably during the evolution of liver fibrosis [17, 85, 87, 89]. Initial in vitro studies indicated that senescent stellate cells expressed a pro-inflammatory phenotype and reduced levels of pro-fibrogenic ECM components, including collagens, tenascin, and fibronectin [162]. These results were later replicated in vivo: it was shown that mutant mice harbouring a disrupted pathway to cell senescence specifically targeted to stellate cells were more prone to develop chemically induced liver fibrosis compared to wild-type controls [89].

However, exceptions to the hypothesis proposing a beneficial effect of stellate cell senescence are now emerging. For example, it has been recently reported that the obesity-associated gut microbiome promotes the growth of HCC via induction of senescence in hepatic stellate cells [188], an observation that parallels results of previous studies relating HCC and hepatocyte senescence [21, 134].

3.3 Oxidative Stress and Liver Regeneration

As with the intricacies of aging and cell senescence, ancient western mythology has also anticipated, conceptually, the remarkable phenomenon of liver regeneration. Both Sumerian [174] and Greek [152] civilizations entertained a similar myth narration of a terrible punishment inflicted by the Gods to the victim: the latter would be tortured each day by a huge eagle or vulture eating his liver, which then grew back each night, ready to be devoured again the next day. While there can be no certainty about the actual knowledge supporting mythological imagination, two fundamental concepts were hidden and implied in this fascinating narrative. First, the liver is endowed with enormous regenerative capacity, in that it is able to recover its mass over and over again after innumerable losses; in addition, liver mass can be restored very rapidly, in that the ritual punishment by the predatory birds could be repeated every day or every few days.

Several thousand years had to elapse before formal and reproducible evidence of the regenerative capacity of the liver could be provided. Scattered observations on this topic appeared in the literature during the nineteenth and early twentieth century [56, 120]. However, sound scientific studies of the process could be really undertaken only after the publication of the now classical report of Higgins and Anderson [71], who described a simple surgical procedure to perform partial hepatectomy in the rat. It was during the 60s and the 70s of the last century that the exploitation of this experimental tool, particularly through the work of Nancy Bucher and her group [22], paved the way for the quantitative analysis of liver regenerative process.

The overall information derived from these studies provided important details on the rate and kinetics of response of parenchymal and non-parenchymal liver cells following partial (usually two thirds) surgical hepatectomy. On the other hand, the regenerative response of the liver turned out to display a few unique and important biological features such that it became a general reference model system for the analysis of cell proliferation and cell cycle regulation in vivo. First, most cell populations in the liver, including hepatocytes, are largely quiescent in the normal adult animal (and human), implying that, for practical purposes, they can entirely be attributed to the G_0 stage of the cell cycle. Second, when two-thirds partial hepatectomy (PH) is performed in young adult animals, at least 80 % of the remaining resident hepatocytes enter the cell cycle and participate in the regenerative response, i.e., a percent fraction comparable to that obtained under optimum *in vitro* conditions. Third, the response of the hepatocytes is remarkably synchronous, at least for the first 24–36 h, such that biochemical and molecular changes associated with the unfolding of cell cycle phases can be analysed with great detail in a homogeneous cell population [119].

On a different level of analysis, the regenerative response of the liver to PH represents one of best examples epitomizing a defining feature of complex multicellular organisms, i.e., their ability to exert a continuous and accurate control over the size of any tissue throughout the body [109]. In fact, hepatocytes and other cell types perceive and respond to "signalling" cues within minutes post-PH, setting in motion a long sequence of events culminating in the restoration of the liver mass; once this end point is achieved, the response subsides. Such a seemingly simple, but in fact highly integrated capacity, lies at the very hearth of the essence of metazoans as societies of cells; and yet, very little is known on the mechanisms enforcing maintenance of this fine homeostatic balance.

In this regard, identification of possible triggers initiating the process of liver regeneration may provide important insights on general strategies overlooking the assembly of multicellular organisms in defined cell and tissue type ratios. Incidentally, it is certainly not by a mere coincidence that the liver is one of the most investigated mammalian organs in the context of the hippo pathway and its possible involvement in the control of organ size [5, 9, 133].

Thus, liver regeneration following PH represents a most useful model system to investigate fundamental questions in basic biology and both basic and clinical hepatology, including the regulation of the cell cycle in vivo, mechanisms of tissue repair, and regeneration and the homeostatic control of organ size. In the next paragraphs, the involvement of ROS in the unfolding of hepatocyte cell cycle and liver regeneration will be discussed in light of the above considerations.

3.3.1 The Cell Cycle as a Redox Cycle

Eukaryotic cell cycle is a highly coordinated process disseminated with check points, restriction points, and licencing steps, in order to ensure that the two daughter cells originating from mitotic division receive an equal complement of genetic material. Complex signalling pathways are set in motion, culminating in the initiation of DNA replication, which is followed, upon completion, by the mitotic division.

The cell cycle is essentially driven by different types of cyclins and related cyclin-dependent kinases (CDKs). The D-type cyclins (D1, D2 and D3) are expressed during G1 and at the G1–S border and are important to traverse the restriction (R) point towards commitment to the S phase [135]; cyclin E is also involved in the G1–S transit, while cyclin A is expressed during S phase and, together with cyclin B, is important during the transit through G2. Finally, progression through mitosis is dependent upon the expression of cyclin B. All cyclins exert part of their regulatory functions via binding and activation of partner CDKs, which in turn phosphorylate critical targets involved in cell cycle progression. Thus, as an example, cyclin D1 binds and activates the CDK4/CDK6 complex, leading to phosphorylation of the retinoblastoma family of proteins (pRb), thereby releasing their inhibitory effect on E2F. The latter are transcription factors controlling the expression of several genes whose products regulate cell cycle progression through multiple mechanisms, including feedback effects on pRb phosphorylation [98].

At least two more levels of complexity add to the intricate wiring of cell cycle regulation through phosphorylation processes. A series of dual specific phosphatases, including CDC25 (A, B and C), can act to remove phosphate groups from threonine or tyrosine residues on CDKs, resulting in activation of cyclin/CDK complexes. Furthermore, CDKs can be targeted by specific inhibitors (CDKIs), including two main families: the INK4 (inhibitors of CDK4) and the CIP/KIP (cyclin/kinase inhibitory protein) proteins. Members of the INK4 family are p15 (INK4B), p16 (INK4A), p18 (INK4C), and p19 (INK4D); they can bind and inhibit the kinase activity of CDK4/CDK6 complexes. On the other hand, the KIP family, including p27 (KIP1) and p57 (KIP2), targets mainly cyclinE/CDK2 complexes, while p21 (CIP1) has the ability to block all CDKs [50, 160].

The increasing relevance that redox mechanisms are being attributed in the regulation of the cell cycle is best exemplified by the emerging concept that the cell division cycle is in fact paralleled and possibly controlled, at least in part, by a redox cycle [26, 36]. To be true, this concept is not exactly new, in that it was introduced over 80 years ago, when it was observed that soluble intracellular thiols displayed a cyclic pattern during mitotic division in sea urchin eggs [147]. It was later shown that variations in the redox status were also present in cycling mammalian cells, notably between mitotic phase and interphase [62, 113]. However, formal evidence for the existence of a redox cycle running in association with the cell division cycle was presented more recently in a landmark study performed in yeast [179]. It was reported that over half of the yeast genome, including genes specifying functions associated with metabolism, tend to be expressed with exceptionally robust periodicity; furthermore, essential cellular events, including cell division cycle, occur in synchrony with the metabolic cycle. Three distinct phases of the yeast cell redox cycle were described, with the cell division cycle beginning during a more oxidative phase, while DNA synthesis and mitosis coincided with a more reductive phase [179].

3.3.2 The Redox Balance and Redox Sensors

The redox status of the cell results from a fine balance between the rate of production of oxidative species and the levels of antioxidant mechanisms. Generation of ROS occurs mainly in mitochondrial electron transport chain and from the activity of other enzymes such as xanthine oxidases, nicotinamide adenine dinucleotide phosphate (NADPH) oxidases, cytochrome P450s, myeloperoxidase, and nitric oxide synthase [114]. On the other hand, antioxidant barriers comprise both enzymatic and non-enzymatic mechanisms. Antioxidant enzymes include superoxide dismutase (SOD), catalase, glutathione peroxidase, glutaredoxin, thioredoxin, and peroxiredoxin. Three different SODs are known to exist in mammalian cells: MnSOD, located in mitochondria; CuZnSOD, expressed in the cytoplasm, in the nucleus as well as in mitochondria; and an extracellular SOD (EcSOD), bound to the external side of the plasma membrane. All SODs convert the superoxide anion (O^{2-}) into hydrogen peroxide (H_2O_2), while H_2O_2 is acted upon by catalase, located mostly in peroxisomes, and glutathione peroxidase, distributed in various subcellular organelles. Non-enzymatic antioxidants include small molecules such as glutathione, cysteine, and vitamins C (ascorbic acid) and E (α-tocopherol). Evidence also indicates that the redox equilibrium is set at different steady state levels in different cellular compartments [61, 73, 76], suggesting that its regulation is specifically important for each cellular function.

There is now ample evidence to indicate that the reversible oxidation-reduction cycle in specific cysteine residues in several proteins is intimately involved in the regulation of virtually every cellular function [26]. Therefore, it makes much biological sense that, in order to maintain and fine tune redox balance, cells are equipped with complex sensing mechanisms, which are able to regulate production and/or degradation of ROS acting on several enzymes and interconnected metabolic pathways [45, 64]. An important role in this context is played by the forkhead box O (FOXO) family of transcription factors, of which there are four in humans (FOXO1, FOXO3a, FOXO4, and FOXO6) [156]. These proteins have been implicated in processes as diverse as cell cycle regulation and arrest, apoptosis, ROS scavenging, DNA damage repair, and are the orthologs of longevity-associated DAF-16 in *C. elegans* [8]. The activity of FOXO is regulated by two main types of input: insulin receptor signalling and ROS. The effect of insulin is to inhibit FOXO transcriptional activity by preventing its shuttling to the nucleus; however, regulation by ROS appears to be far more complex and leads to a wide spectrum of functional consequences, from transcriptional inhibition, to stimulation, to a shift in transcriptional targets [27]. It is precisely this versatility in FOXO responses to ROS that make this protein central to the fine tuning of redox balance [45]. Regulation of FOXO occurs mainly through post-translational modifications, including phosphorylation by mitogen-activated protein kinase (MAPKs) [51, 187], acetylation by p300/cAMP responsive element binding (CREB), binding protein (CBP), and deacetylation by Sirt1 [20, 43] and ubiquitinylation by MDM2 [74]. The picture that is emerging from this multitude of information is that FOXO functions as a central player in the control of redox status and in orchestrating cellular responses to stress. Downstream consequences of FOXO binding to DNA include an increased resistance to stress, which is accomplished through up-regulation of antioxidant enzymes (e.g., MnSOD), withdrawal from the cell cycle (e.g., up-regulation of p27), and metabolic regulation [35, 72, 122].

However, for the message to be conveyed to FOXO transcription factors, the direct cellular sensors of the redox status appear to reside in the cysteine groups of specific proteins involved in cellular pathways regulating FOXO activity. Interestingly, this was found to be the case for both insulin and ROS-regulated pathways impinging on FOXO transcriptional activity.

In fact, many of the proteins involved in the signal cascade downstream of insulin receptor binding appear to be regulated, at least in part, by ROS acting on critical cysteine residues. These include: (1) protein tyrosine phosphatase 1b (PTP1b), which targets both e insulin receptor and insulin receptor substrate 1 (IRS1) [175]; (2) the phosphatase and tensin homolog (PTEN), which dephosphorylates phosphatidylinositol-3-phosphates (PI3P), thereby abrogating the effect of phosphatidylinositol-3-kinase (PI3K) acting downstream of the insulin receptor [101]; (3) protein kinase B (PKB)/Akt), which directly phosphorylates FOXO and inhibits its DNA binding and transcriptional activity [88]. Similarly, the c-Jun N-terminal kinase (JNK), which activates FOXO through phosphorylation, is in turn controlled by other proteins possessing cysteine groups susceptible to modulation by ROS [38]. A cross talk has also been described between insulin and ROS-regulated pathways of FOXO activation [156].

Finally, and most intriguingly, cysteine groups in FOXO protein itself can be a target of ROS. Acetylation by p300/CBP acetylases, which is able to modulate FOXO transcriptional activity, is impaired following oxidation of specific cysteine residues in the molecule, indicating that FOXO can act as a direct redox sensor [43].

Another protein exquisitely designed as a redox sensor in mammalian cells is the NF-E2-related factor 2 (Nrf2). This transcription factor is normally targeted to proteasomal degradation via binding with the E3 Ubiquitin ligase Cullin-3 complex; however, oxidation of critical cysteine residues in Nrf2 prevents formation of the ubiquitin complex, leading to the translocation of Nrf2 to the nucleus, where it binds to the antioxidant responsive element (ARE) of several genes coding for antioxidant enzymes [103, 150].

3.3.3 ROS and the Cell Cycle

The dual role of ROS, as both toxins and signalling molecules, already discussed with reference to cell senescence, becomes even more evident in the context of cell replication. While low levels of H_2O_2 (0.02–0.13μM) were found to stimulate proliferation of NIH3T3 fibroblasts, higher levels (0.25–2μM) were conducive to growth arrest and cell death [97]. Several lines of evidence do indicate that, in mammalian cells, low levels of H_2O_2 are essential to enter the cell cycle. For example, mouse embryo fibroblasts required a transient increase in pro-oxidant levels to proceed from G1 to S phase, and adding *N*-acetyl-cysteine, a known anti-oxidant, was sufficient to block this transition [115]. Furthermore, response to EGF, FGF, PDGF, insulin, and other growth factors is associated with a transient NADPH-dependent increase in H_2O_2 [23, 33]. Similarly, growth in vitro of mouse embryonic stem cells induced by peroxisome proliferator-activated receptor delta (PPARδ)-agonist is critically dependent on ROS generation and it is blocked by antioxidants [78]. The p53R2 small subunit of ribonucleotide reductase, which plays a key role in supplying precursors for DNA and is therefore expressed during S phase of the cell cycle, is also a scavenger for H_2O_2,, and this activity is relevant to its enzymatic function [186].

On the other hand, it is well known that high levels of oxidants lead to a block in the cell cycle, cell senescence, and apoptosis, as discussed in the preceding section. Among the targets mediating this effect is the transcription factor c-Fos, whose binding to DNA is stabilized by excess H_2O_2, thereby impairing cell cycle progression [24].

3.3.4 Triggers of Liver Regeneration

Most molecular pathways involved in the initiation and the step-by-step completion of the cell division cycle have been characterized from in vitro studies. However, the specific features of the proliferative response of hepatocytes after PH, as outlined above, have provided the opportunity to compare the in vitro results with those obtained in a mammalian system in vivo. These studies have highlighted basic commonalities of the cell cycle process between in vitro synchronized cultures, including hepatocyte cultures, and what is observed in vivo in the liver during the regenerative process.

Current knowledge on this topic can be summarized as follows. Hepatocyte cell cycle is mainly controlled by two types of growth factors, i.e., hepatocyte growth factor (HGF), which binds to the c-Met receptor in the plasma membrane, and ligands of the epidermal growth factor receptor (EGF-R), including EGF, transforming growth factor (TGF)-α, amphiregulin, and heparin-binding EGF. Both HGF and EGF-R ligands are considered as *bona fide* growth factors for hepatocytes in that they are potent inducers of DNA synthesis in vitro and are also able to stimulate a significant mitogenic response in vivo [119]. However, several other factors are known to be important to facilitate or optimize the response of hepatocytes to growth stimuli. Included in this list are cytokines such as tumour necrosis factor (TNF) and interleukin 6 (IL6), other growth factors such as vascular endothelial growth factor (VEGF) and fibroblast growth factor (FGF) 1 and FGF2, hormones, and neurotransmitters such as norepinephrine, serotonin, leptin, 27 insulin, 28, 29, 30, Notch/Jagged signalling pathway, complement proteins, and bile acids. The role of these factors should not be overlooked: they do not appear to be essential for liver regeneration to complete, but they have a substantial impact on the timing and coordination of the process [55, 118, 119].

Within minutes after PH, important changes are already detectable in the remnant liver. In fact, over 100 genes, which are not expressed at detectable levels in normal livers, are turned on soon after surgery. One of the earliest such changes is an increase in the activity of urokinase type plasminogen activator (uPA), associated with extensive remodelling of the extracellular matrix (ECM). The latter phenomenon is possibly related to major hemodynamic alterations occurring in the entire organ following removal of 2/3 of its mass, given that portal blood flow remains unchanged and it is now distributed into a reduced tissue volume. Irrespective of the underlying mechanism, matrix remodelling in the pericellulur space represents the main source of biologically active HGF, which is released from its binding to ECM and clived to its active form by uPA [111, 112] and matrix metallo-proteinases (MMPs), particularly MMP9 [116]. The release of HGF results in the activation of c-Met receptor at 30–60 min post-PH [170]. Following the early release of stored HGF, this growth factor is also newly synthesized in response to PH by both stellate cells and sinusoidal endothelium for up to 48 h post-hepactomy [99, 161].

Signalling through the EGF receptor also increases within the first hour post-PH. Its ligand EGF originates mainly from Brunner's glands in the duodenum, which secrete this growth factor constitutively. While the relative increase in portal blood perfusion to the remnant liver could account per se for a more sustained EGF receptor signalling, other factors may be involved, including a possible potentiating role of HGF and norepinephrine on EGF-R sensitivity to its ligands [41] and a direct stimulation of EGF synthesis by norepinephrine [128].

The role of TGFα in the response of hepatocytes to PH is less clear. Parenchymal cells are the main source of this growth factor, whose increase is seen after about 3 h and lasts for over 48 h; given the time-course of its expression, it has been proposed that it might exert a more important role in triggering non-parenchymal cell proliferation, including bile ductular and endothelial cells, possessing receptors for TGFα [118].

Of all factors potentiating and coordinating the regenerative response to PH, not being directly mitogenic, insulin deserves special consideration. This hormone greatly enhances the effect of growth factors in vitro. Adequate supply of insulin to the liver through the portal circulation is essential to maintain normal liver structure and function. In fact, experimental porto-caval shunt leads to profound liver atrophy and this can be reversed by insulin administration [52]. On the other hand, it has long been known that liver regeneration is impaired in diabetic rats and insulin can reverse this defect [80].

The Notch/Jagged signalling pathway is still poorly understood in its potential role during the regenerative response of the liver. It has been reported that binding of Jagged to Notch and subsequent cleavage and nuclear migration of the cytoplasmic portion of the latter occurs within 15 min post-PH in rat liver [86]. On the other hand, activation of this pathway has been associated with both positive and negative effects on cell cycle progression [81, 126]. Its precise role during liver regeneration needs therefore to be further investigated.

3.3.5 ROS and Liver Regeneration

Several lines of evidence have indicated that HGF/c-Met signalling, which emerges as the most critical pathway in the liver regenerative process, also participates in the defence mechanisms against oxidative stress and is a major determinant of cellular redox potential [82, 178]. In addition, c-Met ligation triggers phosphoinositide 3-kinase(PI3K)/Akt activation, which in turn is involved in the control of redox homeostasis [131]. Interestingly, hepatocytes lacking c-Met receptor display overproduction of NADPH-generated ROS, associated with increased basal expression of a distinct set of genes implicated in the ROS defence [39, 84, 172]. Consistent with above results obtained in hepatocytes, it was reported that mice lacking the Prdm16 transcription factor, which promotes HGF expression, had increased levels of ROS and depletion of stem cells throughout the nervous and haematopoietic systems. Adding recombinant HGF to Prdm16-deficient cells cultured in vitro partially rescued the increase in ROS levels [37].

Engagement of the EGF-R pathway has also been reported to impact on redox equilibrium and to act through modulation of ROS production and scavenging. Several studies have documented that EGF receptor signalling in hepatocytes is associated with down-regulation of ROS production and up-regulation of antioxidant mechanisms [34, 57, 130, 157]. For example, rat hepatoma cells showing constitutive overactivation of the EGF-R pathway can grow in the absence of serum and are protected from apoptotic signals; however, inhibition of EGF-R tyrosine kinase activity potently inhibits autocrine growth and induces apoptosis. The apoptotic effect correlates increased ROS production and glutathione depletion coincident with up-regulation of the NADPH oxidase 4 (Nox4) and down-regulation of the γ-glutamylcysteine synthetase (γ-GCS), a key regulatory enzyme of the glutathione synthesis [130]. In another study from the same laboratory, a cell line derived from mouse hepatocytes was cultured in the absence of amino acids, leading to loss in cell viability and apoptosis. However, a subpopulation of cells was able to survive and proliferate, showing resistance to apoptotic stimuli. Interestingly, these cells had increased activation of EGF-R pathway and overexpression of EGFR ligands, such as TGF-α and HB-EGF. While production of ROS was increased during amino acid deprivation, cells that survived showed an enhancement in the levels of reduced glutathione and a higher expression of γ-GCS, suggesting that they had adapted to counteract the oxidative stress [34].

The Nrf2 transcription factor is a crucial regulator of the cellular redox homeostasis through its capacity to induce the expression of antioxidant enzymes, including glutathione S-transferase (GST) glutamate-cysteine ligase catalytic subunit (GCLC) and NAD(P)H quinone oxidoreductase 1 (NQO1) [77]. Interestingly, regeneration after PH was reported to be strongly delayed in the absence of Nrf2. This defect was shown to result from ROS-induced transient resistance to insulin and insulin-like growth factor 1 in hepatocytes, causing impaired phosphorylation of the p38 mitogen-activated kinase (MAPK) and its downstream targets. A cross talk between Nrf2 and NFκB was also suggested, since the activity of the latter was strongly enhanced in the regenerating liver of Nrf2 knockout mice, possibly as a compensatory effect to reduce the extent of oxidative stress [13]. Interestingly, the gene coding for a protein referred to as augmenter of liver regeneration (ALR), which is induced soon after PH, was also reported to contain a promoter sequence regulated by Nrf2 [44].

The role of p38-MAPK in connecting redox status, ROS signalling, and liver regeneration is controversial. In some studies, as with the Nrf2 knockout mouse, a positive correlation between p38-MAPK activation and proliferation has been found. For example, studies on palmitic acid-stimulated cells from a normal human hepatocyte cell line indicated that the mitogenic effect (1) was dependent on ROS production, in that it was inhibited by either *N*-acetylcysteine or catalase; (2) it was associated with increased expression and nuclear translocation of Nrf2; (3) the specific inhibition of p38-MAPK blocked the G1–S transition in these cells [183]. In line with the above results, a recent report indicates that p38-MAPK-knockout mice have reduced levels of hepatocyte proliferation and reduced regeneration after injury [176].

However, other studies suggest that the activation of p38-MAPK inhibits the process of liver regeneration and hepatocyte proliferation [28, 29]. Thus, phosphorylated p38-MAPK was found to be present in normal mouse liver while it was rapidly inactivated within 30 min post-PH, to reappear in its active form only 12 h later. The reduced levels of phosphorylation were attributed to an increase in the relative activity of MAPK phosphatases (MKPs) over kinases (MKK) [28]. Furthermore, increased proliferative activity was reported in hepatocytes in p38-deficient mice [29]. Obviously, additional studies are urgently needed to define the complex role of p38-MAPK in liver regeneration.

Although there is general agreement on the central role played by this signalling pathway in sensing and regulating the redox balance, the exact mechanism(s) mediating these effects are yet to be clarified. As previously mentioned, likely candidates in perceiving and transducing redox balance are specific cysteine residues present in proteins along the signalling cascade whose redox status can impact on their enzymatic and/or interacting properties.

3.4 Concluding Remarks

Several lines of evidence have now converged towards the concept that ROS are intimately integrated in cell metabolic pathways, playing a central role in the modulation of several biological processes. As with molecular oxygen and iron, the bioavailability of ROS needs to be strictly controlled, in that any excess can cause serious harm to cells and tissues. To this end, cells are equipped with both enzymatic and non-enzymatic systems to titrate the redox equilibrium. Interestingly, these control systems appear to operate separately and with different set points in different subcellular compartments, further highlighting their specific functional significance.

Both the emergence of cell senescence and the unfolding of the cell cycle are susceptible to modulation by ROS. Available data suggest that this is also true for hepatocytes. Critical cysteine residues present in specific proteins involved in metabolic and/or signal transduction pathway are capable to translate alterations in redox balance into a biochemical signal. The activity of kinases, phosphatases, as well as acetylation and de-acetylation reaction, glutathionylation, and ubiquitylation processes, among others, are the ultimate likely targets of ROS signalling. However, this intricated and fascinating area of research is just beginning to be explored.

References

1. Acosta JC, O'Loghlen A, Banito A, Guijarro MV, Augert A, Raguz S, Fumagalli M, Da Costa M, Brown C, Popov N, Takatsu Y, Melamed J, D'Adda di Fagagna F, Bernard D, Hernando E, Gil J (2008) Chemokine signaling via the CXCR2 receptor reinforces senescence. Cell 133:1006–1018
2. Allsopp RC, Vaziri H, Patterson C, Goldstein S, Younglai EV, Futcher AB, Greider CW, Harley CB (1992) Telomere length predicts replicative capacity of human fibroblasts. Proc Natl Acad Sci U S A 89:10114–10118
3. Andziak B, O'Connor TP, Qi W, DeWaal EM, Pierce A, Chaudhuri AR, Van Remmen H, Buffenstein R (2006) High oxidative damage levels in the longest-living rodent, the naked mole-rat. Aging Cell 5:463–471
4. Apte RN, Dotan S, Elkabets M, White MR, Reich E, Carmi Y, Song X, Dvozkin T, Krelin Y, Voronov E (2006) The involvement of IL-1 in tumorigenesis, tumor invasiveness, metastasis and tumor–host interactions. Cancer Metastasis Rev 25:387–408
5. Apte U, Gkretsi V, Bowen WC, Mars WM, Luo JH, Donthamsetty S, Orr A, Monga SP, Wu C, Michalopoulos GK (2009) Enhanced liver regeneration following changes induced by hepatocyte-specific genetic ablation of integrin-linked kinase. Hepatology 50:844–851
6. Aravinthan A, Pietrosi G, Hoare M, Jupp J, Marshall A, Verrill C, Davies S, Bateman A, Sheron N, Allison M, Alexander GJ (2013) Hepatocyte expression of the senescence marker p21 is linked to fibrosis and an adverse liver-related outcome in alcohol-related liver disease. PLoS One 8:e72904
7. Aravinthan A, Scarpini C, Tachtatzis P, Verma S, Penrhyn-Lowe S, Harvey R, Davies SE, Allison M, Coleman N, Alexander G (2013) Hepatocyte senescence predicts progression in non-alcohol-related fatty liver disease. J Hepatol 58:549–556
8. Arden KC (2008) FOXO animal models reveal a variety of diverse roles for FOXO transcription factors. Oncogene 27:2345–2350
9. Avruch J, Zhou D, Fitamant J, Bardeesy N (2011) Mst1/2 signalling to Yap: gatekeeper for liver size and tumour development. Br J Cancer 104:24–32
10. Baker DJ, Wijshake T, Tchkonia T, LeBrasseur NK, Childs BG, van de Sluis B, Kirkland JL, van Deursen JM (2011) Clearance of p16Ink4a-positive senescent cells delays ageing-associated disorders. Nature 479:232–236
11. Banito A, Lowe SW (2013) A new development in senescence. Cell 155:977–978

12. Ben-Porath I, Weinberg RA (2004) When cells get stressed: an integrative view of cellular senescence. J Clin Invest 113:8–13
13. Beyer TA, Xu W, Teupser D, auf dem Keller U, Bugnon P, Hildt E, Thiery J, Kan YW, Werner S (2008) Impaired liver regeneration in Nrf2 knockout mice: role of ROS-mediated insulin/IGF-1 resistance. EMBO J 27:212–223
14. Bjelakovic G, Nikolova D, Gluud LL, Simonetti RG, Gluud C (2007) Mortality in randomized trials of antioxidant supplements for primary and secondary prevention: systematic review and meta-analysis. JAMA 297:842–857
15. Blagosklonny MV (2013) Aging is not programmed: genetic pseudo-program is a shadow of developmental growth. Cell Cycle 12:3736–3742
16. Bokov A, Chaudhuri A, Richardson A (2004) The role of oxidative damage and stress in aging. Mech Ageing Dev 125:811–826
17. Borkham-Kamphorst E, Schaffrath C, Van de Leur E, Haas U, Tihaa L, Meurer SK, Nevzorova YA, Liedtke C, Weiskirchen R (2014) The anti-fibrotic effects of CCN1/CYR61 in primary portal myofibroblasts are mediated through induction of reactive oxygen species resulting in cellular senescence, apoptosis and attenuated TGF-β signaling. Biochim Biophys Acta 1843:902–914
18. Braig M, Schmitt C (2006) Tumor development oncogene-induced senescence: putting the brakes on tumor development. Cancer Res 66:2881–2884
19. Brookes S, Rowe J, Gutierrez Del Arroyo A, Bond J, Peters G (2004) Contribution of p16(INK4a) to replicative senescence of human fibroblasts. Exp Cell Res 298:549–559
20. Brunet A, Sweeney LB, Sturgill JF, Chua KF, Greer PL, Lin Y, Tran H, Ross SE, Mostoslavsky R, Cohen HY, Hu LS, Cheng HL, Jedrychowski MP, Gygi SP, Sinclair DA, Alt FW, Greenberg ME (2004) Stress-dependent regulation of FOXO transcription factors by the SIRT1 deacetylase. Science 303:2011–2015
21. Brunt EM, Walsh SN, Hayashi PH, Labundy J, Di Bisceglie AM (2007) Hepatocyte senescence in end-stage chronic liver disease: a study of cyclin-dependent kinase inhibitor p21 in liver biopsies as a marker for progression to hepatocellular carcinoma. Liver Int 27:662–671
22. Bucher NL (1963) Regeneration of the mammalian liver. Int Rev Cytol 15:245–300
23. Burch PM, Heintz NH (2005) Redox regulation of cell-cycle re-entry: cyclin D1 as a primary target for the mitogenic effects of reactive oxygen and nitrogen species. Antioxid Redox Signal 7:741–751
24. Burch PM, Yuan Z, Loonen A, Heintz NH (2004) An extracellular signal-regulated kinase 1- and 2-dependent program of chromatin trafficking of c-Fos and Fra-1 is required for cyclin D1 expression during cell cycle reentry. Mol Cell Biol 24:4696–4709
25. Burgess DJ (2011) Senescence: Tumorigenesis under surveillance. Nat Rev Cancer 12:6
26. Burhans WC, Heintz NH (2009) The cell cycle is a redox cycle: linking phase-specific targets to cell fate. Free Radic Biol Med 47:1282–1293
27. Calnan DR, Brunet A (2008) The FoxO code. Oncogene 27:2276–2288
28. Campbell JS, Argast GM, Yuen SY, Hayes B, Fausto N (2011) Inactivation of p38 MAPK during liver regeneration. Int J Biochem Cell Biol 43:180–188
29. Hui L, Bakiri L, Mairhorfer A, Schweifer N, Haslinger C, Kenner L, Komnenovic V, Scheuch H, Beug H, Wagner EF (2007) p38alpha suppresses normal and cancer cell proliferation by antagonizing the JNK-c-Jun pathway. Nat Genet 39:741–749
30. Campisi J, D'Adda di Fagagna F (2007) Cellular senescence: when bad things happen to good cells. Nat Rev Mol Cell Biol 8:729–740
31. Campisi J, Andersen JK, Kapahi P, Melov S (2011) Cellular senescence: a link between cancer and age-related degenerative disease? Semin Cancer Biol 21:354–359
32. Campisi J (2014) Cell biology: the beginning of the end. Nature 505:35–36
33. Chen K, Kirber MT, Xiao H, Yang Y, Keaney JF Jr (2008) Regulation of ROS signal transduction by NADPH oxidase 4 localization. J Cell Biol 181:1129–1139
34. Chisari AN, Sancho P, Caja L, Bertran E, Fabregat I (2012) Lack of amino acids in mouse hepatocytes in culture induces the selection of preneoplastic cells. Cell Signal 24:325–332
35. Fu Z, Tindall DJ (2008) FOXOs, cancer and regulation of apoptosis. Oncogene 27:2312–2319

36. Chiu J, Dawes IW (2012) Redox control of cell proliferation. Trends Cell Biol 22:592–601
37. Chuikov S, Levi BP, Smith ML, Morrison SJ (2010) Prdm16 promotes stem cell maintenance in multiple tissues, partly by regulating oxidative stress. Nat Cell Biol 12:999–1006
38. Clavel S, Siffroi-Fernandez S, Coldefy AS, Boulukos K, Pisani DF, Derijard B (2010) Regulation of the intracellular localization of Foxo3a by stress-activated protein kinase signaling pathways in skeletal muscle cells. Mol Cell Biol 30:470–480
39. Clavijo-Cornejo D, Enriquez-Cortina C, López-Reyes A, Domínguez-Pérez M, Nuño N, Domínguez-Meraz M, Bucio L, Souza V, Factor VM, Thorgeirsson SS, Gutiérrez-Ruiz MC, Gómez-Quiroz LE (2013) Biphasic regulation of the NADPH oxidase by HGF/c-Met signaling pathway in primary mouse hepatocytes. Biochimie 95:1177–1184
40. Coppé JP, Patil CK, Rodier F, Sun Y, Munoz DP, Goldstein J, Nelson PS, Desprez PY, Campisi J (2008) Senescence-associated secretory phenotypes reveal cell-nonautonomous functions of oncogenic ras and the p53tumor suppressor. PLoS Biol 6:2853–2868
41. Cruise JL, Knechtle SJ, Bollinger RR, Kuhn C, Michalopoulos G (1987) Alpha 1-adrenergic effects and liver regeneration. Hepatology 7:1189–1194
42. d'Adda di Fagagna F, Reaper PM, Clay-Farrace L, Fiegler H, Carr P, Von Zglinicki T, Saretzki G, Carter NP, Jackson SP (2003) A DNA damage checkpoint response in telomere-initiated senescence. Nature 426:194–198
43. Dansen TB, Smits AMM, van Triest M, de Keizer PLJ, van Leenen D, Groot Koerkamp M, Szypowska M, Meppelink A, Brenkman AB, Yodoi J, Holstege FCP, Burgering BMT (2009) Redox-sensitive cysteines bridge p300/CBP-mediated acetylation and FoxO4 activity. Nat Chem Biol 5:664–672
44. Dayoub R, Vogel A, Schuett J, Lupke M, Spieker SM, Kettern N, Hildt E, Melter M, Weiss TS (2013) Nrf2 activates augmenter of liver regeneration (ALR) via antioxidant response element and links oxidative stress to liver regeneration. Mol Med 19:237–244
45. de Keizer PL, Burgering BM, Dansen TB (2011) Forkhead box o as a sensor, mediator, and regulator of redox signaling. Antioxid Redox Signal 14:1093–1106
46. De Marais DJ (2000) Evolution. When did photosynthesis emerge on Earth? Science 289:1703–1705
47. Dimri GP, Lee X, Basile G, Acosta M, Scott G, Roskelley C, Medrano EE, Linskens M, Rubelj I, Pereira-Smith O, Peacocke M, Campisi J (1995) A biomarker that identifies senescent human cells in culture and in aging skin in vivo. Proc Natl Acad Sci U S A 92:9363–9367
48. Dismukes GC, Klimov VV, Baranov SV, Kozlov YN, DasGupta J, Tyryshkin A (2001) The origin of atmospheric oxygen on Earth: the innovation of oxygenic photosynthesis. Proc Natl Acad Sci U S A 98:2170–2175
49. Dulić V, Beney GE, Frebourg G, Drullinger LF, Stein GH (2000) Uncoupling between phenotypic senescence and cell cycle arrest in aging p21-deficient fibroblasts. Mol Cell Biol 20:6741–6754
50. Ekholm SV, Reed SI (2000) Regulation of G(1) cyclin-dependent kinases in the mammalian cell cycle. Curr Opin Cell Biol 12:676–684
51. Essers MA, Weijzen S, de Vries-Smits AM, Saarloos I, de Ruiter ND, Bos JL, Burgering BM (2004) FOXO transcription factor activation by oxidative stress mediated by the small GTPase Ral and JNK. EMBO J 23:4802–4812
52. Evarts RP, Raab M, Marsden E, Thorgeirsson SS (1986) Histochemical changes in livers from portacaval-shunted rats. J Natl Cancer Inst 76:731–738
53. Factor VM, Jensen MR, Thorgeirsson SS (1997) Coexpression of C-myc and transforming growth factor alfa in the liver promotes early replicative senescence and diminishes regenerative capacity after partial hepatectomy in transgenic mice. Hepatology 26:1434–1443
54. Falkowski P (2006) Evolution. Tracing oxygen's imprint on earth's metabolic evolution. Science 311:1724–1725
55. Fausto N, Campbell JS, Riehle KJ (2006) Liver regeneration. Hepatology 43(2 Suppl 1): S45–S53
56. Fishback FC (1929) A morphologic study of regeneration of the liver after partial removal. Arch Pathol 7:955–977

57. Flinder LI, Timofeeva OA, Rosseland CM, Wierød L, Huitfeldt HS, Skarpen E (2011) EGF-induced ERK-activation downstream of FAK requires rac1-NADPH oxidase. J Cell Physiol 226:2267–2278
58. Franceschi C, Capri M, Monti D, Giunta S, Olivieri F, Sevini F, Panourgia MP, Invidia L, Celani L, Scurti M, Cevenini E, Castellani GC, Salvioli S (2007) Inflammaging and anti-inflammaging: a systemic perspective on aging and longevity emerged from studies in humans. Mech Ageing Dev 128:92–105
59. Fumagalli M, Rossiello F, Clerici M, Barozzi S, Cittaro D, Kaplunov JM, Bucci G, Dobreva M, Matti V, Beausejour CM, Herbig U, Longhese MP, d'Adda di Fagagna F (2012) Telomeric DNA damage is irreparable and causes persistent DNA-damage-response activation. Nat Cell Biol 14:355–365
60. Gewirtz DA (2013) Autophagy and senescence: a partnership in search of definition. Autophagy 9:808–812
61. Go YM, Jones DP (2008) Redox compartmentalization in eukaryotic cells. Biochim Biophys Acta 1780:1273–1290
62. Goswami PC, Sheren J, Albee LD, Parsian A, Sim JE, Ridnour LA, Higashikubo R, Gius D, Hunt CR, Spitz DR (2000) Cell cycle-coupled variation in topoisomerase IIalpha mRNA is regulated by the 3′-untranslated region. Possible role of redox-sensitive protein binding in mRNA accumulation. J Biol Chem 275:38384–38392
63. Gutierrez-Reyes G, del Carmen Garcia de Leon M, Varela-Fascinetto G, Valencia P, Pérez Tamayo R, Rosado CG, Labonne BF, Rochilin NM, Garcia RM, Valadez JA, Latour GT, Corona DL, Diaz GR, Zlotnik A, Kershenobich D (2010) Cellular senescence in livers from children with end stage liver disease. PLoS One 5:e10231
64. Haddad JJ, Harb HL (2005) L-gamma-glutamyl-L-cysteinyl-glycine (glutathione; GSH) and GSH-related enzymes in the regulation of pro- and anti-inflammatory cytokines: a signaling transcriptional scenario for redox(y) immunologic sensor(s)? Mol Immunol 42:987–1014
65. Harley CB, Futcher AB, Greider CW (1990) Telomeres shorten during ageing of human fibroblasts. Nature 345:458–460
66. Harman D (1956) Aging: a theory based on free radical and radiation chemistry. J Gerontol 11:298–300
67. Hayflick L (1965) The limited in vitro lifetime of human diploid cell strains. Exp Cell Res 37:614–636
68. Hayflick L, Moorhead PS (1961) The serial cultivation of human diploid cell strains. Exp Cell Res 25:585–621
69. Hekimi S, Lapointe J, Wen Y (2011) Taking a "good" look at free radicals in the aging process. Trends Cell Biol 21:569–576
70. Hewitt G, Jurk D, Marques FD, Correia-Melo C, Hardy T, Gackowska A, Anderson R, Taschuk M, Mann J, Passos JF (2012) Telomeres are favoured targets of a persistent DNA damage response in ageing and stress-induced senescence. Nat Commun 3:708
71. Higgins GM, Anderson RM (1931) Restoration of the liver of the white rat following partial surgical removal. Arch Pathol 12:186–202
72. Ho KK, Myatt SS, Lam EW (2008) Many forks in the path: cycling with FoxO. Oncogene 27:2300–2311
73. Hu J, Dong L, Outten CE (2008) The redox environment in the mitochondrial intermembrane space is maintained separately from the cytosol and matrix. J Biol Chem 283:29126–29134
74. Huang H, Regan KM, Wang F, Wang D, Smith DI, van Deursen JM, Tindall DJ (2005) Skp2 inhibits FOXO1 in tumor suppression through ubiquitin-mediated degradation. Proc Natl Acad Sci U S A 102:1649–1654
75. Huang TT, Carlson EJ, Gillespie AM, Shi Y, Epstein CJ (2000) Ubiquitous overexpression of CuZn superoxide dismutase does not extend life span in mice. J Gerontol A Biol Sci Med Sci 55:B5–B9
76. Hwang C, Sinskey AJ, Lodish HF (1992) Oxidized redox state of glutathione in the endoplasmic reticulum. Science 257:1496–1502
77. Jaiswal A (2004) Nrf2 signaling in coordinated activation of antioxidant gene expression. Free Radic Biol Med 36:1199–1207

78. Jeong AY, Lee MY, Lee SH, Park JH, Han HJ (2009) PPARdelta agonist-mediated ROS stimulates mouse embryonic stem cell proliferation through cooperation of p38 MAPK and Wnt/beta-catenin. Cell Cycle 8:611–619
79. Jiang YY, Kong DX, Qin T, Zhang HY (2010) How does oxygen rise drive evolution? Clues from oxygen-dependent biosynthesis of nuclear receptor ligands. Biochem Biophys Res Commun 391:1158–1160
80. Johnston DG, Johnson GA, Alberti KG, Millward-Sadler GH, Mitchell J, Wright R (1986) Hepatic regeneration and metabolism after partial hepatectomy in diabetic rats: effects of insulin therapy. Eur J Clin Invest 16:384–390
81. Joshi I, Minter LM, Telfer J, Demarest RM, Capobianco AJ, Aster JC, Sicinski P, Fauq A, Golde TE, Osborne BA (2009) Notch signaling mediates G1/S cell-cycle progression in T cells via cyclin D3 and its dependent kinases. Blood 113:1689–1698
82. Kannan R, Jin M, Gamulescu MA, Hinton DR (2004) Ceramide-induced apoptosis: role of catalase and hepatocyte growth factor. Free Radic Biol Med 37:166–175
83. Kang TW, Yevsa T, Woller N, Hoenicke L, Wuestefeld T, Dauch D, Hohmeyer A, Gereke M, Rudalska R, Potapova A, Iken M, Vucur M, Weiss S, Heikenwalder M, Khan S, Gil J, Bruder D, Manns M, Schirmacher P, Tacke F, Ott M, Luedde T, Longerich T, Kubicka S, Zender L (2011) Senescence surveillance of pre-malignant hepatocytes limits liver cancer development. Nature 479:547–551
84. Kaposi-Novak P, Lee JS, Gòmez-Quiroz L, Coulouarn C, Factor VM, Thorgeirsson SS (2006) Met-regulated expression signature defines a subset of human hepatocellular carcinomas with poor prognosis and aggressive phenotype. J Clin Invest 116:1582–1595
85. Kim KH, Chen CC, Monzon RI, Lau LF (2013) Matricellular protein CCN1 promotes regression of liver fibrosis through induction of cellular senescence in hepatic myofibroblasts. Mol Cell Biol 33:2078–2090
86. Köhler C, Bell AW, Bowen WC, Monga SP, Fleig W, Michalopoulos GK (2004) Expression of Notch-1 and its ligand Jagged-1 in rat liver during liver regeneration. Hepatology 39:1056–1065
87. Kong X, Feng D, Wang H, Hong F, Bertola A, Wang FS, Gao B (2012) Interleukin-22 induces hepatic stellate cell senescence and restricts liver fibrosis in mice. Hepatology 56:1150–1159
88. Kops GJ, de Ruiter ND, De Vries-Smits AM, Powell DR, Burgering BM, Bos JL (1999) Direct control of the Forkhead transcription factor AFX by protein kinase B. Nature 398:630–634
89. Krizhanovsky V, Yon M, Dickins RA, Hearn S, Simon J, Miething C, Yee H, Zender L, Lowe SW (2008) Senescence of activated stellate cells limits liver fibrosis. Cell 134:657–667
90. Ksiazek K (2010) Bacterial aging: from mechanistic basis to evolutionary perspective. Cell Mol Life Sci 67:3131–3137
91. Kuilman T, Michaloglou C, Vredeveld LC, Douma S, van Doorn R, Desmet CJ, Aarden LA, Mooi WJ, Peeper DS (2008) Oncogene-induced senescence relayed by an interleukin dependent inflammatory network. Cell 133:1019–1031
92. Kuilman T, Peeper DS (2009) Senescence-messaging secretome: SMS-ing cellular stress. Nat Rev Cancer 9:81–94
93. Kumar S, Millis AJ, Baglioni C (1992) Expression of interleukin 1-inducible genes and production of interleukin 1 by aging human fibroblasts. Proc Natl Acad Sci U S A 89:4683–4687
94. Laconi S, Curreli F, Diana S, Pasciu D, De Filippo G, Sarma DS, Pani P, Laconi E (1999) Liver regeneration in response to partial hepatectomy in rats treated with retrorsine: a kinetic study. J Hepatol 31:1069–1074
95. Laconi S, Pani P, Pillai S, Pasciu D, Sarma DS, Laconi E (2001) A growth-constrained environment drives tumor progression in vivo. Proc Natl Acad Sci U S A 98:7806–7811
96. Lane N (2011) Mitonuclear match: optimizing fitness and fertility over generations drives ageing within generations. Bioessays 33:860–869
97. Laurent A, Nicco C, Chéreau C, Goulvestre C, Alexandre J, Alves A, Lévy E, Goldwasser F, Panis Y, Soubrane O, Weill B, Batteux F (2005) Controlling tumor growth by modulating endogenous production of reactive oxygen species. Cancer Res 65:948–956

98. Lavia P, Jansen-Dürr P (1999) E2F target genes and cell-cycle checkpoint control. Bioessays 21:221–223
99. LeCouter J, Moritz DR, Li B, Phillips GL, Liang XH, Gerber HP, Hillan KJ, Ferrara N (2003) Angiogenesis-independent endothelial protection of liver: role of VEGFR-1. Science 299:890–893
100. Lee AC, Fenster BE, Ito H, Takeda K, Bae NS, Hirai T, Yu ZX, Ferrans VJ, Howard BH, Finkel T (1999) Ras proteins induce senescence by altering the intracellular levels of reactive oxygen species. J Biol Chem 274:7936–7940
101. Lee SR, Yang KS, Kwon J, Lee C, Jeong W, Rhee SG (2002) Reversible inactivation of the tumor suppressor PTEN by H_2O_2. J Biol Chem 277:20336–20342
102. Leontieva OV, Lenzo F, Demidenko ZN, Blagosklonny MV (2012) Hyper-mitogenic drive coexists with mitotic incompetence in senescent cells. Cell Cycle 11:4642–4649
103. Li W, Kong AN (2009) Molecular mechanisms of Nrf2-mediated antioxidant response. Mol Carcinog 48:91–104
104. Liang H, Masoro EJ, Nelson JF, Strong R, McMahan CA, Richardson A (2003) Genetic mouse models of extended lifespan. Exp Gerontol 38:1353–1364
105. Lin WW, Karin M (2007) A cytokine-mediated link between innate immunity, inflammation, and cancer. J Clin Invest 117:1175–1183
106. Macip S, Igarashi M, Fang L, Chen A, Pan ZQ, Lee SW, Aaronson SA (2002) Inhibition of p21-mediated ROS accumulation can rescue p21-induced senescence. EMBO J 21:2180–2188
107. Mantovani A, Bonecchi R, Locati M (2006) Tuning inflammation and immunity by chemokine sequestration: decoys and more. Nat Rev Immunol 6:907–918
108. Marí M, Colell A, Morales A, von Montfort C, Garcia-Ruiz C, Fernández-Checa JC (2010) Redox control of liver function in health and disease. Antioxid Redox Signal 12:1295–1331
109. Marongiu F, Doratiotto S, Sini M, Serra MP, Laconi E (2012) Cancer as a disease of tissue pattern formation. Prog Histochem Cytochem 47:175–207
110. Marongiu F, Serra MP, Sini M, Angius F, Laconi E (2014) Clearance of senescent hepatocytes in a neoplastic-prone microenvironment delays the emergence of hepatocellular carcinoma. Aging (Albany NY) 6:26–34
111. Mars WM, Liu ML, Kitson RP, Goldfarb RH, Gabauer MK, Michalopoulos GK (1995) Immediate early detection of urokinase receptor after partial hepatectomy and its implications for initiation of liver regeneration. Hepatology 21:1695–1701
112. Mars WM, Zarnegar R, Michalopoulos GK (1993) Activation of hepatocyte growth factor by the plasminogen activators uPA and tPA. Am J Pathol 143:949–958
113. Mauro F, Grasso A, Tolmach LJ (1969) Variations in sulfhydryl, disulfide, and protein content during synchronous and asynchronous growth of HeLa cells. Biophys J 9:1377–1397
114. Menon SG, Goswami PC (2007) A redox cycle within the cell cycle: ring in the old with the new. Oncogene 26:1101–1109
115. Menon SG, Sarsour EH, Spitz DR, Higashikubo R, Sturm M, Zhang H, Goswami PC (2003) Redox regulation of the G1 to S phase transition in the mouse embryo fibroblast cell cycle. Cancer Res 63:2109–2117
116. Mohammed FF, Khokha R (2005) Thinking outside the cell: proteases regulate hepatocyte division. Trends Cell Biol 15:555–563
117. Morimoto H, Iwata K, Ogonuki N, Inoue K, Atsuo O, Kanatsu-Shinohara M, Morimoto T, Yabe-Nishimura C, Shinohara T (2013) ROS are required for mouse spermatogonial stem cell self-renewal. Cell Stem Cell 12:774–786
118. Michalopoulos GK (2007) Liver regeneration. J Cell Physiol 213:286–300
119. Michalopoulos GK (2010) Liver regeneration after partial hepatectomy: critical analysis of mechanistic dilemmas. Am J Pathol 176:2–13
120. Milne LS (1909) The histology of liver tissue regeneration. J Pathol Bacteriol 13:127–160
121. Muñoz-Espín D, Cañamero M, Maraver A, Gómez-López G, Contreras J, Murillo-Cuesta S, Rodríguez-Baeza A, Varela-Nieto I, Ruberte J, Collado M, Serrano M (2013) Programmed cell senescence during mammalian embryonic development. Cell 155:1104–1118

122. Nakae J, Oki M, Cao Y (2008) The FoxO transcription factors and metabolic regulation. FEBS Lett 582:54–67
123. Naugler WE, Karin M (2008) NF-kappaB and cancer-identifying targets and mechanisms. Curr Opin Genet Dev 18:19–26
124. Nault JC, Amaddeo G, Zucman-Rossi J (2012) When activated oncogene meets immunity: a fight to prevent liver tumor initiation. Hepatology 56:387–389
125. Nelson G, Wordsworth J, Wang C, Jurk D, Lawless C, Martin-Ruiz C, von Zglinicki T (2012) A senescent cell bystander effect: senescence-induced senescence. Aging Cell 11:345–349
126. Noseda M, Chang L, McLean G, Grim JE, Clurman BE, Smith LL, Karsan A (2004) Notch activation induces endothelial cell cycle arrest and participates in contact inhibition: role of p21Cip1 repression. Mol Cell Biol 24:8813–8822
127. Olovnikov AM (1996) Telomeres, telomerase, and aging: origin of the theory. Exp Gerontol 31:443–448
128. Olsen PS, Poulsen SS, Kirkegaard P (1985) Adrenergic effects on secretion of epidermal growth factor from Brunner's glands. Gut 26:920–927
129. Orjalo AV, Bhaumik D, Gengler BK, Scott GK, Campisi J (2009) Cell surface-bound IL-1α is upstream regulator of the senescence-associated IL-6/IL-8 cytochine network. Proc Natl Acad Sci U S A 106:17031–17036
130. Ortiz C, Caja L, Sancho P, Bertran E, Fabregat I (2008) Inhibition of the EGF receptor blocks autocrine growth and increases the cytotoxic effects of doxorubicin in rat hepatoma cells: role of reactive oxygen species production and glutathione depletion. Biochem Pharmacol 75:1935–1945
131. Ozaki M, Haga S, Zhang HQ, Irani K, Suzuki S (2003) Inhibition of hypoxia/reoxygenation-induced oxidative stress in HGF-stimulated antiapoptotic signaling: role of PI3-K and Akt kinase upon rac1. Cell Death Differ 10:508–515
132. Palm W, de Lange T (2008) How shelterin protects mammalian telomeres. Annu Rev Genet 42:301–334
133. Pan D (2007) Hippo signaling in organ size control. Genes Dev 21:8868–8897
134. Paradis V, Youssef N, Dargère D, Bâ N, Bonvoust F, Deschatrette J, Bedossa P (2001) Replicative senescence in normal liver, chronic hepatitis C, and hepatocellular carcinomas. Hum Pathol 32:327–332
135. Pardee AB (1974) A restriction point for control of normal animal cell proliferation. Proc Natl Acad Sci U S A 71:1286–1290
136. Park EJ, Lee JH, Yu GY, He G, Ali SR, Holzer RG, Osterreicher CH, Takahashi H, Karin M (2010) Dietary and genetic obesity promote liver inflammation and tumorigenesis by enhancing IL-6 and TNF expression. Cell 140:197–208
137. Parola M, Bellomo G, Robino G, Barrera G, Dianzani MU (1999) 4-Hydroxynonenal as a biological signal: molecular basis and pathophysiological implications. Antioxid Redox Signal 1:255–284
138. Passos JF, Nelson G, Wang C, Richter T, Simillion C, Proctor CJ, Miwa S, Olijslagers S, Hallinan J, Wipat A, Saretzki G, Rudolph KL, Kirkwood TB, von Zglinicki T (2010) Feedback between p21 and reactive oxygen production is necessary for cell senescence. Mol Syst Biol 6:347
139. Passos JF, Saretzki G, von Zglinicki T (2007) DNA damage in telomeres and mitochondria during cellular senescence: is there a connection? Nucleic Acids Res 35:7505–7513
140. Pearson KJ, Baur JA, Lewis KN, Peshkin L, Price NL, Labinskyy N, Swindell WR, Kamara D, Minor RK, Perez E, Jamieson HA, Zhang Y, Dunn SR, Sharma K, Pleshko N, Woollett LA, Csiszar A, Ikeno Y, Le Couteur D, Elliott PJ, Becker KG, Navas P, Ingram DK, Wolf NS, Ungvari Z, Sinclair DA, de Cabo R (2008) Resveratrol delays age-related deterioration and mimics transcriptional aspects of dietary restriction without extending life span. Cell Metab 8:157–168
141. Pelicci PG (2004) Do tumor-suppressive mechanisms contribute to organism aging by inducing stem cell senescence? J Clin Invest 113:4–7
142. Pendergrass WR, Gray M, Wold MS, Luo P, Norwood T (1999) Analysis of the capacity of extracts from normal human young and senescent fibroblasts to support DNA synthesis in vitro. J Cell Biochem 73:176–187

143. Pérez VI, Bokov A, Van Remmen H, Mele J, Ran Q, Ikeno Y, Richardson A (2009) Is the oxidative stress theory of aging dead? Biochim Biophys Acta 1790:1005–1014
144. Pérez-Garijo A, Steller H (2014) The benefits of aging: cellular senescence in normal development. EMBO J 33:99–100
145. Rai P (2012) Human Mut T Homolog 1 (MTH1): a roadblock for the tumor-suppressive effects of oncogenic RAS-induced ROS. Small GTPases 3:120–125
146. Rai P, Young JJ, Burton DG, Giribaldi MG, Onder TT, Weinberg RA (2011) Enhanced elimination of oxidized guanine nucleotides inhibits oncogenic RAS-induced DNA damage and premature senescence. Oncogene 30:1489–1496
147. Rapkine L (1931) Su les processus chimiques au cours de la division cellulaire. Ann Physiol Physiochem Biol 7:382–418
148. Raymond J, Blankenship RE (2004) Biosynthetic pathways, gene replacement and the antiquity of life. Geobiology 2:199–203
149. Raymond J, Segrè D (2006) The effect of oxygen on biochemical networks and the evolution of complex life. Science 311:1764–1767
150. Reddy NM, Kleeberger SR, Yamamoto M, Kensler TW, Scollick C, Biswal S, Reddy SP (2007) Genetic dissection of the Nrf2-dependent redox signaling-regulated transcriptional programs of cell proliferation and cytoprotection. Physiol Genomics 32:74–81
151. Rees DC, Howard JB (2003) The interface between the biological and inorganic worlds: iron-sulfur metalloclusters. Science 300:929–931
152. Reuben A (2004) Prometheus and Pandora—together again. Hepatology 39:1460–1463
153. Robles SJ, Adami GR (1998) Agents that cause DNA double strand breaks lead to p16(INK4A) enrichment and the premature senescence of normal fibroblasts. Oncogene 16:1113–1123
154. Rolo AP, Teodoro JS, Palmeira CM (2012) Role of oxidative stress in the pathogenesis of non alcoholic steatohepatitis. Free Radic Biol Med 52:59–69
155. Salama R, Sadaie M, Hoare M, Narita M (2014) Cellular senescence and its effector programs. Genes Dev 28:99–114
156. Salih DA, Brunet A (2008) FoxO transcription factors in the maintenance of cellular homeostasis during aging. Curr Opin Cell Biol 20:126–136
157. Sancho P, Bertran E, Caja L, Carmona-Cuenca I, Murillo MM, Fabregat I (2009) The inhibition of the epidermal growth factor (EGF) pathway enhances TGF-beta-induced apoptosis in rat hepatoma cells through inducing oxidative stress coincident with a change in the expression pattern of the NADPH oxidases (NOX) isoforms. Biochim Biophys Acta 1793:253–263
158. Sang L, Coller HA, Roberts JM (2008) Control of the reversibility of cellular quiescence by the transcriptional repressor HES1. Science 321:1095–1100
159. Sasaki M, Ikeda H, Yamaguchi J, Nakada S, Nakanuma Y (2008) Telomere shortening in the damaged small bile ducts in primary biliary cirrhosis reflects ongoing cellular senescence. Hepatology 48:186–195
160. Satyanarayana A, Kaldis P (2009) Mammalian cell-cycle regulation: several Cdks, numerous cyclins and diverse compensatory mechanisms. Oncogene 28:2925–2939
161. Schirmacher P, Geerts A, Jung W, Pietrangelo A, Rogler CE, Dienes HP (1993) The role of Ito cells in the biosynthesis of HGF-SF in the liver. EXS 65:285–299
162. Schnabl B, Purbeck CA, Choi YH, Hagedorn CH, Brenner D (2003) Replicative senescence of activated human hepatic stellate cells is accompanied by a pronounced inflammatory but less fibrogenic phenotype. Hepatology 37:653–664
163. Serra MP, Marongiu F, Sini M, Laconi E (2012) Hepatocyte senescence in vivo following preconditioning for liver repopulation. Hepatology 56:760–768
164. Serviddio G, Bellanti F, Vendemiale G (2013) Free radical biology for medicine: learning from nonalcoholic fatty liver disease. Free Radic Biol Med 65:952–968
165. Sherr CJ, DePinho RA (2000) Cellular senescence: mitotic clock or culture shock? Cell 102:407–410
166. Sherr CJ, Roberts JM (1999) CDK inhibitors: positive and negative regulators of G1-phase progression. Genes Dev 13:1501–1512

167. Sohal RS, Brunk UT (1992) Mitochondrial production of pro-oxidants and cellular senescence. Mutat Res 275:295–304
168. Sohal RS, Weindruch R (1996) Oxidative stress, caloric restriction, and aging. Science 273:59–63
169. Sozou PD, Seymour RM (2004) To age or not to age. Proc Biol Sci 271:457–463
170. Stolz DB, Mars WM, Petersen BE, Kim TH, Michalopoulos GK (1999) Growth factor signal transduction immediately after two-thirds partial hepatectomy in the rat. Cancer Res 59:3954–3960
171. Storer M, Mas A, Robert-Moreno A, Pecoraro M, Ortells MC, Di Giacomo V, Yosef R, Pilpel N, Krizhanovsky V, Sharpe J, Keyes WM (2013) Senescence is a developmental mechanism that contributes to embryonic growth and patterning. Cell 155:1119–1130
172. Takami T, Kaposi-Novak P, Uchida K, Gomez-Quiroz LE, Conner EA, Factor VM, Thorgeirsson SS (2007) Loss of hepatocyte growth factor/c-Met signaling pathway accelerates early stages of N-nitrosodiethylamine induced hepatocarcinogenesis. Cancer Res 67:9844–9851
173. Terman A, Brunk UT (2006) Oxidative stress, accumulation of biological 'garbage', and aging. Antioxid Redox Signal 8:197–204
174. Tiniakos DG, Kandilis A, Geller SA (2010) Tityus: a forgotten myth of liver regeneration. J Hepatol 53:357–361
175. Tonks NK (2005) Redox redux: revisiting PTPs and the control of cell signalling. Cell 121:667–670
176. Tormos AM, Arduini A, Talens-Visconti R, del Barco Barrantes I, Nebreda AR, Sastre J (2013) Liver-specific p38α deficiency causes reduced cell growth and cytokinesis failure during chronic biliary cirrhosis in mice. Hepatology 57:1950–1961
177. Trusina A (2014) Stress induced telomere shortening: longer life with less mutations? BMC Syst Biol 8:27
178. Tsuboi S (1999) Elevation of glutathione level in rat hepatocytes by hepatocyte growth factor via induction of gamma-glutamylcysteine synthetase. J Biochem 126:815–820
179. Tu BP, Kudlicki A, Rowicka M, McKnight SL (2005) Logic of the yeast metabolic cycle: temporal compartmentalization of cellular processes. Science 310:1152–1158
180. Van Remmen H, Ikeno Y, Hamilton M, Pahlavani M, Wolf N, Thorpe SR, Alderson NL, Baynes JW, Epstein CJ, Huang TT, Nelson J, Strong R, Richardson A (2003) Life-long reduction in MnSOD activity results in increased DNA damage and higher incidence of cancer but does not accelerate aging. Physiol Genomics 16:29–37
181. von Zglinicki T, Pilger R, Sitte N (2000) Accumulation of single-strand breaks is the major cause of telomere shortening in human fibroblasts. Free Radic Biol Med 28:64–74
182. von Zglinicki T, Saretzki G, Ladhoff J, d'Adda di Fagagna F, Jackson SP (2005) Human cell senescence as a DNA damage response. Mech Ageing Dev 126:111–117
183. Wang X, Liu JZ, Hu JX, Wu H, Li YL, Chen HL, Bai H, Hai CX (2011) ROS-activated p38 MAPK/ERK-Akt cascade plays a central role in palmitic acid-stimulated hepatocyte proliferation. Free Radic Biol Med 51:539–551
184. Wiemann SU, Satyanarayana A, Tsahuridu M, Tillmann HL, Zender L, Klempnauer J, Flemming P, Franco S, Blasco MA, Manns MP, Rudolph KL (2002) Hepatocyte telomere shortening and senescence are general markers of human liver cirrhosis. FASEB J 16:935–942
185. Wilson DW, Lamé MW, Dunston SK, Segall HJ (2000) DNA damage cell checkpoint activities are altered in monocrotaline pyrrole-induced cell cycle arrest in human pulmonary artery endothelial cells. Toxicol Appl Pharmacol 166:69–80
186. Xue L, Zhou B, Liu X, Wang T, Shih J, Qi C, Heung Y, Yen Y (2006) Structurally dependent redox property of ribonucleotide reductase subunit p53R2. Cancer Res 66:1900–1905
187. Yang JY, Zong CS, Xia W, Yamaguchi H, Ding Q, Xie X, Lang JY, Lai CC, Chang CJ, Huang WC, Huang H, Kuo HP, Lee DF, Li LY, Lien HC, Cheng X, Chang KJ, Hsiao CD, Tsai FJ, Tsai CH, Sahin AA, Muller WJ, Mills GB, Yu D, Hortobagyi GN, Hung MC (2008) ERK

promotes tumorigenesis by inhibiting FOXO3a via MDM2-mediated degradation. Nat Cell Biol 10:138–148

188. Yoshimoto S, Loo TM, Atarashi K, Kanda H, Sato S, Oyadomari S, Iwakura Y, Oshima K, Morita H, Hattori M, Honda K, Ishikawa Y, Hara E, Ohtani N (2013) Obesity-induced gut microbial metabolite promotes liver cancer through senescence secretome. Nature 499:97–101
189. Zhu H, Jia Z, Misra H, Li YR (2012) Oxidative stress and redox signaling mechanisms of alcoholic liver disease: updated experimental and clinical evidence. J Dig Dis 13:133–142

Part II
Oxidative Stress and Liver Pathophysiology: From Cell Damage and Death to Inflammatory and Fibrogenic Responses

Chapter 4
Oxidative Stress and Hepatocellular Injury

Marjolein H. Tiebosch*, Golnar Karimian*, and Han Moshage

4.1 Oxidative Stress and Hepatocellular Injury

Oxidative stress and inflammation are hallmarks of virtually all (chronic) liver diseases. Not surprisingly, understanding the role of oxidative stress and its deleterious consequences has been and still is an important research topic in hepatology. A thorough understanding of oxidative stress and its impact on liver homeostasis is essential to identify novel targets for the treatment of liver diseases. In this chapter, we will review the current knowledge on the role of oxidative stress in hepatocellular injury. Hepatocytes have received a large share of the attention in research on oxidative stress. That will also be the case in this chapter. Nevertheless, we will also address the role of other liver cell types, such as endothelial cells, hepatic stellate cells, and Kupffer cells in oxidative stress-mediated liver injury.

4.2 Oxidative Stress and Antioxidant Defenses

Oxidative stress is defined as the inappropriate exposure to reactive oxygen species (ROS). It results from the imbalance between pro-oxidants and antioxidants and leads to cell damage. ROS represent a variety of species, including superoxide anions ($O_2^{\bullet-}$), hydrogen peroxide (H_2O_2), and hydroxyl radicals ($HO^{\bullet}$) [1].

*Author contributed equally with all other contributors.

M.H. Tiebosch • G. Karimian • H. Moshage (✉)
Department of Gastroenterology and Hepatology, University Medical Center Groningen, University of Groningen, Groningen, The Netherlands

Department of Laboratory Medicine, University Medical Center Groningen, University of Groningen, Groningen, The Netherlands
e-mail: a.j.moshage@umcg.nl

E. Albano, M. Parola (eds.), *Studies on Hepatic Disorders*,
Oxidative Stress in Applied Basic Research and Clinical Practice,
DOI 10.1007/978-3-319-15539-5_4

The stability and (biological) half-life of these species are variable: the hydroxyl radical is highly reactive, whereas superoxide anions and hydrogen peroxide are more stable. Hydroxyl radicals can be generated through the Haber–Weiss reaction, in which superoxide anions react with hydrogen peroxide, or through the Fenton reaction, in which hydrogen peroxide reacts with iron.

Excessive oxidation of cellular substrates by ROS can lead to cellular injury. Therefore, cells are equipped with various antioxidant defenses, which can significantly delay or prevent the oxidation of substrates [1].

One of these antioxidant defenses are the superoxide dismutases (SODs), a family of enzymes specialized in eliminating superoxide anions. Three distinct SODs have been identified in mammals: Copper-zinc SOD in the cytosol and nucleus (SOD1 or CuZnSOD), manganese SOD in the mitochondria (SOD2 or MnSOD), and extracellular SOD (SOD3 or EC-SOD) [2]. Although they are structurally unrelated and encoded by different genes, they all have the same function, namely converting the superoxide radical to hydrogen peroxide [3]. In general, the generated hydrogen peroxide is less reactive and/or further converted into water and oxygen.

Several enzymes are able to convert hydrogen peroxide into water and oxygen. Catalase, localized in the peroxisomes, can enzymatically decompose hydrogen peroxide. Glutathione peroxidases (GPx) are a family of enzymes, which catalyze the reaction of glutathione (GSH) and H_2O_2 into water and oxidized glutathione (GSSG). Glutathione is highly present in virtually all cells. It is localized in the cytosol as well as in intracellular organelles such as the endoplasmic reticulum, the nucleus, and the mitochondria. Glutathione is exclusively synthesized in the cytosol in a two-step process by the rate-limiting enzyme γ-glutamylcysteine synthetase (γ-GCS or GCL: glutamate-cysteine ligase) followed by γ-glutamyl transpeptidase (γ-GT) [4]. Oxidized glutathione can be reduced again to GSH by the enzyme GSH reductase (GR), a reaction requiring NADPH. This possibility to recycle GSH makes it a crucial antioxidant defense mechanism for cells.

Several redox-sensitive transcription factors are involved in antioxidant responses such as activator protein-1 (AP-1), nuclear factor-κB (NF-κB), and Nrf2.

AP-1 is a heterodimeric transcription factor that binds to the TPA response element (TRE), which is present in the promoter region of many genes, including antioxidant genes, like heme oxygenase-1, genes involved in cell proliferation, but also genes involved in fibrogenesis, e.g., TGF-β and collagen type 1. Many mitogens and pro-inflammatory signals lead to increased AP-1 nuclear binding and corresponding gene transactivation. It is known that lipid peroxidation products (such as 4HNE) and ROS can induce AP-1 and subsequent transcriptional regulation of target genes, including the ones mentioned above.

NF-κB is also a redox-sensitive transcription factor, consisting of homo- or heterodimers of structurally related subunits. NF-κB activity is mostly regulated at the translational level. Under normal conditions, it is present in the cytosol of cells in an inactive form, due to its binding to the inhibitory subunits IκB. The activation of NF-κB consists of the phosphorylation of the inhibitory subunits, followed by its subsequent degradation [5]. Upon dissociation of IκB from NF-κB, NF-κB can translocate to the nucleus where it can bind to promoter regions of genes which have the κB element. These are genes involved in, e.g., inflammation, cell survival (including antioxidant enzymes), and adhesion. Most, if not all, inducers of NF-κB rely on the production of ROS and their reactions with cellular macromolecules.

Nuclear erythroid-related factor 2 (Nrf-2) is a redox-sensitive transcription factor. In response to an oxidative event, Nrf2 dissociates from KEAP (kelch-like ECH-associated protein 1) and translocates to the nucleus. Here it can bind to its response element ARE (antioxidant response element) to transactivate antioxidant genes. Several antioxidant enzymes like glutamate–cysteine ligase (GCL) and heme oxygenase-1 (HO-1) are under the control of AREs via the phosphorylation of the transcription factor Nrf2 [6, 7].

4.3 Oxidative Stress in Liver Diseases

Oxidative stress plays an important role in many types of liver diseases including non-alcoholic fatty liver disease (NAFLD), cholestatic liver diseases, alcohol-induced liver injury, and viral hepatitis.

4.3.1 Non-alcoholic Fatty Liver Disease

Liver injury in NAFLD is characterized by fat accumulation (steatosis), the infiltration of inflammatory cells, and a varying extent of ballooning degeneration of the hepatocytes [8]. NAFLD is the general name for fatty liver disease with steatosis but without severe and overt inflammation, whereas non-alcoholic steatohepatitis (NASH) is the subsequent stage of this disease, characterized by overt inflammation [9]. It is widely accepted that NASH has a two-hit pathogenesis (although some recent reports also imply a "multiple hit" model) [10]. Steatosis is the "first hit": during this stage, free fatty acids (FFAs) accumulate in the cells. FFA accumulation is caused by an increase in fatty acid uptake either from dietary sources or from lipolysis in adipose tissue. The "second hit," leading to NASH, involves oxidative stress, decreased hepatic ATP production, and induction of pro-inflammatory cytokines. Progression of steatosis to NASH is an undesired process because it leads to hepatocellular injury. There are various hypotheses regarding the progression of steatosis to NASH. It has been proposed that the lipotoxicity of FFAs increases the vulnerability of the liver to a "second hit," and that this vulnerability is dependent on environmental and/or genetic factors [10].

Another hypothesis is that the impaired β-oxidation leads to hepatic steatosis and accumulation of lipid intermediates and subsequent impairment of insulin signaling and insulin-resistance, a phenomenon often observed in patients with NASH. In addition, the sustained inflammation and increased levels of TNF may also lead to impaired insulin-signaling via aberrant and sustained phosphorylation of insulin receptor substrate-1 (IRS-1) [11].

As mentioned before, oxidative stress plays a crucial role in the pathogenesis of NASH. Sources for ROS in NASH are mitochondria and peroxisomes, as the β-oxidation occurs in these organelles. Increased β-oxidation in these organelles might be a compensatory mechanism for the excessive FFA accumulation in the hepatocytes and leads to excessive ROS production and oxidative stress [12].

4.3.2 Cholestasis

Cholestasis is the reduction of bile flow which results in elevated serum levels of potentially toxic bile acids. There are several causes of cholestasis in humans. The most common causes are primary biliary cirrhosis and sclerosing cholangitis. In addition to these disorders, cholestasis can result from biliary atresia, familial cholestatic syndromes resulting from defects in bile acid transporters, exposure to certain drugs, pregnancy, graft-versus-host disease, infection, and Alagille syndrome [13–15].

Several animal models as well as human studies suggest a role for oxidative stress in cholestatic liver diseases. In the bile duct ligation (BDL) model, an animal model for cholestatic liver disease, products of lipid peroxidation were present in hepatic mitochondrial membranes [16]. In addition, increased serum and/or tissue levels of 4-hydroxynonenal (4HNE) and malondialdehyde (MDA), both stable end products of lipid peroxidation, were observed after BDL [17]. However, although this was associated with the influx of inflammatory cells, it was not associated with liver damage. Accordingly, several studies proposed that oxidative stress is a consequence of cholestatic liver disease rather than a cause for injury in these diseases [18].

4.3.3 Xenobiotic-Induced Liver Diseases (Alcohol, Paracetamol)

Xenobiotics are generally metabolized by hepatocytes, in particular by members of the enzyme family cytochrome P450 (CYP family). Several cytochrome P450 enzymes are important in metabolizing common xenobiotics. A well-known harmful xenobiotic is ethanol. Alcohol-induced liver disease is one of the most common causes of liver disease. Ethanol is metabolized by CYP2E1, producing ROS through reduction of oxygen to superoxide anions, which are dismutated to hydroxyl radicals, a highly oxidative substance. Consequently, lipid peroxidation occurs that leads to the depletion of cellular antioxidants (especially glutathione). This disturbance of the redox equilibrium leads to oxidative stress.

Another xenobiotic is paracetamol (or acetaminophen, APAP), widely used as an analgesic and antipyretic drug. Upon an APAP overdose, the metabolizing capacity of glucuronidating enzymes is exceeded and the drug is metabolized by CYP2E1 or CYP3A4 into *N*-acetylbenzoquinoneimine (NAPQI), an extremely toxic and oxidizing component. Normally, NAPQI is neutralized by glutathione; however, at high concentrations, the glutathione store is depleted and the excess NAPQI binds to cellular proteins eventually resulting in mitochondrial dysfunction and cell death [19, 20]. Interestingly, acetaminophen toxicity can be attenuated by the addition of glutathione precursors like *N*-acetylcysteine.

4.4 Signaling Pathways in Hepatocellular Injury: Involvement of Different Organelles

4.4.1 Plasma Membrane

In general, the plasma membranes of mammalian cells are exposed to an oxidizing environment, while the cytosol is a reducing environment. If the defense systems are inadequate, oxidative damage will occur to proteins, lipids, and carbohydrates. This damage can be directly, by inhibition of transmembrane transporters and other membrane proteins, or indirectly mediated by lipid peroxidation products.

Lipid peroxidation is the main damaging effect of oxidative stress. Lipids in the cell or organelle membranes are mainly polyunsaturated fatty acids and cholesterol. Upon lipid peroxidation, these lipids are oxidized, which results in the formation of aldehydic by-products such as 4-hydroxy-2-nonenal (4HNE) and malondialdehyde (MDA). This leads to increased cell membrane permeability, decreased cell membrane fluidity, inactivation of membrane proteins, and loss of polarity of mitochondrial membranes. Moreover, 4HNE and MDA have long half-lives, which make them detrimental as they can diffuse over long distances, thereby amplifying the effect of oxidative stress.

As a consequence, oxidative stress can adversely affect the function of cellular organelles such as mitochondria, endoplasmic reticulum, and lysosomes.

4.4.2 Mitochondria

Mitochondrial function is based on membrane potential, which depends on membrane integrity. Mitochondrial DNA (mtDNA) is extremely sensitive to oxidative damage due to its location (close to the ROS producing mitochondrial inner membrane), the absence of protective histones, and incomplete DNA repair mechanisms in mitochondria. Accumulation of mtDNA damage results in mitochondrial dysfunction, leading to increased ROS production.

Oxidative stress leads to activation of JNK (c-Jun N-terminal kinase, a member of the MAP-kinase family) and translocation of phosphorylated (P)-JNK to the mitochondria, which leads to increased mitochondrial ROS production. This process is a critical step in hepatocellular injury, because it can trigger the mitochondrial membrane permeability transition (MPT) and the collapse of membrane potential, mitochondrial swelling, and rupture of the outer mitochondrial membrane [21].

4.4.3 Endoplasmic Reticulum

The membrane of the endoplasmic reticulum can also be damaged by lipid peroxidation. Disruption of the ER membrane leads to calcium disequilibrium, which is considered to be one of the initial and pivotal events of ER stress-mediated cell death. ROS and lipid peroxidation directly disable the sarco-endoplasmic reticulum Ca^{2+}-ATPase (SERCA). The impairment of SERCAs can further lead to Ca^{2+} release from the ER lumen, the accumulation of unfolded proteins in the ER lumen, and ultimately, the disruption of ER homeostasis, i.e., ER stress [22–24]. ER stress causes the activation of the Unfolded Protein Response (UPR). The UPR describes a series of compensatory responses and signaling between organelles to mediate cellular adaptations in order to promote cell survival. However, when ER stress is excessive or chronic, or when the UPR is compromised, pro-apoptotic pathways are activated. A key event in this pro-apoptotic pathway is the induction of CHOP (C/EBP homologous protein) expression [25]. Induced expression of CHOP leads to ROS formation by enhancing the translation of mRNA to stimulate the UPR. Furthermore, ER stress and accompanying CHOP induction is also associated with Ca^{2+} leakage and subsequent cell death in hepatitis C virus-infected cells [26]. All of these events may lead to hepatocellular injury and cell death.

For the proteins to fold into the correct tertiary and quaternary structures, disulphide bonds have to be created. This requires a highly oxidizing environment that is maintained by a low GSH/GSSG ratio. Oxidative protein folding is a source of ROS, predominantly hydrogen peroxide. Normally, the antioxidant defense can cope with this exposure to elevated ROS. However, under conditions of increased protein folding, this system can be exhausted. In addition, protein misfolding can also give rise to ROS [27].

Exogenous oxidants can also activate the UPR. For example, 7-ketocholesterol, an oxidant product of cholesterol, induces the complete UPR, whereas others only mildly or partly stimulate the UPR [28].

During ER stress, mitochondria display increased ROS production: calcium leaking from the ER is taken up by the mitochondria, which results in the opening of the permeability transition pore and subsequent cytochrome c release. This abrogates the normal electron transport chain, and as a result, leads to increased ROS generation. The increased calcium concentrations in the mitochondria stimulate Krebs cycle dehydrogenases, further boosting ROS production. The increased ROS production leads to opening of the calcium channels in the ER, resulting in a vicious circle.

4.4.4 Lysosomes

Lysosomes are involved in necrotic, apoptotic, and autophagic cell death. The key factor in determining the type of cell death is the magnitude of the lysosomal membrane permeabilization (LMP) and the amount of proteolytic enzymes released into the cytosol [29]. Massive breakdown of lysosomes results in unregulated necrosis, whereas selective permeabilization of lysosomes triggers apoptosis. Several mechanisms for the controlled permeabilization of lysosomes have been proposed. One hypothesis includes the accumulation of lysosomotropic detergents such as sphingosine in the lysosomes, facilitating the release of lysosomal enzymes into the cytoplasm [30]. Another hypothesis involves ROS-mediated lysosomal destabilization. In this hypothesis, LMP is suggested to precede mitochondrial dysfunction, thereby creating a feedback loop between mitochondrial-derived ROS and LMP to control cell death. In addition, intralysosomal accumulation of free iron indirectly mediates lysosomal membrane damage via generation of ROS [31, 32].

4.5 Injury to Different Hepatic Cell Type Populations

Liver injury is associated with various responses from different cell types in liver. In response to toxic stimuli, hepatocytes usually become damaged or die, whereas endothelial cells (EC), Kupffer cells (KC), and hepatic stellate cells (HSC) become activated.

4.5.1 Hepatocytes

Hepatocyte injury can result in hepatocyte cell death via apoptosis, necrosis, necroptosis, or autophagy [33, 34]. In a damaged liver all these forms of cell death will be present; however, one may be predominant over the others. They also share some of the same executive pathways and therefore hybrid modes of cell death occur, displaying characteristics of different modes of cell death. For example, cell death may start as apoptosis, but can switch to necroptosis. In the following paragraphs the different forms of cell death are briefly described.

Apoptosis is an ATP-dependent process, also known as programmed cell death. Apoptosis is characterized by DNA condensation, nuclear fragmentation, plasma membrane blebbing, cell shrinkage, and the formation of apoptotic bodies. Two distinct pathways of apoptosis have been described: the organelle-mediated pathway (or intrinsic pathway) and the receptor-mediated pathway (or extrinsic pathway). Both pathways overlap downstream with the activation of effector caspases (caspase-3, -6 and -7), which results in cleavage of intracellular substrates such as lamin A, poly (ADP-ribose) polymerase (PARP), and Inhibitor of Caspase-Activated DNase (ICAD) to induce apoptotic cell death. Apoptotic features include cell shrinkage, nuclear fragmentation, apoptotic DNA fragmentation, and ultimately, cell death [35].

Necrosis is a passive process that is characterized by metabolic disruption, energy depletion (loss of ATP), mitochondrial swelling, and rupture of the plasma membrane. Subsequently, the release of cellular content into the extracellular environment and systemic circulation triggers an inflammatory response.

Necrosis was long thought to be an uncontrolled form of cell death; however, recently it was appreciated that there is an "in between" variant between apoptosis and necrosis, called *necroptosis*. Morphologically, necroptotic cell death resembles necrosis as it is characterized by cell swelling, mitochondrial dysfunction, plasma membrane permeabilization, and release of cytoplasmic content into the extracellular space. However, it is regulated via stimulation of death receptors in a similar manner as in apoptosis. Necroptosis seems to serve as a back-up mechanism for apoptosis in cells, which have become resistant to apoptosis such as virus-infected cells. It has been demonstrated that these cells are highly sensitive to necroptosis [36]. This form of cell death is also associated with high mitochondrial ROS production and, unlike apoptosis, it does not involve DNA fragmentation [37].

Autophagy represents a homeostatic cellular mechanism for the turnover of organelles and proteins through a lysosome-dependent degradation pathway. During starvation, autophagy facilitates cell survival through the recycling of metabolic precursors. Additionally, autophagy can modulate other vital processes such as programmed cell death (e.g., apoptosis), inflammation, and adaptive immune mechanisms and thereby influence the pathogenesis of diseases. Selective pathways can target distinct cargoes (e.g., mitochondria or proteins) for autophagic degradation. At present, the causal relationship between autophagy and various forms of cell death remains unclear, as it has been reported that autophagy can act both as an agonist and as an antagonist of cell death. It has also been described that autophagy is a cell death effector pathway under conditions of impaired apoptosis [38].

Autophagy can occur in association with necroptosis triggered by caspase inhibition. Autophagy and apoptosis have been shown to be coincident or antagonistic, depending on the experimental context, and share cross talk between signal transduction pathways. Autophagy may modulate the outcome of other regulated forms of cell death such as necroptosis [39].

Different ROS can cause different modes of cell death. We have previously demonstrated that the superoxide anions induce mainly apoptosis in primary rat hepatocytes and to a much lesser extent necrosis [40]. Superoxide anion-induced apoptosis was caspase-dependent (caspase-9, -6, and -3 were involved). In contrast, we have shown that hydrogen peroxide is not toxic at low doses, but causes necrosis at higher concentrations (or at low concentrations when hydrogen peroxide detoxification is inhibited).

Exposure to ROS has been shown to activate signal transduction pathways, such as phosphorylation of ERK1/2 and JNK MAP kinases. In hepatocytes, activation of ERK is anti-apoptotic, whereas activation of JNK is pro-apoptotic. Superoxide anions also induce the antioxidant enzyme HO-1. This induction is not sufficient to prevent superoxide-induced hepatocyte cell death. However, when HO-1 is induced before exposure to superoxide anions, the primary hepatocytes are protected against superoxide anion-induced apoptosis [41]. Carbon monoxide, one of the products of the enzymatic reaction of HO-1, is partially responsible for this protective effect as exogenously added CO inhibits superoxide anion-induced caspase activation, PARP cleavage, JNK activation, and apoptosis [41]. CO is suggested to down-regulate the ERK1/2 MAPK pathway and prevent transplant-induced hepatic ischemia/reperfusion injury [42]. In addition, CO is known to activate p38 MAPK. The activation of p38 MAPK protects liver against ischemia/reperfusion injury and has antioxidant effects in systemic liver inflammation [43, 44]. As stated above, superoxide anions also activate pro-apoptotic pathways, e.g., JNK. Apparently, the activation of pro-apoptotic pathways overrules the activation of survival pathways. Inhibition of JNK blocks superoxide anion-induced apoptosis and caspase-9 activation. This could indicate that JNK triggers the mitochondrial apoptotic pathway after exposure to superoxide anions. The well-known anti-apoptotic transcription factor NF-kB is not involved in the protection against superoxide anions-induced apoptosis.

4.5.2 Kupffer Cells

Kupffer cells, the liver-specific macrophages, are generally seen as the main source of oxidative stress in the liver next to neutrophils. ROS generation by Kupffer cells occurs via an oxidant burst. Prolonged production of oxidants involves activated complement factors. Kupffer cells produce hydrogen peroxide, superoxide anions, and hydroxyl radicals. NADPH oxidase is the principal enzyme involved in ROS production.

Kupffer cells are also involved in NASH. Serum leptin levels correlate with NAFLD/NASH severity. Leptin induces NADPH oxidase and inducible nitric oxide synthase (iNOS) in Kupffer cells and, as a result, superoxide anions and nitric oxide are produced. These compounds can react to form peroxynitrite, which is a strong oxidant. Peroxynitrite can also cause posttranslational modifications such as nitrosylation of proteins. Ultimately, this leads to activation of macrophages and production of TNF-α and MCP-1 [45].

DAMPS (damage- or danger-associated molecular patterns) are also known to activate Kupffer cells. For example, ATP is released after necrosis/apoptosis of hepatocytes and is then considered a DAMP. ATP can bind to the P2X7 receptor. This receptor is upstream of NADPH oxidase and can activate NADPH oxidase and subsequent ROS production by increasing the expression of the p47 phox subunit and the binding of p47 phox to the membrane subunit gp91 phox [46].

4.5.3 Endothelial Cells

Kupffer cells and liver sinusoidal endothelial cells (LSECs) together constitute the largest scavenger cell system in the body and they are responsible for the elimination of a wide range of potentially injurious particles, pathogens and molecules from the blood. Particulate matter (>200 nm in diameter) is phagocytosed by the Kupffer cells, whereas LSECs mediate clearance of soluble macromolecules and colloids <200 nm in diameter via receptor-mediated endocytosis [47].

In any injury, the tightly regulated microcirculation in the liver is disrupted as there is a shift towards more vasoconstrictive forces over vasodilating agents.

This causes vasoconstriction, mostly because of an induction of the vasoconstrictor endothelin-1 (ET-1) which cannot be overcome by the vasodilator nitric oxide (NO), generated by endothelial nitric oxide synthase (eNOS) and, in case of inflammation, iNOS. Vasoconstriction leads to hypoxia, which can further aggravate liver damage. ET-1 is induced by oxidative stress, whereas receptor-mediated NO release is reduced by hydrogen peroxide. ET-1 can normally induce NO production by promoting the translocation of eNOS to the plasma membrane. This will restore the balance between vasoconstriction and vasodilation. However, upon acute oxidative stress exposure, eNOS cannot translocate to the plasma membrane, although normal phosphorylation of eNOS can still occur, leading to a reduction of NO production. Subacute oxidative stress also causes reduced NO production [48].

Oxidative stress can further affect these mechanisms. It has been demonstrated that exposure of LSECs to hydrogen peroxide induces the production of eicosanoids such as thromboxane A2 (TXA2), prostaglandin I2 (PGI2), and PGE2 via up-regulation of cyclooxygenase-2 (COX2), thromboxane synthase, and the phosphorylation of cytosolic phospholipase A2 (cPLA2). This causes vasodilatation. However, the combination of high ET-1 and hydrogen peroxide causes a decrease in the production of the vasodilating eicosanoids by LSECs. As a result, this combination leads to vasoconstriction [49].

The synergistic effect of NO and hydrogen peroxide is due to the inhibition of hydrogen peroxide degradation, since NO inhibits catalase (predominantly in hepatocytes) and/or glutathione peroxidases (prevailing in endothelial cells) [50].

In diabetes, there is oxidative injury to the vessel walls. In the liver, there are also ultrastructural changes in the hepatic microcirculation. It has been shown in in vivo models of diabetes that LSEC porosity is increased by ~50 %, and an increase in nitrosative stress has also been shown by demonstrating the presence of nitrosylated proteins in the sinusoidal endothelium, which was reduced by antioxidants [51].

4.5.4 Hepatic Stellate Cells

Hepatic stellate cells become activated in chronically injured liver. Upon activation, they proliferate and produce excessive amounts of extracellular matrix which eventually results in fibrosis and end stage cirrhosis. The exact cause of HSC activation is not known; however, many conditions and mechanisms have been shown to contribute to HSC activation, including oxidative stress. Oxidative stress has both paracrine and autocrine effects on HSC.

Paracrine effects include products that activate HSCs from oxidative-stress-damaged hepatocytes and oxidative stress-activated Kupffer cells.

These paracrine effects were investigated in co-culture experiments of hepatic stellate cells and hepatocytes or Kupffer cells. For example, co-culture of primary hepatic stellate cells with HepG2 cells (a human hepatoma cell line) manipulated to express increased levels of CYP2E1 showed higher rates of proliferation as well as induced levels of the fibrogenic activation markers α-smooth muscle actin (α-SMA) and collagen. This could be inhibited by the addition of antioxidants and CYP2E1 inhibitors [52, 53].

It has also been shown that ROS can induce hepatic stellate cell death. We have shown in hepatic stellate cells, similar to hepatocytes, that different ROS induce different modes of cell death. Superoxide anions induce apoptosis of HSCs in a caspase-independent manner [54]. HSCs were relatively resistant to hydrogen peroxide-induced necrosis, probably related to the induction of glutathione levels during stellate cell activation [55].

Lipid peroxidation products, especially 4HNE, can also activate HSC. A positive correlation between hepatic 4HNE and hepatic hydroxyproline content has been observed. Moreover, HSC are stimulated by 4HNE to produce collagen type 1, via signaling pathways involving kinases that modulate redox-sensitive transcription factors such as AP-1.

Acetaldehyde, the first metabolite of ethanol, also has direct effects on collagen production by HSC [56]: acetaldehyde induces hydrogen peroxide accumulation in stellate cells that in turn activate protein kinase C and phosphatidylinositol 3-kinase pathways [57, 58]. Recently, it was shown that acetaldehyde mediates β-catenin stabilization (by inhibiting its phosphorylation) and nuclear translocation in activated hepatic stellate cells. In addition to hydrogen peroxide, acetaldehyde also induces the production of superoxide anions and 4HNE. Exposure to oxidative stress results in the depletion of cellular glutathione stores and redox imbalance. In this environment redox sensors (NXN, a Wnt pathway regulator) are oxidized, leading to β-catenin stabilization [59].

Direct effects of oxidative stress on HSC activation have also been demonstrated.

Hepatic stellate cells can produce ROS themselves since they contain both the phagocytic as well as the non-phagocytic NADPH oxidase. NOX1, the non-phagocytic form of NADPH oxidase, is induced upon stimulation of HSC with platelet-derived growth factor (PDGF). ROS produced by NOX1 inactivates phosphatase and tensin homolog (PTEN) by oxidizing this protein. This, in turn, leads to activation of the Akt/FOXO4 pathway and, subsequently, downregulation of p27, a cell cycle repressor. Therefore, PDGF leads to HSC proliferation via NOX1-mediated ROS production [60].

NOX2, the phagocytic form of NADPH oxidase, is also present on hepatic stellate cells. It has been demonstrated that stellate cells, like macrophages, have the ability to phagocytose, in particular apoptotic bodies of hepatocytes. The engulfment of apoptotic bodies (from hepatocytes) by HSC was shown to be directly fibrogenic. This is mediated by NOX2, as NOX2-mediated ROS production is directly linked to increased collagen promoter activity.

Both NOX1 and NOX2 are inhibited in the p47 phox knockout model. P47 phox is the subunit known to be involved in the structural organization of the NOX subunits into a functional NADPH complex. The production of ROS by HSC induces PDGF [61]. PDGF signals through the PDGF-receptor to Ras and Raf-1 with subsequent activation of the MAPK pathway. In addition, MAPK activation was also shown to be induced by lipid peroxidation products (4HNE). These results indicate that proliferative pathways for HSC often involve ROS generation.

References

1. Halliwell B, Gutteridge JMC (1989) Free radicals in biology and medicine. Clarendon, Oxford, p 543
2. Miao L, St Clair DK (2009) Regulation of superoxide dismutase genes: implications in disease. Free Radic Biol Med 47:344–356
3. Fridovich I (1995) Superoxide radical and superoxide dismutases. Annu Rev Biochem 64:97–112
4. Rahman I, MacNee W (2000) Oxidative stress and regulation of glutathione in lung inflammation. Eur Respir J 16:534–554
5. Baeuerle PA, Baltimore D (1996) NF-kappa B: ten years after. Cell 87:13–20
6. Keum YS, Owuor ED, Kim BR, Hu R, Kong AN (2003) Involvement of Nrf2 and JNK1 in the activation of antioxidant responsive element (ARE) by chemopreventive agent phenethyl isothiocyanate (PEITC). Pharm Res 20:1351–1356
7. Kobayashi M, Yamamoto M (2005) Molecular mechanisms activating the Nrf2-Keap1 pathway of antioxidant gene regulation. Antioxid Redox Signal 7:385–394
8. Brunt EM, Tiniakos DG (2010) Histopathology of nonalcoholic fatty liver disease. World J Gastroenterol 16:5286–5296
9. Brunt EM (2010) Pathology of nonalcoholic fatty liver disease. Nat Rev Gastroenterol Hepatol 7:195–203
10. Tilg H, Moschen AR (2010) Evolution of inflammation in nonalcoholic fatty liver disease: the multiple parallel hits hypothesis. Hepatology 52:1836–1846
11. Tilg H (2010) The role of cytokines in non-alcoholic fatty liver disease. Dig Dis 28:179–185
12. Dowman JK, Tomlinson JW, Newsome PN (2010) Pathogenesis of non-alcoholic fatty liver disease. QJM 103:71–83
13. Kim WR, Ludwig J, Lindor KD (2000) Variant forms of cholestatic diseases involving small bile ducts in adults. Am J Gastroenterol 95:1130–1138
14. Qureshi WA (1999) Intrahepatic cholestatic syndromes: pathogenesis, clinical features and management. Dig Dis 17:49–59
15. Poupon R, Chazouilleres O, Poupon RE (2000) Chronic cholestatic diseases. J Hepatol 32:129–140
16. Sokol RJ, Devereaux M, Khandwala RA (1991) Effect of dietary lipid and vitamin E on mitochondrial lipid peroxidation and hepatic injury in the bile duct-ligated rat. J Lipid Res 32:1349–1357
17. Parola M, Leonarduzzi G, Robino G, Albano E, Poli G et al (1996) On the role of lipid peroxidation in the pathogenesis of liver damage induced by long-standing cholestasis. Free Radic Biol Med 20:351–359
18. Copple BL, Jaeschke H, Klaassen CD (2010) Oxidative stress and the pathogenesis of cholestasis. Semin Liver Dis 30:195–204
19. Jaeschke H, Bajt ML (2006) Intracellular signaling mechanisms of acetaminophen-induced liver cell death. Toxicol Sci 89:31–41
20. Nelson SD (1990) Molecular mechanisms of the hepatotoxicity caused by acetaminophen. Semin Liver Dis 10:267–278
21. Tsujimoto Y, Nakagawa T, Shimizu S (2006) Mitochondrial membrane permeability transition and cell death. Biochim Biophys Acta 1757:1297–1300
22. Ikesugi K, Mulhern ML, Madson CJ, Hosoya K, Terasaki T et al (2006) Induction of endoplasmic reticulum stress in retinal pericytes by glucose deprivation. Curr Eye Res 31:947–953
23. Badiola N, Penas C, Minano-Molina A, Barneda-Zahonero B, Fado R et al (2011) Induction of ER stress in response to oxygen-glucose deprivation of cortical cultures involves the activation of the PERK and IRE-1 pathways and of caspase-12. Cell Death Dis 2:e149
24. Mekahli D, Bultynck G, Parys JB, De Smedt H, Missiaen L (2011) Endoplasmic-reticulum calcium depletion and disease. Cold Spring Harb Perspect Biol 3(6):pii: a004317. doi:10.1101/cshperspect.a004317

25. Oyadomari S, Mori M (2004) Roles of CHOP/GADD153 in endoplasmic reticulum stress. Cell Death Differ 11:381–389
26. Benali-Furet NL, Chami M, Houel L, De Giorgi F, Vernejoul F et al (2005) Hepatitis C virus core triggers apoptosis in liver cells by inducing ER stress and ER calcium depletion. Oncogene 24:4921–4933
27. Higa A, Chevet E (2012) Redox signaling loops in the unfolded protein response. Cell Signal 24:1548–1555
28. Pedruzzi E, Guichard C, Ollivier V, Driss F, Fay M et al (2004) NAD(P)H oxidase nox-4 mediates 7-ketocholesterol-induced endoplasmic reticulum stress and apoptosis in human aortic smooth muscle cells. Mol Cell Biol 24:10703–10717
29. Li W, Yuan X, Nordgren G, Dalen H, Dubowchik GM et al (2000) Induction of cell death by the lysosomotropic detergent MSDH. FEBS Lett 470:35–39
30. Kagedal K, Zhao M, Svensson I, Brunk UT (2001) Sphingosine-induced apoptosis is dependent on lysosomal proteases. Biochem J 359:335–343
31. Terman A, Gustafsson B, Brunk UT (2006) The lysosomal-mitochondrial axis theory of postmitotic aging and cell death. Chem Biol Interact 163:29–37
32. Terman A, Kurz T, Gustafsson B, Brunk UT (2006) Lysosomal labilization. IUBMB Life 58:531–539
33. Malhi H, Gores GJ, Lemasters JJ (2006) Apoptosis and necrosis in the liver: a tale of two deaths? Hepatology 43:S31–S44
34. Schoemaker M, Moshage H (2004) Defying death: the hepatocyte's survival kit. Clin Sci (Lond) 107:13–25
35. Mahmood Z, Shukla Y (2010) Death receptors: targets for cancer therapy. Exp Cell Res 316:887–899
36. Li M, Beg AA (2000) Induction of necrotic-like cell death by tumor necrosis factor alpha and caspase inhibitors: novel mechanism for killing virus-infected cells. J Virol 74:7470–7477
37. Giampietri C, Starace D, Petrungaro S, Filippini A, Ziparo E (2014) Necroptosis: molecular signalling and translational implications. Int J Cell Biol 2014:490275
38. Maiuri MC, Zalckvar E, Kimchi A, Kroemer G (2007) Self-eating and self-killing: crosstalk between autophagy and apoptosis. Nat Rev Mol Cell Biol 8:741–752
39. Ryter SW, Choi AM (2013) Autophagy: an integral component of the mammalian stress response. J Biochem Pharmacol Res 1:176–188
40. Conde de la Rosa L, Schoemaker M, Vrenken T, Buist-Homan M, Havinga R et al (2006) Superoxide anions and hydrogen peroxide induce hepatocyte death by different mechanisms: involvement of JNK and ERK MAP kinases. J Hepatol 44:918–929
41. Conde de la Rosa L, Vrenken T, Hannivoort R, Buist-Homan M, Havinga R et al (2008) Carbon monoxide blocks oxidative stress-induced hepatocyte apoptosis via inhibition of the p54 JNK isoform. Free Radic Biol Med 44:1323–1333
42. Kaizu T, Ikeda A, Nakao A, Tsung A, Toyokawa H et al (2008) Protection of transplant-induced hepatic ischemia/reperfusion injury with carbon monoxide via MEK/ERK1/2 pathway downregulation. Am J Physiol Gastrointest Liver Physiol 294:G236–G244
43. Amersi F, Shen XD, Anselmo D, Melinek J, Iyer S et al (2002) Ex vivo exposure to carbon monoxide prevents hepatic ischemia/reperfusion injury through p38 MAP kinase pathway. Hepatology 35:815–823
44. Brugger J, Schick MA, Brock RW, Baumann A, Muellenbach RM et al (2010) Carbon monoxide has antioxidative properties in the liver involving p38 MAP kinase pathway in a murine model of systemic inflammation. Microcirculation 17:504–513
45. Chatterjee S, Ganini D, Tokar EJ, Kumar A, Das S et al (2013) Leptin is key to peroxynitrite-mediated oxidative stress and kupffer cell activation in experimental non-alcoholic steatohepatitis. J Hepatol 58:778–784
46. Chatterjee S, Rana R, Corbett J, Kadiiska MB, Goldstein J et al (2012) P2X7 receptor-NADPH oxidase axis mediates protein radical formation and kupffer cell activation in carbon tetrachloride-mediated steatohepatitis in obese mice. Free Radic Biol Med 52:1666–1679

47. Seternes T, Sorensen K, Smedsrod B (2002) Scavenger endothelial cells of vertebrates: a nonperipheral leukocyte system for high-capacity elimination of waste macromolecules. Proc Natl Acad Sci U S A 99:7594–7597
48. Karaa A, Kamoun WS, Clemens MG (2005) Oxidative stress disrupts nitric oxide synthase activation in liver endothelial cells. Free Radic Biol Med 39:1320–1331
49. Karaa A, Kamoun WS, Xu H, Zhang J, Clemens MG (2006) Differential effects of oxidative stress on hepatic endothelial and kupffer cell eicosanoid release in response to endothelin-1. Microcirculation 13:457–466
50. Rauen U, Li T, de Groot H (2007) Inhibitory and enhancing effects of NO on H(2)O(2) toxicity: dependence on the concentrations of NO and H(2)O(2). Free Radic Res 41:402–412
51. McMahon AC, Parry SN, Benson VL, Witting PK, Le Couteur DG (2013) Beneficial effects of the synthetic antioxidant tert-butyl bisphenol on the hepatic microcirculation in a rat model of diabetes mellitus. Acta Diabetol 50:645–649
52. Nieto N, Friedman SL, Cederbaum AI (2002) Stimulation and proliferation of primary rat hepatic stellate cells by cytochrome P450 2E1-derived reactive oxygen species. Hepatology 35:62–73
53. Nieto N, Friedman SL, Cederbaum AI (2002) Cytochrome P450 2E1-derived reactive oxygen species mediate paracrine stimulation of collagen I protein synthesis by hepatic stellate cells. J Biol Chem 277:9853–9864
54. Dunning S, Hannivoort R, de Boer J, Buist-Homan M, Faber K et al (2009) Superoxide anions and hydrogen peroxide inhibit proliferation of activated rat stellate cells and induce different modes of cell death. Liver Int 29:922–932
55. Dunning S, Ur Rehman A, Tiebosch MH, Hannivoort RA, Haijer FW et al (2013) Glutathione and antioxidant enzymes serve complementary roles in protecting activated hepatic stellate cells against hydrogen peroxide-induced cell death. Biochim Biophys Acta 1832:2027–2034
56. Moshage H, Casini A, Lieber CS (1990) Acetaldehyde selectively stimulates collagen production in cultured rat liver fat-storing cells but not in hepatocytes. Hepatology 12:511–518
57. Svegliati-Baroni G, Ridolfi F, Di Sario A, Saccomanno S, Bendia E et al (2001) Intracellular signaling pathways involved in acetaldehyde-induced collagen and fibronectin gene expression in human hepatic stellate cells. Hepatology 33:1130–1140
58. Svegliati-Baroni G, Inagaki Y, Rincon-Sanchez AR, Else C, Saccomanno S et al (2005) Early response of alpha2(I) collagen to acetaldehyde in human hepatic stellate cells is TGF-beta independent. Hepatology 42:343–352
59. Arellanes-Robledo J, Reyes-Gordillo K, Shah R, Dominguez-Rosales JA, Hernandez-Nazara ZH et al (2013) Fibrogenic actions of acetaldehyde are beta-catenin dependent but wingless independent: a critical role of nucleoredoxin and reactive oxygen species in human hepatic stellate cells. Free Radic Biol Med 65:1487–1496
60. Cui W, Matsuno K, Iwata K, Ibi M, Matsumoto M et al (2011) NOX1/nicotinamide adenine dinucleotide phosphate, reduced form (NADPH) oxidase promotes proliferation of stellate cells and aggravates liver fibrosis induced by bile duct ligation. Hepatology 54:949–958
61. Adachi T, Togashi H, Suzuki A, Kasai S, Ito J et al (2005) NAD(P)H oxidase plays a crucial role in PDGF-induced proliferation of hepatic stellate cells. Hepatology 41:1272–1281

Chapter 5
Oxidative Stress and the Unfolded Protein Response in the Liver

William M. Hudson and Michael J. Pagliassotti

5.1 Introduction

Liver and biliary diseases can result from a wide variety of causes, including infectious agents, inherited defects, alcohol, toxins, and environmental insults. Common forms of liver and biliary diseases include viral hepatitis, alcoholic fatty liver disease, non-alcoholic fatty liver disease, and gallstones. Estimates suggest that hepatitis B and C virus are present in ~5 % and ~3 %, respectively, of the world's population. Hepatocellular carcinoma is among the leading causes of cancer-related deaths and non-alcoholic fatty liver disease may affect 10–20 % of the population, largely due to the current world-wide obesity epidemic. Many liver and biliary diseases are characterized by both oxidative and endoplasmic reticulum stress, both of which can lead to inflammation, cell death, and global organ impairment. This chapter will discuss the endoplasmic reticulum (ER) and oxidative stress in the liver.

5.2 The Endoplasmic Reticulum

The ER is the largest continuous organelle in a eukaryotic cell and consists of an array of tubules (cisternae) that form a three-dimensional network (reticulum) stretching from the nuclear envelope to the cell surface. The smooth ER produces structural phospholipids and cholesterol, as well as significant amounts of

W.M. Hudson • M.J. Pagliassotti (✉)
Department of Food Science and Human Nutrition, Colorado State University, Gifford 234, Fort Collins, CO, USA 80523-1571
e-mail: Michael.Pagliassotti@colostate.edu

E. Albano, M. Parola (eds.), *Studies on Hepatic Disorders*, Oxidative Stress in Applied Basic Research and Clinical Practice, DOI 10.1007/978-3-319-15539-5_5

triacylglycerol and cholesterol esters that have non-structural roles [1, 2]. The smooth ER is the main site of cholesterol synthesis, although much of this lipid is transported to other cellular organelles. Thus, the ER membrane is comprised of very low concentrations of cholesterol and complex sphingolipids [1]. It has been suggested that the loose packing of ER membrane lipids may provide an environment conducive to the insertion and transport of newly synthesized lipids and proteins [1]. This specialized lipid environment within the ER may have implications in diseases characterized by abnormal lipid accumulation, such as alcoholic and non-alcoholic fatty liver disease [3].

All eukaryotic cells contain a significant amount of rough ER, the site for protein folding and maturation. Proteins destined for secretion or insertion into membranes require modification, such as glycosylation and disulfide bond formation, which cannot be achieved in the cytosol [4]. The ER lumen provides a specialized environment for protein folding and maturation that is characterized by high concentrations of calcium, a low ratio (1:1–3:1) of reduced glutathione (GSH) to oxidized glutathione (GSSG), and a unique complement of molecular chaperones and folding enzymes [5]. The ER is also equipped with a quality control system that recognizes and degrades improperly folded proteins, termed ER-associated degradation (ERAD). ERAD can target and transport misfolded proteins from the ER lumen to the cytosolic proteasome machinery [6]. The unfolded protein response (UPR) monitors the ability of the ER lumen to match folding and degradation to the rate of entry of newly synthesized proteins and functions to restore ER homeostasis following periods of ER stress (i.e., accumulation of unfolded proteins within the ER lumen).

5.3 The Unfolded Protein Response

In mammalian cells, activation of the UPR (Fig. 5.1) generally involves three ER-localized proteins: inositol-requiring 1α (IRE1α), double-stranded RNA-dependent protein kinase-like ER kinase (PERK), and activating transcription factor-6 (ATF6) [7]. Each of these transmembrane proteins has an ER-luminal domain to sense unfolded proteins, a transmembrane domain for targeting to the ER membrane, and a cytosolic domain to transmit signals to the transcriptional and/or translational apparatus [8]. It is currently thought that in un-stressed cells all three proteins are maintained in an inactive state via their association with the ER protein chaperone glucose-regulated protein 78/immunoglobulin-heavy-chain-binding protein (GRP78). Subsequent to ER stress, GRP78 is released and sequestered on unfolded proteins, thereby allowing activation of PERK, IRE1α, and ATF6 [9]. PERK activation leads to phosphorylation of the α-subunit of the translation initiation factor eIF2 (p-eIF2α) and subsequent attenuation of translation initiation. Attenuation of translation also leads to selective translation of mRNAs containing open reading frames, such as activating transcription factor-4 (ATF4) [10, 11]. Increased expression of GADD34 (which also contains open reading frames), a member of the growth arrest and DNA damage family of proteins, is involved in

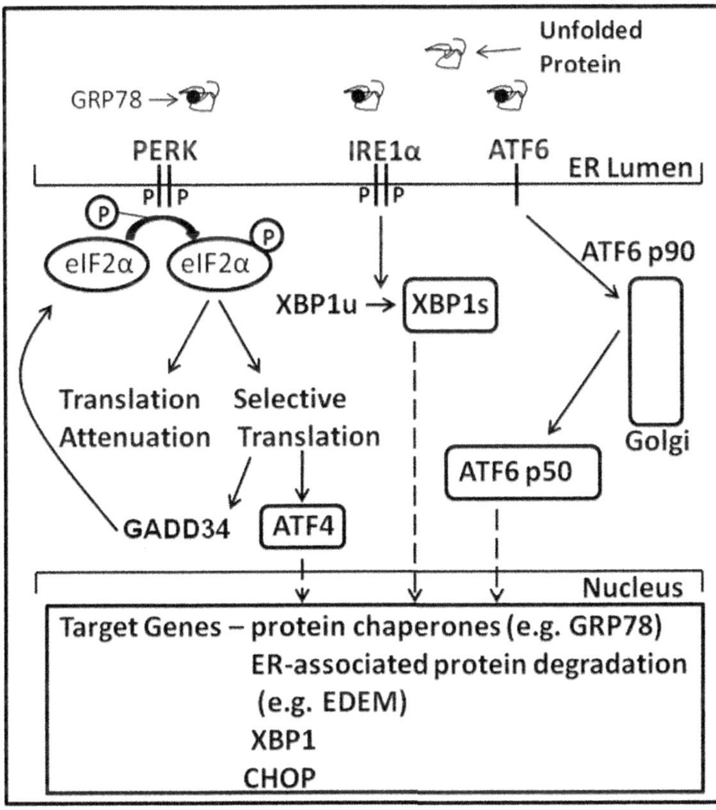

Fig. 5.1 Overview of the mammalian unfolded response. The presence of unfolded proteins in the ER lumen leads to dimerization and autophosphorylation of PERK and IRE1α, and the release and proteolytic cleavage of ATF6 in the Golgi. PERK-mediated phosphorylation of eIF2α leads to transient attenuation of translation, but selective translation of mRNAs containing upstream open reading frames, such as ATF4. Increased transcription and translation of GADD34 subsequently leads to dephosphorylation of eIF2α and resumption of translation. Activation of IRE1α leads to the splicing of XBP1. XBP1s, ATF4, and the cleaved form of ATF6 lead to transcriptional activation of a number of gene targets related to protein folding and ER-associated degradation (see text). Reprinted with permission from ANTIOXIDANT AND REDOX SIGNALING (2011, volume 15, issue 2), published by Mary Ann Liebert, Inc., New Rochelle, NY

dephosphorylation of eIF2α and therefore promotes reversal of translational attenuation [7]. IRE1α activation leads to splicing of X-box-binding protein-1 (XBP1s) mRNA and subsequent transcription of molecular chaperones (e.g., GRP78) and genes involved in ERAD (e.g., ER degradation-enhancing α—like protein (EDEM)) [11]. Activation of ATF6 leads to its release from the ER membrane, processing in the Golgi, and entry into the nucleus. Transcriptional targets of ATF6 include protein chaperones and XBP1 [8]. Thus, activation of the UPR initiates a spectrum of responses that include transient attenuation of global protein synthesis and an increased capacity for protein folding and degradation. This dual response may serve to not only minimize the increase in unfolded proteins in the ER lumen, but also the accumulation of chaperones and may be particularly relevant to cell types that produce a large amount of secreted proteins (e.g., β-cells) [12].

5.3.1 An Expanded View of the UPR

PERK is one of four protein kinases that can phosphorylate eIF2α; the other three are double-stranded RNA-activated protein kinase (PKR) which is activated in response to viral infection, general control non-derepressible 2 kinase (GCN2) which is activated in response to amino acid deprivation, and heme-regulated inhibitor kinase (HRI) which is primarily expressed in reticulocytes and appears to coordinate globin polypeptide synthesis with heme availability [13]. Protein kinase-mediated phosphorylation of eIF2α not only regulates translation, but also the activation of nuclear factor kappa-β (NFκB), via reduction in the abundance of the NFκβ inhibitor Iκβ [11]. PERK can also phosphorylate nuclear erythroid 2 p45-related factor 2 (Nrf2), triggering the dissociation of Nrf2/Keap1 complexes and subsequent nuclear import of Nrf2 [14]. Thus, activation of this branch of the UPR links disruption of ER homeostasis to both inflammation, via NFκβ, and redox balance, via Nrf2 (see below).

IRE1α, in addition to catalyzing XBP1 splicing, has additional functions related to cellular signaling. Activated IRE1α can interact with the adaptor protein TNFR-associated factor 2 and lead to activation of c-Jun-NH_2-terminal kinase and NFκβ via apoptosis signaling-regulating kinase 1 [15]. IRE1α activation has also been linked to the activation of p38 mitogen-activated protein kinase and extracellular-regulated kinase [16–18]. These interactions suggest that the IRE1α branch of the UPR not only regulates adaptation to ER stress and cell survival via XBP1 splicing, but also activation of signaling pathways involved in inflammation, insulin action, and apoptosis. Regulated IRE1α-dependent decay of selected mRNAs can also reduce production of proteins destined for the ER lumen [19, 20].

5.4 The UPR and Antioxidant Defense

Oxidative stress is thought to be an important pathogenic event in many liver diseases. The ER provides a unique oxidizing environment for protein folding and disulfide bond formation. Each disulfide bond formed during oxidative protein folding produces a single reactive oxygen species. It has been estimated that secretory cells produce 3–6 million disulfide bonds per minute, thus protein folding in the ER is intimately linked to the generation of reaction oxygen species and potentially oxidative stress [21, 22]. Conversely, cellular oxidative stress can disrupt ER homeostasis and induce ER stress [23–25]. Therefore, it is not surprising that the UPR engages the antioxidant program via the transcription factor Nrf2 [26]. Nrf2 belongs to the

Cap "n" Collar family of basic leucine zipper transcription factors and regulates the expression of antioxidant response element (ARE)-containing genes [26]. Nrf2 is highly expressed in the liver and kidney and is a substrate of the proximal UPR sensor PERK [27]. Importantly, Nrf2 deletion results in rapid onset and progression of steatohepatitis in mice provided a methionine-choline-deficient diet, often used to model components of non-alcoholic fatty liver disease [28]. In addition, Nrf2-deficient mice were characterized by increased mortality in response to endotoxin- and cecal ligation and puncture-induced septic shock [29]. As noted above, PERK-mediated phosphorylation of eIF2α also leads to the selective translation and upregulation of ATF4. Along with Nrf2, this transcription factor has been linked to the maintenance of cellular glutathione [14]. Thus, the PERK arm of the UPR appears to play a critical role in the defense against oxidative stress and the downstream substrate Nrf2 has been directly linked to steatohepatitis.

In addition to the PERK arm of the UPR, recent evidence has also linked the IRE1α-XBP1 branch of the UPR to the regulation of antioxidant defenses [30]. In this study, hydrogen peroxide-mediated cell death occurred more extensively in mouse embryonic fibroblast cells deficient in XBP1. XBP1 deficiency resulted in reduced catalase expression, and overexpression of XBP1 restored catalase expression in XBP1-deficient cells. Thus, XBP1 may provide protection from oxidative stress; however, whether this regulation occurs in hepatocytes is presently unknown.

5.5 The ER Lumen as a Source of Oxidative Stress

The ER lumen is an oxidizing environment characterized by a GSH:GSSG ratio of 1:1–3:1, much lower than the cytosolic ratio of 30:1–100:1 [31]. This environment is, in part, maintained by ER oxidase 1 (Ero1) and GSSG, and disulfide bond formation in the ER lumen appears to primarily result from electron transfer reactions involving Ero1, protein disulfide isomerase, and molecular oxygen. Hydrogen peroxide is a product of these transfer reactions and therefore disulfide bond formation and protein folding in the ER lumen are associated with the formation of reactive oxygen species [21, 32]. In addition, the luminal NADPH concentration may play an important antioxidant defense role in liver cells in a manner that appears to be independent of the thiol/disulfide redox system [33].

Very few studies have directly examined whether and how protein folding/misfolding influences oxidative stress. In one study, HIP-deficient cells that lack the ability to eliminate misfolded proteins from the ER were employed. Introduction of low levels of a mutant misfolded form of the vacuolar protein carboxypeptidase Y induced ER stress, accumulation of reactive oxygen species, and cell death [34]. Malhotra et al. utilized hydrodynamic delivery of FVIII (coagulation factor VIII, prone to misfolding in the ER lumen) DNA expression vectors into the tail vein of mice [35]. Accumulation of FVIII resulted in oxidative stress (monitored by dihydroethidine staining, malondialdehyde, GSH) and activation of the UPR in the liver. Treatment with butylated hydroxyanisole reduced accumulation of FVIII and attenuated oxidative stress and UPR activation. Taken together, these data are consistent with the notion that protein misfolding in the ER lumen can produce ROS and that ROS and accumulation of misfolded proteins induce ER stress and activate the UPR. In this context, ROS may be generated as a consequence of disulfide bond formation, depletion of cellular GSH, and/or mitochondrial oxidative phosphorylation.

5.6 Oxidative Stress as an Activator of the UPR

ROS can originate from exposure to irradiation and environmental pollutants and enzymatic reactions involving the mitochondrial respiratory chain, arachidonic acid pathway, cytochrome P450 family, glucose, amino acid, xanthine and NADP/NADPH oxidases, and nitric oxide synthase [36]. In cultured liver cells the combination of hydrogen peroxide generation using glucose oxidase and proteasome inhibition resulted in activation of the UPR and formation of inclusion bodies that was reduced by either pretreatment with *N*-acetyl-cysteine or the chemical chaperone, 4-phenylbutyrate [23].

Changes in nutrient flux, particularly fatty acid flux, may influence the functional capacity of the ER, in part, via effects on redox balance. Elevated free fatty acids, in particular saturated fatty acids, have been linked to activation of the UPR in a number of cell types, including hepatocytes [37–40]. Antioxidants, such as taurine, effectively reduce saturated fatty acid-mediated oxidative stress and UPR activation in both H4IIE liver cells and primary hepatocytes [41].

Ethanol impairs protein secretion in hepatocytes and the serum protein deficiency that can lead to clotting disorders, edema, and impaired iron delivery [42]. ER stress and activation of the UPR by ethanol is conserved across vertebrates and reductions in ER stress reduce alcohol-induced liver injury [42, 43]. Using a zebrafish larvae model, Tsedensodnom et al. demonstrated that ethanol exposure induced oxidative stress and that oxidative stress and low doses of ethanol synergize to induce the UPR in the liver [44]. In alcoholic liver disease, ROS is generated by ER-localized cytochrome P450's and, perhaps, interactions between the ER and mitochondria (discussed below) [45]. Regardless of the source of ROS, these data are consistent with the notion that ROS can impair the secretory pathway in hepatocytes and lead to ER stress and UPR activation.

5.7 ER–Mitochondrial Interactions and Oxidative Stress

Protein folding in and clearance of aggregated proteins from the ER lumen requires energy, thus ER homeostasis is linked to mitochondrial bioenergetics and adenosine triphosphate (ATP) supply. Physical interactions between the ER and mitochondria have been documented and involve specific tethering proteins and elements of the cytoskeleton [46, 47]. One function of these physical interactions is to provide efficient calcium transfer from the ER to mitochondria, thereby regulating the activity of matrix dehydrogenases required for mitochondrial respiration and ATP production. In turn, ATP can be supplied to the ER lumen to support the energy requirements for protein folding and clearance [48, 49].

Several proteins have been identified as components of the ER–mitochondria tethering mechanism including mitofusin-1 and -2 [50]. In particular, mitofusin-2 appears to be enriched at ER–mitochondria contact sites, can influence ER morphology, and can directly tether ER and mitochondria via homo- and hetero-typic interactions [50]. Importantly, the absence of mitofusin-2 increased the distance between the ER and mitochondria and impaired mitochondrial calcium uptake [50]. Liver-specific ablation of mitofusin-2 in mice resulted in ER and oxidative stress, increased stress pathway signaling (c-Jun NH_2 Terminal Kinase), and impaired insulin signaling [51]. Chemical chaperones or the antioxidant *N*-acetylcysteine improved insulin signaling and glucose homeostasis in liver-specific mitofusin-2 knockout mice [51]. It should be noted that hydrogen peroxide increased the expression of mitofusin-2 in cardiomyocytes and selective upregulation of mitofusin-2 was sufficient to induce myocyte apoptosis [52]. One interpretation of these data is that both reductions and increases in mitofusin-2 may interfere with mitochondrial fusion/fission, physical interactions between the ER and mitochondria, and potentially calcium and ATP transfer.

5.8 Summary

ER and oxidative stress are typically present in metabolic diseases, such as obesity and diabetes, neurodegenerative diseases, such as Alzheimer and Parkinson disease, and atherosclerosis. Protein folding in the ER lumen generates ROS and in turn, ROS can influence the fidelity of protein folding. The quality control system that responds to ER stress, the UPR, can regulate multiple pathways, including antioxidant defense. Thus, the UPR can be viewed as an adaptive mechanism that ultimately attempts to maintain cell survival and function. The ER is also structurally and functionally linked to mitochondria, and thus mitochondrial ATP and ROS may influence or interact with the ER lumen. Liver diseases are often characterized by both ER and oxidative stress and thus understanding how these two processes interact is critical to understanding the pathogenesis of these diseases.

References

1. van Meer G, Voelker DR, Feigenson GW (2008) Membrane lipids: where they are and how they behave. Nat Rev Mol Cell Biol 9:112–124. doi:10.1038/nrm2330
2. Gentile CL, Frye M, Pagliassotti MJ (2011) Endoplasmic reticulum stress and the unfolded protein response in nonalcoholic fatty liver disease. Antioxid Redox Signal 15:505–521. doi:10.1089/ars.2010.3790

3. Gentile CL, Pagliassotti MJ (2008) The role of fatty acids in the development and progression of nonalcoholic fatty liver disease. J Nutr Biochem 19:567–576. doi:10.1016/j.jnutbio.2007.10.001
4. Marciniak S, Ron D (2006) Endoplasmic reticulum stress signaling in disease. Physiol Rev 86:1133–1149
5. Schroder M, Kaufman RJ (2005) ER stress and the unfolded protein response. Mutat Res 569:29–63. doi:10.1016/j.mrfmmm.2004.06.056
6. Guerriero CJ, Brodsky JL (2012) The delicate balance between secreted protein folding and endoplasmic reticulum-associated degradation in human physiology. Physiol Rev 92:537–576. doi:10.1152/physrev.00027.2011
7. Rutkowski DT, Kaufman RJ (2004) A trip to the ER: coping with stress. Trends Cell Biol 14:20–28
8. Wu J, Kaufman RJ (2006) From acute ER stress to physiological roles of the Unfolded Protein Response. Cell Death Differ 13:374–384
9. Zhang K, Kaufman RJ (2008) From endoplasmic reticulum stress to the inflammatory response. Nature 454:455–462
10. Kaufman RJ (2002) Orchestrating the unfolded protein response in health and disease. J Clin Invest 110:1389–1398
11. Schroder M, Kaufman RJ (2005) The mammalian unfolded protein response. Annu Rev Biochem 74:739–789
12. Trusina A, Papa FR, Tang C (2008) Rationalizing translation attenuation in the network architecture of the unfolded protein response. Proc Natl Acad Sci U S A 105:20280–20285. doi:10.1073/pnas.0803476105
13. Kaufman RJ (1999) Stress signaling from the lumen of the endoplasmic reticulum: coordination of gene transcriptional and translational controls. Genes Dev 13:1211–1233
14. Cullinan SB, Zhang D, Hannink M, Arvisais E, Kaufman RJ, Diehl JA (2003) Nrf2 is a direct PERK substrate and effector of PERK-dependent cell survival. Mol Cell Biol 23:7198–7209
15. Urano F, Wang X, Bertolotti A, Zhang Y, Chung P, Harding HP, Ron D (2000) Coupling of stress in the ER to activation of JNK protein kinases by transmembrane protein kinase IRE1. Science 287:664–666
16. Hu P, Han Z, Couvillon AD, Kaufman RJ, Exton JH (2006) Autocrine tumor necrosis factor alpha links endoplasmic reticulum stress to the membrane death receptor pathway through IRE1alpha-mediated NF-kB activation and down-regulation of TRAF2 expression. Mol Cell Biol 26:3071–3084
17. Nguyen DT, Kebache S, Fazel A, Wong HN, Jenna S, Emadali A, Lee EH, Bergeron JJ, Kaufman RJ, Larose L, Chevet E (2004) Nck-dependent activation of extracellular signal-regulated kinase-1 and regulation of cell survival during endoplasmic reticulum stress. Mol Biol Cell 15:4248–4260
18. Hetz C, Glimcher LH (2009) Fine-tuning of the unfolded protein response: assembling the IRE1alpha interactome. Mol Cell 35:551–561
19. Hollien J, Weissman JS (2006) Decay of endoplasmic reticulum-localized mRNAs during the unfolded protein response. Science 313:104–107. doi:10.1126/science.1129631
20. Hollien J, Lin JH, Li H, Stevens N, Walter P, Weissman JS (2009) Regulated Ire1-dependent decay of messenger RNAs in mammalian cells. J Cell Biol 186:323–331. doi:10.1083/jcb.200903014
21. Santos CX, Tanaka LY, Wosniak J, Laurindo FR (2009) Mechanisms and implications of reactive oxygen species generation during the unfolded protein response: roles of endoplasmic reticulum oxidoreductases, mitochondrial electron transport, and NADPH oxidase. Antioxid Redox Signal 11:2409–2427. doi:10.1089/ARS.2009.2625
22. Shimizu Y, Hendershot LM (2009) Oxidative folding: cellular strategies for dealing with the resultant equimolar production of reactive oxygen species. Antioxid Redox Signal 11: 2317–2331
23. Hanada S, Harada M, Kumemura H, Omary MB, Koga H, Kawaguchi T, Taniguchi E, Yoshida T, Hisamoto T, Yanagimoto C, Maeyama M, Ueno T, Sata M (2007) Oxidative stress induces

the endoplasmic reticulum stress and facilitates inclusion formation in cultured cells. J Hepatol 47:93–102
24. Yokouchi M, Hiramatsu N, Hayakawa K, Okamura M, Du S, Kasai A, Takano Y, Shitamura A, Shimada T, Yao J, Kitamura M (2008) Involvement of selective reactive oxygen species upstream of proapoptotic branches of unfolded protein response. J Biol Chem 283:4252–4260
25. Xue X, Piao JH, Nakajima A, Sakon-Komazawa S, Kojima Y, Mori K, Yagita H, Okumura K, Harding H, Nakano H (2005) Tumor necrosis factor alpha (TNFalpha) induces the unfolded protein response (UPR) in a reactive oxygen species (ROS)-dependent fashion, and the UPR counteracts ROS accumulation by TNFalpha. J Biol Chem 280:33917–33925. doi:10.1074/jbc.M505818200
26. Cullinan SB, Diehl JA (2006) Coordination of ER and oxidative stress signaling: the PERK/Nrf2 signaling pathway. Int J Biochem Cell Biol 38:317–332
27. Moi P, Chan K, Asunis I, Cao A, Kan YW (1994) Isolation of NF-E2-related factor 2 (Nrf2), a NF-E2-like basic leucine zipper transcriptional activator that binds to the tandem NF-E2/AP1 repeat of the beta-globin locus control region. Proc Natl Acad Sci U S A 91:9926–9930
28. Sugimoto H, Okada K, Shoda J, Warabi E, Ishige K, Ueda T, Taguchi K, Yanagawa T, Nakahara A, Hyodo I, Ishii T, Yamamoto M (2010) Deletion of nuclear factor-E2-related factor-2 leads to rapid onset and progression of nutritional steatohepatitis in mice. Am J Physiol Gastrointest Liver Physiol 298:G283–G294
29. Thimmulappa RK, Lee H, Rangasamy T, Reddy SP, Yamamoto M, Kensler TW, Biswal S (2006) Nrf2 is a critical regulator of the innate immune response and survival during experimental sepsis. J Clin Invest 116:984–995
30. Liu Y, Adachi M, Zhao S, Hareyama M, Koong AC, Luo D, Rando TA, Imai K, Shinomura Y (2009) Preventing oxidative stress: a new role for XBP1. Cell Death Differ 16:847–857
31. Hwang C, Sinskey AJ, Lodish HF (1992) Oxidized redox state of glutathione in the endoplasmic reticulum. Science 257:1496–1502
32. Tu BP, Weissman JS (2004) Oxidative protein folding in eukaryotes: mechanisms and consequences. J Cell Biol 164:341–346. doi:10.1083/jcb.200311055
33. Szaraz P, Banhegyi G, Benedetti A (2010) Altered redox state of luminal pyridine nucleotides facilitates the sensitivity towards oxidative injury and leads to endoplasmic reticulum stress dependent autophagy in HepG2 cells. Int J Biochem Cell Biol 42:157–166. doi:10.1016/j.biocel.2009.10.004
34. Haynes CM, Titus EA, Cooper AA (2004) Degradation of misfolded proteins prevents ER-derived oxidative stress and cell death. Mol Cell 15:767–776. doi:10.1016/j.molcel.2004.08.025
35. Malhotra JD, Miao H, Zhang K, Wolfson A, Pennathur S, Pipe SW, Kaufman RJ (2008) Antioxidants reduce endoplasmic reticulum stress and improve protein secretion. Proc Natl Acad Sci U S A 105:18525–18530. doi:10.1073/pnas.0809677105
36. Malhotra JD, Kaufman RJ (2007) Endoplasmic reticulum stress and oxidative stress: a vicious cycle or a double-edged sword? Antioxid Redox Signal 9:2277–2293. doi:10.1089/ars.2007.1782
37. Listenberger LL, Han X, Lewis SE, Cases S, Farese RV Jr, Ory DS, Schaffer JE (2003) Triglyceride accumulation protects against fatty acid-induced lipotoxicity. Proc Natl Acad Sci U S A 100:3077–3082. doi:10.1073/pnas.0630588100
38. Unger RH, Zhou YT (2001) Lipotoxicity of beta-cells in obesity and in other causes of fatty acid spillover. Diabetes 50(Suppl 1):S118–S121
39. Feldstein AE, Werneburg NW, Canbay A, Guicciardi ME, Bronk SF, Rydzewski R, Burgart LJ, Gores GJ (2004) Free fatty acids promote hepatic lipotoxicity by stimulating TNF-alpha expression via a lysosomal pathway. Hepatology 40:185–194. doi:10.1002/hep.20283
40. Wei Y, Wang D, Topczewski F, Pagliassotti MJ (2006) Saturated fatty acids induce endoplasmic reticulum stress and apoptosis independently of ceramide in liver cells. Am J Physiol Endocrinol Metab 291:E275–E281. doi:10.1152/ajpendo.00644.2005
41. Gentile CL, Nivala AM, Gonzales JC, Pfaffenbach KT, Wang D, Wei Y, Jiang H, Orlicky DJ, Petersen DR, Pagliassotti MJ, Maclean KN (2011) Experimental evidence for therapeutic

potential of taurine in the treatment of nonalcoholic fatty liver disease. Am J Physiol Regul Integr Comp Physiol 301:R1710–R1722. doi:10.1152/ajpregu.00677.2010
42. Howarth DL, Vacaru AM, Tsedensodnom O, Mormone E, Nieto N, Costantini LM, Snapp EL, Sadler KC (2012) Alcohol disrupts endoplasmic reticulum function and protein secretion in hepatocytes. Alcohol Clin Exp Res 36:14–23. doi:10.1111/j.1530-0277.2011.01602.x
43. Howarth DL, Passeri M, Sadler KC (2011) Drinks like a fish: using zebrafish to understand alcoholic liver disease. Alcohol Clin Exp Res 35:826–829. doi:10.1111/j.1530-0277.2010.01407.x
44. Tsedensodnom O, Vacaru AM, Howarth DL, Yin C, Sadler KC (2013) Ethanol metabolism and oxidative stress are required for unfolded protein response activation and steatosis in zebrafish with alcoholic liver disease. Dis Model Mech 6:1213–1226. doi:10.1242/dmm.012195
45. Marchi S, Patergnani S, Pinton P (2013) The endoplasmic reticulum-mitochondria connection: one touch, multiple functions. Biochim Biophys Acta. doi:10.1016/j.bbabio.2013.10.015
46. Hayashi T, Rizzuto R, Hajnoczky G, Su TP (2009) MAM: more than just a housekeeper. Trends Cell Biol 19:81–88. doi:10.1016/j.tcb.2008.12.002
47. Pizzo P, Pozzan T (2007) Mitochondria-endoplasmic reticulum choreography: structure and signaling dynamics. Trends Cell Biol 17:511–517. doi:10.1016/j.tcb.2007.07.011
48. Bravo R, Gutierrez T, Paredes F, Gatica D, Rodriguez AE, Pedrozo Z, Chiong M, Parra V, Quest AF, Rothermel BA, Lavandero S (2012) Endoplasmic reticulum: ER stress regulates mitochondrial bioenergetics. Int J Biochem Cell Biol 44:16–20. doi:10.1016/j.biocel.2011.10.012
49. Bravo R, Vicencio JM, Parra V, Troncoso R, Munoz JP, Bui M, Quiroga C, Rodriguez AE, Verdejo HE, Ferreira J, Iglewski M, Chiong M, Simmen T, Zorzano A, Hill JA, Rothermel BA, Szabadkai G, Lavandero S (2011) Increased ER-mitochondrial coupling promotes mitochondrial respiration and bioenergetics during early phases of ER stress. J Cell Sci 124:2143–2152. doi:10.1242/jcs.080762
50. de Brito OM, Scorrano L (2008) Mitofusin 2 tethers endoplasmic reticulum to mitochondria. Nature 456:605–610. doi:10.1038/nature07534
51. Sebastian D, Hernandez-Alvarez MI, Segales J, Sorianello E, Munoz JP, Sala D, Waget A, Liesa M, Paz JC, Gopalacharyulu P, Oresic M, Pich S, Burcelin R, Palacin M, Zorzano A (2012) Mitofusin 2 (Mfn2) links mitochondrial and endoplasmic reticulum function with insulin signaling and is essential for normal glucose homeostasis. Proc Natl Acad Sci U S A 109:5523–5528. doi:10.1073/pnas.1108220109
52. Shen T, Zheng M, Cao C, Chen C, Tang J, Zhang W, Cheng H, Chen KH, Xiao RP (2007) Mitofusin-2 is a major determinant of oxidative stress-mediated heart muscle cell apoptosis. J Biol Chem 282:23354–23361. doi:10.1074/jbc.M702657200

Chapter 6
Oxidative Stress and Liver Inflammation

Stephen N. Greenhalgh, Alexandra I. Thompson, Neil C. Henderson, and John P. Iredale

6.1 Introduction

The inflammatory response has a crucial role in all forms of liver disease, be they of pathogen or sterile origin. Likewise, oxidative stress and the existence of reactive oxygen species (ROS), both in parenchymal and non-parenchymal cell populations, mediate a number of important pathways of injury, including cell recruitment, activation, and function (Fig. 6.1). The importance of ROS in liver inflammation is not simply related to pathogen killing [1, 2] or basic biochemical effects of cell exposure to ROS (such as protein oxidation, DNA damage, and lipid peroxidation (LPO)). In addition, ROS have a range of key roles including redox signaling and the ability to alter protein structure (redox sensor activation), inhibit enzyme function, and modulate transcription [3, 4].

In liver inflammation, ROS have been shown to exert their effects on parenchymal and immune cells, leading to up-regulation of gene transcription and modulation of cytokine and chemokine expression, with recruitment and activation of non-parenchymal cells [3, 5, 6]. These effects have been demonstrated in multiple disease processes and animal models. Although many of the functions performed by ROS in liver inflammation are likely to be conserved across a range of liver pathologies, certain aspects are potentially disease-specific. Indeed, not every effect of ROS has been demonstrated in all forms of liver disease. Therefore, in this chapter, a range of roles for ROS will be considered in relation to the diseases, or relevant animal models, to which they have been shown to contribute.

S.N. Greenhalgh (✉) • A.I. Thompson • N.C. Henderson • J.P. Iredale
The University of Edinburgh/Medical Research Council Centre for Inflammation Research, The Queen's Medical Research Institute, 47 Little France Crescent, Edinburgh EH16 4TJ, United Kingdom
e-mail: Stephen.Greenhalgh@ed.ac.uk

E. Albano, M. Parola (eds.), *Studies on Hepatic Disorders*,
Oxidative Stress in Applied Basic Research and Clinical Practice,
DOI 10.1007/978-3-319-15539-5_6

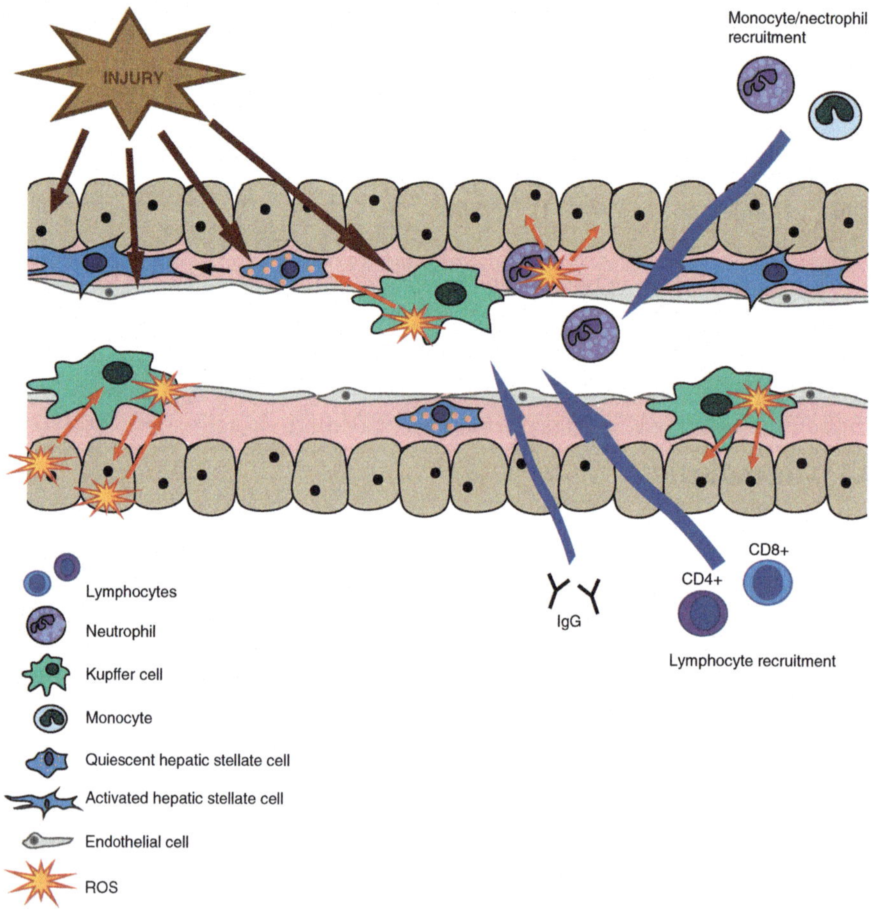

Fig. 6.1 Reactive oxygen species and liver inflammation. Liver injury, of varied etiology, results in ROS generation, predominantly in hepatocytes and Kupffer cells, triggering leukocyte activation and recruitment, stellate cell activation, and further ROS generation

6.2 Mechanisms Underlying ROS Production in Leukocytes

Clearly, ROS production by hepatocytes following injury, the mechanisms of which are discussed in earlier chapters, can have a direct effect on resident and recruited inflammatory cells. Furthermore, leukocytes themselves can produce ROS through a variety of mechanisms (Fig. 6.2). During necrosis of liver parenchymal cells, ROS production in leukocytes is triggered by the release of damage-associated molecular patterns (DAMPs). Key examples of DAMPs include the chromatin protein high mobility group box 1 (HMGB1), heat shock proteins (hsps), DNA, and RNA [7, 8]. Essentially, these are all intracellular molecules whose presence in the extracellular environment alerts the host immune system to the existence of cell damage and disease.

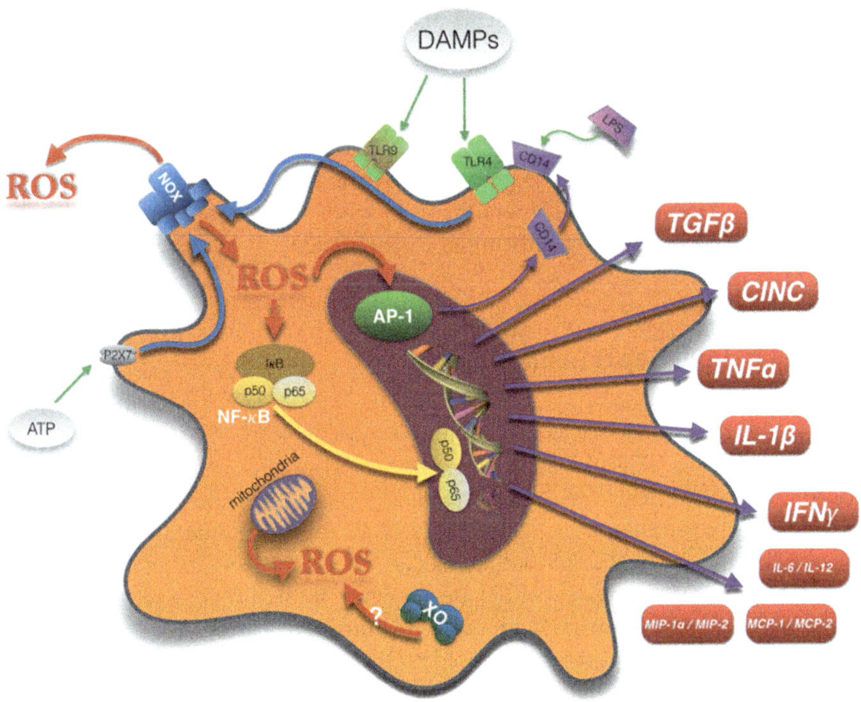

Fig. 6.2 Principle routes for ROS production and action in Kupffer cells

Complexes of DAMPs can be particularly inflammatory and HMGB1 is able to combine with a number of other intracellular molecules to this effect [9].

DAMPs are able to activate liver resident macrophages (Kupffer cells, KCs) and neutrophils via toll-like receptors (TLRs) [10]. For example, during reperfusion injury, HMGB1 can activate TLR4 [11] and signaling via TLR4 leads to activation of nicotinamide adenine dinucleotide phosphate oxidase (NADPH oxidase, Nox) 2, the primary superoxide-producing enzyme in phagocytes [12, 13]. Similarly, DNA fragments are able to activate TLR9 and trigger ROS production in neutrophils, acting in concert with HMGB1 to promote cytokine production and exacerbate injury [14]. As well as Nox, myeloperoxidase (MPO) acts as a major producer of ROS in neutrophils [15]. MPO generates hypochlorous acid, a potent oxidant, from hydrogen peroxide and chloride. This contributes to the formation of chlorotyrosine protein adducts, commonly utilized as a marker of neutrophil-derived oxidative stress [16], although it should be noted that production can also occur in macrophages to a lesser extent [17, 18].

In addition to the direct stimulation of leukocytes, DAMPs also have the potential to trigger ROS production and ROS-mediated signaling through the activation of complement. Both complement and cytokines released upon liver injury can stimulate KCs and neutrophils to produce ROS, as well as recruiting neutrophils into the liver vasculature and up-regulating the neutrophil cell adhesion molecule CD11b/CD18 (Mac-1) [19, 20].

The biology of pathogen-associated inflammation can lead to similar outcomes, since many of the receptors and pathways now shown to be activated by DAMPs were originally identified through the study of pathogen–host interaction [21, 22]. For example, TLR4 is activated by the bacterial endotoxin lipopolysaccharide (LPS) in addition to HMGB1 [10, 22] and the CpG motif in bacterial DNA, similar to mammalian DNA, is a potent activator of TLR9 [14, 23].

6.3 Ischemia-Reperfusion Injury

A substantial number of investigations into the role of ROS in liver inflammation have been carried out to define the cellular processes underlying ischemia-reperfusion injury (IRI). Not only is this due to the great interest in reducing the degree of injury resulting from ischemia-reperfusion (IR), in order to improve outcome following liver transplantation, hepatic tumor resection, and abdominal trauma, but also animal models of IRI are attractive because of their reproducibility and short duration in comparison to models of chronic disease.

Activation of KCs in isolated, perfused rat liver leads to cell damage that is limited by administration of glutathione (GSH) or the antioxidant enzymes superoxide dismutase (SOD) and catalase [24, 25]. GSH and catalase both prevent the lactate dehydrogenase increase observed when KCs are activated by zymosan infusion, suggesting that cell damage can occur due to KC production of hydrogen peroxide, the primary target for catalase [25]. In a rat model of hepatic IRI, both KCs and polymorphonuclear leukocytes (i.e., neutrophils) were shown to have the ability to generate superoxide during the reperfusion phase [26, 27]. They were also primed for additional superoxide generation in response to further stimulation, such as protein kinase C agonists or opsonized zymosan, suggesting that components of the complement cascade are major factors for post-ischemic activation of KCs and polymorphonuclear leukocytes [26, 27].

Inhibition of macrophage function using gadolinium chloride ($GdCl_3$) reduces macrophage ability to generate superoxide and also significantly attenuates, although does not entirely protect against, both early and late IRI [25, 27]. Neutrophil infiltration after 24 h of reperfusion is also decreased by this intervention, although neutrophils retain the ability to generate superoxide [27]. Alternatively, treatment with anti-CD11b or anti-CD18 antibodies, preventing neutrophil adhesion, markedly decreases the amount of necrosis seen after 24 h of reperfusion, but does not alter the extent of initial injury [27]. This is consistent with the conclusion that KCs are of primary importance both in the early phase of injury and for neutrophil recruitment, whereas neutrophils play a major role in the later phase of reperfusion injury and require functional CD11b/CD18 to permit hepatocyte adhesion and to mediate the effects of superoxide generation.

Oxidative stress can also be quantified through measurement of glutathione disulfide (GSSG), formed during the reduction of ROS and oxidation of GSH. Hepatic IRI leads to an increase in plasma GSSG and this is enhanced by activation of KCs, through administration of retinol or *Propionibacterium acnes*, but attenuated by KC inactivation with $GdCl_3$ (or methyl palmitate) [28]. Again, KC inactivation significantly protects the liver from IRI.

This supports the conclusion that, initially, most oxidative stress is present in the hepatic sinusoids, with KCs having a primary role for superoxide production during the early phase of reperfusion [24, 25, 28]. Examination of the proteome of isolated KCs after 8 h of reperfusion confirms up-regulation of the oxidant-antioxidant system [29]. Subsequently, neutrophils are recruited and extravasate, allowing them to adhere to target cells (e.g., hepatocytes) and resulting in significant and long-lasting ROS generation that has the potential to induce cell death [30]. A gradual increase, from 6 to 24 h after reperfusion, in hepatocyte staining for hypochlorite-modified epitopes, a marker for neutrophil-specific oxidative stress, correlates with neutrophil accumulation in the liver parenchyma [31]. Both this marker of oxidative stress and measures of liver injury were attenuated by the Nox inhibitor diphenyleneiodonium chloride (DPI).

There is also a role for ROS in neutrophil recruitment, since pretreatment with SOD attenuates the recruitment of neutrophils and increase in MPO activity that occurs during IRI [32]. A gastrointestinal model of IRI has been used to examine the role of KCs in leukocyte adhesion and oxidative stress in the liver, with NADPH autofluorescence measured to assess redox status [33]. $GdCl_3$ attenuated the leukostasis, oxidative stress, and injury observed in the liver early in reperfusion, findings supported by later investigations assessing microcirculatory failure following liver transplant in rats [34]. Treatment with tumor necrosis factor alpha (TNFα) antibody also reduced the degree of sinusoidal leukostasis and oxidative stress, pointing to a central role for this important leukocyte cytokine in relation to injury and oxidative stress [33].

The same gut IRI model, when performed in mice genetically deficient for a range of cell adhesion molecules, suggested that leukostasis and the effects of neutrophil-derived oxidative stress were dependent on the leukocyte adhesion molecule Mac-1 (CD11b/CD18) and intercellular adhesion molecule (ICAM)-1 [35]. This is in agreement with both an earlier in vitro study of neutrophil adherence and the in vivo use of blocking antibodies [36, 37].

Attempts to understand the mechanisms underlying intrahepatic ROS production during reperfusion have utilized an ex vivo model, with KC inactivation by $GdCl_3$ and activation using schizophyllan glucan [38]. As expected, superoxide production (as measured by real-time chemiluminescence) increased with KC stimulation and decreased with inactivation. The ex vivo nature of the model prevents the investigation of the role of recruited neutrophils during reperfusion; however, the importance of KCs in ROS generation in vivo is reinforced by the finding that whole liver ROS production, as measured by chemiluminescence, is impaired when $GdCl_3$ is administered prior to IRI [39]. Chemiluminescence during reperfusion also decreased when either a Nox inhibitor or xanthine oxidase (XO) inhibitor was administered during the initial perfusion phase at the time of organ harvest [38]. The effect observed with Nox inhibition was more marked, but this experiment does suggest a role for superoxide production by XO during IRI. XO is an enzyme derived from proteolytic cleavage of xanthine dehydrogenase; superoxide production occurs in the course of converting hypoxanthine to xanthine to uric acid. Although the cell-specific effects of these inhibitors were not investigated, it is possible to conclude that KCs are a major source of ROS during reperfusion and that ROS generation likely occurs, at least in part, through the actions of Nox, and potentially XO.

Exogenous administration of XO to rats results in increased oxidative stress (quantified by measurement of LPO using thiobarbituric acid reactive substances, TBARS) and neutrophil infiltration (measured by MPO activity) in the liver [40]. XO levels also increase in plasma and liver after reperfusion of hindlimb ischemia and parallel an increase in TNFα and KC activation [40]. Pretreatment with allopurinol, a XO inhibitor and antioxidant, inhibits KC activation and liver leukocyte infiltration, supporting the conclusion that neutrophil recruitment requires KC activation [40]. XO inhibition has also been shown to reduce the production of cytokine-induced neutrophil chemoattractant (CINC) by KCs, identified through their ED2 expression, during hepatic IRI in rats [41]. This was accompanied by decreased neutrophil number and MPO activity after 24 h of reperfusion. In vitro, CINC production can be triggered in quiescent KCs by exposure to XO and hypoxanthine [41]. The promoter region of the CINC gene contains a nuclear factor (NF)-κB-binding site and transcription of CINC has been shown to be regulated by NF-κB, which itself can be activated by XO superoxide production [42, 43]. This supports a role for ROS production by KCs in neutrophil recruitment following injury; however, the importance and contribution of XO to post-ischemic oxidative stress in clinically relevant situations has been questioned [44].

The relevance of Nox in liver inflammation has been demonstrated using transgenic knockout mice lacking the gene transcribing one of the Nox subunits: $p47^{phox}$ [45]. In IRI, these Nox-deficient mice exhibited greatly reduced liver injury, as assessed by serum alanine aminotransferase (ALT) levels and degree of hepatic necrosis. Neutrophil infiltration (assessed by MPO^+ leukocytes) was also decreased by approximately 50 % and evidence of oxidative stress (positive staining for 4-hydroxynonenal, 4-HNE) was approximately one quarter of that seen in wild-type mice. Although the transgenic mouse utilized in this work has global rather than leukocyte-specific knockout of $p47^{phox}$, phagocytes are key expressers of Nox and thus highly likely to have a significant role in Nox-derived ROS production in IRI. Interestingly, a reduction in nitrotyrosine expression, representative of peroxynitrite formation, was also seen in Nox-deficient mice, supporting the conclusion that the KC is a cell of prime interest due to its ability to produce both the superoxide and nitric oxide required for peroxynitrite generation. Further, there is clear evidence for the role of Nox-derived ROS in neutrophil recruitment.

Classically, one of the most important pathways through which ROS exert an effect on gene transcription is via NF-κB. Intracellular ROS trigger phosphorylation of the IκB subunit which enables dissociation of the p50-p65 κB heterodimer and its subsequent nuclear translocation and promotion of transcription [46]. Short periods of ischemia result in up-regulation of protective mitochondrial Mn-SOD (SOD2), predominantly in hepatocytes [47]. However, with extended ischemia time, activation of NF-κB in KCs results in increasing transcription of the pro-inflammatory genes TNFα and interleukin (IL)-1β [47]. This latter up-regulation is prevented by pretreatment with $GdCl_3$. So although NF-κB is responsible for up-regulation of genes protective against oxidative stress in parenchymal cells during the early course of injury, in the face of persistent and severe injury, its pro-inflammatory role predominates.

6.4 NAFLD and NASH

The evidence presented thus far for the role of oxidative stress in the pathogenesis of IRI has focused on experimental models in rodents. However, shifting attention to non-alcoholic fatty liver disease (NAFLD) and its important potential sequela—non-alcoholic steatohepatitis (NASH)—permits consideration of evidence that oxidative stress has an important role in human inflammatory liver disease.

Comparing steatohepatitis with steatosis and normal samples of human liver reveals a correlation between neutrophil number, degree of liver damage, and oxidative stress markers [48]. Oxidized phosphatidylcholine is the principle oxidized phospholipid present in oxidatively modified low-density lipoprotein, a lipoprotein with cytotoxic and macrophage chemoattractant properties [49, 50]. Oxidized phosphatidylcholine immunoreactivity in human liver correlated positionally and quantitatively with MPO^+ neutrophils, suggesting that neutrophils either produce or are recruited by this oxidatively modified substance [48]. In an unrelated study, liver biopsies from obese subjects with steatosis or NASH were stained for MPO and oxidative stress-related changes, namely hypochlorite-modified proteins [51]. Numbers of both hepatic neutrophils and MPO^+ KCs increased in NASH, accompanied by positive staining for hypochlorite-modified proteins. The increased sequestration of neutrophils was also correlated with increased C-X-C chemokine generation, leading to the conclusion that MPO-derived products of oxidative stress are associated with C-X-C chemokine induction and hepatic neutrophil infiltration.

In a rat model of NASH, the antioxidant *N*-acetylcysteine (NAC) has been shown to ameliorate injury, reduce the increase of inflammatory mediators such as TNFα, and prevent the changes in markers of oxidative stress (decreased GSH and increased TBARS), which are seen in untreated controls [52].

Of interest, the decreased levels of adiponectin seen in obesity [53] may contribute to an enhancement of the inflammatory response in NASH, mediated via oxidative stress. In a NASH model, examined in adiponectin knockout mice, lack of this adipose-produced hormone resulted in an increase in markers of oxidative stress (TBARS) with a decrease in GSH and mRNA expression of both catalase and SOD [54]. This was accompanied by increased infiltration of $F4/80^+$ macrophages into the liver. All these changes were reduced when adiponectin expression was restored using an adenovirus vector.

Conversely, serum levels of the hormone leptin increase in obesity and NASH in humans [55, 56]. Elevated leptin, in a murine model of NASH, was associated with markers of oxidative stress, with leptin-dependent increases in $p47^{phox}$ mRNA and release of both TNFα and monocyte chemotactic protein (MCP)-1 [57]. Transcription of inducible nitric oxide synthase also increased and the combination of increased generation of superoxide and nitric oxide permitted formation of peroxynitrite. Both the use of a peroxynitrite decomposition catalyst and inhibitors of Nox or inducible nitric oxide synthase reduced protein radical formation and expression of TNFα and MCP-1. The role of leptin is confirmed by the reduction in oxidative stress that occurred when leptin antibodies were administered or *ob/ob* mice were used instead of wild type. Macrophages again appear to mediate this pathway, since inactivation (with $GdCl_3$ or liposomal clodronate) leads to reduced protein radical formation and serum levels of TNFα and MCP-1, whereas adoptive transfer of leptin-primed KCs restores levels of protein radicals. Iatrogenic hyperleptinaemia, investigated in an alternative NASH rodent model, led to an increase in Nox-derived hepatic oxidative stress, accompanied by an increase in thromboxane A2, vasoconstriction, and portal hypertension [58]. KC inactivation, by means of $GdCl_3$ or encapsulated clodronate, limits these effects, suggesting that leptin can promote portal hypertension in NASH cirrhosis via oxidative stress and KCs.

As in IRI, ROS generation by KCs in obesity occurs through up-regulation of Nox. Hepatocyte injury can result in release of adenosine triphosphate, which is then able to bind the P2X7 receptor on KCs, up-regulating both expression of the $p47^{phox}$ Nox subunit and membrane binding of Nox [59, 60]. In obese mice, in which liver injury was triggered by carbon tetrachloride (CCl_4) injection, the formation of protein radicals in KCs was evident after 24 h [60]. Further, these mice showed more evidence of hepatic oxidative stress (4-HNE adducts) than lean controls. The oxidant generation, co-localized to KCs, resulted in release of TNFα and MCP-2. KC activation and pro-inflammatory cytokine release were blocked by antagonists of Nox and P2X7, and also in P2X7 receptor knockout mice and $p47^{phox}$-deficient (i.e., Nox knockout) mice. Moreover, $p47^{phox}$ expression was significantly down-regulated in KCs of P2X7 receptor knockout mice. Reduced liver injury was seen in P2X7 receptor knockout mice administered CCl_4. This suggests that in the injured liver of obese mice, stimulation of the P2X7 receptor plays an important role in $p47^{phox}$ expression and ROS generation by KCs.

Neutrophils also have an important role in NASH, through their ability to generate powerful oxidants and promote LPO through MPO [15, 61]. MPO and nitrotyrosine (a marker of MPO activity) levels increase in response to high fat feeding of mice, predictably prevented in MPO knockout, which is also associated with down-regulation of the pro-inflammatory mediators TNFα, IL-1α, and MCP-1 [62].

Recently, the role of the adaptive immune system in response to oxidative stress has also been explored. High titers of IgG against malondialdehyde (MDA) adducts generated by oxidative stress have been measured in both adults and children with NASH [63, 64]. Rats that were protected against NASH by administration of the antioxidant NAC showed lower titers of autoantibodies against MDA adducts [52]. In mice with NASH, the degree of IgG reactivity against MDA adducts correlates with elevations in ALT and hepatic TNFα mRNA and is accompanied by infiltration of both T and B lymphocytes [65]. Stimulation of an immune response against MDA-protein adducts, through injection of MDA-modified bovine serum albumin, exacerbates the parenchymal injury and inflammation that occurs in response to dietary-induced NASH. Immunization also promotes recruitment of T cells and leads to an increase in both $CD4^+$ and $CD8^+$ pools. $CD8^+$ T cells have been linked to NASH in children, [66] while the increase of $CD4^+$ T cells appears to promote macrophage activation through increased expression of CD40 ligand and the CD40 receptor [65]. $CD4^+$ T cell depletion lowered expression of both CD40 ligand and markers of macrophage activation and was accompanied by histological improvements in inflammation and necrosis.

6.5 Alcoholic Liver Disease

Classically, the hepatocyte cytochrome p450 enzyme CYP2E1 has been considered to have a major role in alcohol metabolism and thus ROS generation [67]. However, no differences in injury or ROS production were noted in CYP2E1 knockout mice fed ethanol when compared to wild type, suggesting that, at least in early alcohol-induced liver injury, other mechanisms may be of greater importance [68]. Almost two decades ago, a role for macrophages in oxidative stress and alcohol-induced liver injury was demonstrated by decreased radical adduct formation and reduced histological injury in rats receiving the macrophage inactivator $GdCl_3$ alongside chronic ethanol feeding [69]. Ethanol-induced superoxide production and its reduction by $GdCl_3$ administration were confirmed in the isolated rat liver [70]. Both KCs in situ and those isolated following acute alcohol binge or chronic administration have been demonstrated to produce increased amounts of superoxide, hydrogen peroxide, and TNFα, with increased LPO and Nox activity [70–72]. The hepatic injury, LPO, and TNFα production resultant from ethanol administration can be suppressed by treatment with the antioxidant NAC and also by caffeine [73–75]. Of note, evidence for superoxide production by liver sinusoidal endothelial cells was also shown [70, 71] and chronic alcohol administration leads to increased superoxide release in circulating neutrophils as well [71].

With regard to the mechanism by which chronic ethanol administration has its inflammatory effect, attention again falls upon the NF-κB pathway. NF-κB activation correlates with increased LPO and increased expression of a range of pro-inflammatory cytokines (TNFα, IL-1β, interferon-γ, IL-12) as well as C-C (MCP-1, macrophage inflammatory protein-1α) and C-X-C (macrophage inflammatory protein-2, CINC) chemokines [76]. The role of the chemokine CINC has already been discussed in relation to the inflammatory response that arises following oxidative stress generated in IRI. Exacerbation of oxidative stress, hepatic injury, and pro-inflammatory cytokine up-regulation following alcohol feeding in mice is prevented by loss of MCP-1 [77]. Thus, these investigations support the scenario in which alcohol administration leads to oxidative stress, NF-κB pathway activation, and activation of a range of pro-inflammatory mediators that perpetuate injury through both monocyte and neutrophil recruitment and activation. Indeed, isolated KCs from alcohol-injured liver show increased superoxide production and NF-κB activation, with increased hepatic TNFα [78]. These changes can all be blocked by maintenance of GSH [78]. Further investigation of the mechanisms through which the NF-κB pathway is activated has implicated HSPs and the heat shock transcription factor HSF-1. HSF-1 can be induced by oxidative stress [79] and in vitro treatment of blood monocytes and macrophages with ethanol increases the nuclear translocation of HSF-1 [80]. Extended ethanol exposure in macrophages results in increased expression of the HSF-1 target hsp90, which in turn increases NF-κB activity and TNFα production. Conversely, inhibition of hsp90 prevents the NF-κB elevation seen after prolonged alcohol administration. Therefore, it appears that the role of ROS in redox signaling and inflammatory mediator stimulation may be at least as important as any direct cytotoxicity. Enteral feeding of alcohol to TNF-receptor 1 knockout mice resulted in similar levels of free radical production as seen in wild-type mice [81]. However, liver injury, assessed by ALT increase and histopathology, was markedly attenuated.

ROS production in alcohol-induced liver disease may occur through both the action of Nox and XO [70, 82, 83]. $P47^{phox}$ knockout mice do not manifest the same degree of radical adduct production in bile that is seen in wild-type mice, nor does hepatic NF-κB activation occur to the same extent [82]. A similar outcome was achieved through administration of the Nox inhibitor DPI to rats receiving chronic dietary ethanol, leading to a reduction in free radical adduct formation, NF-κB activation, TNFα mRNA, and overall liver injury [84]. In isolated, ethanol-perfused rat liver, both DPI and the serine threonine kinase inhibitor staurosporine resulted in a decrease in superoxide production, supporting the conclusion that ethanol-mediated stimulation of superoxide production by Nox may involve protein kinase C [70]. Considering alternative means of ROS generation, concomitant administration of the XO inhibitor allopurinol with ethanol in rats also reduces liver injury, free radical adduct formation in bile, and NF-κB activation [83]. However, allopurinol's ability to inhibit XO and independently scavenge free radicals does confound the conclusion that XO generation of ROS is definitely at play in ethanol-induced liver injury.

Irrespective of the route of superoxide generation, the overexpression of the cytoplasmic SOD isoform, SOD1 or Cu/Zn-SOD, in KCs and hepatocytes, through infection of rats with an adenovirus vector, ameliorates liver injury induced by chronic alcohol feeding. In acute and chronic ethanol administration, increased SOD1 results in fewer free radical adducts, decreased activation of NF-κB and activator protein (AP)-1 transcription factor, and reduced pro-inflammatory cytokine production [85, 86]. Thus, oxidative stress is important for NF-κB and AP-1 activation and inflammatory cytokine production. Interestingly, a similar approach using adenovirus-mediated up-regulation of SOD2 (the mitochondrial isoform) in the liver provided protection against alcohol-induced injury, but did not affect TNFα expression or neutrophil infiltration. This suggests that mitochondrial oxidative stress in parenchymal cells of the liver appears to occur downstream of KC activation, inflammatory cytokine production, and neutrophil recruitment [87].

Importantly, the pathogenesis of ethanol-induced liver inflammation does not merely comprise a direct response to ethanol itself. There is clinical and experimental evidence that ethanol exposure induces sensitization to gastrointestinal-derived bacterial endotoxin, as discussed in the following section [88–91].

6.6 Endotoxin-Associated Oxidative Stress and Inflammation

The ability of LPS to trigger superoxide production by human monocytes was originally identified in vitro and was shown, through the use of blocking antibodies, to be dependent on the CD14 receptor [92]. The antioxidants NAC and α-tocopherol succinate can suppress LPS-induced KC activation, and TNFα mRNA and protein expression, via decreased NF-κB activation [93]. Similarly, transfection of KCs with adenovirus containing SOD decreases the superoxide production, NF-κB activation, and TNFα production that occurs following LPS stimulation, confirming that these processes are dependent on oxidative stress [94]. In vivo, LPS leads to liver injury with evidence of superoxide generation, accompanied by NF-κB activation and TNFα production in KCs [95]. Pretreatment with the Nox inhibitor DPI protects against liver injury [96].

The principle reason underlying the interest in endotoxin-mediated liver injury derives from evidence that alcoholic liver disease partly owes its pathogenesis to its ability to sensitize the immune system to bacterial endotoxin. Firstly, acute alcohol-induced liver injury increases plasma endotoxin, whereas neutralization of endotoxin ameliorates alcohol-induced liver injury in a similar manner to antioxidant administration [73]. Also, a single enteral dose of ethanol in mice worsens the increase in ALT that occurs following LPS administration [86]. This sensitization can be blocked by prior overexpression of SOD1 or DPI, suggesting a role for oxidative stress and in particular Nox [86, 97]. In fact, ethanol leads to an increase in Rac1-GTPase activity, a critical component of Nox, and an increase in translocation of the $p67^{phox}$ Nox subunit to the plasma membrane [97]. The downstream consequence of this is increased LPS-induced mitogen-associated protein kinase phosphorylation and TNFα production. As well as inducing pro-inflammatory cytokine production through activation of Nox and the NF-κB and AP-1 signaling pathways, ethanol also increases CD14 mRNA expression in sinusoidal cells identified as KCs [86]. This up-regulation can be inhibited by overexpression of SOD1 and also by inhibition of AP-1 or loss of $p47^{phox}$. Taken together this suggests that ethanol promotes CD14 expression and thus endotoxin sensitivity through ROS production by Nox and activation of the AP-1 pathway, consistent with the finding that the CD14 promoter contains AP-1-responsive elements [98]. Mice deficient in CD14 are resistant to ethanol-induced hepatic injury, with blunting of NF-κB activation and TNFα expression [99].

It also appears that neutrophils play a significant part in endotoxin-induced liver injury and that oxidative stress mediates this process. Plasma and hepatic GSSG levels increase during the neutrophil injury phase of endotoxic shock, providing evidence of oxidative stress [100]. Further, more severe injury occurs in mice lacking the antioxidant enzyme glutathione peroxidase [100], so oxidative stress is critical for this process. As seen in IRI, the leukocyte adhesion molecule Mac-1 (CD11b/CD18) and ICAM-1 are essential to permit neutrophil adhesion to hepatocytes to occur in this model [101]. In addition, neutrophils co-localize to areas where hepatocytes stain positively for neutrophil-derived chlorotyrosine protein adducts [96]. Hence, these data support the conclusion that injury occurs following neutrophil adhesion to hepatocytes, permitting ROS generation and diffusion into parenchymal cells.

6.7 Viral Hepatitis

Clearly, bacterial endotoxin is not the only pathogen-derived stimulant of liver inflammation. Viral hepatitis poses an enormous global disease burden, with over 400 million people estimated to be infected by or have been exposed to hepatitis B or C [102, 103]. Hepatitis C virus (HCV), in particular, has oxidative stress as a key component of its pathology [104]. Individuals with chronic HCV have elevated markers of oxidative stress in liver, peripheral blood mononuclear cells, and serum [105–107]. SOD2 expression is also increased in peripheral blood mononuclear cells of HCV patients, although not in liver [108].

HCV proteins have been clearly demonstrated to increase ROS production, although most work has been carried out using hepatocytes or immortalized cell lines [109, 110]. With regard to inflammatory cells, leukocytes obtained from the buffy coat of patients with HCV show evidence of oxidative stress and a role for phagocyte MPO in HCV has been suggested by a genotypic association between MPO and severity of HCV-induced fibrosis [111, 112]. Superoxide production was triggered in vitro in human monocytes through stimulation with HCV nonstructural protein (NS)-3, whereas stimulation with NS4, NS5A, NS5B, and core proteins of HCV did not lead to ROS production [113].

As well as mononuclear cells, NS3 can induce prolonged release of ROS from neutrophils [114]. HCV-induced ROS production due to NS3 requires Nox, since treatment with DPI abolished superoxide generation [113, 114]. This phagocyte production of ROS can lead to dysfunction and apoptosis in natural killer and T cells, lymphocytes with antiviral properties [114]. Finally, in addition to the direct production of ROS by leukocytes via Nox2, the up-regulation of cytokine production, especially of TNFα, transforming growth factor beta (TGFβ) and IFNγ, that is seen during HCV can induce expression of Nox homologs (namely Nox1 and Nox4) in other cell types, including hepatocytes [115, 116].

6.8 Fibrosis

All chronic liver diseases have fibrosis as their end point, but does oxidative stress in the inflammatory response have a direct fibrogenic effect? When hepatic stellate cells (HSCs) were cultured with human neutrophils that had been stimulated to produce ROS, not only was evidence of oxidative stress apparent (by way of MDA and 4-HNE protein adducts), but procollagen α1(I) mRNA and protein levels increased compared to HSCs in co-culture with unstimulated neutrophils [117]. This increase was inhibited by addition of vitamin E or SOD. Exogenous XO was also able to stimulate procollagen expression. Similarly, co-culture of KCs and HSCs results in activation and proliferation of HSCs, along with increased production of collagen I and hydrogen peroxide [118]. Addition of catalase prevents the increase in HSC-derived collagen I, suggesting up-regulation is mediated by oxidative stress. Increased tissue inhibitor of metalloproteinase 1, IL-6, and decreased matrix metallopeptidase 13 were also detected in co-culture, providing further evidence of progression to a fibrogenic phenotype.

Induction of oxidative stress in macrophages can lead to production of pro-fibrogenic mediators. Treatment of human and murine macrophages, or KCs isolated from cirrhotic rat liver, with 4-HNE resulted in up-regulation of expression of TGFβ, a central cytokine involved in myofibroblast activation [119]. The up-regulation in TGFβ was associated with activation of the AP-1 pathway, but no change in NF-κB activation. The importance of the AP-1 pathway is supported by the in vivo observation that AP-1 DNA binding, accompanied by evidence of oxidative stress, occurs after acute CCl_4 administration [120]. This AP-1 activation can be prevented by macrophage inactivation with $GdCl_3$ or antioxidant administration to quench 4-HNE activity.

Concomitant with the ability of the macrophage to exert a pro-fibrotic effect on HSCs, addition of 4-HNE (or xanthine/XO) to HSCs in vitro up-regulates expression of the chemokine MCP-1, suggesting that oxidative stress in these myofibroblast precursors can promote monocyte recruitment [121]. MCP-1 knockout in acute CCl_4 liver injury reduces monocyte-macrophage infiltration and the degree of both liver injury and pro-inflammatory cytokine production [122]. This is associated with reduced elevations in markers of oxidative stress.

As fibrosis progresses, KCs exhibit changes consistent with oxidative stress, such as NF-κB activation and a decrease in GSH, combined with elevations in mRNA expression of inflammatory and fibrogenic cytokines including TGFβ, IL-6, TNFα, and IL-1β [123, 124]. Cellular expression of Nox homologs also increases: KCs express Nox2 and HSCs express both Nox1 and Nox2 [125]. Genetic knockout of either Nox1 or Nox2 attenuates fibrosis, accompanied by reduced markers of oxidative stress [125]. The importance of Nox production in both KCs and HSCs in fibrogenesis is confirmed by an alternative experimental approach utilizing $p47^{phox}$ knockout and bone marrow transplantation to generate mice in which Nox expression was only present or only deficient in bone marrow-derived cells [126]. Both chimeras show reduced fibrosis and markers of oxidative stress in comparison to wild type. In a CCl_4 model of hepatic fibrogenesis in rats, inhibition of NF-κB activation, through administration of (Z)2-(5-(4-methoxybenzylidene)-2, 4-dioxothiazolidin-3-yl) acetic acid, ameliorates injury, lessens fibrosis, and limits the decrease in GSH [127]. The same compound was shown in vitro in macrophages to inhibit the phosphorylation of IκB that is required for NF-κB activation [127].

It is therefore apparent that, in addition to the direct effects of injury and oxidative stress on HSCs and fibrogenesis that are described elsewhere, oxidative stress in both KCs and neutrophils can promote fibrogenesis through the production of fibrogenic cytokines and activation of HSCs to collagen-producing myofibroblasts.

6.9 Hepatocellular Carcinoma

Fibrosis is an independent risk factor for development of hepatocellular carcinoma (HCC) and over 80 % of patients who develop HCC will have pre-existing fibrotic disease [128]. The mechanisms mediating malignant transformation are not well-understood, neither is the contribution of inflammatory-derived oxidative stress. *N*-nitrosodiethylamine (DEN) is frequently used to model HCC in mice. KCs isolated from mice treated with DEN have enhanced superoxide production and the effect is also seen when DEN is applied directly to KCs in culture [129]. DEN also causes liver inflammation and necrosis in vivo, with evidence of DNA damage. However, in $p47^{phox}$ knockout mice (deficient in Nox) treated with DEN, hepatic injury and TNFα levels are reduced, with no evidence of DNA damage in hepatocytes. Isolated KCs from these mice did not produce superoxide, leading to the conclusion that DEN can stimulate KCs to generate superoxide via Nox which can in turn lead to genotoxic and cytotoxic effects in hepatocytes, with the potential for neoplastic transformation.

6.10 Toxin Injury

The role of oxidative stress in toxin-mediated liver injury is unfortunately of necessary interest due to the fact that drug-induced hepatotoxicity is a leading cause of acute liver failure in the developed world [130]. Further, acetaminophen has served as a predictable and translatable trigger of hepatotoxicity to permit the study of a variety of mechanisms underlying liver injury. As in many of the disease processes considered thus far, a role for macrophages in acetaminophen hepatotoxicity was initially demonstrated using inactivation with $GdCl_3$. Pretreatment prior to acetaminophen administration prevented the formation of nitrotyrosine protein adducts indicative of peroxynitrite [131]. In control mice, acetaminophen adducts co-localized with nitrotyrosine in a centrilobular location. Similar findings of reduced markers of oxidative stress were reported after macrophage inactivation with $GdCl_3$ in thioacetamide-induced liver injury in rats [132].

As well as evidence of oxidative stress, with decreased hepatic GSH and SOD1, acetaminophen administration in mice also leads to increased expression of classical pro-inflammatory mediators TNFα, MCP-1, and IL-1β [133]. Antioxidant treatment with NAC limits oxidative stress and liver injury. Of interest and contrary to findings in many other liver pathologies, disruption of the NF-κB signaling pathway, as found in Nox-deficient mice, does not prevent acetaminophen-induced liver injury, suggesting other oxidative stress-mediated pathways are of greater importance. Indeed, studies of acetaminophen toxicity in mouse macrophages in vitro have shown that ROS production is associated with mitochondrial dysfunction, including a decreased mitochondrial membrane potential and inhibition of complexes I and IV [134]. This suggests that mitochondrial oxidative and metabolic stress may be of central importance in acetaminophen-induced cytotoxicity in macrophages.

Acetaminophen is not the only toxic substance that can lead to oxidative stress and liver inflammation. Silica nanoparticles, one of the most popular nanomaterials, with potential applications in a wide variety of biomedical fields, have been shown to activate KCs in vitro, causing ROS production and release and increased TNFα production [135]. In vivo, they lead to an increase in sinusoidal KC number and signs of oxidative stress (decreased GSH and increased MDA).

6.11 Cholestatic Liver Disease

Cholestatic liver disease is commonly modeled in experimental animals by means of bile duct ligation (BDL). Using this model in rats, gene delivery of SOD2 was shown to reduce injury, assessed by histopathology, measurement of ALT, and TNFα, TGFβ, and collagen I(α)1 expression [136]. Both the oxidative stress marker 4-HNE and NF-κB activation were reduced. There was less or no effect when SOD1 was overexpressed, suggesting that cholestasis leads to liver injury via mechanisms involving mitochondrial rather than cytoplasmic oxidative stress.

When considering the inflammatory response in cholestatic liver injury, neutrophils are the dominant non-parenchymal cell type [16, 137]. Investigations performed using CD18-deficient mice, i.e., with neutrophils lacking functional Mac-1 adhesion molecule, revealed that the number of neutrophils located in the liver parenchyma 3 days after BDL was reduced by 91 % in these CD18-deficient mice [16]. The mice also exhibited less severe hepatocellular injury with little staining for neutrophil-derived oxidative stress in the form of chlorotyrosine adducts. Similar findings were observed in ICAM-1-deficient mice, supporting the conclusion that Mac-1 and ICAM-1 interaction is required for neutrophil adherence to hepatocytes and subsequent oxidative stress-mediated injury [137].

6.12 Hepatoprotective Functions of the Immune System

It is important to remember that the principle role of the host inflammatory response is actually protection as opposed to exacerbation of injury. In the same way that ROS production by the phagocytic cells of the immune system can protect the host through pathogen killing, it is also possible that oxidative stress can have protective functions in certain circumstances. Proteome analysis of isolated KCs following 8 h of reperfusion post-ischemia revealed that both oxidant and antioxidant protein expression (including catalase and SOD1) was up-regulated [29].

Neutrophils producing the ROS that trigger oxidative stress and procollagen expression in HSCs also produce nitric oxide [117]. When nitric oxide formation is inhibited, ROS-stimulated collagen production increases. Conversely, treatment with the nitric oxide donor sodium nitroprusside inhibited procollagen expression. This demonstrates the presence of an in-built system to counter-balance at least some of the negative effects that ensue from ROS production.

Of greater import is the phenomenon of ischemic preconditioning, whereby a short period of ischemia and reperfusion subsequently reduces the injurious effects of a longer period of tissue or organ ischemia and reperfusion. Thus, ROS, of which KCs and neutrophils are major producers in IRI, can serve a protective function, when generated in limited quantities. In a rat model of IRI, preconditioning with 5 min of ischemia and 15 min of reperfusion prior to 45 min of ischemia led to reduced hepatic injury, decreased markers of oxidative stress, and less neutrophil extravasation [31]. This short period of oxidative stress serves to protect against future and excessive ROS generation by KCs and neutrophils [138]. Indeed, antioxidant administration, so often shown to be protective, abolishes the protective effects of either a short period of ischemic preconditioning or hydrogen peroxide infusion prior to prolonged IR [139].

Lower levels of ROS formation may also lead to activation of the Keap1-Nrf2-ARE signaling pathway [140]. Oxidation of Kelch ECH-associating protein 1 (Keap1) cysteine residues leads to dissociation from nuclear factor erythroid 2-related factor 2 (Nrf2), which is phosphorylated by protein kinase C and then able to translocate to the nucleus and bind to antioxidant response elements (AREs). This results in up-regulation of a number of defense genes including heme oxygenase (HO)-1 (protective in all hepatic IRI models [141]), SOD-1, glutathione peroxidase, and glutathione reductase [4]. KCs can act as a major source of HO-1 production in the liver [142], with expression up-regulated in KCs from ethanol-fed rats [143]. Up-regulation can be inhibited by small interfering RNA directed against hypoxia-inducible factor 1α, Nrf2, or Nox. Prevention of HO-1 expression results in increased inflammatory cytokine expression (of TNFα and IL-1β).

The combination of acute alcohol administration followed by IRI revealed a divergent effect of oxidative stress and the inflammatory response [144]. While alcohol prior to IRI worsened injury and superoxide release shortly after reperfusion, NF-κB activation was suppressed, in comparison to animals receiving IRI alone, along with reduced expression of TNFα and CINC. Neutrophil recruitment was blunted and this was accompanied by a reduction in the degree of histopathologic injury.

6.13 Summary

The generation of ROS, oxidative stress, and the inflammatory response are inextricably intertwined in all of the principle causes of liver pathology. Oxidative stress plays a key role in leukocyte activation and recruitment during liver injury and also contributes significantly to intra- and intercellular signaling, gene expression, and cytokine production.

The pervasiveness of ROS within the liver, with production by, and action in, all of the cell types of the organ does pose certain challenges with regard to demonstrating the cells of origin and dissecting cause from effect. Defining the production of ROS and consequences of oxidative stress in specific cell types is particularly relevant since it is evident that ROS can actually be of benefit in certain cells at certain times, so cannot be viewed as a purely negative and detrimental component of injury.

Advances in cell-specific labeling techniques now permit improved identification of cells and fate-mapping of cellular progeny. These same tools can be utilized to target gene deletion to specific cells and even at specific times during development and disease. Thus, there are now elegant approaches available that would allow abrogation of ROS production or signaling specifically in Kupffer cells or any other cell of interest. Through this it will be possible to determine precisely in vivo the relative contributions of various cell types during the response to oxidative stress in the liver, thus paving the way for targeted therapies to limit the propagation of its negative effects during liver inflammation.

References

1. Rosen GM, Pou S, Ramos CL, Cohen MS, Britigan BE (1995) Free radicals and phagocytic cells. FASEB J 9(2):200–209
2. Murray PJ, Wynn TA (2011) Protective and pathogenic functions of macrophage subsets. Nat Rev Immunol 11(11):723–737. doi:10.1038/nri3073
3. Dröge W (2002) Free radicals in the physiological control of cell function. Physiol Rev 82(1):47–95
4. Jaeschke H (2011) Reactive oxygen and mechanisms of inflammatory liver injury: present concepts. J Gastroenterol Hepatol 26(Suppl 1):173–179. doi:10.1111/j.1440-1746.2010.06592.x
5. D'Autréaux B, Toledano MB (2007) ROS as signalling molecules: mechanisms that generate specificity in ROS homeostasis. Nat Rev Mol Cell Biol 8(10):813–824
6. Novo E, Parola M (2008) Redox mechanisms in hepatic chronic wound healing and fibrogenesis. Fibrogenesis Tissue Repair 1(1):5. doi:10.1186/1755-1536-1-5
7. Scaffidi P, Misteli T, Bianchi ME (2002) Release of chromatin protein HMGB1 by necrotic cells triggers inflammation. Nature 418(6894):191–195
8. Rock KL, Latz E, Ontiveros F, Kono H (2010) The sterile inflammatory response. Annu Rev Immunol 28:321–342. doi:10.1146/annurev-immunol-030409-101311
9. Bianchi ME (2009) HMGB1 loves company. J Leukoc Biol 86(3):573–576. doi:10.1189/jlb.1008585

10. Schwabe RF, Seki E, Brenner DA (2006) Toll-like receptor signaling in the liver. Gastroenterology 130(6):1886–1900
11. Tsung A, Sahai R, Tanaka H, Nakao A, Fink MP, Lotze MT, Yang H, Li J, Tracey KJ, Geller DA, Billiar TR (2005) The nuclear factor HMGB1 mediates hepatic injury after murine liver ischemia-reperfusion. J Exp Med 201(7):1135–1143
12. Fan J, Li Y, Levy RM, Fan JJ, Hackam DJ, Vodovotz Y, Yang H, Tracey KJ, Billiar TR, Wilson MA (2007) Hemorrhagic shock induces NAD(P)H oxidase activation in neutrophils: role of HMGB1-TLR4 signaling. J Immunol 178(10):6573–6580
13. Bae YS, Lee JH, Choi SH, Kim S, Almazan F, Witztum JL, Miller YI (2009) Macrophages generate reactive oxygen species in response to minimally oxidized low-density lipoprotein: toll-like receptor 4- and spleen tyrosine kinase-dependent activation of NADPH oxidase 2. Circ Res 104(2):210–8. doi:10.1161/CIRCRESAHA.108.181040
14. Bamboat ZM, Balachandran VP, Ocuin LM, Obaid H, Plitas G, DeMatteo RP (2010) Toll-like receptor 9 inhibition confers protection from liver ischemia-reperfusion injury. Hepatology 51(2):621–632. doi:10.1002/hep.23365
15. Klebanoff SJ (2005) Myeloperoxidase: friend and foe. J Leukoc Biol 77(5):598–625
16. Gujral JS, Farhood A, Bajt ML, Jaeschke H (2003) Neutrophils aggravate acute liver injury during obstructive cholestasis in bile duct-ligated mice. Hepatology 38(2):355–363
17. Brown KE, Brunt EM, Heinecke JW (2001) Immunohistochemical detection of myeloperoxidase and its oxidation products in Kupffer cells of human liver. Am J Pathol 159(6): 2081–2088
18. Sugiyama S, Okada Y, Sukhova GK, Virmani R, Heinecke JW, Libby P (2001) Macrophage myeloperoxidase regulation by granulocyte macrophage colony-stimulating factor in human atherosclerosis and implications in acute coronary syndromes. Am J Pathol 158(3):879–891
19. Jaeschke H, Farhood A, Bautista AP, Spolarics Z, Spitzer JJ (1993) Complement activates Kupffer cells and neutrophils during reperfusion after hepatic ischemia. Am J Physiol 264(4 Pt 1):G801–G809
20. Bajt ML, Farhood A, Jaeschke H (2001) Effects of CXC chemokines on neutrophil activation and sequestration in hepatic vasculature. Am J Physiol Gastrointest Liver Physiol 281(5): G1188–G1195
21. Kubes P, Mehal WZ (2012) Sterile inflammation in the liver. Gastroenterology 143(5):1158–1172. doi:10.1053/j.gastro.2012.09.008
22. Poltorak A, He X, Smirnova I, Liu MY, Van Huffel C, Du X, Birdwell D, Alejos E, Silva M, Galanos C, Freudenberg M, Ricciardi-Castagnoli P, Layton B, Beutler B (1998) Defective LPS signaling in C3H/HeJ and C57BL/10ScCr mice: mutations in Tlr4 gene. Science 282(5396):2085–2088
23. Hemmi H, Takeuchi O, Kawai T, Kaisho T, Sato S, Sanjo H, Matsumoto M, Hoshino K, Wagner H, Takeda K, Akira S (2000) A Toll-like receptor recognizes bacterial DNA. Nature 408(6813):740–745
24. Bilzer M, Jaeschke H, Vollmar AM, Paumgartner G, Gerbes AL (1999) Prevention of Kupffer cell-induced oxidant injury in rat liver by atrial natriuretic peptide. Am J Physiol 276(5 Pt 1):G1137–G1144
25. Bilzer M, Baron A, Schauer R, Steib C, Ebensberger S, Gerbes AL (2002) Glutathione treatment protects the rat liver against injury after warm ischemia and Kupffer cell activation. Digestion 66(1):49–57
26. Jaeschke H, Bautista AP, Spolarics Z, Spitzer JJ (1992) Superoxide generation by neutrophils and Kupffer cells during in vivo reperfusion after hepatic ischemia in rats. J Leukoc Biol 52(4):377–382
27. Liu P, McGuire GM, Fisher MA, Farhood A, Smith CW, Jaeschke H (1995) Activation of Kupffer cells and neutrophils for reactive oxygen formation is responsible for endotoxin-enhanced liver injury after hepatic ischemia. Shock 3(1):56–62
28. Jaeschke H, Farhood A (1991) Neutrophil and Kupffer cell-induced oxidant stress and ischemia-reperfusion injury in rat liver. Am J Physiol 260(3 Pt 1):G355–G362

29. Hirsch J, Hansen KC, Choi S, Noh J, Hirose R, Roberts JP, Matthay MA, Burlingame AL, Maher JJ, Niemann CU (2006) Warm ischemia-induced alterations in oxidative and inflammatory proteins in hepatic Kupffer cells in rats. Mol Cell Proteomics 5(6):979–986
30. Jaeschke H, Smith CW (1997) Mechanisms of neutrophil-induced parenchymal cell injury. J Leukoc Biol 61(6):647–653
31. Hasegawa T, Malle E, Farhood A, Jaeschke H (2005) Generation of hypochlorite-modified proteins by neutrophils during ischemia-reperfusion injury in rat liver: attenuation by ischemic preconditioning. Am J Physiol Gastrointest Liver Physiol 289(4):G760–G767
32. Komatsu H, Koo A, Ghadishah E, Zeng H, Kuhlenkamp JF, Inoue M, Guth PH, Kaplowitz N (1992) Neutrophil accumulation in ischemic reperfused rat liver: evidence for a role for superoxide free radicals. Am J Physiol 262(4 Pt 1):G669–G676
33. Horie Y, Wolf R, Russell J, Shanley TP, Granger DN (1997) Role of Kupffer cells in gut ischemia/reperfusion-induced hepatic microvascular dysfunction in mice. Hepatology 26(6):1499–1505
34. Schauer RJ, Bilzer M, Kalmuk S, Gerbes AL, Leiderer R, Schildberg FW, Messmer K (2001) Microcirculatory failure after rat liver transplantation is related to Kupffer cell-derived oxidant stress but not involved in early graft dysfunction. Transplantation 72(10):1692–1699
35. Horie Y, Wolf R, Anderson DC, Granger DN (1997) Hepatic leukostasis and hypoxic stress in adhesion molecule-deficient mice after gut ischemia/reperfusion. J Clin Invest 99(4): 781–788
36. Shappell SB, Toman C, Anderson DC, Taylor AA, Entman ML, Smith CW (1990) Mac-1(CD11b/CD18) mediates adherence-dependent hydrogen peroxide production by human and canine neutrophils. J Immunol 144(7):2702–2711
37. Jaeschke H, Farhood A, Bautista AP, Spolarics Z, Spitzer JJ, Smith CW (1993) Functional inactivation of neutrophils with a Mac-1 (CD11b/CD18) monoclonal antibody protects against ischemia-reperfusion injury in rat liver. Hepatology 17(5):915–923
38. Shibuya H, Ohkohchi N, Seya K, Satomi S (1997) Kupffer cells generate superoxide anions and modulate reperfusion injury in rat livers after cold preservation. Hepatology 25(2): 356–360
39. Cutrín JC, Llesuy S, Boveris A (1998) Primary role of Kupffer cell-hepatocyte communication in the expression of oxidative stress in the post-ischaemic liver. Cell Biochem Funct 16(1):65–72
40. Vega VL, Mardones L, Maldonado M, Nicovani S, Manríquez V, Roa J, Ward PH (2000) Xanthine oxidase released from reperfused hind limbs mediate kupffer cell activation, neutrophil sequestration, and hepatic oxidative stress in rats subjected to tourniquet shock. Shock 14(5):565–571
41. Matsumura F, Yamaguchi Y, Goto M, Ichiguchi O, Akizuki E, Matsuda T, Okabe K, Liang J, Ohshiro H, Iwamoto T, Yamada S, Mori K, Ogawa M (1998) Xanthine oxidase inhibition attenuates kupffer cell production of neutrophil chemoattractant following ischemia-reperfusion in rat liver. Hepatology 28(6):1578–1587
42. Blackwell TS, Holden EP, Blackwell TR, DeLarco JE, Christman JW (1994) Cytokine-induced neutrophil chemoattractant mediates neutrophilic alveolitis in rats: association with nuclear factor kappa B activation. Am J Respir Cell Mol Biol 11(4):464–472
43. Satriano J, Schlondorff D (1994) Activation and attenuation of transcription factor NF-kB in mouse glomerular mesangial cells in response to tumor necrosis factor-alpha, immunoglobulin G, and adenosine 3′:5′-cyclic monophosphate. Evidence for involvement of reactive oxygen species. J Clin Invest 94(4):1629–1636
44. Jaeschke H, Woolbright BL (2012) Current strategies to minimize hepatic ischemia-reperfusion injury by targeting reactive oxygen species. Transplant Rev (Orlando) 26(2):103–114. doi:10.1016/j.trre.2011.10.006
45. Lehnert M, Arteel GE, Smutney OM, Conzelmann LO, Zhong Z, Thurman RG, Lemasters JJ (2003) Dependence of liver injury after hemorrhage/resuscitation in mice on NADPH oxidase-derived superoxide. Shock 19(4):345–351

46. Fan C, Li Q, Ross D, Engelhardt JF (2003) Tyrosine phosphorylation of I kappa B alpha activates NF kappa B through a redox-regulated and c-Src-dependent mechanism following hypoxia/reoxygenation. J Biol Chem 278(3):2072–2080
47. Llacuna L, Marí M, Lluis JM, García-Ruiz C, Fernández-Checa JC, Morales A (2009) Reactive oxygen species mediate liver injury through parenchymal nuclear factor-kappaB inactivation in prolonged ischemia/reperfusion. Am J Pathol 174(5):1776–1785. doi:10.2353/ajpath.2009.080857
48. Ikura Y, Ohsawa M, Suekane T, Fukushima H, Itabe H, Jomura H, Nishiguchi S, Inoue T, Naruko T, Ehara S, Kawada N, Arakawa T, Ueda M (2006) Localization of oxidized phosphatidylcholine in nonalcoholic fatty liver disease: impact on disease progression. Hepatology 43(3):506–514
49. Guyton JR, Lenz ML, Mathews B, Hughes H, Karsan D, Selinger E, Smith CV (1995) Toxicity of oxidized low density lipoproteins for vascular smooth muscle cells and partial protection by antioxidants. Atherosclerosis 118(2):237–249
50. Wang GP, Deng ZD, Ni J, Qu ZL (1997) Oxidized low density lipoprotein and very low density lipoprotein enhance expression of monocyte chemoattractant protein-1 in rabbit peritoneal exudate macrophages. Atherosclerosis 133(1):31–36
51. Rensen SS, Slaats Y, Nijhuis J, Jans A, Bieghs V, Driessen A, Malle E, Greve JW, Buurman WA (2009) Increased hepatic myeloperoxidase activity in obese subjects with nonalcoholic steatohepatitis. Am J Pathol 175(4):1473–1482. doi:10.2353/ajpath.2009.080999
52. Baumgardner JN, Shankar K, Hennings L, Albano E, Badger TM, Ronis MJ (2008) N-acetylcysteine attenuates progression of liver pathology in a rat model of nonalcoholic steatohepatitis. J Nutr 138(10):1872–1879
53. Hui JM, Hodge A, Farrell GC, Kench JG, Kriketos A, George J (2004) Beyond insulin resistance in NASH: TNF-alpha or adiponectin? Hepatology 40(1):46–54
54. Fukushima J, Kamada Y, Matsumoto H, Yoshida Y, Ezaki H, Takemura T, Saji Y, Igura T, Tsutsui S, Kihara S, Funahashi T, Shimomura I, Tamura S, Kiso S, Hayashi N (2009) Adiponectin prevents progression of steatohepatitis in mice by regulating oxidative stress and Kupffer cell phenotype polarization. Hepatol Res 39(7):724–738. doi:10.1111/j.1872-034X.2009.00509.x
55. Friedman JM (2002) The function of leptin in nutrition, weight, and physiology. Nutr Rev 60(10 Pt 2):S1–S14; discussion S68–S84, 85–87
56. Chitturi S, Farrell G, Frost L, Kriketos A, Lin R, Fung C, Liddle C, Samarasinghe D, George J (2002) Serum leptin in NASH correlates with hepatic steatosis but not fibrosis: a manifestation of lipotoxicity? Hepatology 36(2):403–409
57. Chatterjee S, Ganini D, Tokar EJ, Kumar A, Das S, Corbett J, Kadiiska MB, Waalkes MP, Diehl AM, Mason RP (2013) Leptin is key to peroxynitrite-mediated oxidative stress and Kupffer cell activation in experimental non-alcoholic steatohepatitis. J Hepatol 58(4):778–784. doi:10.1016/j.jhep.2012.11.035
58. Yang YY, Huang YT, Tsai TH, Hou MC, Lee FY, Lee SD, Lin HC (2012) Kupffer cell depletion attenuates leptin-mediated methoxamine-stimulated portal perfusion pressure and thromboxane A2 release in a rodent model of NASH-cirrhosis. Clin Sci (Lond) 123(12):669–680. doi:10.1042/CS20110572
59. Hewinson J, Moore SF, Glover C, Watts AG, MacKenzie AB (2008) A key role for redox signaling in rapid P2X7 receptor-induced IL-1 beta processing in human monocytes. J Immunol 180(12):8410–8420
60. Chatterjee S, Rana R, Corbett J, Kadiiska MB, Goldstein J, Mason RP (2012) P2X7 receptor-NADPH oxidase axis mediates protein radical formation and Kupffer cell activation in carbon tetrachloride-mediated steatohepatitis in obese mice. Free Radic Biol Med 52(9):1666–1679. doi:10.1016/j.freeradbiomed.2012.02.010
61. Malle E, Furtmüller PG, Sattler W, Obinger C (2007) Myeloperoxidase: a target for new drug development? Br J Pharmacol 152(6):838–854
62. Rensen SS, Bieghs V, Xanthoulea S, Arfianti E, Bakker JA, Shiri-Sverdlov R, Hofker MH, Greve JW, Buurman WA (2012) Neutrophil-derived myeloperoxidase aggravates non-

alcoholic steatohepatitis in low-density lipoprotein receptor-deficient mice. PLoS One 7(12):e52411. doi:10.1371/journal.pone.0052411

63. Albano E, Mottaran E, Vidali M, Reale E, Saksena S, Occhino G, Burt AD, Day CP (2005) Immune response towards lipid peroxidation products as a predictor of progression of non-alcoholic fatty liver disease to advanced fibrosis. Gut 54(7):987–993
64. Nobili V, Parola M, Alisi A, Marra F, Piemonte F, Mombello C, Sutti S, Povero D, Maina V, Novo E, Albano E (2010) Oxidative stress parameters in paediatric non-alcoholic fatty liver disease. Int J Mol Med 26(4):471–476
65. Sutti S, Jindal A, Locatelli I, Vacchiano M, Gigliotti L, Bozzola C, Albano E (2014) Adaptive immune responses triggered by oxidative stress contribute to hepatic inflammation in NASH. Hepatology 59(3):886–897. doi:10.1002/hep.26749
66. Ferreyra Solari NE, Inzaugarat ME, Baz P, De Matteo E, Lezama C, Galoppo M, Galoppo C, Cherñavsky AC (2012) The role of innate cells is coupled to a Th1-polarized immune response in pediatric nonalcoholic steatohepatitis. J Clin Immunol 32(3):611–621. doi:10.1007/s10875-011-9635-2
67. Purohit V, Gao B, Song BJ (2009) Molecular mechanisms of alcoholic fatty liver. Alcohol Clin Exp Res 33(2):191–205. doi:10.1111/j.1530-0277.2008.00827.x
68. Kono H, Bradford BU, Yin M, Sulik KK, Koop DR, Peters JM, Gonzalez FJ, McDonald T, Dikalova A, Kadiiska MB, Mason RP, Thurman RG (1999) CYP2E1 is not involved in early alcohol-induced liver injury. Am J Physiol 277(6 Pt 1):G1259–G1267
69. Knecht KT, Adachi Y, Bradford BU, Iimuro Y, Kadiiska M, Xuang QH, Thurman RG (1995) Free radical adducts in the bile of rats treated chronically with intragastric alcohol: inhibition by destruction of Kupffer cells. Mol Pharmacol 47(5):1028–1034
70. Hasegawa T, Kikuyama M, Sakurai K, Kambayashi Y, Adachi M, Saniabadi AR, Kuwano H, Nakano M (2002) Mechanism of superoxide anion production by hepatic sinusoidal endothelial cells and Kupffer cells during short-term ethanol perfusion in the rat. Liver 22(4): 321–329
71. Bautista AP, Spitzer JJ (1999) Role of Kupffer cells in the ethanol-induced oxidative stress in the liver. Front Biosci 4:D589–D595
72. Cubero FJ, Nieto N (2008) Ethanol and arachidonic acid synergize to activate Kupffer cells and modulate the fibrogenic response via tumor necrosis factor alpha, reduced glutathione, and transforming growth factor beta-dependent mechanisms. Hepatology 48(6):2027–2039. doi:10.1002/hep.22592
73. Zhou Z, Wang L, Song Z, Lambert JC, McClain CJ, Kang YJ (2003) A critical involvement of oxidative stress in acute alcohol-induced hepatic TNF-alpha production. Am J Pathol 163(3):1137–1146
74. Ronis MJ, Butura A, Sampey BP, Shankar K, Prior RL, Korourian S, Albano E, Ingelman-Sundberg M, Petersen DR, Badger TM (2005) Effects of N-acetylcysteine on ethanol-induced hepatotoxicity in rats fed via total enteral nutrition. Free Radic Biol Med 39(5):619–630
75. Lv X, Chen Z, Li J, Zhang L, Liu H, Huang C, Zhu P (2010) Caffeine protects against alcoholic liver injury by attenuating inflammatory response and oxidative stress. Inflamm Res 59(8):635–645. doi:10.1007/s00011-010-0176-6
76. Nanji AA, Jokelainen K, Rahemtulla A, Miao L, Fogt F, Matsumoto H, Tahan SR, Su GL (1999) Activation of nuclear factor kappa B and cytokine imbalance in experimental alcoholic liver disease in the rat. Hepatology 30(4):934–943
77. Mandrekar P, Ambade A, Lim A, Szabo G, Catalano D (2011) An essential role for monocyte chemoattractant protein-1 in alcoholic liver injury: regulation of proinflammatory cytokines and hepatic steatosis in mice. Hepatology 54(6):2185–2197. doi:10.1002/hep.24599
78. Iimuro Y, Bradford BU, Yamashina S, Rusyn I, Nakagami M, Enomoto N, Kono H, Frey W, Forman D, Brenner D, Thurman RG (2000) The glutathione precursor L-2-oxothiazolidine-4-carboxylic acid protects against liver injury due to chronic enteral ethanol exposure in the rat. Hepatology 31(2):391–398
79. Ahn SG, Thiele DJ (2003) Redox regulation of mammalian heat shock factor 1 is essential for Hsp gene activation and protection from stress. Genes Dev 17(4):516–528

80. Mandrekar P, Catalano D, Jeliazkova V, Kodys K (2008) Alcohol exposure regulates heat shock transcription factor binding and heat shock proteins 70 and 90 in monocytes and macrophages: implication for TNF-alpha regulation. J Leukoc Biol 84(5):1335–1345. doi:10.1189/jlb.0407256
81. Yin M, Gäbele E, Wheeler MD, Connor H, Bradford BU, Dikalova A, Rusyn I, Mason R, Thurman RG (2001) Alcohol-induced free radicals in mice: direct toxicants or signaling molecules? Hepatology 34(5):935–942
82. Kono H, Rusyn I, Yin M, Gäbele E, Yamashina S, Dikalova A, Kadiiska MB, Connor HD, Mason RP, Segal BH, Bradford BU, Holland SM, Thurman RG (2000) NADPH oxidase-derived free radicals are key oxidants in alcohol-induced liver disease. J Clin Invest 106(7): 867–872
83. Kono H, Rusyn I, Bradford BU, Connor HD, Mason RP, Thurman RG (2000) Allopurinol prevents early alcohol-induced liver injury in rats. J Pharmacol Exp Ther 293(1):296–303
84. Kono H, Rusyn I, Uesugi T, Yamashina S, Connor HD, Dikalova A, Mason RP, Thurman RG (2001) Diphenyleneiodonium sulfate, an NADPH oxidase inhibitor, prevents early alcohol-induced liver injury in the rat. Am J Physiol Gastrointest Liver Physiol 280(5):G1005–G1012
85. Wheeler MD, Kono H, Yin M, Rusyn I, Froh M, Connor HD, Mason RP, Samulski RJ, Thurman RG (2001) Delivery of the Cu/Zn-superoxide dismutase gene with adenovirus reduces early alcohol-induced liver injury in rats. Gastroenterology 120(5):1241–1250
86. Wheeler MD, Thurman RG (2003) Up-regulation of CD14 in liver caused by acute ethanol involves oxidant-dependent AP-1 pathway. J Biol Chem 278(10):8435–8441
87. Wheeler MD, Nakagami M, Bradford BU, Uesugi T, Mason RP, Connor HD, Dikalova A, Kadiiska M, Thurman RG (2001) Overexpression of manganese superoxide dismutase prevents alcohol-induced liver injury in the rat. J Biol Chem 276(39):36664–36672
88. Bode C, Kugler V, Bode JC (1987) Endotoxemia in patients with alcoholic and non-alcoholic cirrhosis and in subjects with no evidence of chronic liver disease following acute alcohol excess. J Hepatol 4(1):8–14
89. Nanji AA, Khettry U, Sadrzadeh SM, Yamanaka T (1993) Severity of liver injury in experimental alcoholic liver disease. Correlation with plasma endotoxin, prostaglandin E2, leukotriene B4, and thromboxane B2. Am J Pathol 142(2):367–373
90. Adachi Y, Moore LE, Bradford BU, Gao W, Thurman RG (1995) Antibiotics prevent liver injury in rats following long-term exposure to ethanol. Gastroenterology 108(1):218–224
91. Parlesak A, Schäfer C, Schütz T, Bode JC, Bode C (2000) Increased intestinal permeability to macromolecules and endotoxemia in patients with chronic alcohol abuse in different stages of alcohol-induced liver disease. J Hepatol 32(5):742–747
92. Landmann R, Scherer F, Schumann R, Link S, Sansano S, Zimmerli W (1995) LPS directly induces oxygen radical production in human monocytes via LPS binding protein and CD14. J Leukoc Biol 57(3):440–449
93. Bellezzo JM, Leingang KA, Bulla GA, Britton RS, Bacon BR, Fox ES (1998) Modulation of lipopolysaccharide-mediated activation in rat Kupffer cells by antioxidants. J Lab Clin Med 131(1):36–44
94. Wheeler MD, Yamashina S, Froh M, Rusyn I, Thurman RG (2001) Adenoviral gene delivery can inactivate Kupffer cells: role of oxidants in NF-kappaB activation and cytokine production. J Leukoc Biol 69(4):622–630
95. Zhou Z, Wang L, Song Z, Saari JT, McClain CJ, Kang YJ (2004) Abrogation of nuclear factor-kappaB activation is involved in zinc inhibition of lipopolysaccharide-induced tumor necrosis factor-alpha production and liver injury. Am J Pathol 164(5):1547–1556
96. Gujral JS, Hinson JA, Farhood A, Jaeschke H (2004) NADPH oxidase-derived oxidant stress is critical for neutrophil cytotoxicity during endotoxemia. Am J Physiol Gastrointest Liver Physiol 287(1):G243–G252
97. Thakur V, Pritchard MT, McMullen MR, Wang Q, Nagy LE (2006) Chronic ethanol feeding increases activation of NADPH oxidase by lipopolysaccharide in rat Kupffer cells: role of

increased reactive oxygen in LPS-stimulated ERK1/2 activation and TNF-alpha production. J Leukoc Biol 79(6):1348–1356
98. Liu S, Shapiro RA, Nie S, Zhu D, Vodovotz Y, Billiar TR (2000) Characterization of rat CD14 promoter and its regulation by transcription factors AP1 and Sp family proteins in hepatocytes. Gene 250(1–2):137–147
99. Yin M, Bradford BU, Wheeler MD, Uesugi T, Froh M, Goyert SM, Thurman RG (2001) Reduced early alcohol-induced liver injury in CD14-deficient mice. J Immunol 166(7): 4737–4742
100. Jaeschke H, Ho YS, Fisher MA, Lawson JA, Farhood A (1999) Glutathione peroxidase-deficient mice are more susceptible to neutrophil-mediated hepatic parenchymal cell injury during endotoxemia: importance of an intracellular oxidant stress. Hepatology 29(2): 443–450
101. Jaeschke H, Farhood A, Smith CW (1991) Neutrophil-induced liver cell injury in endotoxin shock is a CD11b/CD18-dependent mechanism. Am J Physiol 261(6 Pt 1):G1051–G1056
102. Mohd Hanafiah K, Groeger J, Flaxman AD, Wiersma ST (2013) Global epidemiology of hepatitis C virus infection: new estimates of age-specific antibody to HCV seroprevalence. Hepatology 57(4):1333–1342. doi:10.1002/hep.26141
103. WHO (2013) Global policy report on the prevention and control of viral hepatitis in WHO member states. WHO, Geneva. ISBN 978 92 4 156463 2
104. Choi J (2012) Oxidative stress, endogenous antioxidants, alcohol, and hepatitis C: pathogenic interactions and therapeutic considerations. Free Radic Biol Med 52(7):1135–1150. doi:10.1016/j.freeradbiomed.2012.01.008
105. Farinati F, Cardin R, De Maria N, Della Libera G, Marafin C, Lecis E, Burra P, Floreani A, Cecchetto A, Naccarato R (1995) Iron storage, lipid peroxidation and glutathione turnover in chronic anti-HCV positive hepatitis. J Hepatol 22(4):449–456
106. De Maria N, Colantoni A, Fagiuoli S, Liu GJ, Rogers BK, Farinati F, Van Thiel DH, Floyd RA (1996) Association between reactive oxygen species and disease activity in chronic hepatitis C. Free Radic Biol Med 21(3):291–295
107. Boya P, de la Peña A, Beloqui O, Larrea E, Conchillo M, Castelruiz Y, Civeira MP, Prieto J (1999) Antioxidant status and glutathione metabolism in peripheral blood mononuclear cells from patients with chronic hepatitis C. J Hepatol 31(5):808–814
108. Larrea E, Beloqui O, Muñoz-Navas MA, Civeira MP, Prieto J (1998) Superoxide dismutase in patients with chronic hepatitis C virus infection. Free Radic Biol Med 24(7–8): 1235–1241
109. Okuda M, Li K, Beard MR, Showalter LA, Scholle F, Lemon SM, Weinman SA (2002) Mitochondrial injury, oxidative stress, and antioxidant gene expression are induced by hepatitis C virus core protein. Gastroenterology 122(2):366–375
110. Abdalla MY, Ahmad IM, Spitz DR, Schmidt WN, Britigan BE (2005) Hepatitis C virus-core and non structural proteins lead to different effects on cellular antioxidant defenses. J Med Virol 76(4):489–497
111. Farinati F, Cardin R, Degan P, De Maria N, Floyd RA, Van Thiel DH, Naccarato R (1999) Oxidative DNA damage in circulating leukocytes occurs as an early event in chronic HCV infection. Free Radic Biol Med 27(11–12):1284–1291
112. Reynolds WF, Patel K, Pianko S, Blatt LM, Nicholas JJ, McHutchison JG (2002) A genotypic association implicates myeloperoxidase in the progression of hepatic fibrosis in chronic hepatitis C virus infection. Genes Immun 3(6):345–349
113. Bureau C, Bernad J, Chaouche N, Orfila C, Béraud M, Gonindard C, Alric L, Vinel JP, Pipy B (2001) Nonstructural 3 protein of hepatitis C virus triggers an oxidative burst in human monocytes via activation of NADPH oxidase. J Biol Chem 276(25):23077–23083
114. Thorén F, Romero A, Lindh M, Dahlgren C, Hellstrand K (2004) A hepatitis C virus-encoded, nonstructural protein (NS3) triggers dysfunction and apoptosis in lymphocytes: role of NADPH oxidase-derived oxygen radicals. J Leukoc Biol 76(6):1180–1186

115. Kuwano Y, Kawahara T, Yamamoto H, Teshima-Kondo S, Tominaga K, Masuda K, Kishi K, Morita K, Rokutan K (2006) Interferon-gamma activates transcription of NADPH oxidase 1 gene and upregulates production of superoxide anion by human large intestinal epithelial cells. Am J Physiol Cell Physiol 290(2):C433–C443
116. de Mochel NS, Seronello S, Wang SH, Ito C, Zheng JX, Liang TJ, Lambeth JD, Choi J (2010) Hepatocyte NAD(P)H oxidases as an endogenous source of reactive oxygen species during hepatitis C virus infection. Hepatology 52(1):47–59. doi:10.1002/hep.23671
117. Casini A, Ceni E, Salzano R, Biondi P, Parola M, Galli A, Foschi M, Caligiuri A, Pinzani M, Surrenti C (1997) Neutrophil-derived superoxide anion induces lipid peroxidation and stimulates collagen synthesis in human hepatic stellate cells: role of nitric oxide. Hepatology 25(2):361–367
118. Nieto N (2006) Oxidative-stress and IL-6 mediate the fibrogenic effects of Kupffer cells on stellate cells. Hepatology 44(6):1487–1501
119. Leonarduzzi G, Scavazza A, Biasi F, Chiarpotto E, Camandola S, Vogel S, Dargel R, Poli G (1997) The lipid peroxidation end product 4-hydroxy-2,3-nonenal up-regulates transforming growth factor beta1 expression in the macrophage lineage: a link between oxidative injury and fibrosclerosis. FASEB J 11(11):851–857
120. Camandola S, Aragno M, Cutrin JC, Tamagno E, Danni O, Chiarpotto E, Parola M, Leonarduzzi G, Biasi F, Poli G (1999) Liver AP-1 activation due to carbon tetrachloride is potentiated by 1,2-dibromoethane but is inhibited by alpha-tocopherol or gadolinium chloride. Free Radic Biol Med 26(9–10):1108–1116
121. Marra F, DeFranco R, Grappone C, Parola M, Milani S, Leonarduzzi G, Pastacaldi S, Wenzel UO, Pinzani M, Dianzani MU, Laffi G, Gentilini P (1999) Expression of monocyte chemotactic protein-1 precedes monocyte recruitment in a rat model of acute liver injury, and is modulated by vitamin E. J Investig Med 47(1):66–75
122. Zamara E, Galastri S, Aleffi S, Petrai I, Aragno M, Mastrocola R, Novo E, Bertolani C, Milani S, Vizzutti F, Vercelli A, Pinzani M, Laffi G, LaVilla G, Parola M, Marra F (2007) Prevention of severe toxic liver injury and oxidative stress in MCP-1-deficient mice. J Hepatol 46(2):230–238
123. Luckey SW, Petersen DR (2001) Activation of Kupffer cells during the course of carbon tetrachloride-induced liver injury and fibrosis in rats. Exp Mol Pathol 71(3):226–240
124. Wang H, Wei W, Wang NP, Gui SY, Wu L, Sun WY, Xu SY (2005) Melatonin ameliorates carbon tetrachloride-induced hepatic fibrogenesis in rats via inhibition of oxidative stress. Life Sci 77(15):1902–1915
125. Paik YH, Iwaisako K, Seki E, Inokuchi S, Schnabl B, Osterreicher CH, Kisseleva T, Brenner DA (2011) The nicotinamide adenine dinucleotide phosphate oxidase (NOX) homologues NOX1 and NOX2/gp91(phox) mediate hepatic fibrosis in mice. Hepatology 53(5):1730–1741. doi:10.1002/hep.24281
126. De Minicis S, Seki E, Paik YH, Osterreicher CH, Kodama Y, Kluwe J, Torozzi L, Miyai K, Benedetti A, Schwabe RF, Brenner DA (2010) Role and cellular source of nicotinamide adenine dinucleotide phosphate oxidase in hepatic fibrosis. Hepatology 52(4):1420–1430. doi:10.1002/hep.23804
127. Wang ZL, Deng CY, Zheng H, Xie CF, Wang XH, Luo YF, Chen ZZ, Cheng P, Chen LJ (2012) (Z)2-(5-(4-methoxybenzylidene)-2, 4-dioxothiazolidin-3-yl) acetic acid protects rats from CCl(4) -induced liver injury. J Gastroenterol Hepatol 27(5):966–973. doi:10.1111/j.1440-1746.2011.06913.x
128. Simonetti RG, Cammà C, Fiorello F, Politi F, D'Amico G, Pagliaro L (1991) Hepatocellular carcinoma. A worldwide problem and the major risk factors. Dig Dis Sci 36(7):962–972
129. Teufelhofer O, Parzefall W, Kainzbauer E, Ferk F, Freiler C, Knasmüller S, Elbling L, Thurman R, Schulte-Hermann R (2005) Superoxide generation from Kupffer cells contributes to hepatocarcinogenesis: studies on NADPH oxidase knockout mice. Carcinogenesis 26(2):319–329
130. Bernal W, Auzinger G, Dhawan A, Wendon J (2010) Acute liver failure. Lancet 376(9736):190–201. doi:10.1016/S0140-6736(10)60274-7

131. Michael SL, Pumford NR, Mayeux PR, Niesman MR, Hinson JA (1999) Pretreatment of mice with macrophage inactivators decreases acetaminophen hepatotoxicity and the formation of reactive oxygen and nitrogen species. Hepatology 30(1):186–195
132. Andrés D, Sánchez-Reus I, Bautista M, Cascales M (2003) Depletion of Kupffer cell function by gadolinium chloride attenuates thioacetamide-induced hepatotoxicity. Expression of metallothionein and HSP70. Biochem Pharmacol 66(6):917–926
133. Dambach DM, Durham SK, Laskin JD, Laskin DL (2006) Distinct roles of NF-kappaB p50 in the regulation of acetaminophen-induced inflammatory mediator production and hepatotoxicity. Toxicol Appl Pharmacol 211(2):157–165
134. Al-Belooshi T, John A, Tariq S, Al-Otaiba A, Raza H (2010) Increased mitochondrial stress and modulation of mitochondrial respiratory enzyme activities in acetaminophen-induced toxicity in mouse macrophage cells. Food Chem Toxicol 48(10):2624–2632. doi:10.1016/j.fct.2010.06.031
135. Chen Q, Xue Y, Sun J (2013) Kupffer cell-mediated hepatic injury induced by silica nanoparticles in vitro and in vivo. Int J Nanomedicine 8:1129–1140. doi:10.2147/IJN.S42242
136. Zhong Z, Froh M, Wheeler MD, Smutney O, Lehmann TG, Thurman RG (2002) Viral gene delivery of superoxide dismutase attenuates experimental cholestasis-induced liver fibrosis in the rat. Gene Ther 9(3):183–191
137. Gujral JS, Liu J, Farhood A, Hinson JA, Jaeschke H (2004) Functional importance of ICAM-1 in the mechanism of neutrophil-induced liver injury in bile duct-ligated mice. Am J Physiol Gastrointest Liver Physiol 286(3):G499–G507
138. Schauer RJ, Gerbes AL, Vonier D, op den Winkel M, Fraunberger P, Bilzer M (2003) Induction of cellular resistance against Kupffer cell-derived oxidant stress: a novel concept of hepatoprotection by ischemic preconditioning. Hepatology 37(2):286–295
139. Rüdiger HA, Graf R, Clavien PA (2003) Sub-lethal oxidative stress triggers the protective effects of ischemic preconditioning in the mouse liver. J Hepatol 39(6):972–977
140. Kensler TW, Wakabayashi N, Biswal S (2007) Cell survival responses to environmental stresses via the Keap1-Nrf2-ARE pathway. Annu Rev Pharmacol Toxicol 47:89–116
141. Tsuchihashi S, Fondevila C, Kupiec-Weglinski JW (2004) Heme oxygenase system in ischemia and reperfusion injury. Ann Transplant 9(1):84–87
142. Kiemer AK, Gerwig T, Gerbes AL, Meissner H, Bilzer M, Vollmar AM (2003) Kupffer-cell specific induction of heme oxygenase 1 (hsp32) by the atrial natriuretic peptide–role of cGMP. J Hepatol 38(4):490–498
143. Yeligar SM, Machida K, Kalra VK (2010) Ethanol-induced HO-1 and NQO1 are differentially regulated by HIF-1alpha and Nrf2 to attenuate inflammatory cytokine expression. J Biol Chem 285(46):35359–35373. doi:10.1074/jbc.M110.138636
144. Bautista AP (2002) Acute ethanol binge followed by withdrawal regulates production of reactive oxygen species and cytokine-induced neutrophil chemoattractant and liver injury during reperfusion after hepatic ischemia. Antioxid Redox Signal 4(5):721–731

Chapter 7
Oxidative Stress and Liver Ischemia–Reperfusion Injury

Carmen García-Ruíz, Albert Morales, and José C. Fernández-Checa

7.1 Introduction

Liver damage induced by ischemia/reperfusion (I/R) has an important impact in different clinical settings such as liver resection, transplantation, trauma, or hemorrhagic shock. Hepatocyte ischemia during liver transplantation, resection, and shock causes anoxia, depletion of glycolytic substrates, loss of adenosine triphosphate (ATP), and acidosis. When blood flow returns to hepatocytes, the normal oxygen concentration and acid–base balance are restored; however, paradoxically,

C. García-Ruíz (✉)
Department of Cell Death and Proliferation, Institute of Biomedical Research of Barcelona (IIBB), Consejo Superior Investigaciones Científicas (CSIC) and Liver Unit-Hospital Clinic and Networked Biomedical Research Center for Hepatic and Digestive Diseases, Barcelona 08036, Spain

Southern California Research Center for ALPD and Cirrhosis, Keck School of Medicine of the University of Southern California, Los Angeles, CA, USA
e-mail: carmen.garcia@iibb.csic.es

J.C. Fernández-Checa (✉)
Department of Cell Death and Proliferation, Institute of Biomedical Research of Barcelona (IIBB), Spanish National Research Council and Liver Unit-Hospital Clinic and CIBEREHD, Barcelona 08036, Spain

Southern California Research Center for ALPD and Cirrhosis, Keck School of Medicine of the University of Southern California, Los Angeles, CA, USA
e-mail: checa229@yahoo.com

A. Morales
Department of Cell Death and Proliferation, Institute of Biomedical Research of Barcelona - Spanish National Research Council and Liver Unit-Hospital Clinic and Networked Biomedical Research Center for Hepatic and Digestive Diseases, Barcelona 08036, Spain

E. Albano, M. Parola (eds.), *Studies on Hepatic Disorders*, Oxidative Stress in Applied Basic Research and Clinical Practice, DOI 10.1007/978-3-319-15539-5_7

hepatocyte injury is prominent upon reperfusion. Hence, a better understanding of the molecular events contributing to hepatic I/R injury may be of significant relevance to ensure successful graft function. While this process is multifactorial, cumulating evidence indicates that mitochondrial dysfunction and overgeneration of reactive oxygen species (ROS) and oxidants and compromised antioxidant defense are crucial events in liver I/R damage. In addition to the contribution of inflammatory cells in the overgeneration of ROS, recent evidence point to mitochondria as another critical source of ROS generation, which contribute to hepatic I/R injury. This event is important as it opens the possibility to design specific antioxidant strategies to target the mitochondrial contribution of ROS generation as a valid approach to protect the liver against I/R-mediated damage. In the following sections, we will briefly describe the current evidence describing the major reactive species generated, their source, and contribution to hepatic I/R damage and potential strategies aimed to improve function in suboptimal grafts, such as those with steatosis.

7.2 Reactive Oxygen Species and Free Radicals in Hepatic I/R Injury

The toxicity of oxygen remained unclear until the seminal contribution of Rebeca Gerschman in the 1950s proved for the first time the existence of free radicals in her "oxygen poisoning" article published in Science [22]. After her findings of the synergism between radiation and hyperbaric oxygen in decreasing the survival of exposed mice, she proposed that the toxicity of oxygen was due to its partially reduced forms. Free radicals are molecules or atoms (e.g., oxygen, nitrogen) with unpaired electrons. Oxygen atoms are an intrinsic part of most of the molecules that make up our bodies. A molecule of *atmospheric* oxygen is composed of two oxygen atoms bound together chemically (O_2). This molecule is relatively stable, although inhaled oxygen does participate in metabolic reactions within the body's cells. Through several enzymatic and no enzymatic processes that routinely occur in cells, O_2 accepts single electrons to form ROS. These species encompass a variety of molecules and free radicals that derived from molecular oxygen. The consequences of ROS on cellular processes depend on the strength and duration of exposure, as well as the context of the exposure. ROS are highly reactive, because they are chemically or electrically unstable, and are known to be involved in a number of physiological and pathological processes, liver diseases, and hepatic I/R injury. Free radicals are very reactive species, which can derive from oxygen or nitrogen that when generated in a controlled fashion may regulate signaling pathways and control cellular process, but when generated in excess promote cell and tissue injury [56]. In the case of ROS, mitochondria are an important source of these species, which generate in the mitochondrial electron transport chain (ETC) where oxygen consumption is used to generate ROS. The estimation of oxygen consumed in the ETC used for ROS generation is about 2 %. The liver is a metabolically active and highly aerobic organ whose viability depends on the accessibility to oxygen. Oxygen consumption in the liver is 100–150 μmol/h/g of wet weight. Sinusoidal blood flow and hepatic metabolism create gradients of oxygen, metabolites, and hormones between periportal and pericentral regions of the liver lobule contributing to the development of biochemical differences between hepatocytes in different regions. Although the liver is fairly vulnerable to hypoxic injury, its dual vascularization provides some protection against hypoxia. Prolonged liver ischemia is characterized by reducing tissue oxygenation and a lack of ATP in hepatocytes, Kupffer and sinusoidal endothelial cells (SECs) [77], with a transition to activation of anaerobic metabolic pathways, which cannot maintain cellular function for prolonged periods. During anoxia and ischemia, glycolytic ATP formation replaces, in part, ATP lost from oxidative phosphorylation and delays anoxic hepatocellular cell death. In low flow states, pericentral regions of the liver become anoxic, whereas periportal areas remain normoxic.

7.2.1 Major ROS Generation and Defense During Hepatic I/R

Ischemia implies the loss of blood supply, which can be relative or absolute. Tissue injury and stress in ischemia begins as tissue oxygen levels reach very low levels [38]. Depletion on intracellular and extracellular ATP during ischemia results in increased ATP degradation products, including adenosine, hypoxanthine and xanthine, and a shift towards anaerobic metabolism. Although hypoxia produces an overgeneration of ROS [26], and ischemia induces necrotic cell death after 30 min from oxygen deprivation, it is the reperfusion that paradoxically initiates a cascade of pathways that cause further cellular injury after prolonged ischemia. Reoxygenation of hypoxic liver mediates liver injury through different mechanisms in three phases, namely early, intermediate, and late, whose importance is dependent on the length of the ischemia period [47]. In the very early phase, mitochondrial dysfunction leads to hepatocellular cell death and involves mitochondrial permeability transition (MPT). When MPT occurs, mitochondria begin to swell and the mitochondrial membrane potential collapses leading to the ATP depletion, release of cytochrome c and apoptosis. The translocation of iron from lysosomes into mitochondria has been shown to synergistically promote ROS formation, MPT opening, and cell death [88]. Thus, damaged mitochondria produce progressively greater amounts of ROS, which in turn, impairs mitochondrial function, thereby increasing the reduction of ATP production and compromised cell function. In the intermediate phase, which lasts up to 6 h after restoration of the blood flow, Kupffer cells mediate cell death (apoptosis and necrosis), and the release of free radicals, chemokines, and proinflammatory cytokines such as tumor necrosis factor (TNF) and IL-1β. Inflammatory cytokines, e.g., TNF, can also further stimulate generation of further bursts of ROS, thus amplifying the generation of ROS from different sources. TNF is of particular importance by triggering the inflammatory cascade and inducing the expression of adhesion molecules (i.e., ICAM-1 and VCAM-1) and further promote the adhesion, migration, and chemotaxis of neutrophils, resulting in increased liver damage. Activation of neutrophils can directly damage liver cells by the release of oxidants and proteases after reperfusion. TNF may directly cause liver damage by binding receptors on the surface of the liver cells and induce overproduction of ROS, activation of nuclear factor-κB (NF-κB), MAPK, and JNK. In addition, ischemia reperfusion injury causes calcium release from the endoplasmic reticulum [76], leading to the activation of the unfolded protein response (UPR), a signal transduction pathway, that further generates ROS, contributing to apoptosis and necrosis. In this regard, it has been shown that knockout mice deficient in the ER protein BI-1 (Bax inhibitor 1) exhibit increased histological and biochemical liver injury when subjected to transient blood flow occlusion [2]. In the late phase, between 6 and 24 h after reperfusion, injury is caused by neutrophil activation, involving margination and transendothelial migration of neutrophils in response to Kupffer cell-generated chemoattractants and release of toxic mediators such as ROS and proteases (e.g., cathepsins). Oxidative stress occurs when the rate of ROS/RNS generation exceeds the capacity of the cell for their removal. In addition to the concept that an unbalance between oxidants and antioxidants triggers oxidative stress, a strict equilibrium between antioxidants ensures defense against ROS [56]. For instance, depletion of glutathione (GSH) contributes to cell injury in the face of increased superoxide scavenging due to the accumulation of potent oxidants and free radicals [93]. Aerobic metabolism releases ROS, which under normal circumstances are neutralized through diverse antioxidant mechanisms. However, under stress conditions the balance between ROS and antioxidants shifts towards the former, resulting in oxidative stress and cytotoxicity [92]. Mitochondrial proteins are particularly sensitive to oxidative (ROS/RNS) stress. Inactivation of mitochondrial proteins by oxidants has been shown to contribute to dysfunction and injury during hepatic ischemia reperfusion [61]. Superoxide anion, hydrogen peroxide, and peroxynitrite are the main ROS species involved in I/R injury.

7.2.1.1 Superoxide Formation

Mitochondria are a main source of ROS (Fig. 7.1). ROS are produced during oxidative metabolism mainly by the complexes I and III of the ETC [4, 93]. Most of the oxygen consumed by aerobic organisms is reduced to water by the enzyme cytochrome c oxidase in the terminal reaction of the mitochondrial respiratory chain. However, a small amount (1–2 %) of the oxygen molecules is converted to superoxide anion radical by other respiratory components [9]. Superoxide can potentially be generated at a number of different sites, mainly complex I (NADH-ubiquinone oxidoreductase) and complex III (ubisemiquinone), but also by other mitochondrial enzyme complexes including the dihydrolipoamide dehydrogenase containing FAD-linked pyruvate and α-ketoglutarate dehydrogenase complexes, as well as the flavoenzymes α-glycerophosphate dehydrogenase, and the electron-transferring flavoprotein Q oxidoreductase (ETFQOR) of fatty acid β-oxidation [37, 82, 83]. Complex I is an important source of superoxide formation and one of the major sites where premature electron leakage to oxygen occurs. Complex I produces superoxide during forward electron flow (NADH-oxidizing) [30, 44] or during reverse electron transport from the ubiquinone (Q) pool (NAD^+-reducing) at sufficiently high protonmotive force [46]. Inhibitors of electron transfer to ubiquinone (IQ inhibitors), such

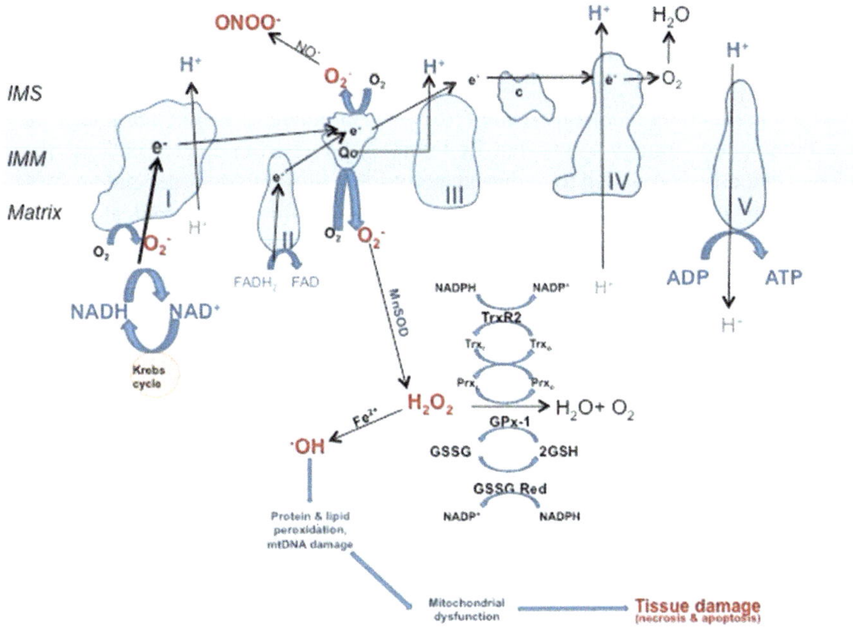

Fig. 7.1 Mitochondrial sites of ROS generation and defense. The mitochondrial respiratory chain is a major source of ROS due to the leakage of electrons to molecular oxygen to form superoxide anion. This species is then dismutated by MnSOD to form hydrogen peroxide, which is detoxified by several antioxidant systems, most predominantly by the GSH redox cycle but also by the peroxiredoxin/thioredoxin couple. The reaction of superoxide with NO generates peroxynitrite, which contributes to mitochondrial dysfunction by targeting tyrosine residues of affected proteins

as rotenone and piericidin, prevent electron escape from the complex and increase superoxide production by complex I under conditions of high NADH/NAD$^+$. During forward ETC, the production of superoxide is very low (about 0.1 % of the total oxygen consumed during electron flow). While in the reverse flow of electron in which electron transport occurs in a reverse direction to reduce NAD$^+$ to NADH, the rate of superoxide formation is high, as much as 5 % of the electron flow is directed towards complex I. The ratio of NADH to NAD$^+$ determines the rate of superoxide formation. Superoxide formation by complexes I and III in the respiratory chain is a nonenzymatic process. This activity should increase by mass action when the electron carriers are highly reduced or in the presence of mitochondrial inhibitors. Superoxide production by mitochondrial electron carriers should also increase with oxygen concentration since cytochrome c oxidase, with a K_m for oxygen in the submicromolar range, is basically saturated under physiological conditions and will not compete for the extra oxygen. Electrons from substrate oxidation are carried to complex III via the reduction of ubiquinone (Q) to ubiquinol (QH2). At complex III, superoxide is produced at the Qo site (the ubisemiquinone located on the outer side of the inner membrane). Inhibitors of complex III that prevent Q binding at the Qo site, such as stigmatellin and myxothiazol, block electron entry into the complex and prevent ROS production even in the presence of Qi inhibitors [64]. A role for complex III in mediating autophagy has been described after its inhibition by myxothiazol [52]. A recent study by Sun et al. [86] suggests that autophagy suppressed ischemic liver injury by reducing ROS-induced necrosis, pointing to new strategies in the pretreatment of liver injury, and restoration or enhancement of autophagy may be a therapeutic strategy to prevent IR injury in livers from elderly patients [94]. O_2^- is efficiently dismutated to hydrogen peroxide (H_2O_2) by Mn^{2+}-dependent superoxide dismutase (Mn-SOD). In fact, superoxide is a mild reactive radical and considered a primary ROS, which can lead to the generation of more reactive species, and secondary ROS, such as hydrogen peroxide.

7.2.1.2 Hydrogen Peroxide

Hydrogen peroxide, a potent oxidant, is produced as a secondary product via superoxide dismutation by Mn-SOD. The stoichiometry between superoxide and hydrogen peroxide is close to 2.0 [10]. Hydrogen peroxide in mitochondria is detoxified to water by several antioxidant strategies, predominantly by peroxiredoxin-III and thioredoxin-2 as well as by GSH peroxidases, which use reduced GSH as a cofactor [56]. GSH is one of the most important intracellular defenses against the deleterious effects of ROS and electrophiles. The antioxidant function of GSH is determined by the redox-active thiol of cysteine that becomes oxidized when GSH reduces target molecules [69]. Upon reaction with ROS or electrophiles, GSH becomes oxidized to GSSG, which can be reduced to GSH by the GSSG reductase (GR). Thus, the GSH/GSSG ratio reflects the oxidative state and can interact with redox couples to maintain appropriate redox balance in the cell. In addition to GSH and the GSH redox cycle there are a number of small molecule antioxidants that play a role in detoxification. Vitamin C or ascorbic acid is a water-soluble molecule capable of reducing ROS, while vitamin E (α-tocopherol) is a lipid-soluble molecule that has been suggested as playing a similar role in membranes. In liver mitochondria, GSH is the only defense against hydrogen peroxide, because this organelle lacks catalase. As described below, the levels of mitochondrial GSH (mGSH) in relation to those of Manganese superoxide dismutase (MnSOD) is an important factor determining the therapeutic benefit of superoxide scavenging by regulating efficient detoxification of hydrogen peroxide. Indeed, hydrogen peroxide accumulation can be the source of the hydroxyl radical, a very reactive radical with a short half-life in vivo ($\sim 10^{-9}$ s), which derives from hydrogen peroxide and superoxide anion in the presence of reduced transition metals (i.e., Fe^{2+} or Cu^+) in the Fenton reaction.

7.2.1.3 Peroxynitrite

Besides ROS, nitric oxide (NO) is the source of other potent reactive nitrogen species (RNS) that also participate in pathophysiology of hepatic I/R injury. NO combines with superoxide to form the RNS peroxynitrite or the radical nitrogen dioxide. Early generation of ROS and RNS during reperfusion is considered to play a pivotal role in initiating a chain of deleterious cellular responses leading to inflammation and cell death, which eventually culminate in target organ dysfunction and failure [63]. An important early mechanism of liver failure is the marked sinusoid vasoconstriction that develops during reperfusion, as a consequence of an imbalance between vasoconstrictors, especially endothelins, and vasodilators, essentially NO. In the liver, NO is normally produced exclusively by eNOS expressed by sinusoid endothelial cells. In liver ischemia reperfusion injury, a significant reduction of eNOS activity has been demonstrated both in humans and animals, precipitating in sinusoidal vasoconstriction [14]. Increased superoxide generation consumes NO, contributing not only to peroxynitrite formation but also to the further aggravation of vasoconstriction. As such, a viable approach to attenuate hepatic I/R injury is the scavenging of superoxide, which is expected to result in the sparing of NO and in the reduction of peroxynitrite, a potent reactive species with a short half-life responsible for many damaging modifications of tyrosine residues in mitochondrial target proteins [93]. As mentioned above, ischemia reperfusion injury also rapidly activates Kupffer cells, which then produce proinflammatory cytokines, free radicals, oxidants, and large amounts of NO due to iNOS expression. These events result in the activation of neutrophils, endothelial cells, and hepatocytes, which further release toxic levels of oxidant species and iNOS-derived NO in the delayed phase of IR injury. The pathogenic role of oxygen-centered radicals in this setting has been well established by the significant reduction of liver damage provided by various antioxidant strategies.

Although NO is crucial for peroxynitrite and other species, whether hepatic mitochondria generate NO or not is controversial. Although the existence of mitochondrial nitric oxide synthase (NOS) has been described in mitochondrial fractions from different organs, recent evidence in rat liver mitochondria has questioned the existence of mtNOS, minimizing the contribution of in situ NO generation within mitochondria to the formation of peroxynitrite [91]. However, since NO is freely diffusible across membranes, it is possible that the mitochondrial production of peroxynitrite may derive from extramitochondrial NO diffusing into mitochondria to react with superoxide generated by ETC.

7.2.2 *Sources of ROS in Hepatic I/R Injury*

Cellular sources for ROS production after IR injury include mitochondrial metabolism, hepatocyte-derived xanthine oxidase activation, and Kupffer- and SEC-associated NADPH oxidase (Fig. 7.2).

7.2.2.1 Mitochondrial Metabolism

Mitochondria are intracellular organelles that play important functions including cell signaling and apoptosis. As mentioned before, the physiological role of mitochondria is energy generation in the form of ATP within the ETC using oxygen as

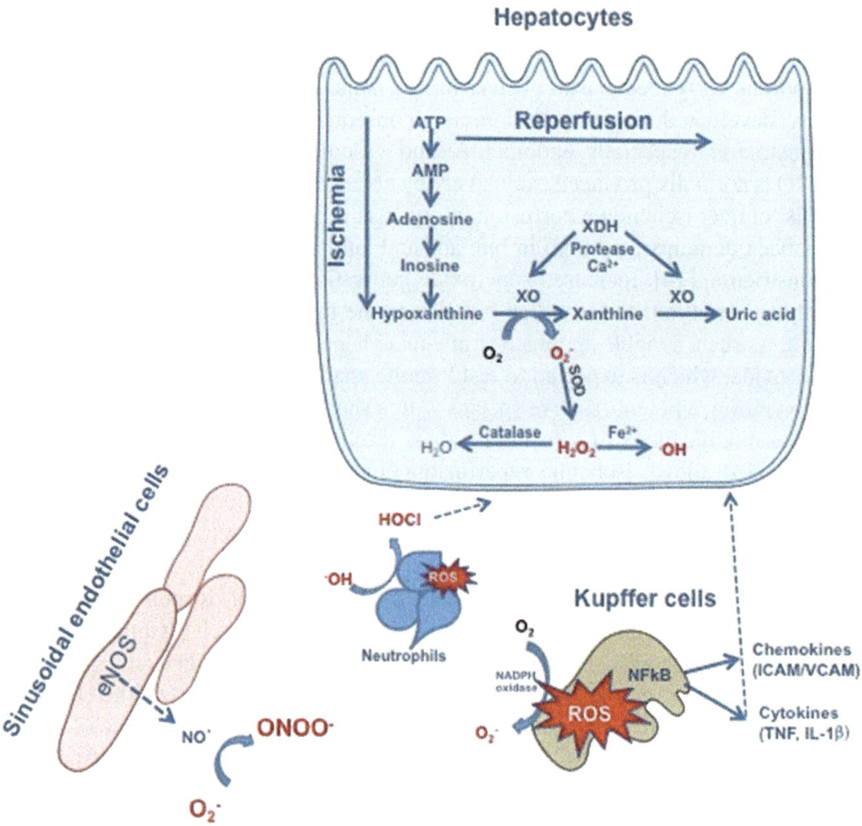

Fig. 7.2 Extramitochondrial sources of ROS production. Besides the ETC of mitochondria, other liver cells contribute to the generation of ROS during hepatic I/R. As depicted, hepatocytes are a major source of ROS generation via XO activation. In addition, endothelial cells, neutrophils, and Kupffer cells generated diverses ROS species, which contribute to liver damage during I/R

the ultimate electron acceptor. The movement of electrons along the ETC generates a protonmotive force, which is transduced to phosphorylate ADP to form ATP. The electron transport is a very well organized process that involves several redox centers assembled in four protein complexes embedded in the inner mitochondrial membrane. The main sites of side-reaction with oxygen are localized at complexes I and III, and as recently shown, also at complex II [66]. Complex I is comprised of 46 different subunits; its complexity contributes to the still unknown exact sites of superoxide generation. Nevertheless, ubisemiquinone and flavin mononucleotide have been suggested as superoxide generators [46, 87]. At complex III, ubisemiquinone or cytochrome b has been considered as the source of superoxide production. The Q-binding pocket of mGPDH has been suggested as a H_2O_2 source at complex II [66]. From the several sites that can generate superoxide in the mitochondrial matrix, only the superoxide produced at complex III appears to be released both into the matrix and the intermembrane space (IMS) [72]. This spatial difference (matrix vs. IMS) may determine whether mitochondrial superoxide reaches the cytosol or not. The anionic nature of superoxide and the fact that it is mostly produced in the mitochondrial matrix determine that the bulk of antioxidant defenses to neutralize superoxide and other ROS reside in the matrix (Fig. 7.1).

Although it is highly recognized that mitochondria are targets for the ROS/RNS generation during I/R, their contribution to ROS generation in hepatic I/R injury has remained elusive [36]. In line with this role, it has been shown that targeting antioxidants to mitochondria, such as Mito-Q or Mito-CP dose-dependently attenuated I/R-induced liver dysfunction, the early and delayed oxidative and nitrative stress response (HNE/carbonyl adducts, malondialdehyde, 8-OHdG, and 3-nitrotyrosine formation), and mitochondrial and histopathological injury/dysfunction, as well as delayed inflammatory cell infiltration and cell death [63]. These findings support the view that these antioxidants can attenuate mitochondrial dysfunction and ROS/RNS generation in vivo under hepatic I/R, most probably by attenuating lipid peroxidation and the subsequent deleterious cascade of events associated with hepatic I/R. Emerging evidence indicates that mitochondrial ROS/RNS formation may directly or indirectly control expression of NAD(P)H oxidases and hence recruit or modulate the degree of the burst of ROS generation caused by Kupffer and inflammatory infiltrates [15]. Moreover, in line with these findings two-photon confocal microscopy analyses indicated increased mitochondrial ROS generation in vivo during hepatic I/R injury that paralleled the loss of mitochondrial membrane potential, thus further indicating that mitochondria are a critical source and target of ROS contributing to liver I/R damage.

7.2.2.2 Extramitochondrial Sources

The xanthine oxidase has been identified as an important ROS source during the reperfusion stage. In fact the protective effect of allopurinol in in vivo and in vitro studies has been reported in hepatic I/R, involving the xanthine oxidase-induced oxidant stress in the pathogenesis of the injury. Liver contains high concentrations of xanthine oxidase (XO) and xanthine dehydrogenase (XDH). During prolonged hepatic ischemia periods, XDH is converted in XO either by reversible oxidation of cysteine residues or irreversibly with the involvement of Ca^{2+} stimulated proteases. In this scenario, XO no longer reduces NAD^{+} but rather transfer electrons to molecular oxygen producing the superoxide radical. Upon reoxygenation of the tissue, oxygen enters the vicious cycle producing a burst of superoxide production.

Hepatocytes and SEC are the two main cell types that are injured after I/R. Hepatocytes are more sensitive to warm (37 °C) ischemic injury while SECs are more sensitive to cold ischemia (4 °C) found in cold preservation of donor liver grafts before transplantation. There is also evidence that under cold ischemia, Kupffer cells contribute to ROS generation. The conversion of XDH to XOD following cold storage is slow in SEC and hepatocytes but much faster in Kupffer cells. Moreover, increased vascular oxidant stress under ischemia is attenuated by inactivation of Kupffer cells but not by allopurinol [34]. In line with this evidence, treatment with gadolinium chloride has been shown to reduce the capacity of Kupffer cells to generate ROS and protect against hepatic I/R injury [12]. A note of caution regarding the contribution of Kupffer cells to ROS generation implies that the stimulatory state of Kupffer cells after I/R depends on the duration of ischemia, and it may also differ between warm and cold ischemia. NADPH oxidase is the main source of superoxide formation by neutrophils, which also include generation of hydrogen peroxide and hydroxyl radical. Activated neutrophils also release elastase, cathepsin G, and hydrolytic enzymes that directly inflict damage to hepatocytes [33]. Neutrophil recruitment into the liver occurs during the first hours after reperfusion, however, at this time period, neutrophils do not generate significant amount of ROS but are primed for their generation, which is maximal between 6 and 24 h after reperfusion, a finding that correlates with a neutrophil-mediated injury phase [33]. Consistent with this role, the main cytotoxic ROS induced by neutrophils include hydrogen peroxide and hypochlorous acid [25, 28].

7.3 Dual Role of ROS in Hepatic I/R Injury: NF-κB Activation and Inflammatory Cytokines Generation

Besides the ability of ROS to inflict cell damage through attack to biomembranes, enzymes, proteins, and nucleic acids, ROS also play a signaling function recruiting the activation of downstream target of relevance for hepatic I/R injury [6, 14, 23, 32, 47, 58, 84]. Some of these events regulated by (mitochondrial) ROS generation include NF-κB activation, JNK activation, and the activation of Toll-like receptors (TLR). This family of receptors is known to play a fundamental role in innate immunity and inflammation, and recent evidence indicated that TLR, particularly TLR4, plays a key role in hepatic I/R injury [98]. Furthermore, TLR4 deficiency in the donor organ reduces I/R injury in a murine liver transplantation model [80]. Using differential proteomic analysis, Scaloni et al. [75], described early protein targets of I/R during liver transplantation. This approach uncovered the overoxidation of the active-site thiol of PrxI into sulfonic acid, indicating that the redox state of Prx may be an early and sensitive marker of oxidative stress in hepatic I/R and liver transplantation [90]. Although these factors are useful in the design of biochemical strategies to modulate hepatic I/R injury, ischemic preconditioning (IPC) is a surgical strategy of therapeutic value during I/R, involving a number of mechanisms, including the redox-sensitive mitochondrial ATP-sensitive K^+ channels that control ROS production [18]. Moreover, TNF has been identified as a key player in hepatic I/R damage [73]. In this regard, IPC with TNF/Fas reduces ischemic injury in the liver through a compensatory mechanism modulating the activation of oxidative stress [35]. Consistent with the role of TNF/Fas in acid sphingomyelinase (ASMase) activation, it was shown that ASMase null mice are resistant to hepatic I/R injury [50]. Indeed, ASMase-induced ceramide generation targeted mitochondria through JNK and BimL activation. Thus, in addition to the previously described mechanisms and consistent with the role of ceramide in recruiting the mitochondrial pathway of cell death, ASMase emerges as a potential target for intervention against hepatic I/R.

The role of NF-κB and ROS in hepatic I/R injury is quite intriguing, with NF-κB playing a controversial role because of its dual action in the induction of survival/inflammatory genes. For instance, hepatic NF-κB activation has been shown to ameliorate hepatic I/R injury, improving orthotopic liver transplantation, whereas NF-κB inactivation protects against hepatic I/R [43, 85]. These opposing results can be explained on the basis of different degrees of residual NF-κB activation. Moreover, ROS and oxidative stress exert a dual effect on NF-κB activation, which involves its release from the inhibitory moiety in the cytosol, its nuclear translocation, and the subsequent transactivation of target genes. Recent data showed that NF-κB transactivation is impaired in hepatocytes after GSH depletion, involving IKK-dependent and -independent mechanisms [51]. In addition, GSSG formation following ROS overgeneration inactivates NF-κB [16]. In addition to this dual role, the activation of NF-κB in different population of cells in the liver also may contribute to the dichotomy of NF-κB in I/R-induced liver injury. Recent findings showed that ROS mediate liver injury selective parenchymal inactivation of NF-κB in I/R [49]. Thus, activation of NF-κB after I/R in hepatocytes occurs through a non-canonical pathway without IκB degradation requiring Src activation, NF-κB activation in Kupffer cells involves a canonical pathway mediated by TLR. These findings have important implications, suggesting that strategies that increase GSH in hepatocytes would maintain NF-κB activation and hence the expression of survival genes without changing the induction of proinflammatory cytokines, which may thus contribute to the protection against I/R damage. In contrast, marked GSH depletion contributes to the downregulation of survival genes and in the enhanced generation of TNF/IL-1β, which translated into exaggerated liver damage after I/R.

7.4 Fatty Liver Sensitizes to I/R-Mediated Injury

Hepatic I/R is an intrinsic condition of diverse clinical settings, including liver transplantation, which is indicated for the treatment for chronic end stage liver disease and acute liver failure. As the number or quality of organs available for transplantation is insufficient to meet the demand for liver transplantation, the use of suboptimal quality grafts, such as steatotic livers, are increasingly being used in clinical practice [78]. However, this approach may compromise the overall success of liver transplantation due to the known sensitization of fatty liver to second hits, including I/R-mediated ROS generation leading to hepatocellular death [81]. Due to the rising prevalence of obesity and type II diabetes among the general population, fatty livers constitute an expanding pool of marginal grafts available for transplantation. Hepatic steatosis is known to impose an additional risk of primary graft dysfunction. Fat accumulation within hepatocytes increases cell volume resulting in a decreased sinusoidal space and impaired microcirculatory blood flow [27], even though the molecular mechanisms, underlying the susceptibility of fatty livers to ischemia/reperfusion injury, remain incompletely understood. In this regard, according to the morphology, size, and percentage of hepatocytes exhibiting cytoplasmic fat deposits, it has been reported that fatty livers with macrovesicular steatosis are more susceptible to I/R injury than those with microsteatosis, correlating with impaired hepatic perfusion and portal vein flow [17, 79]. Moreover, the contribution of microvesicular steatosis to graft dysfunction after transplantation remains unclear. Some studies showed that livers with moderate microsteatosis are nonfunctional after transplantation, while other reports discarded a role for microvesicular steatosis on graft or patient survival [20, 89, 97]. Regardless of the morphological appearance of fat, steatotic livers are more susceptible than nonsteatotic livers to lipid peroxidation [57] and fatty livers tolerate poorly cold and warm ischemia/reperfusion injury and are associated with high mortality after major surgery and transplantation [5]. Interestingly, IL-6 has been shown to protect obesity and alcohol-associated fatty liver against I/R injury [31]. Long-term and short-term IL-6 treatments elevate serum levels of triglyceride and cholesterol in both lean and ob/ob mice, which reflects the stimulation of hepatic triglyceride secretion by IL6 by a poorly understood mechanism. Importantly, IL-6 protected fatty liver from hepatic I/R injury by suppressing TNF expression and serum TNF levels, which is associated by downregulation of proinflammatory cytokines and reduced ROS production and oxidative stress. Furthermore, IL6 upregulated the expression of PPARα, which is known to induce fat oxidation via increased peroxisomal, microsomal, and particularly mitochondrial fatty acid oxidation that prevents steatosis. Overall, uncovering mechanisms that protect fatty livers against hepatic I/R injury may be of clinical relevance by increasing the number of suboptimal grafts available for live transplantation.

7.5 Cholesterol and Hepatic I/R Injury

Hypercholesterolemia is an important risk factor for the development of atherosclerosis and a number of diverse diseases. The chronic inflammatory nature of atherosclerotic lesions has been described as consisting of inflammatory cell infiltrates, enhanced cytokine production, and an increased expression of endothelial cell adhesion molecules. Emerging evidence indicates that in addition to vascular diseases hypercholesterolemia may be a sensitizing factor for hepatic I/R injury. As discussed above, hepatic steatosis is an important factor sensitizing to hepatic I/R injury, although the contribution of macro vs. microsteatosis to hepatic I/R injury is not well established [17, 20, 79, 89, 97] Recent data have examined the specific contribution of the individual lipids accumulating in fatty liver to I/R injury. Using nutritional models of hepatic steatosis characterized by predominant TG/FFA or cholesterol accumulation by feeding mice diet-deficient in choline (CD) or enriched in cholesterol (HC), respectively [55], it has been recently reported that increased cholesterol levels sensitized to hepatic I/R damage [48] In particular, HC-fed but not CD-fed mice exhibit mitochondrial cholesterol loading that correlated with increased mitochondrial depolarization and increased ROS generation due to selective mGSH depletion. Moreover, ob/ob mice, characterized by increased hepatic TG, FFA, and cholesterol levels, are as sensitive to I/R-mediated liver injury as mice fed the HC diet. Livers from ob/ob mice displayed increased StAR expression, a polypeptide responsible for the trafficking of cholesterol to the mitochondrial inner membrane [11, 60], and mitochondrial cholesterol accumulation, resulting in mGSH depletion. Moreover, rats fed a high-cholesterol diet for 12 weeks also exhibited increased susceptibility to 45 min of ischemia followed by 2 h reperfusion [42]. The contribution of cholesterol to normothermic ischemic injury is independent of the qualitative appearance of steatosis (microsteatosis vs. macrosteatosis). Livers with microsteatosis induced by CD diet feeding are less sensitive to I/R damage than those with macrosteatosis observed in ob/ob mice [79]. However, the steatosis of CD-fed mice is relatively microvesicular compared to that of ob/ob mice, but it is more macrovesicular than that of HC-fed mice, and yet HC-fed mice are more sensitive to hepatic I/R injury than CD-fed mice. Thus, these findings indicate that the relative qualitative aspect of steatosis plays a minor role in hepatic I/R susceptibility. The sensitizing effect of cholesterol to liver I/R injury is independent of the approach to induced hypercholesterolemia and of the species examined, as indicated in studies with LDLR null mice and in rabbits fed a high-cholesterol diet [21, 62]. The role of cholesterol in sensitizing to normothermic hepatic I/R injury is due to its trafficking and accumulation in mitochondria, resulting in subsequent mGSH depletion Mitochondrial cholesterol loading is known to perturb membrane dynamics affecting selective carriers, including the GSH mitochondrial transporter but not the *S*-adenosyl-L-methionine (SAM) carrier nor the ANT translocator [13, 19].

Cholesterol is synthesized in the ER from acetate in a multistep cascade that requires oxygen and energy. The precursor acetyl-CoA is first converted to 3-hydroxy-3-methylglutaryl-CoA (HMG-CoA) and then to mevalonate. The rate-limiting step of cholesterol biosynthesis is the conversion of HMG-CoA to mevalonate catalyzed by the HMG-CoA reductase (HMGCR), which is the target of statins (Fig. 7.3). Atorvastatin therefore has been shown to downregulate mitochondrial cholesterol loading and consequently prevents mGSH depletion. However, by targeting HMG-CoA R, statins may have broad effects far from just preventing cholesterol synthesis, as the conversion of mevalonate into farnesyl-PP can branch into non-sterol intermediates including isoprenoids, which can regulate a number of target proteins posttranslationally. For instance, it has been reported that atorvastatin protected obese mice against hepatic I/R damage by suppressing Toll-like receptor 4 and endothelial NOS activation [1]. Moreover, statins have been reported to activate the transcription factor Kruppel-like factor 2 (KLF2), which plays an important role in orchestrating a variety of vasoprotective pathways [24, 68]. Although these findings illustrate the pleiotropic effects of statins in hepatic I/R defense, the beneficial effect of atorvastatin in ob/ob mice against hepatic I/R injury due to the downregulation of cholesterol and its trafficking to mitochondria was further confirmed upon inhibition of squalene synthase, which catalyzes the first committed step in cholesterol synthesis (Fig. 7.3). Squalene synthase inhibition by YM-53601 paralleled the effects observed with atorvastatin with respect to mitochondrial cholesterol depletion and subsequent mGSH restoration, strongly

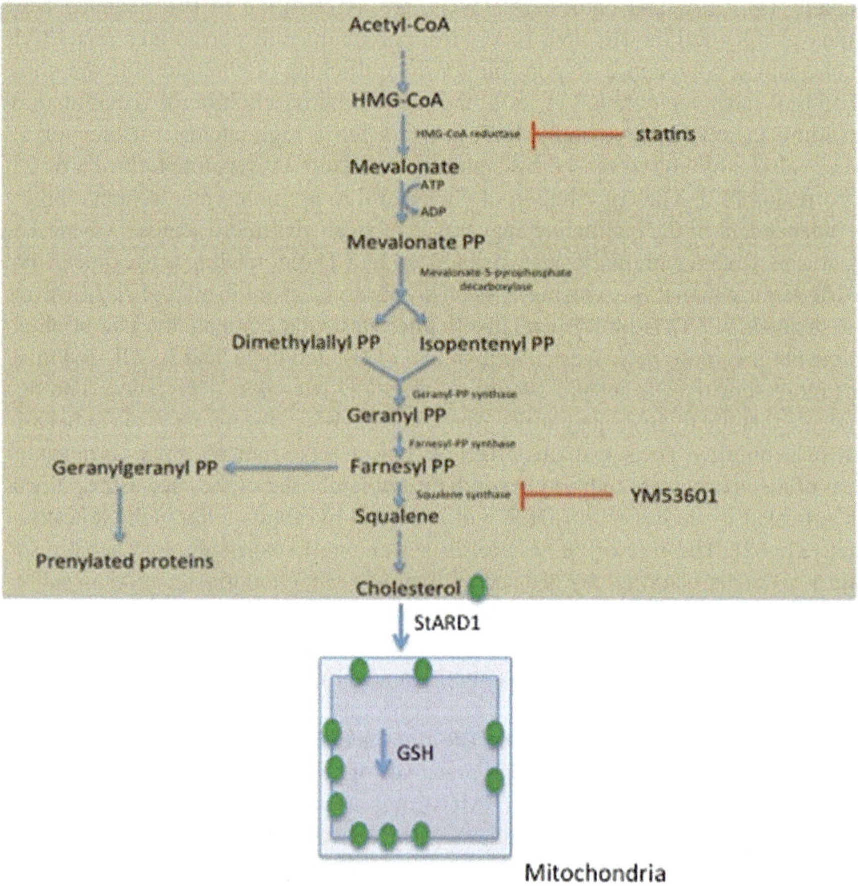

Fig. 7.3 The mevalonate pathway and the synthesis of cholesterol and isoprenoids. Cholesterol is synthesized from acetyl-CoA in a cascade of reactions in which the rate-limiting step is catalyzed by HMG-CoA reductase, the pharmacological target of statins. Generation of farnesyl-PP branches in the generation of geranylgeranyl PP, and squalene catalyzed by squalene synthases, which is the first committed step in the synthesis of cholesterol. Cholesterol then accumulates in mitochondria induced by StARD1 which impairs the transport of GSH into mitochondrial matrix resulting in its depletion

suggesting that the beneficial effects of atorvastatin in hepatic I/R susceptibility is due to cholesterol regulation. In line with the effects of atorvastatin, previous studies reported that simvastatin reduces the mortality and hepatic injury after hemorrhage/resuscitation in rats or during I/R [45, 70]. Interestingly, it has also been reported that ischemia reduces caveolin-1 protein expression, which persisted for 1 h post-reperfusion and recovered after 4 h [39]. While the mechanisms underlying the contribution of impaired expression of caveolin-1 to hepatic I/R injury remains to be understood, previous findings have indicated that hepatocytes from caveolin-1 knockout mice exhibit increased mitochondrial cholesterol loading, mGSH depletion, mitochondrial dysfunction, and sensitization to oxidative stress [8].

Therefore, it is conceivable that ischemia sensitizes to reperfusion in part due to decreased caveolin-1 expression resulting in increased mitochondrial cholesterol loading. Overall, from a translational perspective, statins may be of relevance to increase the pool of potential donors useful for liver transplantation. In this regard, recent data have demonstrated that liver grafts from healthy animals preserved in a commercially available cold storage solution supplemented with simvastatin exhibit reduced hepatocellular damage and improved microcirculatory status in comparison to grafts preserved in nonsupplemented cold storage solution [74].

7.6 Strategies to Protect Against Oxidative Stress During I/R

As described above, production of ROS, including superoxide, hydrogen peroxide, and hydroxyl radical, is an important mediator of hepatic I/R injury. Therefore, targeting these reactive species by antioxidant defense mechanisms may be of clinical relevance in ameliorating hepatic I/R injury. Manganese superoxide dismutase (MnSOD; also known as SOD2) is an essential primary antioxidant enzyme that converts superoxide radical to hydrogen peroxide within the mitochondrial matrix, generated by respiratory chain activity. MnSOD deficiency leads to neonatal death in mice, illustrating the crucial role of superoxide scavenging in development. Several reports have demonstrated a protective role for MnSOD in I/R injury. Its partial deficiency ($MnSOD^{+/-}$) exacerbates cerebral infarction after I/R, while overexpression of Cu/ZnSOD or MnSOD afforded protection from I/R injury with delayed induction of NF-κB and abrogated response of activator protein-1 (AP-1) in rat heart and in mouse liver (Kim 2002, [96, 99]). Recent evidence has shown that MnSOD is upregulated in the liver after I/R and that this response is mediated by Kupffer cell damage and activation, leading to DAMP molecule release particularly HMGB1 [67]. These molecules are secreted by dying Kupffer cells and account for the activation of the mitochondrial antioxidants in the liver, which protect hepatocytes from cell death. Furthermore, recent studies tested the effects of a novel recombinant form of human MnSOD (rMnSOD) in hepatic I/R damage [29]. This rMnSOD form remains stable in solution, has a good biodistribution, and efficiently scavenges intra- and extracellular superoxide, crosses plasma membrane, and is constitutively active in cytoplasm, nuclei, and mitochondria ([7, 53, 54]; Guillaume 2013). Using this formulation, it has been shown that rMnSOD markedly blunted oxidative stress, which was associated with a global improvement in liver damage and microcirculatory derangements in cold storage and warm reperfusion liver injury [29]. In line with these findings using rMnSOD, it has also been shown that SOD mimetic MnTBAP protected hepatocytes against prolonged I/R [49]. This anionic porphyrin MnTBAP has been shown to exhibit scavenging activity towards superoxide anion and peroxynitrite [3]. While MnTBAP was inefficient in supporting the aerobic growth of SOD-deficient *Escherichia coli*, studies in MnSOD null mice demonstrated the therapeutic benefit of MnTBAP administration, including prevention of systemic pathology such as dilated cardiomyopathy, hepatic lipid accumulation, and most importantly, a dramatically extended lifespan [59]. Thus, these data with the MnSOD null mice indicate that this particular SOD mimetic can functionally overcome MnSOD deficiency. Despite these findings, the therapeutic use of SOD mimetics, such as MnTBAP, as a strategic defense against hepatic I/R injury needs to be carefully assessed, especially in steatotic hepatocytes. In conditions of cholesterol accumulation and mitochondrial cholesterol trafficking, which sensitizes to hepatic I/R injury [48], MnSOD impairment can concur with mGSH depletion. In this scenario, the treatment with SOD mimetics may exacerbate liver injury during I/R, as the scavenging of superoxide in the face of mGSH depletion would result in increased hydrogen peroxide, which in the presence of transition metals, gives rise to potent free radical hydroxyl radical via Fenton reaction, which triggers the onset of the MPT pore opening by targeting critical cysteine residues in cyclophilin D [65]. The addition of the iron chelator deferoxamine or cyclosporine A, which have been shown to protect mGSH depleted hepatocytes against superoxide anion scavenging [93], in conjunction with MnTBAP may be a more efficient approach to protect against hepatic I/R than just MnTBAP alone. In addition, the selective replenishment of mGSH in addition to total GSH may be an interesting approach to protect hepatocytes from I/R-mediated injury. Moreover, *N*-acetylcysteine (NAC) is a GSH precursor and hence it is expected to increase cytosol GSH. However, in conditions of mitochondrial cholesterol loading, which depletes mGSH due to impairment of the transport of cytosol GSH into mitochondrial matrix [71], NAC may not be efficient in replenishing mGSH levels as shown in alcohol-fed models. In line with these pitfalls, it has been recently reported that NAC administration does not improve patient outcome after liver resection [72]. Interestingly, however, recent findings have shown that NAC attenuates hepatic I/R injury and apoptosis via regulation of a ROS/JNK/Bcl2-axis [95]. In contrast to NAC, the administration of GSH ethyl ester is able to replenish mGSH despite impairment of the mitochondrial transport of GSH [71], and therefore may emerge as a potential approach in the protection against hepatic I/R-induced injury, especially in combination with superoxide scavenging. Under these circumstances, the elimination of superoxide along with mGSH repletion may offer the advantage of preventing the formation of peroxynitrite, which has a short half-life and is responsible for many damaging modifications of tyrosine residues in mitochondrial target proteins. Quite intriguingly, the addition of MnTBAP in conditions of mGSH depletion exacerbates hepatocellular injury against oxidative stress despite of significantly preventing peroxynitrite generation [93]. In addition to these therapeutic combinations, IPC has been also of relevance in protecting rats fed a HC diet against hepatic oxygenation, microcirculation, and I/R-mediated injury [42]. Whether IPC has a beneficial role in GSH homeostasis and in the replenishment of mGSH remains to be further investigated.

Finally, although the role of autophagy, in hepatic I/R injury has not been thoroughly examined, there has been evidence indicating impaired autophagy during anoxia/reoxygenation in hepatocytes [40] due to increased calpain activity, decreased induction of autophagic regulatory proteins (e.g., Atg7 and Beclin-1), and limited autophagosome formation with consequent onset of the MPT and hepatocyte death [40]. In line with these findings, pharmacological induction of autophagy by carbamazepine, an FDA-approved anticonvulsant and mood stabilizing drug, has been shown to alleviate lethal reperfusion injury both in in vitro and in vivo models of I/R by preventing calcium overload, calpain activation, Atg7 and Beclin-1 depletion, defective autophagy, onset of MPT, mitochondrial dysfunction, ROS generation, and cell death [41].

7.7 Concluding Remarks

Hepatic I/R injury is a serious clinical complication that may compromise liver function because of extensive hepatocellular loss affecting important clinical conditions such as liver surgery, trauma, hemorrhagic shock, or liver transplantation. While ischemia causes anoxia, depletion of glycolytic substrates, loss of ATP and acidosis, hepatocyte injury occurs upon reperfusion when blood flow returns to hepatocytes. Hence, a better understanding of the molecular events contributing to hepatic I/R injury may be of significant relevance to ensure successful graft function. Although ROS can be generated from different sources during hepatic I/R, mitochondrial ROS production is crucial and contributes to liver injury as indicated by the protective effects of mitochondrial-targeted antioxidants. Besides overgeneration, increased ROS levels can also reflect the depletion of mitochondrial antioxidant defenses. In this regard, mGSH depletion underlies the susceptibility of steatotic livers to I/R injury due to the specific accumulation of cholesterol in mitochondria. Thus, targeting mitochondrial cholesterol accumulation or mGSH depletion may be of relevance in protecting graft function during hepatic I/R and may offer the advantage of expanding the pool of grafts available for liver transplantation.

References

1. Ajamieh H, Farrell G, Wong HJ, Yu J, Chu E, Chen J, Teoh N (2012) Atorvastatin protects obese mice against hepatic ischemia-reperfusion injury by Toll-like receptor-4 suppression and endothelial nitric oxide synthase activation. J Gastroenterol Hepatol 27(8):1353–1361
2. Bailly-Maitre B, Fondevila C, Kaldas F, Droin N, Luciano F, Ricci JE, Croxton R et al (2006) Cytoprotective gene bi-1 is required for intrinsic protection from endoplasmic reticulum stress and ischemia-reperfusion injury. Proc Natl Acad Sci U S A 103:2809–2814
3. Batinić-Haberle I, Cuzzocrea S, Rebouças JS, Ferrer-Sueta G, Mazzon E, Di Paola R, Radi R, Spasojević I, Benov L, Salvemini D (2009) Pure MnTBAP selectively scavenges peroxynitrite over superoxide: comparison of pure and commercial MnTBAP samples to MnTE-2-PyP in two models of oxidative stress injury, an SOD-specific Escherichia coli model and carrageenan-induced pleurisy. Free Radic Biol Med 46:192–201
4. Balaban RS, Nemoto S, Finkel T (2005) Mitochondria, oxidants and aging. Cell 120:483–495
5. Belghiti J, Durand F (2000) Living donor liver transplantation: present and future. Br J Surg 87(11):1441–1443
6. Bilzer M, Gerbes AL (2000) Preservation injury of the liver: mechanisms and novel therapeutic strategies. J Hepatol 32:508–515

7. Borrelli A, Schiattarella A, Fasano S, Occhiello A, Pica A, Sehr P, Tommasino M, Nuesch JP, Rommelaere J (2006) Tumor suppressive activity of a variant isoform of manganese superoxide dismutase released by a human liposarcoma cell line. Int J Cancer 119:932–943
8. Bosch M, Marí M, Herms A, Fernández A, Fajardo A, Kassan A, Giralt A, Colell A, Balgoma D, Barbero E, González-Moreno E, Matias N, Tebar F, Balsinde J, Camps M, Enrich C, Gross SP, García-Ruiz C, Pérez-Navarro E, Fernández-Checa JC, Pol A (2011) Caveolin-1 deficiency causes cholesterol-dependent mitochondrial dysfunction and apoptotic susceptibility. Curr Biol 21(8):681–686
9. Boveris A, Chance B (1973) The mitochondrial generation of hydrogen peroxide. General properties and effect of hyperbaric oxygen. Biochem J 134:707–716
10. Cadenas E, Boveris A, Ragan CI, Stoppani AOM (1977) Production of superoxide radicals and hydrogen peroxide by NADH-ubiquinone reductase and ubiquinol-cytochrome c reductase from beef-heart mitochondria. Arch Biochem Biophys 180:248–257
11. Caron KM, Soo SC, Wetsel WC, Stocco DM, Clark BJ, Parker KL (1997) Targeted disruption of the mouse gene encoding steroidogenic acute regulatory protein provides insights into congenital lipoid adrenal hyperplasia. Proc Natl Acad Sci U S A 94(21):11540–11545
12. Casillas-Ramírez A, Mosbah IB, Ramalho F, Roselló-Catafau J, Peralta C (2006) Past and future approaches to ischemia-reperfusion lesion associated with liver transplantation. Life Sci 79(20):1881–1894
13. Colell A, García-Ruiz C, Lluis JM, Coll O, Mari M, Fernández-Checa JC (2003) Cholesterol impairs the adenine nucleotide translocator-mediated mitochondrial permeability transition through altered membrane fluidity. J Biol Chem 278(36):33928–33935
14. Datta G, Fuller BJ, Davidson BR (2013) Molecular mechanisms of liver ischemia reperfusion injury: Insights from transgenic knockout models. World J Gastroenterol 21:1683–1698
15. Dikalov S (2011) Cross talk between mitochondria and NADPH oxidases. Free Radic Biol Med 51(7):1289–1301
16. Dröge W, Schulze-Osthoff K, Mihm S, Galter D, Schenk H, Eck HP, Roth S, Gmünder H (1994) Functions of glutathione and glutathione disulfide in immunology and immunopathology. FASEB J 8(14):1131–1138. Review
17. El-Brady AM, Moritz W, Contaldo C, Tian Y, Graf R, Clavien PA (2007) Prevention of reperfusion injury and microcirculatory failure in macrosteatosis mouse liver by omega-3 fatty acids. Hepatology 45:855–863
18. Facundo HT, de Paula JG, Kowaltowski AJ (2007) Mitochondrial ATP-sensitive K+ channels are redox-sensitive pathways that control reactive oxygen species production. Free Radic Biol Med 42(7):1039–1048
19. Fernández A, Colell A, Caballero F, Matías N, García-Ruiz C, Fernández-Checa JC (2009) Mitochondrial S-adenosyl-L-methionine transport is insensitive to alcohol-mediated changes in membrane dynamics. Alcohol Clin Exp Res 33:1169–1180
20. Fishbein TM, Fiel MI, Emre S, Cubukcu O, Guy SR, Schwartz ME et al (1997) Use of livers with microvesicular fat safely expands the donor pool. Transplantation 64:248–251
21. Fusai G, Glantzounis GK, Hafez T, Yang W, Quaglia A, Sheth H et al (2005) N-acetylcysteine ameliorates the late phase of liver ischemia/reperfusion injury in the rabbit with hepatic steatosis. Clin Sci 109:465–473
22. Gerschman R, Gilbert DL, Nye SW, Dwyer P, Fenn W (1954) Oxygen poisoning and x-irradiation: a mechanism in common. Science 119(3097):623–626
23. Gonzalez-Flecha B, Cutrin JC, Boveris A (1993) Time course and mechanism of oxidative stress and tissue damage in rat liver subjected to in vivo ischemia-reperfusion. J Clin Invest 91:456–464
24. Guillaume M, Rodriguez-Vilarrupla A, Gracia-Sancho J, Rosado E, Mancini A, Bosch J, Garcia-Pagán JC (2013) Recombinant human manganese superoxide dismutase reduces liver fibrosis and portal pressure in CCl4-cirrhotic rats. J Hepatol 58:240–246
25. Gracia-Sancho J, Russo L, García-Calderó H, García-Pagán JC, García-Cardeña G, Bosch J (2011) Endothelial expression of transcription factor Kruppel-like factor 2 and its vasoprotective target genes in the normal and cirrhotic rat liver. Gut 60(4):517–524

26. Gujral JS, Farhood A, Bajt ML, Jaeschke H (2003) Neutrophils aggravate acute liver injury during obstructive cholestasis in bile duct-ligated mice. Hepatology 38(2):355–363
27. Guzy RD, Hoyos B, Robin E, Chen H, Liu L, Mansfield KD, Simon MC, Hammerling U, Schumacker PT (2005) Mitochondrial complex III is required for hypoxia-induced ROS production and cellular oxygen sensing. Cell Metab 1(6):401–408
28. Hakamada K, Sasaki M, Takahashi K, Umehara Y, Konn M (1997) Sinusoidal flow block after warm ischemia in rats with diet induced fatty liver. J Surg Res 70:12–20
29. Hasegawa T, Malle E, Farhood A, Jaeschke H (2005) Generation of hypochlorite-modified proteins by neutrophils during ischemia-reperfusion injury in rat liver: attenuation by ischemic preconditioning. Am J Physiol Gastrointest Liver Physiol 289(4):G760–G767
30. Hide D, Ortega-Ribera M, Fernández-Iglesias A, Fondevila C, Salvadó MJ, Arola L, García-Pagán JC, Mancini A, Bosch J, Gracia-Sancho J (2014) A novel form of the human manganese superoxide dismutase protects rat and human livers undergoing ischaemia and reperfusion injury. Clin Sci (Lond) 127(8):527–537
31. Hirst J, King MS, Pryde KR (2008) The production of reactive oxygen species by complex I. Biochem Soc Trans 36:976–980
32. Hong F, Radaeva S, Pan HN, Tian Z, Veech R, Gao B (2004) Interleukin 6 alleviates hepatic steatosis and ischemia/reperfusion injury in mice with fatty liver disease. Hepatology 40(4): 933–941
33. Jaeschke H (2003) Molecular mechanisms of hepatic ischemia-reperfusion injury and preconditioning. Am J Physiol 284:G15–G26
34. Jaeschke H, Lemasters JJ (2003) Apoptosis versus oncotic necrosis in hepatic ischemia/reperfusion injury. Gastroenterology 125:1246–1257
35. Jaeschke H, Bautista AP, Spolarics Z, Spitzer JJ (1991) Superoxide generation by Kupffer cells and priming of neutrophils during reperfusion after hepatic ischemia. Free Radic Res Commun 15(5):277–284
36. Jang JH, Moritz W, Graf R, Clavien PA (2008) Preconditioning with death ligands FasL and TNF-alpha protects the cirrhotic mouse liver against ischaemic injury. Gut 57(4):492–499
37. Jassem W, Fuggle SV, Rela M, Koo DD, Heaton ND (2002) The role of mitochondria in ischemia/reperfusion injury. Transplantation 73(4):493–499
38. Jastroch M, Divakaruni AJ, Mookerjee S, Treberg JR, Brand MD (2010) Mitochondrial proton and electron leaks. Essays Biochem 47:53–67
39. Jungermann K, Kietzmann T (2000) Oxygen: modulator of metabolic zonation and disease of the liver. Hepatology 31:225–260
40. Kang JW, Lee SM (2014) Impaired expression of caveolin-1 contributes to hepatic ischemia reperfusion injury. Biochem Biophys Res Commun 450:1351–1357
41. Kim GW, Kondo T, Noshita N, Chan PH (2002) Manganese superoxide dismutase deficiency exacerbates cerebral infarction after focal cerebral ischemia/reperfusion in mice: implications for the production and role of superoxide radicals. Stroke 33:809–815
42. Kim JS, Nitta T, Mohuczy D, O'Malley KA, Moldawer LL, Dunn WA Jr, Behrns KE (2008) Impaired autophagy: a mechanism of mitochondrial dysfunction in anoxic rat hepatocytes. Hepatology 47(5):1725–1736
43. Kim JS, Wang JH, Biel TG, Flores-Toro JA, Vijayvargiya R, Zendejas I, Behrns KE (2013) Carbamazepine suppresses calpain-mediated autophagy impairment after ischemia/reperfusion in mouse livers. Toxicol Appl Pharmacol 273(3):600–610
44. Koti RS, Yang W, Glantzounis G, Quaglia A, Davidson BR, Seifalian AM (2005) Effect of ischaemic preconditioning on hepatic oxygenation, microcirculation and function in a rat model of moderate hepatic steatosis. Clin Sci (Lond) 108(1):55–63
45. Kuboki S, Okaya T, Schuster R, Blanchard J, Denenberg A, Wong HR, Lentsch AB (2007) Hepatocyte NF-kappaB activation is hepatoprotective during ischemia-reperfusion injury and is augmented by ischemic hypothermia. Am J Physiol Gastrointest Liver Physiol 292(1): G201–G207
46. Kussmaul L, Hirst J (2006) The mechanism of superoxide production by NADH:ubiquinone oxidoreductase (complex I) from bovine heart mitochondria. Proc Natl Acad Sci U S A 103:7607–7612

47. Lai IR, Chang KJ, Tsai HW, Chen CF (2008) Pharmacological preconditioning with simvastatin protects liver from ischemia-reperfusion injury by heme oxygenase-1 induction. Transplantation 85:732–738
48. Lambert AJ, Brand MD (2004) Superoxide production by NADH:ubiquinone oxidoreductase (complex I) depends on the pH gradient across the mitochondrial inner membrane. Biochem J 382:511–517
49. Lemasters JJ (2001) Hypoxic, ischemic, and reperfusion injury to liver. In: Arias IM (ed) The liver: biology and pathobiology, 4th edn. Lippincott Williams & Wilkins, Philadelphia, pp 257–279
50. Llacuna L, Fernandez A, von Montfort C, Matias N, Martinez L, Caballero F, Rimola A, Elena M, Morales A, Fernandez-Checa JC, Garcia-Ruiz C (2011) Targeting cholesterol at different levels of the mevalonate pathway protects fatty livers against ischemia-reperfusion injury. J Hepatol 54:1002–1010
51. Llacuna L, Mari M, LLuis JM, Garcia-Ruiz C, Fernandez-Checa A, Morales A (2009) Reactive oxygen species mediate liver injury through parenchymal nuclear factor-B inactivation in prolonged ischemia/reperfusion. Am J Pathol 174:1776–1785
52. Llacuna L, Marí M, Garcia-Ruiz C, Fernandez-Checa JC, Morales A (2006) Critical role of acidic sphingomyelinase in murine hepatic ischemia-reperfusion injury. Hepatology 44(3): 561–572
53. Lou H, Kaplowitz N (2007) Glutathione depletion down-regulates tumor necrosis factor alpha-induced NF-kappaB activity via IkappaB kinase-dependent and-independent mechanisms. J Biol Chem 282(40):29470–29481
54. Ma X, Jin M, Cai Y, Xia H, Long K, Liu J, Yu Q, Yuan J (2011) Mitochondrial electron transport chain complex III is required for antimycin A to inhibit autophagy. Chem Biol 18: 1474–1481
55. Mancini A, Borrelli A, Schiattarella A, Fasano S, Occhiello A, Pica A, Sehr P, Tommasino M, Nuesch JP, Rommelaere J (2006) Tumor suppressive activity of a variant isoform of manganese superoxide dismutase released by a human liposarcoma cell line. Int J Cancer 119: 932–943
56. Mancini A, Borrelli A, Schiattarella A, Aloj L, Aurilio M, Morelli F, Pica A, Occhiello A, Lorizio R, Mancini R et al (2008) Biophysical and biochemical characterization of a liposarcoma-derived recombinant MnSOD protein acting as an anticancer agent. Int J Cancer 123:2684–2695
57. Mari M, Caballero F, Colell A, Morales A, Caballeria J, Fernandez A et al (2006) Mitochondrial free cholesterol loading sensitizes to TNF- and Fas-mediated steatohepatitis. Cell Metab 4:185–198
58. Mari M, Colell A, Morales A, von Montfort C, Garcia-Ruiz C, Fernández-Checa JC (2010) Redox control of liver function in health and disease. Antioxid Redox Signal 12(11): 1295–1331
59. Massip-Salcedo M, Roselló-Catafau J, Prieto J, Avíla MA, Peralta C (2007) The response of the hepatocyte to ischemia. Liver Int 27(1):6–16
60. Menger MD, Vollmar B (2000) Role of microcirculation in transplantation. Microcirculation 7:291–306
61. Melov S, Schneider JA, Day BJ, Hinerfeld D, Coskun P, Mirra SS, Crapo JD, Wallace DC (1998) A novel neurological phenotype in mice lacking mitochondrial manganese superoxide dismutase. Nat Genet 18:159–163
62. Miller WL (2007) Steroidogenic acute regulatory protein (StAR), a novel mitochondrial cholesterol transporter. Biochim Biophys Acta 1771(6):663–676
63. Moon KH, Hood BL, Mukhopadhyay P, Rajesh M, Abdelmegeed MA, Kwon YI, Conrads TP, Veenstra TD, Song BJ, Pacher P (2008) Oxidative inactivation of key mitochondrial proteins leads to dysfunction and injury in hepatic ischemia reperfusion. Gastroenterology 135(4):1344–1357
64. Mori N, Horie Y, Yuji N, Wolf R, Granger N (2000) Hepatic microvesicular responses to ischemia-reperfusion in low-density lipoprotein receptor knockout mice. Am J Physiol 279: G1257–G1264

65. Mukhopadhyay P, Horváth B, Zsengellér Z, Zielonka J, Tanchian G, Holovac E, Kechrid M, Patel V, Stillman IE, Parikh SM, Joseph J, Kalyanaraman B, Pacher P (2012) Mitochondrial-targeted antioxidants represent a promising approach for prevention of cisplatin-induced nephropathy. Free Radic Biol Med 52(2):497–506
66. Muller FL, Roberts AG, Bowman MK, Kramer DM (2003) Architecture of the Qo site of the cytochrome bc1 complex probed by superoxide production. Biochemistry 42:6493–6499
67. Nguyen TT, Stevens MV, Kohr M, Steenberg C, Sack MN, Murphy E (2011) Cysteine 203 of cyclophilin D is critical for cyclophilin D activation of the mitochondrial permeability transition pore. J Biol Chem 286:40184–40192
68. Orr AL, Quinlan CL, Perevoshcchikova IV, Brand MD (2012) A refined analysis of superoxide production by mitochondria sn-glycerol 3-phosphate dehydrogenase. J Biol Chem 287: 42921–42935
69. Pardo M, Budick-Harmeln N, Tirosh B, Tirosh O (2008) Antioxidant defense in hepatic ischemia-reperfusion injury is regulated by damage-associated molecular pattern signals molecules. Free Radic Biol Med 45:1073–1083
70. Parmar KM, Nambudiri V, Dai G, Larman HB, Gimbrone MA Jr, García-Cardeña G (2005) Statins exert endothelial atheroprotective effects via the KLF2 transcription factor. J Biol Chem 280(29):26714–26719
71. Pompella A, Visvikis A, Paolicchi A, De Tata V, Casini AF (2003) The changing faces of glutathione, a cellular protagonist. Biochem Pharmacol 66(8):1499–1503
72. Quinlan CL, Perevoshchikova IV, Hey-Mogensen M, Orr AL, Brand MD (2013) Sites of reactive oxygen species generation by mitochondria oxidizing different substrates. Redox Biology 1:304–312
73. Relja B, Lehnert M, Seyboth K, Bormann F, Hohn C, Czerny C et al (2010) Simvastatin reduces mortality and hepatic injury after hemorrhage/resuscitation in rats. Shock 34:46–54
74. Ribas V, García-Ruiz C, Fernández-Checa JC (2014) Glutathione and mitochondria. Front Pharmacol 5:151
75. Robinson SM, Saif R, Sen G, French JJ, Jacques BC, Charnley RM, Manas DM, White SA (2013) N-acetylcysteine administration does not improve patient outcome after liver resection. HPB 15:457–462
76. Rüdiger HA, Clavien PA (2002) Tumor necrosis factor alpha, but not Fas, mediates hepatocellular apoptosis in the murine ischemic liver. Gastroenterology 122(1):202–210
77. Russo L, Gracia-Sancho J, García-Calderó H, Marrone G, García-Pagán JC, García-Cardeña G, Bosch J (2012) Addition of simvastatin to cold storage solution prevents endothelial dysfunction in explanted rat livers. Hepatology 55(3):921–930
78. Scaloni A, Codarin E, Di Maso V, Arena S, Renzone G, Tiribelli C, Quadrifoglio F, Tell G (2009) Modern strategies to identify new molecular targets for the treatment of liver diseases: the promising role of proteomics and redox proteomics investigations. Proteomics Clin Appl 3:242–262
79. Sakon M, Ariyoshi H, Umeshita K, Monden M (2002) Ischemia-reperfusion injury of the liver with special reference to calcium-dependent mechanisms. Surg Today 32:1–12
80. Seizner M, Seizmer N, Jochum W, Graf R, Clavien PA (2007) Increased ischemic injury in old mouse liver: an ATP-dependent mechanism. Liver Transpl 13(3):382–390
81. Selzner M, Clavien PA (2001) Fatty liver in liver transplantation and surgery. Semin Liver Dis 21:105–113
82. Selzner N, Selzner M, Jochum W, Amann-Vesti B, Graf R, Clavien PA (2006) Mouse liver with macrosteatosis are more susceptible to normothermic ischemic injury than those with microsteatosis. J Hepatol 44:694–701
83. Shen XD, Ke B, Zhai Y, Gao F, Tsuchihashi S, Lassman CR, Busuttil RW, Kupiec-Weglinski JW (2007) Absence of toll-like receptor 4 (TLR4) signaling in the donor organ reduces ischemia and reperfusion injury in a murine liver transplantation model. Liver Transpl 13(10):1435–1443
84. Soltys K, Dikdan G, Koneru B (2001) Oxidative stress in fatty livers of obese Zucker rats: rapid amelioration and improved tolerance to warm ischemia with tocopherol. Hepatology 34(1):13–18

85. Starkov AA, Fiskum G, Chinopoulos C, Lorenzo BJ, Browne SE, Patel MS, Beal MF (2004) Mitochondrial alpha-ketoglutarate dehydrogenase complex generates reactive oxygen species. J Neurosci 24:7779–7788
86. St Pierre J, Buckingham JA, Roebuck SJ, Brand MD (2002) Topology of superoxide production from different sites in the mitochondrial electron transport chain. J Biol Chem 277: 44784–44790
87. Strasberg SM (1997) Preservation injury and donor selection: it all starts here. Liver Transpl Surg 5(Suppl 1):S1–S7
88. Suetsugu H, Iimuro Y, Uehara T, Nishio T, Harada N, Yoshida M, Hatano E, Son G, Fujimoto J, Yamaoka Y (2005) Nuclear factor {kappa}B inactivation in the rat liver ameliorates short term total warm ischaemia/reperfusion injury. Gut 54(6):835–842
89. Sun K, Xie X, Liu Y, Han Z, Zhao X, Cai N, Zhang S, Song J, Wei L (2013) Autophagy lessens ischemic liver injury by reducing oxidative damage. Cell Biosci 3:26
90. Treberg JR, Quinlan CL, Brand MD (2011) Evidence for two sites of superoxide production by mitochondrial NADH-ubiquinone oxidoreductase (complex I). J Biol Chem 286: 27103–27110
91. Uchiyama A, Kim JS, Kon K, Jaeschke H, Ikejima K, Watanabe S, Lemasters JJ (2008) Translocation of iron from lysosomes into mitochondria is a key event during oxidative stress-induced hepatocellular injury. Hepatology 48:1644–1654
92. Urena MA, Ruiz-Delgado FC, Gonzalez EM, Segurola CL, Romero CJ, Garcia IG et al (1998) Assessing risk of the use of livers with macro and microsteatosis in a liver transplant program. Transplant Proc 30:3288–3291
93. Vascotto C, Cesaratto L, D'Ambrosio C, Scaloni A, Avellini C, Paron I, Baccarani U, Adani GL, Tiribelli C, Quadrifoglio F, Tell G (2006) Proteomic analysis of liver tissues subjected to early ischemia/reperfusion injury during human orthotopic liver transplantation. Proteomics 6(11):3455–3465
94. Venkatakrishnan P, Nakayasu ES, Almeida IC, Miller RT (2009) Absence of nitric-oxide synthase in sequentially purified rat liver mitochondria. J Biol Chem 284(30):19843–19855
95. Videla LA, Fernandez V (1988) Biochemical aspects of cellular oxidative stress. Arch Biol Med Exp 21:85–92
96. von Montfort C, Matías N, Fernandez A, Fucho R, Conde de la Rosa L, Martinez-Chantar ML, Mato JM, Machida K, Tsukamoto H, Murphy MP, Mansouri A, Kaplowitz N, Garcia-Ruiz C, Fernandez-Checa JC (2012) Mitochondrial GSH determines the toxic or therapeutic potential of superoxide scavenging in steatohepatitis. J Hepatol 57:852–859
97. Wang JH, Behrns KE, Leeuwenburgh C, Kim JS (2012) Critical role of autophage in ischemia/reperfusion injury to aged livers. Autophagy 8:140–141
98. Wang C, Chen K, Xia Y, Dai W, Wang F, Shen M, Cheng P, Wang J, Lu J, Zhang Y, Yang J, Zhu R, Zhang H, Li J, Zhou Y, Guo C (2014) N-acetylcysteine attenuates ischemia-reperfusion induced apoptosis and autophagy in mouse liver via regulation of the ROS/JNK/Bcl-2 pathway. PLoS ONE 9(9):e108855
99. Yang J, Marden JJ, Fan C, Sanlioglu S, Weiss RM, Ritchie TC, Davisson RL, Engelhardt JF (2003) Genetic redox preconditioning differentially modulates AP-1 and NF kappa B responses following cardiac ischemia/reperfusion injury and protects against necrosis and apoptosis. Mol Ther 7:341–353
100. Yoong KF, Gunson BK, Neil DA, Mirza DF, Mayer AD, Buckels JA et al (1999) Impact of donor liver microvesicular steatosis on the outcome of liver retransplantation. Transplant Proc 31:550–551
101. Zhai Y, Shen XD, O'Connell R, Gao F, Lassman C, Busuttil RW, Cheng G, Kupiec-Weglinski JW (2004) Cutting edge: TLR4 activation mediates liver ischemia/reperfusion inflammatory response via IFN regulatory factor 3-dependent MyD88-independent pathway. J Immunol 173(12):7115–7119
102. Zhou W, Zhang Y, Hosch MS, Lang A, Zwacka RM, Engelhardt JF (2001) Subcellular site of superoxide dismutase expression differentially controls AP-1 activity and injury in mouse liver following ischemia/reperfusion. Hepatology 33:902–914

Chapter 8
Oxidative Stress and Liver Fibrogenesis

Erica Novo, Fabio Marra, and Maurizio Parola

8.1 Liver Fibrogenesis, Liver Fibrosis, and the Progression of Chronic Liver Diseases

8.1.1 Basic Definitions and Concepts

Persistent parenchymal injury (i.e., hepatocyte injury) is the hallmark of any chronic liver disease (CLD) of clinical relevance which, at least in industrialized countries, can be induced by chronic infection by hepatotropic viruses (hepatitis B and C viruses), chronic alcohol consumption or exposure to toxins and drugs, persisting autoimmune injury or chronic conditions of altered metabolism. Persistent hepatic parenchymal injury invariably results in the activation of those mechanisms and events able to sustain liver fibrogenesis, including chronic activation of inflammatory and wound healing response as well as persisting oxidative stress and derangement of interactions between epithelial and mesenchymal cells. Liver fibrogenesis is best defined as a dynamic and highly integrated molecular, tissue, and cellular process,

---

E. Novo • M. Parola (✉)
Department of Clinical and Biological Sciences, Unit of Experimental Medicine, University of Torino, C.so Raffaello 30, Torino 10125, Italy

Interuniversity Centre for Liver Pathophysiology, University of Torino, C.so Raffaello 30, Torino 10125, Italy
e-mail: maurizio.parola@unito.it

F. Marra
Department of Experimental and Clinical Medicine, University of Florence, Largo Brambilla 3, Florence 50134, Italy

Center for Research, Transfer and Higher Education DenoTHE, University of Florence, Largo Brambilla 3, Florence 50134, Italy

E. Albano, M. Parola (eds.), *Studies on Hepatic Disorders*,
Oxidative Stress in Applied Basic Research and Clinical Practice,
DOI 10.1007/978-3-319-15539-5_8

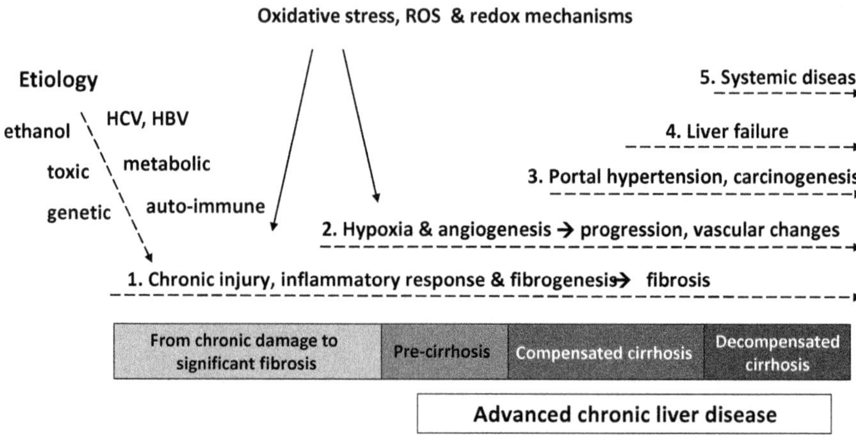

Fig. 8.1 Oxidative stress, ROS, and redox mechanisms in the fibrogenic progression of chronic liver diseases. Whatever the aetiology, oxidative stress, ROS, and more generally redox mechanisms are critical in major pathophysiological events that drive progression of chronic liver diseases from early fibrosis to the most advanced stages of the disease: (1) the scenario of chronic injury and chronic activation of inflammatory response and fibrogenesis, which accompanies a chronic liver disease; (2) the scenario of developing hypoxia and angiogenesis, which in turn is relevant for vascular changes

potentially reversible, that drives the progression of CLD first towards liver fibrosis (i.e., the "result" of fibrogenesis) and then with time towards liver cirrhosis and hepatic failure [1–6]. In this CLD scenario (Fig. 8.1) liver cirrhosis can be defined not just as an advanced stage of CLD, but a condition characterized by alteration in the vascular microanatomy of the liver. Cirrhosis is characterized by the formation of regenerative nodules of parenchyma surrounded and separated by fibrotic septa, and associated with significant changes in organ vascular architecture, development of portal hypertension and related complications, including variceal bleeding, hepatic encephalopathy, ascites, and hepatorenal syndrome as well as the development of hepatocellular carcinoma (HCC) [7].

Fibrogenic progression of CLDs, although difficult to predict, has a very significant clinical impact as indicated by the following epidemiological data [8–10]: (a) 180 millions of patients worldwide are affected by a form of CLD, 25–30 % of which are expected to develop cirrhosis and related complications; (b) approximately one million deaths (2010 data) are due to liver cirrhosis worldwide; (c) cirrhosis is the most common non-neoplastic cause of death in Europe and USA among diseases of the GI tract (overall the seventh most common cause of death in western countries); (d) HCC, which almost invariably develops in a cirrhotic or (as emerging for NASH-related patients) fibrotic liver, is a very aggressive malignant cancer representing the fifth most common cancer and the third most common cause of cancer mortality worldwide; (e) a peak for end-stage CLDs and HCC is predicted for the next decade in parallel with a shortage of donor organs for liver transplantation (at present the best treatment option).

Whatever the aetiology, CLD progression is dominated by an interrelated sequence of persisting chronic parenchymal injury (resulting in chronic necrosis and/or apoptosis), chronic inflammatory response, and chronic activation of fibrogenesis, which is the driving force for excess deposition of ECM component (i.e., fibrosis). A number of related relevant concepts can be offered [1–7, 11, 12]: (1) persisting chronic parenchymal injury is not only aetiology-related, but also depends on chronic injury itself and chronic inflammatory response through the involvement of several mediators that also include reactive oxygen species (ROS) and other oxidative stress-related reactive intermediates; (2) persisting activation of inflammatory response and the related recruitment/activation of cells of the innate or acquired immunity result in a pro-fibrogenic environment characterized by synthesis and release of growth factors, cytokines, chemokines, ROS, and other mediators able to significantly impair liver regeneration but also to lead to chronic activation of wound healing and fibrogenesis; (3) such a pro-fibrogenic environment is critical in inducing persistent activation of myofibroblast (MF)-like cells which are responsible for increased deposition of ECM components paralleled by altered/inefficient remodeling; (4) hypoxia and angiogenesis are proposed to play a major role in sustaining and, possibly, driving fibrogenesis as well as vascular changes during CLD progression; (5) liver fibrosis, at least in its initial stage, is potentially reversible and this reversibility critically depends on the possibility to either remove the exposure to the aetiological agent/condition or the use of an effective therapy.

8.1.2 *Liver Cell Populations Involved in Fibrogenic Progression of CLDs*

All liver cells provide some contribution to the fibrogenic progression of CLDs and liver fibrogenesis can be reasonably envisaged as a dynamic process resulting from the interactions (see Fig. 8.2) between hepatic cell populations. This cross talk is further modulated and/or affected by the presence of hypoxia, and involves the synthesis and release of several mediators, including growth factors, cytokines and chemokines, ROS, adipokines, vasoactive compounds, and plasma proteins, to name just a few [1–7]. Hepatocytes are the obvious major target liver cells in relation to aetiology-related parenchymal cell injury and death, either necrotic or apoptotic. However, as it will be emphasized later, they have a significant role in the fibrogenic scenario as sources of ROS and other oxidative stress-related reactive intermediates as a consequence of injury (i.e., released by damaged hepatocytes) as well as relevant sources of pro-angiogenic cytokines (mainly vascular endothelial growth factor, VEGF) following either exposure to hypoxic conditions or to specific polypeptides. Resident Kupffer cells and macrophages recruited from the bloodstream or of bone marrow origin represent another population of activated cells to be obviously considered in CLD progression. In fact, these cells are able to synthesize and release several pro-fibrogenic and pro-inflammatory mediators, where

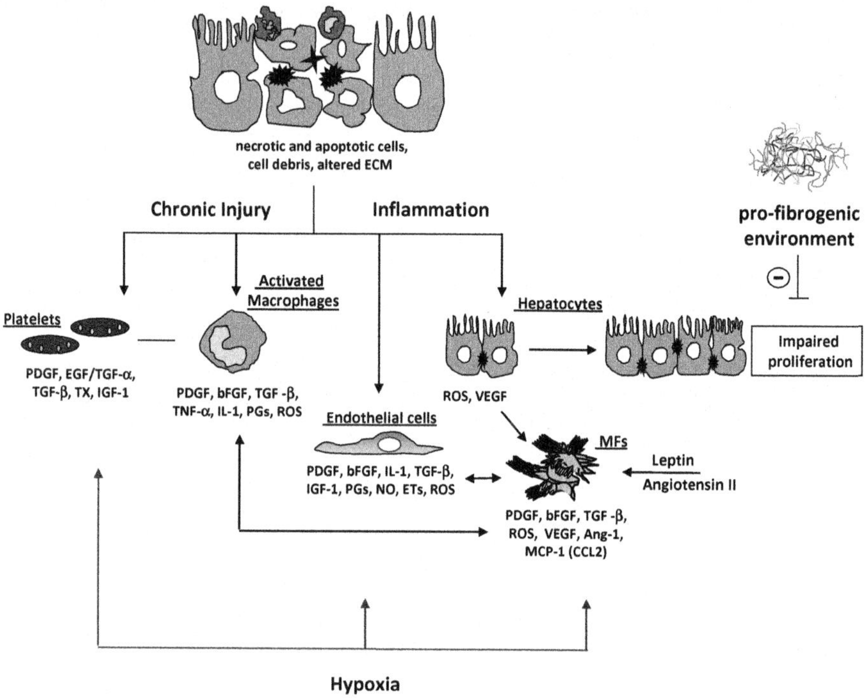

Fig. 8.2 The cross talk between hepatic cell populations in the CLD scenario. The interactions between the different hepatic cell populations under conditions of chronic liver injury are outlined in terms of release of cytokines, growth factors, and other polypeptides or mediators (including ROS and other oxidative stress-related intermediates), with an additional modulating role exerted by both hypoxia and the "pro-fibrogenic" environment, the latter believed to impair proliferation of parenchymal cells

platelet-derived growth factor (PDGF), basic fibroblast growth factor (bFGF), transforming growth factor β1 (TGFβ1), tumor necrosis factor α (TNFα), interleukin-1 (IL-1), monocyte chemoattractant protein-1 (MCP-1 or CCL2, and ROS represent the best characterized and most efficient ones. The chronic scenario is affected also by mediators released by damaged and/or activated endothelial cells (PDGF, basic fibroblast growth factor or bFGF, interleukin 1 or IL-1, TGF-β, insulin-like growth factor-1 or IGF-1, prostaglandins, nitric oxide or NO, as well as endothelins or ETs and ROS), with platelets, sometimes neglected but certainly involved, also contributing to the genesis of the pro-fibrogenic environment. Whatever the aetiology, liver fibrogenesis is mainly sustained by a heterogeneous population of hepatic MFs that are easily recognized in fibrotic or cirrhotic human liver specimens by the immune-positivity for a-smooth-muscle actin (α-SMA), the most reliable in vivo marker for these cells [13]. Because of their relevance, hepatic MFs deserve a dedicated section.

8.1.3 Hepatic Myofibroblasts: Origin, Activation, Phenotypic Responses, and Fate

Hepatic MFs are the major effectors of liver fibrogenesis and their prevalent role is intimately related to their ability to act as a unique cellular crossroad able to integrate and to respond to paracrine or autocrine signals released from the other hepatic cell populations (Fig. 8.2) [2, 6].

Hepatic MFs are considered as a heterogeneous population since α-SMA-positive cells in fibrotic/cirrhotic livers have been reported to originate from different cellular sources. There is consistent evidence that hepatic MFs originate mainly from hepatic stellate cells (HSC/MFs) or, to a lesser extent and in particular pathological conditions, from portal fibroblasts (P/MFs) through a process of activation/trans-differentiation [1–4, 6, 14, 15]. This process is believed to be sustained by the "pro-fibrogenic environment", and then by the related mediators already mentioned in a previous section, as well as by mechanical stimuli related to properties of extracellular matrix components [16]. This second issue is of interest and implies the relevance of mechanical tension and mechanochemical transduction processes as key elements in the induction of MF phenotype. Fibroblasts or other mesenchymal cells would exert forces on the ECM with an intensity proportional to its stiffness, producing α-SMA in response to mechanical stimulation and re-arranging cytoskeletal elements to match the cellular stiffness with that of the substrate. These mechanochemical transduction processes are operated through mechanosensitive receptors, mainly $\alpha1\beta1$ and $\alpha2\beta1$ integrins [16]. Hepatic MFs, although to a lesser extent, have been also reported to originate from cells derived from bone marrow (mesenchymal stem cells or MSC, or fibrocytes) and recruited/activated in the chronically injured liver parenchyma [1–6]. At present, the possible origin of at least part of pro-fibrogenic cells from liver epithelial cells (i.e., hepatocytes and/or cholangiocytes) through a process of epithelial to mesenchymal transition (EMT) is still uncertain and represents the subject of an intense scientific debate [4, 6, 17–19].

There is substantial agreement on the fact that the so-called "fully activated hepatic MF-like phenotype" is highly proliferative, migratory, and contractile. Along these lines, hepatic MFs are not only able to synthesize an excess of ECM components and to contribute to ECM remodelling, but also operate actively in order to sustain angiogenesis and/or inflammatory response as well as to modulate the immune response (Fig. 8.3). Although most of current knowledge on liver MFs comes directly from studies performed on HSC/MFs, the available evidence suggests that the most relevant phenotypic responses of MF-like cells are common to MFs originated from portal fibroblasts and, possibly, from bone marrow-derived cells engrafting chronically injured liver [1–6, 14–19]. As far as the phenotype of hepatic MFs is concerned, one of the most interesting concepts is that during CLD progression MFs enter into a "survival attitude" that is likely to be the consequence of growth factor- and ROS-mediated activation of Nuclear-Factor κB (NF-κB) [20]

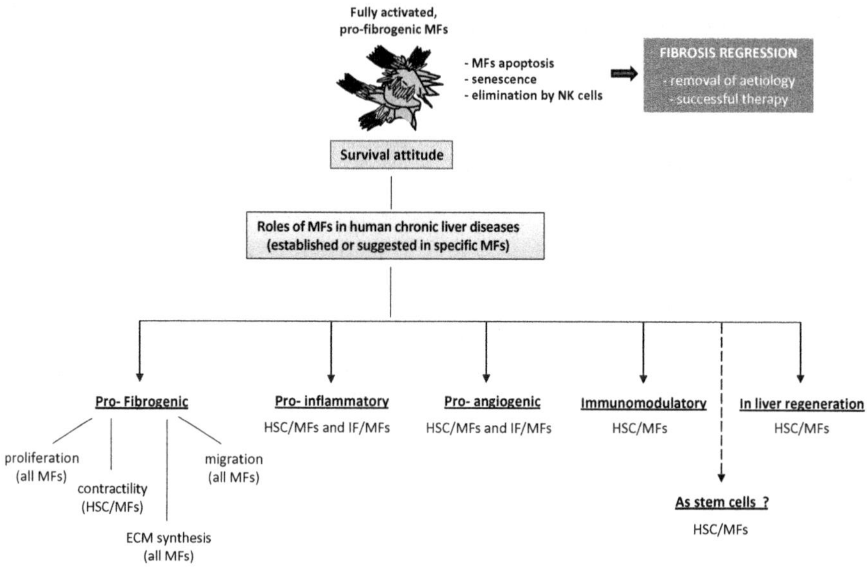

Fig. 8.3 Phenotypic responses of activated MF-like cells. Fully activated and MF-like cells can display a number of established phenotypic responses, which are believed to be shared by almost all hepatic MFs, whatever their origin, and the specific aetiology of a chronic liver disease

and up-regulation of anti-apoptotic Bcl-2 [21] in hepatic MFs. This is relevant and directly connected with the issue of liver fibrosis reversal, mainly described in experimental studies and somewhat confirmed in clinical studies (for example, in patients undergoing successful viral eradication). Indeed, cessation of liver injury can result in a significant and progressive regression in the degree of liver fibrosis which is accompanied by evidence of apoptosis of hepatic MFs. The "rise and fall" of MFs is believed to be modulated by an altered balance between pro-apoptotic and anti-apoptotic signals (for example, NGF versus pro-NGF) deriving from either extracellular or intracellular environment, with liver fibrosis regression being the result of a switch in the TIMP/MMP balance also favoured by infiltration of the hepatic scars by bone-marrow-derived macrophages (i.e., rich source of MMPs) [1–6, 11, 12, 20–22].

8.1.4 Pattern of Fibrosis: Aetiology and Prevailing Pro-fibrogenic Mechanisms

As a preliminary step in introducing the role of oxidative stress, ROS, and redox signalling, one should recall that although cirrhosis may be considered as a common end-point of progressive fibrogenesis, the specific aetiology has a relevant impact on CLD progression. According to the original hypothesis proposed in 2004 [23] and then progressively refined [4, 6], a number of distinct patterns of fibrosis development can be recognized in relation to the specific aetiology which, in turn, can also significantly influence the prevailing pro-fibrogenic mechanisms and the type(s) of pro-fibrogenic cells involved.

1. Post-necrotic or bridging fibrosis. A pattern which is typically identified in patients affected by chronic HCV or HBV infections or by an autoimmune disease which results in the prevailing formation of portal–central septa, which follows portal–central bridging necrosis and is associated with interface hepatitis, as well as of septa connecting different portal areas. This pattern leads to an early involvement of centrilobular vein, formation of neo-vessels (i.e., angiogenesis), and then porto-central shunting. In this pattern the prevalent mechanism is represented by chronic activation of the wound healing response with a significant contribution of oxidative stress, while MF-like cells mainly originating from HSC, portal fibroblasts and, to a minor extent, bone marrow-derived cells.
2. Pericellular/perisinusoidal fibrosis. This characteristic pattern, also defined as intercellular fibrosis or chickenwire fibrosis, is typically detected in the liver of patients suffering of metabolic derangements such as those induced by NASH or alcoholic steato-hepatitis (ASH), with a major pathogenic contribution of oxidative stress. Excess deposition of ECM initiates in the space of Disse and is considered as the result of activation of perisinusoidal HSC which indeed predominates in the early phases of both conditions. With the time this pattern rapidly results in the so-called capillarization of sinusoids, which precedes formation of fibrotic septa and leads to a pattern of fibrosis development that tends to be centro-portal.
3. Biliary fibrosis. This is a rather unique pattern of fibrosis, observed typically in primary biliary cirrhosis (PBC), secondary (i.e., obstructive) biliary fibrosis, and primary sclerosing cholangitis (PSC), which is sustained by concomitant proliferation of reactive bile ductules and periductular MFs. This occur at the interface between portal areas and parenchyma, leading to the characteristic development of portal–portal fibrotic septa, a scenario that leaves unchanged connections with the portal system for long time. Oxidative stress and derangement of the epithelial–stromal equilibrium around bile ducts are believed to play a major role in sustaining fibrogenesis. Moreover, MFs are supposed to derive mainly from portal fibroblasts and HSC.
4. Centrilobular fibrosis. This pattern is secondary to conditions resulting in an alteration of venous outflow (i.e., Budd–Chiari syndrome or chronic heart failure) and is characterized by the development of central–central (vein) fibrotic septa and the so-called "reversed lobulation".

8.2 Oxidative Stress, ROS, and Redox Signalling in Liver Fibrogenesis

8.2.1 *Oxidative Stress in CLD Progression: Involvement, Genesis, and General Significance*

Oxidative stress can be defined as a significant derangement in redox homeostasis resulting in sustained increase of ROS levels (i.e., intracellular) and of other reactive intermediates or mediators which exceed antioxidant defences and significantly affect signal transduction, gene expression, and functional responses of involved cells [2, 6, 24, 25]. Evidence of oxidative stress has been detected in almost all the clinical and experimental conditions of CLD with different aetiology and fibrosis progression rate, and oxidative stress (i.e., with ROS and redox signalling) has been proposed as a major pro-fibrogenic mechanism. Mechanistic involvement of oxidative stress in the fibrogenic progression of CLDs is unequivocally supported by more than two decades of experimental and clinical studies, with major concepts that may be summarized synthetically as follows [1–6, 24, 25]: (a) direct detection of oxidative stress-related parameters and/or of reactive intermediates has been reported in liver specimens from either fibrotic or cirrhotic human patients or rodent models of chronic liver injury and fibrosis; (b) administration of powerful antioxidant molecules or the use of selective therapeutic strategies designed to significantly prevent, inhibit, or interfere with ROS generation has been reported to negatively affect fibrogenic progression (mostly experimental models); (c) fibrogenic progression of clinical and experimental conditions of CLD, whatever the aetiology, is associated with a significant decrease and/or depletion of antioxidant defences.

According to the definition of oxidative stress and the related consequences previously mentioned, a critical point is then represented by the rise of intracellular levels of ROS. In CLD, ROS levels may theoretically increase in any kind of liver cell directly or indirectly involved, including hepatocytes, MFs, sinusoidal endothelial cells (SECs), Kupffer cells, cholangiocytes, T lymphocytes as well as NK and NK-T cells. However, immuno-histochemical analysis for heme-oxygenase 1 (HO-1), a redox-sensitive target gene whose expression is up-regulated in cells undergoing or exposed to oxidative stress [26], suggests that a significant rise in intracellular ROS and/or significant redox change is usually detectable mostly in hepatocytes as well as in activated HSC/MFs or, more generally, hepatic MFs in either human or rodent pathologic liver specimens [27, 28]. As summarized in Fig. 8.4, a rise of intracellular ROS levels as well as of other oxidative stress-related reactive intermediates in liver target cells is believed to represent the consequence of a number of established events [2, 6, 25] that may be synthetized as follows: (a) entry of ROS from extracellular environment as a consequence of ROS release mainly by activated inflammatory cells or by damaged and/or steatotic parenchymal cells; hydrogen peroxide can indeed easily cross biological membranes, a property also shared by nitric oxide; (b) activation of nicotinamide adenine dinucleotide phosphate (NADPH) oxidase in non-phagocytic cells which is known to parallel interaction

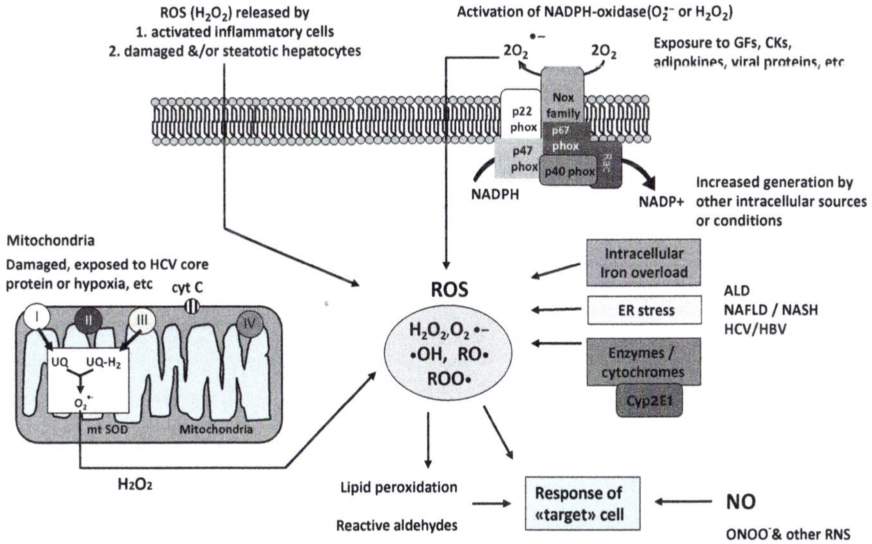

Fig. 8.4 Increase of intracellular ROS and related intermediates in chronic injury scenario. A significant increase in intracellular ROS levels can be the consequence of: (1) release of ROS (mainly H_2O_2) by damaged hepatocytes or activated inflammatory cells that can enter target cells through biological membranes; (2) activation of NADPH-oxidase (NOX) isoforms which parallel ligand-receptor interaction as exerted by several growth factors, cytokines, chemokines, viral proteins, adipokines, etc.; (3) release of ROS by mitochondria either damaged or following exposure to hypoxic conditions; (4) generation of ROS as a consequence of increased intracellular iron overload, endoplasmic reticulum (ER) stress, activation of several ROS-generating enzymes (several isoforms of cytochrome P450, oxidases, etc.)

with specific receptors by several polypeptides which are available in the extracellular environment and known to be involved in fibrogenesis, including growth factors, cytokines, chemokines, plasma proteins, adipokines, viral proteins, etc.; (c) intracellular release of ROS by damaged mitochondria (whatever the nature of the damaging stimulus or condition) or by mitochondria of cells facing exposure to hypoxic conditions, which is very common during CLD progression; (d) increased ROS generation within the cell as a consequence of several events which are known to be common during, for example, ALD, NAFLD/NASH, or HBV/HCV infection, including up-regulation of the expression of specific cytochrome P450 (CYP) isoforms (like Cyp2E1 in ALD as well as NAFLD/NASH), induction of endoplasmic reticulum (ER) stress, or iron overload, just to cite some of the major events that can affect intracellular redox homeostasis.

Irrespective of the specific aetiology or condition, an intracellular elevation of ROS and possibly of other oxidative stress-related reactive intermediates like aldehydic end products of lipid peroxidation, uncommon lipids, nitric oxide, and other reactive nitrogen species is believed to result in [24, 25]: (a) a low or transient increase in intracellular ROS resulting in up-regulation of genes carrying antioxidant-response elements (ARE) in their promoter or enhancer sequences and involved in

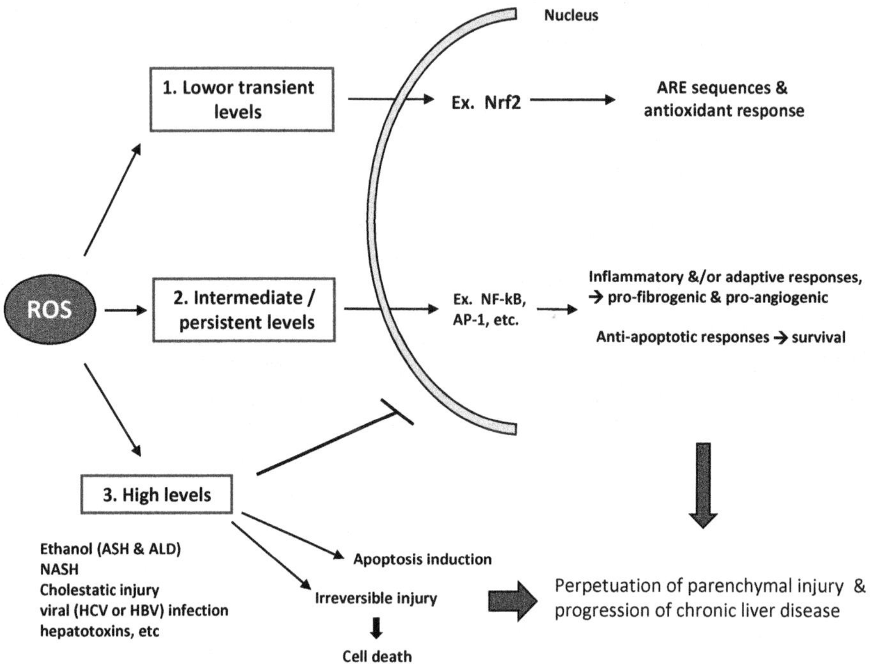

Fig. 8.5 The response of target cells to a raise in intracellular ROS and related intermediates. In the scenario of CLDs, redox-related cellular responses are closely dependent on the overall levels of oxidative stress mediators and on their temporal duration. Low/transient levels of oxidative stress lead to nuclear translocation of redox-dependent transcription factor, activation of antioxidant responsive elements (ARE) or sequences, and up-regulation of the "antioxidant response". High levels of ROS can induce both necrotic or apoptotic cell death. Intermediate and/or persistent levels of ROS and oxidative stress can up-regulate target genes that sustain inflammatory and adaptive responses, including fibrogenesis and angiogenesis

the so-called antioxidant response (i.e., designed to up-regulate antioxidant defences and then counteract oxidative stress); (b) high or very high increase in ROS, pro-oxidants, and potentially damaging intermediates, resulting by means of different mechanisms either in irreversible cell damage and necrotic cell death or induction of apoptosis, mainly of hepatocytes; (c) intermediate and persistent increase in intracellular ROS that, by operating mostly through redox-mediated activation of NF-κB and AP-1 transcription factors, can result into adaptive responses which in turn can sustain inflammatory response and fibrogenesis as well as angiogenesis and survival attitude in target cells (Fig. 8.5).

Most of these critical aspects, particularly those linking oxidative stress to hepatocellular damage and inflammatory response (i.e., major determinants of fibrogenic progression of CLDs), are already covered in detail by other chapters in this volume and then here we will focus the attention mainly on those issues and aspects which are related to redox-mediated modulation/activation of the phenotypic responses of hepatic MFs.

8.2.2 Hepatic MFs as Pro-fibrogenic Functional Target Cells Sensitive to Oxidative Stress-Related Mediators: The Pioneer Studies

As already mentioned elsewhere in this volume, ROS as well as other redox-related reactive intermediates can operate as critical secondary messengers in numerous signalling pathways, including transcriptional regulation, differentiation, proliferation, oncogenic transformation, and activation of programmed cell death [1–6, 24, 25]. ROS contribution to fibrogenic progression of CLDs has been described in chronic liver injury following alcohol abuse, HCV or HBV infection, iron overload, and chronic cholestasis. The hypothesis of a causative involvement of oxidative stress in liver fibrogenesis was initially provided by several experimental studies providing evidence that supplementation with antioxidants like silymarin, vitamin E, silybin, *S*-adenosylmethionine, and several others is able to prevent fibrogenic progression in animal models of CLDs by reducing the extent of oxidative stress and/or lipid peroxidation [2, 6, 24, 25]. In addition, different laboratories were able to describe a temporal sequence of events strongly indicating that oxidative stress and lipid peroxidation were preceding or concomitant with activation of HSC and/or involvement of MFs and collagen deposition [29–31].

The concept that MFs may represent a functional "target" cell for oxidative stress and related reactive mediators generated in hepatic chronic injury comes directly from pioneer studies in this area. Using either rat or human HSC/MFs in culture (at that time the widely accepted major source of ECM components), several groups showed that exposure to extracellularly available oxidative-stress-related mediators was rapidly resulting in up-regulation of pro-collagen type I. This was obtained by exposing HSC/MFs directly to ROS like hydrogen peroxide or superoxide anion (the latter as generated by systems like xanthine/xanthine oxidase), aldehydic end products of lipid peroxidation (mainly 4-hydroxynonenal or HNE, other 4-hydroxyalkenals of different chain length and malonyldialdehyde or MDA) [32–36] as well as to the conditioned medium of normal hepatocytes undergoing oxidative stress [37] or by co-culturing HSC/MFs with hepatocytes transfected to overexpress CYP2E1 and then exposed to ethanol in order to up-regulate ROS generation [38, 39]. Up-regulation of pro-collagen type I by either ROS, 4-hydroxynonenal, or acetaldehyde was reported to depend on activation of a c-Jun-NH_2-kinase (JNK)/AP-1 signaling [40–43], although other ROS-related mechanisms involving parallel signalling pathways have been proposed, including up-regulation of cyclooxygenase 2, activation of the redox-sensitive transcription factor Sp-1 or of Erk1/2 and p38MAPK through involvement of Janus kinases 1 and 2 (JAK1 and JAK2) [reviewed in [25]]. An additional paradigmatic example of response of HSC/MFs to ROS generated in the extracellular environment and then entering the intracellular milieu is represented by the acquisition of the redox-dependent, ERK1/2- and JKN1/2-mediated ability of these cells to migrate [27].

The overall message emerging from these earlier studies was quite clear: relevant oxidative stress-related mediators (hydrogen peroxide, HNE, MDA, F2-isoprostanes, etc.)

released in a paracrine way by damaged hepatocytes, endothelial cells, or activated inflammatory cells can easily cross the membrane of HSC/MFs and induce the up-regulation of procollagen type I (i.e., a critical component of the fibrillar-like ECM in fibrotic and cirrhotic livers).

This early "extracellular" pro-fibrogenic view was rapidly integrated by other studies that started to suggest that generation of ROS within HSC/MFs (i.e., a change in intracellular redox homeostasis) was the critical event to up-regulate pro-collagen type I. As an example, Nieto and coworkers showed the occurrence of this phenomenon in rat HSC/MFs transfected to overexpress CYP2E1 and then challenged with ethanol or arachidonic acid (i.e., conditions leading to CYP2E1-related increased generation of ROS) [44, 45]. However, a turning point in the field was a study showing that TGFβ1, the most potent pro-fibrogenic cytokine, was able to up-regulate collagen type I expression in HSC/MFs by eliciting H_2O_2-dependent signalling involving the binding of p35 C/EBPβ protein to a specific region of the promoter of the collagen α1(I) gene [46]. The relevance of this observation relies in two major points: (1) extracellular mediators, including ROS, HNE, and F2-isoprostanes, are all able to up-regulate TGFβ1 synthesis in either HSC/MFs or mononuclear cells [36, 43, 47], with TGFβ1 that in turn is operating through redox changes within the target cell, then outlining a redox-sensitive amplifying scenario in chronic activation of repair mechanisms; (2) the study prompted several laboratories to investigate the role of redox changes elicited by ligand/receptor interactions in relation to HSC/MFs phenotypic responses.

8.2.3 NADPH-Oxidase Isoforms and Pro-fibrogenic Responses: Of Liver Cells, NOX, and Aetiologies

As mentioned in the previous section, ROS were reported to either directly stimulate the production of the type I collagen or to act as intracellular signaling mediators of the fibrogenic action of TGFβ1. According to current literature and as detailed in other chapters, ROS can be generated in the liver by multiple sources, including the mitochondrial respiratory chain, CYP family members, peroxisomes, xanthine oxidase, and NADPH oxidases (NOXs) [6, 24, 25, 48]. A series of elegant studies, mostly coming from the laboratory of David Brenner, rapidly provided evidence indicating that intracellular generation of ROS mediated by NOXs had a critical role in modulating the phenotypic responses of HSC/MFs and then hepatic fibrogenesis. This area of research has been extremely productive in the last decade, as outlined recently in a nice and exhaustive review by Brenner and coworkers [49], also in the view of the fact that NOX isoforms have been envisaged as potential therapeutic targets for the development of selective anti-fibrotic therapeutic intervention. In this section we will focus the attention on major messages emerged in these years that are more directly related to MFs and their pro-fibrogenic behaviour.

As it is well-known, NOX is a multimeric transmembrane enzymatic complex which generates superoxide anion and hydrogen peroxide from molecular oxygen using NADPH as an electron donor. The complex is formed by one of the seven described isoforms of NOX (NOX 1–5, DUOX1, and DUOX2), usually heterodimeric, and by a number of NOX regulatory proteins ($p47^{phox}$/NOXOA1, $p40^{phox}$, $p67^{phox}$/NOXA1) and a Rac isoform. On stimulation with agonists/ligands, the cytosolic subunits translocate to the membrane-bound NOX complex leading to enzymatic activity, with NOX2 (also known as flavocytochrome b558, a heterodimer of $gp91^{phox}$ and $p22^{phox}$) representing the phagocytic form of NOX found in professional phagocytes, whereas the other isoforms are the non-phagocytic ones. Whether the liver is concerned, only three major NOX isoforms are expressed in the different liver cell populations [49–51]: (1) Kupffer cells as well as other immune-related cells (neutrophils, B and T cells, dendritic cells) only express NOX2 isoform; (2) HSC, endothelial cells, and hepatocytes express NOX1, NOX2, and NOX4 isoforms, with hepatocytes found also to express the non-heterodimeric isoforms DUOX1 and DUOX2. The phagocytic NOX2 is known to release high amount of superoxide anion in response to agonists (no constitutive activity), whereas non-phagocytic NOX1 and NOX4 usually generate low levels of ROS in basal conditions that can be further increased in response to a wide range of polypeptide mediators (TNFα, IL-1β, PDGF, AngII, leptin, MCP-1), with NOX4 generating predominantly hydrogen peroxide instead of superoxide anion.

The emerging scenario in conditions of liver injury, acute or chronic, indicate that ROS derived from specific NOX isoforms in specific liver cell types may affect different signalling and function in target cells. For example, acute alcohol-induced liver injury is mainly mediated by NOX2 and the non-phagocytic isoforms NOX4 and DUOXs, the latter not expressed in physiological conditions [49]. Along these lines, HCV-related proteins can induce ROS generation in different liver cells affecting either their survival or functional response. NS3 protein, for example, can phosphorylate $p47^{phox}$ and induce a prolonged release of ROS through NOX2 in mononuclear cells as well as in neutrophils, resulting in dysfunction and/or apoptosis of T lymphocytes, and negatively affecting HCV clearance [52, 53]. However, HCV core as well as NS3 and NS5 proteins can induce ROS generation in activated HSC (NOX isoforms responsible in HSC being still unidentified), resulting in (1) an increased proliferation through Ras/ERK/PI3K/Akt signalling (HCV core), (2) pro-inflammatory responses through NF-κB and c-Jun-NH2-kinase pathways (NS3/NS5), or (3) increased TGFβ1 and procollagen type I expression (HCV core, NS3 and NS5 proteins) [54]. As an additional relevant example, HCV (structural and non-structural constructs) have been reported to up-regulate TGFβ1 expression, which in turn up-regulates NOX4 (and then ROS generation) in either hepatocytes and HSC [55]. Up-regulation of NOX4 by TGFβ1 and the related increase in intracellular ROS generation (mainly hydrogen peroxide by NOX4) has been suggested as a critical event able to mediate hepatocyte apoptosis during the course of CLDs (by any aetiology, including HCV infection, alcohol, toxic bile acids, and NASH), possibly by hydrogen peroxide-dependent down-regulation of Bcl-X_L and/or by increasing the sensitivity to pro-apoptotic signals like TNF and FasL [49, 56–58].

8.2.4 NADPH-Oxidase Isoforms, Intracellular ROS Generation, and Pro-fibrogenic Responses of HSC and MFs

As previously mentioned, intracellular levels of ROS are critical in affecting the functional response of target cells as well as in inducing irreversible damage and cell death, either necrotic or apoptotic. HSCs and, more generally, hepatic MFs respond to low/intermediate levels of ROS or of other intermediates like aldehydic end products of lipid peroxidation by up-regulating expression of procollagen type I and of TIMPs as well as (ROS only) by increasing either cell proliferation or migration [32–35, 44, 59–61]. Exposure to high levels of the same mediators is followed by induction of cell death, either necrotic or apoptotic [60, 61], with HSC/MFs being in particular extremely resistant to ROS. Indeed, human HSC/MFs can undergo caspase-independent cell death only when exposed to very high concentrations of either ROS (hydrogen peroxide or superoxide anion) or HNE, which are hardly comparable with levels observed in vivo. This feature fits well with the concept that these "activated" and MF-like cells enter in a Bcl-2- and NF-κB-dependent survival attitude [1–6, 62], with an overall message that HSC/MFs may easily survive the conditions of oxidative stress usually operating in CLDs and then are more likely to sustain redox-dependent pro-inflammatory and pro-fibrogenic responses.

According to existing literature data, expression of various NOX isoforms is increased in patients with liver fibrosis [63, 64] as well as in the liver of animal models of CLD [65]. Moreover, NOX isoforms, through increased levels of intracellular ROS, can mediate fibrogenic actions in response to major polypeptide ligands involved in CLDs, including at least PDGF [66], TGFβ1 [56], Ang II [51], and leptin [67] as well as in response to advanced glycation end products [68] or phagocytosis of apoptotic bodies [69]. As mentioned in a previous section, activated HSCs (HSC/MFs) express multiple NOX isoforms and their regulatory components, with NOX1, NOX2, and NOX4 being significantly up-regulated in rodent-activated cells versus quiescent cells [65, 70]. In human-activated HSCs the situation is very similar, with documented up-regulation of NOX1, NOX2, and the cytosolic regulatory factor $p47^{phox}$ either in human HSCs activated in culture or in activated HSCs isolated from the fibrotic liver of human patients [51]. The exact structure of the complex is still uncertain [49]. Experiments performed in animal models of liver fibrosis (bile duct ligation or BDL, MCD model of NAFLD/NASH, chronic CCl_4 administration) either in genetically manipulated mice ($p47^{phox}$ or NOX2-deficient mice) or in the presence of therapeutic approaches designed to clear Kupffer cells (then the contribution of NOX2) clearly indicate that NOX2 is relevant in sustaining activation of HSCs and relevant pro-fibrogenic responses like up-regulation of collagen type I expression [51, 65, 67, 69]. However, experimental evidence has been provided strongly suggesting that also non-phagocytic NOX isoforms can contribute to liver fibrogenesis as is the case in particular for models run in NOX1-deficient mice [65, 71], with NOX1 found to be able to promote HSC proliferation via a mechanism involving ROS-mediated inactivation of phosphatase and tensin homologue (PTEN), leading then to phosphorylation of Akt and forkhead box O 4 and the resulting down-regulation of the cell cycle suppressor $p27^{kip1}$. Accordingly, specifically designed anti-fibrotic strategies in order to target other components of the NOX complex have been shown to be effective as well. The following examples are worth mentioning: (1) the increase in NOX-dependent superoxide production and Rac1 activity in SOD1 G37R-mutant mice results in a significant increase in CCl_4-induced liver fibrosis as compared with wild-type mice [70]; (2) prevention of liver fibrosis obtained by administration of the potent and specific NOX1/NOX4 inhibitor GKT137831, with reported inhibition of collagen α1(I) promoter activity and of ROS generation induced by Ang II in WT and SOD1 G37R-mutant HSCs [70]; (3) attenuation of liver fibrosis (in addition to prevention of hepatocyte apoptosis) in NOX4-deficient mice compared with WT mice [56].

As mentioned, NOX isoforms are relevant because intracellular ROS generated by NOX can contribute to boost the intracellular signalling pathways elicited by a number or relevant polypeptide pro-fibrogenic ligands, including PDGF, TGFβ1, and Ang II [6, 49]. As a direct consequence, NOX isoforms and ROS are then involved in modulating phenotypic responses of activated HSCs (HSC/MFs) or, more generally, of hepatic MFs. In relation to PDGF, this growth factor is known to represent the most potent mitogen for HSCs [72] and exerts at least part of its mitogenic effect through NOX isoforms, as shown in vitro or in vivo (DMN model of liver fibrosis) by employing an antioxidant like Mn-TBAP or the pharmacological inhibitors of the NOX complex diphenylene-iodonium (DPI) or apocynin [70]. However, the NOX isoform that contributes to PDGF-induced mitogenic signaling in HSCs is not yet identified [49]. It should be noted that NOX-derived intracellular ROS have also a role in mediating the migration of human HSC/MFs stimulated by PDGF-BB as well as by other polypeptide chemoattractants effective for these cells like VEGF-A or MCP1 (CCL2) [27]. All these polypeptides are able to induce a NADPH-oxidase-dependent intracellular rise in ROS, resulting in activation of ERK1/2 and JNK1/2 and, in particular, activation of JNK1 isoform was found to be predominant for migration as shown by specific silencing. Moreover, both ROS generation and oriented as well non-oriented migration induced by polypeptide chemoattractants were almost abolished by pre-treatment with the NOX inhibitor apocynin [27]. The potential in vivo relevance of these data was suggested by the fact that activation of ERK1/2 and JNK1/2 was evident in extracts obtained from HSC/MFs isolated during the course of an oxidative stress-mediated model of liver injury and that phosphorylated JNK1/2 isoforms were detected in α-SMA-positive myofibroblasts lining fibrotic septa in human cirrhotic livers [27].

Leptin is another relevant mediator relying on NOX to propagate its downstream signals in activated HSCs [1–6]. Leptin-induced intracellular signaling pathways have been evaluated in primary wild-type (WT), $p47^{phox}$ (−/−), or signal transducer and activator of transcription protein 3 (STAT3)-deleted HSCs [67]. In this study leptin-stimulated ROS production was found to be attenuated in human and mouse HSCs by the NOX inhibitor DPI and in HSCs lacking the NADPH component $p47^{phox}$. Of interest, leptin-induced phosphorylation of ERK and AKT, but not of STAT3, was blocked by NOX inhibition. Moreover, leptin-induced ROS production was inhibited by the Janus kinase (JAK) pharmacological inhibitor AG490, but normal ROS production was observed in STAT3-deleted HSCs. Pharmacologic or genetic inhibition of NADPH in HSCs resulted in the reduction of leptin-mediated HSC proliferation as well as in the inhibition of leptin-mediated up-regulation of collagen type 1, α-SMA, and of inflammatory-related peptides, including MCP-1, macrophage inflammatory protein 1 (MIP-1), and macrophage inflammatory protein 2 (MIP-2). The same study also showed that the leptin-enhanced chemokine expression induced by chemokine (C–C motif) ligand 4, detected in wild-type mice, was significantly blunted in $p47^{phox}$ (−/−) mice [67].

NOX isoforms and ROS have been shown also to mediate the pro-fibrogenic action of Ang II, a key effector of the renin-angiotensin system (RAS). The RAS, including the angiotensin-converting enzyme (ACE) and the angiotensin receptor type 1 (AT1R), has been shown to be markedly up-regulated in the fibrotic liver [73, 74] and several experimental strategies designed to interfere with RAS have shown to result in a significant decrease in hepatic inflammation and fibrosis in animal models of CLD (reviewed in [49]). Of interest, most of the biological actions of Ang II have been reported to involve ROS generated by NOX isoforms [73, 75] and, as mentioned previously, liver injury and fibrosis were attenuated in $p47^{phox}$ (−/−) mice undergoing BDL as compared to wild-type mice [51]. Moreover, activated HSCs express high levels of AT1R and exposure of these pro-fibrogenic cells to Ang II resulted in increased proliferation, migration, and contraction [76] as well as in up-regulation of TGFβ1 and type I collagen expression [73, 77]. NOX has been reported to stimulate, through ROS generation, several signalling pathways including PKC, PI3K/Akt, and MAPKs [49, 51]. A further study has finally indicated that Ang II is able to up-regulate expression of both NOX1 and NOX4 in HSCs [70].

A brief mention should be dedicated also to the role of NOX-generated ROS in mediating the pro-fibrogenic action of TGFβ1 [1–6]. A relevant role of NOX4 isoform in TGFβ1-mediated HSCs activation has been proposed by a study that used NOX4-deficient HSC; the same study also reported that NOX4 siRNA inhibited ROS production in HSCs, and that the activation of HSCs in culture is inhibited in NOX4-deficient mice [56]. Moreover, TGFβ1 was found to up-regulate NOX4 expression in a SMAD3-dependent way. Of interest Ang II-dependent NOX4 up-regulation is suppressed in HSCs silenced for NOX1 but not NOX4 induction by TGFβ1, suggesting that NOX1 is required for NOX4 up-regulation in HSCs stimulated with Ang II, but that TGFβ1 induces NOX4 independently of NOX1 in HSCs [70].

Finally, the role of ROS and NOX isoforms has been investigated in relation to toll-like receptor (TLR)-related responses of activated HSCs in the typical pro-inflammatory scenario of a CLD. A critical point in this area (see for more details [49]) is that both quiescent and activated HSCs express the mRNAs for multiple TLR isoforms [78], with activated HSCs responding mainly to TLR3 ligands by producing interferon β (IFNβ) [78] or to TLR4 ligands through NF-κB and JNK/AP1 activation to produce pro-inflammatory cytokines [79]. This point is critical since it has opened the way to studies that in the end showed a strict relationship, antibiotic-sensitive, between gut microflora and liver fibrosis, indicating an essential role of bacterial products in hepatic fibrogenesis, as emphasized by the finding that TLR4-mutant mice and MyD88-deficient mice are resistant to fibrosis [80] and by several studies (reviewed in [49]) investigating related molecular mechanisms. In this scenario it has been suggested that NOX and ROS might play a role in lipopolysaccharide (LPS)-TLR4-mediated inflammatory or fibrogenic signaling in HSCs. LPS, for example, has been found to induce NOX1 expression through TLR4 in macrophages, suggesting that the interaction between LPS/TLR4 signaling and NOX1 might modulate fibrogenic response in the liver by affecting macrophage response with similar data also reported for neutrophils [49]. However, although a relevant role of ROS in mediating LPS/TLR4 signalling has been proposed [80], we still need specifically designed studies to fully understand the role of NOX4-derived ROS in mediating the LPS–TLR4–NF-κB cascade in HSCs activation, suggested to possibly rely on a direct interaction between NOX4 and TLR4.

8.2.5 Hypoxia, Mitochondrial ROS, and Pro-angiogenic and Pro-fibrogenic Responses of Activated HSCs and MFs

Literature data in the last decade have provided compelling evidence that angiogenesis, a dynamic, hypoxia-stimulated, and growth factor-dependent process, is intimately linked to liver fibrogenesis and CLDs progression, with several authors suggesting that angiogenesis may represent a critical process that favours fibrogenesis [81, 82]. The presence of hypoxic areas within liver parenchyma (i.e., a very common finding) in CLD is the most obvious stimulus able to up-regulate transcription of pro-angiogenic genes through the action of hypoxia-inducible factors or HIFs, although this may happen also through hypoxia-independent mechanisms [81–84]. The progressive increase of tissue hypoxia detected in the scenario of CLDs is directly related to the histopathological changes of liver tissue, particularly the increased deposition of ECM components and formation of fibrotic septa, paralleled by vascular changes. With the time this results in an impairment of the oxygen diffusion and consequent up-regulation of pro-angiogenic pathways [81, 82]. Literature evidence supporting the link between angiogenesis and fibrogenesis has been recently reviewed by several laboratories [6, 81, 82] and major points are: (1) the parallel development of fibrogenesis and angiogenesis in experimental and clinical conditions of CLD in areas which are positive for hypoxia and VEGF, with VEGF expression mainly limited to hepatocytes and MFs [85]; (2) the relevance of HIFs for the development of liver fibrosis, as shown in mechanistic studies employing liver conditional knock out mice for HIF1α [86]; (3) the efficiency of experimental anti-angiogenic therapy in inhibiting not only fibrogenic progression but also in reducing inflammatory infiltrate, the number of MFs as well as the increase in portal pressure [6, 81, 82]; (4) the critical role of MFs as cellular crossroads modulating fibrogenesis, inflammatory response, and angiogenesis in CLDs. Pertinent to this review is the evidence that some hypoxia-dependent or independent features linked to the role of MF in the fibrogenic progression are potentially mediated through the generation of ROS. Indeed, hepatic MFs, in particular HSC/MFs, have been shown not only to respond to the exposure to hypoxic conditions in a HIF1α-dependent way by up-regulating expression of VEGF, Angiopoietin 1, and related receptor VEGFR-2 and Tie2 [52, 83] (i.e., then behaving as pro-angiogenic cells), but also to be a target for VEGF and Angiopoietin 1. VEGF and Angiopoietin 1 indeed can stimulate proliferation and increased deposition of extracellular matrix components [81, 82]), as well as increased migration and chemotaxis [28, 85], the latter response being also significantly elicited in cells just exposed to hypoxia [85]. In particular, oriented migration of MFs in response to either hypoxia or VEGF (as other chemotactic peptides) has been proposed as a biphasic mechanism that is first switched on by ROS released by either mitochondria under hypoxia or through ligand–receptor-related activation of NADPH-oxidase and proceeds through redox-dependent activation of Ras/ERK and c-Jun-NH2-terminal kinase isoforms (JNKs). This early phase is then followed by a delayed and sustained phase of migration depending on HIF-1α-mediated, ROS-stabilized, up-regulation of VEGF expression, resulting in the subsequent chemotactic action of extracellularly released VEGF [87, 88].

The concept of hypoxia-related and ROS-modulated migration of HSC/MFs and/or hepatic MFs, as a critical step for either angiogenesis and fibrogenesis, is of further extreme interest if one considers additional evidence reported in the last decade. First, immunohistochemistry analysis performed on either human or rodent fibrotic/cirrhotic liver has shown that MF-like cells in developing septa and at the border of more mature and larger fibrotic septa stain positively for both HIF-2α and heme-oxygenase 1 (HO-1, a redox-related marker). These data, which suggest that hypoxia-related and ROS-modulated events are likely to occur also in vivo [88], overlap with morphological analysis performed on human and rat fibrotic/cirrhotic livers [85] where α-SMA-positive hepatic MFs expressing VEGF, Ang-1, or the related receptors VEGFR-2 and Tie-2 were detected at the leading edge of small and incomplete developing septa, but not in larger bridging septa. It has been suggested that such a distribution may reflect the existence of two distinct phases of the angiogenic process during CLDs [83]: (a) an early phase, occurring in developing septa, in which fibrogenesis and angiogenesis (with intracellular ROS playing here a relevant role) may be driven/modulated by HSC/MFs; (b) a later phase, occurring in larger and more mature fibrotic septa, in which pro-angiogenic factors and receptors are expressed only by endothelial cells, a scenario that may favour the stabilization of the newly formed vessels.

An additional, and often neglected concept is that MFs can operate their pro-angiogenic role also in a hypoxia-independent manner by responding to a number of polypeptide mediators, including mainly PDGF and leptin, by signalling pathways that are still related to intracellular ROS generation and the involvement of HIFs. A critical point here is that a raise in intracellular ROS, whatever the origin (i.e., even in the absence of hypoxia), can result in recruitment/stabilization of HIFs, particularly HIF-1α, and up-regulation of the transcription of HIFs-related genes. This has been reported to be due to the ability of ROS to inhibit prolyl-hydroxylases, thus preventing hydroxylation of HIF-α subunits and their subsequent binding to von Hippel–Lindau protein, ubiquitylation, and proteasome degradation (i.e., the mechanism commonly operating under conditions of normoxia and preventing HIFs to act as transcription factors on their target genes) [81–84]. As far as PDGF and leptin are concerned, it is interesting to note that both mediators can promote an angiogenic response in HSCs and HSC/MFs. In an elegant study PDGF has been shown to promote an angiogenic phenotype of HSC that regulates HSC-driven vascular tube formation in vitro and enhanced coverage of sinusoids in vivo, supporting the concept that PDGF can sustain the critical role of HSC/MFs in the modulation of microvascular structure and function in liver parenchyma [89]. Concerning leptin, this adipokine can directly up-regulate in human HSC/MFs the expression of VEGF and Ang-1 as well as of monocyte-chemoattractant protein 1 (MCP-1 or CCL2) [90]. Interestingly, leptin was found to operate the pro-angiogenic actions by recruitment/stabilization of HIF-1a and nuclear translocation of HIF-1 in vitro and in vivo the specific leptin receptor ObR co-localized with VEGF and α-SMA after induction of fibrosis in rats [90]. A more recent study performed on human HSC/MFs has revealed that both leptin and PDGF-BB can directly up-regulate VEGF and then the pro-angiogenic role of these MF-like cells by a common mechanism involving both activation of the mammalian target of rapamycin (mTOR) pathway as well as generation of ROS via NADPH-oxidase, the latter being relevant for HIF-1α stabilization but not for mTOR activation [91].

8.2.6 Therapeutic Strategies to Counteract Oxidative Stress in Liver Fibrogenesis?

According to past and present literature data emphasizing the pathogenic role of ROS and redox signalling in liver fibrogenesis, one should expect a significant efficiency of therapeutic strategies designed to counteract ROS generation, oxidative stress, and redox signalling in inhibiting or at least limiting CLD progression. As already pointed out by several authors [1–6, 23–25, 49, 92–94], experimental literature on this topic is indeed impressive and at the same time quite straightforward in offering a number of well-defined take-home messages.

The first message is indeed apparently positive and unequivocal since therapeutic strategies that employed antioxidants were able to significantly prevent liver fibrosis in animal models as well as to inhibit or limit the progression of the disease toward cirrhosis. These positive results were associated with an inhibition of recruitment of inflammatory cells, a decrease in the number of hepatic MFs and, more generally, with a reduction of the levels of pro-inflammatory and pro-fibrogenic cytokines, as confirmed in different animal models of CLDs and by using either naturally occurring or synthetic antioxidants differing for their structure and/or mechanism of action. Accordingly, there is a long list of effective antioxidants employed in experimental animal models of liver fibrosis [1–6, 23–25, 49, 92–94] and includes well-known molecules such as like α-tocopherol (vitamin E), carotenoids, the selenium antioxidant ebselen, hydroxyl radical scavengers (such as dimethyl-sulphoxide of dimethyl-thiourea), *N*-acetyl-cysteine, several flavonoids and polyphenols (such as silymarin, quercetin, curcumin, and epigallocatechin), the Japanese herbal medicine sho-saiko-to, the GSH-replenishing compound *S*-adenosyl methionine, the CYP2E1 inhibitor diallyl sulphide, and the supernutrient polyenyl-phosphatidylcholine.

Unfortunately, as recently reviewed [49], data from clinical trials employing antioxidants in human conditions of CLDs are generally quite disappointing. The use of agents with general antioxidant action such as vitamin E, vitamin C, polyenyl phosphatidyl choline, or urso-deoxycholic acid (UDCA) did not result in significant antifibrotic efficacy in patients carrying different CLDs, including alcoholic hepatitis [95], NASH [96, 97], and chronic hepatitis C [98]. Two only apparent exceptions to this rule have been reported for some trials with NAFLD/NASH patients treated with vitamin E [93, 94]. In addition, these negative results had also to face a serious report indicating that the use of high dosage of the very popular antioxidant vitamin E was even associated with an increase of all-cause mortality [99].

A number of strategies have been suggested to potentially overcome these problems. A first notion was related to the fact that antioxidants failed to provide benefit because these were employed in patients in whom the CLD was already advanced. Theoretically, early diagnosis of the CLD should reasonably allow administration of safe antioxidants as soon as possible during the natural history of the disease, likely being able to slow down fibrogenic progression [23–25, 92–94]. Alternatively, it has been suggested that a more promising strategy may rely in the identification of candidate antioxidants with sufficiently rapid rate constants as to be pharmacologically active (i.e., effective even at low doses). Although it may be theoretically possible to improve the design of trials or identify more powerful antioxidant molecules, it seems clear that a successful strategy can only rely on more specific and selective approaches.

As recently reviewed by Brenner and coworkers [49], more recent literature evidence suggesting the critical role of NOX in hepatic fibrogenesis of course provides a strong rationale for the use also in clinical conditions of CLDs of pharmacological inhibitors of NOX. Promising data have been reported by drugs able to block angiotensin type 1 receptor (AT1-R) in retrospective or pilot studies in human patients with chronic hepatitis C [100–103] and NASH [104]. A relevant retrospective study in which losartan was employed showed that in approx. 50 % of patients with chronic hepatitis C the degree of fibrosis was significantly decreased. This was associated to a significant decrease in the expression of several fibrogenic genes, including procollagen a1(I) and a1(IV), urokinase type plasminogen activator, metalloproteinase type 2, and NOX genes, including NOXO1, NOXA1, and Rac1 [100]. Similar results were obtained with another AT1-R antagonist like candesartan [105], but, to be honest, conflicting and/or more negative results have been reported for this approach with efficacy that may be also variable in relation to the specific aetiology and/or CLD (reviewed in [49]). Novel pharmacological NOX inhibitors have been developed, including in particular the efficient drug

GKT137831, a small molecule able to inhibit both NOX1 and NOX4 and found to attenuate CCl4- or BDL-induced ROS production and hepatic fibrosis in mice [56, 70]. This drug has been found to be well-tolerated in a phase I clinical study, but a specifically designed clinical trial to investigate antifibrotic efficacy in the treatment of CLDs is still lacking.

References

1. Friedman SL (2008) Mechanisms of hepatic fibrogenesis. Gastroenterology 134:1655–1669
2. Parola M, Marra F, Pinzani M (2008) Myofibroblast—like cells and liver fibrogenesis: emerging concepts in a rapidly moving scenario. Mol Aspects Med 29:58–66
3. Dranoff JA, Wells RG (2010) Portal fibroblasts: underappreciated mediators of biliary fibrosis. Hepatology 51:1438–1444
4. Forbes SJ, Parola M (2011) Liver fibrogenic cells. Best Pract Res Clin Gastroenterol 25:207–218
5. Lee UE, Friedman SL (2011) Mechanisms of hepatic fibrogenesis. Best Pract Res Clin Gastroenterol 25:195–206
6. Novo E, Cannito S, Paternostro C, Bocca C, Miglietta A, Parola M (2014) Cellular and molecular mechanisms in liver fibrogenesis. Arch Biochem Biophys 548:20–37
7. Rosselli M, MacNaughtan J, Jalan M, Pinzani M (2013) Liver cirrhosis. Gut 62:1234–1241
8. El-Serag HB (2011) Hepatocellular carcinoma. N Engl J Med 365:1118–1127
9. El-Serag HB (2012) Epidemiology of viral hepatitis and hepatocellular carcinoma. Gastroenterology 142:1264–1273
10. Rehm J, Samokhvalov AV, Shield KD (2013) Global burden of alcoholic liver diseases. J Hepatol 59:160–168
11. Iredale JP, Thompson A, Henderson NC (2013) Extracellular matrix degradation in liver fibrosis: biochemistry and regulation. Biochim Biophys Acta 1832:876–883
12. Povero D, Busletta C, Novo E, Valfrè di Bonzo L, Cannito S, Paternostro C, Parola M (2010) Liver fibrosis: a dynamic and potentially reversible process. Histol Histopathol 25:1075–1091
13. Cassiman D, Libbrecht L, Desmet V, Denef C, Roskams T (2002) Hepatic stellate cell/myofibroblast subpopulations in fibrotic human and rat livers. J Hepatol 36:200–209
14. Novo E, Valfrè di Bonzo L, Cannito S, Colombatto S, Parola M (2008) Hepatic myofibroblasts: a heterogeneous population of multifunctional cells in liver fibrogenesis. Int J Biochem Cell Biol 41:2089–2093
15. Mederacke I, Hsu CC, Troeger JS, Huebener P, Mu X, Dapito DH, Pradere JP, Schwabe RF (2013) Fate tracing reveals hepatic stellate cells as dominant contributors to liver fibrosis independent of its aetiology. Nat Commun 4:2823. doi:10.1038/ncomms3823
16. Siani A, Tirelli N (2014) Myofibroblast differentiation: main features, biomedical relevance, and the role of reactive oxygen species. Antioxid Redox Signal 21(5):768–785. doi:10.1089/ars.2013.5724
17. Pinzani M (2011) Epithelial-mesenchymal transition in chronic liver disease: fibrogenesis or escape from death? J Hepatol 55:459–465
18. Xie G, Diehl AM (2013) Evidence for and against epithelial-to-mesenchymal transition in the liver. Am J Physiol Gastrointest Liver Physiol 305:G881–G890
19. Mederacke I, Hsu CC, Troeger JS, Huebener P, Mu X, Dapito DH, Pradere JP, Schwabe RF (2013) Fate tracing reveals hepatic stellate cells as dominant contributors to liver fibrosis independent of its aetiology. Nat Commun 4:2823. doi:10.1038/ncomms3823
20. Watson MR, Wallace K, Gieling RG, Manas DM, Jaffray E, Hay RT, Mann DA, Oakley F (2008) NF-κB is a critical regulator of the survival of rodent and human hepatic myofibroblasts. J Hepatol 48:589–597

21. Novo E, Marra F, Zamara E, Valfrè di Bonzo L, Monitillo L, Cannito S, Petrai I, Mazzocca A, Bonacchi A, De Franco RS, Colombatto S, Autelli R, Pinzani M, Parola M (2006) Overexpression of Bcl-2 by activated human hepatic stellate cells: resistance to apoptosis as a mechanism of progressive hepatic fibrogenesis in humans. Gut 55:1174–1182
22. Kendall TJ, Hennedige S, Aucott RL, Hartland SN, Vernon MA, Benyon RC, Iredale JP (2009) p75 neurotrophin receptor signaling regulates hepatic myofibroblast proliferation and apoptosis in recovery from rodent liver fibrosis. Hepatology 49:901–910
23. Pinzani M, Rombouts K (2004) Liver fibrosis: from the bench to clinical targets. Dig Liver Dis 36:231–242
24. Robino G, Parola M (2001) Oxidative stress-related molecules and liver fibrosis. J Hepatol 35:297–306
25. Novo E, Parola M (2008) Redox mechanisms in hepatic chronic wound healing and fibrogenesis. Fibrogenesis Tissue Repair 1:5
26. Ryter SW, Choi AM (2005) Heme oxygenase-1: redox regulation of a stress protein in lung and cell culture models. Antioxid Redox Signal 7:80–91
27. Novo E, Busletta C, Valfrè di Bonzo L, Povero D, Paternostro C, Mareschi K, Ferrero I, David E, Bertolani C, Caligiuri A, Cannito S, Tamagno E, Compagnone A, Colombatto S, Marra F, Fagioli F, Pinzani M, Parola M (2011) Intracellular reactive oxygen species are required for directional migration of resident and bone marrow-derived hepatic pro-fibrogenic cells. J Hepatol 54:964–974
28. Novo E, Povero D, Busletta C, Paternostro C, di Bonzo LV, Cannito S, Compagnone A, Bandino A, Marra F, Colombatto S, David E, Pinzani M, Parola M (2012) The biphasic nature of hypoxia-induced directional migration of activated human hepatic stellate cells. J Pathol 226:588–597
29. Bedossa P, Houglum K, Trautwein C, Holstege A, Chojkier M (1994) Stimulation of collagen α 1(I) gene expression is associated with lipid peroxidation in hepatocellular injury. A link to tissue fibrosis? Hepatology 19:1262–1271
30. Niemela O, Parkkila S, Yla-Herttuala S, Villanueva J, Ruebner B, Halsted CH (1995) Sequential acetaldehyde production, lipid peroxidation and fibrogenesis in micropig model of alcohol induced disease. Hepatology 22:1208–1214
31. Tsukamoto H, Horne W, Kamimura S, Niemela O, Parkkila S, Yla-Herttuala S, Brittenham GM (1995) Experimental liver cirrhosis induced by alcohol and iron. J Clin Invest 96:620–630
32. Parola M, Pinzani M, Casini A, Albano E, Poli G, Gentilini A, Gentilini P, Dianzani MU (1993) Stimulation of lipid peroxidation or 4-hydroxynonenal treatment increases procollagen α 1(I) gene expression in human liver fat-storing cells. Biochem Biophys Res Commun 194:1044–1050
33. Casini A, Ceni E, Salzano R, Biondi P, Parola M, Galli A, Foschi M, Caligiuri A, Pinzani M, Surrenti C (1997) Neutrophil-derived superoxide anion induces lipid peroxidation and stimulates collagen synthesis in human hepatic stellate cells: role of nitric oxide. Hepatology 25:361–367
34. Maher JJ, Tzagarakis C, Gimenez A (1994) Malondialdehyde stimulates collagen production by hepatic lipocytes only upon activation in primary culture. Alcohol Alcohol 29:605–610
35. Parola M, Pinzani M, Casini A, Leonarduzzi G, Marra F, Caligiuri A, Ceni E, Biondi P, Poli G, Dianzani MU (1996) Induction of procollagen type I gene expression and synthesis in human hepatic stellate cells by 4-hydroxy-2,3-nonenal and other 4-hydroxy-2,3-alkenals is related to their molecular structure. Biochem Biophys Res Commun 1996(222):261–264
36. Comporti M, Arezzini B, Signorini C, Sgherri C, Monaco B, Gardi C (2005) F2-isoprostanes stimulate collagen synthesis in activated hepatic stellate cells: a link with liver fibrosis? Lab Invest 2005(85):1381–1391
37. Svegliati Baroni G, D'Ambrosio L, Ferretti G, Casini A, Di Sario A, Salzano R, Ridolfi F, Saccomanno S, Jezequel AM, Benedetti A (1998) Fibrogenic effect of oxidative stress on rat hepatic stellate cells. Hepatology 27:720–726

38. Nieto N, Friedman SL, Cederbaum AI (2002) Cytochrome P502E1-derived reactive oxygen species mediate paracrine stimulation of collagen I protein synthesis by hepatic stellate cells. J Biol Chem 277:9853–9864
39. Nieto N, Cederbaum AI (2003) Increased Sp1-dependent transactivation of the LAMã promoter in hepatic stellate cells co-cultured with HepG2 cells overexpressing cytochrome P450 2E1. J Biol Chem 278:15360–15372
40. Parola M, Robino G, Marra F, Pinzani M, Bellomo G, Leonarduzzi G, Chiarugi P, Camandola S, Poli G, Waeg G, Gentilini P, Dianzani MU (1998) HNE interacts directly with JNK isoforms in human hepatic stellate cells. J Clin Invest 102:1942–1950
41. Chen A, Davis BH (1999) UV irradiation activates JNK and increases alpha (I) collagen gene expression in rat hepatic stellate cells. J Biol Chem 274:158–164
42. Greenwel P, Dominguez-Rosales JA, Mavi G, Rivas-Estilla AM, Rojkind M (2000) Hydrogen peroxide: a link between acetaldehyde-elicited alpha1(I) collagen gene up-regulation and oxidative stress in mouse hepatic stellate cells. Hepatology 31:109–116
43. Chen A, Davis BH (2000) The DNA binding protein BTEB mediates acetaldehyde-induced, Jun N-terminal kinase-dependent alphaI(I) collagen gene expression in rat hepatic stellate cells. Mol Cell Biol 20:2818–2826
44. Nieto N, Friedman SL, Greenwel P, Cederbaum AI (1999) Cyp2E1-mediated oxidative stress induces collagen type I expression in rat hepatic stellate cells. Hepatology 30:987–996
45. Nieto N, Greenwel P, Friedman SL, Zhang F, Dannenberg AJ, Cederbaum AI (2000) Ethanol and arachidonic acid increase α 2 (I) collagen expression in rat hepatic stellate cells overexpressing cytochrome P450 2E1. J Biol Chem 26:20136–20145
46. Garcia-Trevijano E, Iraburu MJ, Fontana L, Dominguez-Rosales JA, Auster A, Covarrubias-Pinedo A, Rojkind M (1999) Trasforming growth factor β1 induces the expression of α (I) procollagen mRNA by a hydrogen peroxide-C/EBPβ-dependent mechanism in rat hepatic stellate cells. Hepatology 29:960–970
47. Leonarduzzi G, Scavazza A, Biasi F, Chiarpotto E, Camandola S, Vogel S, Dargel R, Poli G (1997) The lipid peroxidation end product 4-hydroxy-2,3-nonenal up-regulates transforming growth factor β1 expression in the macrophage lineage: a link between oxidative injury and fibrosclerosis. FASEB J 11:851–857
48. Czaja MJ (2007) Cell signaling in oxidative stress-induced liver injury. Semin Liver Dis 27:378–389
49. Paik YH, Kim J, Aoyama T, De Minicis S, Bataller R, Brenner DA (2014) Role of NADPH oxidases in liver fibrosis. Antioxid Redox Signal 20:2854–2872
50. Bedard K, Krause KH (2007) The NOX family of ROS-generating NADPH oxidases: physiology and pathophysiology. Physiol Rev 87:245–313
51. Bataller R, Schwabe RF, Choi YH, Yang L, Paik YH, Lindquist J, Qian T, Schoonhoven R, Hagedorn CH, Lemasters JJ, Brenner DA (2003) NADPH oxidase signal transduces angiotensin II in hepatic stellate cells and is critical in hepatic fibrosis. J Clin Invest 112:1383–1394
52. Bureau C, Bernad J, Chaouche N, Orfila C, Béraud M, Gonindard C, Alric L, Vinel JP, Pipy B (2001) Nonstructural 3 protein of hepatitis C virus triggers an oxidative burst in human monocytes via activation of NADPH oxidase. J Biol Chem 276:23077–23083
53. Thorén F, Romero A, Lindh M, Dahlgren C, Hellstrand K (2004) A hepatitis C virus-encoded, nonstructural protein (NS3) triggers dysfunction and apoptosis in lymphocytes: role of NADPH oxidase-derived oxygen radicals. J Leukoc Biol 76:1180–1186
54. Bataller R, Paik YH, Lindquist JN, Lemasters JJ, Brenner DA (2004) Hepatitis C virus core and nonstructural proteins induce fibrogenic effects in hepatic stellate cells. Gastroenterology 126:529–540
55. Boudreau HE, Emerson SU, Korzeniowska A, Jendrysik MA, Leto TL (2009) Hepatitis C virus (HCV) proteins induce NADPH oxidase 4 expression in a transforming growth factor beta-dependent manner: a new contributor to HCV-induced oxidative stress. J Virol 83:12934–12946

56. Jiang JX, Chen X, Serizawa N, Szyndralewiez C, Page P, Schröder K, Brandes RP, Devaraj S, Török NJ (2012) Liver fibrosis and hepatocyte apoptosis are attenuated by GKT137831, a novel NOX4/NOX1 inhibitor in vivo. Free Radic Biol Med 53(2):289–296
57. Reinehr R, Becker S, Eberle A, Grether-Beck S, Häussinger D (2005) Involvement of NADPH oxidase isoforms and Src family kinases in CD95-dependent hepatocyte apoptosis. J Biol Chem 280:27179–27194
58. Yang L, Roh YS, Song J, Zhang B, Liu C, Loomba R, Seki E (2014) Transforming growth factor beta signaling in hepatocytes participates in steatohepatitis through regulation of cell death and lipid metabolism in mice. Hepatology 59:483–495
59. Lee KS, Buck M, Houglum K, Chojkier M (1995) Activation of hepatic stellate cells by TGF alpha and collagen type I is mediated by oxidative stress through c-myb expression. J Clin Invest 96:2461–2468
60. Novo E, Marra F, Zamara E, Valfrè di Bonzo L, Caligiuri A, Cannito S, Antonaci C, Colombatto S, Pinzani M, Parola M (2006) Dose dependent and divergent effects of superoxide anion on cell death, proliferation, and migration of activated human hepatic stellate cells. Gut 55:90–97
61. Zamara E, Novo E, Marra F, Gentilini A, Romanelli RG, Caligiuri A, Robino G, Tamagno E, Aragno M, Danni O, Autelli R, Colombatto S, Dianzani MU, Pinzani M, Parola M (2004) 4-Hydroxynonenal as a selective pro-fibrogenic stimulus for activated human hepatic stellate cells. J Hepatol 40:60–68
62. Novo E, Marra F, Zamara E, Valfrè di Bonzo L, Monitillo L, Cannito S, Petrai I, Mazzocca A, Bonacchi A, De Franco RS, Colombatto S, Autelli R, Pinzani M, Parola M (2006) Overexpression of Bcl-2 by activated human hepatic stellate cells: resistance to apoptosis as a mechanism of progressive hepatic fibrogenesis in humans. Gut 55:1174–1182
63. Colmenero J, Bataller R, Sancho-Bru P, Bellot P, Miquel R, Moreno M, Jares P, Bosch J, Arroyo V, Caballeria J, Gines P (2007) Hepatic expression of candidate genes in patients with alcoholic hepatitis: correlation with disease severity. Gastroenterology 132:687–697
64. de Mochel NS, Seronello S, Wang SH, Ito C, Zheng JX, Liang TJ, Lambeth JD, Choi J (2010) Hepatocyte NAD(P)H oxidases as an endogenous source of reactive oxygen species during hepatitis C virus infection. Hepatology 52:47–59
65. Paik YH, Iwaisako K, Seki E, Inokuchi S, Schnabl B, Osterreicher CH, Kisseleva T, Brenner DA (2011) The nicotinamide adenine dinucleotide phosphate oxidase (NOX) homologues NOX1 and NOX2/gp91(phox) mediate hepatic fibrosis in mice. Hepatology 53:1730–1741
66. Adachi T, Togashi H, Suzuki A, Kasai S, Ito J, Sugahara K, Kawata S (2005) NAD(P)H oxidase plays a crucial role in PDGF-induced proliferation of hepatic stellate cells. Hepatology 41:1272–1281
67. De Minicis S, Seki E, Oesterreicher C, Schnabl B, Schwabe RF, Brenner DA (2008) Reduced nicotinamide adenine dinucleotide phosphate oxidase mediates fibrotic and inflammatory effects of leptin on hepatic stellate cells. Hepatology 48:2016–2026
68. Guimaraes EL, Empsen C, Geerts A, van Grunsven LA (2010) Advanced glycation end products induce production of reactive oxygen species via the activation of NADPH oxidase in murine hepatic stellate cells. J Hepatol 52:389–397
69. Zhan SS, Jiang JX, Wu J, Halsted C, Friedman SL, Zern MA, Torok NJ (2006) Phagocytosis of apoptotic bodies by hepatic stellate cells induces NADPH oxidase and is associated with liver fibrosis in vivo. Hepatology 43:435–443
70. Aoyama T, Paik YH, Watanabe S, Laleu B, Gaggini F, Fioraso-Cartier L, Molango S, Heitz F, Merlot C, Szyndralewiez C, Page P, Brenner DA (2012) Nicotinamide adenine dinucleotide phosphate oxidase in experimental liver fibrosis: GKT137831 as a novel potential therapeutic agent. Hepatology 56:2316–2327
71. Cui W, Matsuno K, Iwata K, Ibi M, Matsumoto M, Zhang J, Zhu K, Katsuyama M, Torok NJ, Yabe-Nishimura C (2011) NOX1/nicotinamide adenine dinucleotide phosphate, reduced form (NADPH) oxidase promotes proliferation of stellate cells and aggravates liver fibrosis induced by bile duct ligation. Hepatology 54:949–958

72. Pinzani M, Gesualdo L, Sabbah GM, Abboud HE (1989) Effects of platelet-derived growth factor and other polypeptide mitogens on DNA synthesis and growth of cultured rat liver fat-storing cells. J Clin Invest 84:1786–1793
73. Bataller R, Sancho-Bru P, Gines P, Brenner DA (2005) Liver fibrogenesis: a new role for the renin-angiotensin system. Antioxid Redox Signal 7:1346–1355
74. Paizis G, Cooper ME, Schembri JM, Tikellis C, Burrell LM, Angus PW (2002) Up-regulation of components of the renin angiotensin system in the bile duct-ligated rat liver. Gastroenterology 123:1667–1676
75. Griendling KK, Ushio-Fukai M (2000) Reactive oxygen species as mediators of angiotensin II signaling. Regul Pept 91:21–27
76. Bataller R, Gines P, Nicolas JM, Gorbig MN, Garcia-Ramallo E, Gasull X, Bosch J, Arroyo V, Rodes J (2000) Angiotensin II induces contraction and proliferation of human hepatic stellate cells. Gastroenterology 118:1149–1156
77. Yoshiji H, Kuriyama S, Yoshii J, Ikenaka Y, Noguchi R, Nakatani T, Tsujinoue H, Fukui H (2001) Angiotensin-II type 1 receptor interaction is a major regulator for liver fibrosis development in rats. Hepatology 34:745–750
78. Wang B, Trippler M, Pei R, Lu M, Broering R, Gerken G, Schlaak JF (2009) Toll-like receptor activated human and murine hepatic stellate cells are potent regulators of hepatitis C virus replication. J Hepatol 51:1037–1045
79. Paik YH, Schwabe RF, Bataller R, Russo MP, Jobin C, Brenner DA (2003) Toll-like receptor 4 mediates inflammatory signaling by bacterial lipopolysaccharide in human hepatic stellate cells. Hepatology 37:1043–1055
80. Seki E, De Minicis S, Osterreicher CH, Kluwe J, Osawa Y, Brenner DA, Schwabe RF (2007) TLR4 enhances TGF-beta signaling and hepatic fibrosis. Nat Med 13(1324–1332):2007
81. Fernández M, Semela D, Bruix J, Colle I, Pinzani M, Bosch J (2009) Angiogenesis in liver diseases. J Hepatol 50:604–620
82. Valfrè di Bonzo L, Novo E, Cannito S, Busletta C, Paternostro C, Povero D, Parola M (2009) Angiogenesis and liver fibrogenesis. Histol Histopathol 23(2009):1324–1341
83. Semenza GL (2012) Hypoxia-inducible factors in physiology and medicine. Cell 148:399–408
84. Carmeliet P, Jain RK (2011) Molecular mechanisms and clinical applications of angiogenesis. Nature 473:298–307
85. Novo E, Cannito S, Zamara E, Valfrè di Bonzo L, Caligiuri A, Cravanzola C, Compagnone A, Colombatto S, Marra F, Pinzani M, Parola M (2007) Proangiogenic cytokines as hypoxia-dependent factors stimulating migration of human hepatic stellate cells. Am J Pathol 170:1942–1953
86. Moon J-K, Welch TP, Gonzalez FJ, Copple BL (2009) Reduced liver fibrosis in hypoxia-inducible factor-1alpha-deficient mice. Am J Physiol Gastrointest Liver Physiol 296:G582–G592
87. Novo E, Povero D, Busletta C, Paternostro C, Valfrè di Bonzo L, Cannito S, Compagnone A, Bandino A, Marra F, Colombatto S, David E, Pinzani M, Parola M (2012) The biphasic nature of hypoxia-induced directional migration of activated human hepatic stellate cells. J Pathol 226:588–597
88. Novo E, Busletta C, Valfrè di Bonzo L, Povero D, Paternostro C, Mareschi K, Ferrero I, David E, Bertolani C, Caligiuri A, Cannito S, Tamagno E, Compagnone A, Colombatto S, Marra F, Fagioli F, Pinzani M, Parola M (2011) Dissection of the biphasic nature of hypoxia-induced motogenic action in bone marrow-derived human mesenchymal stem cells. J Hepatol 54:964–974
89. Semela D, Das A, Langer D, Kang N, Leof E, Shah V (2008) Platelet-derived growth factor signaling through ephrin-b2 regulates hepatic vascular structure and function. Gastroenterology 135:671–679
90. Aleffi S, Petrai I, Bertolani C, Parola M, Colombatto S, Novo E, Vizzutti F, Anania FA, Milani S, Rombouts K, Laffi G, Pinzani M, Marra F (2005) Upregulation of pro-inflammatory

and pro-angiogenic cytokines by leptin in human hepatic stellate cells. Hepatology 42:1339–1348

91. Aleffi S, Navari N, Delogu W, Galastri S, Novo E, Rombouts K, Pinzani M, Parola M, Marra F (2011) Mammalian target of rapamycin mediates the angiogenic effects of leptin in human hepatic stellate cells. Am J Physiol Gastrointest Liver Physiol 301:G210–G219
92. Pinzani M, Rombouts K, Colagrande S (2005) Fibrosis in chronic liver diseases: diagnosis and management. J Hepatol 42(Suppl 1):S22–S36
93. Chang CY, Argo CK, Al-Osaimi AMS, Caldwell SH (2006) Therapy of NAFLD: antioxidants and cytoprotective agents. J Clin Gastroenterol 40(Suppl 1):S51–S60
94. de Alwis NMW, Day CP (2008) Non-alcoholic fatty liver: the mist gradually clear. J Hepatol 48(Suppl 1):S105–S112
95. Lieber CS, Weiss DG, Groszmann R, Paronetto F, Schenker S (2003) II. Veterans Affairs Cooperative Study of polyenylphosphatidylcholine in alcoholic liver disease. Alcohol Clin Exp Res 27:1765–1772
96. Lindor KD, Kowdley KV, Heathcote EJ, Harrison ME, Jorgensen R, Angulo P, Lymp JF, Burgart L, Colin P (2004) Ursodeoxycholic acid for treatment of nonalcoholic steatohepatitis: results of a randomized trial. Hepatology 39:770–778
97. Sanyal AJ, Chalasani N, Kowdley KV, McCullough A, Diehl AM, Bass NM, Neuschwander-Tetri BA, Lavine JE, Tonascia J, Unalp A, Van Natta M, Clark J, Brunt EM, Kleiner DE, Hoofnagle JH, Robuck PR (2010) Pioglitazone, vitamin E, or placebo for nonalcoholic steatohepatitis. N Engl J Med 362:1675–1685
98. Zarski JP, Sturm N, Desmorat H, Melin P, Raabe JJ, Bonny C, Sogni P, Pinta A, Rouanet S, Babany G, Cheveau A, Chevallier M (2010) Non-invasive assessment of liver fibrosis progression in hepatitis C patients retreated for 96 weeks with antiviral therapy: a randomized study. Liver Int 30:1049–1058
99. Miller ER 3rd, Pastor-Barriuso R, Dalal D, Riemersma RA, Appel LJ, Guallar E (2005) Meta-analysis: high-dosage vitamin E supplementation may increase all-cause mortality. Ann Intern Med 142:37–46
100. Colmenero J, Bataller R, Sancho-Bru P, Dominguez M, Moreno M, Forns X, Bruguera M, Arroyo V, Brenner DA, Gines P (2009) Effects of losartan on hepatic expression of nonphagocytic NADPH oxidase and fibrogenic genes in patients with chronic hepatitis C. Am J Physiol Gastrointest Liver Physiol 297:G726–G734
101. Corey KE, Shah N, Misdraji J, Abu Dayyeh BK, Zheng H, Bhan AK, Chung RT (2009) The effect of angiotensin-blocking agents on liver fibrosis in patients with hepatitis C. Liver Int 29:748–753
102. Rimola A, Londono MC, Guevara G, Bruguera M, Navasa M, Forns X, Garcia-Retortillo M, Garcia-Valdecasas JC, Rodes J (2004) Beneficial effect of angiotensin-blocking agents on graft fibrosis in hepatitis C recurrence after liver transplantation. Transplantation 78:686–691
103. Sookoian S, Fernandez MA, Castano G (2005) Effects of six months losartan administration on liver fibrosis in chronic hepatitis C patients: a pilot study. World J Gastroenterol 11:7560–7563
104. Zarski JP, Sturm N, Desmorat H, Melin P, Raabe JJ, Bonny C, Sogni P, Pinta A, Rouanet S, Babany G, Cheveau A, Chevallier M (2010) Non-invasive assessment of liver fibrosis progression in hepatitis C patients retreated for 96 weeks with antiviral therapy: a randomized study. Liver Int 30:1049–1058
105. Kim MY, Cho MY, Baik SK, Jeong PH, Suk KT, Jang YO, Yea CJ, Kim JW, Kim HS, Kwon SO, Yoo BS, Kim JY, Eom MS, Cha SH, Chang SJ (2012) Beneficial effects of candesartan, an angiotensin-blocking agent, on compensated alcoholic liver fibrosis—a randomized open-label controlled study. Liver Int 32:977–987

Part III
Oxidative Stress in Specific Liver Disease Conditions

Chapter 9
Oxidative Stress in Acute Liver Failure

Mitchell R. McGill and Hartmut Jaeschke

9.1 Introduction

Acute liver failure (ALF) is not itself a disease, but a condition that develops rapidly in response to some forms of liver injury. Clinically, ALF is defined as reduced blood clotting and altered mental function, the latter beginning within just days or weeks of initial symptoms (e.g., jaundice, coagulopathy, etc.), in the absence of previously established liver disease [44, 45]. ALF is sometimes further divided into three categories: hyper-acute (caused almost exclusively by acetaminophen [APAP] overdose), acute, and subacute liver failure are characterized by the development of encephalopathy within 1, 4, or 12 weeks of illness onset, respectively. Generally, patient outcomes are worse in acute and subacute liver failure cases, while hyper-acute ALF patients have the best long-term outlook. Although ALF is relatively rare (approximately 2,000 cases in the US each year), mortality is very high [44, 45].

Importantly, the systemic effects of ALF are the result of hepatocyte injury leading to reduced hepatocyte function. Coagulopathy is caused by impaired synthesis of blood coagulation factors in the damaged liver. Encephalopathy can occur through multiple mechanisms, including decreased blood flow to the brain and possibly an increase in endogenous signaling molecules that are normally removed from circulation by the liver [34]. The reduced cranial blood flow is a result of cerebral edema and increased intracranial pressure, which are unique features of ALF. There is considerable evidence that the brain swelling is due to accumulation of ammonia within astrocytes [72]. Ammonia is produced in the GI tract and skeletal muscles as a result of amino acid catabolism. Normally, the liver converts

M.R. McGill, Ph.D. (✉) • H. Jaeschke, Ph.D.
Department of Pharmacology, Toxicology & Therapeutics, University of Kansas Medical Center, 3901 Rainbow Blvd, MS 1018, Kansas City, KS 66160, USA
e-mail: mmcgill@kumc.edu; hjaeschke@kumc.edu

E. Albano, M. Parola (eds.), *Studies on Hepatic Disorders*,
Oxidative Stress in Applied Basic Research and Clinical Practice,
DOI 10.1007/978-3-319-15539-5_9

ammonia to urea through the urea cycle, and muscle tissue and astrocytes in the brain play a lesser role in detoxification of ammonia. However, when hepatocytes are damaged, the urea cycle in the liver is compromised and astrocytes take up more, which results in brain swelling due to increased intracellular solute concentration. Infection and sepsis can also occur as complications of ALF [43], probably due to reduced complement protein synthesis and reduced phagocytic activity in the liver, and the resulting inflammation may also contribute to brain swelling [72]. Glial cells in the brain also have immunomodulatory functions and can be activated by lipopolysaccharide and cytokines that result from infection and liver damage, resulting in localized neuroinflammation that can exacerbate the swelling [72]. In any case, brain hernia and infection due to loss of liver function are major causes of death in ALF patients.

The causes of ALF are numerous, but the most common are acetaminophen (APAP) hepatotoxicity, non-APAP drug-induced liver injury (DILI), viral hepatitis, various genetic liver diseases, and autoimmune hepatitis. Oxidative stress is thought to occur in several of these conditions. The focus of this chapter is on the role of oxidative stress in the causes and progression of ALF. Because APAP overdose is frequent and because it is well-studied experimentally, emphasis will be placed on APAP-induced liver injury, though other etiologies will also be discussed.

9.2 Oxidative Stress in the Causes of ALF

9.2.1 Drug-Induced Liver Injury

9.2.1.1 Acetaminophen Hepatotoxicity

Acetaminophen is a very popular drug [36]. However, APAP overdose is currently the most common cause of ALF and ALF-related deaths in several countries [6, 24, 44]. Oxidative stress is believed to play a major role in the mechanisms of liver injury caused by APAP (Fig. 9.1). Like many drugs, the major site of APAP metabolism is the liver. At recommended doses, most is glucuronidated or sulfated. The resulting glucuronide and sulfate conjugates of APAP are harmless and are quickly eliminated in urine. The small remaining fraction of a dose is converted to an electrophilic intermediate, commonly believed to be N-acetyl-p-benzoquinone imine (NAPQI) [17], by cytochrome P450 enzymes. Although this reactive metabolite of APAP binds to sulfhydryl groups on proteins even after therapeutic doses [27, 59], most of it reacts with endogenous glutathione (GSH) in the liver and is then transported out of the cell for further processing and excretion by other organs [55]. However, after an overdose, the parent drug depletes GSH stores in the liver and reactive metabolite formation and protein binding become overwhelming [59, 61]. This leads to massive oxidative stress and organelle dysfunction. Eventually, the damaged hepatocytes undergo oncotic necrosis.

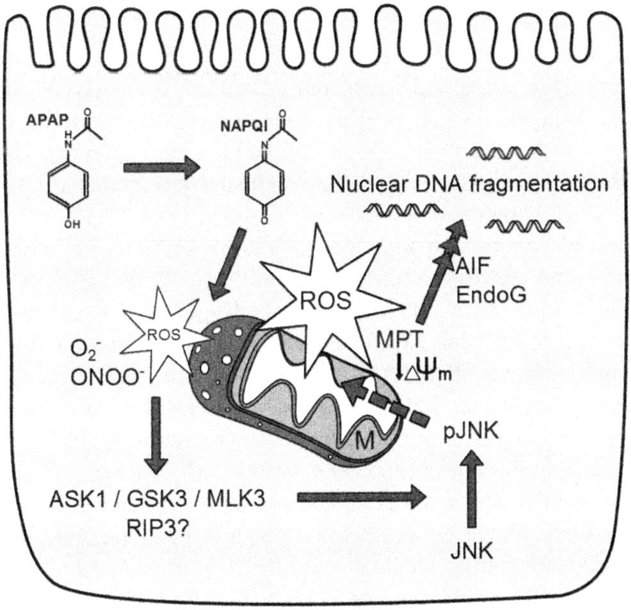

Fig. 9.1 Oxidative stress in acetaminophen hepatotoxicity. After an overdose of acetaminophen, excess reactive metabolite binds to mitochondrial proteins and causes increased reactive oxygen species (ROS) production and oxidative stress. This initial oxidative stress activates kinase signaling pathways that converge on the c-Jun N-terminal kinases (JNK) 1/2. Activated JNK then translocates into mitochondria, where it increases the oxidative stress. The oxidative stress drives the mitochondrial membrane permeability transition (MPT), resulting in loss of mitochondrial membrane potential. Eventually, the MPT and mitochondrial membrane swelling and rupture result in release of mitochondrial proteins, including endonucleases that can cleave nuclear DNA

The source of oxidative stress during APAP hepatotoxicity is controversial. Several possibilities have been suggested. Because it is known that electrons and reactive oxygen species (ROS) can escape from the active site of P450s during drug metabolism, it was once thought that P450-mediated APAP metabolism was the major cause of oxidative stress and that lipid peroxidation was a major pathological consequence [32, 97]. However, this was based in large part on experiments done under special conditions with mice that were primed for lipid peroxidation through a diet that was deficient in vitamin E and in which saturated fats were replaced with polyunsaturated fats from soybean oil [32]. In mice fed a normal control diet, treatment with a toxic dose of APAP does not significantly increase liver levels of malondialdehyde, an indicator of lipid peroxidation [38]. Additionally, treatment with exogenous vitamin E does not protect against APAP [38]. Although lipid peroxidation occurs in cell culture models of APAP hepatotoxicity and vitamin E does protect in vitro, there are strong data which suggest that this is due to the high oxygen concentration at which most cell culture experiments are performed [100]. More direct evidence for the lack of P450 involvement in the oxidative stress caused by APAP is the fact that oxidized GSH (GSSG) content in the liver does not change in the first 2 h after APAP treatment, when P450-mediated metabolism is occurring in mice [84, 85]. This was also shown in vitro. Dichlorofluorescein fluorescence, an indicator of oxidative stress, did not increase until 3.5 h after APAP treatment in primary mouse hepatocytes [2]. By this time, depletion of GSH by the reactive metabolite of APAP was complete, but injury had not yet begun [2]. The data suggest that while oxidative stress is not the result of P450-mediated metabolism, it does occur in APAP hepatotoxicity.

Another possible source of oxidative stress during APAP hepatotoxicity that was once proposed is the enzyme xanthine oxidoreductase (XOR) [92]. XOR has two forms: xanthine oxidase (XO) and xanthine dehydrogenase. XO can produce ROS, and there is a shift toward the XO form in mice after treatment with APAP [31]. Importantly, pretreatment with the XOR inhibitor allopurinol protects against APAP hepatotoxicity [31]. However, it was found that the dose of allopurinol required for this is much higher than the minimum dose needed to inhibit the enzyme, making it unlikely that XOR inhibition was the mechanism by which this compound was protected [31]. It was then assumed that allopurinol served instead as an antioxidant in APAP hepatotoxicity. However, recent work has shown that the protection provided by this drug is dependent upon aldehyde oxidase (AO)-mediated conversion of allopurinol to oxypurinol [98]. AO activity is known to produce ROS. Thus, it is now thought that metabolism of allopurinol by AO results in early oxidative stress and this has a preconditioning effect that results in increased expression of antioxidant genes which can scavenge both ROS and the reactive metabolite of APAP [98].

Mitochondria are a more likely source of oxidative stress in APAP-induced liver injury. In early experiments, it was shown that, although total liver GSSG is increased, biliary GSSG does not change during APAP hepatotoxicity [31, 84]. Because hepatocytes transport GSSG from the cytosol into bile, these data suggested that the oxidative stress is contained within an organelle. Consistent with this, GSSG was found to increase approximately tenfold in mitochondria after treatment with a large dose of APAP [31, 37]. More recent work has revealed that peroxynitrite ($ONOO^-$) forms specifically within mitochondria from the reaction of superoxide (O_2^-) with nitric oxide (NO), and $ONOO^-$ damages both mitochondrial DNA and proteins during APAP-induced liver injury in mice [15]. Importantly, preventing the oxidative stress by replenishing antioxidant levels in the liver protects against APAP [2, 15]. Although the exact cause of the mitochondrial oxidative stress is not yet clear, it may involve protein binding. Binding to mitochondrial proteins seems to be particularly important in APAP-induced liver injury. This is based on the knowledge that the reactive metabolite of a non-hepatotoxic isomer of APAP, 3′-hydroxyacetanilide (AMAP), shows lower affinity for mitochondrial proteins [69, 91]. More recently, it was shown that doses of APAP which cause similar levels of total protein binding in mice and rats result in less mitochondrial protein binding in the latter species, which is much less susceptible to APAP hepatotoxicity [57].

Whatever the cause, it is now clear that there is an initial oxidative stress after APAP overdose which activates mitogen-activated protein kinase (MAPK) signaling pathways that converge on the c-jun N-terminal kinases (JNK) 1/2 [26, 76]. Several MAPKs upstream of JNK have been identified. Normally, the MAP3K apoptosis signal-regulating kinase 1 (ASK1) is held inactive by thioredoxin, but oxidative stress can cause the two proteins to dissociate [77]. This has been shown to occur in APAP hepatotoxicity in mice, and ASK1-deficient mice have reduced JNK activation and reduced liver injury at later time points [64]. Glycogen synthase kinase 3β (GSK3β), another redox-responsive kinase, has also been shown to activate JNK in APAP hepatotoxicity and it was suggested that this protein acts at earlier time points after APAP overdose [82]. Mixed lineage kinase 3 (MLK3) can also be activated by oxidative stress, and MLK3 knockout mice are protected against APAP and have reduced activation of both JNK and GSK3β [81]. Once JNK has been activated, it translocates into mitochondria where it is thought to augment the mitochondrial oxidative stress in a feed-forward loop [33, 76]. The receptor-interacting protein kinase 3 (RIP3) also appears to control JNK activation and can likely regulate the mitochondrial oxidative stress through JNK [71]. Eventually, the mitochondrial membrane permeability transition pore (MPTP) forms and mitochondrial membrane potential is lost [39, 73]. Cyclophilin D (CypD), which regulates the MPTP, controls APAP toxicity. CypD knockout mice are less susceptible to APAP-induced liver injury [70]. Interestingly, however, the classical MPTP may not be necessary in all cases. Oxidative stress, particularly $ONOO^-$, appears to somehow drive the mitochondrial permeability transition at very high doses of APAP [50]. Combined with Bax translocation into the outer membrane, the MPT, mitochondrial swelling, and eventually outer mitochondrial membrane rupture result in the release of mitochondrial contents into the cytosol. The mitochondrial endonucleases apoptosis-inducing factor (AIF) and endonuclease G then enter the nucleus and cleave nuclear DNA [3, 4]. Interestingly, it appears that AIF may also play a role in the early mitochondrial oxidative stress [5]. The result of all this is oncotic cell necrosis. Importantly, while most of our knowledge of the molecular events underlying APAP-induced liver injury has come from rodent studies, there is accumulating evidence that the mechanisms are similar in humans. In particular, GSH turnover and depletion [19, 42, 56] protein binding [18, 56] and mitochondrial damage [56, 58] have been documented in humans and in vitro human models. However, future studies are needed to explore the role of oxidative stress in humans.

9.2.1.2 Non-Acetaminophen Drug-Induced Liver Injury

Most non-acetaminophen cases of DILI are idiosyncratic. Idiosyncratic DILI (iDILI) is defined as hepatotoxicity which develops in less than 1 in 10,000 users of a given drug. Idiosyncratic toxicity is generally not dose-dependent, although there is evidence that drugs with daily doses greater than 50–100 mg are more likely to cause liver injury than others [21]. It is also characterized by delayed onset. Extremely low incidence, variable histopathology, and experimental irreproducibility of iDILI make it very difficult to identify it clinically or to study it scientifically. However, there is evidence from the laboratory that oxidative stress can play a role in idiosyncratic toxicity caused by some drugs.

Troglitazone is a peroxisome proliferator-activated receptor agonist that was briefly used for the treatment of diabetes. It is also arguably the most important drug in the study of iDILI because the scandal surrounding its approval and use boosted awareness of idiosyncratic hepatotoxicity among the public and created pressure for regulatory agencies. Interestingly, it has been suggested that troglitazone toxicity involves mitochondrial oxidative stress. One study found that mice heterozygous for the mitochondrial antioxidant enzyme manganese superoxide dismutase (MnSOD, Sod2), which converts superoxide (O_2^-) to hydrogen peroxide (H_2O_2), developed liver injury after one month of daily treatments, while wild-type mice did not [68]. The same group later showed oxidation of mitochondrial proteins in the Sod2-deficient mice [46] and depletion of mitochondrial GSH [47]. Similar results were shown in these animals after treatment with the non-steroidal anti-inflammatory drug (NSAID) nimesulide [67], the anti-androgen flutamide [35], and the antibiotic trovafloxacin [28], each of which can also cause liver injury in some people. However, the results for troglitazone could not be reproduced in subsequent work [22]. Nevertheless, there is evidence for an association between some forms of iDILI and a particular variant of Sod2 in humans [29, 51]. Surprisingly though, this allele has been shown to result in increased Sod2 activity within mitochondria. It is thought that the increase in mitochondrial Sod2 in these patients may result in increased production of H_2O_2, which, despite being less reactive than O_2^-, is still a strong oxidant [51]. Consistent with this idea is the fact that individuals who have both this variant of Sod2 and a variant of glutathione peroxidase, which detoxifies H_2O_2, with reduced activity have an even higher risk of DILI [51]. Antituberculosis drugs, namely isoniazid and rifampicin, are among the most common causes of DILI throughout the world [1, 74]. Interestingly, there is some, albeit limited, evidence of oxidative stress in both humans and rodents after exposure to these drugs [13, 14]. Both mitochondrial dysfunction and oxidative stress have been implicated in the pathogenesis of a number of other drugs that cause DILI as well [12, 33, 54, 79]. In fact, it has been proposed that measuring oxidative stress in preclinical models could be a useful way to screen drugs for DILI potential, and a method for imaging of ROS in vivo using nanoparticle-based sensors has even been developed [83].

However, although oxidative stress has been observed in both humans and animal models in idiosyncratic hepatotoxicity, it is unlikely to be the primary mechanism of liver injury for most idiosyncratic drugs. The delayed onset of injury, the fact that many of the drugs are hepatotoxic upon re-challenge, the fact that most cases of iDILI involve protein adduct formation, and the association of various human leukocyte antigen (HLA) variants with iDILI suggest that the adaptive immune system drives the injury in most cases [93].

9.2.2 *Viral Hepatitis*

Viral hepatitis is the third most common cause of ALF in the US and the single most common cause in some developing countries [1, 44]. Hepatitis A (HAV) and B (HBV) account for most of these cases, although HCV can also contribute. Infection with HAV and the resulting liver injury are always acute, while HBV and HCV infection can be acute or chronic. The clinical course of the infection is determined by the strength of the immune response; if the initial response is strong enough, either the virus will be cleared and the patient will survive, or the patient will die of acute liver failure. A weak response results in chronic infection. Most of the liver injury caused by viral hepatitis is due to these immune responses to the infection. In the early phase, $CD8^+$ T cells recognize infected hepatocytes and produce perforins and proteases that kill the compromised cells. However, there is evidence that oxidative damage or increased susceptibility to oxidative stress caused by viral proteins expressed within the infected hepatocytes also play a role. Unfortunately, like non-acetaminophen DILI, viral hepatitis is difficult to study using common laboratory animals. Rodents are not natural vehicles for these viruses and are resistant to infection. As a consequence, most researchers rely on human cell culture systems or transgenic mice that are engineered to overexpress viral proteins. The data from such studies must be interpreted with caution. Moreover, much of the work on oxidative stress in humans with viral hepatitis has been done in patients with chronic infection, and the animal models that do exist could be considered chronic models. Thus, it is unclear how data from these studies translate to the early acute liver injury and liver failure caused by viral hepatitis. Nevertheless, it is clear that oxidative stress occurs in these diseases.

Oxidative stress is probably more important in HCV than in other forms of viral hepatitis. Overexpression of the HCV core protein causes mitochondrial damage and oxidative stress both in vitro and in vivo, and mice overexpressing this protein develop more oxidative damage than wild-type mice after exposure to hepatotoxicants [41, 66]. It appears that the core protein associates with mitochondria and facilitates movement of Ca^{2+} from the endoplasmic reticulum into these organelles [41, 96]. The result is mitochondrial swelling and increased ROS production [41, 96]. Similarly, human hepatoma cells transfected with the HBV genome have elevated levels of oxidized GSH and an increase in oxidative stress-response genes early after induction of HBV gene expression when compared with control cells [80]. Oxidation of DNA and a similar induction of antioxidant genes have also been observed in the livers of HBV transgenic mice [102]. Interestingly, the HBV X protein (HBx), like the HCV core protein, is known to localize to mitochondria and appears to be sufficient on its own to induce mitochondrial oxidative stress in human hepatoma cells [48]. Evidence of oxidative stress has also been observed in patients with chronic HBV or HCV and with HAV infection [11, 23, 40, 94]. However, as mentioned, the significance of these results for patients who develop acute liver failure after the early injury is unknown. The role of oxidative stress in chronic viral hepatitis is discussed elsewhere in this volume.

In addition to the intracellular oxidative stress directly caused by viral proteins, there is some evidence that granzyme A, a protease produced by the cytotoxic T cells that are responsible for the early liver injury, can cleave a subunit in complex I of the electron transport chain, resulting in increased production of reactive oxygen species in the mitochondria [53]. Unfortunately, the role of granzyme A in oxidative stress caused by viral hepatitis has not been thoroughly investigated.

9.2.3 Wilson Disease

Wilson disease is caused by a mutation in the transporter *ATP7B* gene [9, 90, 99]. This gene is primarily expressed in hepatocytes and is responsible for the biliary excretion of copper. When ATP7B is defective, copper accumulates within hepatocytes and causes damage. There is considerable variation in patient presentation [75]. In many cases, individuals with Wilson disease present with some form of chronic liver injury or with psychiatric complications in the absence of overt hepatic damage. However, some are asymptomatic until they suddenly develop ALF [46].

It has been known for decades that patients with Wilson disease have altered mitochondrial morphology [88] and that proper morphology can be restored with copper chelation treatment [89]. These findings, coupled with the knowledge that Cu^{2+} is a strong oxidizing agent capable of producing hydroxyl radical through Fenton-like chemistry, led to the hypothesis that the pathogenesis of Wilson's disease involves mitochondrial oxidative stress. This idea is supported by the observation of lipid peroxidation in mitochondria from rats with dietary copper overload [86] and from humans with Wilson disease [87]. Some evidence of DNA damage resulting from oxidative stress and lipid peroxidation products has also been observed in both Wilson patients and in animal models of the disease [10, 62, 63, 101], and some Wilson disease patients appear to have mitochondrial DNA alterations that are consistent with oxidative lesions [52]. More recently, a pattern of inhibition of mitochondrial enzymes consistent with what has been seen in a mouse model of mitochondrial oxidative stress has been reported in humans with Wilson disease [25]. Based on these data, it is commonly thought that the mitochondrial damage in Wilson disease is caused by oxidative stress resulting from accumulation of copper in hepatocytes. However, a difficulty with this hypothesis is that most copper is bound to protein under physiological conditions and is therefore not available to act as a pro-oxidant. An alternative explanation is that the copper ions help to crosslink proteins in mitochondrial membranes, resulting in aggregation of mitochondria and extreme changes in mitochondrial morphology and function [103]. In this scenario, the mitochondrial oxidative stress is a result of these changes in mitochondrial ultrastructure and may or may not be important in the downstream mechanisms of injury [103]. While it's clear that mitochondria are central in the mechanisms of liver injury in Wilson disease, additional work is needed to clarify the role of oxidative stress in this condition.

9.2.4 Other Causes of ALF

Little data are available for oxidative stress in other causes of ALF. Like DILI and viral hepatitis, the liver injury in autoimmune hepatitis is primarily caused by cells of the adaptive immune system [16, 49]. Nevertheless, there is limited evidence to suggest that oxidative stress occurs in patients with this condition [78]. Liver disease occasionally occurs in pregnant women who are near-term. The disease is often aggressive and can sometimes only be cured by birth. Serum arachidonic acid levels are elevated in some patients with acute fatty liver of pregnancy [65], and it has been suggested that the arachidonic acid causes mitochondrial damage and oxidative stress in hepatocytes of these patients [65]. Hepatotoxins found in the death cap mushroom *Amanita phalloides* damage the liver by inhibiting RNA polymerase. Although it has been suggested that these toxins can also cause oxidative stress, limited data suggest that this is not the case in humans [8].

9.3 The Role of Oxidative Stress in Liver Regeneration

There is evidence that the final outcome of acute liver injury is determined by the balance between injury and regeneration, especially after exposure to hepatotoxic chemicals [60]. The liver has a remarkable ability to regenerate. In experimental models, as much as 70 % of liver mass can be restored after surgical removal or acute hepatotoxicity. However, if the damage is so severe that the remaining cells cannot proliferate to replace the liver's original functional mass, the victim will likely enter liver failure. Generally, oxidative stress is detrimental during liver injury. However, there is new evidence emerging that oxidative stress can also encourage regeneration. Nuclear factor (erythroid derived 2)-like 2 (NFE2L2/Nrf2) is a redox-responsive nuclear transcription factor that binds to antioxidant response element (ARE) sequences in DNA that are upstream of antioxidant genes and upregulates expression of these genes. Nrf2 is normally held inactive in the cytosol by the cysteine-rich protein Kelch-like ECH-associated protein 1 (Keap1) [30]. When oxidative stress develops, two critical cysteine residues of Keap1 are oxidized, causing it to dissociate from Nrf2 [95]. Nrf2 can then translocate to the nucleus. It was found that Nrf2-deficient mice had delayed liver regeneration after partial hepatectomy [7]. It was concluded that the loss of Nrf2 and the resulting loss of antioxidant response gene induction resulted in greater oxidative stress in these animals, which led to a reduction in insulin and insulin-like growth factor 1 signaling [7]. While this suggests that extreme oxidative stress actually inhibits regeneration, it is important to keep in mind that this work was done with Nrf2 knockout animals. More recent work has shown that the gene for augmenter of liver regeneration (ALR) is downstream of an ARE in normal wild-type mice and that activation of Nrf2 results in increased expression of ALR, which can enhance liver regeneration [20]. Thus, it seems that oxidative stress has an important role in determining the balance between injury and repair mechanisms. It is possible that pharmacological Nrf2 activators could be used in the treatment of ALF in the future.

9.4 The Importance of Choosing the Right Endpoint

While there is evidence for oxidative stress in a number of liver diseases, the inability to detect it in some cases does not mean that it doesn't occur. It is critical to consider the sources and sites of the oxidative stress, the specific ROS involved, and the mechanisms of protection against the specific ROS. For example, markers of lipid peroxidation are likely not significantly elevated in serum of patients with APAP hepatotoxicity, as lipid peroxidation is not a major contributor to the mechanisms of the injury [38]. It is very important to choose the right end point when studying oxidative stress, and any study proposing a role for oxidative stress in liver injury, particularly those involving humans, should be critically evaluated.

9.5 Conclusions

ALF is a rare, but devastating, condition which is often fatal. Clinically, ALF is characterized by coagulopathy and encephalopathy developing within just days or weeks of the first symptoms of disease. Currently, liver transplantation is the only cure. Importantly, oxidative stress has been implicated in many causes of ALF. It is clear that it has a role in both of the two major causes of ALF throughout the world: a considerable amount of data from rodent studies demonstrate that oxidative stress is critical in the mechanisms of APAP hepatotoxicity, and similar evidence exists for oxidative stress in viral hepatitis from both rodents and humans. Although a role for oxidative stress has been proposed for other causes of ALF (e.g., Wilson disease, pregnancy-related liver disease, etc.), much more work is needed in those areas. Paradoxically, recent studies have also uncovered an important role for oxidative stress in liver regeneration through Nrf2 activation. The latter finding suggests that Nrf2 activators may be clinically useful in the treatment of ALF to promote tissue repair and regrowth. Future work should explore this possibility in greater detail.

References

1. Acharya SK, Batra Y, Hazati S, Choudhury V, Panda SK, Dattagupta S (2002) Etiopathogenesis of acute hepatic failure: Eastern versus Western countries. J Gastroenterol Hepatol 17:S268–S273
2. Bajt ML, Knight TR, Lemasters JJ, Jaeschke H (2004) Acetaminophen-induced oxidant stress and cell injury in cultured mouse hepatocytes: protection by N-acetyl cysteine. Toxicol Sci 80:343–349
3. Bajt ML, Cover C, Lemasters JJ, Jaeschke H (2006) Nuclear translocation of endonuclease G and apoptosis-induced factor during acetaminophen-induced liver cell injury. Toxicol Sci 94:217–225
4. Bajt ML, Farhood A, Lemasters JJ, Jaeschke H (2008) Mitochondrial bax translocation accelerates DNA fragmentation and cell necrosis in a murine model of acetaminophen hepatotoxicity. J Pharmacol Exp Ther 324:8–14

5. Bajt ML, Ramachandran A, Yan HM, Lebofsky M, Farhood A, Lemasters JJ, Jaeschke H (2011) Apoptosis-inducing factor modulates mitochondrial oxidant stress in acetaminophen hepatotoxicity. Toxicol Sci 122:598–605
6. Bernal W (2003) Changing patterns of causation and the use of transplantation in the United Kingdom. Semin Liver Dis 23:227–237
7. Beyer TA, Xu W, Teupser D, auf dem Keller U, Bugnon P, Hildt E, Thiery J, Kan YW, Werner S (2008) Impaired liver regeneration in Nrf2 knockout mice: role of ROS-mediated insulin/IGF-1 resistance. EMBO J 27:212–223
8. Biasi F, Bosco M, Lanfranco G, Poli G (1995) Cytolysis does not per se induce lipid peroxidation: evidence in man. Free Radic Biol Med 18:909–912
9. Bull PC, Thomas GR, Rommens JM, Forbes JR, Cox DW (1993) The Wilson disease gene is a putative copper transporting P-type ATPase similar to the Menkes gene. Nat Genet 5:327–337
10. Carmichael PL, Hewer A, Osborne MR, Strain AJ, Phillips DH (1995) Detection of bulky DNA lesions in the liver of patients with Wilson's disease and primary haemochromatosis. Mutat Res 326:235–243
11. Cemek M, Dede S, Bayiroglu F, Caksen H, Cemek F, Mert N (2006) Relationship between antioxidant capacity and oxidative stress in children with acute hepatitis A. World J Gastroenterol 12:6212–6215
12. Chang TK, Abbott FS (2006) Oxidative stress as a mechanism of valproic acid-associated hepatotoxicity. Drug Metab Rev 38:627–639
13. Chowdhury A, Santra A, Kundu S, Mukherjee A, Pandit A, Chaudhuri S, Dhali GK (2001) Induction of oxidative stress in antitubercular drug-induced hepatotoxicity. Indian J Gastroenterol 20:97–100
14. Chowdhury A, Santra A, Bhattacharjee K, Ghatak S, Saha DR, Dhali GK (2006) Mitochondrial oxidative stress and permeability transition in isoniazid and rifampicin induced liver injury in mice. J Hepatol 45:117–126
15. Cover C, Mansouri A, Knight TR, Bajt ML, Lemasters JJ, Pessayre D, Jaeschke H (2005) Peroxynitrite-induced mitochondrial and endonuclease-mediated nuclear DNA damage in acetaminophen hepatotoxicity. J Pharmacol Exp Ther 315:879–887
16. Czaja AJ, Manns MP (2010) Advances in the diagnosis, pathogenesis, and management of autoimmune hepatitis. Gastroenterology 139:58–72
17. Dahlin DC, Miwa GT, Lu AY, Nelson SD (1984) N-acetyl-p-benzoquinone imine: a cytochrome P-450-mediated oxidation product of acetaminophen. Proc Natl Acad Sci U S A 81:1327–1331
18. Davern TJ, James LP, Hinson JA, Polson J, Larson AM, Fontana RJ, Lalani E, Munoz S, Shakil AO, Lee WM, Acute Liver Failure Study Group (2006) Measurement of serum acetaminophen-protein adducts in patients with acute liver failure. Gastroenterology 130:687–694
19. Davis M, Ideo G, Harrison NG, Williams R (1975) Hepatic glutathione depletion and impaired bromosulphthalein clearance early after paracetamol overdose in man and the rat. Clin Sci Mol Med 49:495–502
20. Dayoub R, Vogel A, Schuett J, Lupke M, Spieker SM, Kettern N, Hildt E, Melter M, Weiss TS (2013) Nrf2 activates augmenter of liver regeneration (ALR) via antioxidant response element and links oxidative stress to liver regeneration. Mol Med 19:237–244
21. Fontana RJ (2014) Pathogenesis of idiosyncratic drug-induced liver injury and clinical perspectives. Gastroenterology 146:914–928
22. Fujimoto K, Kumagai K, Ito K, Arakawa S, Oda S, Yamoto T, Manabe S (2009) Sensitivity of liver injury in heterozygous Sod2 knockout mice treated with troglitazone or acetaminophen. Toxicol Pathol 37:193–200
23. Fujita N, Sugimoto R, Ma N, Tanaka H, Iwasa M, Kobayashi Y, Kawanishi S, Watanabe S, Kaito M, Takei Y (2008) Comparison of hepatic oxidative DNA damage in patients with chronic hepatitis B and C. J Viral Hepat 15:498–507

24. Gow PJ, Jones RM, Dobson JL, Angus PW (2004) Etiology and outcome of fulminant hepatic failure managed at an Australian transplant unit. J Gastroenterol Hepatol 19:154–159
25. Gu M, Cooper JM, Butler P, Walker AP, Mistry PK, Dooley JS, Schapira AH (2000) Oxidative-phosphorylation defects in liver of patients with Wilson's disease. Lancet 356:469–474
26. Gunawan BK, Liu ZX, Han D, Hanawa N, Gaarde WA, Kaplowitz N (2006) c-Jun N-terminal kinase plays a major role in murine acetaminophen hepatotoxicity. Gastroenterology 131:165–178
27. Heard KJ, Green JL, James LP, Judge BS, Zolot L, Rhyee S, Dart RC (2011) Acetaminophen-cysteine adducts during therapeutic dosing and following overdose. BMC Gastroenterol 11:20
28. Hsiao CJ, Younis H, Boelsterli UA (2010) Trovafloxacin, a fluoroquinolone antibiotic with hepatotoxic potential, causes mitochondrial peroxynitrite stress in a mouse model of underlying mitochondrial dysfunction. Chem Biol Interact 188:204–213
29. Huang YS, Su WJ, Huang YH, Chen CY, Chang FY, Lin HC, Lee SD (2007) Genetic polymorphisms of manganese superoxide dismutase, NAD(P)H:quinone oxidoreductase, glutathione S-transferase M1 and T1, and the susceptibility to drug-induced liver injury. J Hepatol 47:128–134
30. Itoh K, Wakabayashi N, Katoh Y, Ishii T, Igarashi K, Engel JD, Yamamoto M (1999) Keap1 represses nuclear activation of antioxidant responsive elements by Nrf2 through binding to the amino-terminal Neh2 domain. Genes Dev 13:76–86
31. Jaeschke H (1990) Glutathione disulfide formation and oxidant stress during acetaminophen-induced hepatotoxicity in mice in vivo: the protective effect of allopurinol. J Pharmacol Exp Ther 255:935–941
32. Jaeschke H, Knight TR, Bajt ML (2003) The role of oxidant stress and reactive nitrogen species in acetaminophen hepatotoxicity. Toxicol Lett 144:279–288
33. Jaeschke H, McGill MR, Ramachandran A (2012) Oxidant stress, mitochondria, and cell death mechanisms in drug-induced liver injury: lessons learned from acetaminophen hepatotoxicity. Drug Metab Rev 44:88–106
34. Jones EA, Mullen KD (2012) Theories of the pathogenesis of hepatic encephalopathy. Clin Liver Dis 16:7–26
35. Kashimshetty R, Desai VG, Kale VM, Lee T, Moland CL, Branham WS, New LS, Chan EC, Younis H, Boelsterli UA (2009) Underlying mitochondrial dysfunction triggers flutamde-induced oxidative liver injury in a mouse model of idiosyncratic drug toxicity. Toxicol Appl Pharmacol 238:150–159
36. Kaufman DW, Kelly JP, Rosenberg L, Anderson TE, Mitchell AA (2002) Recent patterns of medication use in the ambulatory adult population of the United States: the Slone survey. JAMA 287:337–344
37. Knight TR, Kurtz A, Bajt ML, Hinson JA, Jaeschke H (2001) Vascular and hepatocellular peroxynitrite formation during acetaminophen toxicity: role of mitochondrial oxidant stress. Toxicol Sci 62:212–220
38. Knight TR, Fariss MW, Farhood A, Jaeschke H (2003) Role of lipid peroxidation as a mechanism of liver injury after acetaminophen overdose in mice. Toxicol Sci 76:229–236
39. Kon K, Kim JS, Jaeschke H, Lemasters JJ (2004) Mitochondrial permeability transition in acetaminophen-induced necrosis and apoptosis of cultured mouse hepatocytes. Hepatology 40:1170–1179
40. Konishi M, Iwasa M, Araki J, Kobayashi Y, Katsuki A, Sumida Y, Nakagawa N, Kojima Y, Watanabe S, Adachi Y, Kaito M (2006) Increased lipid peroxidation in patients with non-alcoholic fatty liver disease and chronic hepatitis C as measured by the plasma level of 8-isoprostane. J Gastroenterol Hepatol 21:1821–1825
41. Korenaga M, Wang T, Li Y, Showalter LA, Chan T, Sun J, Weinman SA (2005) Hepatitis C virus core protein inhibits mitochondrial electron transport and increases reactive oxygen species (ROS) production. J Biol Chem 280:37481–37488

42. Lauterburg BH, Mitchell JR (1987) Therapeutic doses of acetaminophen stimulate the turnover of cysteine and glutathione in man. J Hepatol 4:206–211
43. Leber B, Spindelboeck W, Stadlbauer V (2012) Infectious complications of acute and chronic liver disease. Semin Respir Crit Care Med 33:80–95
44. Lee WM (2008) Etiologies of acute liver failure. Semin Liver Dis 28:142–152
45. Lee WM (2012) Acute liver failure. Semin Respir Crit Care Med 33:36–45
46. Lee YH, Chung MC, Lin Q, Boelsterli UA (2008) Troglitazone-induced hepatic mitochondrial proteome expression dynamics in heterozygous Sod2(+/-) mice: two-stage oxidative injury. Toxicol Appl Pharmacol 231:43–51
47. Lee YH, Goh WW, Ng CK, Raida M, Wong L, Lin Q, Boelsterli UA, Chung MC (2013) Integrative toxicoproteomics implicates impaired mitochondrial glutathione import as an off-target effect of troglitazone. J Proteome Res 12:2933–2945
48. Lee YI, Hwang JM, Im JH, Lee YI, Kim NS, Kim DG, Yu DY, Moon HB, Park SK (2004) Human hepatitis B virus-x protein alters mitochondrial function and physiology in human liver cells. J Biol Chem 279:15460–15471
49. Liberal R, Longhi MS, Mieli-Vergani G, Vergani D (2011) Pathogenesis of autoimmune hepatitis. Best Pract Res Clin Gastroenterol 25:653–664
50. LoGuidice A, Boelsterli UA (2011) Acetaminophen overdose-induced liver injury in mice is mediated by peroxynitrite independently of the cyclophilin D-regulated permeability transition. Hepatology 54:969–978
51. Lucena MI, García-Martín E, Andrade RJ, Martínez C, Stephens C, Ruiz JD, Ulzurrun E, Fernandez MC, Romero-Gomez M, Castiella A, Planas R, Durán JA, De Dios AM, Guarner C, Soriano G, Borraz Y, Agundez JA (2010) Mitochondrial superoxide dismutase and glutathione peroxidase in idiosyncratic drug-induced liver injury. Hepatology 52:303–312
52. Mansouri A, Gaou I, Fromenty B, Berson A, Lettéron P, Degott C, Erlinger S, Pessayre D (1997) Premature oxidative aging of hepatic mitochondrial DNA in Wilson's disease. Gastroenterology 113:599–605
53. Martinvalet D, Dykxhoorn DM, Ferrini R, Lieberman J (2008) Granzyme A cleaves a mitochondrial complex I protein to initiate caspase-independent cell death. Cell 133:681–692
54. Matsuo K, Sasaki E, Higuchi S, Takai S, Tsuneyama K, Fukami T, Nakajima M, Yokoi T (2014) Involvement of oxidative stress and immune- and inflammation-related factors in azathioprine-induced liver injury. Toxicol Lett 224:215–224
55. McGill MR, Jaeschke H (2013) Metabolism and disposition of acetaminophen: recent advances in relation to hepatotoxicity and diagnosis. Toxicol Sci 131:325–335
56. McGill MR, Yan HM, Ramachandran A, Murray GJ, Rollins DE, Jaeschke H (2011) HepaRG cells: a human model to study mechanisms of acetaminophen hepatotoxicity. Hepatology 53:974–982
57. McGill MR, Williams CD, Xie Y, Ramachandran A, Jaeschke H (2012) Acetaminophen-induced liver injury in rats and mice: comparison of protein adducts, mitochondrial dysfunction, and oxidative stress in the mechanism of toxicity. Toxicol Appl Pharmacol 264:387–394
58. McGill MR, Sharpe MR, Williams CD, Taha M, Curry SC, Jaeschke H (2012) The mechanism underlying acetaminophen-induced hepatotoxicity in humans and mice involves mitochondrial damage and nuclear DNA fragmentation. J Clin Invest 122:1574–1583
59. McGill MR, Lebofsky M, Norris HR, Slawson MH, Bajt ML, Xie Y, Williams CD, Wilkins DG, Rollins DE, Jaeschke H (2013) Plasma and liver acetaminophen-protein adduct levels in mice after acetaminophen treatment: dose-response, mechanisms, and clinical implications. Toxicol Appl Pharmacol 269(3):240–249
60. Mehendale HM (2005) Tissue repair: an important determinant of final outcome of toxicant-induced injury. Toxicol Pathol 33:41–51
61. Mitchell JR, Jollow DJ, Potter WZ, Gillette JR, Brodie BB (1973) Acetaminophen-induced hepatic necrosis. IV. Protective role of glutathione. J Pharmacol Exp Ther 187:211–217

62. Nair J, Sone H, Nagao M, Barbin A, Bartsch H (1996) Copper-dependent formation of miscoding etheno-DNA adducts in the liver of Long Evans cinnamon (LEC) rats developing hereditary hepatitis and hepatocellular carcinoma. Cancer Res 56:1267–1271
63. Nair J, Carmichael PL, Fernando RC, Phillips DH, Strain AJ, Bartsch H (1998) Lipid peroxidation-induced etheno-DNA adducts in the liver of patients with the genetic metal storage disorders Wilson's disease and primary hemochromatosis. Cancer Epidemiol Biomarkers Prev 7:435–440
64. Nakagawa H, Maeda S, Hikiba Y, Ohmae T, Shibata W, Yanai A, Sakamoto K, Ogura K, Noguchi T, Karin M, Ichijo H, Omata M (2008) Deletion of apoptosis signal-regulating kinase 1 attenuates acetaminophen-induced liver injury by inhibiting c-Jun N-terminal kinase activation. Gastroenterology 135:1311–1321
65. Natarajan SK, Thangaraj KR, Eapen CE, Ramachandran A, Mukhopadhya A, Mathai M, Seshadri L, Peedikayil A, Ramakrishna B, Balasubramanian KA (2010) Liver injury in acute fatty liver of pregnancy: possible link to placental mitochondrial dysfunction and oxidative stress. Hepatology 51:191–200
66. Okuda M, Li K, Beard MR, Showalter LA, Scholle F, Lemon SM, Weinman SA (2002) Mitochondrial injury, oxidative stress, and antioxidant gene expression are induced by hepatitis C virus core protein. Gastroenterology 122:366–375
67. Ong MM, Latchoumycandance C, Boelsterli UA (2006) Nimesulid-induced hepatic mitochondrial injury in heterozygous Sod2(+/−) mice. Free Radic Biol Med 40:420–429
68. Ong MM, Latchoumycandance C, Boelsterli UA (2007) Troglitazone-induced hepatic necrosis in an animal model of silent genetic mitochondrial abnormalities. Toxicol Sci 97:205–213
69. Qiu Y, Benet LZ, Burlingame AL (2001) Identification of hepatic protein targets of the reactive metabolites of the non-hepatotoxic regioisomer of acetaminophen, 3′-hydroxyacetanilide, in the mouse in vivo using two-dimensional gel electrophoresis and mass spectrometry. Adv Exp Med Biol 500:663–673
70. Ramachandran A, Lebofsky M, Baines CP, Lemasters JJ, Jaeschke H (2011) Cyclophilin D deficienty protects against acetaminophen-induced oxidant stress and liver injury. Free Radic Res 45:156–164
71. Ramachandran A, McGill MR, Xie Y, Ni HM, Ding WX, Jaeschke H (2013) Receptor interacting protein kinase 3 is a critical early mediator of acetaminophen-induced hepatocyte necrosis in mice. Hepatology 58:2099–2108
72. Rama Rao KV, Jayakumar AR, Norenberg MD (2014) Brain edema in acute liver failure: mechanisms and concepts. Metab Brain Dis 29(4):927–936
73. Reid AB, Kurten RC, McCullough SS, Brock RW, Hinson JA (2005) Mechanisms of acetaminophen-induced hepatotoxicity: role of oxidative stress and mitochondrial permeability transition in freshly isolated mouse hepatocytes. J Pharmacol Exp Ther 312:509–516
74. Reuben A, Koch DG, Lee WM, Group ALFS (2010) Drug-induced acute liver failure: results of a U.S. multicenter prospective study. Hepatology 52:2065–2076
75. Roberts EA, Schilsky ML, AASLD (2008) Diagnosis and treatment of Wilson disease: an update. Hepatology 47:2089–2111
76. Saito C, Lemasters JJ, Jaeschke H (2010) c-Jun N-terminal kinase modulates oxidant stress and peroxynitrite formation independent of inducible nitric oxide synthase in acetaminophen hepatotoxicity. Toxicol Appl Pharmacol 246:8–17
77. Saitoh M, Nishitoh H, Fujii M, Takeda K, Tobiume K, Sawada Y, Kawabata M, Miyazono K, Ichijo H (1998) Mammalian thioredoxin is a direct inhibitor of apoptosis signal-regulating kinase (ASK) 1. EMBO J 17:2596–2606
78. Sanz-Cameno P, Medina J, García-Buey L, García-Sánchez A, Borque MJ, Martín-Vílchez S, Gamallo C, Jones EA, Moreno-Otero R (2002) Enhanced intrahepatic inducible nitric oxide synthase expression and nitrotyrosine accumulation in primary biliary cirrhosis and autoimmune hepatitis. J Hepatol 37:723–729

79. Sasaki E, Matsuo K, Iida A, Tsuneyama K, Fukami T, Nakajima M, Yokoi T (2013) A novel mouse model for phenytoin-induced liver injury: involvement of immune-related factors and P450-mediated metabolism. Toxicol Sci 136:250–263
80. Severi T, Ying C, Vermeesch JR, Cassiman D, Cnops L, Verslype C, Fevery J, Arckens L, Neyts J, van Pelt JF (2006) Hepatitis B virus replication causes oxidative stress in HepAD38 liver cells. Mol Cell Biochem 290:79–85
81. Sharma M, Gadang V, Jaeschke A (2012) Critical role for mixed-lineage kinase 3 in acetaminophen-induced hepatotoxicity. Mol Pharmacol 82:1001–1007
82. Shinohara M, Ybanez MD, Win S, Than TA, Jain S, Gaarde WA, Han D, Kaplowitz N (2010) Silencing glycogen synthase kinase-3beta inhibits acetaminophen hepatotoxicity and attenuates JNK activation and loss of glutamate cysteine ligase and myeloid cell leukemia sequence 1. J Biol Chem 285:8244–8255
83. Shuhendler AJ, Pu K, Cui L, Uetrecht JP, Rao J (2014) Real-time imaging of oxidative and nitrosative stress in the liver of live animals for drug-toxicity testing. Nat Biotechnol 32:373–380
84. Smith CV, Jaeschke H (1989) Effect of acetaminophen on hepatic content and biliary efflux of glutathione disulfide in mice. Chem Biol Interact 70:241–248
85. Smith CV, Mitchell JR (1985) Acetaminophen hepatotoxicity in vivo is not accompanied by oxidant stress. Biochem Biophys Res Commun 133:329–336
86. Sokol RJ, Devereaux M, Mierau GW, Hambridge KM, Shikes RH (1990) Oxidant injury to hepatic mitochondrial lipids in rats with dietary copper overload. Modification by vitamin E deficiency. Gastroenterology 99:1061–1071
87. Sokol RJ, Twedt D, McKim JM Jr, Devereaux MW, Karrer FM, Kam I, von Steigman G, Narkewicz MR, Bacon BR, Britton RS et al (1994) Oxidant injury to hepatic mitochondria in patients with Wilson's disease and Bedlington terriers with copper toxicosis. Gastroenterology 107:1788–1798
88. Sternlieb I (1968) Mitochondrial and fatty changes in hepatocytes of patients with Wilson's disease. Gastroenterology 55:354–367
89. Sternlieb I, Feldmann G (1976) Effects of anticopper therapy on hepatocellular mitochondria in patients with Wilson's disease: an ultrastructural and stereological study. Gastroenterology 71:457–461
90. Tanzi RE, Petrukhin K, Chernov I, Pellequer JL, Wasco W, Ross B, Romano DM, Parano E, Pavone L, Brzustowicz LM et al (1993) The Wilson disease gene is a copper transporting ATPase with homology to the Menkes disease gene. Nat Genet 5:344–350
91. Tirmenstein MA, Nelson SD (1989) Subcellular binding and effects on calcium homeostasis produced by acetaminophen and a nonhepatotoxic regioisomer, 3′-hydroxyacetanilide, in mouse liver. J Biol Chem 264:9814–9819
92. Tirmenstein MA, Nelson SD (1990) Acetaminophen-induced oxidation of protein thiols. Contribution of impaired thiol-metabolizing enzymes and the breakdown of adenine nucleotides. J Biol Chem 265:3059–3065
93. Uetrecht J, Naisbitt DJ (2013) Idiosyncratic adverse drug reactions: current concepts. Pharmacol Rev 65:779–808
94. Valgimigli L, Valgimigli M, Gaiani S, Pedulli GF, Bolondi L (2000) Measurement of oxidative stress in human liver by EPR spin-probe technique. Free Radic Res 33:167–178
95. Wakabayashi N, Dinkova-Kostova AT, Holtzclaw WD, Kang MI, Kobayashi A, Yamamoto M, Kensler TW, Talalay P (2004) Protection against electrophile and oxidant stress by induction of the phase 2 response: fate of cysteines of the Keap1 sensor modified by inducers. Proc Natl Acad Sci U S A 101:2040–2045
96. Wang T, Weinman SA (2013) Interactions between hepatitis C virus and mitochondria: impact on pathogenesis and innate immunity. Curr Pathobiol Rep 1:179–187
97. Wendel A, Feuerstein S (1981) Drug-induced lipid peroxidation in mice: I. Modulation by monooxygenase activity, glutathione and selenium status. Biochem Pharmacol 30:2513–2520

98. Williams CD, McGill MR, Lebofsky M, Bajt ML, Jaeschke H (2014) Protection against acetaminophen-induced liver injury by allopurinol is dependent on aldehyde oxidase-mediated liver preconditioning. Toxicol Appl Pharmacol 274:417–424
99. Yamaguchi Y, Heiny ME, Gitlin JD (1993) Isolation and characterization of a human liver cDNA as a candidate gene for Wilson disease. Biochem Biophys Res Commun 197:271–277
100. Yan HM, Ramachandran A, Bajt ML, Lemasters JJ, Jaeschke H (2010) The oxygen tension modulates acetaminophen-induced mitochondrial oxidant stress and cell injury in cultured hepatocytes. Toxicol Sci 117:515–523
101. Yamamoto F, Kasai H, Togashi Y, Takeichi N, Hori T, Nishimura S (1993) Elevated level of 8-hydroxydeoxyguanosine in DNA of liver, kidneys, and brain of Long-Evans Cinnamon rats. Jpn J Cancer Res 84:508–511
102. Yang F, Yan S, He Y, Wang F, Song S, Guo Y, Zhou Q, Wang Y, Lin Z, Yang Y, Zhang W, Sun S (2008) Expression of hepatitis B virus proteins in transgenic mice alters lipid metabolism and induces oxidative stress in the liver. J Hepatol 48:12–19
103. Zischka H, Lichtmannegger J, Schmitt S, Jägemann N, Schulz S, Wartini D, Jennen L, Rust C, Larochette N, Galluzzi L, Chajes V, Bandow N, Gilles VS, DiSpirito AA, Esposito I, Goettlicher M, Summer KH, Kroemer G (2011) Liver mitochondrial membrane crosslinking and destruction in a rat model of Wilson disease. J Clin Invest 121:1508–1518

Chapter 10
Oxidative Stress in Alcoholic Liver Disease

Emanuele Albano

10.1 Introduction

The excessive consumption of alcoholic beverages is an important cause of death and disease worldwide. In developed countries alcohol-related mortality is estimated to range from 7.9 to 14.3 per 100.000 habitants representing the third cause of death [1]. However, in the recent years alcohol abuse has become a leading cause of disease also in the developing countries of Central and South America and East Asia [2]. Although ethanol can damage several organs, alcoholic liver disease (ALD) is the most common medical consequence of excessive alcohol intake. ALD accounts for 8 % of newly diagnosed liver diseases and for more than 50 % of chronic liver diseases [3]. In addition, heavy alcohol consumption is often a comorbility factor in hepatic injury caused by viral infections or by metabolic disorders [4, 5]. ALD encompasses a broad spectrum of histological features ranging from lipid accumulation within the hepatocytes (fatty liver or steatosis) with minimal parenchymal injury to more advanced liver damage, including steatohepatitis and fibrosis/cirrhosis. Almost all heavy drinkers develop steatosis with variable degree of necro-inflammation; however, overt alcoholic hepatitis is diagnosed in only 10–35 % of cases [5]. The progression of ALD to alcoholic cirrhosis is evident in about 8–20 % of alcohol abusers [5]. Even so, only in US alcohol accounts for about 44 % of the 30,000 deaths caused every year by hepatic cirrhosis [2]. Similar figures hold for Europe, where the prevalence of alcoholic cirrhosis in the different countries correlates quite well to the national per capita alcohol consumption [6, 7]. At the individual level, the risk of developing ALD shows a dose–effect relationship with daily alcohol intake [8]. However, such an association is not linear as only 6 %

E. Albano (✉)
Department of Health Sciences, University of Eastern Piedmont Amedeo Avogadro, Via Solaroli 17, Novara 28100, Italy
e-mail: emanuele.albano@med.unipmn.it

E. Albano, M. Parola (eds.), *Studies on Hepatic Disorders*,
Oxidative Stress in Applied Basic Research and Clinical Practice,
DOI 10.1007/978-3-319-15539-5_10

of the subjects consuming more than 80–90 g of ethanol a day show clinical sings of cirrhosis [9]. Moreover, prospective studies in heavy drinkers suggest that the mortality for alcoholic cirrhosis has a threshold, but not a dose–response effect [10]. It is also noteworthy that chronic alcohol consumption increases by about fivefold the risk of hepatocellular carcinoma (HCC) and that about 15 % of the patients with alcoholic cirrhosis develop HCC [11, 12]. At present, the reasons for the large inter-individual variability in the risk of alcohol-induced liver injury are still poorly characterized. A gender difference is well evident with women being twice more sensitive to alcohol-mediated hepatotoxicity than men. They develop ALD at lower intake of alcohol and with shorter duration of consumption as compared to males [5]. Ethnic differences are also appreciable with African–Americans and Hispanic males showing incidence rates of alcoholic cirrhosis higher than Caucasian males, irrespective of the alcohol intake [5]. Thus, the duration of alcohol abuse, the drinking patterns, the type of alcoholic beverages along with metabolic and genetic factors likely influence the risk of ALD progression independently from the amount of ethanol consumed [5, 13]. Furthermore, in spite of the fact that from the clinical point of view steatosis, steatohepatitis, and fibrosis/cirrhosis represent the evolution of alcohol-induced hepatic injury, it is increasingly clear that the mechanisms leading to each of these lesions can be quite different.

10.2 Ethanol Metabolism in the Liver

Ethanol is largely (90–98 %) metabolized in the liver. Such a specificity justifies why ethanol toxicity mostly involves this organ. In human hepatocytes ethanol is converted to acetaldehyde by the action of alcohol dehydrogenase (ADH), microsomal cytochrome P450, and to minor extent by catalase [14, 15]. The major pathway for hepatic ethanol oxidation involves ADH, a NAD-dependent zinc metalloenzyme present in human tissues in five different iso-enzyme classes [14, 15]. These enzymes originate from the association of eight different subunits into active dimeric molecules. The hepatic alcohol metabolism largely relies on the Class I ADH (ADH1A/B/C) that has high affinity for ethanol. Microsomal ethanol oxidation mainly relies on the action of cytochrome P4502E1 isoenzyme (CYP2E1) with a minor contribution of other isoforms (CYP1A2 and CYP3A4) [14]. CYP2E1 is mainly expressed in centrilobular hepatocytes, but small amounts are detectable also in hepatic macrophages obtained from alcohol-fed rats [16]. CYP2E1 K_m for ethanol is about tenfold higher than that of ADH, and at low alcohol concentration, microsomal ethanol oxidation accounts for about 10 % of the overall alcohol elimination [16]. However, chronic alcohol exposure increases by 5- to 20-fold CYP2E1 activity through both enzyme stabilization and increased gene expression [16]. Thus, CYP2E1 may have a major role in ethanol elimination in alcohol abusers. However, the evaluation of CYP2E1 activity in humans through the measure of the oxidation of the myorelaxant drug chlorzoxazone reveals that an appreciable CYP2E1 up-regulation occurs already following moderate alcohol consumption [17]. There is also a large inter-individual variability in CYP2E1 induction with a minority of heavy drinkers who do not show appreciable increase in CYP2E1 activity [18]. Recent evidence indicates that CYP2E1 induction not only involves the microsomal form of the enzyme, but also two CYP2E1 variants that are present in the mitochondrial matrix (mtCYP2E1s). These mtCYP2E1s consist of a highly phosphorylated form and a truncated form lacking of the hydrophobic part at the NH_2-terminus [16]. At difference from the microsomal CYP2E1 that uses NADPH and cytochrome P450 reductase for electron transfer, the electron suppliers of mtCYP2E1s are adrenodoxin and adrenodoxin reductase [16]. Overall, mtCYP2E1s account for 30–40 % of the microsomal enzyme activity, and because of their localization, have important implications in alcohol hepatotoxicity [16]. Finally, a small fraction of CYP2E1 (about 10 % or the microsomal content) is also expressed on the outer layer of the hepatocyte plasma membranes, where it is transported from the Golgi apparatus via the secretory vesicles [19]. Plasma membrane CYP2E1 is catalytically active, but its importance in alcohol toxicity mainly relies on the fact that it is the target for allo- and auto-immune reactions (see below). Finally, catalase has been implicated in ethanol oxidation [14, 15]. Although such a reaction might have a role in brain alcohol metabolism, its role in human liver appears to be negligible [14].

Acetaldehyde that originates from ethanol oxidation is largely detoxified by the action of NAD-dependent aldehyde dehydrogenases (ALDH) with the formation of acetate. ALDHs are tetramers or dimers formed by the same subunits [20]. Among the 19 ALDHs characterized in humans, only a few are involved in acetaldehyde oxidation, and of these, the mitochondrial ALDH2 and the cytosolic ALDH1A are the most effective. These enzymes are highly efficient (Km values of about 1 M) and detoxify more than 99 % of the acetaldehyde generated within the hepatocytes [20]. However, low amounts of acetaldehyde can escape detoxification interacting with proteins and nucleic acids [20]. Acetaldehyde binding to DNA is regarded as an important factor in the carcinogenic effects of alcohol [12]. It is noteworthy that ALDHs play also an important role in the detoxification of a variety of aldehydic compounds originating from the peroxidation of unsaturated lipids during oxidative stress [21].

During acute ethanol intoxication, large amount of NADH are generated and the increased NADH/NAD ratio can affect hepatic enzymes involved in lipid, carbohydrate, and uric acid metabolism [4]. In particular, the excess of NADH decreases fatty acid oxidation and enhances lipogenesis leading to triglyceride accumulation within the hepatocytes, while the block of pyruvate conversion to glucose increases lactic acid production causing alcoholic hypoglycaemia and acidosis [4]. These metabolic disturbances are, however, rapidly reversible and seem to be attenuated during chronic ethanol ingestion.

10.3 Oxidative Stress in the Pathogenesis of ALD

It is now widely accepted that ALD has a multifactorial pathogenesis and that the disease evolution is the result of the interaction between many factors including oxidative damage, metabolic disturbances, endoplasmic reticulum stress, and inflammatory responses. The role of oxidative stress in alcohol liver injury has received attention since the early 1960s when Di Luzio first reported the detection of lipid peroxidation in the livers of alcohol-treated rats [22]. During the following three decades, the capacity of ethanol to trigger oxidative stress was further supported by the demonstration that ethanol increases the hepatic production of free radical species in both parenchimal and non-parenchimal cells and by the detection of oxidative modifications in hepatic constituents following both acute and chronic ethanol administration to experimental animals [23, 24]. In line with these findings, many reports have demonstrated an increase in lipid peroxidation products, protein carbonyls, and other oxidative stress biomarkers in both the liver and the serum of patients with alcohol abuse, particularly in those with advanced ALD [25–28]. The relevance of these observations to the human disease has also been confirmed by the immunohistochemical detection of proteins adducted by lipid peroxidation products in liver biopsies from alcohol abusers where these adducts are particularly evident in the areas of focal necrosis and around fibrotic septa [29]. However, rodent aversion for alcohol has hampered for many years the possibility to verify experimentally the actual role of oxidative stress in the pathogenesis ALD and only the introduction of rodent enteral nutrition with an alcohol-containing liquid diet has now partially overcome this limitation. By using this experimental approach, it has been possible to show that rats receiving ethanol in combination with a diet rich in highly oxidizable unsaturated fatty acids from corn or fish oil develop extensive hepatic lipid peroxidation and parenchymal damage [30, 31], while minor effects are evident when a similar amount of ethanol is given together with poorly oxidizable medium chain triglycerides [32]. Furthermore, replacing fish oil with less-oxidizable palm oil while continuing ethanol administration improves oxidative stress and ameliorates already established liver damage [33]. On the same line, several other studies have demonstrated that supplementing alcohol-fed rodents with a variety of antioxidants and free radical scavengers reduces oxidative stress and hepatic injury [34–36]. Protection against oxidative stress and hepatotoxicity caused by the enteral alcohol administration is also evident in mice over-expressing antioxidant enzymes such as cytosolic (Cu–Zn)-superoxide dismutase (SOD-1) or mitochondrial (Mn)-superoxide dismutase (SOD-2) [37, 38]. Conversely, SOD-1 knockout mice show increased lipid peroxidation, extensive centrilobular necrosis, and inflammation upon moderate ethanol consumption [39]. In a similar manner, mice deficient for the transcription factor erythroid-2-related factor (Nfr2) which regulates the gene expression of antioxidant enzymes are more sensitive to alcohol hepatotoxicity undergoing extensive liver injury and an increased mortality when exposed to low amounts of ethanol otherwise well-tolerated by wild-type animals [40].

10.4 Mechanisms Responsible for Alcohol-Induced Oxidative Stress in the Liver

The occurrence of oxidative stress during ethanol intake is the result of the combined action of alcohol in increasing the generation of oxidizing species such as reactive oxygen species (ROS) and nitric oxide (NO) and in lowering intracellular antioxidant defenses [12]. In addition, ethanol itself can generate free radical intermediates, known as hydroxyethyl-free radicals.

10.4.1 Reactive Oxygen Species

During alcohol intake the mitochondrial respiratory chain and CYP2E1-dependent microsomal monoxygenase system are the main intracellular sources of ROS in hepatocytes, while extracellular ROS generated by activated inflammatory cells also significantly contribute to the pro-oxidant action.

The capacity of ethanol to promote ROS generation by the mitochondria has emerged from the observation that both acute and chronic ethanol exposure increase the oxidation of mitochondrial proteins and DNA (mtDNA) and deplete the organelle pool of reduced glutathione (mtGSH) [39, 40]. During acute ethanol intoxication, an excess of O_2^- results from an enhanced leakage of electrons from complexes I and III of the mitochondrial respiratory chain in relation to an increased availability of NADH [41, 42]. Conversely, the stimulation in mitochondrial ROS production that characterizes chronic alcohol administration is the consequence of an impaired synthesis of mitochondria-encoded constituents of the respiratory chain [41, 42]. Indeed, single or multiple mtDNA deletions are frequent in the liver of alcohol-treated rats as well as hepatic biopsies from alcoholic patients [42]. On their turn, these mtDNA alterations contribute to the selective lowering of mitochondrial respiratory chain components evident in alcohol-treated rodents [43]. Additional mechanisms that can contribute to mitochondrial ROS generation during alcohol exposure can involve the interaction of the complexes II and III with N-acetylsphingosine (C_2-ceramide) released by hepatocytes in response to tumor necrosis factor α (TNF-α) [44] and the development of hepatic hypoxia.

As mentioned above, CYP2E1-dependent microsomal monoxygenase system represents an important biotransformation pathway of ethanol, particularly in the presence of high alcohol intake [16]. CYP2E1 has a especially high NADPH oxidase activity leading to the production of large quantities of O_2^- and H_2O_2. In liver microsomes from either humans or alcohol-fed rodents, CYP2E1 content positively correlates with NADPH-oxidase activity as well as with the extent of lipid peroxidation. Recent studies have also implicated mitochondrial CYP2E1s (mtCYP2E1s) as additional sources of ROS [16, 45]. In fact, cells over-expressing mtCYP2E1s undergo oxidative damages upon ethanol exposure [46]. Thus, the high efficiency of CYP2E1 in reducing oxygen to O_2^- and H_2O_2 is regarded as key contributor to oxidative stress during chronic exposure to alcohol. Indeed, the addition of ethanol to HepG2 hepatoma cells stably transfected with the *CYP2E1* gene increases ROS production and causes oxidative stress-mediated cell injury [16]. In these cells as well as in the liver of heavy drinkers, CYP2E1 levels also correlate with the amounts of DNA adducts with the lipid peroxidation products 4-hydroxynonenal (4-HNE) [16, 47]. These 4-HNE-DNA adducts involve the codon 249 of human p53 gene, a unique mutational hot spot in hepatocellular carcinoma [48], suggesting CYP2E1-

dependent oxidative injury as a relevant factor in alcohol-induced hepatocarcinogenesis. In line with these findings, experiments performed using enteral alcohol-fed rats have demonstrated that the induction of CYP2E1 by ethanol enhances hepatic lipid peroxidation, while compounds interfering with CYP2E1 expression significantly decrease oxidative stress and hepatic damage [49]. However, CYP2E1 knockout mice are not protected from alcohol toxicity [50] in spite of the fact that protein carbonyls and oxidized DNA products are lower in CYP2E1-null mice than in wild-type animals [51]. Such a discrepancy can be explained considering that, differently from rats and humans, CYP2E1 represents less than 5 % of the total hepatic cytochrome P450 content in ethanol-fed mice making its contribution to alcohol toxicity easily blunted by other factors [51]. On the other hand, ethanol administration to CYP2E1 transgenic mice over-expressing CYP2E1 causes more liver injury than in naïve mice and this effect is associated with a potentiation of oxidative damage and centrilobular hypoxia [52]. Interestingly, following alcohol feeding mice expressing human instead of murine CYP2E1 show higher hepatic CYP2E1 content and increased oxidative stress and hepatic injury than similarly treated wild-type mice [53], confirming the importance of the monoxygenase system in alcohol hepatotoxicity.

As discussed in the next paragraph, inflammation plays an important role in alcohol hepatotoxicity. Lobular inflammation associated to both acute and chronic alcohol intoxication is an important source of extracellular ROS that are generated by Kupffer cells and liver-infiltrating leucocytes through the activation of NADPH oxidase. Phagocyte NADPH oxidase is a multimeric transmembrane enzymatic complex that comprises the catalytic subunit (NOX2), regulatory subunits ($p22^{phox}$, $p40^{phox}$, and $p47^{phox}$), and the Rho small GTPase Rac1/2 [54]. The contribution of phagocyte-derived ROS to alcohol-induced oxidative stress has emerged from the observation that macrophage depletion with gadolinium chloride lowers the hepatic production of O_2^- following ethanol infusion and decreases both liver injury and lipid peroxidation markers in chronic enteral alcohol-fed rats [55]. A similar protection is also evident in mice knockout for NADPH oxidase $p47^{phox}$ [56] or in animals deficient for ICAM-1, an endothelial adhesion molecule required for the recruitment of leucocytes into inflammatory sites [57]. Nonetheless, following alcohol administration, the extent of DNA oxidation induced is comparable in both $p47^{phox}$ knockout and naïve mice, whereas CYP2E1-null mice are instead protected from such damages [51]. Altogether, these results indicate a different contribution of intracellular and extracellular ROS sources in ethanol-induced oxidative stress.

10.4.2 Hydroxyethyl-Free Radicals

One peculiarity of CYP2E1-mediated alcohol metabolism is represented by the conversion of ethanol itself to 1-hydroxyethyl-free radical ($CH_3C^{\bullet}HOH$; HER) [58]. The mechanisms responsible for HER formation involve the presence of ferric-CYP2E1-oxygen complexes as well as of hydroxyl radicals [58]. HERs are produced by rat liver microsomes at a rate 10 times lower than the ethanol conversion to acetaldehyde [59]. Nonetheless, because of their high reactivity, they are mainly responsible for the alkylation of hepatic constituents including CYP2E1 itself [58]. One of the consequences of protein alkylation by HER is the stimulation of an immune response characterized by the generation of antibodies specifically recognizing HER-derived epitopes [59]. These antibodies are detectable in the sera of both chronically ethanol-fed rats and patients with ALD where they strictly correlate with CYP2E1 activity [58, 60], indicating that HER generation actually takes place during "in vivo" alcohol metabolism in humans. Interestingly, heavy drinkers who do not display CYP2E1 induction have titers of anti-HER IgG comparable to non-drinking controls and significantly lower than drinkers with normally induced CYP2E1 activity [18].

10.4.3 Nitric Oxide

An additional contribution to alcohol-induced oxidative stress may originate from the interaction of nitric oxide (NO) with O_2^- to generate highly oxidizing peroxynitrite ($ONOO^-$). During alcohol intoxication, NO is mainly generated in activated phagocytes by inducible NO synthetase (iNOS) that increases by threefold following chronic ethanol exposure [61]. Accordingly, signs of ONOO-mediated nitrosative stress are evident in alcohol-treated rodents [62]. One of the targets of the increased NO formation are mitochondria that shows a NO-dependent decrease in the respiratory chain functions [63]. Consistently, the selective iNOS inhibitor N-(3-aminomethyl)benzyl-acetamindine (1,400 W) or iNOS genetic deficiency reduce oxidative stress, mitochondrial respiratory impairment, and hepatic injury induced by chronic alcohol feeding [34]. Nonetheless, the actual role of NO in the pathogenesis of alcohol-mediated oxidative stress awaits further investigations, as the treatment with the non-selective NO-inhibitor, N-nitro-L-arginine methyl ester (L-NAME), worsens alcohol liver damage [64]. This suggests that NO generated by phagocyte iNOS might be harmful in ALD, while NO produced by endothelial (eNOS) and possibly other NO sources can prevent alcoholic injury. According to this view, the stimulation of NO production within the hepatocytes increases their resistance to ethanol-induced oxidative stress by lowering intracellular low molecular weight iron content [65].

10.4.4 Iron Overload

Alteration of iron homeostasis is a well-recognized cofactor in the promotion of oxidative stress in many liver diseases, in relation to the capacity of this metal to stimulate the breakdown of lipid hydroperoxides and to catalyze the generation of hydroxyl radicals. A common feature of chronic alcohol intake in humans is the accumulation of an excess of iron within the hepatocytes [66]. However, an increased liver iron deposition is also evident even after moderate alcohol consumption [67]. Growing evidence suggests that ethanol induces liver iron overload by interfering with the functions of hepcidin [68], a liver-produced peptide that regulates the circulating iron levels by inhibiting ferroportin-mediated iron release from enterocytes and macrophages [69]. Alcohol-mediated oxidative stress has been proposed to affect hepatocyte hepcidin synthesis by acting on its transcription factor CCAAT/enhancer-binding protein-α (C/EBP-α) [70]. Nonetheless, enhanced activity of hypoxia-inducible factors (HIF-1/2α) as a consequence of hepatocyte hypoxia can also contribute in down-modulating C/EBP-α expression during chronic alcohol intake [71]. This latter effect can be related to CYP2E1 activity as modulation of intrahepatic CYP3E1 content influence in a similar manner liver hypoxia and HIF-1α activity in alcohol-treated mice [72]. In line with these findings, hepatic hepcidin mRNA levels are low in alcoholic patients [73], while ferroportin expression is increased in macrophages from alcohol-exposed rats [74]. In this scenario, low hepatocyte hepcidin production would favor an increased iron release by enterocytes and macrophages and its concomitant storage within the liver. On its turn, an increased hepatic content of intracellular low molecular weight non-protein iron exacerbates ethanol-induced oxidative damage in parenchymal cells of animals receiving alcohol [23] and stimulates macrophage ROS production [74]. Accordingly, iron supplementation of enteral alcohol-fed rats enhances lipid peroxidation and worsens liver pathology [75].

10.4.5 Interference with Liver Antioxidant Defenses

Beside alcohol capacity to stimulate free radical generation, the impairment of liver antioxidant defenses substantially contributes to oxidative injury in ALD. Alcohol action on antioxidant defenses involves a lowering of small molecular weight antioxidants, such as reduced glutathione (GSH) and α-tocopherol (Vitamin E), and an impairment in antioxidant enzymes. Early studies have shown that a decrease in the liver GSH content is common in ethanol-fed animals as well as in alcoholic patients independently from the nutritional status or the degree of liver disease [23]. On the other hand, the stimulation of GSH re-synthesis by rat supplementation with the GSH precursors L-2-oxothiazolidine-4-carboxylic acid or N-acetylcysteine prevents oxidative stress in enteral alcohol-feed rats [36, 76]. Hepatic GSH lowering can be regarded as one of the consequences of the impairment of S-adenosylmethionine (SAMe) production, as SAMe is a precursor of cysteine that is required for GSH synthesis [77]. SAMe is generated in an ATP-dependent reaction catalyzed by the enzyme methionine adenosyltransferase isoenzymes MATI and MATIII [77]. Beside to be a cysteine precursor, SAMe acts as the principal methyl group donor for the methylation reactions involving DNA, RNA biogenic amine, histones, and phospholipids [77]. These latter reactions generate homocysteine that is converted back to methionine by the action of methionine synthetase using methyl groups supplied by the combined action of methyltetrahydrofolate and vitamin B12 [77]. A lowering of hepatic SAMe content is evident in experimental animals chronically treated with alcohol as well as in patients with alcoholic hepatitis [78]. Alcohol affects SAMe formation either directly by impairing MAT and methionine synthetase activities, or indirectly by interfering with the turnover of folate, a key cofactor in methyl group transfer [78]. Consistently, SAMe administration attenuates alcohol-induced GSH depletion and oxidative injury in rats and mini-pigs [78]. It is noteworthy that during chronic alcohol intake, hepatic GSH depletion mainly involves the mitochondrial pool (mtGSH) of centrilobular hepatocytes [79]. Such a selective mtGSH depletion is partially due to an enhanced mtGSH oxidation in response to mitochondrial ROS production as the expression of mtCYP2E1 in HepG2 cell selectively depletes mtGSH [16]. In addition, defects in the GSH transport from cytosol to the mitochondrial matrix further contribute to affect mtGSH homeostasis. In fact, alcohol increases the hepatic synthesis of cholesterol and cholesterol unbalance in mitochondrial membranes interferes with the activity of GSH carrier proteins [79]. The action of ethanol on mtGSH homeostasis favors oxidative mitochondrial damage and enhances hepatocyte susceptibility to TNF-α-mediated cytotoxicity [79, 80]. Beside the effects on GSH, in both humans and rodents chronic alcohol intake decreases the liver and the plasma levels of vitamin E. Vitamin E depletion contributes to oxidative injury as in humans vitamin E levels inversely correlate with the extent of lipid peroxidation [25]. Moreover, vitamin E-deficient rats show an increased susceptibility to alcohol-induced oxidative stress and hepatotoxicity [81], while upon discontinuation of alcohol feeding the administration of vitamin E reduces the severity of hepatic lesions [32].

The effects of alcohol on antioxidant enzymes are less well-characterized. In one hand, ethanol increases the liver expression of the nuclear factor erythroid-2-related factor (Nfr2), which regulates the expression of antioxidant enzyme genes [81]. Consistently, the mRNA levels of liver glutathione peroxidase and catalase are increased following chronic alcohol administrations [82, 83], whereas Nfr2 deficiency greatly enhances mice susceptibility to alcohol-induced oxidative stress [38]. On the other hand, alcohol hepatotoxicity is associated with a significant decline in the hepatic content and enzymatic activity of (Cu–Zn)-superoxide dismutase (SOD-1), catalase, and glutathione peroxidase and the loss of these antioxidant enzymes inversely correlates with the extent of both lipid peroxidation and hepatic injury [31]. Similarly, mice ethanol feeding rapidly reduces the hepatic content of thioredoxin-1 (Trx-1), a redox-sensitive protein implicated in the reduction of oxidized proteins [33]. Little is known about the mechanisms responsible for such contrasting effects, but it is possible that ethanol might stimulate the intracellular degradation of antioxidant enzymes or interfere with their post-transcriptional regulation. On this latter respect, alcohol feeding in mice has been recently shown to modulate several epigenetic systems in the liver including the expression of microRNAs (miRNAs) that might influence the synthesis regulation of a variety of proteins [84]. Although in rodents manipulations of antioxidant enzymes such as Trx-1, SOD-1, and SOD-2 strongly influence alcohol-induced oxidative injury and hepatotoxicity [35, 37–39], human studies investigating the possible role of genetic polymorphisms of antioxidant genes in ALD have been inconclusive [13]. One of the most frequent of those genetic variants is a ^{16}alanine/valine substitution in the leader amino acid sequence (about 25 % prevalence in Caucasians) that is responsible for the mitochondrial localization of SOD-2 [13]. The Ala-SOD-2 variant translocates less efficiently to the mitochondria than the Val-SOD-2 [13], but Ala-SOD-2 does not appear to influence oxidative damage in ALD [85].

10.5 Role of Oxidative Stress in the Pathogenesis of Alcohol-Induced Liver Injury

As previously mentioned, ALD is characterized by a variety of histological lesions including steatosis, hepatocyte death by either necrosis or apoptosis, the presence of Mallory's bodies, lobular inflammation, and fibrosis/cirrhosis that implicate to different extent oxidative stress.

10.5.1 Alcoholic Steatosis

An increase in the hepatic content of triglycerides represents the most common histological and biochemical feature of excessive alcohol intake. Alcoholic steatosis mainly consists in the presence of a medium-sized/large fat droplet in hepatocyte cytoplasm (macrovesicular steatosis) with lateral displacement of the nucleus. Fat accumulation mostly involves all hepatic acinus, but may be prominent in the centrilobular areas [4]. Microvesicular steatosis, consisting in liver cell filling by small fat droplets, is relatively rare in ALD (0.8–2.3 %), but it is often associated with a more severe evolution of the hepatic injury [4]. Although steatosis is mostly asintomatic and often reversible, it is presently regarded as an important contributor in the progression of liver damage to fibrosis [86]. Alcoholic steatosis results from the combination of increased triglyceride synthesis, lowered fatty acids oxidation, and impaired lipoprotein secretion. These effects are the consequence of ethanol interferences with the nuclear transcription factors, sterol regulatory element-binding protein-1 (SREBP-1) and peroxisome proliferator-activated transcription factor-α (PPAR-α) controlling lipid metabolizing enzymes, as well as from direct damage of mitochondria and endoplasmic reticulum [87, 88]. Oxidative stress might contribute to the pathogenesis of alcoholic steatosis through actions on transcription factors regulating lipid metabolism, mitochondrial injury, and endoplasmic reticulum stress.

Mitochondrial β-oxidation of fatty acids represents a key pathway in hepatic lipid metabolism. During chronic alcohol intake, an important cause of steatosis relays on the lowering in fatty acid oxidation due the impairment of mitochondrial functions as consequence of oxidative stress-mediated mtDNA mutations or oxidation of mitochondrial proteins [42, 89]. The prevalence of mtDNA deletions is particularly high in alcoholics with microvesicular steatosis (about 85 % of the cases), indicating that the loss of mitochondrial respiratory capacity may represent the main cause of this lesion [90].

Hepatocyte endoplasmic reticulum (ER) is the site of protein folding, lipid and sterol synthesis, and intracellular calcium storage. The perturbation of ER functions activates several sensor proteins that trigger a signal network collectively termed the unfolded protein response (UPR) [91]. UPR counteracts ER alterations by reducing protein synthesis and promoting protein re-folding and/or degradation. Furthermore, ER stress stimulates the transcription of cytoprotective genes under the control of Nfr2 transcription factor and activates autophagy [91]. Increasing evidence indicates that alcohol causes ER stress in the liver. Oxidative stress is implicated in this process, as sulphydril redox unbalance in the ER, CYP2E1-mediated ROS production, and protein alkylation by lipid peroxidation products cause protein unfolding [92]. In particular, 4-HNE binding to heat shock proteins 70 and 90, protein sulphide isomerase, and the fatty acid-binding protein L-FABP have been recently detected in the liver of alcohol-fed rodents [93]. Moreover, ethanol can also contribute to hepatic ER stress by affecting protein degradation by the ubiquitin-proteasome system and autophagy [94, 95]. On its turn, ER stress is a stimulus for the proteolytical cleavage of SREBP-1 that translocates to the nucleus inducing the expression of genes encoding for enzymes involved in fatty acid synthesis, namely fatty acid synthetase, acyl-CoA carboxylase, and ATP citrate lysase [87, 89]. An increased SREBP-1 activation is evident in mice receiving an ethanol-containing diet, while SREBP-1 knockout mice are protected against ethanol-induced steatosis [87]. Consistently, betaine supplementation of alcohol-fed mice prevents ER stress, SREBP-1 activation, and fatty liver [96]. These observations suggest that alcohol-induced ER may stimulate intrahepatic lipid accumulation by triggering SREBP-1-dependent lipogenetic enzymes. In addition, ethanol interferes with PPAR-α action [87, 89] and oxidative stress mediated by CYP2E1 induction has been proposed to contribute to this effect [97]. PPAR-α regulates several genes responsible for the mitochondrial transport of fatty acids and both mitochondrial and peroxisomal fatty acid oxidation [89], making its down-modulation an important contributor in the development of alcoholic steatosis.

10.5.2 Lobular Inflammation

The transition from steatosis to steato-hepatitis is characterized by the appearance of mixed lobular inflammation featuring scattered infiltration by polymorphonuclear leucocytes and mononucleated cells [4, 88]. The persistence of mild-moderate parenchymal injury and inflammation is now recognized to be the main factor in the progression of ALD to cirrhosis. Furthermore, extensive lobular inflammation characterizes alcoholic steatohepatitis and contributes to impair liver functions [4, 88]. As previously discussed, inflammatory cell activation represents a relevant source of ROS and NO, leading to oxidative and nitrosative injury in ALD. In its turn, oxidative stress significantly contributes to the mechanisms that promote hepatic inflammation during alcohol abuse. In fact, growing evidence indicates that oxidized lipids and the adducts between proteins and end products of lipid peroxidation can act as damage-associated molecular patterns (DAMPs), promoting the activation of inflammatory cells through the interaction with soluble and cell-associated pattern recognition receptors such as Toll-like receptors 2 and 4 (TLR-2, TLR-4) and lectin-like oxidized LDL receptor-1 (LOX-1) [98, 99]. Recent studies have also implicated oxidative mechanisms in inflammasome activation, as intracellular ROS formation stimulates Nod-like receptor protein 3 (NLPR3) to trigger caspase-1-mediated release of interleukins (IL-1β and IL-18) [100]. Interestingly, caspase-1-dependent release of IL-1β has been recently shown to be required for the onset of hepatic inflammation in alcohol-treated mice, while the supplementation with IL-1β receptor antagonist ameliorated ethanol hepatotoxicity [101]. Furthermore, ethanol-triggered redox signals in liver macrophages amplify pro-inflammatory responses by modulating TLR-4 transduction pathway, ERK1/2 and p38 MAPK kinase signal cascades and TNF-α transcriptional control [102].

An increased translocation of bacterial lipopolysaccarides (LPS) to the portal circulation is presently recognized as an important cause for hepatic inflammation during chronic alcohol intake [88, 103]. In chronic alcohol-fed rats as well as in ALD patients, plasma LPS content increases several fold over physiological levels and correlates with the circulating TNF-α levels and the severity of alcoholic hepatitis [103]. Interestingly, CYP2E1-mediated oxidative stress in the gut and increased intestinal NO production have been shown to enhance the permeability of enteral mucosa to LPS in chronic alcohol-treated rats [104, 105], while prevention of oxidative damage ameliorates endotoxemia associated with ethanol intake [105, 106].

It is increasingly evident that, beside the effect on innate immunity, oxidative stress induces adaptive immune responses [107]. In fact, oxidized phospholipids and proteins modified by lipid peroxidation are recognized as important antigens in autoimmune disease and in atherosclerosis [108, 109]. In line with these observations, we have reported that advanced ALD is characterized by the detection of elevated titres of circulating antibodies recognizing epitopes derived from protein modification by malonyldialdehyde (MDA), 4-hydroxynonenal (4-HNE), and lipid hydroperoxides [107] as well as by malonyldialdehyde-acetaldehyde (MAA) condensation products [110]. In about 35 % of the ALD patients, the presence of these antibodies is also associated with the detection of $CD4^+$ T-lymphocytes recognizing malonyldialdehyde-derived antigens, indicating that oxidative stress promotes both humoral and cellular immune responses [111]. Finally, patients with alcoholic hepatitis or cirrhosis often have high titers of anti-phospholipid antibodies targeting oxidized phospholipids, namely oxidized cardiolipin and phosphatidylserine [112]. This latter observation is consistent with a recent report demonstrating increased levels of oxidized phospholipids in the plasma of alcohol-fed mice and patients with alcoholic hepatitis [28]. These clinical observations are supported by data generated in enteral alcohol-fed rats in which the development of lipid peroxidation-derived antibodies is associated with a sustained increase of TNF-α and IL-12 and histological evidence of necro-inflammation [113]. In these animals the supplementation with the antioxidant N-acetylcysteine ameliorates oxidative stress, hepatic inflammation, and the immune response triggered by lipid peroxidation [36]. Consistently, heavy drinkers with elevated titers of IgG targeting lipid peroxidation-induced antigens have a fivefold higher prevalence of elevated plasma TNF-α levels than the subjects with these antibodies within the control range irrespective of alcohol intake [114]. The risk of advanced ALD also increases by 11-fold in the heavy drinkers with the combination of high TNF-α and lipid peroxidation-induced antibodies as compared to the subjects with high TNF-α, but no immune responses [114]. Furthermore, the combination of steatosis and high titers of antibodies against lipid peroxidation-derived adducts is an independent predictor of advanced fibrosis/cirrhosis in alcohol-consuming patients with chronic hepatitis C [115]. Interestingly, antibodies towards oxidative stress-derived antigens are also a risk factor for severe lobular inflammation and fibrosis in both children and adults with non-alcoholic steatohepatitis (NASH) [116, 117]. In this latter respect, using an experimental model of NASH, it has been possible to show that liver injury and lobular inflammation parallel with the development of IgG against MDA- and 4-HNE-derived antigens and the hepatic recruitment of $CD4^+$ and $CD8^+$ T-lymphocytes responsive to the same antigens [117]. In this setting, mice immunization against MDA-adducted proteins further stimulates transaminase release and lobular inflammation and increases hepatic macrophage activation by promoting the Th-1 activation of $CD4^+$ T-lymphocytes [118]. These findings are in line with early studies showing that, in alcoholic hepatitis and active alcoholic cirrhosis, liver inflammatory infiltrates contain $CD8^+$ and $CD4^+$ T lymphocytes [119] and that $CD4^+$ T-cells respond to T-cell receptor stimulation by producing Th-1 cytokines such as interferon-γ (IFN-γ) and TNF-α [120]. More recently, IL-17-producing T helper (Th-17) lymphocytes have also been found to increase in hepatic inflammatory infiltrates of patients with alcoholic hepatitis/cirrhosis [121]. Although still indirect, these evidences suggest that the development of humoral and cellular immunity against oxidative stress-derived antigens may contribute to hepatic inflammation during the evolution of ALD.

A further contributor to adaptive immunity involves the antibody responses against HER-derived antigens. Human anti-HER IgG selectively recognize HER-CYP2E1 adducts [60]. Furthermore, CYP2E1 alkylation by HER favors the breaking of the self-tolerance towards CYP2E1, leading to the development of anti-CYP2E1 auto-antibodies that are detectable in about 40 % of patients with advanced ALD [122]. Anti-CYP2E1 autoimmune responses are more frequent in individuals carrying a genetic polymorphism in the locus coding for the cytotoxic T lymphocyte-associated antigen-4 (CTLA-4), a membrane receptor that down-modulates T-cell activation [122]. Thus, the combination of the antigenic stimulation by HER-modified CYP2E1 and an impaired T cell control due to CTLA-4 mutation act synergically in promoting the development of anti-CYP2E1 auto-antibodies. Both allo- and auto-reactivity involving CYP2E1 may contribute to alcohol hepatotoxicity, as anti-HER IgG recognize HER-CYP2E1 adducts present on the outer layer of the plasma membranes of ethanol-treated hepatocytes and activate antibody-dependent cell-mediated cytotoxicity [60].

The involvement of B- and T-cell responses in ALD is not in contrast with the observations concerning the role of innate immunity in alcohol-induced hepatic inflammation. Indeed, the activation of the innate immunity is an important stimulus for lymphocyte activation towards oxidative stress-derived antigens. On their turn, lymphocyte-derived cyto/chemokine might provide a stimulus for phagocyte recruitment and activation maintaining chronic hepatic inflammation in ALD.

10.5.3 Hepatocellular Injury

Hepatocyte death by either necrosis or apoptosis characterizes parenchymal injury in ALD and significantly contributes to the progression of the disease. In particular, an increase in hepatocyte apoptosis is associated with the severity of liver injury in patients with alcoholic steatohepatitis [123]. At histology, hepatocyte injury is characterized by cell swelling (ballooning) and by the presence of Mallory's bodies, consisting in intracellular protein aggregates mainly containing cytokeratins (CK) 8 and 18 [5]. Serum levels of CK18 and of caspase-cleaved CK18 have been recently proposed as specific markers of hepatocyte death [124]. These markers are increased in heavy drinkers and prominently in patients with alcoholic hepatitis in whom they correlate with the prevalence of Mallory's bodies, hepatocyte ballooning, and fibrosis [124]. Several mechanisms account for the cytotoxic action of ethanol. Alcohol-induced oxidative damage of mitochondria causes the collapse of mitochondrial membrane potential and the onset of mitochondria permeability transition (MPT) [42]. In this contest, the mitochondrial translocation of the pro-apoptotic factor Bax favors MTP by complexing the voltage-dependent anion channel, (VDAC) [125]. According to this view, ethanol addition to HepG2 over-expressing human CYP2E1 collapses mitochondrial membrane potential causing MPT and apoptosis that can be prevented by CYP2E1 inhibitors, antioxidants as well as by the over-expression of the anti-apoptotic protein Bcl-2 [16]. Similarly, antioxidants block ethanol-induced apoptosis in isolated hepatocytes [125]. On the other hand, ethanol-fed SOD-1-deficient mice show extensive hepatocellular damage, mitochondrial depolarization, and increase in MTP [126]. A further mechanism contributing to alcohol-induced hepatocyte apoptosis involves the stimulation ER stress and the subsequent production of the pro-apoptotic CHOP protein [127]. Intracellular accumulation and aggregation of oxidized proteins is also implicated in the formation of Mallory's bodies [94]. In fact, CYP2E1-expressing HepG2 cells exposed to alcohol form insoluble protein aggregates containing CK8 and 18 as a result of oxidative proteasome impairment [128].

A further important factor in hepatocellular damage concerns altered hepatocyte responses to inflammation. As discussed above, ALD is characterized by an increased production of pro-inflammatory cytokines including TNF-α and elevated plasma TNF-α levels correlate with the severity and the mortality of alcoholic hepatitis [88]. Experiments using enteral alcohol feed rodents have shown that the treatment with anti-TNF-α antibodies as well as TNF-α receptor 1 (TNF-R1) deficiency protect against liver damage [88]. At difference of other parenchymal cells, hepatocytes are resistant to the pro-apoptotic action of TNF-α, because TNF-α interaction with TNF-R1 stimulates pro-survival responses through the activation of anti-apoptotic NF-κB-dependent genes and PI3K/PKB/Akt (protein kinase B)-mediated signals [129]. However, CYP2E1-expressing HepG2 cells or CYP2E1 induction in mouse liver sensitize hepatocytes to the cytotoxic action of TNF-α [15], suggesting that alcohol may alter the balance between pro- and anti-apoptotic signals. More detailed investigations in hepatocytes from chronically ethanol-fed rats have shown that the selective depletion of mtGSH enhances the susceptibility to TNF-α-induced killing, without interfering with NF-κB response [79]. Such an effect involves cardiolipin oxidation in the mitochondrial membranes that favors Bax-induced MPT and cytochrome c release [130]. Finally, oxidative stress-dependent mechanisms can interfere at different levels with the signal network controlling hepatocyte pro-survival responses. In fact, ethanol activates the apoptosis signaling kinase-1 (ASK-1) by oxidizing its binding proteins thioredoxin [131], while lipid peroxidation products affect hepatocyte ERK1/2, a kinase responsible for transducing anti-apoptotic signals [132]. Furthermore, chronic alcohol treatment inhibits hepatocyte AMP-activated protein kinase (AMPK) that is an important regulator of cell responses to pro-apoptotic signals consequent to ER stress, oxidative injury, and mitochondrial damage [133]. This latter effect appears to be mediated by the direct interaction of AMPK with 4-HNE generated by lipid peroxidation [134]. Thus, by inducing mitochondrial damage, ER stress, and by interfering with hepatocyte pro-survival signals, oxidative stress contributes to hepatocyte killing in alcohol-exposed livers.

Beside its hepatotoxic action, oxidative stress is also implicated in causing liver cell senescence, a condition characterized by the block of proliferative capacity, morphological changes, the expression of senescence-associated (SA) ß-galactosidase, the up-regulation of p53-dependent cycline-dependent kinase inhibitors p21, p16, and the production of inflammatory cytokines [135]. This action has been implicated in reducing the regenerative capacity of mature hepatocytes by activating p38 MAPK and p21 [136]. Concomitantly, oxidative damage promotes the proliferation and accumulation of hepatic progenitor cells (HPCs) that are the precursors of both hepatocytes and biliary duct epithelial cells [137]. It has been observed that in both ethanol-treated mice and patients with severe alcoholic hepatitis, the increase in HPCs within liver midzonal and perivenular areas is associated with a higher mortality [138]. Conversely, a recent report has shown that acute alcohol intoxication promotes liver regeneration after partial hepatectomy. This apparent paradox is due to the induction of aldehyde dehydrogenase 2 (ALDH2) that can effectively metabolizes lipid peroxidation-derived aldehydes such as 4-HNE that are generated by oxidative injury associated to liver surgical resection [139].

Altogether these data indicate that in response to chronic ethanol hepatotoxicity, HPCs can sustain hepatocyte turnover in an attempt to compensate for the impaired proliferation of mature cells. This reaction disturbs the normal parenchymal structure contributing to hepatic dysfunction associated with the evolution of cirrhosis. It is also possible that the expansion of immature cell populations, together with the mutagenic effects induced by oxidative stress products [25], might be important in the development of hepatocellular carcinomas in alcoholic cirrhotics.

10.5.4 Alcohol-Induced Fibrosis and Cirrhosis

Liver cirrhosis represents the terminal stage of ALD and is one of the main causes of death among patients with alcohol abuse. Fibrosis develops primarily in the pericentral areas, where thin bundles of fibrotic tissue surround groups of hepatocytes and thicken the space of Disse, in a "chicken wire" fashion [4]. However, during the progression of alcohol liver damage, parenchymal injury along with unresolved inflammation promotes further deposition of collagen-rich extracellular matrix (ECM), causing the extension of fibrosis to periportal areas and the formation of fibrous septa that encircle islets of hepatic parenchyma, eventually leading to micronodular cirrhosis [5]. As in other chronic liver diseases, alcohol-induced fibrosis is the consequence of a wound healing response mediated by hepatic stellate cells (HSCs), which under the local influence of transforming growth factor β1 (TGF-β1), platelet-derived growth factor (PDGF), and the chemochine CCL2 (MCP-1), trans-differentiate into myofibroblast-like cells (HSC/MSs) producing collagen and ECM components [140]. Furthermore, decreased hepatic matrix degradation due to a reduced production of matrix metalloproteases (MMPs) and/or an increased production of matrix metalloprotease inhibitors might also contribute to collagen accumulation [140]. Although hepatic macrophages are responsible for the secretion of HSC-activating cyto/chemokines, HSCs themselves can sustain through paracrine mechanisms liver wound-healing responses [140, 141]. In this contest, ROS and aldehydic end products of lipid peroxidation are well-recognized pro-fibrogenic stimuli triggering intracellular signals, leading to myofibroblast transition [54, 142]. In particular, redox changes activate multiple signal pathways involving the kinases INK/AP-1 and Janus kinases 1/2 as well as the transcription factors NF-κB and C/EBPß, which promote pro-collagen α1(I) gene expression, stimulate HSC proliferation, and enhance HSC migration [142]. Similarly, NADPH-oxidase generation of intracellular ROS in HSC/MFs sustains pro-fibrogenic signals in response to PDGF-BB, angiotensin II, and leptin [143]. In this latter respect, microarray analysis has shown that the gene expression of non-phagocitic components of NADPH-oxidase (NOX4, DUOX1 and DUOX2) is markedly elevated in the livers of patients with alcoholic hepatitis as compared to healthy controls where these mRNAs are barely detectable [144]. From the functional point of view, the relevance of oxidative stress-related mechanisms in the development of alcohol-induced hepatic fibrosis is suggested by several observations. Studies *in vitro* have shown that ROS directly triggers collagen synthesis in ethanol-treated HSCs over-expressing human CYP2E1 [145] as well as in HSCs isolated from ethanol-fed mice livers [146]. Interestingly, such an effect is also evident when ethanol is added to cocultures of HSC and CYP2E1-expressing HepG2 cells [147], indicating that oxidative stress-derived mediators generated within alcohol-metabolizing hepatocytes may diffuse out and signal to neighboring HSC. Consistently, *in vivo* experiments reveal that alcohol-induced lipid peroxidation is associated with macrophage production of TGF-1ß and precedes the appearance of fibrosis [148]. Ethanol-stimulated TGF-ß1 synthesis is particularly evident in the perivenous regions and is abolished by CYP2E1 inhibition with chlormethiazole [149]. Furthermore, ethanol-induced liver fibrosis is exacerbated by the combined administration of carbonyl iron that greatly enhances oxidative stress as well as TGF-ß1 and procollagen-α1 mRNA expression in either the whole liver and freshly isolated HSCs [75]. Animal experiments have also shown that binge whisky administration to rats causes oxidative stress, p38 MAP kinase activation, and collagen deposition only in the livers with moderate steatosis induced by a choline-deficient diet [150]. On the same line, the combination of oxidative stress and steatosis is a risk factor for severe fibrosis in hepatitis C patients consuming alcohol [115], suggesting that alcoholic steatosis might worsen oxidative stress-dependent mechanisms of fibrogesis. Nonetheless, it should be considered that the pathogenesis of alcoholic fibrosis/cirrhosis is complex, and beside oxidative stress, chronic inflammation, impaired NK cell functions, and alterations in adipokine and endocannabinoid secretion might also specifically contribute to the disease evolution [140].

10.6 Conclusions

A large body of experimental and clinical data strongly implicates free radical formation and oxidative injury in several aspects of alcohol hepatotoxicity such as mitochondrial failure, hepatocyte apoptosis, amplification of alcohol-dependent pro-inflammatory signals, stimulation of immune responses, liver cell sensitization to the cytotoxic action of TNF-α, and activation of hepatic stellate cells to matrix production. Altogether, these observations give a rationale for the possible clinical use of antioxidant compounds to reduce the progression of ALD. However, despite experimental studies have shown that antioxidants can ameliorate alcohol liver injury in rodents [3, 32], the available clinical reports on the use of antioxidants in the therapy of human ALD give negative or inconclusive results [151–154]. On the same line, meta-analyses concerning the effects of milk thistle and S-adenosyl-L-methionine in alcoholic hepatitis were rather inconclusive [155, 156]. There are several possible explanations for such discrepancies: (1) alcohol-dependent liver injury in humans has a complex multifactorial pathogenesis and oxidative stress represents one of the factors influencing the disease evolution; (2) large inter-individual variability might influence the actual contribution of oxidative stress-dependent factors; (3) most of the clinical studies were performed in patients with severe alcoholic steatohepatitis or cirrhosis in which the lowering of oxidative stress would likely minimally modify already advanced hepatic lesions. Furthermore, to our knowledge, in none of the studies evaluating antioxidant therapy in ALD the actual effects on oxidative damage were monitored. Nonetheless, new hopes on the possible use of antioxidants in ameliorating hepatic damage in ALD come from recent evidences indicating that antioxidant treatments might be effective in improving hepatic damage in non-alcoholic steatohepatitis (NASH) [154, 157], which also implicate oxidative stress in the disease pathogenesis [158].

References

1. Mandayam S, Jamal MM, Morgan TR (2004) Epidemiology of alcoholic liver disease. Semin Liver Dis 24:217–232
2. Tzukamoto H (2007) Conceptual importance of identifying alcoholic liver disease as a lifestyle disease. J Gastroenterol 42:603–609
3. Bell BP, Manos MM, Zaman A, Terrault N, Thomas A, Navarro VJ, Dhotre KB, Murphy RC, Van Ness GR, Stabach N, Robert ME, Bower WA, Bialik SR, Sofair AN (2008) The epidemiology of newly diagnosed chronic liver disease in gastroenterology practices in the United States: results from population-based surveillance. Am J Gastroenterol 103:1–10
4. Zakhari S, Kai LT (2007) Determinants of alcohol use and abuse: impact of quantity and frequency patterns on the liver. Hepatology 46:2032–2039
5. O'Shea R, Dasarathy S, McCullough AJ, Practice Guideline Committee of the American Association for the Study of Liver Diseases and the Practice Committee of the American College of Gastroenterology (2010) Alcoholic liver disease. Hepatology 51:307–328
6. Sheron N, Oslen N, Gilmore I (2008) An evidence based alcohol reduction policy. Gut 57:1341–1344
7. Blachier M, Leleu H, Peck-Radosavljevic M, Valla DC, Roudot-Thoraval F (2013) The burden of liver disease in Europe: a review of available epidemiological data. J Hepatol 58:593–608
8. Corrao G, Bagnardi V, Zambon A, La Vecchia C (2004) A meta-analysis of alcohol consumption and the risk of 15 diseases. Prev Med 38:613–619
9. Bellentani S, Saccoccio G, Costa G, Tiribelli C, Manenti F, Sodde M, Saveria Crocè L, Sasso F, Pozzato G, Cristianini G, Brandi G (1977) Drinking habitus as co factor of the risk of alcohol-induced liver damage. Gut 41:845–850
10. Kamper-Jorgensen M, Gronback M, Tolstrup J, Becker U (2004) Alcohol and cirrhosis: dose-response or threshold effect? J Hepatol 41:25–30
11. Morgan TR, Mandayam S, Jamal MM (2004) Alcohol and hepatocellular carcinoma. Gastroenterology 127:587–596
12. Seitz HK, Stickel F (2007) Molecular mechanisms of alcohol-mediated carcinogenesis. Nat Rev Cancer 7:599–612
13. Stickel F, Hampe J (2012) Genetic determinants of alcoholic liver disease. Gut 61:150–159
14. Crabb DW, Liangpunsakul S (2007) Acetaldehyde generating enzymes: roles of alcohol dehydrogenase, CYP2E1 and catalase, and speculation on the role of the others enzymes and processes. In: Chadwick DJ, Goode J (eds) Acetaldehyde-related pathology; bridging the transdisciplinary divide. Wiley, Chichister, pp 4–16
15. Cederbaum AL (2012) Alcohol metabolism. Clin Liver Dis 16:667–685
16. Lu Y, Cederbaum AL (2008) CYP2E1 and oxidative liver injury by alcohol. Free Radic Biol Med 44:723–738
17. Liangpunsakul S, Kolwankar D, Pinto A, Gorski CJ, Hall SD, Chalasani N (2005) Activity of CYP2E1 and CYP3A enzymes in adults with moderate alcohol consumption: a comparison with nonalcoholics. Hepatology 41:1144–1150
18. Dupont I, Lucas D, Clot P, Ménez C, Albano E (1998) Cytochrome P4502E1 inducibility and hydroxyethyl radical formation among alcoholics. J Hepatol 28:564–571
19. Neve EP, Ingelman-Sundberg M (2008) Intracellular transport and localization of microsomal cytochrome P450. Anal Bioanal Chem 392:1075–1084
20. Deitrich RA, Petersen D, Vasiluou V (2007) Removal of acetaldehyde from the body. In: Chadwick DJ, Goode J (eds) Acetaldehyde-related pathology; bridging the transdisciplinary divide. Wiley, Chichister, pp 23–40
21. Singh S, Brocker C, Koppaka V, Chen Y, Jackson BC, Matsumoto A, Thompson DC, Vasiliou V (2013) Aldehyde dehydrogenases in cellular responses to oxidative/electrophilic stress. Free Radic Biol Med 56:89–101

22. Di Luzio NR (1963) Prevention of acute ethanol-induced fatty liver by antioxidants. Physiologist 6:169–173
23. Nordmann R, Ribière C, Rouach H (1992) Implication of free radical mechanisms in ethanol induced cellular injury. Free Radic Biol Med 12:219–240
24. Albano E (2008) New concept in the pathogenesis of alcoholic liver disease. Expert Rev Gastroenterol Hepatol 2:749–759
25. Clot P, Tabone M, Aricò S, Albano E (1994) Monitoring oxidative damage in patients with liver cirrhosis and different daily alcohol intake. Gut 35:1637–1643
26. Aleynik SI, Leo MA, Aleynik MK, Lieber CS (1998) Increased circulating products of lipid peroxidation in patients with alcoholic liver disease. Alcohol Clin Exp Res 22:192–196
27. Meager EA, Barry OP, Burke A, Lucey MR, Lawson JA, Rokach J, FitzGerald GA (1999) Alcohol-induced generation of lipid peroxidation products in humans. J Clin Invest 104:805–813
28. Yang L, Latchoumycandane C, McMullen MR, Pratt BT, Zhang R, Papouchado BG, Nagy LE, Feldstein AE, McIntyre TM (2010) Chronic alcohol exposure increases circulating bioactive oxidized phospholipids. J Biol Chem 285:22211–22220
29. Niemelä O, Parkkila S, Ylä-Herttuala S, Halsted C, Witztum JL, Lanca A, Israel Y (1994) Covalent protein adducts in the liver as a result of ethanol metabolism and lipid peroxidation. Lab Invest 70:537–546
30. Nanji AA, Zhao S, Sadrzadeh SMH, Dannenberg AJ, Tahan SR, Waxman DJ (1994) Markedly enhanced cytochrome P4502E1 induction and lipid peroxidation is associated with severe liver injury in fish oil-treated ethanol-fed rats. Alcohol Clin Exp Res 18:1280–1285
31. Polavarapu R, Spitz DR, Sim JE, Follansbee MH, Oberley LW, Rahemtulla A, Nanji AA (1998) Increased lipid peroxidation and impaired antioxidant enzyme function is associated with pathological liver injury in experimental alcoholic liver disease in rats fed diets high in corn oil and fish oil. Hepatology 27:1317–1323
32. Nanji AA, Yang EK, Fogt F, Sadrzadeh SMH, Dannenberg AJ (1996) Medium chain triglycerides and vitamin E reduce the severity of established experimental alcoholic liver disease. J Pharmacol Exper Ther 277:1694–1700
33. Nanji AA, Sadrzadeh SMH, Yang EK, Fogt F, Maydani M, Dannenberg AJ (1995) Dietary saturated fatty acids: a novel treatment for alcoholic liver disease. Gastroenterology 109:547–554
34. Arteel GE (2003) Oxidants and antioxidants in alcohol-induced liver disease. Gastroenterology 124:778–790
35. Cohen JI, Roychowdhury S, DiBello PM, Jacobsen DW, Nagy LE (2009) Exogenous thioredoxin prevents ethanol-induced oxidative damage and apoptosis in mouse liver. Hepatology 49(5):1709–1717
36. Ronis MJJ, Butura A, Sampey BP, Prior RL, Korourian S, Albano E, Ingelman-Sundberg M, Petersen DR, Badger TM (2005) Effects of N-acetyl cysteine on ethanol-induced hepatotoxicity in rats fed via total enteral nutrition. Free Radic Biol Med 39:619–630
37. Wheeler MD, Kono H, Yin M, Rusyn I, Froh M, Connor HD, Mason RP, Samulski RJ, Thurman RG (2001) Delivery of Cu/Zn-superoxide dismutase gene with adenovirus reduces early alcohol-induced liver injury in rats. Gastroenterology 120:1241–1250
38. Wheeler MD, Nakagami M, Bradford BU, Uesugi T, Mason RP, Connor HD, Dikalova A, Kadiiska M, Thurman RG (2001) Overexpression of manganese superoxide dismutase prevents alcohol-induced liver injury in the rat. J Biol Chem 276:36664–36672
39. Kessova IG, Ho YS, Thung S, Cederbaum AI (2003) Alcohol-induced liver injury in mice lacking Cu, Zn-superoxide dismutase. Hepatology 38:1136–1145
40. Lamlé J, Marhenke S, Borlak J, von Wasielewski R, Eriksson PCJ, Geffers R, Manns MP, Yamamoto M, Vogel A (2008) Nuclear factor-eythroid 2-related factor prevents alcohol-induced fulminant liver injury. Gastroenterology 134:1159–1168
41. Bailey SM, Cunningham CC (2002) Contribution of mitochondria to oxidative stress associated with alcohol liver disease. Free Radic Biol Med 32:11–16

42. Hoek JB, Cahill A, Pastorino JG (2002) Alcohol and mitochondria: a dysfunctional relationship. Gastroenterology 122:2049–2063
43. Venkatraman A, Landar A, Davis AJ, Chamlee L, Sanderson T, Kim H, Page G, Pompilius M, Ballinger S, Darley-Usmar V, Bailey SM (2004) Modification of the mitochondrial proteome in response to the stress of ethanol-dependent hepatotoxicity. J Biol Chem 279:22092–22101
44. Garcia-Ruiz C, Colell A, Paris R, Fernandez-Checa JC (2000) Direct interaction of GD3 ganglioside with mitochondria generates reactive oxygen species followed by mitochondrial permeability transition, cytochrome c release and caspase activation. FASEB J 14:847–850
45. Konckaert L, Fromenty B, Robin MA (2011) Mechanisms of mitochondrial targeting of cytochrome P4502E1: physiopathological role in liver injury and obesity. FEBS J 278:4252–4280
46. Bansal S, Liu CP, Sepuri NB, Anandatheerthavarada HK, Selvaraj V, Hoek J, Milne GL, Guengerich FP, Avadhani NG (2010) Mitochondria-targeted cytochrome P450 2E1 induces oxidative damage and augments alcohol-mediated oxidative stress. J Biol Chem 285: 24609–24619
47. Frank A, Seitz HK, Bartsch H, Frank N, Nair J (2004) Immunohistochemical detection of 1, N6-ethenodeoxyadenosine in nuclei of human liver affected by diseases predisposing to hepatocarcinogenesis. Carcinogenesis 25:1027–1031
48. Hu W, Feng Z, Eveleigh J, Iyer G, Pan J, Amin S, Chung FL, Tang MS (2002) The major lipid peroxidation product, trans-4-hydroxy-2-nonenal, preferentially forms DNA adducts at codon 249 of human p53 gene, a unique mutational hot spot in hepatocellular carcinoma. Carcinogenesis 23:1781–1789
49. Albano E, Clot P, Morimoto M, Tomasi A, Ingelman-Sundberg M, French SW (1996) Role of cytochrome P4502E1-dependent formation of hydroxyethyl free radicals in the development of liver damage in rats intragastrically fed with ethanol. Hepatology 23:155–163
50. Kono H, Bradford BU, Yin M, Sulik KK, Koop DR, Peters JM, Gonzalez FJ, McDonald T, Dikalova A, Kadiiska MB, Mason RP, Thurman RG (1999) CYP2E1 is not involved in early alcohol-induced liver injury. Am J Physiol 277:G1259–G1267
51. Bradford BU, Kona H, Isayama F, Kosyk O, Wheeler MD, Akiyama TE, Bleye L, Krausz KW, Gonzalez FJ, Koop DR, Rusyn I (2005) Cytochrome P450 CYP2E1, but not nicotinamide adenine dinucleotide phosphate oxidase is required for ethanol-induced oxidative DNA damage in rodent liver. Hepatology 41:336–344
52. Morgan K, French SW, Morgan TR (2002) Production of a cytochrome P450 2E1 transgenic mouse and initial evaluation of alcoholic liver damage. Hepatology 36:122–134
53. Lu Y, Wu D, Wang X, Ward SC, Cederbaum AI (2010) Chronic alcohol-induced liver injury and oxidant stress are decreased in cytochrome P4502E1 knockout mice and restored in humanized cytochrome P4502E1 knock-in mice. Free Radic Biol Med 49:1406–1416
54. Pail YH, Kim J, Aoyama T, De Minicis S, Battalier R, Brenner DA (2014) Role of NADPH oxidase in liver fibrosis. Antioxid Redox Signal 20:2854–2872
55. Hines IN, Wheeler MD (2004) Recent advances in alcoholic liver disease III. Role of the innate immune response in alcoholic hepatitis. Am J Physiol Gastrointest Liver Physiol 287:G310–G314
56. Kono H, Rusyn I, Yin M, Gabele E, Yamashina S, Dikalova A, Kadiiska MB, Connor HD, Mason RP, Segal BH, Bradford BU, Holland SM, Thurman RG (2000) NADPH oxidase-derived free radicals are key oxidants in alcohol-induced liver disease. J Clin Invest 106:867–872
57. Kono H, Uesugi T, Froh M, Rusyn I, Bradford BU, Thurman RG (2001) ICAM-1 is involved in the mechanism of alcohol-induced liver injury: studies with knockout mice. Am J Physiol 280:G1289–G1295
58. Albano E, French SW, Ingelman-Sundberg M (1999) Hydroxyethyl radicals in ethanol hepatotoxicity. Front Biosci 4:533–540

59. Stoyanovsky DA, Cederbaum AI (1998) ESR and HPLC-EC analysis of ethanol oxidation to 1-hydroxyethyl radical: rapid reduction and quantification of POBN and PBN nitroxides. Free Radic Biol Med 25:536–545
60. Clot P, Parola M, Bellomo G, Dianzani U, Carini R, Tabone M, Aricò S, Ingelman-Sundberg M, Albano E (1997) Plasma membrane hydroxyethyl radical adducts cause antibody-dependent cytotoxicity in rat hepatocytes exposed to alcohol. Gastroenterology 113:265–276
61. Chamulitrat W, Spitzer JJ (1996) Nitric oxide and liver injury in alcohol-fed rats after lipopolysaccaride administration. Alcohol Clin Exp Res 20:1065–1070
62. Loguercuio C, Federico A (2003) Oxidative stress in viral and alcoholic hepatitis. Free Radic Biol Med 34:1–10
63. Venkatraman A, Shiva S, Davis AJ, Bailey SM, Brookes PS, Darley-Usmar VM (2003) Chronic alcohol consumption increases the sensitivity of rat liver mitochondrial respiration to inhibition by nitric oxide. Hepatology 38:141–147
64. Nanji AA, Greenberg SS, Tahan SR, Fogt F, Loscalzo J, Sadrzadeh SMH, Xie J, Stamler JS (1995) Nitric oxide production in experimental alcoholic liver disease in the rat: role in protection from injury. Gastroenterology 109:899–907
65. Sergent O, Griffon B, Morel I, Chevanne M, Dubus MP, Cillard P, Cillard J (1997) Effect of nitric oxide on iron-mediated oxidative stress in primary hepatocyte culture. Hepatology 23:122–127
66. Irving MG, Halliday JW, Powell LW (1988) Association between alcoholism and increased hepatic iron store. Alcohol Clin Exp Res 12:7–12
67. Ioannou GN, Dominitz JA, Weiss NS, Haegerty PJ, Wowdley KV (2004) The effect of alcohol consumption on the prevalence of iron overload, iron deficiency and ion deficiency anemia. Gastroenterology 126:1293–1301
68. Harrison-Findik DD, Klein E, Crist C, Evans J, Timchenko N, Gollan J (2007) Iron-mediated regulation of liver hepcidin expression in rats and mice is abolished by alcohol. Hepatology 46:1979–1985
69. Ganz T, Nemeth E (2011) Hepcidin and disorders of iron metabolism. Ann Rev Med 62:347–360
70. Harrison-Findik DD, Schafer D, Klein E, Timchenko NA, Kulaksiz H, Clemens D, Fein E, Andriopoulos B, Pantopoulos K, Gollan J (2006) Alcohol metabolism-mediated oxidative stress down-regulates hepcidin transcription and leads to increased duodenal iron transporter expression. J Biol Chem 281:22974–22982
71. Anderson ER, Taylor M, Xue X, Martin A, Moons DS, Omary MB, Shah YM (2012) The hypoxia-inducible factor-C/EBPα axis controls ethanol-mediated hepcidin repression. Mol Cell Biol 32:4068–4077
72. Wang X, Wu D, Yang L, Gan L, Cederbaum AI (2013) Cytochrome P450 2E1 potentiates ethanol induction of hypoxia and HIF-1α in vivo. Free Radic Biol Med 63:175–186
73. Costa-Matos L, Batista P, Monteiro N, Simões M, Egas C, Pereira J, Pinho H, Santos N, Ribeiro J, Cipriano MA, Henriques P, Girão F, Rodrigues A, Carvalho A (2012) Liver hepcidin mRNA expression is inappropriately low in alcoholic patients compared with healthy controls. Eur J Gastroenterol Hepatol 24:1158–1165
74. Xiong S, She H, Zhang AS, Wang J, Mkrtchyan H, Dynnyk A, Gordeuk VR, French SW, Enns CA, Tsukamoto H (2008) Hepatic macrophage iron aggravates experimental alcoholic steatohepatitis. Am J Physiol Gastrointest Liver Physiol 295:G512–G521
75. Tsukamoto H, Horne W, Kamimura S, Niemelä O, Parkkila S, Ylä-Herttuala S, Brittenham GM (1995) Experimental liver cirrhosis induced by alcohol and iron. J Clin Invest 96:620–630
76. Iimuro Y, Bradford BU, Yamashina S, Rusyn I, Nakagami M, Enomoto N, Kono H, Frey W, Forman D, Brenner D, Thurman RG (2000) The glutathione precursor L-2-oxothiazolidine-4-carboxylic acid protects against liver injury due to chronic enteral ethanol exposure in the rat. Hepatology 31:391–398

77. Lu SC, Tzukamoto H, Mato JM (2002) Role of abnormal methionine metabolism in alcoholic liver injury. Alcohol 27:155–162
78. Purohit V, Abdelmalek MF, Barve S, Benevenga NJ, Halsted CH, Kaplowitz N, Kharbanda KK (2007) Role of S-adenosylmethionine, folate and betaine in the treatment of alcoholic liver disease: summary of a symposium. Am J Clin Nutr 86:14–24
79. Fernandez-Checa JC, Kaplowitz N (2005) Hepatic mitochondrial glutathione: transport and role in disease and toxicity. Toxicol Appl Pharmacol 204:263–273
80. Fernandez A, Colell A, Garcia-Ruiz C, Fernandez-Checa JC (2008) Cholesterol and sphingolipids in alcohol-induced liver injury. J Gastroenterol Hepatol 23(Suppl 1):S9–S15
81. Sadrazadeh SMH, Nanji AA, Meydani M (1994) Effect of chronic ethanol feeding on plasma and liver α- and γ-tocophetol levels in normal and vitamin E-deficient rats. Biochem Pharmacol 47:2005–2010
82. Gong P, Cederbaum AI (2006) Nrf2 is increased by CYP2E1 in rodent liver HepG2 cells and protects against oxidative stress caused by CYP2E1. Hepatology 43:144–153
83. Nanji AA, Griniuviene B, Sadrzadeh SMH, Levitsky S, McCully JD (1995) Effect of dietary fat and ethanol on antioxidant enzyme mRNA induction in rat liver. J Lipid Res 36:736–744
84. Mandrekar P (2011) Epigenetic regulation in alcoholic liver disease. World J Gastroenterol 17:2456–2464
85. Stewart SF, Leathart JB, Chen Y, Daly AK, Rolla R, Mottaran E, Vay D, Vidali M, Day CP, Albano E (2002) Valine-alanine manganese superoxide dismutase polymorphism is not associated with alcohol-induced oxidative stress or fibrosis. Hepatology 36:1355–1360
86. Powel EE, Jonsson JR, Clouston AD (2005) Steatosis: co-factor in other liver diseases. Hepatology 42:5–13
87. Purohit V, Gao B, Song BJ (2009) Molecular mechanisms of alcoholic fatty liver. Alcohol Clin Exp Res 33:1–15
88. Gao B, Bataller R (2011) Alcoholic liver disease: pathogenesis and new therapeutic targets. Gastroenterology 141:1572–1585
89. Sozio M, Crabb DW (2008) Alcohol and lipid metabolism. Am J Physiol Endocrinol Metab 295:E10–E16
90. Fromenty B, Grimbert S, Mansouri A, Beaugrand M, Erlinger S, Röting A, Pessayre D (1995) Hepatic mitochondrial DNA deletion in alcoholics: association with microvesicular steatosis. Gastroenterology 108:193–200
91. Malhi H, Kaufman RJ (2011) Endoplasmic reticulum stress in liver disease. J Hepatol 54:795–809
92. Malhotra JD, Kaufman RJ (2007) Endoplasmic reticulum stress and oxidative stress: a vicious circle or a double-edged sword? Antiox Redox Signal 9:2277–2293
93. Smalthers RL, Galligan JJ, Stewart BJ, Pettersen DR (2011) Overview of lipid peroxidation products and hepatic protein modification in alcoholic liver disease. Chem Biol Interact 192:107–112
94. Donohue TM, Cederbaum AI, French SW, Barve S, Gao B, Osna NA (2007) Role of the proteasome in ethanol-induced liver pathology. Alcohol Clin Exp Res 31:1446–1459
95. Donohue TM (2009) Autophagy and ethanol-induced liver injury. World J Gastroenterol 15:1178–1185
96. Ji C, Kaplowitz N (2003) Betaine decreases hyperhomocysteinemia, endoplasmic reticulum stress, and liver injury in alcohol-fed mice. Gastroenterology 124:1488–1499
97. Lu Y, Zhuge J, Wang X, Bai J, Cederbaum AI (2008) Cytochrome P450 2E1 contributes to ethanol-induced fatty liver in mice. Hepatology 47:1483–1494
98. Miller YI, Choi SH, Weisner P, Fang L, Harkewicz R, Hartvigsen K, Buillier A, Gonen A, Diehl CJ, Que X, Montano E, Shaw PX, Tsimikas S, Binder CJ, Witztum JL (2011) Oxidation-specific epitopes are danger-associated molecular patterns recognized by pattern recognition receptors of innate immunity. Circ Res 108:235–248
99. Weismann D, Binder CJ (2012) The innate immune response to products of phospholipid peroxidation. Biochim Biophys Acta 1818:2465–2475

100. Jin C, Flavell RA (2010) Molecular mechanisms of NLPR3 inflammosome activation. J Clin Immunol 30:628–631
101. Petrasek J, Bala S, Csak T, Lippai D, Kodys K, Menashy V, Barrieau M, Min SY, Kurt-Jones EA, Szabo G (2012) IL-1 receptor antagonist ameliorates inflammasome-dependent alcoholic steatohepatitis in mice. J Clin Invest 122:3476–3489
102. Cohen JI, Chen X, Nagy LE (2011) Redox signaling and the innate immune system in alcoholic liver disease. Antioxid Redox Signal 16:523–534
103. Wang JH, Gao B, Zakhari S, Nagi LE (2012) Inflammation in alcoholic liver disease. Ann Rev Nutr 32:343–368
104. Keshavarzian A, Farhadi A, Forsyth CB, Rangan J, Jakate S, Shaikh M, Banan A, Fields JZ (2009) Evidence that chronic alcohol exposure promotes intestinal oxidative stress, intestinal hyperpermeability and endotoxemia prior to development of alcoholic steatohepatitis in rats. J Hepatol 50:538–547
105. Abdelmegeed MA, Banerjee A, Jang S, Yoo SH, Yun JW, Gonzalez FJ, Keshavarzian A, Song BJ (2013) CYP2E1 potentiates binge alcohol-induced gut leakiness steatohepatitis and apoptosis. Free Radic Biol Med 65:1238–1245
106. Tang Y, Forsyth CB, Banan A, Fields JZ, Keshavarzian A (2009) Oats supplementation prevents alcohol-induced gut leakiness in rats by preventing alcohol-induced oxidative tissue damage. J Pharmacol Exp Ther 329:952–958
107. Vidali M, Stewart SF, Albano E (2008) Interplay between oxidative stress and immunity in the progression of alcohol-mediated liver injury. Trends Mol Med 14:63–71
108. Witztum JL, Lichtman AH (2014) Influence of innate and adaptive immune responses on atherosclerosis. Annu Rev Pathol 9:73–102
109. Kurien BT, Scofield RH (2008) Autoimmunity and oxidatively modified antigens. Autoimmun Rev 7:567–573
110. Thiele GM, Freeman TK, Klassen LW (2004) Immunological mechanisms of alcoholic liver disease. Sem Liver Dis 24:273–287
111. Stewart SF, Vidali M, Day CP, Albano E, Jones DEJ (2004) Oxidative stress as a trigger for cellular immune response in patients with alcoholic liver disease. Hepatology 39:197–203
112. Vay D, Rigamonti C, Vidali M, Mottaran E, Alchera E, Occhino G, Sartori M, Albano E (2006) Anti-phospholipid antibodies associated with alcoholic liver disease target oxidized phosphatidylserine on apoptotic cell plasma membranes. J Hepatol 44:183–189
113. Ronis MJ, Butura A, Korourian S, Shankar K, Simpson P, Badeaux J, Albano E, Ingelman-Sundberg M, Badger TM (2008) Cytokine and chemokine expression associated with steatohepatitis and hepatocyte proliferation in rats fed ethanol via total enteral nutrition. Exp Biol Med 233:344–355
114. Vidali M, Vietala J, Occhino G, Ivaldi A, Sutti S, Niemelä O, Albano E (2008) Immune responses against oxidative stress-derived antigens are associated with increased circulating tumor necrosis factor-α and accelerated liver damage in heavy drinkers. Free Radic Biol Med 45:306–311
115. Vidali M, Occhino G, Ivaldi A, Rigamonti C, Sartori M, Albano E (2008) Combination of oxidative stress and steatosis is a risk factor for fibrosis in alcohol-drinking patients with chronic hepatitis C. Am J Gastroenterol 103:147–153
116. Albano E, Mottaran E, Vidali M, Reale E, Saksena S, Occhino D, Burt AD, Day CP (2005) Immune response towards lipid peroxidation products as a predictor of progression of non-alcoholic fatty liver disease to advanced fibrosis. Gut 54:987–993
117. Nobili V, Parola M, Alisi A, Marra F, Piemonte F, Mombello C, Sutti S, Povero D, Maina V, Novo E, Albano E (2010) Oxidative stress papameters in paediatric non-alcoholic fatty liver disease. Int J Mol Med 26:471–476
118. Sutti S, Jindal A, Locatelli I, Vacchiano M, Gigliotti L, Bozzola C, Albano E (2014) Adaptive immune responses triggered by oxidative stress contribute to hepatic inflammation in NASH. Hepatology 59:886–897
119. Colombat M, Charlotte F, Ratziu V, Poyard T (2002) Portal lymphocytic infiltrate in alcoholic liver disease. Hum Pathol 33:1170–1174

120. Batey RG, Cao Q, Gould B (2002) Lymphocyte-mediated liver injury in alcohol-related hepatitis. Alcohol 27:37–41
121. Lemmers A, Moreno C, Gustot T, Maréchal R, Degreé D, Demetter P, de Nadal P, Geerts A, Quertinmont E, Vercruysse V, Le Moine O, Devière J (2009) The interleukin-17 pathway is involved in human alcoholic liver disease. Hepatology 49:646–657
122. Vidali M, Stewart SF, Rolla R, Daly AK, Chen Y, Mottaran E, Jones DEJ, Leathart JB, Day CP, Albano E (2003) Genetic and epigenetic factors in autoimmune reactions toward cytochrome P4502E1 in alcoholic liver disease. Hepatology 37:277–285
123. Natori S, Rust C, Stadheim LM, Srinivasan A, Burgart LJ, Gores GJ (2001) Hepatocyte apoptosis is a pathological feature of human alcoholic hepatitis. J Hepatol 34:248–253
124. Eguchi A, Wree A, Feldstein AE (2014) Biomarkers of cells death. J Hepatol 60:1063–1074
125. Adachi M, Ishii H (2002) Role of mitochondria in alcoholic liver injury. Free Radic Bio Med 32:487–491
126. Kessova IG, Cederbaum AI (2007) Mitochondrial alterations in livers of SOD1–/– mice fed alcohol. Free Radic Biol Med 42:1470–1480
127. Ji C, Mehrian-Shai R, Chan C, Hsu YH, Kaplowitz N (2005) Role of CHOP in hepatic apoptosis in the murine model of intragastric ethanol feeding. Alcohol Clin Exp Res 29:1496–1503
128. Bardag-Gorce F, French BA, Nan L, Song H, Nguyen SK, Yong H, Dede J, French SW (2006) CYP2E1 induced by ethanol causes oxidative stress, proteasome inhibition and cytokeratin aggresome (Mallory body-like) formation. Exp Mol Pathol 81:191–201
129. Malhi H, Guicciardi ME, Gores GJ (2010) Hepatocyte death: a clear and present danger. Physiol Rev 90:1165–1194
130. Mari M, Colell A, Morales A, Caballero F, Moles A, Fernández A, Terrones O, Basañez G, Antonsson B, García-Ruiz C, Fernández-Checa JC (2008) Mechanism of mitochondrial glutathione-dependent hepatocellular susceptibility to TNF despite NF-kB activation. Gastroenterology 134:1507–1520
131. Hoek JB, Pastorino JG (2004) Cellular signaling mechanisms in alcoholic liver damage. Sem Liver Dis 24:257–272
132. Sampey BP, Stewart BJ, Petersen DR (2007) Ethanol-induced modulation of hepatocellular extracellular signal-regulated kinase 1/2 activity via 4-hydroxynonenal. J Biol Chem 282:1925–1937
133. Sid B, Verrax J, Calderon PB (2013) Role of AMPK activation in oxidative cell damage: implication for alcohol-induced liver disease. Biochem Pharmacol 86:200–209
134. Shearn CT, Backos DS, Orlicky DJ, Smathers-McCullough RL, Petersen DR (2014) Identification of 5′ AMP activated kinase as a target of reactive aldehydes during chronic ingestion of high concentrations of ethanol. J Biol Chem 289:15449–15462
135. Hoare M, Das T, Alexander G (2010) Ageing, telomeres, senescence and liver injury. J Hepatol 53:950–961
136. Diehl AM (2005) Recent events in alcoholic liver disease V. Effect of ethanol on liver regeneration. Am J Physiol Gastrointest Liver Physiol 288:G1–G6
137. Roskams T, Yang SQ, Koteish A, Durnez A, DeVos R, Huang X, Achten R, Verslype C, Diehl AM (2003) Oxidative stress and oval cell accumulation in mice and humans with alcoholic and nonalcoholic fatty liver disease. Am J Pathol 163:1301–1311
138. Jung Y, Brown KD, Witek RP, Omenetti A, Yang L, Vandongen M, Milton RJ, Hines IN, Rippe RA, Spahr L, Rubbia-Brandt L, Diehl AM (2008) Accumulation of hedgehog-responsive progenitors paralles alcoholic liver disease in mice and humans. Gastroenterology 134:1532–1543
139. Ding X, Beier JI, Baldauf KJ, Jokinen JD, Zhong H, Arlteel GE (2014) Acute ethanol preexposure promotes liver regeneration after partiel hepatectomy by activating ALDH2. Am J Physiol 306:G37–G47
140. Friedman SL (2008) Mechanism of hepatic fibrogenesis. Gastroenterology 134:1655–1669

141. Pellicoro A, Ramachandran P, Iredale JP, Fallowfield JA (2014) Liver fibrosis and repair: immune regulation of wound healing in a solid organ. Nat Rev Immunol 14:181–194
142. Novo E, Parola M (2008) Redox mechanisms in hepatic chronic wound healing and fibrogenesis. Fibrogenesis Tissue Repair 1:5. doi:10.1186/1755-1536-1-5
143. Novo E, Busletta C, Bonzo LV, Povero D, Paternostro C, Mareschi K, Ferrero I, David E, Bertolani C, Caligiuri A, Cannito S, Tamagno E, Compagnone A, Colombatto S, Marra F, Fagioli F, Pinzani M, Parola M (2011) Intracellular reactive oxygen species are required for directional migration of resident and bone marrow-derived hepatic pro-fibrogenic cells. J Hepatol 54:964–974
144. Colmenero J, Balaller R, Sancho-Bru P, Bellot P, Miguel R, Moreno M, Jares P, Bosch J, Arroyo V, Caballeira J, Gines P (2007) Hepatic expression of candidate genes in patients with alcoholic hepatitis: correlations with disease severity. Gastroenterology 132:687–697
145. Nieto N, Greenwel P, Friedman SL, Zhang F, Dannenberg AJ, Cederbaum AJ (2000) Ethanol and arachidonic acid increase α2(1) collagen expression in rat hepatic stellate cells overexpression cytochrome P450 2E1. J Biol Chem 26:20136–20145
146. Nieto N (2007) Ethanol and fish oil induce NFkB transactivation of the collagen α2(1) promoter through lipid peroxidation-driven activation of PKC-PI3K-Akt pathway. Hepatology 45:1433–1445
147. Nieto N, Friedman SL, Cederbaum AJ (2002) Cytochrome P4502E1-derived reactive oxygen species mediate paracrine of collagen 1 protein synthesis by hepatic stellate cells. J Biol Chem 277:9853–9864
148. Kamimura S, Tsukamoto H (1995) Cytokine gene expression by kupffer cells in experimental alcoholic liver disease. Hepatology 21:1304–1309
149. Fang C, Lindros KO, Badger TM, Ronis JJ, Ingelman-Sundberg M (1998) Zonated expression of cytokines in rat liver: effect of chronic ethanol and cytochrome P4502E1 inhibitor, chlormethiazole. Hepatology 27:1304–1310
150. Nieto N, Rojkind M (2007) Repeated whiskey binges promote liver injury in rats fed a choline-deficient diet. J Hepatol 46:330–339
151. Mezey E, Potter JJ, Rennie-Tankersley L, Caballeire J, Pares A (2004) A randomized placebo controlled trial of vitamin E for alcoholic hepatitis. J Hepatol 40(1):40–46
152. Phillips M, Curtis H, Portmann B, Donaldson N, Bomford A, O'Grady J (2006) Antioxidants versus corticosteroids in the treatment of severe alcoholic hepatitis: a randomised clinical trial. J Hepatol 44:784–790
153. Stewart S, Prince M, Bassendine M, Hudson M, James O, Jones D, Record C, Day CP (2007) A randomized trial of antioxidant therapy alone or with corticosteroids in acute alcoholic hepatitis. J Hepatol 47:277–283
154. Singal AK, Jampana SC, Weinman SA (2011) Antioxidants as therapeutic agents for liver disease. Liver Int 31:1432–1448
155. Rambaldi A, Jacobs BP, Iaquinto G, Gluud C (2005) Milk thistle for alcoholic and/or hepatitis B or C liver diseases–a systematic Cochrane hepato-biliary group review with meta-analyses of randomized clinical trials. Am J Gastroenterol 100:2583–2591
156. Rambaldi A, Gluud C (2006) S-adenosyl-L-methionine for alcoholic liver diseases. Cochrane Database Syst Rev 19, CD002235
157. Pacana T, Sanyal AJ (2012) Vitamin E and nonalcoholic fatty liver disease. Curr Opin Clin Nutr Metab Care 15:641–648
158. Gambino R, Musso G, Cassader M (2011) Redox balance in the pathogenesis of nonalcoholic fatty liver disease: mechanisms and therapeutic opportunities. Antioxid Redox Signal 15:1325–1365

Chapter 11
Oxidative Stress in Chronic Viral Hepatitis

Nicole L.B. Corder, Bhargav Koduru, Seung Bum Park, Yiyan Wang, and Jinah Choi

11.1 Introduction

11.1.1 Chronic Viral Hepatitis

Chronic viral hepatitis (CVH) is a persistent state of hepatic inflammation lasting greater than 6 months that is of viral origin. The primary etiological agents of CVH are hepatitis B virus (HBV) and hepatitis C virus (HCV) which, together, are responsible for 78 % of hepatocellular carcinoma (HCC) worldwide [1]. An estimated 3.2 million persons in the United States have chronic hepatitis C (CHC), and approximately 730,000 have chronic hepatitis B (CHB) [2, 3]. Globally, approximately 170 million, or 3 % of the world population, are chronically infected with HCV, and 378 million people are experiencing CHB [4, 5]. Although significant advances have improved the overall survival and quality of life in patients with CVH, the incidence of HCC continues to rise [6]. In contrast to HBV, there is no vaccine to prevent HCV infection, and CHC develops in approximately 75–85 % of those infected with HCV compared to CHB which develops in approximately 25 % of those infected with HBV [2, 3]. Disease progression occurs slowly, over several decades. Many HCV-and HBV-infected individuals often experience only mild flu-like symptoms or are asymptomatic and do not know that they are infected [7, 8]. Also, by the time they are diagnosed, many patients have already progressed to an advanced disease state, which can complicate the treatment.

N.L.B. Corder • B. Koduru • S.B. Park • Y. Wang • J. Choi (✉)
School of Natural Sciences, University of California, Merced,
5200 N. Lake Rd, SE248, Merced, CA 95343, USA
e-mail: jchoi@ucmerced.edu

E. Albano, M. Parola (eds.), *Studies on Hepatic Disorders*,
Oxidative Stress in Applied Basic Research and Clinical Practice,
DOI 10.1007/978-3-319-15539-5_11

A key pathogenic factor, present in both HCV and HBV infections, is increased oxidative stress. In particular, studies indicate that HCV induces oxidative stress by multiple mechanisms, which is likely to contribute to the pathogenesis. The aim of this chapter is to address the significance of oxidative stress in chronic viral hepatitis, with emphasis on HCV. A follow-up discussion on HBV and its interactions with oxidative stress can be found in Sect. 11.4. below.

11.1.2 Hepatitis C Virus

11.1.2.1 Prevalence

Hepatitis C virus was originally isolated from the serum of a patient with non-A, non-B hepatitis (NANBH). The agent was successfully cloned in 1989 and subsequently found to be responsible for 90 % of posttransfusion NANBH in the United States [9, 10]. Since its discovery, HCV remains a leading cause of chronic liver disease worldwide, with an estimated 170 million persons chronically infected globally. Regions with the highest prevalence (>3.5 %) include Central and East Asia, North Africa, and the Middle East. The prevalence of HCV in the United States is estimated to be 1.6 %, corresponding to 3–4 million individuals [2]. In 2011, the number of acute HCV cases reported to the Centers for Disease Control (CDC) totaled 1,229, representing a 44 % increase over the previous year. This corresponds to an increase from 0.3 cases per 100,000 in 2010 to 0.4 cases per 100,000 in 2011 within the United States. A total of 185,979 cases of CHC were reported to the National Notifiable Diseases Surveillance System (NNDSS) by 34 states in 2011. Furthermore, among viral hepatitis A, B, and C, hepatitis C accounted for the highest proportion of deaths in 2011, and an increase in mortality rate was observed from 4.4 deaths per 100,000 in 2006 to 4.7 deaths per 100,000 in 2010 in the United States. Despite the recent statistics pointing to an increase in HCV prevalence, it is still estimated that approximately 50 % of chronic HCV infections remain undiagnosed and unreported in the United States [11]. Worldwide, the estimated prevalence of HCV is highest in Africa, Asia, and Eastern Europe, ranging from approximately 2.0–2.9 % [12]. It is also estimated that there are between 3 and 4 million new cases of HCV every year globally [13].

11.1.2.2 Transmission

HCV is a blood-borne pathogen and is spread through contact with contaminated blood or blood products. Several risk factors have been identified in the transmission of HCV from person to person. The first risk factor involves receipt of contaminated tissues, organs, blood, and/or blood products. Before the implementation of routine screening of donor blood in the early 1990s, receipt of contaminated blood products during blood transfusion and/or organ transplantation served as a significant source of infection. Although healthcare-associated contraction of HCV still exists, the frequency is minimal; eight confirmed cases of healthcare-related exposure to HCV were investigated by the CDC in 2011. The second risk factor is exposure to the virus through shared and repeated needle sticks/punctures during injection drug use or body modification. The third risk factor involves close contact that is primarily sexual, but can include nonsexual, with individuals who have hepatitis C or are at risk of contracting the virus. Several subpopulations that have been identified as "at risk" groups include current or former injection drug users, individuals receiving chronic hemodialysis or those who received blood transfusions/organ transplants before July of 1992 or clotting factors before 1987, those with previous HBV or human immunodeficiency virus (HIV) infection, children born to HCV-positive mothers, and persons with known exposures, such as health care workers and blood or organ recipients from which the donor later tested positive.

11.1.2.3 Diseases and Therapy

Infection with hepatitis C virus induces a state of hepatic inflammation that can be characterized by two phases: acute infection and chronic infection. As previously mentioned, the chronic phase involves all intra- and extrahepatic manifestations occurring after the initial 6 months of infection, or the acute phase. Despite this simplification, physiological events that lead to the various disease states associated with HCV infection are complex. Acute hepatitis, the initial phase of the illness, is characterized by a mild, and often asymptomatic, illness. Symptoms of acute infection, which typically develops in 20–30 % of patients, are nonspecific and can include mild fever, fatigue, muscle aches, loss of appetite, nausea/vomiting, joint pain, dark urine, or jaundice. Similarly, chronic hepatitis often presents as an asymptomatic illness, but underlying liver injury is common and occurs in a progressive manner. The typical cascade of progressive liver damage begins with chronic inflammation of the liver leading to fibrosis and the development of cirrhosis, followed by HCC. About 20–30 % of chronic hepatitis C patients develop cirrhosis in 20–30 years, and 4–7 % of these individuals progress to HCC and end-stage liver disease each year [8]. Reports indicate that from 2000 to 2005, the incidence rates of HCC significantly increased in Hispanic, black, and white middle-aged men and 1-year survival rates remain below 50 % in the United States [14]. The presence of HCV infection has also been associated with an increased risk of extrahepatic manifestations. For example, HCV infection has been correlated with a higher incidence of diabetes mellitus, glomerulonephritis, essential mixed cryoglobulinemia, porphyria cutanea tarda, and Non-Hodgkin's lymphoma [15, 16]. Of these manifestations, diabetes mellitus is three times more likely to occur with HCV infection [17].

Previous standard of care treatment of hepatitis C involved administration of pegylated interferon (peg-IFN) and ribavirin. The factors that dictate whether a patient will achieve sustained virological response (SVR), defined as having an undetectable HCV RNA level 24 weeks after the end of treatment, are complex and include viral genotype, presence of coinfection with HIV or HBV, host-genetic factors, and consumption of alcohol. Typically, 76–80 % of patients infected with HCV genotype 2 or 3 achieve SVR with standard IFN-based therapy, compared to 42–46 % of those with genotype 1 [18]. Genotype 1 is the most prevalent genotype in North America and Europe, accounting for approximately 70 % of cases in the United States [18].

Direct acting antiviral (DAA) therapies, developed within the last decade, have certainly enhanced treatment outcomes in CHC patients. Common viral targets of DAAs include nonstructural (NS) 3/4A protease, NS5A, and HCV RNA-dependent RNA polymerase (RdRp), and DAAs provide patients with promising alternatives in the treatment of HCV infection. For example, 61 % of treatment-naïve subjects administered telaprevir, a protease inhibitor, for 12 weeks and peg-IFN/ribavirin regimen for 48 weeks achieved SVR, compared to 41 % treated with 12 weeks of placebo and 48 weeks of peg-IFN/ribavirin [19]. A nucleotide polymerase inhibitor, sofosbuvir, was recently approved by the FDA for use in conjunction with IFN, and as an IFN-free treatment option [20]. A clinical phase III trial in which treatment-naïve participants with genotype 2 or 3 HCV were administered sofosbuvir and ribavirin for 12 weeks ($n=256$), or 24 weeks of peg-IFN and ribavirin treatment ($n=243$), resulted in 67 % of participants from both groups achieving SVR at 12 weeks after discontinuing therapy [21]. Once-daily oral daclatasvir, an inhibitor of NS5A, and sofosbuvir was associated with high rates of SVR among patients infected with HCV genotype 1, 2, or 3 [18].

11.1.2.4 Viral Genome and Life Cycle

HCV is a single-stranded, positive-sense RNA virus from the genus Hepacivirus of the family *Flaviviridae*. The viral genome is monopartite with a length of 9.6 kb, containing coding and noncoding sequences (see Fig. 11.1a). The noncoding sequences or untranslated regions (UTR) flank either side of the coding sequence. The 5′ UTR forms secondary structures that comprise an internal ribosomal entry site (IRES) essential for cap-independent translation of viral RNA. The coding sequence codes for both structural and nonstructural proteins. Structural regions include, in order of appearance, core that also codes for F protein/Alternative Reading Frame Protein (F/ARFP), and envelope glycoproteins 1 and 2 (E1 and E2). Nonstructural regions include, in order of appearance, NS1, NS2, NS3, NS4A, NS4B, NS5A, and NS5B. The functions of each structural and nonstructural region can be found in Table 11.1.

Hepatitis C virus enters host cells through a complex set of interactions with various cellular membrane proteins (see Fig. 11.1b). Initial attachment of free virions to hepatocytes occurs through association with the cell surface receptors low-density lipoprotein receptor (LDLR), glycosaminoglycan (GAG), and heparan sulfate proteoglycan (HSPG). Viral particle entry, occurring through clathrin-mediated endocytosis, also requires interaction with the cell surface molecules, cluster of differentiation 81 (CD81), scavenger receptor B type 1 (SRB1), claudin

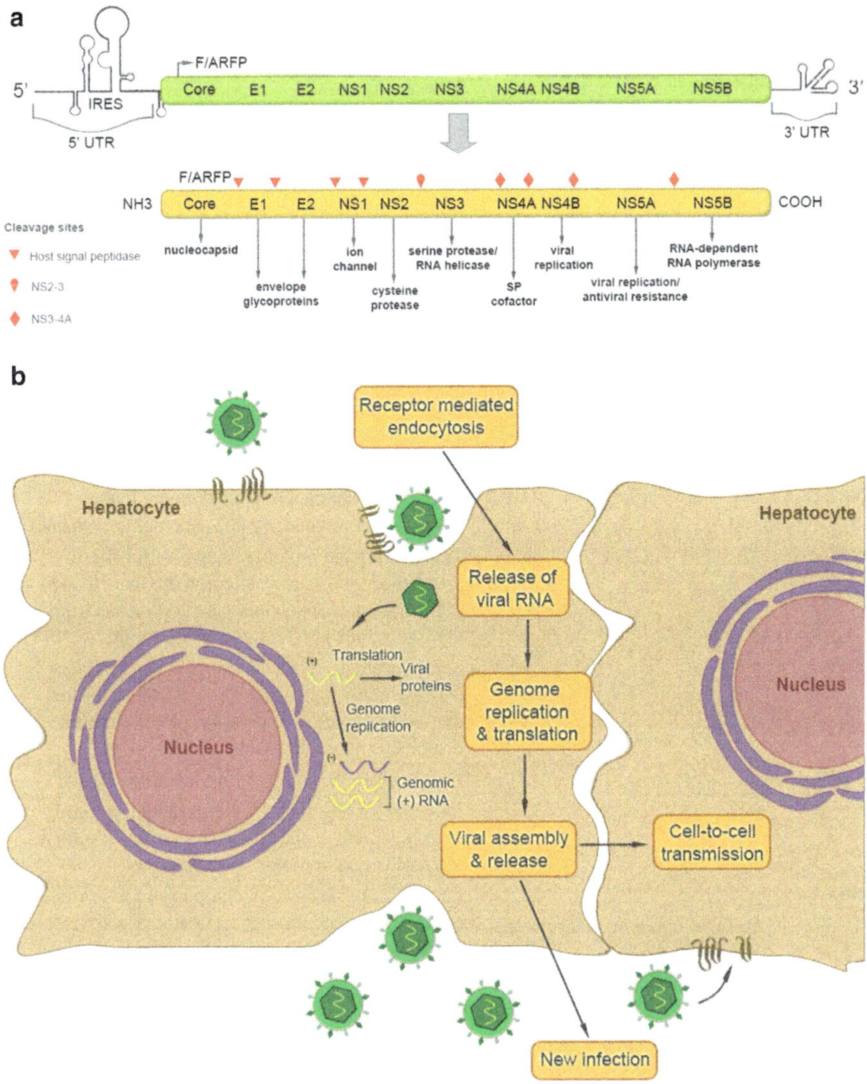

Fig. 11.1 Genome and replication cycle of hepatitis C virus. (**a**) HCV genome is a single-stranded (+) sense RNA of approximately 9.6 kb in length. Both the 5′ and 3′ UTRs contain secondary structures. HCV RNA contains a single open reading frame encoding for structural and nonstructural proteins. Translation of HCV RNA produces a single polyprotein sequence cleaved at the indicated sites. Primary functions of the cleaved protein segments are indicated below the polyprotein sequence (also, see Table 11.1). (**b**) HCV replication, initiated by receptor-mediated endocytosis, involves release of viral RNA into the cytoplasm, and subsequent translation of viral proteins and genome replication. Viral assembly follows whereupon new virions can cause new infection by direct transmission to adjacent cells, or through exocytic release and reentry

Table 11.1 Hepatitis C virus structural and nonstructural proteins

Common name	Alternate designation and structural considerations	Primary functions
Structural proteins		
Core	p22 21 kDa (mature form) 23 kDa (immature form)	Nucleocapsid formation, pro- and antiapoptotic functions, induces lipid droplet formation important for viral assembly and release; oxidative stress; pathogenesis
F/ARFP	−2/+1 frameshift protein	Not clear as of yet; may play a role in modulating the host cell immune response, regulation of iron metabolism, and/or apoptosis
E1	gp35	Envelope glycoprotein 1, possibly acts as a fusion domain
E2	gp70	Envelope glycoprotein 2, receptor binding, possibly acts as a fusion domain
Nonstructural proteins		
NS1	p7	Integral membrane protein belonging to the viroporin family making up a calcium ion channel
NS2	p23	Transmembrane protein responsible for ER membrane association; autoprotease that cleaves the NS2-3 zinc-dependent metalloprotease from the C terminus of NS2 and the N terminus of NS3
NS3	p70	Serine protease (NS3/4A protease) and helicase/NTPase (NS3 helicase/NTPase); important for the viral lifecycle, suppression of host immune response, and pathogenesis
NS4A	p8	Cofactor of NS3/4A proteinase; pathogenesis
NS4B	p27	Induces formation of the membranous web by rearrangement of the ER, acts as an anchor for the replication complex; pathogenesis
NS5A	p56/58 56 kDa (hypophosphorylated) 58 kDa (hyperphosphorylated)	Formation of replication complexes for genome replication; regulation of replication cycle; zinc metalloprotein; ER stress; pathogenesis
NS5B	p68	RNA-dependent RNA polymerase (RdRp)

(CLN1), and occludin (OCLN) [22]. After endocytosis, acidification of vesicle effectively uncoats and releases the viral RNA into the cytoplasm.

Translation of viral RNA by host cell machinery is initiated in a cap-independent and an IRES-dependent manner. In addition to the IRES, several viral proteins are involved in the translation of viral RNA. Translation of viral RNA occurs in vesicular structures known as membranous webs that are derived from the endoplasmic reticulum (ER) [23]. Briefly, the 40S ribosomal subunit first associates directly with the IRES site forming a complex near the start codon [24]. Next, a 48S-like ribosomal initiation complex is formed by the adjoining eukaryotic initiation factors (eIF) 3 and eIF2, along with a methionine initiator tRNA (Met-tRNA$_i$) and associated GTP [25]. The structural conformation of the IRES and the 48S-like complex allows dissociation of phosphorylated eIF2 upon initiation of translation [26].

Translation occurs in the canonical 5′–3′ direction to produce a single long polyprotein sequence of approximately 3,000 amino acids that will undergo co- and posttranslational processing by host and viral proteases. Nine clearly defined proteolytic cleavage sites are located within the polyprotein sequence, generating at least ten non-overlapping proteins. Structural proteins are cleaved by host cellular signal peptidases (SP), whereas nonstructural proteins are cleaved by the viral proteases NS2-3 and NS3-4A (see Fig. 11.1a). Upon cleavage of the various viral proteins, replication of viral RNA occurs through the generation of a negative-sense RNA that serves as a template for mass production of positive-sense genomic RNA by viral RdRp. Newly formed positive-sense RNA may be translated into viral proteins or encapsidated by core protein during the assembly of virions.

The complete roles of individual viral proteins are still under investigation, but some of the main events that facilitate assembly and packaging of new virions include trafficking of envelope glycoproteins 1 and 2 to the lumen of the ER where they assemble into heterodimers. Core protein becomes associated with cytoplasmic lipid droplets (cLD) with the assistance of cytosolic phospholipase A2 (cPLA2). Then, interactions between the ER transmembrane protein NS1 (p7), cytosolic NS2, and NS3/4A in the cytoplasm facilitates the recruitment of core. During this process, viral RNA generated in the membranous web associates with core protein that together migrate across the cytoplasmic space and bud into the lumen of the ER. The action of budding into the ER lumen is, in part, mediated by the interaction between core protein and envelope glycoproteins.

The process of viral maturation and exit appears to follow the pathway of very low-density lipoprotein (VLDL) secretion, which is believed to occur through a series of budding and fission events involving the endosomal sorting complex required for transport (ESCRT) [27]. Posttranslational modification of E1 and E2 occurs as virions transport through the Golgi apparatus preparing the proteins for subsequent pH-dependent uncoating of virions in a newly infected cell. Viral p7 protein acts as an ion channel stabilizing the pH during maturation [28–31]. Finally, virions bud out through the plasma membrane.

11.1.2.5 In Vitro and In Vivo HCV Models

HCV shows a narrow host range of infectivity toward humans and chimpanzees. Chimpanzees have 98 % genetic identity to humans, and until the development of infectious cell culture systems, chimpanzees served as an important model for studying the HCV replication cycle, pathogenesis, and therapy [32]. HCV infection in chimpanzees can be initiated by inoculation of clinical isolates, cell culture-generated virus, or intrahepatic injection of in vitro transcribed HCV RNA [33–35]. Nevertheless, this model has several drawbacks such as high cost, ethical issues, and genetic heterogeneity. In addition, HCV-induced pathogenesis is milder in chimpanzees than in humans, and severe liver diseases such as fibrosis, cirrhosis, and hepatocellular carcinoma are rarely seen in chimpanzees [33]. Other nonhuman primate models appear to be resistant to HCV infection, which may be ascribed, at least in part, to resistance to NS3/4A protease-mediated proteolytic cleavage of mitochondrial antiviral signal (MAVS, also known as IPS-1, VISA, or Cardiff) protein [36]. Several approaches have been used to develop in vitro replication models for HCV in order to study its viral life cycle and develop antiviral drugs.

Subgenomic replicon consisted of HCV genotype 1b NS2-3′UTR/NS3-3′UTR and a neomycin phosphotransferase gene under two different IRES sequences of encephalomyocarditis virus (EMCV) and HCV, respectively. This system show autonomous and persistent HCV RNA replication after transfection of in vitro-transcribed viral RNA and selection using neomycin phosphate (G418) [37, 38]. In addition to a genotype 1b system, genotype 1a, 2a systems and different full-length replicon RNAs were also developed [38–41]. Although the subgenomic and full-length replicon systems provided valuable knowledge about HCV replication in the cell, these systems were not effective to study the impact of complete HCV replication in cell culture.

Successful isolation and cloning of Japanese Fulminant Hepatitis-1 (JFH-1), an HCV genotype 2a sequence, led to the development of a complete HCV replication system. Transfection of human hepatoma (Huh-7) cells with in vitro transcribed JFH-1 full-length RNA resulted in robust HCV replication and production of infectious virus particles [32, 42]. Heller et al., on the other hand, developed a ribozyme-based HCV replication system that was infectious in vivo [43]. Recently, a genotype 1a (H77S virus) replication system was developed [44]. The H77S RNA, rendered after five cell culture-adaptive mutations, is capable of producing moderate levels of cell culture-infectious virus. Together, these systems have provided the opportunity to study the life cycle of HCV and develop various antiviral drugs. Although several infectious HCV replication models were developed, development of additional replication models is still under investigation.

In addition, several different approaches have been used to develop small animal models to study HCV using mice, despite lack of natural infection of mice. For example, human entry factors (CD81, SCARB1, CLDN1, and OCLN) have been introduced into mice to facilitate HCV entry [45]. This murine model exhibits complete host immunity and can be infected with HCV, and cytokine responses to HCV could be observed. However, the innate immune response may be stronger in the murine model, which may not support high levels of viral replication; protein kinase R (PKR), interferon regulatory factor 3 (IRF3), or STAT1 disruption could enhance HCV replication in mouse embryonic fibroblasts or in mice [46–48]. Whether these animal models develop liver fibrosis or HCC has not yet been shown.

Another method involves xenotransplantation of human cells or human liver tissues into the livers of immunocompromised mice, with or without introducing the human immune system. This model utilizes the proliferative and regenerative capabilities of hepatocytes in response to liver injury to "humanize" the mouse liver. An example of the employment of this technique is to engraft human hepatocytes into a severe combined immunodeficient (SCID) mouse expressing a urokinase plasminogen activator (uPA) under the control of an albumin promoter to induce liver injury and allow the human hepatocytes to proliferate. This mouse model (SCID/Alb-uPA) can subsequently be infected with HCV [26]. A recent study further described a dual humanized mouse model in which human hepatocytes and CD34$^+$ hematopoietic stem cells were introduced into AFC8 mice; and successful establishment of human hepatocytes and various immune cells was achieved [49]. This model does not contain a complete human immune system but supported HCV infection in the liver and generated a human T-cell response against HCV. Furthermore, HCV infection in this model produced liver inflammation, hepatitis, and fibrosis [49].

An additional approach for developing a mouse model utilizes the high capacity of HCV to mutate. For example, with a few adaptive mutations, HCV may gain the ability to enter murine hepatocytes. One recent study showed that adaptive mutations in E1 and E2 proteins conferred HCV the ability to enter mouse cells, but whether these HCV can infect mouse hepatocytes in vivo remains to be tested [50]. Another method that has been commonly used in various pathogenesis studies, despite the apparent drawback that these animals do not generate infectious HCV or can be infected with HCV, are HCV transgenic mice. These animals generally harbor either full-length or partial HCV coding sequence under the control of murine albumin or other promoters and enhancers. Such transgenic mouse models have been shown to develop steatosis and hepatic tumors including HCC. Lastly, there is some evidence that tree shrews, or tupaias, are susceptible to HCV infection and can support viral entry into hepatocytes [51]. The advantage of this model is that tree shrews develop severe liver diseases such as steatosis, fibrosis, and cirrhosis [52]. However, limited availability and difficulty of genetic manipulation remain barriers to this model.

11.2 HCV Increases Oxidative Stress

Oxidative stress is a state of imbalance between the production and dismutation, or detoxification, of reactive oxygen species (ROS) by cellular mechanisms that can disrupt normal cell functions or cause cell damage. An important distinction should be made between the beneficial and detrimental effects of ROS. For example, while ROS are frequently associated with toxicity, ROS are essential in the phagocytic respiratory burst (PRB) that functions in the destruction of microbes within phagosomes [53]. In addition, ROS can regulate a variety of cellular processes through redox signaling [54, 55].

Several hepatocyte and nonhepatocyte sources of oxidative stress have been identified in HCV infection (see Fig. 11.2). Both irreversible reactions culminating in cell damage and more subtle modifications of cellular factors by ROS are likely to contribute to HCV-associated liver disease, which are discussed in Sect. 11.3.. In this section, the primary mechanisms of oxidative stress in CHC, including iron overload, mitochondrial ROS, hepatocyte and phagocytic NADPH oxidase (Nox) proteins, and inducible nitric oxide synthase (iNOS), are discussed.

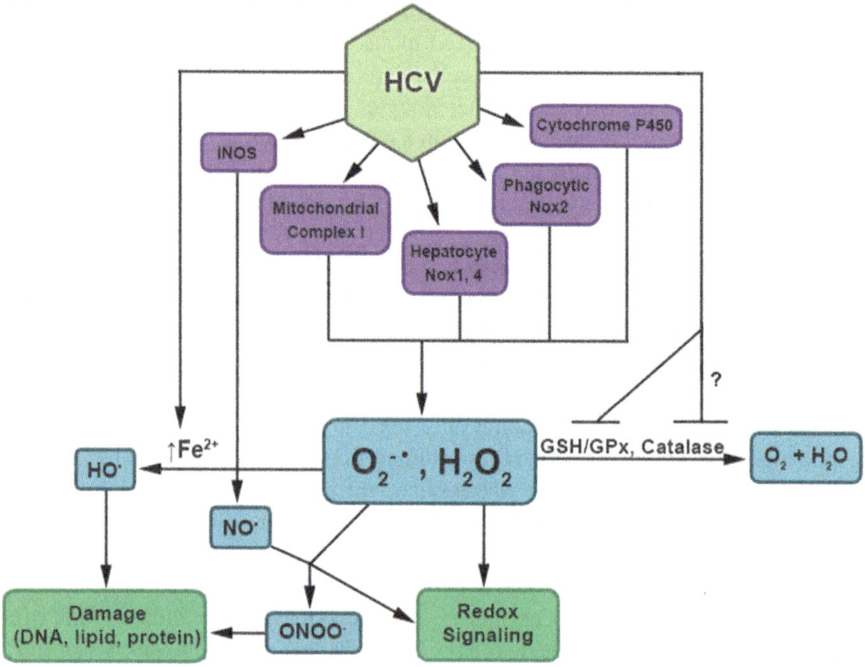

Fig. 11.2 Sources and signaling pathways of oxidative stress with hepatitis C virus infection. Sources of superoxide ($O_2^{-\bullet}$) and/or hydrogen peroxide (H_2O_2) with HCV include mitochondrial complex I, hepatocyte Nox1 and 4, phagocytic Nox2, and cytochrome P450. Elevation of iNOS signaling by HCV induces nitric oxide ($NO^\bullet$) that can induce redox signaling, or react with $O_2^{-\bullet}$ forming peroxynitrite ($ONOO^-$) causing cell damage. Increased ferrous iron (Fe^{2+}) reacts with H_2O_2 to produce hydroxyl radical ($HO^\bullet$) that can induce cellular damage. Alterations in redox signaling would also result from H_2O_2 production observed with HCV. Detoxification of H_2O_2 by glutathione (GSH) and glutathione peroxidase (GPx) are inhibited by HCV. The role of HCV in the inhibition of catalase activity may contribute to decreases in antioxidant defense

11.2.1 Iron Overload

Hepatocytes and liver macrophages act as a substantial reservoir of iron, where approximately 20–30 % remains bound to ferritin and hemosiderin [56]. The role of iron in both the induction and propagation of oxidative stress is due, in part, to its oxidation and reduction properties. Ferrous iron (Fe^{2+}) can become oxidized by hydrogen peroxide, producing the ferric form (Fe^{3+}) and hydroxyl radical ($HO^\bullet$). This reaction, known as the Fenton reaction, is shown below in (11.1).

$$Fe^{2+} + H_2O_2 + H^+ \rightarrow Fe^{3+} + HO^\bullet + H_2O \tag{11.1}$$

Hydroxyl radical is highly reactive and reacts with whatever molecule is next to its site of generation. Thus, iron, in the presence of elevated ROS, can lead to lipid, protein, and DNA damage. It has been reported that patients with CHC exhibit elevated iron accumulation and decreased expression of hepcidin, a negative regulator of iron absorption [57]. Iron depletion therapy reduced oxidative damage in CHC patients [58–61]. Furthermore, IFN was shown to decrease accumulation of 4-hydroxy-2-nonenal (HNE) protein adduct, a marker for lipid peroxidation, and iron deposits in responders [62].

11.2.2 Mitochondrial ROS

HCV also increases ROS generation. Increased levels of superoxide and/or H_2O_2 have been demonstrated in hepatocytes expressing HCV genes or full-length HCV that produces infectious virions in cell culture, and in hepatitis C patient liver [34, 63–68]. HCV infection has been associated with significant mitochondrial abnormalities, and the mitochondrial electron transport chain was identified as a key source of ROS in various HCV protein-expressing cells [69–71]. It has been reported that HCV core protein decreased complex I (NADH-ubiquinone reductase) activity, increased production of superoxide, and decreased mitochondrial glutathione (GSH) and NADPH concentrations [72]. HCV increased the sensitivity to mitochondrial ROS generation occurring at Complex I and membrane permeability transition (MPT) by elevating Ca^{2+} uptake by mitochondrial Ca^{2+} uniporters [63, 71]. Ca^{2+} loading can elevate mitochondrial ROS production in hepatocytes [73]. HCV core also decreased the levels of prohibitin, a mitochondrial protein chaperone, as well as cytochrome c oxidase activity that may further contribute to mitochondrial dysfunction [74]. Furthermore, NS4A co-localized with the mitochondria causing alterations in mitochondrial transmembrane potential and release of cytochrome c into the cytoplasm [75].

11.2.3 Nox2 of Phagocytes and Nox1 and 4 of Hepatocytes

Nox family enzymes consist of Nox and Dual oxidase (Duox) proteins (Nox1 through 5 and Duox 1 and 2) that catalyze the transfer of electrons from NADPH to O_2 to generate superoxide and H_2O_2 [76]. Unlike many other sources of ROS, such as mitochondria that produce ROS as a byproduct of aerobic metabolism, the primary function of Nox/Duox proteins is superoxide and H_2O_2 generation.

Nox enzymes perform diverse physiological roles, ranging from thyroid hormone synthesis, maintenance of equilibrioception, signaling in the vascular system, and antimicrobial processes of the immune system. Nox2, also referred to as gp91^{phox} or simply phox (phagocytic oxidase), functions in the PRB. When activated in response to pathogens, transmembrane Nox2 coupled with its catalytic subunit p22^{phox}, associate with several activating cytosolic factors, p67^{phox}, p40^{phox}, p47^{phox}, and Rac. Upon their association, large quantities of oxygen are reduced to generate superoxide anion ($O_2^{-\bullet}$). The significance of Nox2 in innate immune function is evident in patients with chronic granulomatous disease (CGD), in which mutations in one or more of the catalytic subunits of Nox2, or its cytosolic factors, produce an insufficient respiratory burst resulting in aberrant inflammation and increased susceptibility to infections [77].

Evidence points to the activation of Nox2 in phagocytes and monocytes with HCV infection. Specifically, the NS3 protease was shown to activate phagocytic Nox2 leading to apoptosis in T cells, natural killer (NK) cells, and cytotoxic T cells [78, 79]. In addition, extracellular secreted core protein of HCV was shown to induce STAT3 activation in an IL-6 autocrine-dependent manner. As STAT3 is an activator of the p47^{phox} cytosolic factor of Nox2, authors report that induction of myeloid-derived suppressor cells by HCV core protein may suppress T cells in an ROS-dependent manner via Nox2 [80, 81].

In addition, hepatocytes express Nox isoforms. Both Nox1 and 4 are transmembrane proteins that remain bound to the p22^{phox} catalytic subunit. Similar to Nox2, Nox1 produces ROS when activated by the association of the cytosolic subunits NoxO1, NoxA1, and Rac. Like other Nox enzymes, Nox4 is composed of six predicted transmembrane domains and an intracellular carboxyl-terminal region, which contains FAD- and NADPH-binding sites [82]. Truncated Nox4 proteins without the C-terminal FAD and NADPH binding sites are unable to transport electrons and show negative effects on the endogenous Nox4 activity [83]. Furthermore, the Nox4 intracellular B-loop, located between transmembrane domains 2 and 3, is essential for the enzyme activity, and the intracellular D-loop is necessary for the structural interaction between Nox4 and the p22^{phox} subunit [82]. To date, five alternatively spliced isoforms of Nox4 have been reported [84]. Compared to full-length Nox4 (~67 kDa), variant B (~63 kDa) lacks the first NADPH-binding site while variant C (26 kDa) does not contain any FAD- or NADPH-binding sites due to a frameshift-derived stop codon after transmembrane domain 5; variant D (32 kDa) contains transmembrane domain 1 and all FAD- and NADPH-binding sites; and interestingly, variant E (28 kDa) is similar to variant D, but is missing the first NADPH-binding domain [84]. Nox4 localizes in the ER, mitochondria, plasma membrane, and nucleus in various cells which may be isoform-dependent [65, 85]. In hepatocytes, Nox4 was shown to contribute significantly to nuclear superoxide production [66, 86]. Recently, Anilkumar et al. observed a functional 28 kDa splice variant of Nox4 within the nucleus and nucleolus of vascular cells that was able to generate ROS in an NADPH-dependent manner [85].

Furthermore, HCV has been shown to elevate superoxide and hydrogen peroxide production via an upregulation of Nox1 and Nox4 protein expression in hepatocytes, in addition to nuclear and/or perinuclear localization [66]. Dominant negative Nox4 and Nox4 knockdown significantly decreased the HCV-associated ROS production in hepatocytes [65, 66]. Increases in p22^{phox}, p67^{phox}, NOXA1, and NOXA2 mRNAs could also be found [66]. In addition, Nox4 and Nox1 protein levels and

Nox activity were elevated in HCV-positive human liver [77]. Nox4 elevation by HCV and transcriptional activation of Nox4 promoter by HCV core protein were shown to be mediated by transforming growth factor beta (TGF-β) [65].

11.2.4 Cytochrome P450 Enzymes

Cytochrome p450 2E1 (CYP2E1) is involved in the detoxification of alcohol in the liver to acetaldehyde. This reaction occurs in an NADPH-dependent manner and produces ROS as a byproduct. It has been reported that HCV core protein and alcohol increase the level of CYP2E1 in human hepatoma (Huh-7) cells [87]. Liver tissue biopsies from patients with CHC showed elevations in CYP2E1 mRNA, in addition to tumor necrosis factor alpha (TNF-α) and collagen 1 [88]. It has also been suggested that CYP2C9 contributed to HCV-induced increase in ROS that was associated with an increased level of c-Myc [89].

11.2.5 iNOS

Hepatocytes upregulate inducible nitric oxide synthase (iNOS, NOS2) in response to proinflammatory cytokines, such as TNF-α and interleukin-1β (IL-1β), interferon γ (IFN-γ), and/or lipopolysaccharide (LPS) [90]. iNOS catalyzes the synthesis of nitric oxide ($NO^{\bullet}$) by the following reaction (11.2):

$$\text{L-citrulline} + 2O_2 + \frac{3}{2}\text{NADPH} + H^+ \xrightarrow{\text{iNOS}} \text{L-arginine} + NO^{\bullet} + \frac{3}{2}\text{NADP}^+ \quad (11.2)$$

Nitric oxide is a key mediator of immune and inflammatory responses in a variety of cell types, including macrophages, hepatocytes, and Kupffer cells [90, 91]. iNOS has been shown to be upregulated in response to CVH, and such elevation contributed to the DNA damage observed in HCV-infected cells [92–95]. Furthermore, the role of c-Jun/AP-1 in activation of iNOS was shown in HCV core transgenic mice [96]. Additionally, increased levels of nitrotyrosine were found in liver biopsies from patients with chronic viral hepatitis, including hepatitis C, that co-localized with iNOS [97].

11.2.6 Decreased Antioxidant Defense

Decreases in the body's ability to counter the effects of ROS through its antioxidant defense can increase susceptibility to oxidative stress and promote oxidative damage. Several key antioxidants are decreased in CHC, which include glutathione (GSH) and vitamins C and E [98]. GSH is the most abundant non-protein endogenous thiol that is involved in the detoxification of ROOH in a reaction catalyzed by glutathione peroxidase (GSH-Px) that results in the formation of glutathione disulfide (GSSG) (11.3). Subsequent reduction of GSSG back to GSH is dependent on glutathione disulfide reductase (GSR) that uses NADPH as an electron donor (11.4).

$$2GSH + H_2O_2 \xrightarrow{\text{GSH-Px}} GSSG + 2H_2O \quad (11.3)$$

$$GSSG + NADPH + H^+ \xrightarrow{\text{GSR}} 2GSH + NADP^+ \quad (11.4)$$

Structural proteins of HCV, specifically core protein, were shown to decrease reduced GSH and increase GSSG levels in hepatocytes in vitro. Conversely, hepatocytes expressing nonstructural proteins increased GSH levels [99]. In addition, CHC patients showed higher GSH-Px levels posttreatment with interferon α-2b and ribavirin compared to baseline values [100]. It has been reported that HCV causes delocalization of small Maf (sMaf) proteins from the nucleus to the viral replication complex in the ER, thereby, inhibiting the ability of NF-E2-related factor 2 (Nrf2) to activate electrophile response element (EpRE)-dependent gene expression [101]. While GSH is clearly suppressed in patients, however, the extent to which HCV interferes with Nrf2 and the upregulation of antioxidant defense in response to oxidative stress is still unclear [102, 103].

In addition, various ROS scavengers show perturbed levels with HCV infection, some of which include superoxide dismutase (SOD), catalase (CAT), and thioredoxin (Trx). Patients with HCV-related HCC presented with higher serum MnSOD and Trx levels compared to healthy patients, suggesting upregulation of antioxidant defense in response to oxidative stress [104–106]. In vitro data by Abdalla et al. showed upregulation of CAT in hepatocytes expressing NS proteins of HCV [99]. In contrast, a recent study conducted in Egypt observed significant decreases in CAT, SOD, and zinc in patients with HCC [107].

11.3 Consequences of Oxidative Stress in Chronic Hepatitis C

11.3.1 Hepatic Steatosis

Hepatic steatosis, or fatty liver disease, results from the accumulation of lipid droplets in the liver, and is a common histological manifestation found in almost 50 % of HCV infected patients [108, 109]. Individuals with genotype 3 HCV exhibited a higher incidence of steatosis compared to the other genotypes [110]. Development of hepatic fibrosis, cirrhosis, and further progression to HCC is strongly associated with the incidence of steatosis in HCV patients [111–113]. Furthermore, the severity of steatosis has been shown to affect the likelihood of achieving SVR with

IFN-based therapy, and the response to therapy also coincided with the resolution of fatty liver [36, 114, 115].

HCV affects the host hepatic cellular physiology by multiple mechanisms, leading to the buildup of lipids, mainly composed of triglyceride and cholesterol ester. HCV core expressing cells showed an increased expression of mature sterol regulatory element-binding protein-1 (SREBP-1) and peroxisome proliferators-activated receptor γ (PPARγ), which are known to increase the transcription of genes involved in hepatic fatty acid synthesis [116]. Increased hepatic expression of SREBP-1 and transcription of SREBP-1-related lipogenic genes were also observed in HCV-infected patients [117]. HCV-induced SREBP-1 and -2 activities were sensitive to pyrrolidine dithiocarbamate (PDTC), Ca^{2+} chelator, and inhibitor of phosphatidylinositol 3-kinase (PI3-K) [118]. Also, HCV core, NS2, and NS4B proteins induced accumulation of lipid droplets in vitro [118–120].

HCV core protein expression was strongly associated with steatosis in vivo, in transgenic mice, and both PPARα- and PPARγ-mediated hepatic steatosis and HCC induction by core [121–124]. NS5A protein also triggered steatosis in vivo in another study [125]. Microarray analysis of liver biopsy samples from HCV-infected chimpanzees showed an induction of various genes involved in lipid metabolism and cholesterol/fatty acid biosynthesis [126, 127].

HCV has also been associated with insulin resistance (see Sect. 11.3.4. below) [128]. Thus, the combined influence of viral factors and other metabolic factors, including insulin resistance and hyperglycemia, may facilitate the development of hepatic steatosis in CHC patients. Increased accumulation of polyunsaturated fatty acids is often observed in steatosis, which makes the liver more susceptible to peroxidation compared to normal liver. Lipid peroxidation products are commonly detected in the liver of CHC patients, and the increased hepatic iron reserves observed in steatosis, together with polysaturated fatty acids, are likely to induce lipid peroxidation in these patients [129, 130].

11.3.2 Hepatic Fibrosis and Cirrhosis

Fibrosis is characterized by an excessive accumulation of extracellular matrix (ECM) proteins, including collagen, leading to fibrotic scarring and/or cirrhosis. A key source of ECM in the fibrotic liver is perisinusoidal hepatic stellate cells (HSC) capable of transforming into proliferative myofibroblasts [131]. Activation of HSCs occurs in response to proinflammatory cytokines, ROS, and lipid peroxidation [132–136]. Increased iron levels have also been shown to induce inflammatory cytokines and subsequent activation of HSCs leading to fibrosis [137]. During hepatic fibrosis, activated Kupffer cells increase oxidative stress leading to the production of type I collagen by HSCs in the liver [135, 136, 138, 139]. In CHC, HCV peptides derived from NS5A have been associated with the migration and activation of Kupffer cells in the liver [79, 140]. Evidence also suggests that fibrosis occurs in response to hepatocyte apoptosis [141]. Activation of Kupffer cells and HSCs occurs in response to the release of apoptotic bodies during chronic liver injury [142, 143]. HCV increases apoptosis in hepatocytes, and apoptosis mediated by elevated Fas, a death ligand, was observed in HCV-infected hepatocytes [64, 144–147]. Upregulation of Fas ligand was observed in CHC patients [148]. In addition, HSC activation by hepatocyte apoptotic bodies could be amplified when the apoptotic bodies contained HCV nonstructural proteins [149]. Therefore, chronic inflammation with activated Kupffer cells, tissue injury, and apoptosis may play an important role in HCV-associated liver fibrosis through mechanism(s) involving ROS.

TGF-β, a major fibrogenic cytokine, was shown to be involved in hepatocyte Nox4 elevation with HCV [66]. TGF-β stimulates type I collagen transcription and regulates the expression of MMPs and their inhibitors during fibrogenesis [150]. In addition, TGF-β has been shown to increase Nox4 in fibroblasts and HSCs, and to activate Nox4 and Nox1/Rac in hepatocytes [135, 151–159]. ROS generated by Nox4 and mitochondrial electron transport chain in turn mediate many fibrogenic effects of TGF-β [135]. TGF-β levels are elevated in CHC patients and in HCV cell culture [65, 66, 160]. The discovery that HCV induces Nox4 in a TGF-β-dependent manner, therefore, is significant and suggests a role of oxidative stress in HCV-associated fibrosis. HCV-infected patient liver showed elevated Nox4 expression that correlated with the degree of fibrosis [66, 159]. DPI, a nonspecific inhibitor of flavoproteins such as Nox enzymes, decreased changes in fibrogenic gene expression induced by HCV and human immunodeficiency virus in HSCs and hepatoma cells [161]. Indeed, recent studies strongly suggest a role of Nox enzymes in hepatic fibrosis. Nox1, 2, and 4 were upregulated in activated HSCs, and both Nox1 and 2 were shown to be required for fibrosis [133, 162, 163]. Furthermore, GKT137831, a pharmacological inhibitor of Nox1 and 4, significantly reduced bile duct ligation- or CCl_4-induced liver fibrosis in vivo in mice [132, 164]. Supplemental treatment of CHC patients with Cu/Zn SOD also resulted in a decrease in TGF-β-mediated liver fibrosis [165].

Other enhancers of fibrosis increased in response to HCV include various growth factors, chemokines, and cytokines, such as TNF-α [166]. Elevated levels of TNF-α have been documented in CHC patients as well as in vitro and animal models of HCV [88, 167]. Nox2-dependent activation of TNF-α-converting enzyme (TACE) was recently associated with fibrogenic markers in an experimental model of non-alcoholic steatohepatitis [168]. Also, TNF-α treatment of hepatocytes increased mitochondrial ROS generation at complex I and complex III [169]. Together, this evidence points to the role that oxidative stress may play in HCV-associated fibrosis, which typically precedes HCC.

11.3.3 Hepatocellular Carcinoma

HCC is a primary malignancy of the hepatocytes, and as mentioned previously, viral hepatitis is the major risk factor in the development of liver cancer. Other risk factors include alcohol abuse, aflatoxin exposure, obesity, and nonalcoholic fatty liver disease. The odds ratio for developing HCC is 11.5 in HCV infected patients and it has been estimated that by the year 2020, the incidence of HCC will increase by 81 % [170, 171]. Various in vivo studies suggest that oxidative stress is a key mediator of HCC formation. For example, mice deficient in CuZn SOD exhibited a reduced lifespan, and an increased incidence of nodular hyperplasia and HCC, while liver-specific knockout of Nrf-1 resulted in spontaneous liver cancer, increased ROS, and oxidative damage [172, 173]. Several other studies highlighted the role of oxidative stress in HCV-induced HCC: oxidative DNA damage is elevated in CHC, and iron depletion therapy reduced inflammation and progression of fibrosis, improved alanine aminotransferase (ALT) levels, decreased oxidative damage, and reduced the risk of HCC in patients with HCV [58–61, 174]. Furthermore, increased hepatic iron elevated 8-hydroxydeoxyguanosine (8-OHdG), liver steatosis, and/or HCC formation in HCV polyprotein expressing transgenic mice [175, 176]. In addition, the tumor rates in HCV core mice could be decreased with butylated hydroxyanisole (BHA) treatment [96].

Oxidative DNA damage occurs when oxidants attack bases and the deoxyribosyl backbone of DNA, resulting in DNA strand breaks and other DNA modifications [177, 178]. The mechanism of DNA damage by nitric oxide can involve base deamination causing G:C to A:T transitions, oxidation of DNA by peroxynitrite ($ONOO^-$), and formation of N-nitroso compounds such as nitrosamines and nitrosamides [179]. NO-dependent inhibition of DNA damage repair mechanisms have been observed with HCV [92, 180, 181]. The HCV-associated DNA damage could also be decreased with *N*-acetylcysteine (NAC) and by decreasing iron, suggesting a role of hydroxyl radical in the damage [58, 181]. Together, these findings suggest that HCV induces DNA damage through ROS, iron, and nitric oxide-dependent mechanisms. Note that interactions between reactive species (RS) and their target depend on chemical reactivity of the RS, rate of RS generation, distance between the source and target, and efficiency of their removal by cellular processes [103, 182]. Thus, the nuclear location of Nox4 in hepatocytes and HCV-associated increases in nuclear Nox4 are likely to further increase the probability of redox-mediated DNA damage in CHC [66].

Oxidative DNA damage results in the random mutation of critical cellular genes, as well as strand breaks, and is one of the key mechanisms by which ROS leads to carcinogenesis. Mutation, or loss, of tumor suppressor genes such as p53 and/or retinoblastoma (Rb), important cell cycle regulators, is a common feature of HCC [183–185]. For example, 8-OHdG is increased in HCC livers where it has been shown to induce G:C to T:A transversions frequently observed in cancers [186]. Elevated 8-OHdG found in the noncancerous regions of HCC livers at the time of hepatectomy correlated with an increased risk of tumor recurrence [187]. HCV increased mutations in p53, β-catenin, B-cell lymphoma-6, and immunoglobulin heavy chain genes in B cells, peripheral blood mononuclear cells, and lymphomas [92, 188]. Elimination of 8-OHdG typically occurs by base excision repair usually initiated by the DNA glycosylase, 8-oxoguanine glycosylase (OGG1). Human OGG1 (hOGG1) polymorphism was associated with some cancers and could also predict survival of HCC patients [189, 190]. hOGG1 activity can be modulated by thiol modification and has been shown to be inhibited by nitric oxide [191]. Interestingly, HCV was found to decrease DNA repair, in part, by decreasing DNA glycosylase activity in a nitric oxide-dependent manner [92, 180]. As mentioned in Sect. 11.3.1. above, steatosis correlates with an increased risk of HCC in HCV-infected individuals, and increased lipid peroxidation can result in the formation of DNA adducts [192].

Chronic inflammation associated with HCV also potentially increases ROS production by activation of the host immune cells, which would contribute to chronic tissue damage and cell proliferation in HCC formation. Various studies also establish the significant role of Nox2 in chronic inflammation-mediated pathogenesis. Mice deficient in $p47^{phox}$ showed significant decrease in inflammation, apoptosis, and regenerative DNA synthesis in response to DEN-induced liver damage. DEN mediated activation of Kupffer cells, leading to production of superoxide [193]. These studies suggest that the superoxide generation by phagocytes during inflammation will exacerbate genotoxic and cytotoxic effects of HCV in hepatocytes that may further contribute to tumor initiation and promotion.

Increases in blood oxidative stress markers preceding ALT flare-ups in HCV patients support a role of oxidative stress in inflammation and tissue damage [194]. TNF-α, a proinflammatory cytokine, is elevated during inflammation and HCV infection. TNF-α was shown to upregulate Nox2, Nox4, Nox1, $p67^{phox}$, $p47^{phox}$, $p22^{phox}$, and/or Duox2 gene expression in nonhepatocytes, and was attenuated in $p47^{phox}$ knockout mice, but whether TNF-α participates in the regulation of hepatocyte Nox enzyme expression is still unclear [193, 195–197]. TNF-α also activated NF-κB through increased mitochondrial ROS generation, and upregulated hypoxia-inducible factor 1 (HIF-1)-mediated FoxMI expression that could be inhibited by BHA [169, 198]. FoxMI, a cell cycle regulator vital for hepatocyte survival and proliferation, is upregulated in HCC tissues, and its overexpression has been correlated with poor prognosis in HCC patients [199, 200]. Also, FoxMI conditional knockout mice did not develop HCC in response to DEN, and primary hepatocytes isolated from these mice were unable to proliferate in culture [201]. Another study reported that transcriptional inhibition of FoxMI with $p19^{ARF}$ in FoxMI overexpressing transgenic mice reduced hepatic tumor proliferation and angiogenesis while inducing apoptosis [202].

Toll-like receptor 4 (TLR4), a pattern recognition receptor activated by endotoxins such as LPS present in gram negative bacteria, is upregulated in HCV patients [203–205]. TLR4 has been shown to increase cell proliferation and survival in response to DEN, and gut sterilization reduced the incidence of HCC [206, 207]. Alcohol is a key mediator in HCV-associated pathogenesis, and alcohol-induced endotoxemia may further contribute to TLR4-mediated pathogenesis [208]. Alcohol and LPS, in conjugation with HCV NS5A, increased the rate of hepatic tumors through Nanog, a progenitor cell marker [209]. Furthermore, the rs4986791 T allele of TLR4 was associated with decreased risk of HCC in CHC patients [210]. LPS stimulation was shown to upregulate Nox1 transcription in bone marrow-derived macrophages [211]. Another study showed that the LPS-stimulated Nox1 expression was associated with accumulation and nuclear translocation of β-catenin in macrophages [212]. TLR4 signaling, which can increase TNF-α levels, required Nox4 in Raw264.7 murine macrophages, HEK293 cells, and human aortic endothelial cells

[213–215]. However, the role of Nox enzymes with respect to TLR4-mediated pathogenesis in the liver is still not understood. Altogether, the influence of oxidative stress in the initiation and development of HCC is a downstream result of the various pathogenic mechanisms involved in steatosis, fibrosis, and inflammation.

11.3.4 Extrahepatic Manifestations: Lymphoproliferative Disorders, Diabetes

Metabolic syndrome is a global pandemic of the modern world, with a prevalence of 25 % in the United States [216]. Metabolic syndrome is defined as an array of interrelated metabolic risk factors for diabetes and cardiovascular disease, including elevated blood pressure, dysglycemia, increased triglyceride levels, and obesity [217]. Type 2 diabetes (T2DM), characterized by insulin resistance, accounts for 90–95 % of all diabetes cases. According to the third National Health and Nutrition Examination Survey (NHANES), among 9,841 patients who were aged 20 years or older, 8.4 % had diabetes, and 2.1 % had hepatitis C, with an estimated fourfold increase in risk of developing T2DM among HCV-infected patients [218]. In another study, 21 % of HCV-infected patients and 12 % of HBV-infected patients were found to have diabetes [219, 220].

Studies indicate that HCV-induced insulin resistance is viremia-dependent, genotype-specific, and inversely correlated with SVR [219–223]. Downregulation of GLUT2 receptors and promotion of gluconeogenesis at the transcriptional level can further augment hyperglycemia [224]. Mitochondrial ROS-mediated Janus kinase (JNK) activation has been shown to play a vital role in upregulating hepatic gluconeogenesis in an HCV NS4A-mediated and FoxO1-dependent manner [224]. Chronic inflammation associated with HCV may also induce insulin resistance through TNF-α and IL-6 [225, 226]. The resulting hyperglycemia may, in turn, affect the insulin signaling cascade in an ROS-dependent manner [227].

Mixed cryoglobulinemia (MC), a B-cell lymphoproliferative disorder is the most common HCV-related extrahepatic disorder. HCV has also been associated with Non-Hodgkin's lymphoma [228, 229]. These lymphoproliferative disorders may result either from direct infection of B cells with HCV or through antigen stimulation [230]. HCV-mediated oxidative stress has been shown to inhibit DNA repair mechanisms in B cells, potentially leading to chromosome instability and the development of lymphoma in CHC [180].

11.3.5 Effects on HCV

ROS can also affect HCV replication. Previously, exogenous H_2O_2, GSH depletion, lipid peroxidation products, and arsenite were found to decrease the replication capacity of HCV replicons and infectious HCV in hepatocytes; the suppression was mediated by elevated cytosolic Ca^{2+} [231–235]. These findings support a role of ROS in innate immunity [236] Alcohol (ethanol), on the other hand, enhanced HCV replication through the lipogenic effects of acetaldehyde, acetate, and elevated $NADH/NAD^{+}$; alcohol also enhanced HCV replication through mechanisms involving miR-122 and decreased cyclin G1, cyclooxygenase-2 (COX2), and the endogenous opioid system [237–245].

Furthermore, HCV shows a high level of sequence heterogeneity and exists as a mixture of related but distinct genome sequences, or "quasispecies," that can complicate antiviral therapy [246–249]. RdRp is notoriously error-prone, contributing greatly to the sequence variability commonly observed in HCV. Recently, ethanol and GSH depletion were found to increase nucleotide and amino acid substitution rates of genotype 1b HCV replicon in vitro, suggesting a role of oxidative stress in HCV sequence variability; alcohol consumption was also associated with increased sequence heterogeneity of HCV in patients [250, 251]. The percentage of sites undergoing positive selection increased with ethanol and GSH depletion, increasing the detection of NS5A variants that have been associated with antiviral resistance [250]. Other amino acid substitutions were also found that may affect the ability of HCV to replicate. These findings are consistent with the adverse effects that alcohol can have on antiviral therapy, and suggest that ethanol and GSH depletion may affect HCV's ability to replicate during antiviral treatment.

Antioxidants have also been tested for their effects on virological endpoints in patients. Interestingly, certain antioxidant strategies could enhance the virological outcome of antiviral therapy, whereas others showed either no effect or inconsistent benefit; possible explanations are limited bioavailability, efficacy, and nonspecific effects of traditional antioxidants [170]. A majority of clinical studies testing the effects of antioxidant strategies in the absence of concurrent antiviral therapy, on the other hand, reported no significant effect [103].

11.4 Oxidative Stress in Chronic Hepatitis B

11.4.1 HBV-Induced Pathogenesis

Infection with HBV results in acute infection, fulminant hepatitis failure, inactive carrier state, and chronic hepatitis, often progressing to liver cirrhosis or HCC [7]. In spite of the availability of an effective vaccine, the World Health Organization (WHO) has estimated that nearly two billion people have been infected with HBV worldwide. Of these individuals, 350 million are chronically infected, and from this group, it has been projected that 65 million people will die from liver disease due to HBV infection [252]. Currently, IFN-α and nucleoside/nucleotide analogues are used to reduce the severity of HBV-associated diseases; these antivirals have shown limited efficacy [253, 254]. Eight genotypes of HBV have been defined (forms A to H) which are distributed across various geographical areas [255].

HBV is a small, enveloped, partially double-stranded deoxyribonucleic acid (DNA) virus with an incomplete double-stranded circular genome of 3.2 kb in length. The HBV genome is comprised of four overlapping open reading frames (ORFs): the surface (S) ORF that encodes for small, middle, and large envelope proteins, the polymerase (P) ORF encoding a multifunctional protein necessary for genome replication and encapsidation, the core (C) ORF encoding nucleocapsid proteins hepatitis B core antigen (HBcAg), and hepatitis B e antigen (HBeAg), and the X ORF encoding hepatitis B x protein (HBx) which is essential for in vivo replication [256].

HBV replication is mediated by a hepatocyte-specific receptor [257]. Recent studies identified sodium taurocholate cotransporting polypeptide (NTCP, also known as SLC10A1) as an HBV receptor; ferritin light chain (FTL), and squamous cell carcinoma antigen 1 (SCCA1) were also found to serve as co-receptors for HBV infection of hepatocytes [258, 259]. After receptor binding, viral entry occurs through endocytosis mediated by caveolin-1. Involvement of endogenous GTPases of the Ras family, Rab5 and Rab7, were recently shown to mediate trafficking of HBV particles postentry [260]. Capsid accumulation occurs at the nuclear membrane, where interaction with the nuclear pore complex may favor release of the HBV genome into the nucleus [261].

Following release of the viral genome, the incomplete double-stranded circular viral DNA is converted to a covalently closed circular DNA (cccDNA) by host DNA repair proteins. The cccDNA serves as a template for production of viral RNAs required for replication and protein synthesis, and its incorporation into the host DNA and its long half-life may play a role in persistent infection [262, 263]. Host RNA polymerase II mediates transcription of subgenomic (sg) and pregenomic (pg) RNAs from the cccDNA. Subsequent export of sg- and pgRNA to the cytoplasm initiates translation of viral proteins from sgRNA. Viral polymerase protein and pgRNA are subsequently packaged into immature nucleocapsids. Reverse transcription of pgRNA into incomplete circular DNA occurs in the nucleocapsid during maturation. Mature nucleocapsids containing genomic DNA may be trafficked back to the nucleus to further augment cccDNA formation, or be enveloped and excreted by budding [264]. Large quantities of viral particles appear in the serum, in two forms, during active replication; the complete virion 42 nm in diameter, and a 22 nm empty viral envelope containing HBsAg. HBeAg also appears in the serum during the highly replicative phase of HBV infection.

HBx sequences with C-terminal deletions and point mutations have been detected in CHB and HCC, suggesting that the integrated fragments of subviral DNA could play a role in hepatocarcinogenesis [265–267]. The HBx gene product has also been shown to affect many cellular genes associated with growth control [268, 269]. Furthermore, HBx protein interacts and interferes with various transcription factors including Ca^{2+}/cAMP-response element-binding protein (CREB), activating transcription factor (ATF) 3, nuclear factor interleukin-6 (NFIL-6), octamer-binding protein (Oct1), and retinoid x receptor (RXR). In addition, interactions between DNA repair proteins and HBx have been described, specifically p53 and UV-damaged DNA-binding protein (UVDDB) [270].

Various transgenic mouse models carrying full-length HBV coding sequence, or subgenomic HBV coding sequences such as the large, middle, and small surface proteins, x protein, or core protein, either under the control of an HBV promoter or a host liver-specific albumin promoter, have been studied [271]. Interestingly, only the large surface envelope protein and HBx protein were found to be carcinogenic [272, 273]. However, HBx-induced carcinogenesis lacked characteristic inflammation, or cirrhosis observed in clinical HCC progression, and hepatic tumors developed within 11–15 months, with male mice reaching mortality due to HCC before females [272]. Synergistic effects of HBx in combination with the oncogenic effects of *c-myc* shortened the time to tumor development by as much as 4 months in another study [274]. Unlike the HBx mouse model, HBsAg expressing mice showed inflammation associated with HCC that was particularly prominent in male mice [275].

Oxidative stress in chronic HBV infection plays a vital role in carcinogenesis, as shown in a study where 8-OHdG levels correlated with the initiation and progression of HCC in transgenic mice with liver-specific expression of HBV large envelope protein [276]. In addition, it has been reported that the accumulation of the pre-S mutant HBsAg in the ER may result in oxidative DNA damage that may lead to genomic instability and mutagenesis [277]. Disruption of iron metabolism leading to iron deposition has also been observed in HBV-infected patients and can act as a cofactor in HCC initiation and progression [278]. HBx-associated increase in oxidative stress was associated with the modulation of iron regulatory protein 1 (IRP1), a key regulator of the iron uptake and storage system, by HBx in hepatocytes, resulting in increased iron deposition [279].

Mitochondrial association of HBx has been shown to mediate STAT3 activation and further induce NF-κB activation [280]. HBx is known to regulate the Ras- and Raf-induced mitogen-activated protein kinase (MAPK) pathway [281]. Raf serine/threonine kinases act downstream of Ras and are involved in carcinogenesis [282]. The HBx protein induced mitochondrial translocation of cytoplasmic Raf-1, and also formed a protein–protein complex with Raf-1 kinase that may lead to HBV-induced carcinogenesis [283]. STAT3 activation and mitochondrial translocation of Raf-1 mediated by HBx could be inhibited by NAC and PDTC. Thus, HBV is associated with oxidative stress, which may play an important role in HBV-mediated pathogenesis in multiple ways. On the other hand, H_2O_2 treatment decreased HBV virion production in cell culture by affecting viral gene transcription, which supports a role of oxidative stress in antiviral response [284].

11.5 Conclusions

CVH is a global health concern. Accumulating evidence suggests an important role of oxidative stress in HCV as well as HBV-associated pathogenesis. For example, oxidative stress is a direct consequence of HCV infection of hepatocytes and involves specific induction of ROS generation from sources that include NADPH oxidases and mitochondria. Inflammation is also likely to exacerbate oxidative stress and tissue damage which, together with virus-induced oxidative stress in hepatocytes, would lead to pathogenesis. These findings suggest that strategies designed to target oxidative stress may be beneficial in the treatment of CVH; indeed, CHC patients can still progress to liver disease even after achieving SVR, suggesting a need for continued development of therapy in addition to a vaccine [285–287]. However, further studies will be necessary. Possible effects on antiviral host immunity will also need to be evaluated further.

References

1. Perz JF, Armstrong GL, Farrington LA et al (2006) The contributions of hepatitis B virus and hepatitis C virus infections to cirrhosis and primary liver cancer worldwide. J Hepatol 45:529–538. doi:10.1016/j.jhep.2006.05.013
2. Armstrong GL, Wasley A, Simard EP et al (2006) The prevalence of hepatitis C virus infection in the United States, 1999 through 2002. Ann Intern Med 144:705–714
3. Wasley A, Kruszon-Moran D, Kuhnert W et al (2010) The prevalence of hepatitis B virus infection in the United States in the era of vaccination. J Infect Dis 202:192–201. doi:10.1086/653622
4. Global Burden of Hepatitis C Working Group (2004) Global burden of disease (GBD) for hepatitis C. J Clin Pharmacol 44:20–29. doi:10.1177/0091270003258669
5. Franco E, Bagnato B, Marino MG et al (2012) Hepatitis B: epidemiology and prevention in developing countries. World J Hepatol 4:74. doi:10.4254/wjh.v4.i3.74
6. Jemal A, Siegel R, Ward E et al (2009) Cancer statistics, 2009. CA Cancer J Clin 59:225–249. doi:10.3322/caac.20006
7. McMahon BJ (2009) The natural history of chronic hepatitis B virus infection. Hepatology 49:S45–S55. doi:10.1002/hep.22898
8. Tong MJ, Farra NS, Reikes AR, Co RL (1995) Clinical outcomes after transfusion-associated hepatitis. C N Engl J Med 332:1463–1466. doi:10.1056/NEJM199506013322202
9. Choo QL, Kuo G, Weiner AJ et al (1989) Isolation of a cDNA clone derived from a blood-borne non-A, non-B viral hepatitis genome. Science 244:359–362
10. Tobler LH, Busch MP (1997) History of posttransfusion hepatitis. Clin Chem 43:1487–1493
11. El-Serag HB (2012) Epidemiology of viral hepatitis and hepatocellular carcinoma. Gastroenterology 142:1264–1273. doi:10.1053/j.gastro.2011.12.061, e1
12. Averhoff FM, Glass N, Holtzman D (2012) Global burden of hepatitis C: considerations for healthcare providers in the United States. Clin Infect Dis 55:S10–S15. doi:10.1093/cid/cis361
13. Brown RS, Gaglio PJ (2003) Scope of worldwide hepatitis C problem. Liver Transpl 9:S10–S13. doi:10.1053/jlts.2003.50244
14. Altekruse SF, McGlynn KA, Reichman ME (2009) Hepatocellular carcinoma incidence, mortality, and survival trends in the United States from 1975 to 2005. J Clin Oncol 27:1485–1491. doi:10.1200/JCO.2008.20.7753
15. Dammacco F, Gatti P, Sansonno D (1998) Hepatitis C virus infection, mixed cryoglobulinemia, and non-Hodgkin's lymphoma: an emerging picture. Leuk Lymphoma 31:463–476. doi:10.3109/10428199809057606
16. DeCastro M, Sánchez J, Herrera JF et al (1993) Hepatitis C virus antibodies and liver disease in patients with porphyria cutanea tarda. Hepatology 17:551–557
17. el-Zayadi AR, Selim OE, Hamdy H et al (1998) Association of chronic hepatitis C infection and diabetes mellitus. Trop Gastroenterol 19:141–144
18. Feld JJ, Hoofnagle JH (2005) Mechanism of action of interferon and ribavirin in treatment of hepatitis C. Nature 436:967–972. doi:10.1038/nature04082

19. Cunningham M, Foster GR (2012) Efficacy and safety of telaprevir in patients with genotype 1 hepatitis C infection. Therap Adv Gastroenterol 5:139–151. doi:10.1177/1756283X11426895
20. Marino Z, van Bommel F, Forns X, Berg T (2013) New concepts of sofosbuvir-based treatment regimens in patients with hepatitis C. Gut 63(2):207–215. doi:10.1136/gutjnl-2013-305771
21. Lawitz E, Mangia A, Wyles D et al (2013) Sofosbuvir for previously untreated chronic hepatitis C infection. N Engl J Med 368:1878–1887. doi:10.1056/NEJMoa1214853
22. Zeisel MB, Fofana I, Fafi-Kremer S, Baumert TF (2011) Hepatitis C virus entry into hepatocytes: molecular mechanisms and targets for antiviral therapies. J Hepatol 54:566–576. doi:10.1016/j.jhep.2010.10.014
23. Romero-Brey I, Merz A, Chiramel A et al (2012) Three-dimensional architecture and biogenesis of membrane structures associated with hepatitis C virus replication. PLoS Pathog 8:e1003056. doi:10.1371/journal.ppat.1003056
24. Lukavsky PJ (2009) Structure and function of HCV IRES domains. Virus Res 139:166–171. doi:10.1016/j.virusres.2008.06.004
25. Pestova TV, Shatsky IN, Fletcher SP et al (1998) A prokaryotic-like mode of cytoplasmic eukaryotic ribosome binding to the initiation codon during internal translation initiation of hepatitis C and classical swine fever virus RNAs. Genes Dev 12:67–83
26. Lindenbach BD, Meuleman P, Ploss A et al (2006) Cell culture-grown hepatitis C virus is infectious in vivo and can be recultured in vitro. Proc Natl Acad Sci U S A 103:3805–3809. doi:10.1073/pnas.0511218103
27. Lindenbach BD, Rice CM (2013) The ins and outs of hepatitis C virus entry and assembly. Nat Rev Microbiol 11:688–700. doi:10.1038/nrmicro3098
28. Griffin SDC (2009) Plugging the holes in hepatitis C virus antiviral therapy. Proc Natl Acad Sci U S A 106:12567–12568. doi:10.1073/pnas.0906760106
29. Atkins E, Tatineni R, Li H et al (2014) The stability of secreted, acid-labile H77/JFH-1 hepatitis C virus (HCV) particles is altered by patient isolate genotype 1a p7 sequences. Virology 448:117–124. doi:10.1016/j.virol.2013.10.003
30. Whitfield T, Miles AJ, Scheinost JC et al (2011) The influence of different lipid environments on the structure and function of the hepatitis C virus p7 ion channel protein. Mol Membr Biol 28:254–264. doi:10.3109/09687688.2011.581253
31. Chew CF, Vijayan R, Chang J et al (2009) Determination of pore-lining residues in the hepatitis C virus p7 protein. Biophys J 96:L10–L12. doi:10.1016/j.bpj.2008.10.004
32. Lindenbach BD, Evans MJ, Syder AJ et al (2005) Complete replication of hepatitis C virus in cell culture. Science 309:623–626. doi:10.1126/science.1114016
33. Bassett SE, Brasky KM, Lanford RE (1998) Analysis of hepatitis C virus-inoculated chimpanzees reveals unexpected clinical profiles. J Virol 72:2589–2599
34. Wakita T, Pietschmann T, Kato T et al (2005) Production of infectious hepatitis C virus in tissue culture from a cloned viral genome. Nat Med 11:791–796. doi:10.1038/nm1268
35. Kolykhalov AA, Agapov EV, Blight KJ et al (1997) Transmission of hepatitis C by intrahepatic inoculation with transcribed RNA. Science 277:570–574
36. Patton HM, Patel K, Behling C et al (2004) The impact of steatosis on disease progression and early and sustained treatment response in chronic hepatitis C patients. J Hepatol 40:484–490. doi:10.1016/j.jhep.2003.11.004
37. Lohmann V, Körner F, Koch J et al (1999) Replication of subgenomic hepatitis C virus RNAs in a hepatoma cell line. Science 285:110–113
38. Blight KJ, Kolykhalov AA, Rice CM (2000) Efficient initiation of HCV RNA replication in cell culture. Science 290:1972–1974
39. Kato T, Date T, Miyamoto M et al (2003) Efficient replication of the genotype 2a hepatitis C virus subgenomic replicon. Gastroenterology 125:1808–1817
40. Blight KJ, McKeating JA, Marcotrigiano J, Rice CM (2003) Efficient replication of hepatitis C virus genotype 1a RNAs in cell culture. J Virol 77:3181–3190
41. Pietschmann T, Lohmann V, Kaul A et al (2002) Persistent and transient replication of full-length hepatitis C virus genomes in cell culture. J Virol 76:4008–4021. doi:10.1128/JVI.76.8.4008-4021.2002

42. Zhong J, Gastaminza P, Cheng G et al (2005) Robust hepatitis C virus infection in vitro. Proc Natl Acad Sci 102:9294–9299. doi:10.1073/pnas.0503596102
43. Kato T, Matsumura T, Heller T et al (2007) Production of infectious hepatitis C virus of various genotypes in cell cultures. J Virol 81:4405–4411. doi:10.1128/JVI. 02334-06
44. Yi M, Lemon SM (2009) Genotype 1a HCV (H77S) infection system. Methods Mol Biol 510:337–346
45. Billerbeck E, de Jong Y, Dorner M et al (2013) Animal models for hepatitis C. Curr Top Microbiol Immunol 369:49–86. doi:10.1007/978-3-642-27340-7_3
46. Chang K-S, Cai Z, Zhang C et al (2006) Replication of hepatitis C virus (HCV) RNA in mouse embryonic fibroblasts: protein kinase R (PKR)-dependent and PKR-independent mechanisms for controlling HCV RNA replication and mediating interferon activities. J Virol 80:7364–7374. doi:10.1128/JVI. 00586-06
47. Lin L-T, Noyce RS, Pham TNQ et al (2010) Replication of subgenomic hepatitis C virus replicons in mouse fibroblasts is facilitated by deletion of interferon regulatory factor 3 and expression of liver-specific microRNA 122. J Virol 84:9170–9180. doi:10.1128/JVI. 00559-10
48. Dorner M, Horwitz JA, Donovan BM et al (2013) Completion of the entire hepatitis C virus life cycle in genetically humanized mice. Nature 501:237–241. doi:10.1038/nature12427
49. Washburn ML, Bility MT, Zhang L et al (2011) A humanized mouse model to study hepatitis C virus infection, immune response, and liver disease. Gastroenterology 140:1334–1344. doi:10.1053/j.gastro.2011.01.001
50. Bitzegeio J, Bankwitz D, Hueging K et al (2010) Adaptation of hepatitis C virus to mouse CD81 permits infection of mouse cells in the absence of human entry factors. PLoS Pathog 6:e1000978. doi:10.1371/journal.ppat.1000978
51. Tong Y, Zhu Y, Xia X et al (2011) Tupaia CD81, SR-BI, claudin-1, and occludin support hepatitis C virus infection. J Virol 85:2793–2802. doi:10.1128/JVI. 01818-10
52. Amako Y, Tsukiyama-Kohara K, Katsume A et al (2010) Pathogenesis of hepatitis C virus infection in Tupaia belangeri. J Virol 84:303–311. doi:10.1128/JVI. 01448-09
53. Hurtado-Nedelec M, Makni-Maalej K, Gougerot-Pocidalo M-A et al (2014) Assessment of priming of the human neutrophil respiratory burst. Methods Mol Biol 1124:405–412. doi:10.1007/978-1-62703-845-4_23
54. Overstreet JM, Samarakoon R, Meldrum KK, Higgins PJ (2014) Redox control of p53 in the transcriptional regulation of TGF-β1 target genes through SMAD cooperativity. Cell Signal; 26(7):1427–1436. doi:10.1016/j.cellsig.2014.02.017
55. Brandes RP, Weissmann N, Schröder K (2014) Redox-mediated signal transduction by cardiovascular Nox NADPH oxidases. J Mol Cell Cardiol 73:70–79. doi:10.1016/j.yjmcc.2014.02.006
56. Tandara L, Salamunic I (2012) Iron metabolism: current facts and future directions. Biochem Med 22:311–328
57. Miura K, Taura K, Kodama Y et al (2008) Hepatitis C virus-induced oxidative stress suppresses hepcidin expression through increased histone deacetylase activity. Hepatology 48:1420–1429. doi:10.1002/hep.22486
58. Kato J, Kobune M, Nakamura T et al (2001) Normalization of elevated hepatic 8-hydroxy-2′-deoxyguanosine levels in chronic hepatitis C patients by phlebotomy and low iron diet. Cancer Res 61:8697–8702
59. Hayashi H, Takikawa T, Nishimura N et al (1994) Improvement of serum aminotransferase levels after phlebotomy in patients with chronic active hepatitis C and excess hepatic iron. Am J Gastroenterol 89:986–988
60. Yano M, Hayashi H, Wakusawa S et al (2002) Long term effects of phlebotomy on biochemical and histological parameters of chronic hepatitis C. Am J Gastroenterol 97:133–137. doi:10.1111/j.1572-0241.2002.05436.x
61. Kato J, Miyanishi K, Kobune M et al (2007) Long-term phlebotomy with low-iron diet therapy lowers risk of development of hepatocellular carcinoma from chronic hepatitis C. J Gastroenterol 42:830–836. doi:10.1007/s00535-007-2095-z

62. Kageyama F, Kobayashi Y, Kawasaki T et al (2000) Successful interferon therapy reverses enhanced hepatic iron accumulation and lipid peroxidation in chronic hepatitis C. Am J Gastroenterol 95:1041–1050. doi:10.1111/j.1572-0241.2000.01979.x
63. Piccoli C, Scrima R, Quarato G et al (2007) Hepatitis C virus protein expression causes calcium-mediated mitochondrial bioenergetic dysfunction and nitro-oxidative stress. Hepatology 46:58–65. doi:10.1002/hep.21679
64. Deng L, Adachi T, Kitayama K et al (2008) Hepatitis C virus infection induces apoptosis through a Bax-triggered, mitochondrion-mediated, caspase 3-dependent pathway. J Virol 82:10375–10385. doi:10.1128/JVI. 00395-08
65. Boudreau HE, Emerson SU, Korzeniowska A et al (2009) Hepatitis C virus (HCV) proteins induce NADPH oxidase 4 expression in a transforming growth factor-dependent manner: a new contributor to HCV-induced oxidative stress. J Virol 83:12934–12946. doi:10.1128/JVI. 01059-09
66. De Mochel NSR, Seronello S, Wang SH et al (2010) Hepatocyte NAD(P)H oxidases as an endogenous source of reactive oxygen species during hepatitis C virus infection. Hepatology 52:47–59. doi:10.1002/hep.23671
67. Wang T, Campbell RV, Yi MK et al (2010) Role of Hepatitis C virus core protein in viral-induced mitochondrial dysfunction. J Viral Hepat 17:784–793. doi:10.1111/j.1365-2893.2009.01238.x
68. Valgimigli M, Valgimigli L, Trerè D et al (2002) Oxidative stress EPR measurement in human liver by radical-probe technique. Correlation with etiology, histology and cell proliferation. Free Radic Res 36:939–948
69. Okuda M, Li K, Beard M et al (2002) Mitochondrial injury, oxidative stress, and antioxidant gene expression are induced by hepatitis C virus core protein. Gastroenterology 122:366–375. doi:10.1053/gast.2002.30983
70. Hara Y, Hino K, Okuda M et al (2006) Hepatitis C virus core protein inhibits deoxycholic acid-mediated apoptosis despite generating mitochondrial reactive oxygen species. J Gastroenterol 41:257–268. doi:10.1007/s00535-005-1738-1
71. Li Y, Boehning DF, Qian T et al (2007) Hepatitis C virus core protein increases mitochondrial ROS production by stimulation of Ca2+ uniporter activity. FASEB J 21:2474–2485. doi:10.1096/fj.06-7345com
72. Korenaga M, Wang T, Li Y et al (2005) Hepatitis C virus core protein inhibits mitochondrial electron transport and increases reactive oxygen species (ROS) production. J Biol Chem 280:37481–37488. doi:10.1074/jbc.M506412200
73. Gaspers LD, Mémin E, Thomas AP (2012) Calcium-dependent physiologic and pathologic stimulus-metabolic response coupling in hepatocytes. Cell Calcium 52:93–102. doi:10.1016/j.ceca.2012.04.009
74. Tsutsumi T, Matsuda M, Aizaki H et al (2009) Proteomics analysis of mitochondrial proteins reveals overexpression of a mitochondrial protein chaperon, prohibition, in cells expressing hepatitis C virus core protein. Hepatology 50:378–386. doi:10.1002/hep.22998
75. Nomura-Takigawa Y, Nagano-Fujii M, Deng L et al (2006) Non-structural protein 4A of Hepatitis C virus accumulates on mitochondria and renders the cells prone to undergoing mitochondria-mediated apoptosis. J Gen Virol 87:1935–1945. doi:10.1099/vir. 0.81701-0
76. Lambeth JD (2004) NOX enzymes and the biology of reactive oxygen. Nat Rev Immunol 4:181–189. doi:10.1038/nri1312
77. Pollock JD, Williams DA, Gifford MA et al (1995) Mouse model of X-linked chronic granulomatous disease, an inherited defect in phagocyte superoxide production. Nat Genet 9:202–209. doi:10.1038/ng0295-202
78. Bureau C (2001) Nonstructural 3 protein of hepatitis C virus triggers an oxidative burst in human monocytes via activation of NADPH oxidase. J Biol Chem 276:23077–23083. doi:10.1074/jbc.M100698200
79. Thoren F (2004) A hepatitis C virus-encoded, nonstructural protein (NS3) triggers dysfunction and apoptosis in lymphocytes: role of NADPH oxidase-derived oxygen radicals. J Leukoc Biol 76:1180–1186. doi:10.1189/jlb.0704387

80. Tacke RS, Tosello-Trampont A, Nguyen V et al (2011) Extracellular hepatitis C virus core protein activates STAT3 in human monocytes/macrophages/dendritic cells via an IL-6 autocrine pathway. J Biol Chem 286:10847–10855. doi:10.1074/jbc.M110.217653
81. Corzo CA, Cotter MJ, Cheng P et al (2009) Mechanism regulating reactive oxygen species in tumor-induced myeloid-derived suppressor cells. J Immunol 182:5693–5701. doi:10.4049/jimmunol.0900092
82. Von Lohneysen K, Noack D, Wood MR et al (2009) Structural insights into Nox4 and Nox2: motifs involved in function and cellular localization. Mol Cell Biol 30:961–975. doi:10.1128/MCB. 01393-09
83. Von Lohneysen K, Noack D, Hayes P et al (2012) Constitutive NADPH oxidase 4 activity resides in the composition of the B-loop and the penultimate C terminus. J Biol Chem 287:8737–8745. doi:10.1074/jbc.M111.332494
84. Goyal P, Weissmann N, Rose F et al (2005) Identification of novel Nox4 splice variants with impact on ROS levels in A549 cells. Biochem Biophys Res Commun 329:32–39. doi:10.1016/j.bbrc.2005.01.089
85. Anilkumar N, Jose GS, Sawyer I et al (2013) A 28-kDa splice variant of NADPH oxidase-4 is nuclear-localized and involved in redox signaling in vascular cells. Arterioscler Thromb Vasc Biol 33:e104–e112. doi:10.1161/ATVBAHA.112.300956
86. Spencer NY, Yan Z, Boudreau RL et al (2011) Control of hepatic nuclear superoxide production by glucose 6-phosphate dehydrogenase and NADPH oxidase-4. J Biol Chem 286:8977–8987. doi:10.1074/jbc.M110.193821
87. Otani K, Korenaga M, Beard MR et al (2005) Hepatitis C virus core protein, cytochrome P450 2E1, and alcohol produce combined mitochondrial injury and cytotoxicity in hepatoma cells. Gastroenterology 128:96–107
88. Gochee PA, Jonsson JR, Clouston AD et al (2003) Steatosis in chronic hepatitis C: association with increased messenger RNA expression of collagen I, tumor necrosis factor-alpha and cytochrome P450 2E1. J Gastroenterol Hepatol 18:386–392
89. Higgs MR, Lerat H, Pawlotsky J-M (2013) Hepatitis C virus-induced activation of β-catenin promotes c-Myc expression and a cascade of pro-carcinogenetic events. Oncogene 32:4683–4693. doi:10.1038/onc.2012.484
90. Geller DA, Lowenstein CJ, Shapiro RA et al (1993) Molecular cloning and expression of inducible nitric oxide synthase from human hepatocytes. Proc Natl Acad Sci U S A 90:3491–3495
91. Xie QW, Cho HJ, Calaycay J et al (1992) Cloning and characterization of inducible nitric oxide synthase from mouse macrophages. Science 256:225–228
92. Machida K, Cheng KT-H, Sung VM-H et al (2004) Hepatitis C virus infection activates the immunologic (type II) isoform of nitric oxide synthase and thereby enhances DNA damage and mutations of cellular genes. J Virol 78:8835–8843. doi:10.1128/JVI. 78.16.8835-8843.2004
93. Majano PL, García-Monzón C, López-Cabrera M et al (1998) Inducible nitric oxide synthase expression in chronic viral hepatitis. Evidence for a virus-induced gene upregulation. J Clin Invest 101:1343–1352. doi:10.1172/JCI774
94. Kane JM 3rd, Shears LL 2nd, Hierholzer C et al (1997) Chronic hepatitis C virus infection in humans: induction of hepatic nitric oxide synthase and proposed mechanisms for carcinogenesis. J Surg Res 69:321–324. doi:10.1006/jsre.1997.5057
95. Vodovotz Y, Kim PKM, Bagci EZ et al (2004) Inflammatory modulation of hepatocyte apoptosis by nitric oxide: in vivo, in vitro, and in silico studies. Curr Mol Med 4:753–762
96. Machida K, Tsukamoto H, Liu J-C et al (2010) c-Jun mediates hepatitis C virus hepatocarcinogenesis through signal transducer and activator of transcription 3 and nitric oxide-dependent impairment of oxidative DNA repair. Hepatology 52:480–492. doi:10.1002/hep.23697
97. García-Monzón C, Majano PL, Zubia I et al (2000) Intrahepatic accumulation of nitrotyrosine in chronic viral hepatitis is associated with histological severity of liver disease. J Hepatol 32:331–338

98. El-Kannishy G, Arafa M, Abdelaal I et al (2012) Persistent oxidative stress in patients with chronic active hepatitis-C infection after antiviral therapy failure. Saudi J Gastroenterol 18:375–379. doi:10.4103/1319-3767.103429
99. Abdalla MY, Ahmad IM, Spitz DR et al (2005) Hepatitis C virus-core and non structural proteins lead to different effects on cellular antioxidant defenses. J Med Virol 76:489–497. doi:10.1002/jmv.20388
100. Levent G, Ali A, Ahmet A et al (2006) Oxidative stress and antioxidant defense in patients with chronic hepatitis C patients before and after pegylated interferon alfa-2b plus ribavirin therapy. J Transl Med 4:25. doi:10.1186/1479-5876-4-25
101. Carvajal-Yepes M, Himmelsbach K, Schaedler S et al (2011) Hepatitis C virus impairs the induction of cytoprotective Nrf2 target genes by delocalization of small maf proteins. J Biol Chem 286:8941–8951. doi:10.1074/jbc.M110.186684
102. Burdette D, Olivarez M, Waris G (2010) Activation of transcription factor Nrf2 by hepatitis C virus induces the cell-survival pathway. J Gen Virol 91:681–690. doi:10.1099/vir.0.014340-0
103. Choi J (2012) Oxidative stress, endogenous antioxidants, alcohol, and hepatitis C: pathogenic interactions and therapeutic considerations. Free Radic Biol Med 52:1135–1150. doi:10.1016/j.freeradbiomed.2012.01.008
104. Tamai T, Uto H, Takami Y et al (2011) Serum manganese superoxide dismutase and thioredoxin are potential prognostic markers for hepatitis C virus-related hepatocellular carcinoma. World J Gastroenterol 17:4890–4898. doi:10.3748/wjg.v17.i44.4890
105. Marotta F, Chui DH, Jain S et al (2011) Effect of a fermented nutraceutical on thioredoxin level and TNF-alpha signalling in cirrhotic patients. J Biol Regul Homeost Agents 25:37–45
106. Mitsuyoshi H, Itoh Y, Sumida Y et al (2008) Evidence of oxidative stress as a cofactor in the development of insulin resistance in patients with chronic hepatitis C. Hepatol Res 38:348–353. doi:10.1111/j.1872-034X.2007.00280.x
107. Yahya RS, Ghanem OH, Foyouh A-AA et al (2013) Role of interleukin-8 and oxidative stress in patients with hepatocellular carcinoma. Clin Lab 59:969–976
108. Ramesh S, Sanyal AJ (2004) Hepatitis C and nonalcoholic fatty liver disease. Semin Liver Dis 24:399–413. doi:10.1055/s-2004-860869
109. Lonardo A, Loria P, Adinolfi LE et al (2006) Hepatitis C and steatosis: a reappraisal. J Viral Hepat 13:73–80. doi:10.1111/j.1365-2893.2005.00669.x
110. Hui JM, Kench J, Farrell GC et al (2002) Genotype-specific mechanisms for hepatic steatosis in chronic hepatitis C infection. J Gastroenterol Hepatol 17:873–881
111. Ohata K, Hamasaki K, Toriyama K et al (2003) Hepatic steatosis is a risk factor for hepatocellular carcinoma in patients with chronic hepatitis C virus infection. Cancer 97:3036–3043. doi:10.1002/cncr.11427
112. Leandro G, Mangia A, Hui J et al (2006) Relationship between steatosis, inflammation, and fibrosis in chronic hepatitis C: a meta-analysis of individual patient data. Gastroenterology 130:1636–1642. doi:10.1053/j.gastro.2006.03.014
113. Pekow JR, Bhan AK, Zheng H, Chung RT (2007) Hepatic steatosis is associated with increased frequency of hepatocellular carcinoma in patients with hepatitis C-related cirrhosis. Cancer 109:2490–2496. doi:10.1002/cncr.22701
114. Poynard T (2003) Effect of treatment with peginterferon or interferon alfa-2b and ribavirin on steatosis in patients infected with hepatitis C. Hepatology 38:75–85. doi:10.1053/jhep.2003.50267
115. Westin J, Lagging M, Dhillon AP et al (2007) Impact of hepatic steatosis on viral kinetics and treatment outcome during antiviral treatment of chronic HCV infection. J Viral Hepat 14:29–35. doi:10.1111/j.1365-2893.2006.00777.x
116. Kim KH, Hong SP, Kim K et al (2007) HCV core protein induces hepatic lipid accumulation by activating SREBP1 and PPARgamma. Biochem Biophys Res Commun 355:883–888. doi:10.1016/j.bbrc.2007.02.044

117. Fujino T, Nakamuta M, Yada R et al (2010) Expression profile of lipid metabolism-associated genes in hepatitis C virus-infected human liver: Lipid metabolism in HCV-infected liver. Hepatol Res 40:923–929. doi:10.1111/j.1872-034X.2010.00700.x
118. Waris G, Felmlee DJ, Negro F, Siddiqui A (2007) Hepatitis C virus induces proteolytic cleavage of sterol regulatory element binding proteins and stimulates their phosphorylation via oxidative stress. J Virol 81:8122–8130. doi:10.1128/JVI. 00125-07
119. Oem J-K, Jackel-Cram C, Li Y-P et al (2008) Activation of sterol regulatory element-binding protein 1c and fatty acid synthase transcription by hepatitis C virus non-structural protein 2. J Gen Virol 89:1225–1230. doi:10.1099/vir. 0.83491-0
120. Harris C, Herker E, Farese RV Jr, Ott M (2011) Hepatitis C virus core protein decreases lipid droplet turnover: a mechanism for core-induced steatosis. J Biol Chem 286:42615–42625. doi:10.1074/jbc.M111.285148
121. Lerat H, Honda M, Beard M et al (2002) Steatosis and liver cancer in transgenic mice expressing the structural and nonstructural proteins of hepatitis c virus. Gastroenterology 122:352–365. doi:10.1053/gast.2002.31001
122. Naas T (2005) Characterization of liver histopathology in a transgenic mouse model expressing genotype 1a hepatitis C virus core and envelope proteins 1 and 2. J Gen Virol 86:2185–2196. doi:10.1099/vir. 0.80969-0
123. Moriishi K, Mochizuki R, Moriya K et al (2007) Critical role of PA28 in hepatitis C virus-associated steatogenesis and hepatocarcinogenesis. Proc Natl Acad Sci 104:1661–1666. doi:10.1073/pnas.0607312104
124. Tanaka N, Moriya K, Kiyosawa K et al (2008) PPARα activation is essential for HCV core protein–induced hepatic steatosis and hepatocellular carcinoma in mice. J Clin Invest. doi:10.1172/JCI33594
125. Wang A-G, Lee D-S, Moon H-B et al (2009) Non-structural 5A protein of hepatitis C virus induces a range of liver pathology in transgenic mice. J Pathol 219:253–262. doi:10.1002/path.2592
126. Su AI, Pezacki JP, Wodicka L et al (2002) Genomic analysis of the host response to hepatitis C virus infection. Proc Natl Acad Sci U S A 99:15669–15674. doi:10.1073/pnas.202608199
127. Bigger CB, Guerra B, Brasky KM et al (2004) Intrahepatic gene expression during chronic hepatitis C virus infection in chimpanzees. J Virol 78:13779–13792. doi:10.1128/JVI.78.24.13779-13792.2004
128. Kaddai V, Negro F (2011) Current understanding of insulin resistance in hepatitis C. Expert Rev Gastroenterol Hepatol 5:503–516. doi:10.1586/egh.11.43
129. Farinati F, Cardin R, De Maria N et al (1995) Iron storage, lipid peroxidation and glutathione turnover in chronic anti-HCV positive hepatitis. J Hepatol 22:449–456. doi:10.1016/0168-8278(95)80108-1
130. Macdonald GA, Bridle KR, Ward PJ et al (2001) Lipid peroxidation in hepatic steatosis in humans is associated with hepatic fibrosis and occurs predominately in acinar zone 3. J Gastroenterol Hepatol 16:599–606. doi:10.1046/j.1440-1746.2001.02445.x
131. Bataller R, Brenner DA (2005) Liver fibrosis. J Clin Invest 115:209–218. doi:10.1172/JCI24282
132. Jiang JX, Venugopal S, Serizawa N et al (2010) Reduced nicotinamide adenine dinucleotide phosphate oxidase 2 plays a key role in stellate cell activation and liver fibrogenesis in vivo. Gastroenterology 139:1375–1384. doi:10.1053/j.gastro.2010.05.074
133. Paik Y-H, Iwaisako K, Seki E et al (2011) The nicotinamide adenine dinucleotide phosphate oxidase (NOX) homologues NOX1 and NOX2/gp91(phox) mediate hepatic fibrosis in mice. Hepatology 53:1730–1741. doi:10.1002/hep.24281
134. Calvisi DF, Pascale RM, Feo F (2007) Dissection of signal transduction pathways as a tool for the development of targeted therapies of hepatocellular carcinoma. Rev Recent Clin Trials 2:217–236
135. Poli G (2000) Pathogenesis of liver fibrosis: role of oxidative stress. Mol Aspects Med 21:49–98

136. Sánchez-Valle V, Chávez-Tapia NC, Uribe M, Méndez-Sánchez N (2012) Role of oxidative stress and molecular changes in liver fibrosis: a review. Curr Med Chem 19:4850–4860
137. Rigamonti C, Andorno S, Maduli E et al (2002) Iron, hepatic stellate cells and fibrosis in chronic hepatitis C. Eur J Clin Invest 32(Suppl 1):28–35
138. Nieto N, Greenwel P, Friedman SL et al (2000) Ethanol and arachidonic acid increase alpha 2(I) collagen expression in rat hepatic stellate cells overexpressing cytochrome P450 2E1. Role of H_2O_2 and cyclooxygenase-2. J Biol Chem 275:20136–20145. doi:10.1074/jbc.M001422200
139. Nieto N (2007) Ethanol and fish oil induce NFkappaB transactivation of the collagen alpha2(I) promoter through lipid peroxidation-driven activation of the PKC-PI3K-Akt pathway. Hepatology 45:1433–1445. doi:10.1002/hep.21659
140. Lin Q, Fang D, Hou X et al (2011) HCV peptide (C5A), an amphipathic α-helical peptide of hepatitis virus C, is an activator of N-formyl peptide receptor in human phagocytes. J Immunol 186:2087–2094. doi:10.4049/jimmunol.1002340
141. Canbay A, Friedman S, Gores GJ (2004) Apoptosis: the nexus of liver injury and fibrosis. Hepatology 39:273–278. doi:10.1002/hep.20051
142. Canbay A, Higuchi H, Bronk SF et al (2002) Fas enhances fibrogenesis in the bile duct ligated mouse: a link between apoptosis and fibrosis. Gastroenterology 123:1323–1330
143. Canbay A, Feldstein AE, Higuchi H et al (2003) Kupffer cell engulfment of apoptotic bodies stimulates death ligand and cytokine expression. Hepatology 38:1188–1198. doi:10.1053/jhep.2003.50472
144. Pianko S, Patella S, Ostapowicz G et al (2001) Fas-mediated hepatocyte apoptosis is increased by hepatitis C virus infection and alcohol consumption, and may be associated with hepatic fibrosis: mechanisms of liver cell injury in chronic hepatitis C virus infection. J Viral Hepat 8:406–413
145. Mateu G, Donis RO, Wakita T et al (2008) Intragenotypic JFH1 based recombinant hepatitis C virus produces high levels of infectious particles but causes increased cell death. Virology 376:397–407. doi:10.1016/j.virol.2008.03.027
146. Walters K-A, Syder AJ, Lederer SL et al (2009) Genomic analysis reveals a potential role for cell cycle perturbation in HCV-mediated apoptosis of cultured hepatocytes. PLoS Pathog 5:e1000269. doi:10.1371/journal.ppat.1000269
147. Joyce MA, Walters K-A, Lamb S-E et al (2009) HCV induces oxidative and ER stress, and sensitizes infected cells to apoptosis in SCID/Alb-uPA mice. PLoS Pathog 5:e1000291. doi:10.1371/journal.ppat.1000291
148. Hiramatsu N, Hayashi N, Katayama K et al (1994) Immunohistochemical detection of Fas antigen in liver tissue of patients with chronic hepatitis C. Hepatology 19:1354–1359. doi:10.1002/hep.1840190606
149. Gieseler RK, Marquitan G, Schlattjan M et al (2011) Hepatocyte apoptotic bodies encasing nonstructural HCV proteins amplify hepatic stellate cell activation: implications for chronic hepatitis C: HCV-positive apoptotic bodies amplify hepatic stellate cell activation. J Viral Hepat 18:760–767. doi:10.1111/j.1365-2893.2010.01362.x
150. Inagaki Y, Okazaki I (2007) Emerging insights into transforming growth factor smad signal in hepatic fibrogenesis. Gut 56:284–292. doi:10.1136/gut.2005.088690
151. Senturk S, Mumcuoglu M, Gursoy-Yuzugullu O et al (2010) Transforming growth factor-beta induces senescence in hepatocellular carcinoma cells and inhibits tumor growth. Hepatology 52:966–974. doi:10.1002/hep.23769
152. Jiang JX, Chen X, Serizawa N et al (2012) Liver fibrosis and hepatocyte apoptosis are attenuated by GKT137831, a novel NOX4/NOX1 inhibitor in vivo. Free Radic Biol Med 53:289–296. doi:10.1016/j.freeradbiomed.2012.05.007
153. Hu T, Ramachandrarao SP, Siva S et al (2005) Reactive oxygen species production via NADPH oxidase mediates TGF-beta-induced cytoskeletal alterations in endothelial cells. Am J Physiol Renal Physiol 289:F816–F825. doi:10.1152/ajprenal.00024.2005

154. Hecker L, Vittal R, Jones T et al (2009) NADPH oxidase-4 mediates myofibroblast activation and fibrogenic responses to lung injury. Nat Med 15:1077–1081. doi:10.1038/nm.2005
155. Liu R-M, Choi J, Wu J-H et al (2010) Oxidative modification of nuclear mitogen-activated protein kinase phosphatase 1 is involved in transforming growth factor beta1-induced expression of plasminogen activator inhibitor 1 in fibroblasts. J Biol Chem 285:16239–16247. doi:10.1074/jbc.M110.111732
156. Carmona-Cuenca I, Roncero C, Sancho P et al (2008) Upregulation of the NADPH oxidase NOX4 by TGF-beta in hepatocytes is required for its pro-apoptotic activity. J Hepatol 49:965–976. doi:10.1016/j.jhep.2008.07.021
157. Sancho P, Bertran E, Caja L et al (2009) The inhibition of the epidermal growth factor (EGF) pathway enhances TGF-beta-induced apoptosis in rat hepatoma cells through inducing oxidative stress coincident with a change in the expression pattern of the NADPH oxidases (NOX) isoforms. Biochim Biophys Acta 1793:253–263. doi:10.1016/j.bbamcr.2008.09.003
158. Murillo MM, Carmona-Cuenca I, Del Castillo G et al (2007) Activation of NADPH oxidase by transforming growth factor-beta in hepatocytes mediates up-regulation of epidermal growth factor receptor ligands through a nuclear factor-kappaB-dependent mechanism. Biochem J 405:251–259. doi:10.1042/BJ20061846
159. Sancho P, Mainez J, Crosas-Molist E et al (2012) NADPH oxidase NOX4 mediates stellate cell activation and hepatocyte cell death during liver fibrosis development. PLoS One 7:e45285. doi:10.1371/journal.pone.0045285
160. Nelson DR, Gonzalez-Peralta RP, Qian K et al (1997) Transforming growth factor-beta 1 in chronic hepatitis C. J Viral Hepat 4:29–35
161. Lin W, Wu G, Li S et al (2011) HIV and HCV cooperatively promote hepatic fibrogenesis via induction of reactive oxygen species and NFkappaB. J Biol Chem 286:2665–2674. doi:10.1074/jbc.M110.168286
162. Bataller R, Schwabe RF, Choi YH et al (2003) NADPH oxidase signal transduces angiotensin II in hepatic stellate cells and is critical in hepatic fibrosis. J Clin Invest 112:1383–1394. doi:10.1172/JCI18212
163. De Minicis S, Seki E, Paik Y-H et al (2010) Role and cellular source of nicotinamide adenine dinucleotide phosphate oxidase in hepatic fibrosis. Hepatology 52:1420–1430. doi:10.1002/hep.23804
164. Aoyama T, Paik Y-H, Watanabe S et al (2012) Nicotinamide adenine dinucleotide phosphate oxidase in experimental liver fibrosis: GKT137831 as a novel potential therapeutic agent. Hepatology 56:2316–2327. doi:10.1002/hep.25938
165. Emerit J, Samuel D, Pavio N (2006) Cu-Zn super oxide dismutase as a potential antifibrotic drug for hepatitis C related fibrosis. Biomed Pharmacother 60:1–4. doi:10.1016/j.biopha.2005.09.002
166. Hernandez-Gea V, Friedman SL (2011) Pathogenesis of liver fibrosis. Annu Rev Pathol Mech Dis 6:425–456. doi:10.1146/annurev-pathol-011110-130246
167. Capone F, Costantini S, Guerriero E et al (2010) Serum cytokine levels in patients with hepatocellular carcinoma. Eur Cytokine Netw 21:99–104. doi:10.1684/ecn.2010.0192
168. Jiang JX, Chen X, Fukada H et al (2013) Advanced glycation endproducts induce fibrogenic activity in nonalcoholic steatohepatitis by modulating TNF-α-converting enzyme activity in mice. Hepatology 58:1339–1348. doi:10.1002/hep.26491
169. Kastl L, Sauer SW, Ruppert T et al (2014) TNF-α mediates mitochondrial uncoupling and enhances ROS-dependent cell migration via NF-κB activation in liver cells. FEBS Lett 588:175–183. doi:10.1016/j.febslet.2013.11.033
170. Donato F, Boffetta P, Puoti M (1998) A meta-analysis of epidemiological studies on the combined effect of hepatitis B and C virus infections in causing hepatocellular carcinoma. Int J Cancer 75:347–354. doi:10.1002/(SICI)1097-0215(19980130)75:3<347::AID-IJC4>3.0.CO;2-2
171. Davis G (2003) Projecting future complications of chronic hepatitis C in the United States. Liver Transpl 9:331–338. doi:10.1053/jlts.2003.50073

172. Elchuri S, Oberley TD, Qi W et al (2005) CuZnSOD deficiency leads to persistent and widespread oxidative damage and hepatocarcinogenesis later in life. Oncogene 24:367–380. doi:10.1038/sj.onc.1208207
173. Xu Z, Chen L, Leung L et al (2005) Liver-specific inactivation of the Nrf1 gene in adult mouse leads to nonalcoholic steatohepatitis and hepatic neoplasia. Proc Natl Acad Sci U S A 102:4120–4125. doi:10.1073/pnas.0500660102
174. Farinati F, Cardin R, Degan P et al (1999) Oxidative DNA damage in circulating leukocytes occurs as an early event in chronic HCV infection. Free Radic Biol Med 27:1284–1291
175. Furutani T, Hino K, Okuda M et al (2006) Hepatic iron overload induces hepatocellular carcinoma in transgenic mice expressing the hepatitis C virus polyprotein. Gastroenterology 130:2087–2098. doi:10.1053/j.gastro.2006.02.060
176. Nishina S, Korenaga M, Hidaka I et al (2010) Hepatitis C virus protein and iron overload induce hepatic steatosis through the unfolded protein response in mice. Liver Int 30:683–692. doi:10.1111/j.1478-3231.2010.02210.x
177. Burney S, Caulfield JL, Niles JC et al (1999) The chemistry of DNA damage from nitric oxide and peroxynitrite. Mutat Res 424:37–49. doi:10.1016/S0027-5107(99)00006-8
178. Valko M, Izakovic M, Mazur M et al (2004) Role of oxygen radicals in DNA damage and cancer incidence. Mol Cell Biochem 266:37–56
179. deRojas-Walker T, Tamir S, Ji H et al (1995) Nitric oxide induces oxidative damage in addition to deamination in macrophage DNA. Chem Res Toxicol 8:473–477
180. Machida K, McNamara G, Cheng KT-H et al (1950) (2010) Hepatitis C virus inhibits DNA damage repair through reactive oxygen and nitrogen species and by interfering with the ATM-NBS1/Mre11/Rad50 DNA repair pathway in monocytes and hepatocytes. J Immunol 185:6985–6998. doi:10.4049/jimmunol.1000618
181. Machida K, Cheng KT-H, Lai C-K et al (2006) Hepatitis C virus triggers mitochondrial permeability transition with production of reactive oxygen species, leading to DNA damage and STAT3 activation. J Virol 80:7199–7207. doi:10.1128/JVI. 00321-06
182. Beckman JS (1996) Oxidative damage and tyrosine nitration from peroxynitrite. Chem Res Toxicol 9:836–844. doi:10.1021/tx9501445
183. Hussain SP, Schwank J, Staib F et al (2007) TP53 mutations and hepatocellular carcinoma: insights into the etiology and pathogenesis of liver cancer. Oncogene 26:2166–2176. doi:10.1038/sj.onc.1210279
184. Edamoto Y, Hara A, Biernat W et al (2003) Alterations of RB1, p53 and Wnt pathways in hepatocellular carcinomas associated with hepatitis C, hepatitis B and alcoholic liver cirrhosis. Int J Cancer 106:334–341. doi:10.1002/ijc.11254
185. Knudsen ES, Knudsen KE (2008) Tailoring to RB: tumour suppressor status and therapeutic response. Nat Rev Cancer 8:714–724. doi:10.1038/nrc2401
186. Schwarz KB, Kew M, Klein A et al (2001) Increased hepatic oxidative DNA damage in patients with hepatocellular carcinoma. Dig Dis Sci 46:2173–2178
187. Matsumoto K (2003) Immunohistochemical study of the relationship between 8-hydroxy-2′-deoxyguanosine levels in noncancerous region and postoperative recurrence of hepatocellular carcinoma in remnant liver. Hepatol Res 25:435–441. doi:10.1016/S1386-6346(02)00313-3
188. Machida K, Cheng KT-N, Sung VM-H et al (2004) Hepatitis C virus induces a mutator phenotype: Enhanced mutations of immunoglobulin and protooncogenes. Proc Natl Acad Sci 101:4262–4267. doi:10.1073/pnas.0303971101
189. Jung SW, Park NH, Shin JW et al (2012) Polymorphisms of DNA repair genes in Korean hepatocellular carcinoma patients with chronic hepatitis B: possible implications on survival. J Hepatol 57:621–627. doi:10.1016/j.jhep.2012.04.039
190. Kershaw RM, Hodges NJ (2012) Repair of oxidative DNA damage is delayed in the Ser326Cys polymorphic variant of the base excision repair protein OGG1. Mutagenesis 27:501–510. doi:10.1093/mutage/ges012
191. Jaiswal M, LaRusso NF, Nishioka N et al (2001) Human Ogg1, a protein involved in the repair of 8-oxoguanine, is inhibited by nitric oxide. Cancer Res 61:6388–6393

192. Bartsch H, Nair J (2006) Chronic inflammation and oxidative stress in the genesis and perpetuation of cancer: role of lipid peroxidation, DNA damage, and repair. Langenbecks Arch Surg 391:499–510. doi:10.1007/s00423-006-0073-1
193. Teufelhofer O, Parzefall W, Kainzbauer E et al (2005) Superoxide generation from Kupffer cells contributes to hepatocarcinogenesis: studies on NADPH oxidase knockout mice. Carcinogenesis 26:319–329. doi:10.1093/carcin/bgh320
194. Vendemiale G, Grattagliano I, Portincasa P et al (2001) Oxidative stress in symptom-free HCV carriers: relation with ALT flare-up. Eur J Clin Invest 31:54–63. doi:10.1046/j.1365-2362.2001.00747.x
195. Moe KT, Aulia S, Jiang F et al (2006) Differential upregulation of Nox homologues of NADPH oxidase by tumor necrosis factor-alpha in human aortic smooth muscle and embryonic kidney cells. J Cell Mol Med 10:231–239
196. Kim Y-S, Morgan MJ, Choksi S, Liu Z-G (2007) TNF-induced activation of the Nox1 NADPH oxidase and its role in the induction of necrotic cell death. Mol Cell 26:675–687. doi:10.1016/j.molcel.2007.04.021
197. Manea A, Manea SA, Gafencu AV et al (2008) AP-1-dependent transcriptional regulation of NADPH oxidase in human aortic smooth muscle cells: role of p22phox subunit. Arterioscler Thromb Vasc Biol 28:878–885. doi:10.1161/ATVBAHA.108.163592
198. Xia L, Mo P, Huang W et al (2012) The TNF-α/ROS/HIF-1-induced upregulation of FoxMI expression promotes HCC proliferation and resistance to apoptosis. Carcinogenesis 33:2250–2259. doi:10.1093/carcin/bgs249
199. Sun H-C, Li M, Lu J-L et al (2011) Overexpression of Forkhead box M1 protein associates with aggressive tumor features and poor prognosis of hepatocellular carcinoma. Oncol Rep 25:1533–1539. doi:10.3892/or.2011.1230
200. Sun H, Teng M, Liu J et al (2011) FOXM1 expression predicts the prognosis in hepatocellular carcinoma patients after orthotopic liver transplantation combined with the Milan criteria. Cancer Lett 306:214–222. doi:10.1016/j.canlet.2011.03.009
201. Kalinichenko VV (2004) Foxm1b transcription factor is essential for development of hepatocellular carcinomas and is negatively regulated by the p19ARF tumor suppressor. Genes Dev 18:830–850. doi:10.1101/gad.1200704
202. Gusarova GA, Wang I-C, Major ML et al (2007) A cell-penetrating ARF peptide inhibitor of FoxM1 in mouse hepatocellular carcinoma treatment. J Clin Invest 117:99–111. doi:10.1172/JCI27527
203. Caradonna L, Mastronardi ML, Magrone T et al (2002) Biological and clinical significance of endotoxemia in the course of hepatitis C virus infection. Curr Pharm Des 8:995–1005
204. Vespasiani-Gentilucci U, Carotti S, Onetti-Muda A et al (2012) Toll-like receptor-4 expression by hepatic progenitor cells and biliary epithelial cells in HCV-related chronic liver disease. Mod Pathol 25:576–589. doi:10.1038/modpathol.2011.197
205. Brownell J, Polyak SJ (2013) Molecular pathways: hepatitis C virus, CXCL10, and the inflammatory road to liver cancer. Clin Cancer Res 19:1347–1352. doi:10.1158/1078-0432.CCR-12-0928
206. Dapito DH, Mencin A, Gwak G-Y et al (2012) Promotion of hepatocellular carcinoma by the intestinal microbiota and TLR4. Cancer Cell 21:504–516. doi:10.1016/j.ccr.2012.02.007
207. Yu L-X, Yan H-X, Liu Q et al (2010) Endotoxin accumulation prevents carcinogen-induced apoptosis and promotes liver tumorigenesis in rodents. Hepatology 52:1322–1333. doi:10.1002/hep.23845
208. Szabo G, Wands JR, Eken A et al (2010) Alcohol and hepatitis C virus–interactions in immune dysfunctions and liver damage. Alcohol Clin Exp Res 34:1675–1686. doi:10.1111/j.1530-0277.2010.01255.x
209. Machida K, Tsukamoto H, Mkrtchyan H et al (2009) Toll-like receptor 4 mediates synergism between alcohol and HCV in hepatic oncogenesis involving stem cell marker Nanog. Proc Natl Acad Sci U S A 106:1548–1553. doi:10.1073/pnas.0807390106

210. Agúndez JA, García-Martín E, Devesa MJ et al (2012) Polymorphism of the TLR4 gene reduces the risk of hepatitis C virus-induced hepatocellular carcinoma. Oncology 82:35–40. doi:10.1159/000335606
211. Maitra U, Singh N, Gan L et al (2009) IRAK-1 contributes to lipopolysaccharide-induced reactive oxygen species generation in macrophages by inducing NOX-1 transcription and Rac1 activation and suppressing the expression of antioxidative enzymes. J Biol Chem 284:35403–35411. doi:10.1074/jbc.M109.059501
212. Kim J-S, Yeo S, Shin D-G et al (2010) Glycogen synthase kinase 3beta and beta-catenin pathway is involved in toll-like receptor 4-mediated NADPH oxidase 1 expression in macrophages. FEBS J 277:2830–2837. doi:10.1111/j.1742-4658.2010.07700.x
213. Park HS, Jung HY, Park EY et al (2004) Cutting edge: direct interaction of TLR4 with NAD(P)H oxidase 4 isozyme is essential for lipopolysaccharide-induced production of reactive oxygen species and activation of NF-kappa B. J Immunol 173:3589–3593
214. Chiang E, Dang O, Anderson K et al (2006) Cutting edge: apoptosis-regulating signal kinase 1 is required for reactive oxygen species-mediated activation of IFN regulatory factor 3 by lipopolysaccharide. J Immunol 176:5720–5724
215. Park HS, Chun JN, Jung HY et al (2006) Role of NADPH oxidase 4 in lipopolysaccharide-induced proinflammatory responses by human aortic endothelial cells. Cardiovasc Res 72:447–455. doi:10.1016/j.cardiores.2006.09.012
216. McCullough AJ (2011) Epidemiology of the metabolic syndrome in the USA: the metabolic syndrome. J Dig Dis 12:333–340. doi:10.1111/j.1751-2980.2010.00469.x
217. Alberti KGMM, Eckel RH, Grundy SM et al (2009) Harmonizing the metabolic syndrome: a joint interim statement of the International Diabetes Federation Task Force on Epidemiology and Prevention; National Heart, Lung, and Blood Institute; American Heart Association; World Heart Federation; International Atherosclerosis Society; and International Association for the Study of Obesity. Circulation 120:1640–1645. doi:10.1161/CIRCULATIONAHA.109.192644
218. Williams I (1999) Epidemiology of hepatitis C in the United States. Am J Med 107:2S–9S
219. Mehta SH (2000) Prevalence of type 2 diabetes mellitus among persons with hepatitis C virus infection in the United States. Ann Intern Med 133:592. doi:10.7326/0003-4819-133-8-200010170-00009
220. Sangiorgio L, Attardo T, Gangemi R et al (2000) Increased frequency of HCV and HBV infection in type 2 diabetic patients. Diabetes Res Clin Pract 48:147–151
221. Moucari R, Asselah T, Cazals–Hatem D et al (2008) Insulin resistance in chronic hepatitis C: association with genotypes 1 and 4, serum HCV RNA level, and liver fibrosis. Gastroenterology 134:416–423. doi:10.1053/j.gastro.2007.11.010
222. Moucari R, Asselah T, Paradis V et al (2007) Insulin resistance in 424 chronic hepatitis C patients: association with HCV replication and advanced fibrosis. J Hepatol 46:S11. doi:10.1016/S0168-8278(07)61619-5
223. Romero-Gómez M, Del Mar VM, Andrade RJ et al (2005) Insulin resistance impairs sustained response rate to peginterferon plus ribavirin in chronic hepatitis C patients. Gastroenterology 128:636–641
224. Deng L, Shoji I, Ogawa W et al (2011) Hepatitis C virus infection promotes hepatic gluconeogenesis through an NS5A-mediated, FoxO1-dependent pathway. J Virol 85:8556–8568. doi:10.1128/JVI. 00146-11
225. Rotter V (2003) Interleukin-6 (IL-6) induces insulin resistance in 3T3–L1 adipocytes and is, like IL-8 and tumor necrosis factor-, overexpressed in human fat cells from insulin-resistant subjects. J Biol Chem 278:45777–45784. doi:10.1074/jbc.M301977200
226. Borst SE (2004) The role of TNF-α in insulin resistance. Endocrine 23:177–182. doi:10.1385/ENDO:23:2-3:177
227. Evans JL, Maddux BA, Goldfine ID (2005) The molecular basis for oxidative stress-induced insulin resistance. Antioxidants Redox Signal 7:1040–1052. doi:10.1089/ars.2005.7.1040
228. Ferri C, Greco F, Longombardo G et al (1991) Association between hepatitis C virus and mixed cryoglobulinemia. Clin Exp Rheumatol 9:621–624

229. Zignego AL, Ferri C, Giannini C et al (1997) Hepatitis C virus infection in mixed cryoglobulinemia and B-cell non-Hodgkin's lymphoma: evidence for a pathogenetic role. Arch Virol 142:545–555
230. Conca P, Tarantino G (2009) Hepatitis C virus lymphotropism and peculiar immunological phenotype: effects on natural history and antiviral therapy. World J Gastroenterol 15:2305–2308
231. Seronello S, Ito C, Wakita T, Choi J (2010) Ethanol enhances hepatitis C virus replication through lipid metabolism and elevated NADH/NAD+. J Biol Chem 285:845–854. doi:10.1074/jbc.M109.045740
232. Choi J, Lee KJ, Zheng Y et al (2004) Reactive oxygen species suppress hepatitis C virus RNA replication in human hepatoma cells. Hepatology 39:81–89. doi:10.1002/hep.20001
233. Choi J, Forman HJ, Ou J et al (2006) Redox modulation of the hepatitis C virus replication complex is calcium dependent. Free Radic Biol Med 41:1488–1498. doi:10.1016/j.freeradbiomed.2006.08.008
234. Huang H, Chen Y, Ye J (2007) Inhibition of hepatitis C virus replication by peroxidation of arachidonate and restoration by vitamin E. Proc Natl Acad Sci U S A 104:18666–18670. doi:10.1073/pnas.0708423104
235. Kuroki M, Ariumi Y, Ikeda M et al (2009) Arsenic trioxide inhibits hepatitis C virus RNA replication through modulation of the glutathione redox system and oxidative stress. J Virol 83:2338–2348. doi:10.1128/JVI. 01840-08
236. Rada B, Leto TL (2008) Oxidative innate immune defenses by Nox/Duox family NADPH oxidases. Contrib Microbiol 15:164–187
237. Seronello S, Sheikh MY, Choi J (2007) Redox regulation of hepatitis C in nonalcoholic and alcoholic liver. Free Radic Biol Med 43:869–882. doi:10.1016/j.freeradbiomed.2007.05.036
238. Sata M, Fukuizumi K, Uchimura Y et al (1996) Hepatitis C virus infection in patients with clinically diagnosed alcoholic liver diseases. J Viral Hepat 3:143–148
239. Pessione F, Degos F, Marcellin P et al (1998) Effect of alcohol consumption on serum hepatitis C virus RNA and histological lesions in chronic hepatitis C. Hepatology 27:1717–1722. doi:10.1002/hep.510270635
240. Romero-Gómez M, Grande L, Nogales MC et al (2001) Intrahepatic hepatitis C virus replication is increased in patients with regular alcohol consumption. Dig Liver Dis 33:698–702
241. Carrière M, Rosenberg AR, Conti F et al (2006) Low density lipoprotein receptor transcripts correlates with liver hepatitis C virus RNA in patients with alcohol consumption. J Viral Hepat 13:633–642. doi:10.1111/j.1365-2893.2006.00737.x
242. Cromie SL, Jenkins PJ, Bowden DS, Dudley FJ (1996) Chronic hepatitis C: effect of alcohol on hepatitic activity and viral titre. J Hepatol 25:821–826
243. Zhang T, Li Y, Lai J-P et al (2003) Alcohol potentiates hepatitis C virus replicon expression. Hepatology 38:57–65. doi:10.1053/jhep.2003.50295
244. Trujillo-Murillo K, Alvarez-Martínez O, Garza-Rodríguez L et al (2007) Additive effect of ethanol and HCV subgenomic replicon expression on COX-2 protein levels and activity. J Viral Hepat 14:608–617. doi:10.1111/j.1365-2893.2006.00837.x
245. Hou W, Bukong TN, Kodys K, Szabo G (2013) Alcohol facilitates HCV RNA replication Via up-regulation of miR-122 expression and inhibition of cyclin G1 in human hepatoma cells. Alcohol Clin Exp Res 37:599–608. doi:10.1111/acer.12005
246. Halfon P, Locarnini S (2011) Hepatitis C virus resistance to protease inhibitors. J Hepatol 55:192–206. doi:10.1016/j.jhep.2011.01.011
247. Fridell RA, Qiu D, Wang C et al (2010) Resistance analysis of the hepatitis C virus NS5A inhibitor BMS-790052 in an in vitro replicon system. Antimicrob Agents Chemother 54:3641–3650. doi:10.1128/AAC. 00556-10
248. Scheel TKH, Gottwein JM, Mikkelsen LS et al (2011) Recombinant HCV variants with NS5A from genotypes 1–7 have different sensitivities to an NS5A inhibitor but not interferon-α. Gastroenterology 140:1032–1042. doi:10.1053/j.gastro.2010.11.036, e6
249. Sun J-H, O'Boyle DR II, Zhang Y et al (2012) Impact of a baseline polymorphism on the emergence of resistance to the hepatitis C virus nonstructural protein 5a replication complex inhibitor, BMS-790052. Hepatology 55:1692–1699. doi:10.1002/hep.25581

250. Seronello S, Montanez J, Presleigh K et al (2011) Ethanol and reactive species increase basal sequence heterogeneity of hepatitis C virus and produce variants with reduced susceptibility to antivirals. PLoS One 6:e27436. doi:10.1371/journal.pone.0027436
251. Sherman KE, Rouster SD, Mendenhall C, Thee D (1999) Hepatitis cRNA quasispecies complexity in patients with alcoholic liver disease. Hepatology 30:265–270. doi:10.1002/hep.510300131
252. Kane M (1995) Global programme for control of hepatitis B infection. Vaccine 13(Suppl 1):S47–S49
253. Zoulim F, Locarnini S (2009) Hepatitis B virus resistance to nucleos(t)ide analogues. Gastroenterology 137:1593–1608. doi:10.1053/j.gastro.2009.08.063, e1–6
254. Zoulim F, Locarnini S (2012) Management of treatment failure in chronic hepatitis B. J Hepatol 56(Suppl 1):S112–S122. doi:10.1016/S0168-8278(12)60012-9
255. Ni Y-H, Chang M-H, Wang K-J et al (2004) Clinical relevance of hepatitis B virus genotype in children with chronic infection and hepatocellular carcinoma. Gastroenterology 127:1733–1738
256. Ezzikouri S, Ozawa M, Kohara M et al (2014) Recent insights into hepatitis B virus-host interactions. J Med Virol 86(6):925–932. doi:10.1002/jmv.23916
257. Glebe D, Urban S (2007) Viral and cellular determinants involved in hepadnaviral entry. World J Gastroenterol 13:22–38
258. Tong S, Li J (2014) Identification of NTCP as an HBV receptor: the beginning of the end or the end of the beginning? Gastroenterology 146(4):902–905. doi:10.1053/j.gastro.2014.02.024
259. Hao Z, Zheng L, Kluwe L, Huang W (2012) Ferritin light chain and squamous cell carcinoma antigen 1 are coreceptors for cellular attachment and entry of hepatitis B virus. Int J Nanomedicine 7:827–834. doi:10.2147/IJN.S27803
260. Macovei A, Petrareanu C, Lazar C et al (2013) Regulation of hepatitis B virus infection by Rab5, Rab7, and the endolysosomal compartment. J Virol 87:6415–6427. doi:10.1128/JVI.00393-13
261. Schmitz A, Schwarz A, Foss M et al (2010) Nucleoporin 153 arrests the nuclear import of hepatitis B virus capsids in the nuclear basket. PLoS Pathog 6:e1000741. doi:10.1371/journal.ppat.1000741
262. Yang W, Summers J (1999) Integration of hepadnavirus DNA in infected liver: evidence for a linear precursor. J Virol 73:9710–9717
263. Zoulim F (2004) Mechanism of viral persistence and resistance to nucleoside and nucleotide analogs in chronic hepatitis B virus infection. Antiviral Res 64:1–15. doi:10.1016/j.antiviral.2004.07.003
264. Nassal M (2008) Hepatitis B viruses: reverse transcription a different way. Virus Res 134:235–249. doi:10.1016/j.virusres.2007.12.024
265. Koike K, Moriya K, Iino S et al (1994) High-level expression of hepatitis B virus HBx gene and hepatocarcinogenesis in transgenic mice. Hepatology 19:810–819
266. Iavarone M, Trabut J-B, Delpuech O et al (2003) Characterisation of hepatitis B virus X protein mutants in tumour and non-tumour liver cells using laser capture microdissection. J Hepatol 39:253–261
267. León B, Taylor L, Vargas M et al (2005) HBx M130K and V131I (T-A) mutations in HBV genotype F during a follow-up study in chronic carriers. Virol J 2:60. doi:10.1186/1743-422X-2-60
268. Maguire HF, Hoeffler JP, Siddiqui A (1991) HBV X protein alters the DNA binding specificity of CREB and ATF-2 by protein-protein interactions. Science 252:842–844
269. Kekulé AS, Lauer U, Weiss L et al (1993) Hepatitis B virus transactivator HBx uses a tumour promoter signalling pathway. Nature 361:742–745. doi:10.1038/361742a0
270. Kremsdorf D, Soussan P, Paterlini-Brechot P, Brechot C (2006) Hepatitis B virus-related hepatocellular carcinoma: paradigms for viral-related human carcinogenesis. Oncogene 25:3823–3833. doi:10.1038/sj.onc.1209559
271. Li Y, Tang Z-Y, Hou J-X (2012) Hepatocellular carcinoma: insight from animal models. Nat Rev Gastroenterol Hepatol 9:32–43. doi:10.1038/nrgastro.2011.196
272. Kim CM, Koike K, Saito I et al (1991) HBx gene of hepatitis B virus induces liver cancer in transgenic mice. Nature 351:317–320. doi:10.1038/351317a0

273. Chisari FV, Klopchin K, Moriyama T et al (1989) Molecular pathogenesis of hepatocellular carcinoma in hepatitis B virus transgenic mice. Cell 59:1145–1156
274. Lakhtakia R, Kumar V, Reddi H et al (2003) Hepatocellular carcinoma in a hepatitis B "x" transgenic mouse model: a sequential pathological evaluation. J Gastroenterol Hepatol 18:80–91. doi:10.1046/j.1440-1746.2003.02902.x
275. Wang Y, Cui F, Lv Y et al (2004) HBsAg and HBx knocked into the p21 locus causes hepatocellular carcinoma in mice. Hepatology 39:318–324. doi:10.1002/hep.20076
276. Hagen TM, Huang S, Curnutte J et al (1994) Extensive oxidative DNA damage in hepatocytes of transgenic mice with chronic active hepatitis destined to develop hepatocellular carcinoma. Proc Natl Acad Sci U S A 91:12808–12812
277. Hsieh Y-H, Su I-J, Wang H-C et al (2004) Pre-S mutant surface antigens in chronic hepatitis B virus infection induce oxidative stress and DNA damage. Carcinogenesis 25:2023–2032. doi:10.1093/carcin/bgh207
278. Huang X (2003) Iron overload and its association with cancer risk in humans: evidence for iron as a carcinogenic metal. Mutat Res 533:153–171
279. Gu J-M, Lim SO, Oh SJ et al (2008) HBx modulates iron regulatory protein 1-mediated iron metabolism via reactive oxygen species. Virus Res 133:167–177. doi:10.1016/j.virusres.2007.12.014
280. Waris G, Huh K-W, Siddiqui A (2001) Mitochondrially associated hepatitis B virus X protein constitutively activates transcription factors STAT-3 and NF-kappa B via oxidative stress. Mol Cell Biol 21:7721–7730. doi:10.1128/MCB. 21.22.7721-7730.2001
281. Bouchard MJ, Schneider RJ (2004) The enigmatic X gene of hepatitis B virus. J Virol 78:12725–12734. doi:10.1128/JVI. 78.23.12725-12734.2004
282. Gollob JA, Wilhelm S, Carter C, Kelley SL (2006) Role of Raf kinase in cancer: therapeutic potential of targeting the Raf/MEK/ERK signal transduction pathway. Semin Oncol 33:392–406. doi:10.1053/j.seminoncol.2006.04.002
283. Chen J, Siddiqui A (2007) Hepatitis B virus X protein stimulates the mitochondrial translocation of Raf-1 via oxidative stress. J Virol 81:6757–6760. doi:10.1128/JVI. 00172-07
284. Zheng YW, Yen TS (1994) Negative regulation of hepatitis B virus gene expression and replication by oxidative stress. J Biol Chem 269:8857–8862
285. Chang K-C, Wu Y-Y, Hung C-H et al (2013) Clinical-guide risk prediction of hepatocellular carcinoma development in chronic hepatitis C patients after interferon-based therapy. Br J Cancer 109:2481–2488. doi:10.1038/bjc.2013.564
286. George SL, Bacon BR, Brunt EM et al (2009) Clinical, virologic, histologic, and biochemical outcomes after successful HCV therapy: A 5-year follow-up of 150 patients. Hepatology 49:729–738. doi:10.1002/hep.22694
287. Nojiri K, Sugimoto K, Shiraki K (2010) Development of hepatocellular carcinoma in patients with chronic hepatitis C more than 10 years after sustained virological response to interferon therapy. Oncol Lett 1(3):427–430. doi:10.3892/ol_00000075

Chapter 12
Oxidative Stress in Nonalcoholic Fatty Liver Disease

Montserrat Marí, Albert Morales, Anna Colell, Carmen García-Ruiz, and José C. Fernandez-Checa

12.1 Nonalcoholic Fatty Liver Disease

Nonalcoholic fatty liver disease (NAFLD) represents a spectrum of disease ranging from simple fatty liver (hepatic steatosis), with benign prognosis, to a potentially progressive form, nonalcoholic steatohepatitis (NASH), which may lead to fibrosis and even irreversible cirrhosis. While this spectrum of liver pathology resembles

M. Marí (✉)
Department of Cell Death and Proliferation, Institute of Biomedical Research of Barcelona - Spanish National Research Council, IDIBAPS, Liver Unit-Hospital Clinic, Networked Biomedical Research Center for Hepatic and Digestive Disease, Barcelona 08036, Spain

Centre Esther Koplowitz, C/Rosselló 149-153, Barcelona 08036, Spain
e-mail: monmari@clinic.ub.es

A. Morales
Department of Cell Death and Proliferation, Institute of Biomedical Research of Barcelona - Spanish National Research Council and Liver Unit-Hospital Clinic, Networked Biomedical Research Center for Hepatic and Digestive Diseases, Barcelona 08036, Spain

A. Colell
Department of Cell Death and Proliferation, Institute of Biomedical Research of Barcelona - Spanish National Research Council, The August Pi i Sunyer Biomedical Research Institute, Liver Unit-Hospital Clínic, Networked Biomedical Research Center for Hepatic and Digestive Diseases, Barcelona, Spain

C. García-Ruiz
Department of Cell Death and Proliferation, Institute of Biomedical Research of Barcelona (IIBB), Consejo Superior Investigaciones Científicas (CSIC) and Liver Unit-Hospital Clinic and Networked Biomedical Research Center for Hepatic and Digestive Diseases, Barcelona 08036, Spain

Southern California Research Center for ALPD and Cirrhosis, Keck School of Medicine of the University of Southern California, Los Angeles, CA, USA

E. Albano, M. Parola (eds.), *Studies on Hepatic Disorders*,
Oxidative Stress in Applied Basic Research and Clinical Practice,
DOI 10.1007/978-3-319-15539-5_12

alcohol-induced fatty liver damage, it occurs in individuals who do not abuse alcohol. Accumulation of triglycerides (TG) within hepatocytes (that is, hepatic steatosis) is the hallmark of NAFLD. This accumulation is strongly associated with obesity and the metabolic syndrome [93, 136]. In fact, all features of the metabolic syndrome, including obesity, type-2 diabetes, arterial hypertension, and hyperlipidemia (in the form of elevated TG levels) are associated with NAFLD/NASH. Similarly to these other conditions, NAFLD is becoming epidemic. Not only the presence of excess weight and obesity, but also the location of fat storage plays a role in NAFLD pathogenesis. Visceral fat stores increase the risk for NAFLD in both obese and non-obese individuals [154]. NAFLD is divided into two major subtypes: nonalcoholic fatty liver (NAFL, also termed simple steatosis), the nonprogressive form of NAFLD that rarely develops into cirrhosis, and NASH, the progressive form of NAFLD that can lead to cirrhosis, hepatocellular carcinoma (HCC), and liver-related mortality [3, 89, 100]. NASH, characterized not only by the presence of hepatocellular lipid accumulation (steatosis), but also of ballooning degeneration, along with lobular inflammation and varying degrees of fibrosis [166], is a potentially serious condition, and ~10–25 % of patients with NASH might progress to cirrhosis [3, 100]. Cirrhosis can result in liver failure, portal hypertension, and HCC and patients with cirrhosis are at a high risk for developing cardiovascular disease [136]. Moreover, although the majority of HCC cases occur in the setting of cirrhosis, NAFLD can also progress to hepatocellular carcinoma even in the absence of apparent cirrhosis [40].

However, the general prevalence of NASH is difficult to determine, as invasive liver biopsy remains the diagnostic gold-standard approach [166]. NAFLD is usually clinically silent, and its impact has most likely been underestimated. Symptoms, if present, are minimal and nonspecific, such as fatigue and right upper quadrant discomfort. Most findings on physical examination are also normal. Most patients seek care because of an incidental finding of elevated aminotransferase levels or radiographic studies suggesting the liver is fatty [76, 123].

A review of the epidemiology of NAFLD found that in studies of patients who had undergone bariatric surgery (16 individual studies, including 2,956 patients), 76 % had steatosis (range reported in the studies 33–99 %), 37 % had NASH (9.8–72.5 %), 23 % had fibrosis (7.3–49 %), and 5.8 % had cirrhosis (1.6–10 %) [81, 166]. The prevalence of NAFLD in the general population suggests that up to 30 % of adults in the United States may have NAFLD (defined as 5 % or more of liver fat) [81]. However, estimates from several other countries are quite variable, but depict a highly prevalent condition with ~20 % of the adult population. The

J.C. Fernandez-Checa
Department of Cell Death and Proliferation, Institute of Biomedical Research of Barcelona (IIBB), Spanish National Research Council and Liver Unit-Hospital Clinic and CIBEREHD, Barcelona, Spain

Southern California Research Center for ALPD and Cirrhosis, Keck School of Medicine of the University of Southern California, Los Angeles, CA, USA incidence and prevalence of NAFLD is rising globally owing to increasing rates of obesity and diabetes [37, 89, 160]. In this sense, it has been estimated that NAFLD prevalence increases to 57 % in obese patients and to 70 % in diabetic subjects [21, 49]. Therefore, an increasing number of features of the metabolic syndrome, particularly type-2 diabetes, and a family history of diabetes are risk factors for the presence of steatohepatitis in patients with NAFLD [88, 162].

Thus, NAFLD has become the most prevalent cause of liver disease in Western countries and, over the next decade or so, is set to replace viral hepatitis as the primary cause of end-stage liver disease and it is projected to become the leading cause of liver transplantation in the USA by 2020 [125, 166].

The clinical importance of NAFLD and the present lack of effective treatment to stop disease progression have generated great interest and investigation into the basic mechanisms involved in the development and progression of the disease. The current, and most accepted, concept outlining the pathogenesis of NAFLD has evolved from an initial "two-hit" hypothesis to a "multiple-hit" hypothesis [150]. In the original "two-hit" hypothesis, the "first-hit," hepatic TG accumulation, or steatosis, increases susceptibility of the liver to injury mediated by "second-hits." The number of potential second-hits has greatly increased over the last decades being the most relevant mitochondrial dysfunction and oxidative stress, gut-derived endotoxin, inflammatory cytokines/adipokines, which in turn lead to steatohepatitis and/or fibrosis [28, 29, 34]. The presence of oxidative stress limits the ability of mature hepatocytes to proliferate, resulting in reduced endogenous liver repair. However, as there is an increasing recognition on the simultaneous roles that these second-hits have in directly promoting liver injury, the multiple-hit hypothesis seems much more appropriate since it pictures a complex scenario where many parallel hits derived not only from the liver, but also by inflammatory mediators, especially those deriving from adipose tissue and the gut (see Fig. 12.1), act in promoting liver inflammation, fibrosis, and eventually tumorigenesis [150].

12.2 Metabolic Pathways Leading to Hepatic Steatosis

As commented above, excessive accumulation of TG in hepatocytes is the hallmark of NAFLD, both histologically and metabolically, which is strongly associated with insulin resistance [1, 2]. A fatty liver is the result of the accumulation of various lipids and different sources of fatty acids contribute to its development. In particular, hepatic accumulation of TG arises from an imbalance between lipid acquisition (i.e., fatty acid uptake and de novo lipogenesis) and removal (i.e., mitochondrial fatty acid oxidation and export as a component of very low-density lipoprotein particles. Therefore, hepatic fat accumulation [34] can occur as a result of (1) Increased free fatty acids uptake, (2) decreased or saturated free fatty oxidation, (3) increased de novo hepatic lipogenesis (DNL), and (4) alteration in very low-density lipoprotein–triglyceride secretion, as discussed below.

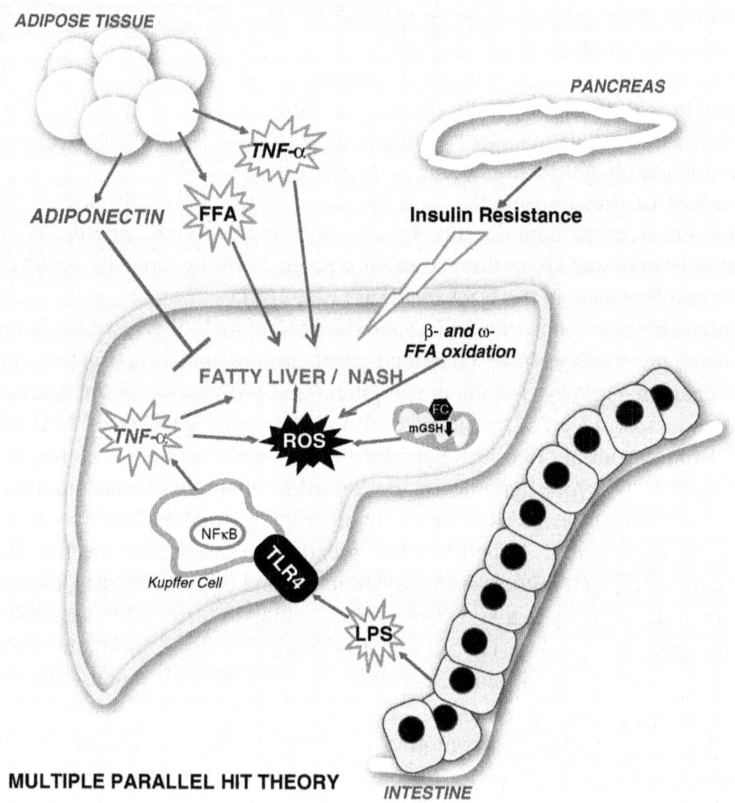

Fig. 12.1 Multiple parallel hit theory of NAFLD progression. Several hits can contribute to the progression from fatty liver to NASH. Among them it is worth highlighting the contribution of gut-derived endotoxin (LPS) acting through TLR4, especially on Kupffer cells; insulin resistance; generation of inflammatory cytokines and adipokines by the adipose tissue; lipotoxicity induced not only by free fatty acids (FFA) but also by free cholesterol (FC) by acting on the mitochondrial glutathione (mGSH) transport therefore resulting in decreased mGSH and enhanced ROS generation

12.2.1 Increased Fatty Acid Uptake

Among these potential components, to establish the relative contribution of lipid accumulation in patients with NAFLD, Donnelly et al. used a multiple-stable-isotope method, demonstrating that approximately 60 % of liver TG content derived from FFA influx from adipose tissue, 26 % from DNL, and 15 % from diet [33, 34]. This contrasts with healthy individuals in whom DNL contributes to less than 5 % of hepatic TG formation [64, 116]. The increased fatty acid uptake can be due to increased lipolysis from both visceral/subcutaneous adipose tissue and/or increased intake of dietary fat.

Given that insulin has a potent action to suppress adipose tissue lipolysis; fatty acid uptake in the liver is in many instances indirectly regulated by insulin. Insulin allows free fatty esterification and TG fat storage in adipose tissues. In this context, several studies have highlighted that insulin resistance is a characteristic feature of

NAFLD even when subjects are not obese [15, 41, 49, 133, 170]. When insulin resistance develops, the adipose tissue becomes resistant to the antilipolytic effect of insulin and the release of fatty acids is increased. These free fatty acids (FFA) are inappropriately shifted to nonadipose tissues, including the liver. Insulin resistance is accompanied by increased insulin levels that, in the presence of increased lipolysis and/or increased fat intake, promote hepatic triglyceride synthesis [49, 153]. In addition, insulin resistance and visceral obesity are commonly associated to the decreased levels of the "protective adipokine," adiponectin, an adipose tissue-specific hormone. Depletion of this anti-inflammatory and antidiabetic adipokines has been linked to the progression of NASH [155]. Adiponectin inhibits liver gluconeogenesis and suppresses lipogenesis. Thus, decreased adiponectin hinders fatty acid oxidation and increases fat accumulation in the liver [58, 76].

In addition, another important source of FFA under postprandial conditions is due to increased spillover from chylomicrons [104]. FFA are taken up by organs saturating their oxidative capacity [15] and accumulated as ectopic fat, mainly not only as intramyocellular and hepatic lipids [68, 90] but also as cardiac and pancreatic fat. It has been hypothesized that ectopic fat could be a defense mechanism against lipotoxicity [26, 109] and that subjects with NAFLD develop NASH and cirrhosis only in consequence of a second hit due to increased inflammation and reactive oxygen species [27, 49].

12.2.2 Decreased or Saturated FFA Oxidation

The steady state balance of hepatic triglycerides is also controlled by the consumption of catalysis of FFA. Oxidation of fatty acids occurs in three subcellular organelles, with β-oxidation confined to mitochondria and peroxisomes and CYP2E1/CYP4A-catalyzed ω-oxidation occurring in the endoplasmic reticulum. Some of the key enzymes of these three fatty acid oxidation systems in liver are regulated by PPAR-α [127]. However, the central organelle for oxidation of fatty acids is mitochondria [36], which is critical for producing ATP and ketone bodies.

Regarding NAFLD, there has been the controversy regarding whether mitochondrial β-oxidation was decreased or enhanced to saturation, with either outcome reinforcing the potential accumulation of triglycerides in the liver. Studies with human NAFLD patients have yielded also mixed results with respect to alterations in rates of fatty acid oxidation [75, 132]. Impaired ATP production has been described in patients with NAFLD and insulin resistance [18]. Others have reported evidence for increased rates of fatty acid oxidation in NAFLD [133], thus suggesting that increased mitochondrial activity could promote oxidative stress within the liver and contribute to the development of NASH [119]. An interesting study evaluated the expression of several fatty acid metabolism-related genes in larger groups of NAFLD ($n = 26$) and normal liver ($n = 10$) samples shedding some light on the role of mitochondrial β-oxidation in NAFLD [78]. In this study, the authors found that while carnitine-palmitoyltransferase 1 (CPT1) (which localizes to the outer mitochondrial membrane and catalyzes the formation of acyl-carnitine from acyl-CoA and carnitine that are subsequently shuttled to the inner mitochondrial membrane) was decreased probably due to an increase in malonyl-CoA, all the other components of the mitochondrial β-oxidation were upregulated. When ß-oxidation reaches maximal levels, a lack of unesterified CoA could inhibit CPT1 and thereby prevent further entry of acyl groups to the mitochondria, thus serving as an intramitochondrial control on β-oxidation. β-oxidation in mitochondria was at maximal levels, thereby resulting in decreased expression of CTP1. In addition, in order to complement the function of mitochondria, peroxisomal (β-oxidation) and microsomal (Ω-oxidation) oxidation were also enhanced to decrease fatty acid accumulation [78].

12.2.3 Increased De Novo Hepatic Lipogenesis

The nonesterified fatty acids that are incorporated into TG within the liver may be derived from the plasma or be newly synthesized from glucose in the so-called *de novo lipogenesis* (DNL) process. DNL is an integrated metabolic pathway that comprises of glycolysis (conversion of glucose to acetyl-CoA), biosynthesis of saturated fatty acid followed by desaturation, and the formation of TG. Key rate-limiting enzymes in the process include glucokinase and liver-type pyruvate kinase in the glycolysis, acetyl-CoA carboxylase (ACC), and fatty acid synthase (FAS) in the fatty acid synthesis, long-chain fatty acid elongase 6 (ELOVL6) and stearoyl-CoA desaturase (SCD) in the formation of monounsaturated fatty acids, and glycerol-3-phosphate acyltransferase (GPAT), lipins, and acyl-CoA: diacylglycerol acyltransferase (DGAT) in the formation of TG [80, 151].

As highlighted before, the study by Donnelly et al. [33], estimated that DNL was responsible for almost 30 % of hepatic TG in NAFLD patients, compared to the modest <5 % under nonpathological conditions. DNL is controlled primarily at the transcriptional level. Two major transcription factors for lipogenesis, sterol regulatory element-binding protein 1c (SREBP-1c) and carbohydrate response element-binding protein (ChREBP), are involved in the transcriptional activation of genes encoding aforementioned rate-limiting enzymes in lipogenesis, and have been associated with increased DNL in NAFLD [121]. Glucose and insulin promote DNL by activation of these two transcription factors. The expression of SREBP-1c is also induced by saturated fatty acids [23]. In obese, insulin-resistant ob/ob mice, both SREBP-1c and ChREBP are highly abundant in the liver, and reduction of either factor was shown to be beneficial in relieving hepatic steatosis in mice, underscoring the importance of these lipogenic transcription factors in de novo lipogenesis and TG accumulation in the liver [31, 80, 140]. TG are the main lipids stored in the liver of patients with NAFLD. Although large epidemiological studies suggest TG-mediated pathways might negatively affect disease [152], studies have shown that cellular TG accumulation per se is not initially toxic [86, 121, 167]. In fact, the accumulation of excess fatty acids into TG pools may divert fatty acids from pathways that could be cytotoxic, such as the generation of ROS [79] or ceramides, and this leads to subsequent alteration of mitochondrial function [52, 118]. Moreover, FFA and cholesterol, especially when accumulated in mitochondria, are considered the “aggressive” lipids leading to tumor necrosis factor-alpha (TNF-α)-mediated liver damage and reactive oxygen species (ROS) formation.

Of note, the term "lipotoxicity" and the concept of "Good Fat /Bad Fat" was introduced almost 10 years ago to define the role that excess FFA or other lipid metabolites play in inducing cellular toxicity [46, 87, 92, 94, 101], as we will discuss later.

12.2.4 *Alteration in Very Low-Density Lipoprotein-Triglyceride Secretion*

TG can also be exported from the liver in VLDL particles, that consist of hydrophobic core lipids containing TG and cholesterol esters, which are formed by the incorporation of TG into apolipoprotein B (apoB) by microsomal transfer protein (MTP) [67, 80]. The VLDL secretion rate appears to depend not only on the availability of hepatic TG, but also on the overall capacity for VLDL assembly [75]. Upon insulin resistance, perturbation of this process results in hypertriglyceridemia via increased TG secretion. However, the rate of TG secretion cannot keep up with the increased rate of TG synthesis in this condition, resulting in hepatic steatosis despite increased VLDL secretion. Similar phenotype is observed in NAFLD patients, which exhibit both hypertriglyceridemia and hepatic steatosis [25]. In these subjects, lipid availability for VLDL assembly is increased, and this is combined with the failure of insulin to suppress VLDL production, thus resulting in hypertriglyceridemia. During NAFLD progression, prolonged exposure of the liver to FFA would promote ER stress and other oxidative stress in the liver, leading to the degradation of apoB, decrease in the VLDL secretion, and worsening of hepatic steatosis [80, 112].

12.3 Multiple-Hits Leading to NAFLD Progression: Transition from Steatosis to Steatohepatitis

As stressed before, fatty liver is considered a benign condition. However, it is predisposed to forms of injury that involve oxidant stress. Steatosis could therefore provide the setting (or "first-hit") for NASH, but a "second hit" or "multiple-hits" may be needed to cause cellular injury or recruit inflammation [29, 150]. In addition to the relevant contribution of mitochondrial dysfunction and oxidative stress, other potential hits responsible for NAFLD progression are (1) gut-derived endotoxin and Toll-like receptors, (2) inflammatory cytokines/adipokines, and (3) lipotoxicity (Fig. 12.1).

We will briefly discuss the participation of "hits" i–iii to next focus on mitochondrial dysfunction and oxidative stress, which is the main topic of this chapter.

12.3.1 *Gut-Derived Endotoxin and Toll-Like Receptors*

Recent evidence suggests that overgrowth of intestinal bacteria and increased intestinal permeability are associated with the development of NASH [14, 102, 103, 130]. It has been suggested that in healthy individuals, maintenance of homeostasis between the gut microbiome and the host plays an important role in health while disruption of such homeostasis occurs in disease [147]. In chronic liver disease, including NASH, intestinal permeability is increased due to bacterial overgrowth or altered composition of bacterial microflora. The liver, receiving approximately 70 % of its blood supply from the intestinal venous outflow, represents the first line of defense against gut-derived antigens and is equipped with a broad array of immune cells (i.e., macrophages, lymphocytes, natural killer cells, and dendritic cells) to accomplish this function. Gut-derived microbial components represent danger signals for the host in the liver, induce inflammatory cascade activation in immune cells and modulate functions and responses of liver parenchymal cells [143]. Indeed, plasma levels of lipopolysaccharide (LPS) are elevated in patients with chronic liver diseases, including NASH [43]. The host immune system recognizes pathogen-associated molecular patterns (PAMPS) as danger signals via pattern recognition receptors including families of the Toll-like receptors (TLRs). These findings suggest that hepatic immune cells might be exposed to high levels of TLR ligands derived from gut bacterial products, which might trigger liver injury. Ligand engagement of TLR induces downstream signaling resulting in NF-κB activation, phosphorylation, nuclear translocation, and activation of proinflammatory cytokines that contribute to hepatic inflammation [147]. In fact, several reports demonstrate the importance of TLR4 and intestine-derived LPS in the animal model of NASH [20, 129]. LPS, a component of the cell wall of Gram-negative bacteria, is the most studied PAMP in relation to gut-derived pathogenic signals. LPS is sensed by TLR4. Remarkably, a key role of TLR4 in Kupffer cells in mediating the progression of simple steatosis to NASH, by also inducing ROS-dependent activation of X-box binding protein-1 (XBP-1) has been suggested [169]. Gut microbiome manipulation through antibiotics, prebiotics, and probiotics yields encouraging results for the treatment of obesity, diabetes, and NAFLD in animal models; however data in humans are currently scarce [108].

Of interest, nonpathogenic substances, such as FFA, and denatured host DNA may activate TLR2, TLR4, and TLR9 [139, 157]. Saturated FFA such as palmitic and stearic acids, potential TLR4 ligands, are abundant in dietary fat, and high levels of circulating FFA have been observed in patients with NAFLD [30]

12.3.2 Inflammatory Cytokine/Adipokines

Cytokines and adipokines (i.e., mediators mainly derived from adipose tissue) play a major role in the orchestration of inflammatory processes throughout the body [149]. In addition, many of these mediators are able to regulate very diverse functions including inflammatory, immune, and metabolic processes. The proinflammatory cytokines tumor necrosis factor-alpha (TNF-α) and interleukin-6 (IL-6) are critically involved in the pathophysiology of various aspects of human NAFLD. Kupffer cells are the main producers of TNF-α in the liver, particularly when they are activated by LPS [145]. Regarding the participation of TNF-α on the metabolic syndrome, characteristic of most NAFLD subjects, TNF-α interferes with insulin signaling, thereby favoring steatosis, and neutralization of TNF-α activity improves insulin resistance and fatty liver disease in animal beings [62, 65]. The lipogenic action of TNF-α is in part accomplished by the upregulation of SREBP-1c [39]. Of importance, a recent report described an essential role for TNF-receptor 1 (TNFR1) as TNFR1 knockout mice are resistant to high-fat diet-induced steatosis and development of hepatocellular carcinoma [115]. Furthermore, TNF-α plays a critical role not only in the production of inflammatory mediators, apoptosis, and necrosis of hepatocytes but also in cholestasis and fibrosis [135].

IL-6 is derived from many cells throughout the body including adipocytes. Serum levels of this cytokine correlate remarkably well with the presence of insulin resistance (IR), and adipose tissue-derived IL-6 has been shown to regulate hepatic IR via upregulation of suppressor of cytokine signaling 3 (SOCS3) [149]. Initial reports supported a hepatoprotective action of IL-6 in steatotic livers, by suppressing oxidative stress and preventing mitochondrial dysfunction [19, 38]. Moreover, IL-6 has a hepatoprotective effect in ischemic preconditioning models [148] and is important for survival after partial hepatectomy in mice [12]. However, this seems to be a paradoxical effect of short- and long-term IL-6 exposure, as the latter may sensitize the liver to injury and apoptotic cell death [71]. It remains to be elucidated whether elevated IL-6 levels in chronic liver injury contribute to inflammation or represent an anti-inflammatory response.

Adiponectin is exclusively secreted by adipocytes and is considered as an anti-inflammatory adipokines [114]. Adiponectin is a potent TNF-α-neutralizing adipokine, and in vitro and experimental animal studies have proven the importance of this mediator in counteracting inflammation and insulin resistance. In addition, the conditions most associated with the development of NAFLD, namely, obesity [9], IR [161], type-2 diabetes [63], and dyslipidemia [99], all have reduced adiponectin levels. Also, the expression of adiponectin and its receptors, AdipoR1 and AdipoR2, is low in the liver of patients with NAFLD [74]. Serum levels of adiponectin correlate negatively with hepatic steatosis [6]. An imbalance between serum TNF-α and adiponectin is considered to be important for the development of NASH [65]. In the liver, adiponectin prevents lipid accumulation by increasing -oxidation of FFA and/or by decreasing DNL within hepatocytes [7, 171]. Part of this action is accomplished through the downregulation of SREBP-1c, a key regulator of fatty acid synthesis [141]. Anti-inflammatory effects of adiponectin are mediated via suppression of TNF-α synthesis as well as induction of anti-inflammatory cytokines such as IL-10 or IL-1 receptor antagonist. Therefore, the balance between pro- and anti-inflammatory acting cytokines/adipokines appears to play a key role in hepatic and systemic insulin action, and they are supposed to have important functions in the development of NAFLD [149].

12.3.3 Lipotoxicity

Although the majority of hepatic lipids in NAFLD are stored in the form of TG, several other lipid metabolites such as different FFA, diacylglycerol, free cholesterol (FC), cholesterol ester (CE), ceramide, and phospholipids also accumulate [5, 24, 122].

Although the amount of fat in the liver is often considered as an index of disease severity, it is unlikely that the neutral TG are lipotoxic. Rather, the magnitude of neutral TG in the liver likely serves as a biomarker for the flux of FFA delivered to the liver [69]. A growing body of evidence, mainly from experimental studies, suggests that lipid compartmentalization in hepatocytes and in particular the type or "quality" as opposed to the "quantity" of lipids accumulating may play a central role in the risk for progressive disease [101]. Herein, lipids such as FFA are the noxious agents, and TG actually serve as "sink" or protective pathway in lipid metabolism. There are multiple definitions for lipotoxicity; a frequently used one is adverse effects of fatty acid accumulation in nonadipose tissues [101]. There is an increased body of evidence indicating that saturated FFA are more hepatotoxic than unsaturated FFA [91, 158, 159]. This difference in toxicity between saturated and unsaturated FFA supports a model of cellular lipid metabolism in which unsaturated FFA serve a protective function against lipotoxicity though promotion of TG accumulation [85, 86]. Of note, the typical dietary pattern of western populations leads to high intake of saturated FFA derived from animal sources, which could explain its high prevalence on NAFLD progression. Moreover, it has been proposed that the accumulation of FFA, especially saturated fatty acids, in the hepatocytes can promote apoptosis by diverse pathways. These may include ROS-induced stress that affects the mitochondrial membranes and lysosomes. Lipid peroxidation increases the levels of reactive oxygen species, which may be partially responsible for hepatocyte dysfunction [70], as we will discuss in detail.

Free cholesterol (FC) has also emerged as a toxic lipid derivative in the liver, whereas cholesterol esters are either stored in lipid droplets in macrophages or incorporated into apoB-containing lipoproteins such as VLDL in hepatocytes. In an experimental study designed to evaluate the differential role of FC loading vs. TG loading, using nutritional and genetic models of hepatic steatosis, we showed that FC, but not FFA or TG, sensitizes to TNF- and Fas-induced progression from fatty liver to NASH. FC distribution in endoplasmic reticulum (ER) and plasma membrane did not cause ER stress or alter TNF signaling. Rather, mitochondrial FC loading accounted for the hepatocellular sensitivity to TNF due to mitochondrial glutathione (mGSH) depletion (Fig. 12.2).

Selective mGSH depletion in primary hepatocytes recapitulated the susceptibility to TNF and Fas seen in FC-loaded hepatocytes; and its repletion with GSH-ethyl ester rescued FC-loaded livers from TNF-mediated steatohepatitis [94]. In addition, the treatment of obese mice with atorvastatin prevented FC accumulation in mitochondria and the subsequent mGSH depletion, thus abolishing the susceptibility to cytokine-induced liver damage [94]. A progressive increase in hepatic FC from controls with normal histology to subjects with simple steatosis and NASH has been reported, together with an increase of serum total cholesterol; in contrast to FC, liver cholesterol ester contents in subjects and controls were comparable [122]. We have also examined the correlation between FC accumulation in patients with NAFLD and the expression of enzymes that regulate cholesterol homeostasis finding that FC increases in human NASH and correlates with sterol regulatory element-binding protein-2 (SREBP-2) and steroidogenic acute regulatory protein (StAR) expression. SREBP-2, a transcription factor that plays an important role in cholesterol synthesis

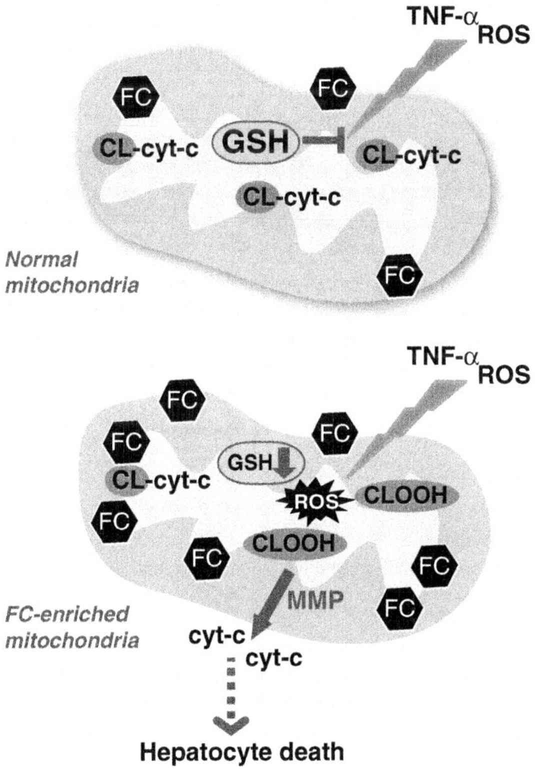

Fig. 12.2 Free cholesterol (FC) deposition in mitochondria sensitizes to other insults, such as TNF-α or ROS, by altering mitochondrial GSH transport resulting in mGSH depletion. mGSH can pose a critical threat to the cell by sensitizing mitochondrial components and lipids such as cardiolipin (CL) to oxidative modifications by ROS that could compromise vital mitochondrial function, and easily engage mitochondria in cell death pathways by inducing mitochondrial permeability transition (MMP)

through hydroxymethylglutaryl-CoA (HMG-CoA) reductase, and StAR, a mitochondrial-cholesterol transporting polypeptide, were overexpressed in patients with NASH compared with those with simple steatosis [16]. In fact, statins have ameliorated surrogate markers of steatosis in several randomized controlled trials, but their impact on liver histology is unknown. Nonetheless, statins have been found to be the only class of lipid-lowering drugs that reduce cardiovascular risk in NAFLD patients [107]. These observations on the role of statins, together with the observation that atherosclerotic cardiovascular disease in adults is often associated to NAFLD, and that animal models of atherosclerosis may also manifest NAFLD [164, 165]; endorse our findings on the role of mitochondrial FC in disease progression from steatosis to steatohepatitis.

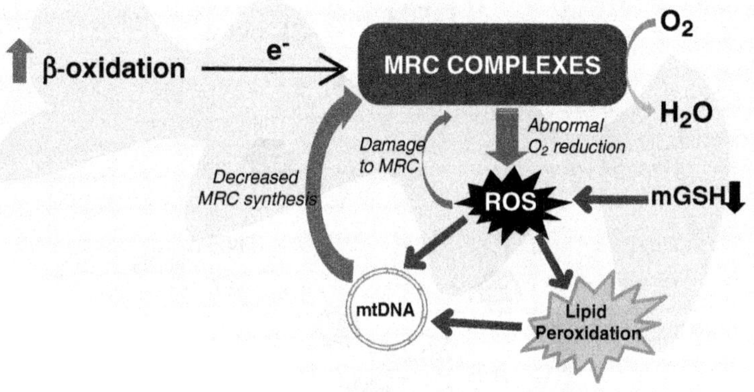

Fig. 12.3 The vicious cycle of mitochondrial ROS generation and mitochondrial respiratory chain (MRC) dysfunction

Another lipid that has been linked to NAFLD progression is ceramide, a dynamic sphingolipid that contributes to stress and death ligand-induced hepatocellular death due to its rapid generation from sphingomyelin by acid sphingomyelinase after the engagement of death ligands such as TNF and Fas with their receptors [97]. In addition, de novo synthesis of ceramide requires sphingosine and a saturated fatty acid moiety, usually palmitoyl CoA. Thus, since long-chain saturated fatty acid availability is the rate-limiting step in ceramide synthesis; in nutritional obesity with associated elevations of palmitic acid and stearic acid, excess ceramide synthesis is possible. In fact, it has been described that ceramide plays a role in insulin resistance by inhibiting insulin-induced glucose uptake, GLUT4 translocation, and glycogen synthesis [146]. These results suggest that ceramide is a candidate metabolite linking lipid oversupply to the development of metabolic syndrome and NAFLD that deserves further attention.

12.4 Oxidative Stress and Mitochondrial Dysfunction in NAFLD

Reactive oxygen species (ROS) including superoxide ($O_2^{\bullet-}$), hydrogen peroxide (H_2O_2), and hydroxyl radical ($OH^{\bullet}$), and reactive nitrogen species (RNS), including nitric oxide (NO) and peroxynitrite ($ONOO^-$), are generated as by-products of biochemical reactions within cells and, hence, at low concentration, considered as inherent intermediates of many physiologic processes. However, when produced in large amounts or in an uncontrolled fashion, free radicals inflict tissue damage and are implicated in many pathologic processes [96].

As originally defined in 1985, oxidative stress is an imbalance between oxidants and antioxidants, in favor of the former [142]. This implies that either the overgeneration of free radicals and ROS and/or the limitation in the function of antioxidants results in the net accumulation of ROS, that exceeds the capacity of normal antioxidant systems, and exerts deleterious effects in cell functions, ultimately contributing to aging and major disease processes, including cardiovascular disorders, pulmonary diseases, diabetes, neurodegeneration, and liver diseases [96, 142].

In the latest years, altered oxidative metabolism of TG at both hepatic and extrahepatic level, an imbalance in pro/antioxidant systems, and an increased oxidative stress are increasingly recognized as central for the development of hepatic steatosis, necroinflammation, and fibrosis that characterize NASH. Many of these alterations can also be found in NAFLD-associated metabolic disorders, including obesity and diabetes, and in the inflammatory process at the basis of atherosclerosis, thus suggesting that common molecular mechanisms act at both hepatic and extrahepatic levels to produce liver and cardiometabolic disease [29, 50].

ROS are toxic to cells because they can react with many cellular macromolecules, denaturing proteins, or inactivating enzymes and causing DNA damage. Moreover, ROS can attack polyunsaturated fatty acids (PUFAs) and initiate lipid peroxidation within the cell. Lipid peroxidation can lead to destruction of biological membranes and to production of aldehydic products such as malondialdehyde (MDA) or 4-hydroxynonenal (4-HNE), that have longer half-life than ROS, and are highly reactive and generate adducts or cross-linking on proteins, which progressively lead to impaired protein function, causing alterations in signal transduction that affect cell dysfunction, inflammatory response, and cell death [96, 120].

In fact, by analyzing the circulating profile of lipid oxidation products in the blood of NAFLD patients it has been demonstrated that specific oxidized fatty acid products, mainly derived from linoleic and arachidonic acids, are markedly increased and are largely the result of free radical-mediated processes [45]. Moreover, using a learning cohort and subsequent validation cohort, and based on these oxidized fatty acids, an OxNASH noninvasive score was developed and a strong positive correlation was revealed between systemic levels of specific oxidation products and liver histology that included inflammation, degree of steatosis, and stage of fibrosis [45]. These promising results have been further validated in an ample cohort of patients thus highlighting the novel utility of the OxNASH score for the noninvasive diagnosis of NASH and demonstrating that OxNASH correlates with individual histologic features of disease severity in NASH, in particular with the degree of inflammatory activity in the liver [4]. These findings provide more evidence on the role that oxidative stress plays in NAFLD progression.

In the fatty liver, different sources of ROS and RNS contribute to the ensuing oxidative stress found during NAFLD progression and will be discussed in detail.

12.4.1 Mitochondria

Several lines of evidence suggest that mitochondrial dysfunction is critical on the progression from simple steatosis to steatohepatitis in NAFLD. Mitochondrial dysfunction is characterized by mitochondrial membrane permeabilization (MMP). MMP causes the release of multiple proteins, especially cytochrome c, from the mitochondrial intermembrane space into the cytosol. This leads to caspase activation in the cytosol, disruption of the mitochondrial respiratory chain (MRC), loss of mitochondrial transmembrane potential, ROS production, and loss of mitochondrial structural integrity. In fact, hepatic mitochondria from patients with NASH show both morphological and functional changes. They present mega-mitochondria that have ultrastructural lesions, loss of cristae, and with paracrystalline inclusion bodies that suggest either an adaptive process or mitochondrial injury [17], and have a decrease in ATP resynthesis after a fructose challenge, leading to a transient ATP depletion [18].

Mitochondria are the most important energy suppliers of the cell. Mitochondrial FFA β-oxidation and other oxidative reactions (e.g., Krebs cycle) produce NADH and $FADH_2$, which are then reoxidized by the MRC, thus regenerating nicotinamide adenine dinucleotide (NAD^+) and flavin adenine dinucleotide (FAD) required for other cycles of oxidation. NADH and $FADH_2$ oxidation is coupled to ATP synthesis through the process. Most of the electrons provided to the MRC migrate along this chain, to finally reach cytochrome c oxidase (COX, or complex IV), where they safely combine with molecular oxygen and protons to form water. A fraction, however, of these electrons leaks from complexes I and III to form superoxide [106]. This radical can be then dismutated by manganese superoxide dismutase (MnSOD) into hydrogen peroxide (H_2O_2), which is normally detoxified into water by glutathione peroxidase (GPx) and reduced glutathione (GSH). Thus, most mitochondrial reactive oxygen species (ROS) are usually detoxified and residual ROS serve as signaling molecules.

Nonetheless, any significant reduction of MRC activity can induce ROS overproduction, thus triggering oxidative stress. The primary factor governing mitochondrial ROS generation is the redox state of the respiratory chain. Importantly, GPx needs an adequate amount of GSH within the mitochondrial matrix to detoxify H_2O_2, so that the depletion of mitochondrial GSH (mGSH) below a crucial level can lead to, or favor, mitochondrial dysfunction and cell death [98, 131]. Thus, mGSH plays a central role in the control of mitochondrial ROS generation.

Along similar lines, it has been described that mGSH depletion due to alcohol-mediated alteration in mitochondrial membrane dynamics underlies the susceptibility of hepatocytes from alcohol-fed models to TNF-α, and in nutritional and genetic models of hepatic steatosis, mGSH depletion occurs due to the enrichment of mitochondria in FC, resulting in decreased mitochondrial membrane fluidity [96, 98]. Altered lipid homeostasis in the liver is a pathophysiologic hallmark of NAFLD, and, as commented above, a striking finding in lipodomic studies of NAFLD biopsies is the progressive increase in FC from NAFLD to NASH [122]. Cholesterol is known to control membrane organization. In particular, cholesterol modulates the coexistence within membranes of lipid-disordered and lipid-ordered phases, regulating membrane permeability and function of resident proteins. In this regard, it has been shown that cholesterol loading in mitochondria results in increased membrane order parameter which impacts negatively in specific membrane carriers, such as the GSH transport system, but without effect in others, including the S-adenosyl-l-methionine transporter [48, 105]. Since GSH is synthesized de novo exclusively in the cytosol but not in mitochondria, the mitochondrial source of GSH depends on its transport from cytosol. Thus, one of the functional consequences of mitochondrial-cholesterol enrichment, as found in NAFLD patients, is the impairment of mGSH transport which results in mGSH depletion in the mitochondrial matrix, promoting the stimulation of mitochondrial ROS.

An important mitochondrial lipid that can be affected by oxidative damage if mGSH levels are compromised is cardiolipin. Cardiolipin, an anionic phospholipid found only in mitochondria, plays a key role in mitochondrial physiology and cell death regulation. Because of its unique structure among phospholipids, cardiolipin confers fluidity and stability to the mitochondrial membrane. Cardiolipin is present almost exclusively in the mitochondrial inner membrane (MIM) but can be also found at low levels in the mitochondrial outer membrane (MOM). Cytochrome c is normally bound to the MIM by association with cardiolipin, and it has been postulated that cardiolipin loss or peroxidation lessens the binding of cytochrome c to the MIM, favoring the permeabilization of isolated liver mitochondria, release of cytochrome c from the mitochondria to the cytosol, causing cell death [73, 98, 113]. Thus, again, the depletion of mGSH can pose a critical threat to the cell by sensitizing mitochondrial components and lipids such as cardiolipin to oxidative modifications by ROS and other reactive species that could compromise the vital mitochondrial function, and easily engage mitochondria in cell death pathways (Fig. 12.2). In addition, ROS per se are also potent inductors of mitochondrial permeabilization that contribute to different forms of apoptotic and necrotic cell death [57, 82]

In addition to enhanced fat content, cytokine overexpression, particularly TNF-α and membrane receptors, has been shown to contribute to steatohepatitis and subsequent progression to hepatocellular apoptosis in NASH [44]. In this regard, we have previously showed that selective depletion of mGSH after incubation with 3-hydroxy-4-pentenoate (HP) sensitizes hepatocytes to TNF-α without interfering with NF-κB activation. By using this strategy, mGSH becomes depleted by 70–80 % while sparing the cytosol pool of GSH [51]. Moreover, we observed that acidic sphingomyelinase (ASMase)-induced ceramide generation plays a key role in TNF-induced hepatocellular death and liver injury through stimulation of mitochondrial ROS [51, 95]. In further analyzing the role of mGSH on the sensitization of hepatocytes to TNF-α, we observed that MOM-permeabilization, cytochrome c release, and procaspase-3 activation take place only after mGSH depletion [95]. These observations suggest that the premitochondrial TNF-α signaling, including tBid generation and mitochondrial Bax activation, are independent of the mGSH status, yet their predicted consequences on mitochondria (MOM-permeabilization) and downstream events (cytochrome c release and apoptosome assembly) are controlled by the levels of mGSH. In other words, the oxidative changes occurring in mGSH depleted hepatocytes precede the upregulation of the NF-kB survival program, committing hepatocytes to TNF-α -mediated cell death [98]. In fact, it has been acknowledged that hepatocytes initially react to fat deposition with an early increase in GSH and thioredoxin, likely to oppose lipid and protein oxidation. However, with ongoing steatosis major redox changes occur in mitochondria and mGSH declines more rapidly than cytosolic GSH, indicating that mitochondria are particularly affected by oxidative damage during NAFLD [53, 55, 117, 168]. Thus, mGSH is a critical factor in the development of steatohepatitis through sensitization of hepatocytes to inflammatory cytokines.

The accumulation of lipids in the cytoplasm of hepatocytes, mostly in the form of FFA and TG, induce several metabolic adaptations in NAFLD mitochondria to restrain fat accretion. In NAFLD mitochondria, there is an increase in FFA oxidation (in position β, also known as β-oxidation), that in the end results in a major burden to mitochondrial and cell function, as follows. Mitochondria catalyze the β-oxidation of the bulk of short-, medium-, and long-chain fatty acids derived from diet, and this pathway constitutes the major process by which fatty acids are oxidized to generate energy. Electrons accumulation along the respiratory chain is further promoted by a simultaneously enhanced mitochondrial β-oxidation, which increases both the formation of NADH and $FADH_2$, and by the flow of electrons to the respiratory chain [50, 119]. Electrons can freely react with oxygen to form superoxide anion radical and other ROS [168], thus enhancing mitochondrial ROS. Further, inducible nitric oxide synthase (iNOS) transforms superoxide into peroxynitrite, which can damage DNA, and in the presence of iron, more powerful oxidants, including hydroxyl radical, and ferryl species, are produced. This creates a vicious cycle of oxidative damage in which damaged mitochondria produce larger amounts of ROS that can damage other mitochondria and other cellular components (Fig. 12.2), and result in less ATP production [131], among other outcomes.

Mitochondrial β-oxidation of FFA is still functional in NAFLD patients presenting with simple steatosis or steatohepatitis, presenting an increased rate of mitochondrial substrate oxidation during the early setting of liver injury in NASH models, as described before to cope with the high demand imposed by the accumulation of FFA. An adaptive mechanism adopted by hepatocyte mitochondria to limit ROS production during NAFLD progression is the induction of uncoupling protein-2 (UCP2). UCP2 is a MIM protein whose function may include heat generation, mitochondrial uncoupling, and protection against oxidative stress by regulating the coupling between the electrons transport chain and ATP synthesis (increasing the rate of proton leak, which partially dissipates the membrane potential). In physiological conditions, UCP2 is not expressed in hepatocytes, and its expression is primarily limited to Kupffer cells. However, UCP2 becomes strikingly abundant in hepatocytes from fatty liver, increasing the proton conductance observed during NASH progression, thus limiting mitochondrial ROS production [22, 137]. This mechanism permits mitochondria to elevate substrate oxidation and reduce redox pressure during fat accumulation, acting as a protective mechanism against damage progression. Despite these favorable effects, the chronic lack of ATP exposes the hepatocytes to increased susceptibility to noxious stimuli when the hepatic energy requirement increases, which emphasizes the key role of the hepatic redox environment in NASH progression [137]. However, as disease progresses and despite the uncoupling mechanism there is a progressive impairment of mitochondrial function and increased mitochondrial ROS [137, 138].

Other significant components greatly affected by enhanced ROS generation and that can suffer oxidative damage in the hepatocyte are mitochondrial MRC complexes and mitochondrial DNA (mtDNA) [128]. Recent evidence indicates that redox reactions are also involved in regulating mitochondrial function via redox-modification of specific cysteine-thiol groups in subunits of respiratory chain complexes [35, 66]. The proteins affected are predominantly involved in fatty acid oxidation or in the regulation of pyruvate dehydrogenase. This finding indicates that the reversible modification of a small subset of mitochondrial protein thiols occurs during low endogenous ROS fluxes, suggesting that protein thiol modification may be an important but neglected aspect of mitochondrial redox signaling [66]. As mitochondria generate more ROS oxidizing fatty acids than when oxidizing carbohydrates [144], thiol regulation could act as a feedback mechanism enabling ROS production during fatty acid oxidation to reciprocally modulate fatty acid and carbohydrate metabolism. Although this proposal is clearly speculative, a link between mitochondrial ROS production and the balance between carbohydrate and fatty acid utilization in NAFLD mitochondria is worth investigating.

Regarding the involvement of mtDNA damage on NAFLD, it is important to emphasize that mtDNA is extremely sensitive to oxidative damage because of its proximity to the MIM (the main cellular source of ROS), the absence of protective histones, and incomplete repair mechanisms in mitochondria [13]. Therefore, oxidative damage to mtDNA by ROS may lead to DNA strand breaks and the occurrence of somatic mtDNA mutations. mtDNA encodes 13 proteins required for normal functioning of the MRC, as well as the two mitochondrial ribosomal RNAs and all the mitochondrial transfer RNAs that are necessary for the synthesis of mtDNA-encoded polypeptides in the mitochondrial matrix [156]. Thus, the accumulation of mtDNA mutations may result in dysfunction of the MRC, leading to increased ROS production in mitochondria and subsequent accumulation of more mtDNA mutations further aggravating the vicious cycle already created by the enhanced β-oxidation of fatty acids, ROS generation, depletion of mGSH, and so forth (Fig. 12.2).

12.4.2 Endoplasmic Reticulum: Microsomes

Fatty acids can also be ω-hydroxylated in the endoplasmic reticulum (ER), by cytochrome P450 enzymes, and the resulting ω-hydroxy fatty acid is then dehydrogenated to a dicarboxylic acid in the cytosol. Although ω-oxidation is a minor pathway of fatty acid metabolism, its importance is dramatically increased during starvation, by ethanol, hypolipidemic drugs, peroxisome proliferators, in situations where mitochondrial β-oxidation becomes impaired and fatty acids accumulate in the cytosol, and in different metabolic diseases (e.g., diabetes, metabolic syndrome, NAFLD) [10, 59]. Microsomal ω-oxidation of fatty acids, catalyzed primarily by cytochrome P450 2E1, and 4A1 generates ROS through a flavoprotein-mediated incomplete transfer of electrons to molecular oxygen [50]. However, and in terms of oxidative stress inducers, the contribution to the formation of ROS by CYP2E1 is much significant than that by CYP4A1 [10].

CYP2E1 is mainly expressed in the liver, with hepatocytes showing the highest expression, but it is also located in other organs such as the brain and intestine. CYP2E1 is mainly located within the endoplasmic reticulum (ER) although it is also expressed in the mitochondria [110]. CYP2E1 metabolizes a variety of substances including multiple drugs, polyunsaturated fatty acids, ethanol, acetaminophen, and most organic solvents. Multiple factors such as insulin, acetone, leptin, adiponectin, and cytokines regulate CYP2E1 mRNA and protein expression [83]. In fact, insulin is known to decrease CYP2E1 expression [163]. Thus, insulin resistance may increase CYP2E1 expression and activity via the high concentration of ketone bodies produced from persistent mitochondrial fatty acid oxidation. Ketone bodies stabilize CYP2E1 and prevent its degradation. The increase in CYP2E1 and enhanced insulin resistance, both factors very common in NAFLD, seem to promote each other by creating a positive feedback loop that may eventually make steatosis progress to steatohepatitis as oxidant stress increases [83]. In addition to generate ROS, CYP2E1 forms dicarboxylic acids that when protonated cycle from the MIM to the MOM resulting in dissipation of the mitochondrial proton gradient without simultaneous ATP production; thus impairing mitochondrial function by uncoupling oxidative phosphorylation [60]. A key feature of CYP2E1 is its inducibility, and consequently liver CYP2E1 levels can be enhanced in different pathophysiological conditions. In NASH, CYP2E1 is inducible by ketones and by a high-fat diet, especially when coupled with low dietary carbohydrate intake [11]. Moreover, CYP2E1 protein content and activity correlate with the development of liver injury in obese NAFLD patients [111], and with higher TNF-α activity, oxidative and nitrosative stress, and liver injury in obese mice [32].

12.4.3 Peroxisomes

Peroxisomes are single membrane-bound subcellular organelles that contain a variety of enzymes involved in a number of metabolic processes. The most well-characterized reactions carried out by peroxisomes are those that catalyze fatty acid β-oxidation. The peroxisomal fatty acid β-oxidation pathway produces hydrogen peroxide through the activity of acyl-CoA oxidase, thus historically accounting for the name "peroxisomes," and fatty acids are converted into acetyl-CoA. Acetyl-CoA is then transported to the mitochondria, where further degradation takes place.

Peroxisomal β-oxidation is streamlined exclusively toward the metabolism of less abundant and relatively more toxic and biologically active very-long-chain fatty acids (containing 20 or more carbon atoms), 2-methyl-branched fatty acids, dicarboxylic acids, prostanoids, and C_{27} bile acid intermediates. Very-long-chain fatty acids ($>C_{20}$) are not processed by the mitochondrial β-oxidation system, and they require peroxisomal β-oxidation to shorten the chain length for further completion of oxidation in mitochondria. Long-chain dicarboxylic acids generated by the microsomal ω-oxidation of fatty acids are metabolized by the peroxisomal β-oxidation system [124, 126, 127]. Dicarboxylic acids are generally more toxic than very-long-chain fatty acids and are known to inhibit the mitochondrial fatty acid oxidation system. Thus, an effective peroxisomal β-oxidation system is needed to minimize the deleterious effects of dicarboxylic and other toxic fatty acids to prevent hepatic steatosis [127].

Moreover, an interesting feature of peroxisomes is their ability to proliferate and multiply. In rodents, for example, the number and size of peroxisomes as well as the expression of peroxisomal β-oxidation enzymes are highly increased when activators of the peroxisome proliferator-activated receptors (PPARs), like fibrates, or when free fatty acids are present [42, 134]. The peroxisomal β-oxidation system is composed of two pathways, one of which is inducible by peroxisome proliferators. The peroxisome proliferator-inducible β-oxidation pathway metabolizes straight-chain saturated fatty acyl-CoAs, whereas the second non-inducible pathway metabolizes 2-methyl-branched fatty acyl-CoAs. The genes encoding the classical β-oxidation pathway in liver are transcriptionally regulated by peroxisome proliferator-activated receptor alpha (PPARα). The first step in peroxisomal FA oxidation is directly coupled to the molecular oxygen, thus generating H_2O_2, while mitochondrial β-oxidation produces energy. Unlike the mitochondrial system, peroxisomal β-oxidation is carnitine-independent, and it does not go to completion as the chain-shortened acyl-CoAs (medium-chain acyl-CoAs) are exported to the mitochondria for the completion of β-oxidation. Evidence derived from mice deficient in PPARα points to the critical importance of PPARα and of the classical peroxisomal fatty acyl-CoA oxidase in energy metabolism, and in the development of hepatic steatosis, steatohepatitis, and liver cancer [126]. Of interest, in hepatic lipid overload a sustained PPARα stimulus manifests in hepatic peroxisome proliferation since PPARα encodes for hepatic peroxisomal acyl-CoA-oxidase, the first and rate-limiting enzyme for peroxisomal β-oxidation. Human genetic deficiencies in peroxisome biogenesis and individual peroxisomal enzymes have been described that result in accumulation of long-chain fatty acids [8, 54].

Therefore, peroxisomes are involved in the β-oxidation chain shortening of long-chain and very-long-chain fatty acyl-coenzyme (CoAs), and they do so more rapidly than mitochondria [61]. In addition, they also oxidize long-chain dicarboxylyl-CoAs, 2-methyl-branched fatty acyl-CoAs, but then again all these oxidations take its toll since in the process they generate H_2O_2, thus adding to oxidative stress [77]. Such conditions, as in NAFLD, are considered to generate peroxisome-induced oxidative stress, which may overwhelm the antioxidant capacity. Furthermore, transition metal ions like iron and copper are abundant in peroxisomes mainly in a complexed form. Under certain conditions, these metal ions can be released (e.g., by xenobiotics) and catalyze the formation of hydroxyl radical in the Fenton reaction, thus leading to lipid peroxidation, damage of the peroxisomal membrane, and loss of peroxisomal functions [134].

12.4.4 Lysosomes

Experimental evidence suggests that lysosomes play an important part in mediating FFA-induced generation of ROS and mitochondrial dysfunction in NASH by activating the lysosomal pathway of apoptosis [46, 47, 84]. Lysosomes are membrane-bound organelles that maintain an acidic intravesicular pH and contain hydrolytic enzymes that are active at an acidic to neutral range of pH. Cathepsin B belongs to the family of lysosomal proteases known as cathepsins that is abundant and active at neutral pH. It is released into the cytosol on lysosomal permeabilization and can mediate downstream mitochondrial permeabilization and caspase activation [56].

In this regard, treatment of hepatocytes with saturated FFA, e.g., palmitic acid, activates the lysosomal pathway of apoptosis through Bax-induced lysosomal membrane permeabilization (LMP), and release of cathepsin B into the cytosol. The mechanism underlying LMP is still incompletely understood; however, a number of factors have been described to affect the stability of the lysosomal membrane, including the level of lysosome-associated membrane proteins (LAMP) and cholesterol [72]. The release of cathepsin B is an early event that occurs hours prior to mitochondrial depolarization and cytochrome c redistribution into the cytosol. Cytosolic cathepsin B therefore enhances mitochondrial dysfunction, ROS generation, and cell death, likely via cathepsin B-mediated tBid generation [84]. Evidence for the importance of cathepsin B in this context is supported by the fact that cathepsin B inhibition in vitro protected against FFA-induced mitochondrial dysfunction and oxidative stress; and cathepsin B deletion in vivo protected against hepatic steatosis, oxidative stress, liver injury, and insulin resistance after cathepsin B inhibition in a dietary murine model of NAFLD [84]. Additionally, cathepsin B release from the lysosome and its redistribution in the cytoplasm has been observed in human liver tissues from patients with NAFLD and correlated with the degree of inflammatory activity [46]. Therefore, factors that prevent LMP or stabilize lysosomal membranes may be a novel mechanism to prevent FFA-mediated mitochondrial dysfunction and oxidative stress in metabolic diseases and NAFLD.

12.5 Concluding Remarks

NAFLD is a complex disease where a number of parallel factors contribute to its progression. There is increasing evidence indicating that mitochondrial dysfunction and oxidative stress play a central role in NAFLD development. Not only mitochondrial dysfunction impacts on fat deposition but it also leads to ROS formation. Hence, several vicious cycles encompassing mitochondrial damage, lipid peroxidation, and imbalance in the antioxidant defense, ROS/RNS generation, and cytokine release cause inflammation and fibrogenesis in a subset of susceptible patients, but not in all patients with fatty liver. Thus, understanding the mechanisms responsible for this differential susceptibility may help us to better design effective therapies.

Of interest, since mitochondrial dysfunction is key feature in the physiopathology of NASH, drugs or other factors improving directly or indirectly mitochondrial function may prove useful in its prevention. Antioxidant molecules specifically targeted to the mitochondria could also be valuable, not only in NASH but also in type-2 diabetes. Moreover, taking into account the epidemic of obesity and related diseases in affluent countries, reducing the incidence of NASH will be a major task to accomplish over the next years. In this regard, since these diseases are not inevitable consequences of a modern society, substantial efforts and long-term clinical intervention are needed to correct the sedentary lifestyle and nutritional habits in western countries.

References

1. Abdelmalek MF, Diehl AM (2007) Nonalcoholic fatty liver disease as a complication of insulin resistance. Med Clin North Am 91(6):1125–1149. doi:10.1016/j.mcna.2007.06.001, ix
2. Adams LA, Lindor KD (2007) Nonalcoholic fatty liver disease. Ann Epidemiol 17(11):863–869. doi:10.1016/j.annepidem.2007.05.013
3. Adams LA, Lymp JF, St Sauver J, Sanderson SO, Lindor KD, Feldstein A, Angulo P (2005) The natural history of nonalcoholic fatty liver disease: a population-based cohort study. Gastroenterology 129(1):113–121
4. Alkhouri N, Berk M, Yerian L, Lopez R, Chung YM, Zhang R, McIntyre TM, Feldstein AE, Hazen SL (2014) OxNASH score correlates with Histologic features and severity of nonalcoholic fatty liver disease. Dig Dis Sci 59(7):1617–1624. doi:10.1007/s10620-014-3031-8
5. Alkhouri N, Dixon LJ, Feldstein AE (2009) Lipotoxicity in nonalcoholic fatty liver disease: not all lipids are created equal. Expert Rev Gastroenterol Hepatol 3(4):445–451. doi:10.1586/egh.09.32
6. Aller R, de Luis DA, Fernandez L, Calle F, Velayos B, Olcoz JL, Izaola O, Sagrado MG, Conde R, Gonzalez JM (2008) Influence of insulin resistance and adipokines in the grade of steatosis of nonalcoholic fatty liver disease. Dig Dis Sci 53(4):1088–1092. doi:10.1007/s10620-007-9981-3
7. Anania FA (2005) Adiponectin and alcoholic fatty liver: is it, after all, about what you eat? Hepatology 42(3):530–532. doi:10.1002/hep.20861
8. Aoyama T, Peters JM, Iritani N, Nakajima T, Furihata K, Hashimoto T, Gonzalez FJ (1998) Altered constitutive expression of fatty acid-metabolizing enzymes in mice lacking the peroxisome proliferator-activated receptor alpha (PPARalpha). J Biol Chem 273(10):5678–5684
9. Arita Y, Kihara S, Ouchi N, Takahashi M, Maeda K, Miyagawa J, Hotta K, Shimomura I, Nakamura T, Miyaoka K, Kuriyama H, Nishida M, Yamashita S, Okubo K, Matsubara K, Muraguchi M, Ohmoto Y, Funahashi T, Matsuzawa Y (2012) Paradoxical decrease of an adipose-specific protein, adiponectin, in obesity. 1999. Biochem Biophys Res Commun 425(3):560–564. doi:10.1016/j.bbrc.2012.08.024
10. Aubert J, Begriche K, Knockaert L, Robin MA, Fromenty B (2011) Increased expression of cytochrome P450 2E1 in nonalcoholic fatty liver disease: mechanisms and pathophysiological role. Clin Res Hepatol Gastroenterol 35(10):630–637. doi:10.1016/j.clinre.2011.04.015
11. Bidlack WR, Brown RC, Mohan C (1986) Nutritional parameters that alter hepatic drug metabolism, conjugation, and toxicity. Fed Proc 45(2):142–148
12. Blindenbacher A, Wang X, Langer I, Savino R, Terracciano L, Heim MH (2003) Interleukin 6 is important for survival after partial hepatectomy in mice. Hepatology 38(3):674–682. doi:10.1053/jhep.2003.50378

13. Bogenhagen DF (1999) Repair of mtDNA in vertebrates. Am J Hum Genet 64(5):1276–1281. doi:10.1086/302392
14. Brun P, Castagliuolo I, Di Leo V, Buda A, Pinzani M, Palu G, Martines D (2007) Increased intestinal permeability in obese mice: new evidence in the pathogenesis of nonalcoholic steatohepatitis. Am J Physiol Gastrointest Liver Physiol 292(2):G518–G525. doi:10.1152/ajpgi.00024.2006
15. Bugianesi E, Gastaldelli A, Vanni E, Gambino R, Cassader M, Baldi S, Ponti V, Pagano G, Ferrannini E, Rizzetto M (2005) Insulin resistance in non-diabetic patients with non-alcoholic fatty liver disease: sites and mechanisms. Diabetologia 48(4):634–642. doi:10.1007/s00125-005-1682-x
16. Caballero F, Fernandez A, De Lacy AM, Fernandez-Checa JC, Caballeria J, Garcia-Ruiz C (2009) Enhanced free cholesterol, SREBP-2 and StAR expression in human NASH. J Hepatol 50(4):789–796. doi:10.1016/j.jhep.2008.12.016
17. Caldwell SH, Swerdlow RH, Khan EM, Iezzoni JC, Hespenheide EE, Parks JK, Parker WD Jr (1999) Mitochondrial abnormalities in non-alcoholic steatohepatitis. J Hepatol 31(3):430–434
18. Cortez-Pinto H, Chatham J, Chacko VP, Arnold C, Rashid A, Diehl AM (1999) Alterations in liver ATP homeostasis in human nonalcoholic steatohepatitis: a pilot study. JAMA 282(17):1659–1664
19. Cressman DE, Greenbaum LE, DeAngelis RA, Ciliberto G, Furth EE, Poli V, Taub R (1996) Liver failure and defective hepatocyte regeneration in interleukin-6-deficient mice. Science 274(5291):1379–1383
20. Csak T, Velayudham A, Hritz I, Petrasek J, Levin I, Lippai D, Catalano D, Mandrekar P, Dolganiuc A, Kurt-Jones E, Szabo G (2011) Deficiency in myeloid differentiation factor-2 and toll-like receptor 4 expression attenuates nonalcoholic steatohepatitis and fibrosis in mice. Am J Physiol Gastrointest Liver Physiol 300(3):G433–G441. doi:10.1152/ajpgi.00163.2009
21. Chalasani N, Younossi Z, Lavine JE, Diehl AM, Brunt EM, Cusi K, Charlton M, Sanyal AJ, American Association for the Study of Liver Diseases, American College of Gastroenterology, American Gastroenterological Association (2012) The diagnosis and management of non-alcoholic fatty liver disease: Practice guideline by the American Association for the Study of Liver Diseases, American College of Gastroenterology, and the American Gastroenterological Association. Am J Gastroenterol 107(6):811–826. doi:10.1038/ajg.2012.128
22. Chavin KD, Yang S, Lin HZ, Chatham J, Chacko VP, Hoek JB, Walajtys-Rode E, Rashid A, Chen CH, Huang CC, Wu TC, Lane MD, Diehl AM (1999) Obesity induces expression of uncoupling protein-2 in hepatocytes and promotes liver ATP depletion. J Biol Chem 274(9):5692–5700
23. Chen G, Liang G, Ou J, Goldstein JL, Brown MS (2004) Central role for liver X receptor in insulin-mediated activation of Srebp-1c transcription and stimulation of fatty acid synthesis in liver. Proc Natl Acad Sci U S A 101(31):11245–11250. doi:10.1073/pnas.0404297101
24. Cheung O, Sanyal AJ (2008) Abnormalities of lipid metabolism in nonalcoholic fatty liver disease. Semin Liver Dis 28(4):351–359. doi:10.1055/s-0028-1091979
25. Choi SH, Ginsberg HN (2011) Increased very low density lipoprotein (VLDL) secretion, hepatic steatosis, and insulin resistance. Trends Endocrinol Metab 22(9):353–363. doi:10.1016/j.tem.2011.04.007
26. Choi SS, Diehl AM (2008) Hepatic triglyceride synthesis and nonalcoholic fatty liver disease. Curr Opin Lipidol 19(3):295–300. doi:10.1097/MOL.0b013e3282ff5e55
27. Day CP (2002) Pathogenesis of steatohepatitis. Best Pract Res Clin Gastroenterol 16(5):663–678
28. Day CP (2006) From fat to inflammation. Gastroenterology 130(1):207–210. doi:10.1053/j.gastro.2005.11.017
29. Day CP, James OF (1998) Steatohepatitis: a tale of two "hits"? Gastroenterology 114(4):842–845
30. de Almeida IT, Cortez-Pinto H, Fidalgo G, Rodrigues D, Camilo ME (2002) Plasma total and free fatty acids composition in human non-alcoholic steatohepatitis. Clin Nutr 21(3):219–223

31. Dentin R, Benhamed F, Hainault I, Fauveau V, Foufelle F, Dyck JR, Girard J, Postic C (2006) Liver-specific inhibition of ChREBP improves hepatic steatosis and insulin resistance in ob/ob mice. Diabetes 55(8):2159–2170. doi:10.2337/db06-0200
32. Dey A, Cederbaum AI (2007) Induction of cytochrome P450 2E1 [corrected] promotes liver injury in ob/ob mice. Hepatology 45(6):1355–1365. doi:10.1002/hep.21603
33. Donnelly KL, Smith CI, Schwarzenberg SJ, Jessurun J, Boldt MD, Parks EJ (2005) Sources of fatty acids stored in liver and secreted via lipoproteins in patients with nonalcoholic fatty liver disease. J Clin Invest 115(5):1343–1351. doi:10.1172/JCI23621
34. Dowman JK, Tomlinson JW, Newsome PN (2010) Pathogenesis of non-alcoholic fatty liver disease. QJM 103(2):71–83. doi:10.1093/qjmed/hcp158
35. Drose S, Brandt U, Wittig I (2014) Mitochondrial respiratory chain complexes as sources and targets of thiol-based redox-regulation. Biochim Biophys Acta 1844(8):1344–1354. doi:10.1016/j.bbapap.2014.02.006
36. Eaton S, Bartlett K, Pourfarzam M (1996) Mammalian mitochondrial beta-oxidation. Biochem J 320(Pt 2):345–357
37. Eguchi Y, Hyogo H, Ono M, Mizuta T, Ono N, Fujimoto K, Chayama K, Saibara T, Jsg N (2012) Prevalence and associated metabolic factors of nonalcoholic fatty liver disease in the general population from 2009 to 2010 in Japan: a multicenter large retrospective study. J Gastroenterol 47(5):586–595. doi:10.1007/s00535-012-0533-z
38. El-Assal O, Hong F, Kim WH, Radaeva S, Gao B (2004) IL-6-deficient mice are susceptible to ethanol-induced hepatic steatosis: IL-6 protects against ethanol-induced oxidative stress and mitochondrial permeability transition in the liver. Cell Mol Immunol 1(3):205–211
39. Endo M, Masaki T, Seike M, Yoshimatsu H (2007) TNF-alpha induces hepatic steatosis in mice by enhancing gene expression of sterol regulatory element binding protein-1c (SREBP-1c). Exp Biol Med 232(5):614–621
40. Ertle J, Dechene A, Sowa JP, Penndorf V, Herzer K, Kaiser G, Schlaak JF, Gerken G, Syn WK, Canbay A (2011) Non-alcoholic fatty liver disease progresses to hepatocellular carcinoma in the absence of apparent cirrhosis. Int J Cancer 128(10):2436–2443. doi:10.1002/ijc.25797
41. Fabbrini E, Magkos F, Mohammed BS, Pietka T, Abumrad NA, Patterson BW, Okunade A, Klein S (2009) Intrahepatic fat, not visceral fat, is linked with metabolic complications of obesity. Proc Natl Acad Sci U S A 106(36):15430–15435. doi:10.1073/pnas.0904944106
42. Fahimi HD, Reinicke A, Sujatta M, Yokota S, Ozel M, Hartig F, Stegmeier K (1982) The short- and long-term effects of bezafibrate in the rat. Ann N Y Acad Sci 386:111–135
43. Farhadi A, Gundlapalli S, Shaikh M, Frantzides C, Harrell L, Kwasny MM, Keshavarzian A (2008) Susceptibility to gut leakiness: a possible mechanism for endotoxaemia in non-alcoholic steatohepatitis. Liver Int 28(7):1026–1033. doi:10.1111/j.1478-3231.2008.01723.x
44. Feldstein AE, Canbay A, Angulo P, Taniai M, Burgart LJ, Lindor KD, Gores GJ (2003) Hepatocyte apoptosis and fas expression are prominent features of human nonalcoholic steatohepatitis. Gastroenterology 125(2):437–443
45. Feldstein AE, Lopez R, Tamimi TA, Yerian L, Chung YM, Berk M, Zhang R, McIntyre TM, Hazen SL (2010) Mass spectrometric profiling of oxidized lipid products in human nonalcoholic fatty liver disease and nonalcoholic steatohepatitis. J Lipid Res 51(10):3046–3054. doi:10.1194/jlr.M007096
46. Feldstein AE, Werneburg NW, Canbay A, Guicciardi ME, Bronk SF, Rydzewski R, Burgart LJ, Gores GJ (2004) Free fatty acids promote hepatic lipotoxicity by stimulating TNF-alpha expression via a lysosomal pathway. Hepatology 40(1):185–194. doi:10.1002/hep.20283
47. Feldstein AE, Werneburg NW, Li Z, Bronk SF, Gores GJ (2006) Bax inhibition protects against free fatty acid-induced lysosomal permeabilization. Am J Physiol Gastrointest Liver Physiol 290(6):G1339–G1346. doi:10.1152/ajpgi.00509.2005
48. Fernandez A, Colell A, Caballero F, Matias N, Garcia-Ruiz C, Fernandez-Checa JC (2009) Mitochondrial S-adenosyl-L-methionine transport is insensitive to alcohol-mediated changes in membrane dynamics. Alcohol Clin Exp Res 33(7):1169–1180. doi:10.1111/j.1530-0277.2009.00940.x

49. Gaggini M, Morelli M, Buzzigoli E, DeFronzo RA, Bugianesi E, Gastaldelli A (2013) Non-alcoholic fatty liver disease (NAFLD) and its connection with insulin resistance, dyslipidemia, atherosclerosis and coronary heart disease. Nutrients 5(5):1544–1560. doi:10.3390/nu5051544
50. Gambino R, Musso G, Cassader M (2011) Redox balance in the pathogenesis of nonalcoholic fatty liver disease: mechanisms and therapeutic opportunities. Antioxid Redox Signal 15(5):1325–1365. doi:10.1089/ars.2009.3058
51. Garcia-Ruiz C, Colell A, Mari M, Morales A, Calvo M, Enrich C, Fernandez-Checa JC (2003) Defective TNF-alpha-mediated hepatocellular apoptosis and liver damage in acidic sphingomyelinase knockout mice. J Clin Invest 111(2):197–208. doi:10.1172/JCI16010
52. Garcia-Ruiz C, Colell A, Mari M, Morales A, Fernandez-Checa JC (1997) Direct effect of ceramide on the mitochondrial electron transport chain leads to generation of reactive oxygen species. Role of mitochondrial glutathione. J Biol Chem 272(17):11369–11377
53. Garcia-Ruiz C, Fernandez-Checa JC (2006) Mitochondrial glutathione: hepatocellular survival-death switch. J Gastroenterol Hepatol 21(Suppl 3):S3–S6. doi:10.1111/j.1440-1746.2006.04570.x
54. Goldfischer SL (1988) Peroxisomal diseases. Prog Clin Biol Res 282:117–137
55. Grattagliano I, de Bari O, Bernardo TC, Oliveira PJ, Wang DQ, Portincasa P (2012) Role of mitochondria in nonalcoholic fatty liver disease–from origin to propagation. Clin Biochem 45(9):610–618. doi:10.1016/j.clinbiochem.2012.03.024
56. Guicciardi ME, Leist M, Gores GJ (2004) Lysosomes in cell death. Oncogene 23(16):2881–2890. doi:10.1038/sj.onc.1207512
57. Halestrap AP, Woodfield KY, Connern CP (1997) Oxidative stress, thiol reagents, and membrane potential modulate the mitochondrial permeability transition by affecting nucleotide binding to the adenine nucleotide translocase. J Biol Chem 272(6):3346–3354
58. Handa P, Maliken BD, Nelson JE, Morgan-Stevenson V, Messner DJ, Dhillon BK, Klintworth HM, Beauchamp M, Yeh MM, Elfers CT, Roth CL, Kowdley KV (2013) Reduced adiponectin signaling due to weight gain results in nonalcoholic steatohepatitis through impaired mitochondrial biogenesis. Hepatology. doi:10.1002/hep.26946
59. Hardwick JP (2008) Cytochrome P450 omega hydroxylase (CYP4) function in fatty acid metabolism and metabolic diseases. Biochem Pharmacol 75(12):2263–2275. doi:10.1016/j.bcp.2008.03.004
60. Hermesh O, Kalderon B, Bar-Tana J (1998) Mitochondria uncoupling by a long chain fatty acyl analogue. J Biol Chem 273(7):3937–3942
61. Hiltunen JK, Karki T, Hassinen IE, Osmundsen H (1986) beta-Oxidation of polyunsaturated fatty acids by rat liver peroxisomes. A role for 2,4-dienoyl-coenzyme A reductase in peroxisomal beta-oxidation. J Biol Chem 261(35):16484–16493
62. Hotamisligil GS, Spiegelman BM (1994) Tumor necrosis factor alpha: a key component of the obesity-diabetes link. Diabetes 43(11):1271–1278
63. Hotta K, Funahashi T, Arita Y, Takahashi M, Matsuda M, Okamoto Y, Iwahashi H, Kuriyama H, Ouchi N, Maeda K, Nishida M, Kihara S, Sakai N, Nakajima T, Hasegawa K, Muraguchi M, Ohmoto Y, Nakamura T, Yamashita S, Hanafusa T, Matsuzawa Y (2000) Plasma concentrations of a novel, adipose-specific protein, adiponectin, in type 2 diabetic patients. Arterioscler Thromb Vasc Biol 20(6):1595–1599
64. Hudgins LC, Hellerstein MK, Seidman CE, Neese RA, Tremaroli JD, Hirsch J (2000) Relationship between carbohydrate-induced hypertriglyceridemia and fatty acid synthesis in lean and obese subjects. J Lipid Res 41(4):595–604
65. Hui JM, Hodge A, Farrell GC, Kench JG, Kriketos A, George J (2004) Beyond insulin resistance in NASH: TNF-alpha or adiponectin? Hepatology 40(1):46–54. doi:10.1002/hep.20280
66. Hurd TR, Prime TA, Harbour ME, Lilley KS, Murphy MP (2007) Detection of reactive oxygen species-sensitive thiol proteins by redox difference gel electrophoresis: implications for mitochondrial redox signaling. J Biol Chem 282(30):22040–22051. doi:10.1074/jbc.M703591200

67. Hussain MM, Shi J, Dreizen P (2003) Microsomal triglyceride transfer protein and its role in apoB-lipoprotein assembly. J Lipid Res 44(1):22–32
68. Hwang JH, Stein DT, Barzilai N, Cui MH, Tonelli J, Kishore P, Hawkins M (2007) Increased intrahepatic triglyceride is associated with peripheral insulin resistance: in vivo MR imaging and spectroscopy studies. Am J Physiol Endocrinol Metab 293(6):E1663–E1669. doi:10.1152/ajpendo.00590.2006
69. Ibrahim SH, Kohli R, Gores GJ (2011) Mechanisms of lipotoxicity in NAFLD and clinical implications. J Pediatr Gastroenterol Nutr 53(2):131–140. doi:10.1097/MPG.0b013e31822578db
70. Jaeschke H (2011) Reactive oxygen and mechanisms of inflammatory liver injury: present concepts. J Gastroenterol Hepatol 26(Suppl 1):173–179. doi:10.1111/j.1440-1746.2010.06592.x
71. Jin X, Zimmers TA, Perez EA, Pierce RH, Zhang Z, Koniaris LG (2006) Paradoxical effects of short- and long-term interleukin-6 exposure on liver injury and repair. Hepatology 43(3):474–484. doi:10.1002/hep.21087
72. Johansson AC, Appelqvist H, Nilsson C, Kagedal K, Roberg K, Ollinger K (2010) Regulation of apoptosis-associated lysosomal membrane permeabilization. Apoptosis 15(5):527–540. doi:10.1007/s10495-009-0452-5
73. Kagan VE, Tyurin VA, Jiang J, Tyurina YY, Ritov VB, Amoscato AA, Osipov AN, Belikova NA, Kapralov AA, Kini V, Vlasova II, Zhao Q, Zou M, Di P, Svistunenko DA, Kurnikov IV, Borisenko GG (2005) Cytochrome c acts as a cardiolipin oxygenase required for release of proapoptotic factors. Nat Chem Biol 1(4):223–232. doi:10.1038/nchembio727
74. Kaser S, Moschen A, Cayon A, Kaser A, Crespo J, Pons-Romero F, Ebenbichler CF, Patsch JR, Tilg H (2005) Adiponectin and its receptors in non-alcoholic steatohepatitis. Gut 54(1):117–121. doi:10.1136/gut.2003.037010
75. Kawano Y, Cohen DE (2013) Mechanisms of hepatic triglyceride accumulation in non-alcoholic fatty liver disease. J Gastroenterol 48(4):434–441. doi:10.1007/s00535-013-0758-5
76. Kim CH, Younossi ZM (2008) Nonalcoholic fatty liver disease: a manifestation of the metabolic syndrome. Cleve Clin J Med 75(10):721–728
77. Koek GH, Liedorp PR, Bast A (2011) The role of oxidative stress in non-alcoholic steatohepatitis. Clin chim Acta 412(15–16):1297–1305. doi:10.1016/j.cca.2011.04.013
78. Kohjima M, Enjoji M, Higuchi N, Kato M, Kotoh K, Yoshimoto T, Fujino T, Yada M, Yada R, Harada N, Takayanagi R, Nakamuta M (2007) Re-evaluation of fatty acid metabolism-related gene expression in nonalcoholic fatty liver disease. Int J Mol Med 20(3):351–358
79. Kohli R, Pan X, Malladi P, Wainwright MS, Whitington PF (2007) Mitochondrial reactive oxygen species signal hepatocyte steatosis by regulating the phosphatidylinositol 3-kinase cell survival pathway. J Biol Chem 282(29):21327–21336. doi:10.1074/jbc.M701759200
80. Koo SH (2013) Nonalcoholic fatty liver disease: molecular mechanisms for the hepatic steatosis. Clin Mol Hepatol 19(3):210–215. doi:10.3350/cmh.2013.19.3.210
81. Lazo M, Clark JM (2008) The epidemiology of nonalcoholic fatty liver disease: a global perspective. Semin Liver Dis 28(4):339–350. doi:10.1055/s-0028-1091978
82. Leung AW, Halestrap AP (2008) Recent progress in elucidating the molecular mechanism of the mitochondrial permeability transition pore. Biochim Biophys Acta 1777(7–8):946–952. doi:10.1016/j.bbabio.2008.03.009
83. Leung TM, Nieto N (2013) CYP2E1 and oxidant stress in alcoholic and non-alcoholic fatty liver disease. J Hepatol 58(2):395–398. doi:10.1016/j.jhep.2012.08.018
84. Li Z, Berk M, McIntyre TM, Gores GJ, Feldstein AE (2008) The lysosomal-mitochondrial axis in free fatty acid-induced hepatic lipotoxicity. Hepatology 47(5):1495–1503. doi:10.1002/hep.22183
85. Li ZZ, Berk M, McIntyre TM, Feldstein AE (2009) Hepatic lipid partitioning and liver damage in nonalcoholic fatty liver disease: role of stearoyl-CoA desaturase. J Biol Chem 284(9):5637–5644. doi:10.1074/jbc.M807616200
86. Listenberger LL, Han X, Lewis SE, Cases S, Farese RV Jr, Ory DS, Schaffer JE (2003) Triglyceride accumulation protects against fatty acid-induced lipotoxicity. Proc Natl Acad Sci U S A 100(6):3077–3082. doi:10.1073/pnas.0630588100

87. Lonardo A, Lombardini S, Ricchi M, Scaglioni F, Loria P (2005) Review article: hepatic steatosis and insulin resistance. Aliment Pharmacol Ther 22(Suppl 2):64–70. doi:10.1111/j.1365-2036.2005.02600.x
88. Loomba R, Abraham M, Unalp A, Wilson L, Lavine J, Doo E, Bass NM, Nonalcoholic Steatohepatitis Clinical Research N (2012) Association between diabetes, family history of diabetes, and risk of nonalcoholic steatohepatitis and fibrosis. Hepatology 56(3):943–951. doi:10.1002/hep.25772
89. Loomba R, Sanyal AJ (2013) The global NAFLD epidemic. Nat Rev Gastroenterol Hepatol 10(11):686–690. doi:10.1038/nrgastro.2013.171
90. Machado MV, Ferreira DM, Castro RE, Silvestre AR, Evangelista T, Coutinho J, Carepa F, Costa A, Rodrigues CM, Cortez-Pinto H (2012) Liver and muscle in morbid obesity: the interplay of fatty liver and insulin resistance. PLoS One 7(2):e31738. doi:10.1371/journal.pone.0031738
91. Malhi H, Bronk SF, Werneburg NW, Gores GJ (2006) Free fatty acids induce JNK-dependent hepatocyte lipoapoptosis. J Biol Chem 281(17):12093–12101. doi:10.1074/jbc.M510660200
92. Malhi H, Gores GJ (2008) Molecular mechanisms of lipotoxicity in nonalcoholic fatty liver disease. Semin Liver Dis 28(4):360–369. doi:10.1055/s-0028-1091980
93. Marchesini G, Bugianesi E, Forlani G, Cerrelli F, Lenzi M, Manini R, Natale S, Vanni E, Villanova N, Melchionda N, Rizzetto M (2003) Nonalcoholic fatty liver, steatohepatitis, and the metabolic syndrome. Hepatology 37(4):917–923. doi:10.1053/jhep.2003.50161
94. Mari M, Caballero F, Colell A, Morales A, Caballeria J, Fernandez A, Enrich C, Fernandez-Checa JC, Garcia-Ruiz C (2006) Mitochondrial free cholesterol loading sensitizes to TNF- and Fas-mediated steatohepatitis. Cell Metab 4(3):185–198. doi:10.1016/j.cmet.2006.07.006
95. Mari M, Colell A, Morales A, Caballero F, Moles A, Fernandez A, Terrones O, Basanez G, Antonsson B, Garcia-Ruiz C, Fernandez-Checa JC (2008) Mechanism of mitochondrial glutathione-dependent hepatocellular susceptibility to TNF despite NF-kappaB activation. Gastroenterology 134(5):1507–1520. doi:10.1053/j.gastro.2008.01.073
96. Mari M, Colell A, Morales A, von Montfort C, Garcia-Ruiz C, Fernandez-Checa JC (2010) Redox control of liver function in health and disease. Antioxid Redox Signal 12(11):1295–1331. doi:10.1089/ars.2009.2634
97. Mari M, Fernandez-Checa JC (2007) Sphingolipid signalling and liver diseases. Liver Int 27(4):440–450. doi:10.1111/j.1478-3231.2007.01475.x
98. Mari M, Morales A, Colell A, Garcia-Ruiz C, Fernandez-Checa JC (2009) Mitochondrial glutathione, a key survival antioxidant. Antioxid Redox Signal 11(11):2685–2700. doi:10.1089/ARS.2009.2695
99. Matsubara M, Maruoka S, Katayose S (2002) Decreased plasma adiponectin concentrations in women with dyslipidemia. J Clin Endocrinol Metab 87(6):2764–2769. doi:10.1210/jcem.87.6.8550
100. Matteoni CA, Younossi ZM, Gramlich T, Boparai N, Liu YC, McCullough AJ (1999) Nonalcoholic fatty liver disease: a spectrum of clinical and pathological severity. Gastroenterology 116(6):1413–1419
101. McClain CJ, Barve S, Deaciuc I (2007) Good fat/bad fat. Hepatology 45(6):1343–1346. doi:10.1002/hep.21788
102. Mehal WZ (2013) The Gordian Knot of dysbiosis, obesity and NAFLD. Nat Rev Gastroenterol Hepatol 10(11):637–644. doi:10.1038/nrgastro.2013.146
103. Miele L, Valenza V, La Torre G, Montalto M, Cammarota G, Ricci R, Masciana R, Forgione A, Gabrieli ML, Perotti G, Vecchio FM, Rapaccini G, Gasbarrini G, Day CP, Grieco A (2009) Increased intestinal permeability and tight junction alterations in nonalcoholic fatty liver disease. Hepatology 49(6):1877–1887. doi:10.1002/hep.22848
104. Miles JM, Nelson RH (2007) Contribution of triglyceride-rich lipoproteins to plasma free fatty acids. Horm Metab Res 39(10):726–729. doi:10.1055/s-2007-990273
105. Montero J, Mari M, Colell A, Morales A, Basanez G, Garcia-Ruiz C, Fernandez-Checa JC (2010) Cholesterol and peroxidized cardiolipin in mitochondrial membrane properties,

permeabilization and cell death. Biochim Biophys Acta 1797(6–7):1217–1224. doi:10.1016/j.bbabio.2010.02.010

106. Murphy MP (2009) How mitochondria produce reactive oxygen species. Biochem J 417(1):1–13. doi:10.1042/BJ20081386
107. Musso G, Anty R, Petta S (2013) Antioxidant therapy and drugs interfering with lipid metabolism: could they be effective in NAFLD patients? Curr Pharm Des 19(29):5297–5313
108. Musso G, Gambino R, Cassader M (2010) Gut microbiota as a regulator of energy homeostasis and ectopic fat deposition: mechanisms and implications for metabolic disorders. Curr Opin Lipidol 21(1):76–83. doi:10.1097/MOL.0b013e3283347ebb
109. Neuschwander-Tetri BA (2010) Nontriglyceride hepatic lipotoxicity: the new paradigm for the pathogenesis of NASH. Curr Gastroenterol Rep 12(1):49–56. doi:10.1007/s11894-009-0083-6
110. Neve EP, Ingelman-Sundberg M (2000) Molecular basis for the transport of cytochrome P450 2E1 to the plasma membrane. J Biol Chem 275(22):17130–17135. doi:10.1074/jbc.M000957200
111. Orellana M, Rodrigo R, Varela N, Araya J, Poniachik J, Csendes A, Smok G, Videla LA (2006) Relationship between in vivo chlorzoxazone hydroxylation, hepatic cytochrome P450 2E1 content and liver injury in obese non-alcoholic fatty liver disease patients. Hepatol Res 34(1):57–63. doi:10.1016/j.hepres.2005.10.001
112. Ota T, Gayet C, Ginsberg HN (2008) Inhibition of apolipoprotein B100 secretion by lipid-induced hepatic endoplasmic reticulum stress in rodents. J Clin Invest 118(1):316–332. doi:10.1172/JCI32752
113. Ott M, Robertson JD, Gogvadze V, Zhivotovsky B, Orrenius S (2002) Cytochrome c release from mitochondria proceeds by a two-step process. Proc Natl Acad Sci U S A 99(3):1259–1263. doi:10.1073/pnas.241655498
114. Pajvani UB, Du X, Combs TP, Berg AH, Rajala MW, Schulthess T, Engel J, Brownlee M, Scherer PE (2003) Structure-function studies of the adipocyte-secreted hormone Acrp30/adiponectin. Implications for metabolic regulation and bioactivity. J Biol Chem 278(11):9073–9085. doi:10.1074/jbc.M207198200
115. Park EJ, Lee JH, Yu GY, He G, Ali SR, Holzer RG, Osterreicher CH, Takahashi H, Karin M (2010) Dietary and genetic obesity promote liver inflammation and tumorigenesis by enhancing IL-6 and TNF expression. Cell 140(2):197–208. doi:10.1016/j.cell.2009.12.052
116. Parks EJ (2002) Dietary carbohydrate's effects on lipogenesis and the relationship of lipogenesis to blood insulin and glucose concentrations. Br J Nutr 87(Suppl 2):S247–S253. doi:10.1079/BJNBJN/2002544
117. Perez-Carreras M, Del Hoyo P, Martin MA, Rubio JC, Martin A, Castellano G, Colina F, Arenas J, Solis-Herruzo JA (2003) Defective hepatic mitochondrial respiratory chain in patients with nonalcoholic steatohepatitis. Hepatology 38(4):999–1007. doi:10.1053/jhep.2003.50398
118. Pessayre D (2007) Role of mitochondria in non-alcoholic fatty liver disease. J Gastroenterol Hepatol 22(Suppl 1):S20–S27. doi:10.1111/j.1440-1746.2006.04640.x
119. Pessayre D, Fromenty B (2005) NASH: a mitochondrial disease. J Hepatol 42(6):928–940. doi:10.1016/j.jhep.2005.03.004
120. Petersen DR, Doorn JA (2004) Reactions of 4-hydroxynonenal with proteins and cellular targets. Free Radic Biol Med 37(7):937–945. doi:10.1016/j.freeradbiomed.2004.06.012
121. Postic C, Girard J (2008) Contribution of de novo fatty acid synthesis to hepatic steatosis and insulin resistance: lessons from genetically engineered mice. J Clin Invest 118(3):829–838. doi:10.1172/JCI34275
122. Puri P, Baillie RA, Wiest MM, Mirshahi F, Choudhury J, Cheung O, Sargeant C, Contos MJ, Sanyal AJ (2007) A lipidomic analysis of nonalcoholic fatty liver disease. Hepatology 46(4):1081–1090. doi:10.1002/hep.21763
123. Ramesh S, Sanyal AJ (2005) Evaluation and management of non-alcoholic steatohepatitis. J Hepatol 42(Suppl (1)):S2–S12. doi:10.1016/j.jhep.2004.11.022

124. Rao MS, Reddy JK (2004) PPAR alpha in the pathogenesis of fatty liver disease. Hepatology 40(4):783–786. doi:10.1002/hep.20453
125. Ray K (2013) NAFLD-the next global epidemic. Nat Rev Gastroenterol Hepatol 10(11):621. doi:10.1038/nrgastro.2013.197
126. Reddy JK, Hashimoto T (2001) Peroxisomal beta-oxidation and peroxisome proliferator-activated receptor alpha: an adaptive metabolic system. Annu Rev Nutr 21:193–230. doi:10.1146/annurev.nutr.21.1.193
127. Reddy JK, Rao MS (2006) Lipid metabolism and liver inflammation. II. Fatty liver disease and fatty acid oxidation. Am J Physiol Gastrointest Liver Physiol 290(5):G852–G858. doi:10.1152/ajpgi.00521.2005
128. Ricci C, Pastukh V, Leonard J, Turrens J, Wilson G, Schaffer D, Schaffer SW (2008) Mitochondrial DNA damage triggers mitochondrial-superoxide generation and apoptosis. Am J Physiol Cell Physiol 294(2):C413–C422. doi:10.1152/ajpcell.00362.2007
129. Rivera CA, Adegboyega P, van Rooijen N, Tagalicud A, Allman M, Wallace M (2007) Toll-like receptor-4 signaling and Kupffer cells play pivotal roles in the pathogenesis of non-alcoholic steatohepatitis. J Hepatol 47(4):571–579. doi:10.1016/j.jhep.2007.04.019
130. Roh YS, Seki E (2013) Toll-like receptors in alcoholic liver disease, non-alcoholic steatohepatitis and carcinogenesis. J Gastroenterol Hepatol 28(Suppl 1):38–42. doi:10.1111/jgh.12019
131. Rolo AP, Teodoro JS, Palmeira CM (2012) Role of oxidative stress in the pathogenesis of nonalcoholic steatohepatitis. Free Radic Biol Med 52(1):59–69. doi:10.1016/j.freeradbiomed.2011.10.003
132. Samuel VT, Shulman GI (2012) Mechanisms for insulin resistance: common threads and missing links. Cell 148(5):852–871. doi:10.1016/j.cell.2012.02.017
133. Sanyal AJ, Campbell-Sargent C, Mirshahi F, Rizzo WB, Contos MJ, Sterling RK, Luketic VA, Shiffman ML, Clore JN (2001) Nonalcoholic steatohepatitis: association of insulin resistance and mitochondrial abnormalities. Gastroenterology 120(5):1183–1192. doi:10.1053/gast.2001.23256
134. Schrader M, Fahimi HD (2006) Peroxisomes and oxidative stress. Biochim Biophys Acta 1763(12):1755–1766. doi:10.1016/j.bbamcr.2006.09.006
135. Schwabe RF, Brenner DA (2006) Mechanisms of Liver Injury. I. TNF-alpha-induced liver injury: role of IKK, JNK, and ROS pathways. Am J Physiol Gastrointest Liver Physiol 290(4):G583–G589. doi:10.1152/ajpgi.00422.2005
136. Schwenger KJ, Allard JP (2014) Clinical approaches to non-alcoholic fatty liver disease. World J Gastroenterol 20(7):1712–1723. doi:10.3748/wjg.v20.i7.1712
137. Serviddio G, Bellanti F, Tamborra R, Rollo T, Capitanio N, Romano AD, Sastre J, Vendemiale G, Altomare E (2008) Uncoupling protein-2 (UCP2) induces mitochondrial proton leak and increases susceptibility of non-alcoholic steatohepatitis (NASH) liver to ischaemia-reperfusion injury. Gut 57(7):957–965. doi:10.1136/gut.2007.147496
138. Serviddio G, Bellanti F, Tamborra R, Rollo T, Romano AD, Giudetti AM, Capitanio N, Petrella A, Vendemiale G, Altomare E (2008) Alterations of hepatic ATP homeostasis and respiratory chain during development of non-alcoholic steatohepatitis in a rodent model. Eur J Clin Invest 38(4):245–252. doi:10.1111/j.1365-2362.2008.01936.x
139. Shi H, Kokoeva MV, Inouye K, Tzameli I, Yin H, Flier JS (2006) TLR4 links innate immunity and fatty acid-induced insulin resistance. J Clin Invest 116(11):3015–3025. doi:10.1172/JCI28898
140. Shimomura I, Bashmakov Y, Horton JD (1999) Increased levels of nuclear SREBP-1c associated with fatty livers in two mouse models of diabetes mellitus. J Biol Chem 274(42):30028–30032
141. Shklyaev S, Aslanidi G, Tennant M, Prima V, Kohlbrenner E, Kroutov V, Campbell-Thompson M, Crawford J, Shek EW, Scarpace PJ, Zolotukhin S (2003) Sustained peripheral expression of transgene adiponectin offsets the development of diet-induced obesity in rats. Proc Natl Acad Sci U S A 100(24):14217–14222. doi:10.1073/pnas.2333912100

142. Sies H (ed) (1985) Oxidative stress: introductory remarks. In: Oxidative stress. Academic, London
143. Son G, Kremer M, Hines IN (2010) Contribution of gut bacteria to liver pathobiology. Gastroenterol Res Pract 2010:453563. doi:10.1155/2010/453563
144. St-Pierre J, Buckingham JA, Roebuck SJ, Brand MD (2002) Topology of superoxide production from different sites in the mitochondrial electron transport chain. J Biol Chem 277(47):44784–44790. doi:10.1074/jbc.M207217200
145. Su GL (2002) Lipopolysaccharides in liver injury: molecular mechanisms of Kupffer cell activation. Am J Physiol Gastrointest Liver Physiol 283(2):G256–G265. doi:10.1152/ajpgi.00550.2001
146. Summers SA (2006) Ceramides in insulin resistance and lipotoxicity. Prog Lipid Res 45(1):42–72. doi:10.1016/j.plipres.2005.11.002
147. Szabo G, Bala S, Petrasek J, Gattu A (2010) Gut-liver axis and sensing microbes. Dig Dis 28(6):737–744. doi:10.1159/000324281
148. Teoh N, Field J, Farrell G (2006) Interleukin-6 is a key mediator of the hepatoprotective and pro-proliferative effects of ischaemic preconditioning in mice. J Hepatol 45(1):20–27. doi:10.1016/j.jhep.2006.01.039
149. Tilg H (2010) The role of cytokines in non-alcoholic fatty liver disease. Dig Dis 28(1): 179–185. doi:10.1159/000282083
150. Tilg H, Moschen AR (2010) Evolution of inflammation in nonalcoholic fatty liver disease: the multiple parallel hits hypothesis. Hepatology 52(5):1836–1846. doi:10.1002/hep.24001
151. Towle HC, Kaytor EN, Shih HM (1997) Regulation of the expression of lipogenic enzyme genes by carbohydrate. Annu Rev Nutr 17:405–433. doi:10.1146/annurev.nutr.17.1.405
152. Triglyceride Coronary Disease Genetics C, Emerging Risk Factors C, Sarwar N, Sandhu MS, Ricketts SL, Butterworth AS, Di Angelantonio E, Boekholdt SM, Ouwehand W, Watkins H, Samani NJ, Saleheen D, Lawlor D, Reilly MP, Hingorani AD, Talmud PJ, Danesh J (2010) Triglyceride-mediated pathways and coronary disease: collaborative analysis of 101 studies. Lancet 375(9726):1634–1639. doi:10.1016/S0140-6736(10)60545-4
153. Utzschneider KM, Kahn SE (2006) Review: the role of insulin resistance in nonalcoholic fatty liver disease. J Clin Endocrinol Metab 91(12):4753–4761. doi:10.1210/jc.2006-0587
154. van der Poorten D, Milner KL, Hui J, Hodge A, Trenell MI, Kench JG, London R, Peduto T, Chisholm DJ, George J (2008) Visceral fat: a key mediator of steatohepatitis in metabolic liver disease. Hepatology 48(2):449–457. doi:10.1002/hep.22350
155. Vonghia L, Michielsen P, Francque S (2013) Immunological mechanisms in the pathophysiology of non-alcoholic steatohepatitis. Int J Mol Sci 14(10):19867–19890. doi:10.3390/ijms141019867
156. Wallace DC, Fan W (2010) Energetics, epigenetics, mitochondrial genetics. Mitochondrion 10(1):12–31. doi:10.1016/j.mito.2009.09.006
157. Watanabe A, Hashmi A, Gomes DA, Town T, Badou A, Flavell RA, Mehal WZ (2007) Apoptotic hepatocyte DNA inhibits hepatic stellate cell chemotaxis via toll-like receptor 9. Hepatology 46(5):1509–1518. doi:10.1002/hep.21867
158. Wei Y, Wang D, Gentile CL, Pagliassotti MJ (2009) Reduced endoplasmic reticulum luminal calcium links saturated fatty acid-mediated endoplasmic reticulum stress and cell death in liver cells. Mol Cell Biochem 331(1–2):31–40. doi:10.1007/s11010-009-0142-1
159. Wei Y, Wang D, Topczewski F, Pagliassotti MJ (2006) Saturated fatty acids induce endoplasmic reticulum stress and apoptosis independently of ceramide in liver cells. Am J Physiol Endocrinol Metab 291(2):E275–E281. doi:10.1152/ajpendo.00644.2005
160. Welsh JA, Karpen S, Vos MB (2013) Increasing prevalence of nonalcoholic fatty liver disease among United States adolescents, 1988-1994 to 2007-2010. J Pediatr 162(3):496–500. doi:10.1016/j.jpeds.2012.08.043
161. Weyer C, Funahashi T, Tanaka S, Hotta K, Matsuzawa Y, Pratley RE, Tataranni PA (2001) Hypoadiponectinemia in obesity and type 2 diabetes: close association with insulin resistance and hyperinsulinemia. J Clin Endocrinol Metab 86(5):1930–1935. doi:10.1210/jcem.86.5.7463

162. Williams CD, Stengel J, Asike MI, Torres DM, Shaw J, Contreras M, Landt CL, Harrison SA (2011) Prevalence of nonalcoholic fatty liver disease and nonalcoholic steatohepatitis among a largely middle-aged population utilizing ultrasound and liver biopsy: a prospective study. Gastroenterology 140(1):124–131. doi:10.1053/j.gastro.2010.09.038
163. Woodcroft KJ, Hafner MS, Novak RF (2002) Insulin signaling in the transcriptional and posttranscriptional regulation of CYP2E1 expression. Hepatology 35(2):263–273. doi:10.1053/jhep.2002.30691
164. Wouters K, van Bilsen M, van Gorp PJ, Bieghs V, Lutjohann D, Kerksiek A, Staels B, Hofker MH, Shiri-Sverdlov R (2010) Intrahepatic cholesterol influences progression, inhibition and reversal of non-alcoholic steatohepatitis in hyperlipidemic mice. FEBS Lett 584(5):1001–1005. doi:10.1016/j.febslet.2010.01.046
165. Wouters K, van Gorp PJ, Bieghs V, Gijbels MJ, Duimel H, Lutjohann D, Kerksiek A, van Kruchten R, Maeda N, Staels B, van Bilsen M, Shiri-Sverdlov R, Hofker MH (2008) Dietary cholesterol, rather than liver steatosis, leads to hepatic inflammation in hyperlipidemic mouse models of nonalcoholic steatohepatitis. Hepatology 48(2):474–486. doi:10.1002/hep.22363
166. Wree A, Broderick L, Canbay A, Hoffman HM, Feldstein AE (2013) From NAFLD to NASH to cirrhosis-new insights into disease mechanisms. Nat Rev Gastroenterol Hepatol 10(11):627–636. doi:10.1038/nrgastro.2013.149
167. Yamaguchi K, Yang L, McCall S, Huang J, Yu XX, Pandey SK, Bhanot S, Monia BP, Li YX, Diehl AM (2007) Inhibiting triglyceride synthesis improves hepatic steatosis but exacerbates liver damage and fibrosis in obese mice with nonalcoholic steatohepatitis. Hepatology 45(6):1366–1374. doi:10.1002/hep.21655
168. Yang S, Zhu H, Li Y, Lin H, Gabrielson K, Trush MA, Diehl AM (2000) Mitochondrial adaptations to obesity-related oxidant stress. Arch Biochem Biophys 378(2):259–268. doi:10.1006/abbi.2000.1829
169. Ye D, Li FY, Lam KS, Li H, Jia W, Wang Y, Man K, Lo CM, Li X, Xu A (2012) Toll-like receptor-4 mediates obesity-induced non-alcoholic steatohepatitis through activation of X-box binding protein-1 in mice. Gut 61(7):1058–1067. doi:10.1136/gutjnl-2011-300269
170. Yki-Jarvinen H (2010) Liver fat in the pathogenesis of insulin resistance and type 2 diabetes. Dig Dis 28(1):203–209. doi:10.1159/000282087
171. You M, Considine RV, Leone TC, Kelly DP, Crabb DW (2005) Role of adiponectin in the protective action of dietary saturated fat against alcoholic fatty liver in mice. Hepatology 42(3):568–577. doi:10.1002/hep.20821

Chapter 13
Oxidative Stress in Nonautoimmune Biliary Diseases

M. Cadamuro, L. Fabris, and M. Strazzabosco

13.1 Cholangiopathies

The diseases primarily affecting the biliary epithelium are collectively termed as cholangiopathies. They comprise a wide group of liver pathologies, most often characterized by a chronic progressive and disabling clinical course. Depending upon the nature of the primary insult, cholangiopathies can be categorized as drug-induced, idiopathic, immunomediated, infectious, inherited/genetic, ischemic, and tumoral (Table 13.1). Unfortunately, pathogenetic mechanisms underlying the progression of these diseases are still uncertain.

If considered as single entity, these pathologies are usually rare, but, as a whole, their prevalence is quite common, accounting for up to 20 % of liver transplants; notably, biliary atresia accounts for the half of liver transplants performed in the pediatric population. All ages can be involved, from neonatal (biliary atresia) to elderly (cholangiocarcinoma), although prevalence is higher in young adults (autoimmune and genetic etiologies).

M. Cadamuro
Department of Surgery and Translational Medicine, University of Milan-Bicocca, Milan, Italy

Department of Molecular Medicine, University of Padova School of Medicine, Padova, Italy

L. Fabris
Department of Molecular Medicine, University of Padova School of Medicine, Padova, Italy

Digestive Disease Section, Yale University School of Medicine, New Haven, CT, USA

M. Strazzabosco (✉)
Department of Surgery and Translational Medicine, University of Milan-Bicocca, Milan, Italy

Digestive Disease Section, Yale University School of Medicine, New Haven, CT, USA
e-mail: mario.strazzabosco@yale.edu; mario.strazzabosco@unimib.it

E. Albano, M. Parola (eds.), *Studies on Hepatic Disorders*,
Oxidative Stress in Applied Basic Research and Clinical Practice,
DOI 10.1007/978-3-319-15539-5_13

Table 13.1 Classification of the cholangiopathies

Drug-induced	Clavulanic acid, floxuridine (by hepatic arterial infusion), non-steroid anti-inflammatory drugs (NSAIDs), statins
Idiopathic	Idiopathic ductopenia, intra- and extrahepatic biliary atresia (BA)
Immunomediated	Chronic allograft rejection, graft vs. host disease, primary biliary cirrhosis (PBC) and its variant autoimmune cholangitis (AIC), primary sclerosing cholangitis (PSC)
Infectious	Bacterial, parasitic, protozoan, and viral
Inherited/genetic	Alagille syndrome, cystic fibrosis, MDR3 deficiency (PFIC-3), fibropolycystic liver diseases [autosomal recessive polycystic kidney disease (ARPKD), Caroli's disease, congenital hepatic fibrosis (CHF)], and polycystic liver diseases [autosomal dominant polycystic kidney disease (ADPKD), isolated polycystic liver disease (PCLD)]
Ischemic	Posttransplant hepatic artery stenosis, systemic vasculitis
Tumoral	Cholangiocarcinoma (CCA)

Despite the differences in etiology and age of onset, cholangiopathies share some common pathological features. These include an intense ductular reaction, associated with the activation of the hepatic progenitor cell (HPC) compartment, progressive portal inflammation resulting in ductopenia, or conversely, bile duct enlargement, and peribiliary fibrosis, that can lead to portal hypertension and its severe complications (ascites, splenomegaly, variceal bleeding) without necessarily evolving to cirrhosis.

Because of the limited knowledge on their pathogenesis, the current curative options for cholangiopathies are quite limited. This chapter will deal with the possible role of oxidative stress, a topic that has hitherto received little attention. Before discussing the specific topic, it seems appropriate to briefly introduce the anatomy and physiology of the biliary system.

13.2 Cholangiocytes and Bile Ducts: Anatomy and Functions

Cholangiocytes are the epithelial cells lining the biliary tree, a complex and highly integrated ductal system responsible, not only for the drainage of the bile, but also for the modification of the primary bile produced by hepatocytes. Intrahepatic biliary tree starts from the canals of Hering, the smallest ductular structure located at the periphery of the portal area and composed partly by hepatocytes and partly by cholangiocytes that links the hepatic canaliculi with the biliary tract system. Canals of Hering represent the putative niche whereby the HPC compartment resides [1–4]. Cholangiocytes may exert both secretory and absorptive functions. In fact, biliary epithelium is involved in the reabsorption of glucose and conjugated bile acids, and plays fundamental secretive functions related to alkalinization and hydration of the primary bile; noteworthy, in humans, 40 % of the bile is secreted by cholangiocytes.

Morphology and functional specialization of cholangiocytes markedly differ along the biliary tree, thus reflecting regionally distinct biological properties. The smallest bile ducts at the interface with the canals of Hering are lined by cuboidal epithelial cells; typically, they express a plastic and reactive phenotype, enabling them to direct the activation of the HPC compartment and the generation of the liver repair response (ductular reaction). Instead, cholangiocytes lining the major ducts are columnar, and mainly possess secretory functions.

13.2.1 Cholangiocyte and Bile Production

Bile released by hepatocytes into the canaliculi (primary bile) is composed by water (more than 90 %), electrolytes, and bile salts. Through the canals of Hering, primary bile is delivered into the distal ramifications of the biliary tree, where it is modified by cholangiocytes, whose activity is regulated by a complex interplay of pro-secretory (acetylcholine, glucagon, secretin, vasoactive intestinal peptide) and anti-secretory signals (somatostatin) [5–8].

Modification of the primary bile depends on the action of a number of carriers and channels expressed on either the apical or the basolateral membrane of cholangiocytes. On the basal side, the import of HCO_3^- is mediated by two different cargo systems, the Na^+-dependent Cl^-/HCO_3^- exchanger (NCHE), and the Na^+/HCO_3^- cotransporter (NCB1). Na^+ turnover is necessary to maintain the correct difference in potential levels among the intra- and extracellular environment, and it is regulated by the Na^+/K^+ adenosine triphosphate (ATP)-ase and the $Na^+/K^+/2Cl^-$ cotransporter (NKCC1), which is also fundamental for the Cl^- uptake by the cell. On the apical side, the cystic fibrosis transmembrane conductance regulator (CFTR), a channel activated by cyclic-adenosine monophosphate (cAMP), regulates Cl^- efflux, that, in turn, is reabsorbed by the action of the Na^+-independent Cl^-/HCO_3^- antiport (AE2), which extrudes bicarbonate into the bile, thus providing biliary alkalinization. Furthermore, Cl^- egress through the CFTR channel generates a negative potential into the ductal lumen that drives the flow of water into the bile duct lumen through Aquaporin-1 and -4; the water efflux is responsible for the bile fluidification. Several other channels responsible for the reabsorption of glutathione, glucose and conjugated bile acids are expressed on the basolateral membrane of cholangiocytes. Other molecules, directly secreted by hepatocytes into the bile, exert pro- or anti-secretory effects in a paracrine fashion, such as ursodeoxycholic acid and ATP, both acting through the ATP-specific receptor P2Y2 [5, 8, 9]. All these mechanisms are ultimately driven and integrated by the adenylyl cyclases (ACs), in particular AC4, 5, 6, 7, 8, 9, and soluble AC [10]. They form a complex enzymatic membrane system expressed on the basolateral side of cholangiocytes, which is activated in response to stimulation of the secretin receptors. By modulating the cAMP intracellular levels, ACs stimulate the extrusion of Cl^- by CFTR, and consequently, of HCO_3^- by the AE2 anion exchanger.

13.2.2 Cholangiocyte and Liver Repair

Cholangiocytes located at the smallest ramifications of the biliary tree possess abilities that are fundamental for directing repair mechanisms to liver damage. In fact, they may acquire some properties of the mesenchymal phenotype (*plasticity*), they may secrete a vast array of proinflammatory mediators, including cytokines and chemokines (*reactivity*, hence the term of "reactive cholangiocytes" or "reactive ductular cells," RDC), and they may activate the HPC compartment (*staminality*) [8, 11]. Thanks to these functions, cholangiocytes react to the damage by increasing their number, and reaching the site of injury where they establish an intense cross talk with other liver nonepithelial cell populations, such as portal fibroblasts, Kupffer cells, lymphocytes, neutrophils, and endothelial cells. Altogether, these interactions lead to the development of the "ductular reaction," a multicellular complex that, though behaving as an adaptive response, may ultimately be responsible for pathologic repair during chronic liver damage. Recent studies indicate a strong correlation between the extent of ductular reaction and the severity of liver fibrosis [12–14].

RDC are believed to originate from the proliferation of preexisting bile ducts and/or from HPC activation or from transdifferentiation of hepatocytes. Originally, activation of the HPC compartment was reported in the mouse model of CCl_4 intoxication, followed by administration of acetaminophen (AAF), to inhibit hepatocyte proliferation [15]. However, following studies showed that HPC expansion is common also in human chronic liver diseases, as in nonalcoholic and alcoholic steatohepatitis, chronic viral hepatitis, and biliary atresia (BA) [1, 12, 14, 16, 17]. HPC can be identified by the co-expression of both cholangiocyte and hepatocyte markers, including Cytokeratins (K) 7, K8, K18, and K19, EpCAM, albumin, Hepar-1 [3, 18, 19], and of stem cell markers, such as CD34, OV-6, and c-Kit [20, 21]. RDC are organized into dysmorphic, highly anastomosed branching structures, devoid of a discernible lumen, aligning along the portal tract interface. RDC have a flattened or cuboidal shape, and in addition to the classic biliary markers, i.e., K7, K8, K18, K19 [2, 3], and EpCAM [22], they express neuroendocrine features, including Chromogranin-A [23] and the neural cell adhesion molecule (NCAM) [2, 22, 23], as well as stem cell markers, such as c-Kit [20] and OV-6 [19], whereas they do not express EMA, a marker of mature biliary phenotype [22, 24]. Expression of features of immaturity is associated with a reduction in the luminal secretory abilities (water, ions, and other solutes), and with a loss of the cell polarity. Given the capability to secrete a number of proinflammatory and chemotactic mediators and growth factors, RDC orchestrate the recruitment and activation of mesenchymal and inflammatory cells, which cooperate in the generation of the ductular reaction. The secretory repertoire of RDC is wide and encompasses vascular endothelial growth factor (VEGF), Angiopoietin-1 and -2 (Ang-1, -2), endothelin-1 (ET-1), platelet-derived growth factor-BB (PDGF-BB), and -DD (PDGF-DD), transforming growth factor-β2 (TGFβ2) and connective tissue growth factor (CTGF), thus making RDC as the "pace-maker" of liver repair [25]. Once activated, fibroblasts are instead the classic effectors of fibrogenesis given their distinctive ability to release several components of the extracellular matrix (fibronectin, collagen, elastin, tenascin). Furthermore, they secrete a large number of soluble factors, such as interleukin-6 (IL-6), IL-8, tumor necrosis factor-α (TNF-α), interferon-γ (IFNγ), CCL2 (or monocyte chemotactic protein-1, MCP-1), cytokine-induced neutrophil chemoattractant (CINC), and nitric oxide (NO), which variably modulate the evolution of liver repair mechanisms by regulating the activity of lymphocytes, macrophages, and neutrophils [5].

13.3 Reactive Oxidative Species in Cholangiopathies

In chronic cholangiopathies, activation of the ductular reaction results in the generation of a brisk proinflammatory microenvironment in the portal area, which induces profound metabolic changes in the adjacent liver cells. Cytokines secreted by inflammatory cells can affect the function of multiple cell types involved in liver repair, including cholangiocytes, hepatocytes, endothelial cells, and hepatic stellate cells. In particular, TNF-α, IFNγ, and IL-6 stimulate the activation of the inducible form of the nitric oxide synthase (iNOS) in cholangiocytes [26]. Inducible-NOS synthesizes high amounts of nitric oxide (NO), a gaseous molecule derived from the biotransformation of L-arginine, which exerts pleiotropic effects, in both physiologic and pathologic contexts [27–30] (Fig. 13.1). In addition, NO is generated, at lower concentrations, by two other synthases, the neuronal NOS (nNOS), and the endothelial NOS (eNOS). At low concentrations, NO mostly possesses antimicrobial and antiviral activities as well as signaling functions. In the biliary epithelium, low concentration NO stimulates choleresis, inducing the efflux of Cl^- and HCO_3^- and of biliary glutathione. On the contrary, at higher concentrations, NO stimulates inflammatory cells to release several proinflammatory and pro-fibrotic cytokines, among which Chemokine (C-C motif) ligand (CCL) 1, IL-6, IL-8, interferon β, MCP-1, Macrophage Inflammatory Protein (MIP)-1, MIP-2, TGFβ, and TNF-α [31–34], which in turn, induce cholestasis and bile duct proliferation. These are harmful effects at potential risk for neoplastic transformation, related to the generation of reactive oxidative species (ROS), as peroxynitrites ($ONOO^-$); they are induced by the reaction of NO_2^- produced by iNOS with O_2^-, formed by superoxide dismutase (SOD) [35]. In fact, peroxynitrites modify tyrosine residues in membrane and structural proteins, including cytoskeletal filaments, to form nitrotyrosine, which interfere with their correct assembly. Protein nitrosylation is currently used as a footprint of the NO accumulation that can be detected in vivo.

Another product of oxidative metabolism relevant in bile duct injury is the arachidonic acid (AA), which derives from the degradation of membrane glycerophospholipids due to the activation of phospholipases A_2 (PLA_2s), in particular $PLA_2\alpha$. Biotransformation of AA is closely related to NO: increased local concentration of NO produced by iNOS stimulates the production and activation of cyclooxygenase 2 (COX-2), the enzyme responsible for the conversion of AA to prostaglandin H_2 (PGH_2). Afterwards, a number of specific PG synthase, such as PGE_2, PGD_2, $PGF_2\alpha$, PGI_2, and thromboxane A_2, convert PGH_2 in different, highly reactive prostaglandins [36, 37].

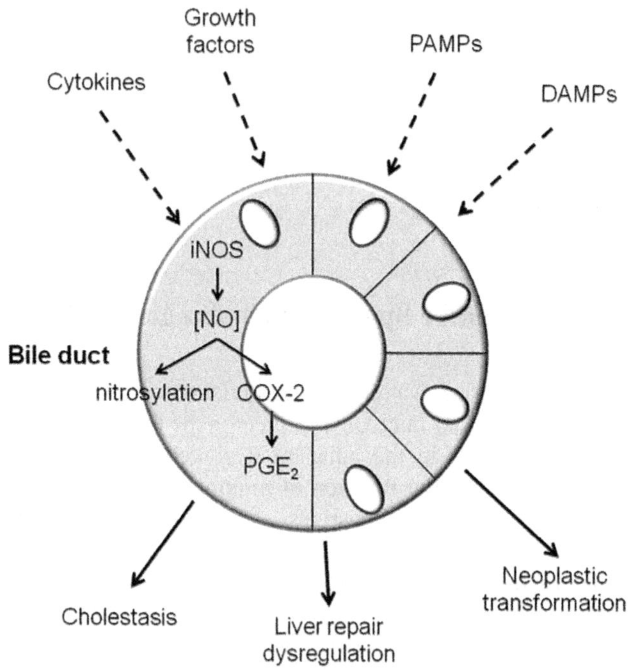

Fig. 13.1 The sustained activation of the enzymes promoting the oxidative stress response (i.e., iNOS and COX-2) caused by chronic inflammation or other biliary injuries, leads to the most severe consequences of cholangiopathies, including cholestasis, dysregulation of liver repair mechanisms, and neoplastic biliary epithelial transformation

PGE2 can bind to different cognate receptors, EP_1, EP_2, EP_3, and EP_4 subtypes, belonging to the G-proteins-coupled receptor family. PGE_2 exerts different effects on epithelial cell proliferation, motility, and invasiveness, depending upon the specific receptor binding.

Both iNOS and COX-2 have been found to be overexpressed by RDC in BA. iNOS immunoreactivity positively correlated with the degree of apoptosis of cholangiocytes, with the content of NO metabolites and with the expression of NF-κB, a transcription factor modulating inflammatory responses which is significantly activated in BA [38–40]. On the other hand, expression of COX-2 by RDC closely correlated with the severity of cholestasis in BA patients [41]. Altogether, these results confirmed that iNOS and COX-2 are strongly involved in the pathogenesis of BA.

13.3.1 *Effects of Inflammation on ROS Generation*

In cholangiopathies, ROS are generated following a dual mechanism, with or without the involvement of pathogens. In case of sepsis or endotoxemia, biliary epithelium may react to different stimuli of bacterial origin, represented by lipopolysaccharides (LPS), DNA, or RNA fragments, and flagellin, collectively recognized as pathogen-associated molecular patterns (PAMPs), by activating the Toll-like receptor (TLR) pathways. In particular, TLR4 mediates signal transduction in response to LPS. Following the binding with LPS, TLR4 dimerizes and recruits the TIR-domain containing adaptor molecules (MyD88, Mal, TRIF, and TRAM) [42–44]. The TLR4-activated intracellular signaling generates ROS acting through the transcription factor NF-κB that directly stimulates iNOS expression on cholangiocytes [42]. Alternatively, ROS may originate from the dysfunction or necrosis of resident liver epithelial cells in absence of pathogens ("sterile inflammation"). This mechanism is driven by alarmins. Alarmins are a group of cell components also known as damage-associated molecular patterns (DAMPs), including DNA fragments, purine nucleotides (ATP), S100 proteins (S100A8-9-12), several heat shock proteins (HSP), and the high mobility box-1 (HMGB-1) protein. HMGB-1 in particular, plays a pivotal role in several liver diseases of metabolic, ischemic, and alcoholic etiology [45, 46]. DAMP-induced cytokine/chemokine secretion includes CCL2, CXCL1, CXCL10, IL-1, LIX, IL-6, TNF-α, and IFNγ, and occurs in a NF-κB-dependent fashion; it leads to the recruitment of immune cells to the site of injury, generating a positive loop that further recruits inflammatory cells and stimulates NO biosynthesis in the site of inflammation.

13.3.2 NO Effects on Bile Secretion

Whereas local nanomolar concentrations of NO exert choleretic effects in cholangiocytes, higher concentrations (micromolar) of NO induce cholestasis and interfere with the mechanisms of bile fluidification and alkalinization. The ambivalent role of local NO was originally shown in vitro in rat cholangiocytes [26, 47]. Among the different membrane transporters, AE2 and CFTR are particularly susceptible to NO effects. In physiological conditions, low concentrations of NO generated by eNOS, are necessary for the normal bile secretion. In cholangiopathies, proinflammatory cytokines (IL-1, IL-6, TNF-α, and IFNγ) released by inflammatory cells and RDC negatively affect the function of AE2 and CFTR [48]. In addition, TNF-α, IL-6 and IFNγ synergistically stimulate the cholangiocyte expression of iNOS that generates a sustained increase in the intracellular levels of NO and consequently of peroxynitrite. ROS interfere with the activity of ACs, thereby deactivating the cAMP-PKA-dependent secretion of HCO_3^- operated by the AE2 transporter [26] (Fig. 13.2). This halts the correct turnover of Cl^- usually maintained by the coordinated action of AE2 and CFTR. The failure of HCO_3^- and Cl^- secretion affects the preservation of the osmolarity gradient, which impairs the water transport performed by aquaporins. Inhibitors of peroxynitrite formation and NO scavengers, such as uric acid, manganese(III)-tetrakis(4-benzoic acid)-porphyrin (MnTBAP), or trolox, are able to restore in vitro, the normal bile secretion in rat bile duct units following exposure to proinflammatory cytokines by normalizing the intracellular levels of cAMP [26].

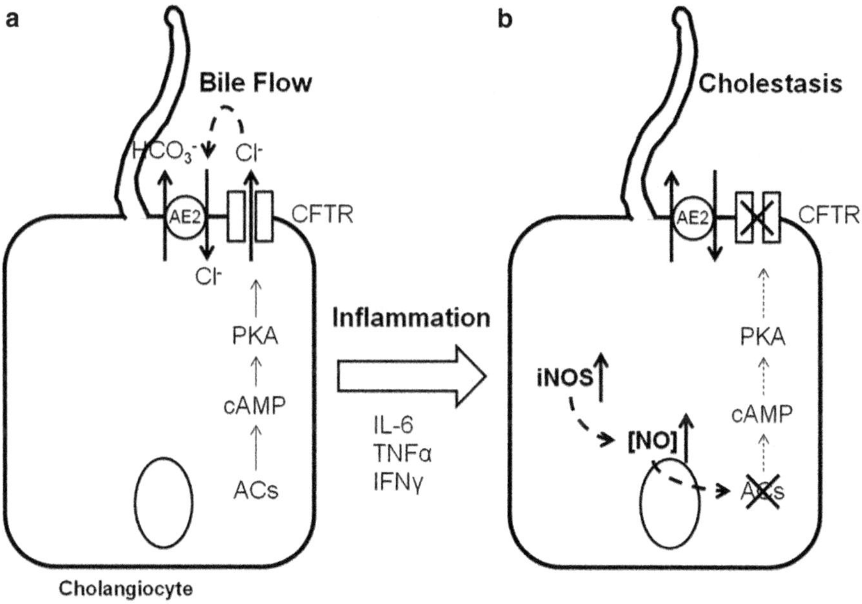

Fig. 13.2 NO effects on bile secretion. (**a**) In normal conditions, cholangiocytes secrete and modify the primary "hepatocellular" bile through a complex interplay of pumps and ion exchangers located on either the apical or the basolateral side. Bile alkalinization is provided by the initial extrusion of Cl^- anions through the CFTR, a cAMP-dependent channel. Then, Cl^- is reabsorbed by the AE2 antiport pump that in turn, extrudes HCO_3^-, a mechanism governed by several AC isoforms. (**b**) In pathologic conditions, NO generated by the upregulation of iNOS, interferes with the activity of ACs, and inhibits the cAMP-PKA-dependent, AE2-mediated secretion of HCO_3. The failure of HCO_3^- and Cl^- turnover affects the osmolarity gradient between the ductal lumen and the cholangiocyte cytoplasm, thereby halting the water transport operated by aquaporins, finally resulting in the bile stagnation into the biliary tree

13.3.3 NO Effects on Liver Repair

Recent evidence indicates that COX2 can activate the HPC compartment. Studies on a mouse model fed with choline-deficient, ethionine-supplemented diet (CDE), showed that COX-2 mRNA expression in total liver correlates with the number of HPCs, whilst selective COX-2 inhibition reduces in vivo the number of HPC, and in vitro, the phosphorylation of AKT in an immortalized HPC-derived cell line [49]. The ability of COX-2 to stimulate HPC proliferation was confirmed further in a rat model of 2-acetylaminofluorene (2-AAF) intoxication associated with 70 % partial hepatectomy. In this model, during liver regeneration, HPCs proliferate in response to COX-2, activated by several cytokines and growth factors released from inflammatory cells, hepatocytes, and RDC. Even in this model, HPC activation by COX-2 is mediated by an AKT-dependent, extracellular-signal-regulated kinases (ERK) 1/2-independent pathway. Specific chemical inhibition of COX-2 reduces HPC proliferation, as well as AKT, but not ERK1/2 phosphorylation in vitro [50].

13.3.4 NO and Cholangiocarcinoma

Several chronic cholangiopathies, such as PSC and fibropolycystic liver diseases (CHF and CD), are characterized by an increased risk to developing cholangiocarcinoma (CCA), a highly aggressive malignancy originating from the biliary epithelium. The pro-carcinogenic effects of long-lasting inflammatory processes are well documented in several organs, and the relative role played by oxidative stress in this context is largely recognized [51–53]. In fact, large amounts of NO produced by iNOS stimulated by proinflammatory mediators may induce malignant transformation of epithelial cells directly, by inducing DNA damage and by repressing the activity of DNA repair mechanisms or indirectly, by affecting the cell proliferation/apoptosis ratio.

DNA damage and inhibition of the DNA reparative systems. A persistent portal inflammatory state leads to a sustained upregulation of iNOS and relatively, to NO accumulation in cholangiocytes upon the effect of proinflammatory cytokines, including TNF-α, IFNγ, TGFβ, and different ILs. ROS are responsible for the lipid peroxidation of membrane proteins. As previously mentioned, peroxynitrite is a highly instable, reactive molecule able to induce multiple single-stranded and double-stranded DNA lesions and ruptures [54]. Furthermore, NO and peroxynitrites are responsible for the formation of 8-oxo-7,8-dihydro-2′-deoxyguanosine (8-oxodG), and 8-Nitro guanine, two potentially mutagenic compounds promoting the G:C → T:A conversion, and inducing the formation of nitrotyrosine, as shown in epithelial cells of chronic gastritis related to *H. pylori* infection [55]. NO and ROS may also inhibit the activity of reparative systems of DNA damage, necessary to maintain the genetic stability of the cells; these are provided by enzymatic proteins containing a catalytic subunit rich in cysteine residues which is extremely sensitive to nitrosylation. In particular, the 8-oxo-deoxiguanine DNA glycosylase 1 (hOGG1), the most important DNA repairing enzyme, can be inactivated by NO and peroxynitrites, thus favoring the progressive accumulation of DNA damages in CCA [56].

Effects on the regulation of the balance epithelial cell proliferation/apoptosis. Recent studies have shown that in CCA, iNOS upregulation leads to COX-2 activation, which in tumoral cells, stimulates proliferation and escape from apoptosis. Studies on immortalized cholangiocytes demonstrate that COX-2 expression is modulated by the activation of p38-mitogen-activated protein kinases (MAPK) and c-Jun N-terminal kinases (JNK) pathways [57, 58]. Murine cholangiocytes challenged with a mix of LPS + TNF-α (reproducing the inflammatory microenvironment of cholangiopathies), stimulated in vitro the expression of both iNOS and COX-2, whose activation was confirmed by the increased production of PGE_2. These effects were reverted by the specific inhibition of either p38/MAPK or JNK, but not of the ERK1/2 pathway, in contrast with what reported in other cell types [58, 59].

In CCA, EP_1 activation by PGE_2 induces a tumorigenic effect by enhancing tumor cell proliferation and invasiveness. In CCA cells, EP_1 activation by selective agonists leads to a proliferative effect through a Src-mediated transactivation of

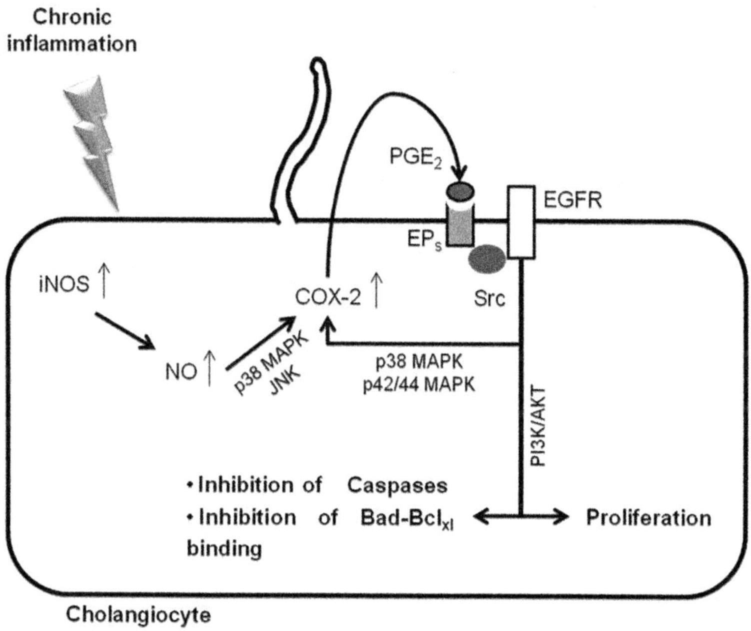

Fig. 13.3 NO modulation of the balance cell proliferation/apoptosis in CCA. In CCA, the NO-induced COX-2 activation, mediated by the p38 MAPK/JNK axis, generates high amounts of PGE_2. PGE_2, released into the luminal space, binds to the specific EP receptors and transactivates EGFR, by a Src binding protein-mediated mechanism. The PI3K/AKT pathway responsible for the proliferative and antiapoptotic responses in CCA cells then transduces the EP/EGFR signal. Moreover, EGFR activation, also activating the p38 MAPK/p42/44 MAPK pathways, further co-operates to the upregulation of COX-2

epidermal growth factor receptor (EGFR) (Fig. 13.3). This increases the phosphorylation of ERKs and AKT [60, 61], in conjunction with an invasive effect dependent upon the cAMP-induced activation of the transcription factor cAMP response element-binding protein (CREB) stimulating the metalloproteinase-2 (MMP2) production [62]. These effects can be counteracted by both EP_1 and EGFR antagonism [60–62].

PGE_2 specific binding to the receptor subtype EP4 increases cell motility and proliferation by activating the phosphatidylinositol-3 kinase/AKT (PI3K/AKT) pathway, but this effect has been demonstrated only in colon cancer cells, a malignancy where relevance of COX2 activation is largely recognized [63, 64]. On the other way, COX2 synthesis and activation can be stimulated in cultured human CCA cell lines by bile acids, through mechanisms dependent on EGFR-p42/44 and EGFR-p38 signaling (Fig. 13.3). Cholangiocytes exposed to a range of bile acids, including deoxycholate, taurodeoxycholate, taurochenodeoxycholate, glycodeoxycholate, and glycochenodeoxycholate, variably upregulate COX-2 expression, an effect counteracted by specific inhibitors of EGFR, p38 MAPK, and rapidly

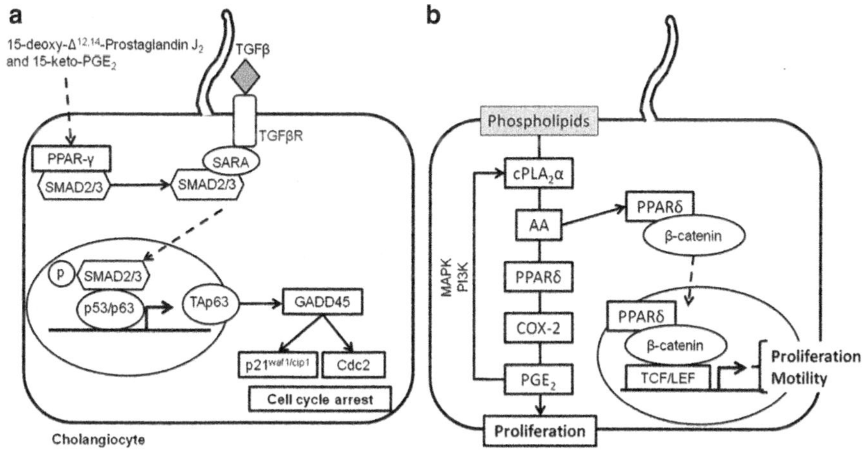

Fig. 13.4 Opposite effects of PPAR-γ and PPAR-δ in CCA. (**a**) Following activation, PPAR-γ detaches from SMAD2/3, that is phosphorylated by TGFβ; pSMAD2/3 translocates into the nucleus where it binds to p53 and activates the transcription of TAp63 and its downstream effectors GADD45, $p21^{waf1/cip1}$, and Cdc2, ultimately blocking the cellular proliferation. (**b**) In contrast, PPAR-δ stimulates proliferation and motility of CCA cells: on one side, by upregulating the expression of COX-2, and on the other, by forming a multimer with β-catenin that translocates to the nucleus, whereby it activates TCF/LEF, thus inducing the transcription of genes regulating cell motility and proliferation

accelerated fibrosarcoma (Raf) and MEK1 mediating the P42/44-MAPK pathway [65]. Notably, both signaling cascades activating COX-2 act in concert to increase the PGE_2 biosynthesis from arachidonic acid, thus stimulating the EGFR-mediated tumor proliferation, invasion, and metastatization [64, 65].

In CCA cells, COX-2 is typically expressed in close proximity to the nucleus where PGI_2 and PGE_2 can bind to different members of the peroxisome proliferator-activated receptor family (PPAR), in particular PPAR-γ and PPAR-δ, thereby activating them. Several studies outlined the opposing effects of PPAR-γ and PPAR-δ activation in CCA, where they behave as tumor suppressor and pro-oncogenic, respectively (Fig. 13.4). CCA cell lines treated with PPAR-γ ligands showed a strong inhibition of cell proliferation. In CCA cells, 15-deoxy-$\Delta^{12,14}$-prostaglandin J_2 inhibits the activation of the G2/M cyclin-dependent kinase, Cdc2, through a p53-dependent mechanism mediated by two downstream effectors, growth arrest and DNA damage 45 (GADD45) and $p21^{WAF1/Cip1}$, an effect erased by a dominant negative inhibition of p53 [66]. Similarly, in vitro treatment of CCA cell lines with 15-keto-prostaglandin E_2 (15-keto-PGE_2), a metabolite of PGE_2 formed by the 15-hydroxyprostaglandin dehydrogenase, results in inhibition of cell proliferation and migration by activating PPAR-γ. Following the binding of 15-keto-PGE_2, SMAD2/3 detaches from PPAR-γ to form a multimer with SMAD anchor for receptor activation (SARA) and TGFβR1. This complex induces the phosphorylation of SMAD2/3 that enters the nucleus where it binds to p53/p63 in the promoter region of the *TAp63* gene, inducing the transcription of the TAp63 protein and inhibiting cancer cell proliferation [67] (Fig. 13.4a). In contrast to PPAR-γ, PPAR-δ exerts a proliferative effect in CCA. In vitro studies showed that AA binding to PPAR-δ upregulates COX-2 [68]. Nevertheless, by binding to PPAR-δ, PGE_2 regulates the biotransformation of phospholipids in AA through a MAPK- and PI3K-mediated pathway that activates $cPLA_2\alpha$ [68]. Furthermore, recent papers demonstrate the presence of a cross talk among PPARs (in particular PPAR-δ), β-catenin, and $cPLA_2\alpha$ signaling. In addition to behave as adhesion protein, β-catenin regulates the Wnt signaling. In absence of Wnt stimulation, β-catenin is phosphorylated in the cytoplasm, where it undergoes proteosomal degradation by ubiquitination. In conditions of Wnt activation, phosphorylation of β-catenin is inhibited: therefore, β-catenin can accumulate in the cytoplasm and translocate into the nucleus where it activates T cell factor (TCF) and lymphoid enhancer-binding factor-1 (LEF-1), which induce the transcription of target genes involved in cell motility, proliferation, and cell reprogramming. Overexpression of $cPLA_2\alpha$ and of its bioactive metabolic product AA, induce the binding of the β-catenin/PPAR-δ complex to the TCF/LEF transcription factors, which become transcriptionally active [69] (Fig. 13.4b). As a positive loop, the activation of the β-catenin signaling can stimulate the expression of PPAR-δ, an effect able to sustain the persistent activation of the β-catenin/PPAR-δ-mediated TCF/LEF complex, as demonstrated in vitro in several cell lines of human colon, lung, and liver carcinomas [70].

In condition of persistent inflammation, accumulation of NO and its derivatives are responsible not only for the increased cell proliferation, but even for the inhibition of pro-apoptotic mechanisms. In hepatocytes and in other epithelial cells, NO can directly inhibit the activation of caspases 3, 8, and 9, through nitrosylation of the cysteine residues present in the active site of the caspases [71, 72]. The inactivation of the caspase-induced apoptosis may also derive from the cross-activation of EGFR by PGE_2 and its receptor EP_1. EGFR activation in rat CCA cells, is responsible for the activation of the PI3K/AKT pathway that inhibits the Bad-Bcl_{xl} binding [37] and the following translocation of Bax to the mitochondria, a fundamental step for activating the caspase cascade. This antiapoptotic effect induced by EGFR activation is counteracted by the treatment with celecoxib, a selective inhibitor of COX-2 [73]. NO may also promote antiapoptotic mechanisms by interacting with the Notch pathway. Notch (1–4) and its ligands Delta-like and Jagged are fundamental players in cell fate determination and in the survival of several cell types, including cholangiocytes. In mouse CCA cell lines, Ishimura and colleagues [74] showed that NO is able to induce the cleavage of the intracellular domain of Notch1 (NICD), which translocates into the nucleus to form a transcriptional complex with recombination signal-binding protein 1 for J-kappa (RBPj), switching on several target genes, among which hairy and enhancer of split-1 (Hes-1). This mechanism, mediated by the activation of the JNK1/2 signaling, is responsible for an increased cell survival and a reduced apoptosis of CCA cells.

13.4 Conclusions

In chronic cholangiopathies with persistent inflammation, the release of proinflammatory mediators including cytokines, chemokines, and growth factors by either RDC or inflammatory cells recruited into the portal area, is the main mechanism responsible for the upregulation of iNOS and COX2, leading in turn, to the generation of NO and its derivatives, and of PGE2, respectively. NO and PGE_2 may profoundly affect critical functions of the biliary epithelium, such as liver repair and bile secretion, and may also predispose it to malignant transformation by interfering with the DNA reparative systems and with several intracellular pathways governing the balance between cell proliferation and apoptosis [29, 30, 53]. Unraveling the molecular mechanisms underlying the generation of ROS and the consequences of their accumulation is a fundamental step for the development of new therapeutic approaches, in both inflammatory cholangiopathies and CCA.

References

1. Sell S (2001) Heterogeneity and plasticity of hepatocyte lineage cells. Hepatology 33:738–750
2. Libbrecht L, Roskams T (2002) Hepatic progenitor cells in human liver diseases. Semin Cell Dev Biol 13:389–396
3. Roskams TA, Libbrecht L, Desmet VJ (2003) Progenitor cells in diseased human liver. Semin Liver Dis 23:385–396
4. Roskams TA, Theise ND, Balabaud C et al (2004) Nomenclature of the finer branches of the biliary tree: canals, ductules, and ductular reactions in human livers. Hepatology 39:1739–1745
5. Lazaridis KN, Strazzabosco M, Larusso NF (2004) The cholangiopathies: disorders of biliary epithelia. Gastroenterology 127:1565–1577
6. Strazzabosco M (1997) New insights into cholangiocyte physiology. J Hepatol 27:945–952
7. Strazzabosco M, Spirli C, Okolicsanyi L (2000) Pathophysiology of the intrahepatic biliary epithelium. J Gastroenterol Hepatol 15:244–253
8. Strazzabosco M, Fabris L, Spirli C (2005) Pathophysiology of cholangiopathies. J Clin Gastroenterol 39:S90–S102
9. Strazzabosco M, Fabris L (2008) Functional anatomy of the normal bile ducts. Anat Rec 291:653–660
10. Strazzabosco M, Fiorotto R, Melero S et al (2009) Differentially expressed adenylyl cyclase isoforms mediate secretory functions in cholangiocyte subpopulation. Hepatology 50:244–252
11. Fabris L, Strazzabosco M (2011) Epithelial-mesenchymal interactions in biliary diseases. Semin Liver Dis 31:11–32
12. Clouston AD, Powell EE, Walsh MJ et al (2005) Fibrosis correlates with a ductular reaction in hepatitis C: roles of impaired replication, progenitor cells and steatosis. Hepatology 41:809–818
13. Clouston AD, Jonsson JR, Powell EE (2007) Steatosis as a cofactor in other liver diseases: hepatitis C virus, alcohol, hemochromatosis, and others. Clin Liver Dis 11:173–189
14. Fabris L, Cadamuro M, Guido M et al (2007) Analysis of liver repair mechanisms in Alagille syndrome and biliary atresia reveals a role for notch signaling. Am J Pathol 171:641–653
15. Ghoshal AK, Mullen B, Medline A et al (1983) Sequential analysis of hepatic carcinogenesis. Regeneration of liver after carbon tetrachloride-induced liver necrosis when hepatocyte proliferation is inhibited by 2-cetylaminofluorene. Lab Invest 48:224–230

16. Lowes KN, Brennan BA, Yeoh GC et al (1999) Oval cell numbers in human chronic liver diseases are directly related to disease severity. Am J Pathol 154:537–541
17. Eleazar JA, Memeo L, Jhang JS et al (2004) Progenitor cell expansion: an important source of hepatocyte regeneration in chronic hepatitis. J Hepatol 41:983–991
18. Ruck P, Xiao JC, Pietsch T et al (1997) Hepatic stem-like cells in hepatoblastoma: expression of cytokeratin 7, albumin and oval cell associated antigens detected by OV-1 and OV-6. Histopathology 31:324–329
19. van den Heuvel MC, Slooff MJ, Visser L et al (2001) Expression of anti-OV6 antibody and anti-N-CAM antibody along the biliary line of normal and diseased human livers. Hepatology 33:1387–1393
20. Baumann U, Crosby HA, Ramani P et al (1999) Expression of the stem cell factor receptor c-kit in normal and diseased pediatric liver: identification of a human hepatic progenitor cell? Hepatology 30:112–117
21. Crosby HA, Kelly DA, Strain AJ (2001) Human hepatic stem-like cells isolated using c-kit or CD34 can differentiate into biliary epithelium. Gastroenterology 120:534–544
22. Fabris L, Strazzabosco M, Crosby HA et al (2000) Characterization and isolation of ductular cells coexpressing neural cell adhesion molecule and Bcl-2 from primary cholangiopathies and ductal plate malformations. Am J Pathol 156:1599–1612
23. Roskams T, van den Oord JJ, De Vos R et al (1990) Neuroendocrine features of reactive bile ductules in cholestatic liver disease. Am J Pathol 137:1019–1025
24. Momburg F, Moldenhauer G, Hämmerling GJ et al (1987) Immunohistochemical study of the expression of a Mr 34,000 human epithelium-specific surface glycoprotein in normal and malignant tissues. Cancer Res 47:2883–2891
25. Desmet VJ (1998) Ludwig symposium on biliary disorders—part I. Pathogenesis of ductal plate abnormalities. Mayo Clin Proc 73:80–89
26. Spirlì C, Fabris L, Duner E et al (2003) Cytokine-stimulated nitric oxide production inhibits adenylyl cyclase and cAMP-dependent secretion in cholangiocytes. Gastroenterology 124:737–753
27. Moncada S, Palmer RMJ, Higgs EA (1991) Nitric oxide: physiology, pathophysiology and pharmacology. Pharmacol Rev 43:109–141
28. Clemens MG (1999) Nitric oxide in liver injury. Hepatology 30:1–5
29. Novo E, Parola M (2008) Redox mechanisms in hepatic chronic wound healing and fibrogenesis. Fibrogenesis Tissue Repair 1:5
30. Novo E, Parola M (2012) The role of redox mechanisms in hepatic chronic wound healing and fibrogenesis. Fibrogenesis Tissue Repair 5(Suppl 1):S4, eCollection 2012
31. Walley KR, McDonald TE, Higashimoto Y et al (1999) Modulation of proinflammatory cytokines by nitric oxide in murine acute lung injury. Am J Respir Crit Care Med 160:698–704
32. Shihab FS, Yi H, Bennett WM et al (2000) Effect of nitric oxide modulation on TGFbeta1 and matrix proteins in chronic cyclosporine nephrotoxicity. Kidney Int 58:1174–1185
33. Schwentker A, Vodovotz Y, Weller R et al (2002) Nitric oxide and wound repair: role of cytokines? Nitric Oxide 7:1–10
34. Kobayashi Y (2010) The regulatory role of nitric oxide in proinflammatory cytokine expression during the induction and resolution of inflammation. J Leukoc Biol 88:1157–1162
35. Fang FC (2004) Antimicrobial reactive oxygen and nitrogen species: concepts and controversies. Nat Rev Microbiol 2:820–832
36. Gupta RA, Dubois RN (2001) Colorectal cancer prevention and treatment by inhibition of cyclooxygenase-2. Nat Rev Cancer 1:11–21
37. Wu T (2005) Cyclooxygenase-2 and prostaglandin signaling in cholangiocarcinoma. Biochim Biophys Acta 1755:135–150
38. Vejchapipat P, Poomsawat S, Imvised T et al (2008) Overexpression of hepatic inducible nitric oxide synthase in biliary atresia. Hepatol Res 38:1018–1025
39. Huang L, Si XM, Feng JX (2010) NF-kappaB related abnormal hyper-expression of iNOS and NO correlates with the inflammation procedure in biliary atresia livers. Pediatr Surg Int 26:899–905

40. Harada K, Sato Y, Itatsu K et al (2007) Innate immune response to double-stranded RNA in biliary epithelial cells is associated with the pathogenesis of biliary atresia. Hepatology 46:1146–1154
41. Honsawek S, Klaikeaw N, Vejchapipat P et al (2010) Cyclooxygenase-2 overexpression is associated with clinical outcome in biliary atresia. Eur J Pediatr Surg 20:164–168
42. Barton GM, Medzhitov R (2002) Toll-like receptors and their ligands. Curr Top Microbiol Immunol 270:81–92
43. Iwasaki A, Medzhitov R (2004) Toll-like receptor control of the adaptive immune responses. Nat Immunol 5:987–995
44. Seki E, Brenner DA (2008) Toll-like receptors and adaptor molecules in liver disease: update. Hepatology 48:322–335
45. Tsan MF, Gao B (2004) Endogenous ligands of Toll-like receptors. J Leukoc Biol 76(3):514–519
46. Kubes P, Mehal WZ (2012) Sterile inflammation in the liver. Gastroenterology 143(5):1158–1172
47. Trauner M, Mennone A, Gigliozzi A et al (1998) Nitric oxide and guanosine 3′,5′-cyclic monophosphate stimulate bile secretion in isolated rat hepatocyte couplets, but not in isolated bile duct units. Hepatology 28:1621–1628
48. Spirlì C, Nathanson MH, Fiorotto R et al (2001) Proinflammatory cytokines inhibit secretion in rat bile duct epithelium. Gastroenterology 121:156–169
49. Davies RA, Knight B, Tian YW et al (2006) Hepatic oval cell response to the choline-deficient, ethionine supplemented model of murine liver injury is attenuated by the administration of a cyclo-oxygenase 2 inhibitor. Carcinogenesis 27:1607–1616
50. Bae SH, Oh SH, Yoon SK et al (2011) Proliferation of hepatic oval cells via cyclooxygenase-2 and extracellular matrix protein signaling during liver regeneration following 2-AAF/partial hepatectomy in rats. Gut Liver 5:367–376
51. Lala PK, Chakraborty C (2001) Role of nitric oxide in carcinogenesis and tumour progression. Lancet Oncol 2:149–156
52. Wise C, Pilanthananond M, Perry BF et al (2008) Mechanisms of biliary carcinogenesis and growth. World J Gastroenterol 14:2986–2989
53. Burke AJ, Sullivan FJ, Giles FJ et al (2013) The yin and yang of nitric oxide in cancer progression. Carcinogenesis 34:503–512
54. Jaiswal M, LaRusso NF, Burgart LJ et al (2000) Inflammatory cytokines induce DNA damage and inhibit DNA repair in cholangiocarcinoma cells by a nitric oxide-dependent mechanism. Cancer Res 60:184–190
55. Mannick EE, Bravo LE, Zarama G et al (1996) Inducible nitric oxide synthase, nitrotyrosine, and apoptosis in Helicobacter pylori gastritis: effect of antibiotics and antioxidants. Cancer Res 56:3238–3243
56. Jaiswal M, LaRusso NF, Shapiro RA et al (2001) Nitric oxide-mediated inhibition of DNA repair potentiates oxidative DNA damage in cholangiocytes. Gastroenterology 120:190–199
57. Yoon JH, Canbay AE, Werneburg NW et al (2004) Oxysterols induce cyclooxygenase-2 expression in cholangiocytes: implications for biliary tract carcinogenesis. Hepatology 39:732–738
58. Ishimura N, Bronk SF, Gores GJ (2004) Inducible nitric oxide synthase upregulates cyclooxygenase-2 in mouse cholangiocytes promoting cell growth. Am J Physiol Gastrointest Liver Physiol 287:G88–G95
59. Eligini S, Habib A, Lebret M, Créminon C, Lévy-Toledano S, Maclouf J (2001) Induction of cyclo-oxygenase-2 in human endothelial cells by SIN-1 in the absence of prostaglandin production. Br J Pharmacol 133:1163–1171
60. Han C, Demetris AJ, Stolz DB et al (2006) Modulation of Stat3 activation by the cytosolic phospholipase A2alpha and cyclooxygenase-2-controlled prostaglandin E2 signaling pathway. J Biol Chem 281:24831–24846
61. Zhang L, Jiang L, Sun Q et al (2007) Prostaglandin E2 enhances mitogen-activated protein kinase/Erk pathway in human cholangiocarcinoma cells: involvement of EP1 receptor, calcium and EGF receptors signaling. Mol Cell Biochem 305:19–26

62. Sun B, Rong R, Jiang H et al (2013) Prostaglandin E2 receptor EP1 phosphorylate CREB and mediates MMP2 expression in human cholangiocarcinoma cells. Mol Cell Biochem 378:195–203
63. Sheng H, Shao J, Washington MK, DuBois RN (2001) Prostaglandin E2 increases growth and motility of colorectal carcinoma cells. J Biol Chem 276:18075–18081
64. Buchanan FG, Wang D, Bargiacchi F, DuBois RN (2003) Prostaglandin E2 regulates cell migration via the intracellular activation of the epidermal growth factor receptor. J Biol Chem 278:35451–35457
65. Yoon JH, Higuchi H, Werneburg NW et al (2002) Bile acids induce cyclooxygenase-2 expression via the epidermal growth factor receptor in a human cholangiocarcinoma cell line. Gastroenterology 122:985–993
66. Han C, Demetris AJ, Michalopoulos GK et al (2003) PPARgamma ligands inhibit cholangiocarcinoma cell growth through p53-dependent GADD45 and p21 pathway. Hepatology 38:167–177
67. Lu D, Han C, Wu T (2013) 15-Hydroxyprostaglandin dehydrogenase-derived 15-keto-prostaglandin E2 inhibits cholangiocarcinoma cell growth through interaction with peroxisome proliferator-activated receptor-γ, SMAD2/3, and TAP63 proteins. J Biol Chem 288:19484–19502
68. Xu L, Han C, Wu T (2006) A novel positive feedback loop between peroxisome proliferator-activated receptor-delta and prostaglandin E2 signaling pathways for human cholangiocarcinoma cell growth. J Biol Chem 281:33982–33996
69. Han C, Lim K, Xu L et al (2008) Regulation of Wnt/beta-catenin pathway by cPLA2alpha and PPARdelta. J Cell Biochem 105:534–545
70. Handeli S, Simon JA (2008) A small-molecule inhibitor of Tcf/beta-catenin signaling down-regulates PPARgamma and PPARdelta activities. Mol Cancer Ther 7:521–529
71. Li J, Billiar TR, Talanian RV et al (1997) Nitric oxide reversibly inhibits seven members of the caspase family via S-nitrosylation. Biochem Biophys Res Commun 240:419–424
72. Jaiswal M, LaRusso NF, Gores GJ (2001) Nitric oxide in gastrointestinal epithelial cell carcinogenesis: linking inflammation to oncogenesis. Am J Physiol Gastrointest Liver Physiol 281:G626–G634
73. Zhang Z, Lai GH, Sirica AE (2004) Celecoxib-induced apoptosis in rat cholangiocarcinoma cells mediated by AKT inactivation and Bax translocation. Hepatology 39:1028–1037
74. Ishimura N, Bronk SF, Gores GJ (2005) Inducible nitric oxide synthase up-regulates Notch-1 in mouse cholangiocytes: implications for carcinogenesis. Gastroenterology 128:1354–1368

Chapter 14
Oxidative Stress in Autoimmune Liver Disease

Verónica Martín-Domínguez, Eduardo Vilar-Gómez, Luisa García-Buey, and Ricardo Moreno-Otero

14.1 Oxidative Stress in Autoimmune Liver Diseases

Oxidative stress (OS) is a commonly used term that refers to a state in which tissue and cellular redox balance is altered towards a more oxidizing environment and an ensuing adaptation of cellular functions occurs. The degree of liver damage caused by OS depends on two factors: (1) the type and amount of oxidant stimuli and their persistence in time and space; and (2) the availability of suitable cellular defense mechanisms. It is the balance between the two opposing processes that ultimately determines the contribution of OS to the pathogenesis of a particular liver disease. The available experimental or clinical evidence of the involvement of OS as a pathophysiological mechanism in alcoholic liver disease (ALD), viral hepatitis, autoimmune liver disease (AILD), nonalcoholic steatohepatitis (NASH), and cirrhosis, as well as the available data with respect to antioxidant therapy in liver disease, has been reviewed in increasing numbers of reports. The majority of reports on this topic are case reports, case series, and pilot studies, but only a few randomized controlled trials, including a placebo control or comparator group, have been carried out [1, 2].

OS begins with the generation of reactive oxygen species (ROS) and reactive nitrogen species (RNS) as a part of normal cellular function; these are highly reactive compounds capable of interacting with most biologic molecules in the absence of enzymes [1]. There are multiple cellular sources of ROS generation but the most significant ones are the mitochondrial electron transport complexes I and III, P450

V. Martín-Domínguez • E. Vilar-Gómez • L. García-Buey • R. Moreno-Otero, M.D. (✉)
Digestive Diseases Service and Networked Biomedical Research Center for Hepatic and Digestive Diseases, Institute of Health Carlos III, Research Institute at La Princesa University Hospital, Autonomous university of Madrid, Diego de León 62, Madrid 28006, Spain
e-mail: rmoreno.hlpr@salud.madrid.org

E. Albano, M. Parola (eds.), *Studies on Hepatic Disorders*,
Oxidative Stress in Applied Basic Research and Clinical Practice,
DOI 10.1007/978-3-319-15539-5_14

enzymes within the endoplasmic reticulum, membrane-bound nicotinamide adenine dinucleotide phosphate (NADPH) oxidase, and peroxisomes. ROS production by each of these sources can be stimulated by cytokines, inflammation, viral proteins, and other mechanisms. These processes initially generate superoxide anion (O_2^-), which is sequentially reduced to form hydrogen peroxide (H_2O_2), hydroxyl radical (HO^-), and ultimately water (H_2O) [3]. These reactive intermediates, however, interact with other molecules to form secondary ROS, such as lipid peroxidation products, sulfenic acid, and disulphides. RNS includes nitric oxide (NO^-) and peroxynitrite ($ONOO^-$), and reactive hydrogen species (RHS) include hypochlorous acid (HOCl) and other halide species. With the exception of NO, all other species can originate from O_2^- by several reactions.

ROS such as H_2O_2 can also be formed in a number of enzymatic reactions not linked to O_2^-, such as those catalyzed by a class of enzymes known as oxidases (acyl-CoA oxidase, xanthine oxidase, amino acid oxidases, etc.). Superoxide anion is primarily produced in cells that have mitochondria and can generate other more powerful reactants. Many reactive species can be made by the same cell type, whereas some can only be made by a specific type of cell. Hepatocytes, for example, can make ROS and RNS, but not RHS. ROS can inactivate a number of signaling protein molecules, such as insulin receptor, protein tyrosine phosphatases, and a number of transcription factors, including nuclear factor kappa β activity (NFκB), and others [4].

Excessive intracellular levels of reactive species are clearly deleterious for cells. However, there are natural mechanisms to remove some of the reactive species. These include either enzyme-catalyzed reactions or nonenzymatic reactions of natural compounds acting as traps for the reactive species. Enzymes that dissipate reactive species include superoxide dismutase (converts O_2^- to H_2O_2), catalases, and peroxidases (convert H_2O_2 into oxygen and water). Therefore, the level of naturally produced reactive species results from a balance between their rate of production and neutralization. OS occurs when the balance is shifted toward excessive formation and is deleterious for the cell. Due to their natural reactivity, these species interact with a plethora of biologic molecules and denature their structure and, therefore, their function [4].

When ROS and RNS are abundant, they cause alternations in mitochondrial functions, modulate cytokine expression, alter immune responses, and activate signaling cascades resulting in hepatocellular injury, apoptosis or cell death, and liver fibrosis [1, 5, 6]. The increase of intrahepatic inducible nitric oxide synthase (iNOS) expression and its correlation with the histological severity of liver disease suggests that NO-mediated nitration of hepatocellular proteins might directly contribute to liver damage in AILD. Some experimental and clinical evidences indicate that OS plays a role in lesions in AILD. An increased intrahepatic iNOS expression and nitrotyrosine (NTY) accumulation in patients with primary biliary cirrhosis (PBC) and autoimmune hepatitis (AIH) has recently been reported; however, the initial trigger for iNOS induction in PBC and AIH remains unidentified [1, 7]. OS is a significant feature of AIH and provides a probable mechanism linking hepatic necroinflammation to fibrogenesis and disease progression [8].

Given the differences between these two liver diseases, it could be suggested that this nitration process represents a nonspecific pathogenic mechanism, common to different autoimmune chronic diseases and secondary to the elicitation of inflammation, rather than its cause. This is supported by the observation of increased NTY localization in tissues affected by other chronic inflammatory disorder [1]. Formation of NTY may result, among other possible mechanisms, from the nitration at the ortho position of aromatic amino acids by peroxynitrite, a potent oxidant agent produced by the reaction of NO and superoxide [7].

Patients with PBC might be particularly sensitive to ROS and/or RNS-mediated damage because a predictable consequence of cholestasis is malabsorption of fat-soluble vitamins and other free radical scavengers, such as carotenoids, leading to an impaired antioxidant capacity; therefore, therapeutic approaches that potentiate the generation of free radical scavengers may be efficacious in both diseases [1]. An additional mechanism whereby OS contributes to liver damage in AILDs has been described. In patients with primary sclerosing cholangitis (PSC), catalase has been characterized as one of the possible autoantigens eliciting an autoimmune reaction. Due to catalase being an important antioxidant enzyme that prevents cell damage induced by highly reactive oxygen-derived free radicals, an impairment of the redox status of cells by catalase autoantibodies has been suggested as a possible pathogenic mechanism in PSC [9].

14.2 Immunopathogenesis of Autoimmune Liver Disease

The causes of AILDs, which are reported to be on the increase all over the world, are unknown. AILD is diagnosed when a series of tests suggest that the body's immune system is attacking the liver. AIH, PBC, and PSC are the major pathogenic mechanisms which provoke an autoimmune reaction. Cellular and humoral-mediated immune reactions against self-antigens participate in the development of liver pathology, but differ according to the focus of autoimmune injury, the pattern of inflammation, and the clinical phenotype [1]. Although these findings are important, the pathogenic mechanisms that trigger the onset of these diseases, and the precise role of the characteristic autoantibodies associated with them, remain unknown [10]. AIH, PBC, and PSC represent complex disorders in that they result from the interaction of multiple factors. Several risk factors for certain AILDs have been determined, such as immunogenic background, sex, and geographical region. All three disorders have a progressive course that, if untreated, develops into cirrhosis and liver failure requiring liver transplantation (LT). The aim of treatment is to abolish or reduce inflammation, cholestasis, and the progression of fibrosis [11]. A number of factors have been considered as possible triggers for the chronicity of the disease process in AILDs, such as viruses, bacteria, chemicals, drugs, and genes [10].

Shared serological, immunological, and histological patterns exist across the spectrum of AIH, PBC, and PSC. Conditions exhibiting features of two different AILDs are commonly referred to as "overlap syndromes." These are represented by variant forms of AIH, in which there are characteristics of both AIH and PBC (*AIH–PBC overlap syndrome*) or AIH and PSC ("AIH/PSC overlap") [11]. *AIH–PBC overlap syndrome* is diagnosed when two or three of the criteria for PBC and AIH are met. A diagnosis of AIH/PSC overlap syndrome is arrived upon when the following criteria are met: (1) revised AIH score > 15; (2) characterized by antinuclear antibodies (ANA) or anti-smooth muscle antibody (ASMA) present in a titer of at least 1:40; and (3) liver histology with piecemeal necrosis, lymphocyte rosetting, moderate or severe periportal, or periseptal inflammation [12].

Recently, a new population of effector CD4+ T (Th17 and T-regs) has been shown to have opposing effects in the immune response and may be involved in the pathogenesis of many diseases, including AILDs [13].

14.2.1 Autoimmune Hepatitis

AIH is a chronic liver disease with unknown etiological factors and is associated with aberrant autoreactivity and a genetic predisposition [14]. It is characterized by a strong prevalence in females and concurrent autoimmune disorders occur in approximately 40 % of patients [15]. The target antigens on the hepatocyte membrane are not known, but it is possible that cytotoxic T lymphocytes produce a specific membrane lesion, via the production of tumor necrosis factor-α (TNF-α), interleukins (IL)-2 and IL-12, leading to a progressive necroinflammatory and fibrotic process [13].

Two types of AIH are recognized: type 1, with ANA and/or ASMA positivity, and type 2, characterized by an anti-liver kidney microsomal type 1 antibody (anti-LKM-1) or by an anti-liver cytosol type 1 antibody (anti-LC-1) [11].

The diagnosis is based on histologic abnormalities (interface hepatitis and typical plasma cell infiltration), characteristic clinical and biochemical findings (AST/ALT high levels), hypergammaglobulinemia, presence of one or more characteristic autoantibodies, and serological and histological response to corticosteroids. Histological findings of bridging necrosis or multilobular necrosis at presentation progress to cirrhosis in 82 % of untreated patients and are associated with a 5-year mortality of 45 %.

Two treatment regimens are equally effective in severe AIH: prednisone alone (60 mg daily) in descending doses, or a lower dose of prednisone (30 mg daily) in conjunction with azathioprine (50 mg or 1–2 mg/kg body weight), for a period of 6 months. Treatment improved symptoms, laboratory tests, histological findings, and survival [15]. Different studies indicate that remission rates with conventional therapy have a high variability of 43–80 % [16]. Overall, 10–20 % of patients don't respond to, or are intolerant of, conventional corticosteroid therapy, with or without azathioprine use (a nonselective immunosuppressant that acts by inhibition of several enzymes involved in purine synthesis). Therefore, the standard therapy for AIH is far from ideal. LT has also evolved as an effective treatment for the decompensated patient, and the 5-year patient and graft survival rate now exceeds 80 %, although the risk for graft loss because of disease recurrence has to be considered [15].

A few studies have reported AIH cases with infiltrated IgG4-positive plasma cells in the liver, suggesting the involvement of IgG4 in the pathogenesis of AIH. This feature was called IgG4-associated AIH. However, the diagnostic criteria of IgG4-associated AIH have not been defined and the epidemiology and clinical features remain uncertain. Corticosteroid therapy normalized liver enzymes in both studies [17].

14.2.2 Primary Biliary Cirrhosis

In PBC, the autoimmune injury affects the small- and medium-sized interlobular bile ducts, causing the typical appearance of non-suppurative, destructive chronic cholangitis [11]. Although several studies have examined the autoimmune mechanisms underlying biliary damage in PBC, the etiology of the disease remains largely unknown [13].

PBC has a strong female predominance (F:M ratio, 10:1) and an association in 30 % of patients with other autoimmune disease. Autoimmunity is also common in families of PBC patients; however, unlike AIH, no immunosuppressive agent to date has been shown to be effective in PBC [11]. Recent data has revealed epidemiologic associations; environmental and genetic risks of disease such as having a first-degree relative with PBC, a history of urinary tract infections, being a past smoker, and using hormone-replacement therapies are all associated with the risk of developing disease [18].

PBC exhibits a number of autoimmune features, including the presence of regulatory T cells (T-regs), autoreactive CD8+ cells, B cells, and innate immune mechanism responses against mitochondrial self-antigens, in particular the E2-domain of pyruvate dehydrogenase complex; the almost universal presence of auto-antibodies reactive with mitochondrial self-antigens links IL-12 signaling pathways to disease pathogenesis [11].

Bile acid therapy with ursodeoxycholic acid (UDCA) remains the only available medical intervention. The choleretic effect of UDCA, together with its ability to cause a marked shift in the composition of the bile acid pool toward hydrophilicity, may account for the beneficial properties of UDCA; there is also evidence that it has direct protective effects at the cellular and molecular levels, including the stabilization of hepatocyte membranes, the enhancement of defenses against oxidative stress, and the inhibition of apoptosis [18].

Transplantation is the only effective treatment for those with decompensated cirrhosis and/or liver failure, making PBC progression more favorable than other liver diseases. About 20–25 % of patients with PBC who undergo LT have recurrent disease over 10 years. Fortunately, recurrent PBC does not affect patient or graft survival [19].

14.2.3 Primary Sclerosing Cholangitis

PSC is a chronic, idiopathic, cholestasis liver disease that affects the medium-sized intra- and extrahepatic bile ducts, causing concentric and obliterative inflammation, fibrosis, and multifocal bile duct structuring. PSC is considered an "autoimmune disease with atypical features" because it displays several differences when compared with the classical autoimmune diseases: these include male predominance (M:F ratio, 2:1), the absence of disease-specific autoantibodies, and the poor response to immunosuppression. However, features that suggest an immune-mediated origin include the major contribution of risk variants within the human leukocyte antigen (HLA) complex, the presence of nonspecific autoantibodies, including atypical anti-neutrophil cytoplasmic antibodies (ANCA), and a strong association with other autoimmune or immune-mediated disorders which occur in approximately 70 % of patients. Most notable is a form of inflammatory bowel disease (IBD), sometimes termed IBD–PSC, which affects up to 90 % of patients [11].

The biliary epithelium appears to be the target for immune-mediated injury characterized by the presence of a mixed predominantly lymphocytic inflammatory infiltrate surrounding bile ducts in liver specimens [20].

Recently, it has been reported that patients with PSC show increased numbers of Th17 cells in response to heat-inactivated pathogens (present in the bile duct of the majority of patients with PSC) relative to healthy controls and patients with PBC. In addition, IL-17+ lymphocytes were detected within the periductal areas of patients with PSC by immunohistochemical analysis. Th17 cells may contribute to fibrosis through production of IL-17A and other cytokines; however, it has not yet been clarified [13].

Despite numerous clinical trials over several decades, safe and effective pharmacotherapy remains to be established. Several pharmacotherapies with highly varying mechanisms of action will likely be investigated over the next 5 years, including novel bile acids and antibacterial agents, among others. Progress in therapeutic developments can be enhanced by a multifaceted, collaborative approach, including better understanding of PSC pathogenesis through basic research, more efficient translation of in vitro and animal model findings to human studies, adequately powered clinical trials with well-defined patient characteristics/subtypes and consistent end points, and consideration of combination pharmacotherapy or combined pharmacologic–endoscopic treatment studies.

IgG4-associated cholangiopathy is an entity often indistinguishable from PSC. It has been reported that approximately 10 % of patients with PSC have elevated serum IgG4 levels, and an even higher proportion of explanted PSC livers have elevated IgG4 deposition in tissue. The significance of this finding is not clear. It has been reported that such patients have a more severe form of the disease, but a significant proportion respond to therapy with corticoids [17].

14.3 Antioxidant Therapy in Autoimmune Liver Disease

OS present in almost all inflammatory and autoimmune liver processes has led to enthusiasm on the possibility of antioxidant therapy in liver diseases.

Therefore, therapeutic approaches that potentiate the generation of free radical scavengers may be efficacious in this disease. For example, in experimental models of cholestatic disease, UDCA has been shown to partially prevent hepatic and mitochondrial glutathione depletion and oxidation, suggesting that the efficacy of the drug in human patients might also be associated, at least in part, with its antioxidant actions [21].

Similarly, H_2O_2 derived from Kupffer cells upregulates collagen expression and prevents collagen degradation by hepatic stellate cells (HSC). The O_2 affects human HSC biology by simulating cell migration and collagen deposition when non-cytotoxic levels, similar to those detectable chronic liver inflammations, are used. Conversely, a higher level of O_2, mimicking conditions of acute liver injury, leads to induction of cell death. It should be noted that inflammation and OS are tightly linked together. In a recent study we demonstrated that interfering with the mechanisms of inflammatory cell recruitment, as observed in mice, limits the generation of intrahepatic OS. According to these results, administration of antioxidants, such as α-tocopherol, silymarin, or resveratrol, may prevent fibrosis progression [22].

Strong evidence exists that several agents mediate their action primarily based on changes in ROS and the redox state of the cell. These agents include vitamins E and C, *N*-acetylcysteine (NAC), mitoquinone (MitoQ), and polyenylphosphatidylcholine (PPC). Other compounds, such as silimaryn, *s*-adenosyl methionine (SAMe), and betaine, have additional prominent effects that may be responsible for their clinical effects, but there is controversy surrounding the non-antioxidant compounds [23].

Conversely, in a recent meta-analysis that included 20 randomized trials with 1,225 participants to compare antioxidant supplements (beta-carotene, vitamin A, C, E, and selenium) with a placebo to assess the impact of antioxidant supplements on AILDs, viral hepatitis, ALD, and cirrhosis (any etiology), no evidence was found to support or refute antioxidant supplements in patients with liver disease [24]. Similarly, Prince et al. tested the effects of 12 weeks of oral antioxidant supplementation (a combination of vitamins A, C and E, selenium, methionine, and ubiquinone) on fatigue or other liver-related symptoms in PBC patients in a double-blind, placebo-controlled crossover trial and, although the drug was well-tolerated, no improvement in the disease was noted after treatment [1, 25].

The natural and pharmacological antioxidant mechanism, together with the evidence found so far with all the antioxidants tested in treating AILDs, is discussed below (Fig. 14.1)

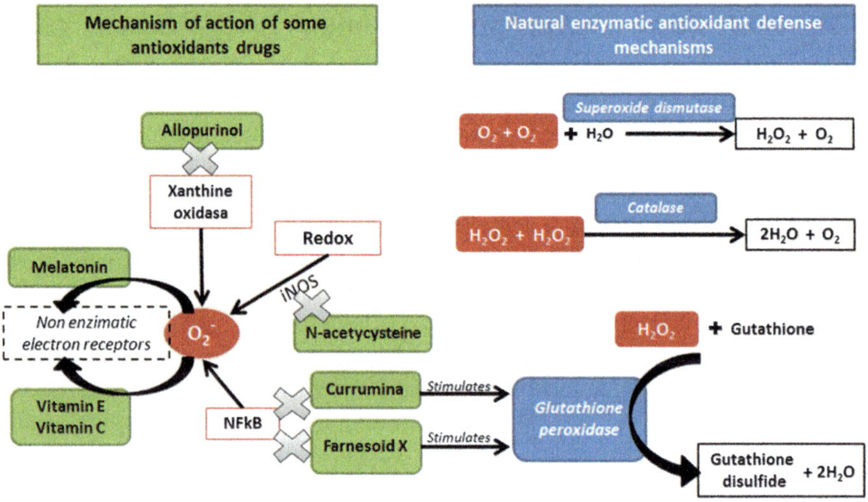

Fig. 14.1 Natural and pharmacological antioxidant defense mechanisms

14.3.1 *Antioxidative Therapy in Autoimmune Hepatitis*

1. *UDCA*: The 7-beta epimer of chenodeoxycholic acid is comprised of 2 % human bile acid and has several interrelated functions including direct choleretic, anti-inflammatory, and antiapoptotic properties. In the first randomized controlled trials, a clinical and biochemical improvement, with a reduction in the histological inflammation but no effect on the seriousness of the fibrosis, was observed [26]. The combination of specific therapeutics for HAI (corticosteroids and azathioprine) with UDCA has provided a clear improvement of abnormal liver enzymes, with a reduction in histological inflammation. However, the severity of fibrosis is unaffected and UDCA does not permit a reduction in corticosteroid dose.
2. *Allopurinol*: It acts through the inhibition of xanthine oxidase and inducing preferential azathioprine breakdown leading to higher 6-thioguanine nucleotides (6-TGN) and lower yielding 6-methilmercaptopurine (6-MMP). High concentrations of 6-TGN are associated with remission in AIH. It may be safely and effectively optimized in AIH patients with intolerance and/or nonresponse due to an unfavorable thiopurine metabolism [27]. The addition of allopurinol with appropriate AZA dose reduction may correct AZA-induced hepatoxicity and induce remission as demonstrated in a study by Al-Shamma et al. [28, 29].

14.3.2 *Antioxidant Therapy in Primary Biliary Cirrhosis*

1. *UDCA*: It is the only drug approved for PBC treatment by the FDA. It decreases serum levels of bilirubin, alkaline phosphatase (ALP), aminotransferase, cholesterol, AMA, and IgM; additionally improves liver histology and prolongs

survival free of liver transplantation [30]. Its effect is dependent on dosage and it is most effective in the initial stages of the illness. A complete response defined as normalized biochemical tests and stabilized or improved liver histology occurs in about 30 % of treated patients with PBC. However, the disease progresses in many patients who do not present a complete response during UDCA therapy. These patients require additional medical treatment [19].

Studies carried out on rats treated with UDCA showed a reduction in the portal pressure and intrahepatic resistance, protein expression of iNOS and thromboxane synthase (TXS), and production of TXA2. Other effects included an increase in hepatic glutathione levels and superoxide dismutase activity. In UDCA-treated rats, the reduction in portal pressure and intrahepatic pressure and increase in antioxidative defense led to the prevention of hepatic fibrosis [31].

2. *Colchicine*: Colchicine has a historical role in the treatment of PBC and several double-blind prospective trials have found that colchicine decreased serum levels of ALP, and aminotransferases (ALT and AST), although not as effectively as UDCA [32]. The addition of colchicine to UDCA has also been reported to improve the liver histological necroinflammatory score over a 10-year follow-up, but with little or no effect on histological stage or predicted patient survival and no additional benefits compared with UDCA alone [33–35]. In many of these preliminary studies, patients were not receiving the optimum dose of UDCA, thus making it difficult to draw firm conclusions regarding the true efficacy of combination therapy. There, therefore, aren't enough tests to support the use of colchicine in the treatment of PBC [36, 37].
3. *Bezafibrate and fenofibrate*: Fibric acid derivatives are used to treat hypertriglyceridemia. Their proposed action mechanism in the treatment of PBC involves: the regulation of immunomodulatory proteins and lipids expression, down-regulation of cholesterol 7 alpha-hydroxylase (an enzyme involved in the synthesis of bile acids), and the reduction of multi-drug resistance genes through the activation of the peroxisome proliferator-activated receptor alpha. They assist the normalization of the liver enzymes in around 45 % of PBC patients who don't respond to UDCA [38]. One pilot study showed that the combination of fibrics acid and AUDC is also effective in maintaining the normalization of liver enzymes [39]. Despite an improvement in long-term prognosis indicators, the effect on histological progression is not clear; an improvement is reported in some, while others show deterioration [40, 41].
4. *Farnesoid X receptor (FXR)*: FXR is a bile acid-activated nuclear receptor, expressed in high levels in both the liver and gastrointestinal tract, which regulates bile acid homeostasis (limits hepatic bile acid accumulation and upregulates bilirubin and phospholipid export pumps). It is also involved in antibacterial defense and protects against inflammation. FXR agonists, such as INT-747, may point towards a new therapeutic option in UDCA-refractory PBC. Obeticholic acid (OCA) is an FXR agonist under investigation for PBC in patients who do not respond to UDCA; a multi-center phase III trial is currently on-going [19, 42, 43].
5. *Modification of UDCA (norUDCA)*: A side chain-modified UDCA derivative has had therapeutic effects in experimental cholestasis. The relative resistance of norUDCA to amidation seems to explain its properties which include the ability

to undergo cholehepatic shunting and to stimulate cholangiocyte secretion directly. Both properties result in a HCO_3-rich hypercholeresis that protects the liver from cholestatic injury. For this reason, it is being studied in the context of PBC treatment [44].

14.3.3 Antioxidative Therapy in Primary Sclerosing Cholangitis

1. *UDCA*: Most published studies have demonstrated an improvement in liver function tests following treatment with UDCA. High doses (25–30 mg/kg/day) also improved the Mayo risk score and are effective in reducing the histological inflammatory stage, although in some cases the treatment should be combined with endoscopic dilations to increase life expectancy. A multi-center study is required to ascertain whether liver transplant indicators are reduced [36]. However, higher doses have a negative long-term effect (including death and transplantation). It is therefore prudent to use doses no higher than 15–20 mg/kg/day, although at this dose, transplant-free survival is not affected [45]. In a pilot study, the combination of UDCA with azathioprine plus prednisone reduced transaminase levels and improved the histological damage. UDCA appears to prevent the development of colorectal carcinoma and dysplasia in patients with UC and PSC.

   The American Association for the Study of Liver Diseases practice guidelines recommend against the use of UDCA in PSC (for treatment of liver disease or chemoprevention of colorectal cancer) [46, 47]. Nevertheless, a follow-up study recently published a clinical trial of UDCA (17-23 mg/kg/day) in the CEP and reported no difference in the frequency of colorectal dysplasia, cancer, or dysplasia/cancer-free survival for a period of 5 years [48].
2. *Colchicine*: Some therapeutic effects were seen in studies in combination with prednisone and there was a trend toward less clinical deterioration and death in PSC patients [49]. However, in a randomized test, there was no evidence of colchicine having a favorable effect on survival, symptoms, serum biochemistry, or liver histology in patients with PSC [50, 51].
3. *Modification of UDCA (norUDCA)*: The protective effects of norUDCA are the result of an increased cholehepatic circulation of norUDCA and stimulation of cholangiocyte secretion resulting in cholangiocyto-protective, bicarbonate-rich hypercholeresis and less toxic bile acid accumulation. In addition, norUDCA has been found to increase expression of efflux transporter proteins (e.g., multidrug resistance protein 4, expressed on the cholangiocyte and hepatocyte membranes), which may have important detoxifying effects [52]. Phase II clinical trials are under way to determine the efficacy of this agent in patients with PSC [53].

4. *Farnesoid X receptor (FXR)*: FXR agonists have potent choleretic activity; so, in patients with downstream obstructions (biliary strictures), they might worsen the liver disease by increasing biliary pressure. Studies of FXR agonists in patients with PSC should therefore be performed with great care [52]. During a phase II clinical trial, the administration of a semi-synthetic derivative of BA 6-ethyl-chenodeoxycholic (6-ECDCA) for patients with diabetes, non-alcoholic fatty liver disease (NAFLD), and PBC led to encouraging results, despite side effects being observed in preclinical studies. Chemical manipulations of the side chain and the steroid nucleus of BAs may yield new semisynthetic BA derivatives that are more specific and selective FXR activators [53, 54].

14.3.4 Antioxidant Therapy in Overlap Syndromes in Autoimmune Liver Disease

AIH–PBC overlap syndrome requires immunosuppressive treatment in addition to UDCA. It appears appropriate to start treatment with UDCA (13–15 mg/kg daily). However, if this therapy does not induce an adequate biochemical response in patients with predominantly hepatitis serum liver tests, a corticosteroid should be added. For corticosteroid-resistant patients, intermediate treatment with another immunosuppressant, such as cyclosporine A, has been considered. LT is regarded as the treatment of choice for end-stage disease [12].

AIH–PSC overlap syndrome: UDCA has been used in combination with immunosuppressive drugs and the long-term course was considered favorable. Thus, UDCA in combination with an immunosuppressive regimen may be an adequate medical treatment for most patients with AIH–PSC overlap syndrome, although more data from controlled trials is required. LT should be considered in late-stage diseases. The only licensed therapy for *PBC and PSC* is UDCA [11, 12].

14.3.5 IgG4-Related Sclerosing Cholangitis

IgG4-related sclerosing cholangitis (IgG4-SC) is a recently described biliary disease that has unknown etiological features, presents with biochemical and cholangiographic features similar to those of PSC, and is often associated with autoimmune pancreatitis and other fibrotic conditions. In this condition, the patient's IgG4 serum level is elevated and IgG4-positive plasma cells infiltrate into the bile ducts and liver tissue. Th2-dominant immune responses or T-reg cells appear to be involved in the underlying immune reaction. Therefore, the immunopathogenesis of IgG4-SC appears to be distinct from that of PBC and PSC. However, the role of Th17 cells in the pathogenesis of IgG4-SC has not yet been clarified, and further studies are required [13].

14.4 Other Antioxidants Therapy in Investigation in AILD

- *Silimaryn*: Silimaryn is an active component of the milk thistle plant and has been used in the treatment of liver diseases, mainly in ALD [55]. The active component of the licorice root, glycyrrhizin (the same as silimaryn), has also been shown to reduce serum ALT and AST values. Silimaryn inhibits immune-mediated cytotoxicity against hepatocytes and *NFκB*, which activates genes encoding inflammatory cytokines in the liver [56]. It is effective in reducing OS (protects against lipid peroxidation) and exerts immunomodulatory and antifibrotic properties [57]. Unfortunately, low-dose studies did not provide a benefit to patients [1]. In the treatment of PBC, the addition of silimaryn to UDCA had no additional benefits compared with UDCA alone [58].

  In PSC, despite 34 % of patients having a positive response to silimaryn (as defined by ≥50 % improvement in liver tests), silimaryn has not been tested in randomized controlled trials and has not been the subject of further research [59].
- *Melatonin*: Melatonin is well-known to exhibit strong direct and indirect antioxidant properties. The powerful antioxidant capacity of melatonin is usually attributed to its potential to eliminate free radicals by the donation of electrons. Melatonin may neutralize hydroxyl radicals by forming 3-hydroxymelatonin, which is excreted in the urine. Furthermore, melatonin was demonstrated to interact with toxic reactants like peroxyl radicals, singlet oxygen species, and hydrogen peroxide. Metabolites of melatonin, including the major hepatic metabolite 6-hydroxymelatonin, as well as *N*-acetyl-*N*-formyl-5-methoxykynuramine and *N*-acetyl-5-methoxykynuramine, have been shown to detoxify radicals themselves. This powerful pyramid scheme of radical scavenging has been named "the antioxidant cascade of melatonin."

  In addition to these direct interactions with ROS, melatonin may induce upregulation of the activity of antioxidants and antioxidant enzymes, in the environment of OS. In addition, the pineal hormone may induce downregulation of pro-oxidant enzymes like iNOS and lipoxygenases, thus reducing the formation of NO, superoxide anions, and subsequently peroxynitrite anions [60]. Through its pro-inflammatory action, it may play an adverse role in autoimmune diseases, although it has not yet been tested in AILDs [61].
- *Docosahexaenoic acid (DHA)*: Cystic fibrosis transmembrane conductance regulator (CFTR) dysfunction is associated with PSC in both children and adults. CFTR dysfunction leads to altered fatty acid metabolism, specifically reduced DHA. A pilot study has investigated the hypothesis that DHA supplementation might be an effective therapy for patients with PSC. It found that oral DHA supplementation is associated with an increase in serum DHA levels and a significant decline in ALP levels in PSC patients. This data supports the need for a rigorous trial of DHA therapy in PSC.

  DHA was shown to reverse the development of bile duct injury. In a recent pilot study ($n=23$), the mean ALP significantly improved after 12 months of DHA treatment in PSC [62].

- *Betaine*: Betaine is a naturally occurring metabolite of choline which has been shown to raise the level of SAMe. SAMe has antioxidant, anti-inflammatory, and anti-fibrotic properties and may in turn play a role in decreasing hepatic steatosis [63]. Its possible action mechanism is to attenuate the hepatic production of NO, through the modulation of nitric iNOS [64].

  This antioxidant is being increasingly used. SAMe levels decrease in most forms of liver disease, and SAMe plays a vital metabolic role by acting as a major methylating agent, as a precursor for glutathione, and in modulating cytokine metabolism, to name only a few of its functions. SAMe therapy has theoretic potential benefits in ALD, NASH, and hepatitis C [4].
- *N-acetylcysteine (NAC)*: NAC modulates the expression of iNOS in human hepatocytes stimulated with proinflammatory cytokines. This effect occurs by blocking the activation of the iNOS promoter and is associated with modulation of NFκB, a central transcription factor for induction of iNOS expression [64]. NAC dose-dependently attenuated the cytokine-induced activation of the iNOS promoter, indicating that the drug exerts its inhibitory effects at the transcriptional level [65]. As the only member of this group that is itself a reduced thiol, it has unique potential to augment glutathione levels and drive protein thiol redox reactions to the reduced form [23].

  NAC represents a promising antioxidant category that has received only limited use in a clinical setting (with the exception of acetaminophen overdose). The role of antioxidant therapy as an adjunct to standard antiviral therapy for hepatitis C requires further investigation [4].
- D-*Penicillamine*: A-penicillamine is an amino acid modified with anti-inflammatory activity and metal chelating properties. The rationale for this cupruretic agent was based on the finding of increased hepatic copper concentrations in patients with PSC and other chronic cholestatic conditions; in addition, aside from its cupruretic properties, penicillamine was known to have immunosuppressive effects based on experience in rheumatoid arthritis. In a review of Cochrane, it was found that D-penicillamine did not appear to reduce the risk of mortality or morbidity and led to more adverse events in patients with PCB [51, 66, 67]. Toxicity includes mucocutaneous, gastrointestinal, renal, hematological, pulmonary, and autoimmune complications. A recent study showed no notable evidence of efficacy in patients with AIH or PBC [68].
- *Pentoxifylline*: In a pilot study for the treatment of PSC, a 400 mg of pentoxifylline alone did not significantly improve symptoms or liver tests in patients with PSC. In addition, side effects included severe nausea and there was no significant alteration to symptoms of fatigue or pruritus, serum liver tests, serum TNF-alpha, or TNF receptor levels [69].
- *PPC*: An extract of soybeans, it is another compound that has been evaluated as a therapy for ALD. It also has lipid peroxidation chain breaking activity and may inhibit ROS generating enzymes [23, 70].
- *Mitoquinone (Mite)*: A mitochondrially targeted antioxidant consisting of a quinone moiety linked to triphenylphosphonium by a carbon alkyl chain. Its resulting positive charge and lipophilic nature allow it to accumulate in the mitochondrial matrix and inner mitochondrial membrane [23, 71].

- *Zinc or polaprezinc (combination of zinc and* l*-carnosine)*: An antioxidant that has been studied as an adjunct to interferon-γ (IFN-γ) in the treatment of chronic hepatitis C. The potential beneficial effects of zinc include: a reduction of hepatic fibrosis, decreasing ferritin, antioxidant activity, and an improvement in hepatic encephalopathy [72].
- *Calcineurin inhibitors*: has been studied in de novo AIH after LT, but its effects are unclear [15].
- *Selenium (selenocysteine)*: The 21st genetically encoded amino acid and the major form of the antioxidant trace element selenium in the human body. The final stage of selenocysteine formation is catalyzed by *O*-phosphoseryl-tRNA: selenocysteinyl-tRNA synthase (SepSecS), also known as soluble liver antigen/liver pancreas (SLA/LP), which represents one of the antigens of AIH [73]. A study comparing levels in blood selenium, glutathione peroxidase, and glutathione observed that whole-blood, plasma selenium, and glutathione concentrations were significantly lower in patients with chronic liver disease (including AILD) compared with the controls [74].
- *Curcumin*: Curcumin is a polyphenol extracted from the plant curcuma longa which exhibits a number of pharmacological properties and has been used for the treatment of inflammatory diseases via its potential to modulate the biological activity of a number of signaling molecules. However, the potential protective effects of curcumin in inflammatory liver diseases have not been clearly elucidated. In animal models, it was found that curcumin pretreatment reduced hepatic OS and pro-inflammatory cytokines, including TNF-α, IFN-γ, and T cell-mediated hepatitis, and decreased serum ALT levels, associated with reduced hepatic necrosis, apoptosis, and mortality [75].

  The curcumin treatment significantly improved endothelial function and reduced OS (malondialdehyde) and inflammatory markers (IL-6, TNF-α, endothelin-1) in type 2 AIH patients [76].
- *Vitamin E (alpha-tocopherol)*: It serves as an antioxidant by complexing with unpaired electrons and therefore stabilizing free radical compounds and preventing lipid peroxidation [77]. Effects observed included a decrease in the production of tumor necrosis factor-β (TNF-β) in alcoholic hepatitis and the prevention of HSC activation in chronic hepatitis C [23].

  Vitamin E is possibly the most widely used antioxidant. Initial pilot studies demonstrated that vitamin E decreased liver enzymes, improved histology in some patients, and downregulated transforming growth factor β-1, a profibrotic cytokine. However, small randomized trials produced more variable results. In the treatment of NASH, vitamin E plus pioglitazone produced a significant decrease in steatosis and improved pericellular fibrosis, Mallory's hyalin, and glucose clearance as well as decreasing fasting free fatty acid and insulin [78]. Vitamin E has also been used in alcoholic hepatitis and hepatitis C and is widely used by many liver patients as an anti-inflammatory and anti-fibrotic agent [79]. However, a recent meta-analysis showed a modest increase in mortality in patients receiving high-dose vitamin E supplementation and further investigation on clinical outcomes is needed [4].

- *Vitamin C or ascorbic acid*: Vitamin C serves as an electron donor and can thus terminate free radical chain reactions (similar to vitamin E). However, its ability to serve as an electron donor makes it possible for it to actually generate free radicals when at high concentrations in the presence of metal ions [23, 80].
- *Sirolimus*: Sirolimus acts by inhibiting the mammalian target of rapamycin (mTOR), a protein that modulates the proliferation and survival of activated lymphocytes. Its antiproliferative effects were recently shown to improve liver fibrosis and inflammation in bile duct ligated rats and represent an avenue to be explored further in human studies [81]. It was reported to have been effective as a salvage treatment in five patients with post-transplant AIH (de novo AIH) who were nonresponders to conventional therapies with prednisone, azathioprine, mycophenolate mofeltil, and calcineurin inhibitors [82]. Consequently, this non-standard off-label medication is a theoretical salvage therapy for refractory AIH in non-transplant patients.
- *Fibric acid derivates*: These stimulate biliary phospholipid secretion via the peroxisome proliferator-activated receptor α-multi-drug resistance protein 3 pathways and may also represent putative therapeutic agents in carefully selected patients with PSC [36, 83].
- *Vitamin D*: This vitamin exhibits immune-modulatory and anti-proliferative effects through vitamin D receptor (VDR) in diseases. Calcitriol has been shown to downregulate intracellular Toll-like receptors (TLRs) expression in human monocytes. TLRs are a group of glycoproteins that function as surface transmembrane receptors, are involved in the innate immune responses to exogenous pathogenic microorganisms, and have an important role in liver inflammation and injury. Calcitriol suppresses major histocompatibility complex (MHC) class II antigen expression in human mononuclear phagocytes and decreases interferon-γ-induced human leukocyte antigen (HLA-DR) antigen expression in normal and transformed human keratinocytes. Calcitriol also stimulated the expression of high levels of cytotoxic T-lymphocyte antigen 4 (CTLA-4) in human CD4+ CD25− T cells. CTLA-4 polymorphisms are associated with type 1 AIH in northern Caucasian European, Chinese patients, and Argentina children. A meta-analysis study demonstrated that CTLA-4 gene +49A/G polymorphisms may be associated with the susceptibility of patients with type 1 AIH, while A/A genotype may be protective against type 1 AIH [84].

  A deficiency in vitamin D levels or VDR is common in autoimmune diseases, with multiple sclerosis being one of the best-studied and well-known examples [85]. Many studies have evaluated the association between many functional polymorphisms in the VDR gene and PBC or AHI risk [36]. Further evaluation of these findings is needed in larger studies. Although low levels of vitamin D, as well as mutations in the VDR genes, have been found to be a feature of PBC patients, it should also be noted that deficiencies in vitamin D are a feature of chronic liver disease in general [86].

References

1. Medina J, Moreno-Otero R (2005) Pathophysiological basis for antioxidant therapy in chronic liver disease. Drugs 65(17):2445–2461
2. Albano E (2006) Alcohol, oxidative stress and free radical damage. Proc Nutr Soc 65(3): 278–290
3. Muriel P (2009) Role of free radicals in liver diseases. Hepatol Int 3(4):526–536
4. Dryden GW, Deaciuc L, Arteel G, McClain CJ (2005) Clinical implications of oxidative stress and antioxidant therapy. Cur Gastroenterol Rep 7(4):308–316
5. Schulze-Osthoff K, Haussinger D (2007) Apoptosis in the liver: a matter of ion fluxes and oxidative stress. Liver Int 27(8):1039–1044
6. Guimaraes EL, Franceschi MF, Grivicich I, Dal-Pizzol F, Moreira JC, Guaragna RM et al (2006) Relationship between oxidative stress levels and activation state on a hepatic stellate cell line. Liver Int 26:477–485
7. Sanz-Cameno P, Medina J, García-Buey L, García-Sánchez A, Borque MJ, Martín-Vílchez S et al (2002) Enhanced intrahepatic inducible nitric oxide synthase expression and nitrotyrosine accumulation in primary biliary cirrhosis and autoimmune hepatitis. J Hepatol 37(6): 723–729
8. Pemberton PW, Aboutwerat A, Smith A, Burrows PC, McMahon RF, Warnes TW (2004) Oxidant stress in type I autoimmune hepatitis: the link between necroinflammation and fibrogenesis? Biochim Biophys Acta 1689(3):182–189
9. Floreani A, Baragiotta A, Martines D, Naccarato R, D'odorica A et al (2000) Plasma antioxidants levels in chronic cholestasis liver disease. Aliment Pharmacol Ther 14(3):353–358
10. Medina J, Jones EA, García-Monzón C, Moreno-Otero R (2001) Immunopathogenesis of cholestatic autoimmune liver diseases. Eur J Clin Invest 31(1):64–71
11. Carbone M, Neuberger JM (2014) Autoimmune liver disease, autoimmunity and liver transplantation. J Hepatol 60(1):210–223
12. Rust C, Beuers U (2008) Overlap syndromes among autoimmune liver diseases. World J Gastroenterol 14(21):3368–3373
13. Abe M, Hiasa Y, Onji M (2013) T helper 17 cells in autoimmune liver diseases. Clin Dev Immunol 2013:607073. doi:10.1155/2013/607073
14. Krawitt EL (2006) Autoimmune hepatitis. N Engl J Med 354(1):54–66
15. Manns MP, Czaja AJ, Gorham JD, Krawitt EL, Mieli-Vergani G, Vergani D et al (2010) Diagnosis and management of autoimmune hepatitis. Hepatology 51(6):2193–2213
16. Mieli-Vergani T, Vergani D (2011) Autoimmune liver disease: news and perspectives. Transl Med UniSa 1:195–212
17. Yada N, Kudo M, Chung H, Watanabe T (2013) Autoimmune hepatitis and immunoglobulin G4-associated autoimmune hepatitis. Dig Dis 31(5–6):415–420
18. Hirschfield GM, Gershwin ME (2013) The immunobiology and pathophysiology of primary biliary cirrhosis. Annu Rev Pathol 8:303–330
19. Kim KA, Jeong SH (2011) The diagnosis and treatment of primary biliary cirrhosis. Korean J Hepatol 17(3):173–179
20. Silveira MG, Lindor KD (2008) Primary sclerosing cholangitis. Can J Gastroenterol 22(8):689–698
21. Serviddio G, Pereda J, Pallardó FV, Carretero J, Borras C, Cutrin J et al (2004) Ursodeoxycholic acid protects against secondary biliary cirrhosis in rats by preventing mitochondrial oxidative stress. Hepatology 39(3):711–720
22. Marra F, Bertolini C (2008) Cellular and molecular mechanisms of hepatic fibrosis. Hematol Rev 5:10–17
23. Singal AK, Jampana SC, Weinman SA (2011) Antioxidants as therapeutic agents for liver disease. Liver Int 31(10):1432–1448
24. Bjelakovic G, Gluud LL, Nikolova D, Bjelakovic M, Nagorni A, Gluud C (2011) Antioxidant supplements for liver diseases. Cochrane Database Syst Rev (3):CD007749. doi:10.1002/14651858.CD007749.pub2

25. Prince MI, Mitchison HC, Ashley D, Burke DA, Edwards N, Bramble MG et al (2003) Oral antioxidant supplementation for fatigue associated with primary biliary cirrhosis: results of a multicentre, randomized, placebo controlled, cross-over trial. Aliment Pharmacol Ther 17(1):137–143
26. Czaja AJ, Carpenter HA, Lindor KD (1999) Ursodeoxycholic acid as adjunctive therapy for problematic type 1 autoimmune hepatitis: a randomized placebo-controlled treatment trial. Hepatology 30(6):1381–1386
27. Zachou K, Muratori P, Koukoulis GK, Granito A, Gatselis N, Fabbri A et al (2013) Review article: autoimmune hepatitis—current management and challenges. Aliment Pharmacol Ther 38(8):887–913
28. de Boer YS, van Gerven NM, de Boer NK, Mulder CJ, Bouma G, van Nieuwkerk CM (2013) Allopurinol safely and effectively optimizes thiopurine metabolites in patients with autoimmune hepatitis. Aliment Pharmacol Ther 37(6):640–646
29. Al-Shamma S, Eross B, Mclaughlin S (2013) Use of a xanthine oxidase inhibitor in autoimmune hepatitis. Hepatology 57(3):1281–1282
30. Lindor KD, Therneau TM, Jorgensen RA, Malinchoc M, Dickson ER (1996) Effects of ursodeoxycholic acid on survival in patients with primary biliary cirrhosis. Gastroenterology 110(5):1515–1518
31. Yang YY, Huang YT, Lee KC, Lee FY, Lee TY, Hou MC (2009) Chronic administration of ursodeoxycholic acid decreases portal pressure in rats with biliary cirrhosis. Clin Sci (Lond) 116(1):71–79
32. Kaplan MM, Alling DW, Zimmerman HJ, Wolfe HJ, Sepersky RA, Hirsch GS (1986) A prospective trial of colchicine for primary biliary cirrhosis. N Engl J Med 315(23):1448–1454
33. Kaplan MM, Cheng S, Price LL, Bonis PA (2004) A randomized controlled trial of colchicine plus ursodiol versus methotrexate plus ursodiol in primary biliary cirrhosis: ten-year results. Hepatology 39(4):915–923
34. Leung J, Bonis P, Kaplan MM (2011) Colchicine or methotrexate, with ursodiol, are effective after 20 years in a subset of patients with primary biliary cirrhosis. Clin Gastroenterol Hepatol 9(9):776–780
35. Battezzati PM, Zuin M, Crosignani A, Allocca M, Invernizzi P, Selmi C (2001) Ten-year combination treatment with colchicine and ursodeoxycholic acid for primary biliary cirrhosis: a double-blind, placebo-controlled trial on symptomatic patients. Aliment Pharmacol Ther 15(9):1427–1434
36. Trivedi PJ, Hirschfield GM (2013) Treatment of autoimmune liver disease: current and future therapeutic options. Ther Adv Chronic Dis 4(3):119–141
37. Gong Y, Gluud C (2005) Colchicine for primary biliary cirrhosis: a Cochrane Hepato-Biliary Group systematic review of randomized clinical trials. Am J Gastroenterol 100(8):1876–1885
38. Kanda T, Yokosuka O, Imazeki F, Saisho H (2003) Bezafibrate treatment: a new medical approach for PBC patients? J Gastroenterol 38(6):573–578
39. Levy C, Peter JA, Nelson DR, Keach J, Petz J, Cabrera R (2011) Pilot study: fenofibrate for patients with primary biliary cirrhosis and an incomplete response to ursodeoxycholic acid. Aliment Pharmacol Ther 33(2):235–242
40. Yano K, Kato H, Morita S, Takahara O, Ishibashi H, Furukawa R (2002) Is bezafibrate histologically effective for primary biliary cirrhosis? Am J Gastroenterol 97(4):1075–1077
41. Kurihara T, Maeda A, Shigemoto M, Yamashita K, Hashimoto E (2002) Investigation into the efficacy of bezafibrate against primary biliary cirrhosis, with histological references from cases receiving long term monotherapy. Am J Gastroenterol 97(1):212–214
42. Lindor KD (2011) Farnesoid X receptor agonists for primary biliary cirrhosis. Curr Opin Gastroenterol 27(3):285–288
43. Kowdley K, Jones D, Luketic V, Chapman R, Burroughs A, Hirschfield G et al (2011) An international study evaluating the farnesoid X receptor agonist obeticholic acid as monotherapy in PBC. J Hepatol 54(Suppl):S13
44. Halilbasic E, Fiorotto R, Fickert P, Marschall HU, Moustafa T (2009) Side chain structure determines unique physiologic and therapeutic properties of norursodeoxycholic acid in Mdr2−/− mice. Hepatology 49(6):1972–1981

45. Lindor KD, Enders FB, Schmoll JA, Hoskin TL, Jorgensen RA, Petz JL et al (2008) Randomized, double-blind controlled trial of high-dose ursodeoxycholic acid (UDCA) for primary sclerosing cholangitis. Hepatology 48(Suppl):94A
46. Triantos CK, Koukias NM, Nikolopoulou VN, Burroughs AK (2011) Meta-analysis: ursodeoxycholic acid for primary sclerosing cholangitis. Aliment Pharmacol Ther 34(8):901–910
47. Pardi DS, Loftus EV Jr, Kremers WK, Keach J, Lindor KD (2003) Ursodeoxycholic acid as a chemopreventive agent in patients with ulcerative colitis and primary sclerosing cholangitis. Gastroenterology 124(4):889–893
48. Lindström L, Boberg KM, Wikman O, Friis-Liby I, Hultcrantz R, Prytz H (2012) High dose ursodeoxycholic acid in primary sclerosing cholangitis does not prevent colorectal neoplasia. Aliment Pharmacol Ther 35(4):451–457
49. Lindor KD, Wiesner RH, Colwell LJ, Steiner B, Beaver S, LaRusso NF (1991) The combination of prednisone and colchicine In patients with primary sclerosing cholangitis. Am J Gastroenterol 86(1):57–61
50. Olsson R, Broomé U, Danielsson A, Hägerstrand I, Järnerot G, Lööf L et al (1995) Colchicine treatment of primary sclerosing cholangitis. Gastroenterology 108(4):1199–1203
51. Tabibian JH, Lindor KD (2013) Primary sclerosing cholangitis: a review and update on therapeutic developments. Expert Rev Gastroenterol Hepatol 7(2):103–114
52. Eaton JE, Talwalkar JA, Lazaridis KN, Gores GJ, Lindor KD (2013) Pathogenesis of primary sclerosing cholangitis and advances in diagnosis and management. Gastroenterology 145(3): 521–536
53. Fickert P, Wagner M, Marschall HU, Fuchbichler A, Zollner G, Tsybrovskyy O (2006) 24-norUrsodeoxycholic acid is superior to ursodeoxycholic acid in the treatment of sclerosing cholangitis in Mdr2 (Abcb4) knockout mice. Gastroenterology 130(2):465–481
54. Stanimirov B, Stankov K, Mikov M (2012) Pleiotropic functions of bile acids mediated by the farnesoid X receptor. Acta Gastroenterol Belg 75(4):389–398
55. Lucena MI, Andrade RJ, de la Cruz JP, Rodriguez-Mendizabal M, Blanco E, Sanchez de la Cuesta F (2002) Effects of silymarin MZ-80 on oxidative stress in patients with alcoholic cirrhosis. Results of a randomized, double-blind, placebo-controlled clinical study. Int J Clin Pharmacol Ther 40(1):2–8
56. Bean P (2002) The use of alternative medicine in the treatment of hepatitis C. Am Clin Lab 21(4):19–21
57. Wagoner J, Negash A, Kane OJ, Martinez LE, Nahmias Y, Bourne N (2010) Multiple effects of silymarin on the hepatitis C virus life cycle. Hepatology 51(6):1912–1921
58. Angulo P, Patel T, Jorgensen RA, Therneau TM, Lindor KD (2000) Silymarin in the treatment of patients with primary biliary cirrhosis with a suboptimal response to ursodeoxycholic acid. Hepatology 32(5):897–900
59. Angulo P, Jorgensen RA, Kowdley KV, Lindor KD (2008) Silymarin in the treatment of patients with primary sclerosing cholangitis: an open-label pilot study. Dig Dis Sci 53(6):1716–1720
60. Mathes AM (2010) Hepatoprotective actions of melatonin: possible mediation by melatonin receptors. World J Gastroenterol 16(48):6087–6097
61. Maestroni GJ (2001) The immunotherapeutic potential of melatonin. Expert Opin Investig Drugs 10(3):467–476
62. Martin C, Blanco P, Keach J, Petz J, Zaman M, Bhaskar K et al (2012) The safety and efficacy of oral docosahexaenoic acid supplementation for the treatment of primary sclerosing cholangitis—a pilot study. Aliment Pharmacol Ther 35(2):255–265
63. Abdelmalek MF, Angulo P, Jorgensen RA, Sylvestre PB, Lindor KD (2001) Betaine, a promising new agent for patients with nonalcoholic steatohepatitis: results of a pilot study. Am J Gastroenterol 96(9):2711–2717
64. Moreno-Otero R (2013) May oxidative stress contribute to autoimmune hepatitis pathogenesis, and can antioxidants be of value as adjuvant therapy for refractory patients? Dig Dis Sci 58(5):1440–1441
65. Majano PL, Medina J, Zubía I, Sunyer L, Lara-Pezzi E, Maldonado-Rodríguez A et al (2004) N-Acetyl-cysteine modulates inducible nitric oxide synthase gene expression in human hepatocytes. J Hepatol 40(4):632–637

66. LaRusso NF, Wiesner RH, Ludwig J, MacCarty RL, Beaver SJ, Zinsmeister AR (1998) Prospective trial of penicillamine in primary sclerosing cholangitis. Gastroenterology 95(4): 1036–1042
67. Gong Y, Klingenberg SL, Gluud C (2006) Systematic review and meta-analysis: D-penicillamine vs. placebo/no intervention in patients with primary biliary cirrhosis—Cochrane Hepato-Biliary Group. Aliment Pharmacol Ther 24(11–12):1535–1544
68. Oo YH, Hubscher SG, Adams DH (2010) Autoimmune hepatitis: new paradigms in the pathogenesis, diagnosis, and management. Hepatol Int 4(2):475–493
69. Bharucha AE, Jorgensen R, Lichtman SN, LaRusso NF, Lindor KD (2010) A pilot study of pentoxifylline for the treatment of primary sclerosing cholangitis. Am J Gastroenterol 95(9): 2338–2342
70. Okiyama W, Tanaka N, Nakajima T, Tanaka E, Kiyosawa K, Gonzalez FJ et al (2009) Polyenephosphatidylcholine prevents alcoholic liver disease in PPAR alpha-null mice through attenuation of increases in oxidative stress. J Hepatol 50(6):1236–1246
71. Smith RA, Murphy MP (2010) Animal and human studies with the mitochondria-targeted antioxidant MitoQ. Ann N Y Acad Sci 1201:96–103
72. Yuasa K, Naganuma A, Sato K, Ikeda M, Kato N, Takagi H et al (2006) Zinc is a negative regulator of hepatitis C virus RNA replication. Liver Int 26(9):1111–1118
73. Palioura S, Herkel J, Simonović M, Lohse AW, Söll D (2010) Human SepSecS or SLA/LP: selenocysteine formation and autoimmune hepatitis. Biol Chem 391(7):771–776
74. Czuczejko J, Zachara BA, Staubach-Topczewska E, Halota W, Kedziora J (2003) Selenium, glutathione and glutathione peroxidases in blood of patients with chronic liver diseases. Acta Biochim Pol 50(4):1147–1154
75. Wang C, Nie H, Li K, Zhang YX, Yang F, Li CB et al (2012) Curcumin inhibits HMGB1 releasing and attenuates concanavalin A-induced hepatitis in mice. Eur J Pharmacol 697(1–3):152–157
76. Gupta SC, Kismali G, Aggarwal BB (2013) Curcumin, a component of turmeric: from farm to pharmacy. Biofactors 39(1):2–13
77. Zingg JM (2007) Vitamin E: an overview of major research directions. Mol Aspects Med 28(5–6):400–422
78. Sanyal AJ, Mofrad PS, Contos MJ, Sargeant C, Luketic VA, Sterling RK et al (2004) A pilot study of vitamin E versus vitamin E and pioglitazone for the treatment of nonalcoholic steatohepatitis. Clin Gastroenterol Hepatol 2(12):1107–1115
79. von Herbay A, Stahl W, Niederau C, Sies H (1997) Vitamin E improves the aminotransferase status of patients suffering from viral hepatitis C: a randomized, double-blind, placebo-controlled study. Free Radic Res 27(6):599–605
80. Abudu N, Miller JJ, Attaelmannan M, Levinson SS (2004) Vitamins in human arteriosclerosis with emphasis on vitamin C and vitamin E. Clin Chim Acta 339(1–2):11–25
81. Patsenker E, Schneider V, Ledermann M, Saegesser H, Dorn C, Hellerbrand C et al (2011) Potent antifibrotic activity of mTOR inhibitors sirolimus and everolimus but not of cyclosporine A and tacrolimus in experimental liver fibrosis. J Hepatol 55(2):388–398
82. Kerkar N, Dugan C, Rumbo C, Morotti RA, Gondolesi G, Shneider BL et al (2005) Rapamycin successfully treats post-transplant autoimmune hepatitis. Am J Transplant 5(5): 1085–1089
83. Kita R, Kita-Sasai Y, Hanaoka I, Kimura T, Kokuryu H, Takamatsu S et al (2002) Beneficial effect of bezafibrate on primary sclerosing cholangitis (three case reports). Am J Gastroenterol 97(7):1849–1851
84. Luong KV, Nguyen LT (2013) The role of vitamin D in autoimmune hepatitis. J Clin Med Res 5(6):407–415
85. Smyk DS, Orfanidou T, Invernizzi P, Bogdanos DP, Lenzi M (2013) Vitamin D in autoimmune liver disease. Clin Res Hepatol Gastroenterol 37(5):535–545
86. Vogel A, Strassburg CP, Manns MP (2002) Genetic association of vitamin D receptor polymorphisms with primary biliary cirrhosis and autoimmune hepatitis. Hepatology 35(1):126–131

Chapter 15
Oxidative Stress and Hepatic Iron Overload

S. Pelusi, L. Valenti, and Silvia Fargion

15.1 Iron Metabolism

15.1.1 Iron Distribution and Absorption

The human body contains approximately 3–5 g of iron. Approximately 60–70 % of body iron is employed within hemoglobin; other iron-rich organs are the liver and muscles. Approximately 20–30 % of body iron is stored in hepatocytes and in macrophages, to a large extent within polymers of ferritin. The remaining is primarily localized in myoglobin, cytochromes, and iron-containing enzymes. A healthy individual absorbs daily 1–2 mg of iron from the diet, which is utilized to compensate non-specific iron losses by desquamation of enterocytes and epidermis and, in childbearing aged women, by periods.

An average daily Western diet contains approximately 15 mg of iron, from which only 1–2 mg is absorbed in the duodenum and jejunum. Two thirds from absorbed iron derives from heme, mainly from myoglobin and hemoglobin of animal origin. The inorganic iron is not efficiently absorbed. The low gastric and duodenal pH dissolves ingested inorganic iron and facilitates its enzymatic reduction to the ferrous form by the ferrireductase on the enterocyte brush border (DCYTB, duodenal cytochrome b). Ferrous iron is transported across the apical membrane by DMT1 (divalent metal transporter-1). Inside absorptive enterocytes, heme iron is enzymatically

S. Pelusi • L. Valenti • S. Fargion (✉)
IRCCS Ca' Granda Foundation Policlinico Hospital Milan and Department of Pathophysiology and Transplantation, University of Milan, via F Sforza 35, Milan 20122, Italy
e-mail: serenapelusi@libero.it; luca.valenti@unimi.it; silvia.fargion@unimi.it

E. Albano, M. Parola (eds.), *Studies on Hepatic Disorders*,
Oxidative Stress in Applied Basic Research and Clinical Practice,
DOI 10.1007/978-3-319-15539-5_15

released by heme oxygenase-1 (HO-1) and follows the fate of inorganic iron: it is either stored in ferritin or transported across the basolateral membrane to plasma transferrin (TF). The transport across the basolateral membrane is mediated by Ferroportin-1 (Fp-1 or SLC40A1), a protein expressed in Kupffer cells and on the basolateral membrane of enterocytes, macrophages, placental cells, and hepatocytes. Then, ceruloplasmin or hephaestin converts ferrous iron to ferric state while it is still associated to TF. Ferric iron binds to plasmatic apotransferrin to form ferric iron–TF complex, which is the major type of iron present in blood. A small amount of iron circulates bound to albumin or other small molecular weight ligands, such as citrate salts. The TF complex facilitates the transport of iron to cells that express transferrin receptors, including erythroid progenitors, and limits the ability of iron to generate toxic radicals.

15.1.2 Iron Storage and Recycling

Not all absorbed iron is utilized in metabolic processes, but it is partly stored as reserve, both for use when iron levels are low, and to prevent toxic effects of free iron in the cell. Stored iron accounts for 20–30 % of body iron, and the major part of it is bound to ferritin. Ferritin is an ubiquitous multimer made up of two different isoforms: H and L ferritin. H ferritin possesses ferroxidase activity that facilitates the oxidation of the cytosolic ferrous iron to the ferric state. Iron is also stored in an insoluble form into a protein, named haemosiderin, likely derived from the lysosomal degradation of ferritin. Under iron overload conditions, ferritin levels increase dramatically, particularly in liver, pancreas, and heart.

Intestinal absorption accounts for only a fraction of TF-bound iron in the circulation. Recovery of iron from senescent erythrocytes plays a key role in iron homeostasis. At the end of their lifespan, human erythrocytes undergo surface alterations that make them to be recognized, phagocytosed, and digested by reticulo-endothelial macrophages in the spleen and liver.

15.1.3 Hepcidin and Systemic Iron Homeostasis

Systemic iron homeostasis is achieved by modulation of the amount of iron absorbed in response to "storage regulators," "erythroid regulators," "inflammatory regulators," and hypoxia regulatory signals. The amount of body storage modulates iron uptake: it is well-established that in iron-deficient conditions, iron absorption is significantly stimulated by two- to threefold compared to basal conditions, which are restored when iron storage are reconstituted. The erythropoietic regulation participates when iron demand for hemoglobin synthesis increases, independently of body iron stores. This mechanism can explain the pathological iron accumulation observed in disorders characterized by ineffective erythropoiesis (such as thalassemia syndromes, congenital dyserythropoietic anemias, sideroblastic anemias). Moreover, a new hormone, erythroferrone (ERFE), mediating hepcidin suppression during stress erythropoiesis and produced by erythroblasts in response to erythropoietin, has been recently identified. It has been shown that ERFE production is increased in mice with thalassemia intermedia, where it causes hepcidin suppression and the systemic iron overload characteristic of this disease [20].

The inflammatory regulators as well as the storage regulators are mediated by hepcidin [32].

Hepcidin-mediated regulation of iron metabolism has been demonstrated to depend upon its ability to bind Fp-1 on cellular surface blocking its iron transport activity and to increase Fp-1 endocytosis and consequently its degradation by lysosomes [28]. In enterocytes, Fp-1 internalization on the basolateral surface causes the retention of absorbed iron with subsequent loss by desquamation, while the same process in macrophages causes the failure to release iron [31]. The final effect is the reduction of plasma iron availability. Indeed, mutations of *HAMP*, the hepcidin gene, cause severe early-onset hereditary hemocromatosis (HH) in humans, whereas deletion of *Hamp* in animal models brings to severe iron accumulation [21].

It is still not clear how systemic iron demand modulates hepcidin release by the liver. A number of molecules have been implicated in different ways in the transduction of the signal: the alteration of each of these molecules causes an insufficient release of hepcidin resulting in a deregulated iron flux from macrophages and enterocytes, which finally brings to iron overload in blood circulation and finally in tissues. Hepcidin release may be impaired by genetic factors, i.e., mutations inactivating *hepcidin* (*HAMP*), *hemojuvelin* (*HJV*), *HFE* or *transferrin receptor-2* (*TFR-2*), or by non-genetic factors: alcohol abuse (that inhibits hepcidin transcription), viral infections such as chronic HCV hepatitis, acute liver insufficiency and cirrhosis (because of the reduced hepcidin synthesis by hepatocytes). In addition, mutations in the *FPN1* gene, encoding for Fp-1, which bring to insensitivity to hepcidin action, are as well responsible for hereditary forms of overload [22].

15.2 Chemical Properties, Biological Functions, and Toxicity of Iron

The biological functions of iron are based on its chemical properties, in particular on the capacity to form a variety of coordination complexes with organic ligands in a dynamic and flexible mode and by its favorable redox potential to switch between the ferrous, Fe(II), and ferric, Fe(III), states. Iron is associated to heme and non-heme complexes. The hemoproteins are hemoglobin and myoglobin, cytochromes, and some enzymes, such as oxygenases, peroxidases, nitric oxide (NO) synthases, or guanylate cyclase. All these proteins contain heme as prosthetic group, which is composed of protoporphyrin IX and a ferrous ion and is involved in oxygen transport to muscles and tissues (hemoglobin and myoglobin), in the respiratory chain (e.g., cytochromes a, b, and c), in the activation of substrates by oxygen (e.g., cytochrome oxidase, cytochrome P450, catalase), and as a NO sensor (guanylate cyclase). The most prevalent forms of non-heme iron are metallo-proteins with iron–sulfur clusters, which are involved in the respiratory chain (e.g., in complex III), DNA synthesis (e.g., ribonucleotide reductase), and in the inflammatory response (e.g., cyclo-oxygenases and lipoxygenase). It should also be noted that non-heme iron has a central function in a recently discovered mechanism for oxygen sensing, via the hypoxia-inducible factor (HIF), which controls the transcription of a wide array of genes involved in erythropoiesis, angiogenesis, cell proliferation/survival, glycolysis, and iron metabolism in response to oxygen availability.

The efficiency of the redox reaction between ferrous and ferric ions is a fundamental feature for many biochemical reactions. However, this property turns iron into a potential biohazard in the cell, because under aerobic conditions, iron can readily catalyze the generation of noxious radicals. Iron toxicity is largely based on the Fenton reaction, where catalytic amounts of iron are sufficient to yield hydroxyl radicals, collectively known as reactive oxygen intermediates (ROIs), from superoxide and hydrogen peroxide. ROIs are inevitable products of aerobic respiration in mitochondria and can be also generated during enzymatic reactions in peroxisomes, in the endoplasmic reticulum, or in the cytoplasm. ROIs are also produced by the membrane-bound NADPH oxidase complex, mainly expressed in phagocytic neutrophils and macrophages during inflammation and involved in the antimicrobial defense. Reactive oxygen species (ROS) include a variety of molecular species, such as hydrogen peroxide (H_2O_2), singlet molecular oxygen ($O^•$), hydroxyl ($OH^•$), superoxide ($O_2^•$), alkoxyl ($RO^•$), peroxyl ($ROO^•$), and nitric oxide ($NO^•$) radicals. Free radicals can promote the oxidation of proteins, peroxidation of membrane lipids, and modification of nucleic acids. An increase in the steady state levels of ROS beyond the antioxidant capacity of the organism, so-called oxidative stress, is observed in many pathological conditions, such as chronic inflammation, ischemia–reperfusion injury, or neurodegeneration.

In limited controlled amounts, ROS play a role in physiologic pathways, such as signal transduction and redox regulation of cell proliferation and apoptosis. Because of their short half-life, small size, and highly diffusible nature, ROS are excellent candidates for second messenger function. Many events of cell regulation, such as protein phosphorylation and binding of transcription factors to DNA, are driven by physiologic oxidant–antioxidant homeostasis. As to the regulation of gene expression, several transcription factors have been identified as being regulated by the intracellular redox state. Binding sites of the redox-regulated transcription factors, nuclear factor-kappa B (NF-κB) and c-*fos*/c-*jun,* are located in the promoter region of a variety of genes such as cytokines, adhesion molecules, chemokines, and major histocompatibility complex proteins, which are directly involved in immune response, apoptosis, inflammation, and fibrogenesis [19].

Under normal conditions, the concentration of ROS in the cell is kept fairly constant by enzymatic (e.g., superoxide dismutases, glutathione peroxidases, and catalase) and nonenzymic (e.g., α-tocopherol, β-carotene, and glutathione) activities that are able to get rid of unwanted ROS and generate nontoxic by-products. In addition to cellular or plasma antioxidants, the modulation of iron availability is the main mean, by which cells keep ROS levels under strict control. This is because the appropriate sequestration of iron may allow the physiologic roles of the relatively safe O_2 and H_2O_2 to take place without production of the highly reactive $OH^•$ by Fenton chemistry [16]. On the other hand, excess of redox active iron aggravates oxidative stress and leads to an accelerated tissue degeneration such as liver fibrosis. Thus, under physiological conditions, extracellular iron is exclusively bound to TF, which maintains iron soluble and nontoxic.

15.3 Primary Iron Overload

The transcription and secretion of hepcidin by the liver is regulated by a mechanism of body iron sensing and by a group of proteins, including the hereditary hemochromatosis protein called HFE, TFR-2, HJV, bone morphogenetic protein 6 (BMP6), matriptase-2 (TMPRSS6), and TF. Mutations in *HFE*, *TFR-2*, *HJV*, and the hepcidin gene (*HAMP*) are responsible for hereditary hemochromatosis (HH), a common iron overload disorder characterized by a deficit of hepcidin release or activity [5, 31].

HFE mutations represent the most frequent cause of HH in Caucasian adults [12]. The most common *HFE* mutation responsible for HH is a single nucleotide substitution that causes the substitution of a cysteine with a tyrosine at position 282 (C282Y). The homozygous genotype is very frequent in Caucasians, particularly in people from Northern Europe (frequency 1/300–400), whereas the prevalence decreases towards Southern Europe. The C282Y substitution brings to HFE mysfolding, abrogating cell surface expression. A second and most frequent mutation is a substitution at position 63 of a histidine with an aspartate (H63D), which likely interferes with the ability of HFE to interact with transferrin receptor-1 (TFR-1). This is a very common polymorphism in the general population, as 25–30 % of the population carries the H63D variant, but its contribution to the pathogenesis of HH and iron overload syndromes is negligible, with the exception of compound heterozygosity for the C282Y/H63D variants. The penetrance of HH depends on age, gender, environmental factors, and on the role of the so-called modifier genes [46]. In fact, only 30 % of C282Y homozygotes will develop a phenotype compatible with HH, thus implying that genetic and environmental factors play a major role.

For example, patients with HH and beta-thalassemia trait develop a more serious iron overload and clinical phenotype [33, 43, 44].

Mutations of *TFR-2* cause a rare recessive form of HH, clinically similar to HFE HH.

Many other molecules have been implicated in the regulation of hepcidin secretion. Binding of the iron-regulated BMP6 ligand, a bone morphogenetic protein of the TGFβ superfamily, to its threonine/serine kinase receptors (BMPR-I and BMPR-II) activates a signaling cascade leading to hepcidin transcription. HJV, a GPI-linked membrane protein synthesized by the hepatocytes, is a BMP6 coreceptor, which is required for its regulatory functions on iron metabolism. The critical role of the BMP6/HJV pathway in iron homeostasis is supported by the loss of hepcidin expression and massive parenchymal iron overload observed in *BMP6−/−* and *HJV−/−* mice [1] and by the fact that *HJV* mutations represent the major cause of juvenile HH. Furthermore, iron overload and increased transferrin saturation have been associated with specific upregulation of BMP6 in hepatocytes and in vivo.

Recently, the serine protease matriptase-2 has been connected to this iron regulatory pathway because of its ability to cleave HJV [25]. Matriptase-2 is a type 2 transmembrane serine protease that is predominately expressed in the liver and was characterized as a negative regulator of hepcidin gene expression [34]. Matriptase-2-deficient mice have very high levels of hepcidin, which lead to the inhibition of dietary iron absorption and cause a severe iron-deficiency anemia phenotype. This anemic phenotype is mirrored in patients with matriptase-2 mutations, who present with iron-refractory iron-deficiency anemia (IRIDA). Indeed, patients with IRIDA show inappropriately high hepcidin levels, which explain the lack of dietary iron absorption and the incomplete response to parenteral iron treatment [13].

HH, due to the C282Y HFE mutation, is the typical model of iron-induced liver damage in which progressive iron overload is responsible for liver damage through enhanced oxidative stress, whose markers decrease with phlebotomy [30].

Severe iron overload in HH involves several organs, mainly the liver, endocrine glands, and heart. However, the involvement of a specific organ varies on the entity of iron accumulation, which depends in part also on the specific mutation at the base of the disease. Usually, the more severe phenotype is observed in the juvenile forms versus adult disease [31]. Besides, also genetic traits such as variations in the activity of antioxidant enzymes (e.g., superoxide-dismutase and myeloperoxidase) are believed to influence the course of HH and the severity of cardiac involvement [27, 41].

15.4 Secondary Iron Overload and Oxidative Stress

A mild increase in body iron stores in the general population is more frequently related to secondary causes such as alcohol abuse, metabolic syndrome, chronic hepatitis C, porphyria cutanea tarda, hemolytic and dyserythropoietic anemias, chronic transfusional therapy, chronic inflammation, and end-stage renal disease. There is growing evidence that the mildly increased amounts of liver iron associated to these conditions can cause liver injury via oxidative stress induction. Experimental data have recently suggested that ferritin, behaving like a cytokine, may also directly induce the fibrogenic process by activating hepatic stellate cells [35].

15.4.1 Chronic Hepatitis C

Mild to moderate iron overload is present in 30–40 % of chronic hepatitis C patients [6, 42]; in contrast, an association between chronic hepatitis B and iron overload has rarely been reported.

Interestingly, recent data in experimental animal models suggest that HCV influences the transcription of hepcidin [29] and oxidative stress has been proposed as a major mechanism of liver injury in chronic HCV hepatitis. It has been demonstrated that HCV-induced ROS raise hepatic iron levels in mice by reducing hepcidin transcription [23]. It has also been hypothesized that the derangement in iron metabolism and possibly of hepcidin transcription in these patients could be related to dysregulation of HFE expression and activity due to the direct action of the virus on MHC-like molecule (which include HFE), as a mean to escape cellular immune response [4].

Since only about one third of HCV-positive patients have iron overload, it is very likely that additional factors are involved, and among the possible candidates are genetic factors able to interfere with iron metabolism. Several studies have focused on the possible role of *HFE* mutations, but contrasting results have been reported [40, 42]. More recently, the presence of beta-globin mutations, a common genetic factor in the Mediterranean area, was independently associated with increased hepatic iron concentrations in patients affected by chronic hepatitis C and *beta-globin* genes mutations being independent predictors of the severity of fibrosis [36, 42].

Whatever the exact mechanisms responsible for increased iron in chronic hepatitis C patients are, once present, they have been reported to exert a negative influence on liver fibrosis progression, particularly in conditions of long-lasting iron overload [40].

15.4.2 Non-alcoholic Fatty Liver Disease

Increased serum ferritin is detected in more than 30 % of the patients with NAFLD and a syndrome termed insulin resistance or dysmetabolic hepatic iron overload has been described in patients with mild to moderate iron overload associated with features of insulin resistance [24]. Iron overload has been hypothesized to induce insulin resistance by catalyzing oxidative stress in the liver and metabolic active tissues [8, 18].

The mechanisms implicated in increased susceptibility to iron accumulation in NAFLD likely involve, similarly to what observed in chronic hepatitis C, dysregulation of hepcidin transcription mediated by inflammation and micronutrient imbalance [3].

In an attempt to explain the cause of increased ferritin in a proportion of patients with NAFLD, several studies analyzed whether *HFE* gene mutations may be responsible for the increased iron tests, but conflicting data have been reported. Hepatocellular iron accumulation has been demonstrated to be associated to a higher risk of moderate/severe fibrosis compared to the absence of siderosis, whereas non-parenchymal siderosis was not associated with fibrosis [43, 44]. Hepatocellular siderosis was more prevalent in patients carrying HFE mutations, but only one third of patients carrying HFE mutations had hepatocellular iron accumulation. Moreover, a prominent role is played by beta-globin mutations [43, 44] and *TMPRSS6* A736V polymorphism [45].

Some evidences have been shown that iron depletion may be beneficial in NAFLD and acts by increasing insulin sensitivity [9]. Furthermore, ferritin levels and body iron stores have also been recently associated with the severity of atherosclerosis, the leading cause of death in patients with NAFLD [43, 44], suggesting that iron depletion may also influence the impact of extrahepatic cardiovascular complications of the metabolic syndrome.

15.4.3 Alcoholic Liver Disease

Subjects with alcohol abuse often present with increased transferrin saturation, serum ferritin levels, and hepatic siderosis. Alcohol abstinence is usually followed by a fast decrease in iron values. The mechanisms by which alcohol is able to interfere with iron metabolism are not fully clear, but oxidative stress certainly plays a key role in the pathogenesis of alcoholic liver disease. Iron itself can induce oxidative stress in this condition by catalyzing the conversion of superoxide and hydrogen peroxide to more potent oxidants such as hydroxyl radicals with consequent damage to different cell structures and leading to the dysregulation of hepcidin release [17]. In addition to hepatocytes, non-parenchymal cells are capable of releasing ROS and may thereby contribute to parenchymal cell necrosis or disease progression. In particular, Kupffer cells are activated by a variety of stimuli to release ROS, such as opsonized zymosan, phorbol esters, and lipopolysaccharide [37]. The effect of lipopolysaccharide on Kupffer cell biology is a key event in the pathogenesis of alcoholic liver fibrogenesis. Thus, iron can play a major role in the pathogenesis of alcohol liver disease by exacerbating oxidative stress and acting synergistically with alcohol to cause liver injury and fibrosis.

15.5 Hepatocellular Carcinoma and Oxidative Stress

Hepatocellular iron accumulation, whether due to HH or acquired factors, is a frequent cause of liver damage and has been associated with increased risk of developing hepatocellular carcinoma (HCC) [10]. Moreover, increased iron store has been reported to be a risk factor for other neoplasia, such as breast and colon cancer [14], whereas iron reduction therapy has been associated with a decreased incidence of cancer [47].

A direct role of iron in hepatic carcinogenesis has been shown in an animal model where pre-neoplastic nodules and HCC developed in rats fed with an iron-rich diet [2].

The molecular pathways underpinning iron-driven carcinogenesis involve both neoplasia initiation by the mutagenic activity of oxidant species and promotion by concurrent stimulation of hepatocellular proliferation. Iron-related carcinogenesis is physiologically contrasted by the oncogene MDM2 (murine double minute 2), by regulation of p53, and consequently of antioxidant response and apoptosis [7]. It was shown that a MDM2 promoter polymorphism associated with higher MDM2 expression and inhibition of p53 influenced the risk of developing HCC in patients with liver cirrhosis and iron overload [7].

Recent evidences, moreover, show that liver iron overload may be associated with development of HCC in NASH-related cirrhosis also in vivo [38].

Interestingly in a recent prospective study in which the relation between liver iron, HFE gene mutations, and HCC was analyzed, it was demonstrated that liver iron accumulation and the C282Y mutation are independent risk factors for HCC in patients with alcoholic cirrhosis and/or NASH [26]. A possible explanation was that iron and alcohol act synergistically in inducing free radicals formation which trigger the carcinogenic process. In the presence of polymorphisms of genes involved in free radicals detoxification, patients with cirrhosis secondary to alcohol, in the presence of a minor increase of iron, are at high risk of developing HCC [11, 39]. Similarly, an interaction has been reported between HFE mutations and viral infection in HCC pathogenesis [15].

15.6 Conclusions

In conclusion, genetic and environmental factors such as diet, alcohol intake, and chronic viral hepatitis can all facilitate iron overload.

Because of iron capability to catalyze redox reactions, ROS are produced leading to oxidative stress and consequent cell damage. In particular, in the liver, ROS-induced cytokines secretion activates hepatic stellate cells which initiate fibrogenesis. Together with ROS-mediated DNA damage, this has been demonstrated to increase the risk of HCC and other neoplasms.

A working model illustrating the role of iron in the pathogenesis of progressive liver disease via oxidative stress is shown in Fig. 15.1.

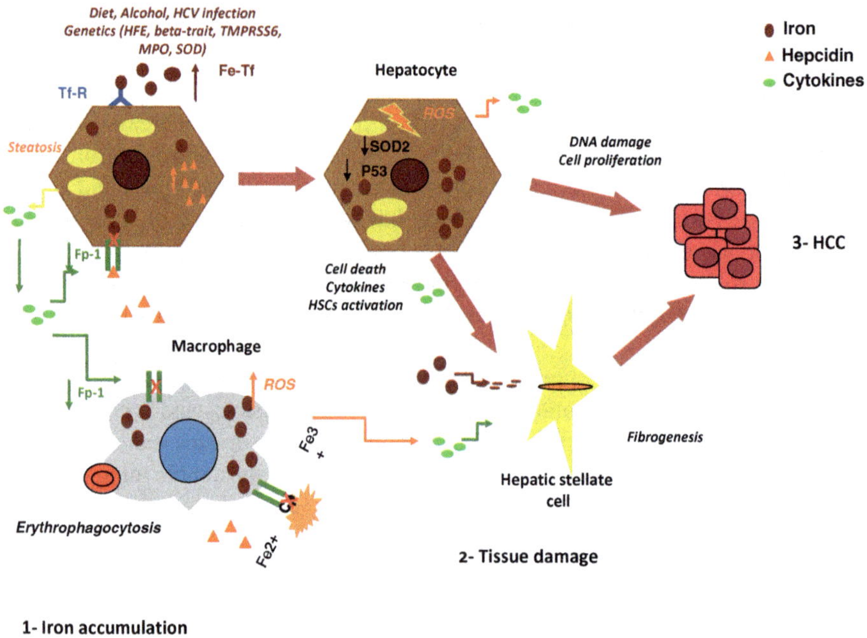

Fig. 15.1 *Mpo* myeloperoxidase, *Sod* superoxide-dismutase, *Tf-R* transferrin receptor, *Fe-Tf* transferrin saturated with iron, *Fp-1* ferroportin-1, *ROS* reactive oxygen species, *Cp* ceruloplasmin, *HCC* hepatocellular carcinoma, *HSC* hepatic stellate cells

References

1. Andriopoulos B Jr, Corradini E et al (2009) BMP6 is a key endogenous regulator of hepcidin expression and iron metabolism. Nat Genet 41(4):482–487
2. Asare GA, Paterson AC, Kew MC et al (2006) Iron-free neoplastic nodules and hepatocellular carcinoma without cirrhosis in Wistar rats fed a diet high in iron. J Pathol 208:82–90
3. Barisani D, Pelucchi S, Mariani R et al (2008) Hepcidin and iron-related gene expression in subjects with dysmetabolic hepatic iron overload. J Hepatol 49:123–133
4. Ben-Arieh SV, Zimerman B, Smorodinsky NI et al (2001) Human cytomegalovirus protein US2 interferes with the expression of human HFE, a nonclassical class I major histocompatibility complex molecule that regulates iron homeostasis. J Virol 75:10557–10562
5. Camaschella C (2005) Understanding iron homeostasis through genetic analysis of hemochromatosis and related disorders. Blood 106(12):3710–3717
6. Corengia C, Galimberti S, Bovo G et al (2005) Iron accumulation in chronic hepatitis C: relation of hepatic iron distribution, HFE genotype, and disease course. Am J Clin Pathol 124:846–853
7. Dongiovanni P, Fracanzani AL et al (2010) Iron-dependent regulation of MDM2 influences p53 activity and hepatic carcinogenesis. Am J Pathol 176:1006–1017
8. Dongiovanni P, Fracanzani AL et al (2011) Iron in fatty liver and in the metabolic syndrome: a promising therapeutic target. J Hepatol 55(4):920–932

9. Facchini FS, Hua NW, Stoohs RA (2002) Effect of iron depletion in carbohydrate intolerant patients with clinical evidence of nonalcoholic fatty liver disease. Gastroenterology 122:931–939
10. Fargion S, Valenti L et al (2010) Hemochromatosis gene (HFE) mutations and cancer risk: expanding the clinical manifestations of hereditary iron overload. Hepatology 51(4): 1119–1121
11. Fargion S, Valenti L, Fracanzani AL (2011) Beyond hereditary hemochromatosis: new insights into the relationship between iron overload and chronic liver disease. Dig Liver Dis 43:89–95
12. Feder JN, Gnirke A et al (1996) A novel MHC class I-like gene is mutated in patients with hereditary haemochromatosis. Nat Genet 13(4):399–408
13. Finberg KE (2009) Iron-refractory iron deficiency anemia. Semin Hematol 46(4):378–386
14. Fracanzani AL, Conte D, Fraquelli M et al (2001) Increased cancer risk in a cohort of 230 patients with hereditary hemochromatosis in comparison to matched control patients with non-iron-related chronic liver disease. Hepatology 33:647–651
15. Fracanzani AL, Fargion S et al (2005) Association between heterozygosity for HFE gene mutations and hepatitis viruses in hepatocellular carcinoma. Blood Cells Mol Dis 35:27–32
16. Halliwell B, Gutteridge JMC (1992) Biologically relevant metal ion-dependent hydroxyl radical generation. An update. FEBS Lett 307:108–112
17. Harrison-Findik DD, Klein E, Crist C et al (2007) Iron-mediated regulation of liver hepcidin expression in rats and mice is abolished by alcohol. Hepatology 46:1979–1985
18. Houstis N, Rosen ED, Lander ES (2006) Reactive oxygen species have a causal role in multiple forms of insulin resistance. Nature 440:944–948
19. Karin M, Takahashi T, Kapahi P, Delhase M, Chen Y, Makris C, Rothwarf D, Baud V, Natoli G, Guido F, Li N (2001) Oxidative stress and gene expression: the AP-1 and NF-kappaB connections. Biofactors 15:87–89
20. Kautz L, Jung G, Valore EV, Rivella S, Nemeth E, Ganz T (2014) Identification of erythroferrone as an erythroid regulator of iron metabolism. Nat Genet 46(7):678–684
21. Lesbordes-Brion JC, Viatte L et al (2006) Targeted disruption of the hepcidin 1 gene results in severe hemochromatosis. Blood 108(4):1402–1405
22. Létocart E, Le Gac G, Majore S et al (2009) A novel missense mutation in SLC40A1 results in resistance to hepcidin and confirms the existence of two ferroportin-associated iron overload diseases. Br J Haematol 147(3):379–385
23. Miura K, Taura K, Kodama Y et al (2008) Hepatitis C virus-induced oxidative stress suppresses hepcidin expression through increased histone deacetylase activity. Hepatology 48:1420–1429
24. Moirand R, Mendler MH, Guillygomarc'h A et al (2000) Non-alcoholic steatohepatitis with iron: part of insulin resistance-associated hepatic iron overload? J Hepatol 33:1024–1026
25. Muckenthaler MU (2008) Fine tuning of hepcidin expression by positive and negative regulators. Cell Metab 8(1):1–3
26. Nahon P, Sutton A, Rufat P et al (2008) Liver iron, HFE gene mutations, and hepatocellular carcinoma occurrence in patients with cirrhosis. Gastroenterology 134:102–110
27. Nahon P, Sutton A, Pessayre D et al (2011) Do genetic variations in antioxidant enzymes influence the course of hereditary hemochromatosis? Antioxid Redox Signal 15(1):31–38
28. Nemeth E, Tuttle MS et al (2004) Hepcidin regulates cellular iron efflux by binding to ferroportin and inducing its internalization. Science 306(5704):2090–2093
29. Nishina S, Hino K, Korenaga M et al (2008) Hepatitis C virus-induced reactive oxygen species raise hepatic iron level in mice by reducing hepcidin transcription. Gastroenterology 134:226–238
30. Pietrangelo A (2004) Hereditary hemochromatosis—a new look at an old disease. N Engl J Med 350:2383–2397
31. Pietrangelo A (2007) Hemochromatosis: an endocrine liver disease. Hepatology 46(4):1291–1301

32. Pigeon C, Ilyin G et al (2001) A new mouse liver-specific gene, encoding a protein homologous to human antimicrobial peptide hepcidin, is overexpressed during iron overload. J Biol Chem 276(11):7811–7819
33. Piperno A, Mariani R et al (2000) Haemochromatosis in patients with beta-thalassaemia trait. Br J Haematol 111(3):908–914
34. Ramsay AJ, Hooper JD et al (2009) Matriptase-2 (TMPRSS6): a proteolytic regulator of iron homeostasis. Haematologica 94(6):840–849
35. Ruddell RG, Hoang-Le D et al (2009) Ferritin functions as a proinflammatory cytokine via iron-independent protein kinase C zeta/nuclear factor kappaB-regulated signaling in rat hepatic stellate cells. Hepatology 49(3):887–900
36. Sartori M, Andorno S et al (2007) Heterozygous beta-globin gene mutations as a risk factor for iron accumulation and liver fibrosis in chronic hepatitis C. Gut 56(5):693–698
37. Smedsrod B, De Bleser PJ, Braet F, Lovisetti P, Vanderkerken K, Wisse E, Geerts A (1994) Cell biology of liver endothelial and Kupffer cells. Gut 35:1509–1516
38. Sorretino P, D'Angelo S et al (2009) Liver iron excess in patients with hepatocellular carcinoma developed on non-alcoholic steato-hepatitis. J Hepatol 50(2):351–357
39. Sutton A, Nahon P, Pessayre D et al (2006) Genetic polymorphisms in antioxidant enzymes modulate hepatic iron accumulation and hepatocellular carcinoma development in patients with alcohol-induced cirrhosis. Cancer Res 66:2844–2852
40. Tung BY, Emond MJ, Bronner MP et al (2003) Hepatitis C, iron status, and disease severity: relationship with HFE mutations. Gastroenterology 124:318–326
41. Valenti L, Conte D et al (2004) The mitochondrial superoxide dismutase A16V polymorphism in the cardiomyopathy associated with hereditary haemochromatosis. J Med Genet 41:946–950
42. Valenti L, Pulixi EA, Arosio P et al (2007) Relative contribution of iron genes, dysmetabolism and hepatitis C virus (HCV) in the pathogenesis of altered iron regulation in HCV chronic hepatitis. Haematologica 92:1037–1042
43. Valenti L, Canavesi E et al (2010) Beta-globin mutations are associated with parenchymal siderosis and fibrosis in patients with non-alcoholic fatty liver disease. J Hepatol 53(5): 927–933
44. Valenti L, Fracanzani A, Bugianesi E et al (2010) HFE genotype, parenchymal iron accumulation, and liver fibrosis in patients with nonalcoholic fatty liver disease. Gastroenterology 138:905–912
45. Valenti L, Rametta R et al (2012) The A736V TMPRSS6 polymorphism influences hepatic iron overload in nonalcoholic fatty liver disease. PLoS One 7(11):e48804
46. Wood MJ, Powell LW et al (2008) Environmental and genetic modifiers of the progression to fibrosis and cirrhosis in hemochromatosis. Blood 111(9):4456–4462
47. Zacharski LR, Chow BK et al (2008) Decreased cancer risk after iron reduction in patients with peripheral arterial disease: results from a randomized trial. J Natl Cancer Inst 100(14):996–1002

Chapter 16
Oxidative Stress in the Central Nervous System Complications of Chronic Liver Failure

Chantal Bemeur

16.1 Introduction

Chronic liver failure is the process of progressive destruction and regeneration of liver parenchyma leading to fibrosis and cirrhosis and would be responsible for an estimated 10.4, 7.3 and 5.3 deaths per 100,000 population in Europe, United States, and Canada, respectively [1–3]. The end-stage process of liver degeneration and failure, or cirrhosis, is the final common pathway of most forms of liver disease. Chronic liver failure may be caused by several conditions including, among others, alcohol, virus, non-alcoholic fatty liver disease (NAFLD) and non-alcoholic steatohepatitis (NASH), biliary disease as well as metabolic disorders. Many complications may arise from chronic liver failure, including portal hypertension, impaired metabolic capacity, synthesis dysfunction, malnutrition, ascites, hepatorenal syndrome, increased risk for the development of hepatocellular carcinoma as well as hepatic encephalopathy (Fig. 16.1).

Hepatic encephalopathy, which is observed in approximately 80 % of patients with chronic liver failure [4–6], is a debilitating neuropsychiatric complication of liver disease. Characterized by a constellation of symptoms, including cognitive, psychiatric, and motor disturbances, hepatic encephalopathy can progress to coma and death. Hepatic encephalopathy encompasses several clinical signs such as asterixis, stupor, seizures, and coma; its severity is usually graded with the West Haven Criteria (Table 16.1) [7]. However, in order to address the universal

C. Bemeur, Dt.P., Ph.D. (✉)
Department of Nutrition, Montréal University Health Center (CHUM), Université de Montréal, 900, rue Saint-Denis, Tour Viger, R12.458, Montréal, QC, Canada H2X 0A9
e-mail: chantal.bemeur@umontreal.ca

E. Albano, M. Parola (eds.), *Studies on Hepatic Disorders*, Oxidative Stress in Applied Basic Research and Clinical Practice, DOI 10.1007/978-3-319-15539-5_16

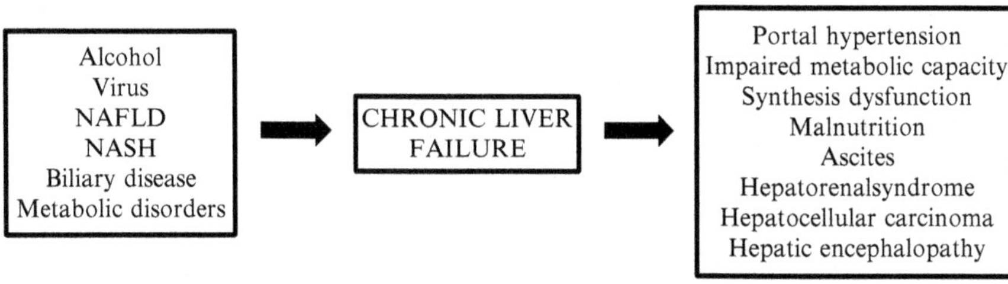

Fig. 16.1 Etiologic factors of chronic liver failure and resulting complications (*NAFLD* non-alcoholic fatty liver disease, *NASH* non-alcoholic steato-hepatitis)

Table 16.1 Grading of hepatic encephalopathy according to the West Haven criteria

Grade	Symptom
0	No signs or symptoms
I	Disturbed sleep-wake rhythm, restless, euphoria, anxiety, aimless, shortened attention span, trivial lack of awareness, impaired performance of addition
II	Lethargia or apathy, overt personality changes, disorientation for time/space, flapping, memory weakness, impaired performance for subtraction
III	Somnolence, stupor, confusion, disturbed articulation, responsive to verbal stimuli
IV	Coma

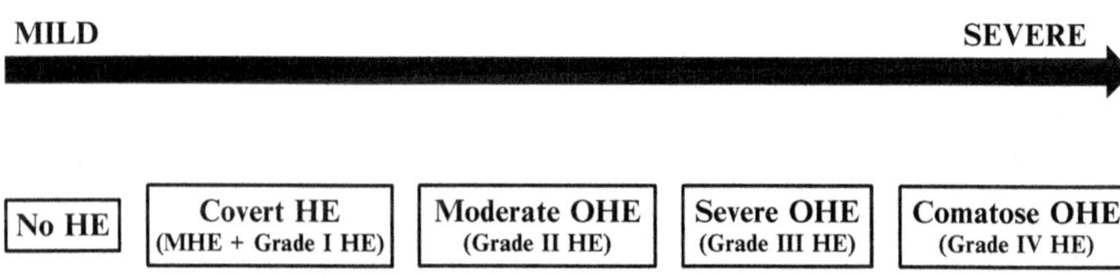

Fig. 16.2 Proposed ISHEN classification [8]. (*HE* hepatic encephalopathy, *MHE* minimal HE, *OHE* overt HE)

concern about the accuracy of the West Haven scale in diagnosing the severity of overt hepatic encephalopathy, the International Society for Hepatic Encephalopathy and Nitrogen metabolism (ISHEN), the official authority for issuing updates on terminology regarding hepatic encephalopathy, has recently proposed a revised classification (Fig. 16.2) [8]. Overall, hepatic encephalopathy has a significant impact on patients' quality of life and on their ability to function daily. Furthermore, hepatic encephalopathy leads to a poor prognosis and a greater risk of mortality [9]. This chapter will summarize the involvement of oxidative stress in the pathogenesis of hepatic encephalopathy, its consequences, and its potential role in therapeutic strategy.

16.2 Pathogenesis of Hepatic Encephalopathy

The pathophysiologic basis of hepatic encephalopathy is multifactorial and remains unclear. However, there is general agreement that ammonia plays a key role [10]. Ammonia accumulates in the brain in chronic liver failure leading to impaired bioenergetics [10, 11], altered neurotransmission [11–13], activation of peripheral benzodiazepine receptors [12, 14], leading to the synthesis of neurosteroids [15] as well as glutamate-mediated excitotoxicity [16, 17] and excessive production of glutamine [18, 19]. However, in the setting of chronic liver failure, the correlation between ammonia and severity of hepatic encephalopathy remains inconclusive [20, 21], suggesting that other pathogenic factors may be implicated. In recent years, oxidative stress has also been suggested to be part of the pathophysiologic cascade in hepatic encephalopathy. Among the factors responsible for oxidative stress development in the setting of hepatic encephalopathy are ammonia, manganese, intracellular calcium, mitochondrial permeability transition (MPT), electron transport chain, *N*-methyl D-aspartate (NMDA) receptors, peripheral benzodiazepine receptor, nuclear factor-Kappa B, inflammation, and glutamine. Data from the literature suggests that the relationship between these factors and oxidative stress in the pathogenesis of hepatic encephalopathy is complex. However, a detailed review of these factors in relation to hepatic encephalopathy is beyond the scope of this chapter. In the present chapter, these factors will be tackled emphasizing the implication of ammonia and manganese-induced oxidative stress in relation with hepatic encephalopathy.

16.3 Oxidative Stress in Hepatic Encephalopathy

Oxidative stress, a condition in which the production of free radicals is far in excess of their rate of detoxification by endogenous mechanisms [22], refers to a state in which tissue and cellular redox balance is altered towards a more oxidizing environment [23, 24]. Precisely, oxidative stress results from an imbalance between the generation of reactive oxygen species and the cellular antioxidant defense capacity, potentially able to affect molecular structure and function. Reactive oxygen species include, among other, hydrogen peroxide (H_2O_2), hydroxyl radical ($^{\bullet}OH$), superoxide anion ($O_2^{\bullet-}$), and peroxynitrite ($ONOO^-$). These reactive oxygen intermediates, which play important roles in cell signaling [25], are highly reactive due to the presence of unpaired valence shell electrons and are constantly produced during oxygen metabolism. However, in excess, reactive oxygen species are very harmful to the cell, for their reaction with cellular structures and macromolecules, and lead to cellular dysfunction. Indeed, oxidative stress affects major cellular components, including lipids, proteins, and DNA. In addition, the brain is absolutely dependent upon oxidative metabolism for cell survival and, being a highly aerobic tissue accounting for 20 % of total oxygen consumed by the body, is prone to dysfunction due to oxidative stress.

The neuropathology of hepatic encephalopathy in chronic liver failure reveals primarily changes in astrocytes (glial cells of the central nervous system) including cell swelling which consequently leads to brain edema [26–28]. Specifically, hepatic encephalopathy resulting from chronic liver failure reflects the clinical manifestation of a low-grade cerebral edema that develops after exhaustion of the volume-regulatory capacity of the astrocytes in response to ammonia and other hepatic encephalopathy-precipitating factors (e.g., inflammation, infection, hyponatremia) [29, 30]. As a consequence, astrocyte swelling triggers a complex signaling cascade which relies on NMDA receptor activation and elevation of intracellular calcium concentration, which result in increased formation of reactive oxygen species through activation of nicotinamide adenine dinucleotide phosphate (NADPH) oxidase and nitric oxide synthase (NOS) [29, 31]. Since oxidative stress in turn promotes astrocyte swelling, a self-amplifying signaling loop between osmotic- and oxidative stress is proposed [31, 32]. Among the consequences of this oxidative stress response are protein-tyrosine nitration, oxidation of RNA, and activation of Zn^{2+}-dependent transcription [31–35].

16.3.1 Ammonia-Induced Oxidative Stress in Hepatic Encephalopathy

Evidence for the involvement of oxidative stress in hepatic encephalopathy initially arose from the observation that Alzheimer type II astrocytes, a distinctive neuropathologic finding in brains of patient with hepatic encephalopathy, contain large amounts of lipofuscin pigments [36], consisting of peroxidized lipids [37]. Excessive amounts of lipofuscin pigments were also detected in ammonia-treated astrocyte cultures [37, 38]. Subsequently, more than 20 years ago, O'Connor and Costell [39] postulated that oxidative stress is implicated in the pathophysiologic mechanisms responsible for hepatic encephalopathy. They reported that hyperammonemic mice displayed evidence of lipid peroxidation in the brain [39]. Since these findings have been reported, substantial evidence from cell cultures and animal studies for an important role of hyperammonemia and oxidative stress in the pathogenesis of hepatic encephalopathy has been reported.

16.3.1.1 In Vitro Evidence of Ammonia-Induced Oxidative Stress in Hepatic Encephalopathy

Much of the evidence for a role of oxidative stress in ammonia neurotoxicity has been derived from cell culture studies [40]. Evidence suggests a close interrelation between astrocyte swelling, a characteristic feature of hepatic encephalopathy, and the production of reactive oxygen species leading to oxidative stress [41]. Specifically, astrocytes have been suggested to be a major source of reactive oxygen species under simulated hepatic encephalopathy conditions [41, 42]. Indeed, astrocytes exposed to a pathophysiological concentration of ammonia were found to stimulate the production of free radicals and reactive oxygen species. In hypoosmotically treated cultured rat astrocytes, a NMDA receptor-dependent elevation of the intracellular calcium con centration was identified to be essential to swelling-dependent reactive oxygen species generation [43]. Ammonia-induced free radical generation via the activation of NADPH oxidase was observed in cultured astrocytes [42–44], while astrocyte swelling triggers a p47(phox)-dependent NADPH oxidase-catalyzed reactive oxygen species production [31]. In addition, in astrocyte cultures exposed to ammonia, increased heme oxygenase-1 (HO-1) [44] and inducible NOS (iNOS) expression as well as nitric oxide production were identified [34, 45]. Ammonia has also been shown to significantly increase soluble guanylyl cyclase [46], a source of nitric oxide, and decrease cellular glutathione (GSH) level [47], a major endogenous antioxidant, in cultured astrocytes. Furthermore, natriuretic peptides, which are known to attenuate the production of reactive oxygen species in other systems [47, 48], were shown to reduce the accumulation of reactive oxygen species in ammonia-treated cultured astrocytes [49]. Altogether, these studies suggest that oxidative stress is induced by astrocytes in conditions associated with increased levels of ammonia.

Another factor by which ammonia and oxidative stress have been related to hepatic encephalopathy is the MPT, a calcium-dependent process characterized by the opening of the permeability transition pore (PTP) in the inner mitochondrial membrane. Oxidative stress triggers the induction of the MPT and, as a consequence, there is an increased permeability to protons, ions, and other solutes <1,500 Da [50], leading to a collapse of the mitochondrial inner membrane potential. Loss of the mitochondrial membrane potential results in osmotic swelling of the mitochondrial matrix, movement of metabolites across the inner membrane, defective oxidative phosphorylation, cessation of ATP synthesis, and the generation of reactive oxygen species. It was reported that, in cultured astrocytes, oxidative stress would be involved in the induction of the MPT by ammonia, suggesting that oxidative stress and the subsequent induction of the MPT contribute to the pathogenesis of hepatic encephalopathy [51, 52]. Also, treatment of cultured astrocytes with ammonia caused a significant dissipation of the mitochondrial membrane potential as well as an increase in the mitochondrial permeability to 2-deoxyglucose. Both of these changes were blocked by cyclosporin A, a MPT inhibitor. Similarly, ammonia caused a decrease in the mitochondrial calcein fluorescence (an index of the MPT), which was also blocked by cyclosporin A [53].

16.3.1.2 In Vivo Evidence of Ammonia-Induced Oxidative Stress in Hepatic Encephalopathy

Evidence for the implication of oxidative stress in the pathogenesis of hepatic encephalopathy has also been reported in studies in experimental animals. For example, NOS activity has been shown to be increased in the brains of portacaval-shunted rats [54]. iNOS and neuronal NOS (nNOS) protein expression is also increased in the brains of these animals [22, 55, 56], whereas brain endothelial NOS (eNOS) protein expression is increased in thioacetamide (TAA)-induced cirrhosis in the rat [57]. It was also reported that astrocyte swelling stimulated the production of cerebral nitric oxide in ammonia-treated rats [58]. An increase in HO-1 mRNA expression as well as a decrease in copper/zinc-superoxide dismutase (Cu/Zn SOD) gene expression have been reported in the brains of portacaval-shunted rats [59]. Protein tyrosine nitration, a consequence of oxidative stress, was also demonstrated in the cerebral cortex of these animals [34, 56].

In a rat model of chronic liver failure (portacaval anastomosis), neurons were immunoreactive to nNOS, whereas iNOS was expressed in pyramidal-like cortical neurons and perivascular astrocytes [56]. In the same animals, nitrotyrosine immunoreactivity was found in pyramidal-like cortical neurons and in perivascular astrocytes. It was also demonstrated that nNOS and iNOS are produced in the Purkinje (neuronal) cells and Bergmann glial cells in rats following portacaval anastomosis [55]. TAA-cirrhotic rats showed nNOS immunoreactivity in stellate and basket neurons and eNOS immunoreactivity in perivascular glial cells of the white matter [57]. In the same experimental model of chronic liver failure, eNOS was located in Purkinje cell bodies and vessels endothelial cells [57]. Taken together, these data suggest that neuronal, glial, and endothelial cells are all sources of free radicals and nitric oxide in hepatic encephalopathy indicating that oxidative stress in hepatic encephalopathy is a multicellular phenomenon.

Systemically, the relationship between hyperammonemia and oxidative stress differs from that depicted in the brain. Indeed, in another model of chronic liver failure and hyperammonemia/hepatic encephalopathy, the bile-duct ligated rat, Bosoi et al. [60] observed the presence of systemic oxidative stress and cerebral edema. The authors suggested that systemic oxidative stress might be an important "first hit," which, followed by increases in ammonia, leads to the onset of brain edema [60]. In a similar model of cirrhosis, an increase in lipid peroxidation and reduction in antioxidant enzymes in the cerebral cortex and cerebellum were reported [61]. Interestingly, the administration of *N*-acetylcysteine exerted a protective effect through the attenuation of oxidative stress [61].

16.3.1.3 Human Studies

While considerable evidence indicates the presence of oxidative stress markers in experimental models of hepatic encephalopathy, documentation of oxidative stress in humans is limited. Increased amount of lipofuscin pigments was found in brains of patients with hepatic encephalopathy [36, 40]. Elevated blood levels of reactive oxygen species were also identified in patients with hepatic encephalopathy resulting from chronic alcohol consumption, which was associated with decrease antioxidant capacity [62]. Increased SOD activity, thiobarbituric acid reactive substances, and decreased catalase activities were observed in cirrhotic children [63]. Increased nitric oxide after transjugular intrahepatic portosystemic shunt insertion in patients with cirrhosis was also reported [64]. Furthermore, postmortem cortical brain tissue samples from patients with cirrhosis dying with or without hepatic encephalopathy were analyzed and compared with brains from patients without liver disease [35]. The results indicate that hepatic encephalopathy in patients with cirrhosis is associated with oxidative stress, protein tyrosine nitration, and RNA oxidation [35].

In postmortem human brain tissue obtained from autopsies of patients with cirrhosis and hepatic encephalopathy, a whole human genome microarray approach revealed altered expression of genes related to oxidative stress [65]. Specifically, expression levels of genes involved in oxidative stress defense, such as HO-1, selenoprotein-V, peroxiredoxin-4, and peroxisome proliferator-activated receptor α (PPARα), were elevated in patients with cirrhosis with hepatic encephalopathy but not in patients with cirrhosis without hepatic encephalopathy, when compared with controls [65]. Taken together, these data strongly suggest a role for oxidative stress in the pathogenesis of hepatic encephalopathy in patients with chronic liver failure and indicate that cerebral oxidative stress is a hallmark of hepatic encephalopathy in patients with liver failure.

16.3.2 Manganese-Induced Oxidative Stress in Hepatic Encephalopathy

Manganese is an essential trace element found in a variety of biological tissues and is necessary for normal functioning of several physiological processes including amino acid, lipid, protein, and carbohydrate metabolism [66]. Manganese is also an important component of a number of cerebral enzymes, in particular, glutamine synthetase, an ammonia detoxifying enzyme. At low levels, manganese binds with superoxide dismutase to form MnSOD, an important mitochondrial antioxidant enzyme [67]. However, when excessive, manganese contributes to neurological abnormalities such as parkinsonism and dystonia [68]. Occupational exposure to excessive manganese levels leads to neurotoxicity, referred to as manganism, which resembles Parkinson's disease [69]. Chronic exposure of various cell types to manganese was shown to induce oxidative stress [70–72].

Manganese has also been implicated in the pathogenesis of hepatic encephalopathy [73]. Manganese highly accumulates in astrocytes [74, 75], which renders these cells more vulnerable to its toxicity. Consistent with this vulnerability, manganese has been shown to decrease antioxidant capacity [72] and generate oxidative stress [72, 76, 77], which are prevented by pre-treatment with *N*-acetylcysteine [78]. Manganese also brings about mitochondrial dysfunction [79, 80], including decreased energy production [72] and the induction of the MPT [81], and causes histopathological changes in astrocytes (Alzheimer type II change) [82]. Interestingly, morphologic and functional changes after exposure of astrocytes to manganese are similar to those observed after ammonia treatment. Cultured astrocytes exposed to ammonia (5 mM) or manganese acetate (100 mM) were shown to increase both free radicals production and L-arginine uptake (a precursor of nitric oxide), and such effects were synergized when manganese was co-treated with ammonia [77, 83]. Similarly, exposure of primary cortical astrocytes to a low concentration of manganese (10 μM) was shown to potentiate interferon-gamma and tumor necrosis factor-alpha-induced expression of iNOS mRNA and protein along with an increased production of nitric oxide [84]. The potentiating effect was a consequence of the activation of soluble guanylate cyclase and mitogen-activated protein kinase (MAPK) signaling pathways [84]. Cultured astrocytes exposed to manganese were also shown to inhibit glutamate uptake by a process involving oxidative stress [85, 86]. Additionally, it was demonstrated that treatment of rats with manganese chloride led to an increase in manganese level in brain that was accompanied by the development of pathological changes similar to those seen in hepatic encephalopathy (Alzheimer type II astrocytosis), and such changes were significantly reduced when rats were treated with antioxidant *N*-actetylcysteine. In primary rat cortical neurons exposed to manganese, an increase in biomarkers of oxidative damage (F(2)-isoprostanes), which was prevented by pretreatment with the antioxidant Trolox (hydrophilic analog of vitamin E), was reported [87]. These results were confirmed in mice exposed to manganese [87]. Finally, it was recently demonstrated that manganese leads to an increase in markers of oxidative stress in rat brain chronically exposed to manganese [88]. These studies suggest that manganese contributes to oxidative stress in hepatic encephalopathy and that such effect is exacerbated in the presence of ammonia.

Patients with chronic liver failure and those who had surgically created portal-systemic shunts have elevated plasma and brain manganese levels, most likely reflecting the combined effects of hepatocellular failure, impaired biliary excretion, and the presence of portal-systemic shunting of blood [89]. This may lead to selective manganese accumulation in the globus pallidus, caudate nucleus, and putamen, and the adjacent areas of the basal ganglia manifest as hyperintensity of these brains areas on T1-Magnetic Resonance Imaging [89, 90]. Indeed, elevated manganese levels were found in the globus pallidus obtained at autopsy from patients with chronic liver failure [90, 91]. Disturbances of manganese homeostasis may partly account for the cognitive impairment associated with chronic liver failure [92].

16.4 Antioxidant Strategies for the Treatment of Hepatic Encephalopathy

Treatment of experimental animals suffering from hyperammonemia and hepatic encephalopathy with antioxidants (e.g., ascorbate, alpha-tocopherol, dimethylsulfoxide) was shown to have beneficial effects by improving antioxidant status as well as their clinical condition [93, 94]. The antioxidant melatonin was shown to reduce blood and brain ammonia level as well as attenuate brain lipid peroxidation in rats after TAA injection [95]. Additionally, increased malondialdehyde levels and decreased glutathione peroxidase, catalase, and SOD activities were found in the hippocampal tissue of rats with portal hypertension (a model of low-grade hepatic encephalopathy), and such effects were reversed when rats were treated with curcumin, a known antioxidant [96]. Rats treated with morin (3,4,7,2′,4′-pentahydroxyflavone), a flavonol, were shown to be protected against oxidative stress in brains of chronic hyperammonemic rats [97]. It was recently demonstrated that guanosine, a nucleoside exhibiting antioxidant properties [98, 99], was neuroprotective in a rat model of chronic hepatic encephalopathy by reducing oxidative stress markers in the brain [100].

The antioxidant *N*-acetylcysteine has proven useful in reducing brain edema in acute liver failure [101] and in the management of patients with acute liver failure [102–105]. In addition, *N*-acetylcysteine was shown to delay the progression of encephalopathy in azoxymethane-induced acute liver failure in mice, as well as to reduce brain water content [101]. Finally, hypothermia, which has been shown to improve brain edema in animals and humans with acute liver failure, is also known to reduce free radical production [106]. Interestingly, *N*-acetylcysteine was able to ameliorate spatial memory and motor coordination deficits observed experimental chronic liver failure (bile-duct ligated rats) [61]. *N*-Acetylcysteine supplementation decreased lipid peroxidation and was also able to restore the activity of antioxidant enzymes as well as structural deficits observed in the cortex and cerebellum of cirrhotic animals with hepatic encephalopathy. Together, these data clearly demonstrate that the protective effect of *N*-acetylcysteine in experimental hepatic encephalopathy is mediated through attenuation of oxidative stress, suggesting a therapeutic role for *N*-acetylcysteine in patients afflicted with hepatic encephalopathy.

16.5 Conclusion

Oxidative stress has evolved in recent years as a major pathogenetic factor in hepatic encephalopathy and experimental evidence for oxidative stress in brain in experimental models of hepatic encephalopathy due to chronic liver failure is increasing. Indeed, several reports suggest that oxidative stress participates in the pathophysiologic cascade responsible for hepatic encephalopathy. While the factors responsible for oxidative stress formation in hepatic encephalopathy remain incompletely understood, it appears that ammonia and manganese would be partly responsible for the production of reactive oxygen species. Although increased oxidative stress in hepatic encephalopathy resulting from chronic liver failure has been demonstrated by some groups, its consequences are not fully established. Additional studies on the role of oxidative stress in chronic hepatic encephalopathy are warranted. Increased oxidative stress has been documented in several studies and antioxidants were shown to be protective against ammonia-induced astrocyte swelling and cerebral edema in liver failure. Antioxidant therapy such as *N*-acetylcysteine is already being used in the management of acute liver failure and its complications. Other antioxidants could prove to be valuable adjuncts to traditional hepatic encephalopathy therapies, such as ammonia-lowering strategies, in the context of chronic liver failure. Further studies are needed in order to assess these possibilities.

References

1. Zatoński WA, Sulkowska U, Mańczuk M, Rehm J, Boffetta P, Lowenfels AB et al (2010) Liver cirrhosis mortality in Europe, with special attention to Central and Eastern Europe. Eur Addict Res 16:193–201
2. FASTSTATS—chronic liver disease or cirrhosis [Internet]. [cited 2014 May 28]. Available from http://www.cdc.gov/nchs/fastats/liver-disease.htm
3. Canadian Liver Foundation (2013) Liver disease in Canada—a crisis in the making. Canadian Liver Foundation
4. Selberg O, Böttcher J, Tusch G, Pichlmayr R, Henkel E, Müller MJ (1997) Identification of high- and low-risk patients before liver transplantation: a prospective cohort study of nutritional and metabolic parameters in 150 patients. Hepatology 25:652–657
5. Hartmann IJ, Groeneweg M, Quero JC, Beijeman SJ, de Man RA, Hop WC et al (2000) The prognostic significance of subclinical hepatic encephalopathy. Am J Gastroenterol 95: 2029–2034
6. Bismuth M, Funakoshi N, Cadranel J-F, Blanc P (2011) Hepatic encephalopathy: from pathophysiology to therapeutic management. Eur J Gastroenterol Hepatol 23:8–22
7. Ferenci P, Lockwood A, Mullen K, Tarter R, Weissenborn K, Blei AT (2002) Hepatic encephalopathy—definition, nomenclature, diagnosis, and quantification: final report of the working party at the 11th World Congresses of Gastroenterology, Vienna, 1998. Hepatology 35:716–721
8. Mullen KD, Prakash RK (2012) Hepatic encephalopathy. Humana Press, Totowa
9. Stewart CA, Smith GE (2007) Minimal hepatic encephalopathy. Nat Clin Pract Gastroenterol Hepatol 4:677–685
10. Butterworth RF (2002) Pathophysiology of hepatic encephalopathy: a new look at ammonia. Metab Brain Dis 17:221–227
11. Mousseau DD, Butterworth RF (1994) Current theories on the pathogenesis of hepatic encephalopathy. Proc Soc Exp Biol Med 206:329–344
12. Norenberg MD, Huo Z, Neary JT, Roig-Cantesano A (1997) The glial glutamate transporter in hyperammonemia and hepatic encephalopathy: relation to energy metabolism and glutamatergic neurotransmission. Glia 21:124–133
13. Szerb JC, Butterworth RF (1992) Effect of ammonium ions on synaptic transmission in the mammalian central nervous system. Prog Neurobiol 39:135–153
14. Desjardins P, Butterworth RF (2002) The "peripheral-type" benzodiazepine (omega 3) receptor in hyperammonemic disorders. Neurochem Int 41:109–114
15. Ahboucha S, Pomier-Layrargues G, Mamer O, Butterworth RF (2006) Increased levels of pregnenolone and its neuroactive metabolite allopregnanolone in autopsied brain tissue from cirrhotic patients who died in hepatic coma. Neurochem Int 49:372–378
16. Hermenegildo C, Marcaida G, Montoliu C, Grisolía S, Miñana MD, Felipo V (1996) NMDA receptor antagonists prevent acute ammonia toxicity in mice. Neurochem Res 21:1237–1244
17. Marcaida G, Felipo V, Hermenegildo C, Miñana MD, Grisolía S (1992) Acute ammonia toxicity is mediated by the NMDA type of glutamate receptors. FEBS Lett 296:67–68
18. Hawkins RA, Jessy J, Mans AM, De Joseph MR (1993) Effect of reducing brain glutamine synthesis on metabolic symptoms of hepatic encephalopathy. J Neurochem 60:1000–1006
19. Takahashi H, Koehler RC, Brusilow SW, Traystman RJ (1991) Inhibition of brain glutamine accumulation prevents cerebral edema in hyperammonemic rats. Am J Physiol 261: H825–H829
20. Ong JP, Aggarwal A, Krieger D, Easley KA, Karafa MT, Van Lente F et al (2003) Correlation between ammonia levels and the severity of hepatic encephalopathy. Am J Med 114: 188–193
21. Kundra A, Jain A, Banga A, Bajaj G, Kar P (2005) Evaluation of plasma ammonia levels in patients with acute liver failure and chronic liver disease and its correlation with the severity of hepatic encephalopathy and clinical features of raised intracranial tension. Clin Biochem 38:696–699

22. Rao VLR (2002) Nitric oxide in hepatic encephalopathy and hyperammonemia. Neurochem Int 41:161–170
23. Medina J, Moreno-Otero R (2005) Pathophysiological basis for antioxidant therapy in chronic liver disease. Drugs 65:2445–2461
24. Albano E (2006) Alcohol, oxidative stress and free radical damage. Proc Nutr Soc 65: 278–290
25. Valko M, Leibfritz D, Moncol J, Cronin MTD, Mazur M, Telser J (2007) Free radicals and antioxidants in normal physiological functions and human disease. Int J Biochem Cell Biol 39:44–84
26. Bosoi CR, Parent-Robitaille C, Anderson K, Tremblay M, Rose CF (2011) AST-120 (spherical carbon adsorbent) lowers ammonia levels and attenuates brain edema in bile duct-ligated rats. Hepatology 53:1995–2002
27. Wright G, Soper R, Brooks HF, Stadlbauer V, Vairappan B, Davies NA et al (2010) Role of aquaporin-4 in the development of brain oedema in liver failure. J Hepatol 53:91–97
28. Davies NA, Wright G, Ytrebø LM, Stadlbauer V, Fuskevåg O-M, Zwingmann C et al (2009) L-ornithine and phenylacetate synergistically produce sustained reduction in ammonia and brain water in cirrhotic rats. Hepatology 50:155–164
29. Häussinger D, Görg B (2010) Interaction of oxidative stress, astrocyte swelling and cerebral ammonia toxicity. Curr Opin Clin Nutr Metab Care 13:87–92
30. Shah NJ, Neeb H, Kircheis G, Engels P, Häussinger D, Zilles K (2008) Quantitative cerebral water content mapping in hepatic encephalopathy. Neuroimage 41:706–717
31. Reinehr R, Görg B, Becker S, Qvartskhava N, Bidmon HJ, Selbach O et al (2007) Hypoosmotic swelling and ammonia increase oxidative stress by NADPH oxidase in cultured astrocytes and vital brain slices. Glia 55:758–771
32. Häussinger D, Schliess F (2008) Pathogenetic mechanisms of hepatic encephalopathy. Gut 57:1156–1165
33. Kruczek C, Görg B, Keitel V, Pirev E, Kröncke KD, Schliess F et al (2009) Hypoosmotic swelling affects zinc homeostasis in cultured rat astrocytes. Glia 57:79–92
34. Schliess F, Görg B, Fischer R, Desjardins P, Bidmon HJ, Herrmann A et al (2002) Ammonia induces MK-801-sensitive nitration and phosphorylation of protein tyrosine residues in rat astrocytes. FASEB J 16:739–741
35. Görg B, Qvartskhava N, Bidmon H-J, Palomero-Gallagher N, Kircheis G, Zilles K et al (2010) Oxidative stress markers in the brain of patients with cirrhosis and hepatic encephalopathy. Hepatology 52:256–265
36. Brunk UT (1989) On the origin of lipofuscin; the iron content of residual bodies, and the relation of these organelles to the lysosomal vacuome. A study on cultured human glial cells. Adv Exp Med Biol 266:313–320; discussion 321–322
37. Gregorios JB, Mozes LW, Norenberg MD (1985) Morphologic effects of ammonia on primary astrocyte cultures. II. Electron microscopic studies. J Neuropathol Exp Neurol 44:404–414
38. Gregorios JB, Mozes LW, Norenberg LO, Norenberg MD (1985) Morphologic effects of ammonia on primary astrocyte cultures. I. Light microscopic studies. J Neuropathol Exp Neurol 44:397–403
39. O'Connor JE, Costell M (1990) New roles of carnitine metabolism in ammonia cytotoxicity. Adv Exp Med Biol 272:183–195
40. Norenberg MD (1987) The role of astrocytes in hepatic encephalopathy. Neurochem Pathol 6:13–33
41. Schliess F, Görg B, Häussinger D (2006) Pathogenetic interplay between osmotic and oxidative stress: the hepatic encephalopathy paradigm. Biol Chem 387:1363–1370
42. Norenberg MD, Rao KVR, Jayakumar AR (2005) Mechanisms of ammonia-induced astrocyte swelling. Metab Brain Dis 20:303–318
43. Schliess F, Foster N, Görg B, Reinehr R, Häussinger D (2004) Hypoosmotic swelling increases protein tyrosine nitration in cultured rat astrocytes. Glia 47:21–29
44. Warskulat U, Görg B, Bidmon H-J, Müller HW, Schliess F, Häussinger D (2002) Ammonia-induced heme oxygenase-1 expression in cultured rat astrocytes and rat brain in vivo. Glia 40:324–336

45. Sinke AP, Jayakumar AR, Panickar KS, Moriyama M, Reddy PVB, Norenberg MD (2008) NFkappaB in the mechanism of ammonia-induced astrocyte swelling in culture. J Neurochem 106:2302–2311
46. Konopacka A, Zielińska M, Albrecht J (2008) Ammonia inhibits the C-type natriuretic peptide-dependent cyclic GMP synthesis and calcium accumulation in a rat brain endothelial cell line. Neurochem Int 52:1160–1166
47. Woodard GE, Rosado JA (2008) Natriuretic peptides in vascular physiology and pathology. Int Rev Cell Mol Biol 268:59–93
48. De Vito P, Incerpi S, Pedersen JZ, Luly P (2010) Atrial natriuretic peptide and oxidative stress. Peptides 31:1412–1419
49. Skowrońska M, Zielińska M, Albrecht J (2010) Stimulation of natriuretic peptide receptor C attenuates accumulation of reactive oxygen species and nitric oxide synthesis in ammonia-treated astrocytes. J Neurochem 115:1068–1076
50. Zoratti M, Szabò I (1995) The mitochondrial permeability transition. Biochim Biophys Acta 1241:139–176
51. Bai M, Qi X, Yang Z, Yin Z, Nie Y, Yuan S et al (2011) Predictors of hepatic encephalopathy after transjugular intrahepatic portosystemic shunt in cirrhotic patients: a systematic review. J Gastroenterol Hepatol 26:943–951
52. Rao KV, Panickar KS, Jayakumar AR, Norenberg MD (2005) Astrocytes protect neurons from ammonia toxicity. Neurochem Res 30:1311–1318
53. Bai G, Rao KVR, Murthy CR, Panickar KS, Jayakumar AR, Norenberg MD (2001) Ammonia induces the mitochondrial permeability transition in primary cultures of rat astrocytes. J Neurosci Res 66:981–991
54. Rao VL, Audet RM, Butterworth RF (1995) Selective alterations of extracellular brain amino acids in relation to function in experimental portal-systemic encephalopathy: results of an in vivo microdialysis study. J Neurochem 65:1221–1228
55. Suárez I, Bodega G, Rubio M, Fernández B (2005) Down-regulation of astroglial proteins in the rat cerebellum after portacaval anastomosis. Neuropathol Appl Neurobiol 31:163–169
56. Suárez I, Bodega G, Arilla E, Felipo V, Fernández B (2006) The expression of nNOS, iNOS and nitrotyrosine is increased in the rat cerebral cortex in experimental hepatic encephalopathy. Neuropathol Appl Neurobiol 32:594–604
57. Hernández R, Martínez-Lara E, Moral MLD, Blanco S, Cañuelo A, Siles E et al (2004) Upregulation of endothelial nitric oxide synthase maintains nitric oxide production in the cerebellum of thioacetamide cirrhotic rats. Neuroscience 126:879–887
58. Master S, Gottstein J, Blei AT (1999) Cerebral blood flow and the development of ammonia-induced brain edema in rats after portacaval anastomosis. Hepatology 30:876–880
59. Song G, Dhodda VK, Blei AT, Dempsey RJ, Rao VLR (2002) GeneChip analysis shows altered mRNA expression of transcripts of neurotransmitter and signal transduction pathways in the cerebral cortex of portacaval shunted rats. J Neurosci Res 68:730–737
60. Bosoi CR, Yang X, Huynh J, Parent-Robitaille C, Jiang W, Tremblay M et al (2012) Systemic oxidative stress is implicated in the pathogenesis of brain edema in rats with chronic liver failure. Free Radic Biol Med 52:1228–1235
61. Dhanda S, Kaur S, Sandhir R (2013) Preventive effect of N-acetyl-L-cysteine on oxidative stress and cognitive impairment in hepatic encephalopathy following bile duct ligation. Free Radic Biol Med 56:204–215
62. Negru T, Ghiea V, Păsărică D (1999) Oxidative injury and other metabolic disorders in hepatic encephalopathy. Rom J Physiol 36:29–36
63. Ribeiro L, Andreazza AC, Salvador M, da Silveira TR, Vieira S, Nora DB et al (2007) Oxidative stress and S100B protein in cirrhotic children. Neurochem Res 32:1600–1603
64. Jalan R, Olde Damink SWM, Ter Steege JC, Redhead DN, Lee A, Hayes PC et al (2011) Acute endotoxemia following transjugular intrahepatic stent-shunt insertion is associated with systemic and cerebral vasodilatation with increased whole body nitric oxide production in critically ill cirrhotic patients. J Hepatol 54:265–271

65. Görg B, Bidmon H-J, Häussinger D (2013) Gene expression profiling in the cerebral cortex of patients with cirrhosis with and without hepatic encephalopathy. Hepatology 57:2436–2447
66. Aschner M, Erikson KM, Dorman DC (2005) Manganese dosimetry: species differences and implications for neurotoxicity. Crit Rev Toxicol 35:1–32
67. Holley AK, Dhar SK, St Clair DK (2010) Manganese superoxide dismutase versus p53: the mitochondrial center. Ann N Y Acad Sci 1201:72–78
68. Pal PK, Samii A, Calne DB (1999) Manganese neurotoxicity: a review of clinical features, imaging and pathology. Neurotoxicology 20:227–238
69. Olanow CW (2004) Manganese-induced parkinsonism and Parkinson's disease. Ann N Y Acad Sci 1012:209–223
70. Sun AY, Yang WL, Kim HD (1993) Free radical and lipid peroxidation in manganese-induced neuronal cell injury. Ann N Y Acad Sci 679:358–363
71. Desole MS, Esposito G, Migheli R, Fresu L, Sircana S, Zangani D et al (1995) Cellular defence mechanisms in the striatum of young and aged rats subchronically exposed to manganese. Neuropharmacology 34:289–295
72. Chen C-J, Liao S-L (2002) Oxidative stress involves in astrocytic alterations induced by manganese. Exp Neurol 175:216–225
73. Butterworth RF (2010) Metal toxicity, liver disease and neurodegeneration. Neurotox Res 18:100–105
74. Aschner M, Vrana KE, Zheng W (1999) Manganese uptake and distribution in the central nervous system (CNS). Neurotoxicology 20:173–180
75. Aschner M, Gannon M, Kimelberg HK (1992) Manganese uptake and efflux in cultured rat astrocytes. J Neurochem 58:730–735
76. Erikson KM, Dorman DC, Lash LH, Aschner M (2005) Persistent alterations in biomarkers of oxidative stress resulting from combined in utero and neonatal manganese inhalation. Biol Trace Elem Res 104:151–163
77. Jayakumar AR, Rao KVR, Kalaiselvi P, Norenberg MD (2004) Combined effects of ammonia and manganese on astrocytes in culture. Neurochem Res 29:2051–2056
78. Zhang F, Xu Z, Gao J, Xu B, Deng Y (2008) In vitro effect of manganese chloride exposure on energy metabolism and oxidative damage of mitochondria isolated from rat brain. Environ Toxicol Pharmacol 26:232–236
79. Gavin CE, Gunter KK, Gunter TE (1990) Manganese and calcium efflux kinetics in brain mitochondria. Relevance to manganese toxicity. Biochem J 266:329–334
80. Heron P, Cousins K, Boyd C, Daya S (2001) Paradoxical effects of copper and manganese on brain mitochondrial function. Life Sci 68:1575–1583
81. Rao KVR, Norenberg MD (2004) Manganese induces the mitochondrial permeability transition in cultured astrocytes. J Biol Chem 279:32333–32338
82. Olanow CW, Good PF, Shinotoh H, Hewitt KA, Vingerhoets F, Snow BJ et al (1996) Manganese intoxication in the rhesus monkey: a clinical, imaging, pathologic, and biochemical study. Neurology 46:492–498
83. Hazell AS, Norenberg MD (1998) Ammonia and manganese increase arginine uptake in cultured astrocytes. Neurochem Res 23:869–873
84. Moreno JA, Sullivan KA, Carbone DL, Hanneman WH, Tjalkens RB (2008) Manganese potentiates nuclear factor-kappaB-dependent expression of nitric oxide synthase 2 in astrocytes by activating soluble guanylate cyclase and extracellular responsive kinase signaling pathways. J Neurosci Res 86:2028–2038
85. Volterra A, Trotti D, Tromba C, Floridi S, Racagni G (1994) Glutamate uptake inhibition by oxygen free radicals in rat cortical astrocytes. J Neurosci 14:2924–2932
86. Hazell AS, Norenberg MD (1997) Manganese decreases glutamate uptake in cultured astrocytes. Neurochem Res 22:1443–1447
87. Milatovic D, Gupta RC, Yu Y, Zaja-Milatovic S, Aschner M (2011) Protective effects of antioxidants and anti-inflammatory agents against manganese-induced oxidative damage and neuronal injury. Toxicol Appl Pharmacol 256:219–226

88. Fernsebner K, Zorn J, Kanawati B, Walker A, Michalke B (2014) Manganese leads to an increase in markers of oxidative stress as well as to a shift in the ratio of Fe(II)/(III) in rat brain tissue. Metallomics 6:921–931
89. Inoue E, Hori S, Narumi Y, Fujita M, Kuriyama K, Kadota T et al (1991) Portal-systemic encephalopathy: presence of basal ganglia lesions with high signal intensity on MR images. Radiology 179:551–555
90. Layrargues GP, Shapcott D, Spahr L, Butterworth RF (1995) Accumulation of manganese and copper in pallidum of cirrhotic patients: role in the pathogenesis of hepatic encephalopathy? Metab Brain Dis 10:353–356
91. Krieger D, Krieger S, Jansen O, Gass P, Theilmann L, Lichtnecker H (1995) Manganese and chronic hepatic encephalopathy. Lancet 346:270–274
92. Tarter RE, Hegedus AM, Van Thiel DH, Schade RR, Gavaler JS, Starzl TE (1984) Nonalcoholic cirrhosis associated with neuropsychological dysfunction in the absence of overt evidence of hepatic encephalopathy. Gastroenterology 86:1421–1427
93. Bruck R, Aeed H, Shirin H, Matas Z, Zaidel L, Avni Y et al (1999) The hydroxyl radical scavengers dimethylsulfoxide and dimethylthiourea protect rats against thioacetamide-induced fulminant hepatic failure. J Hepatol 31:27–38
94. Guerrini VH (1994) Effect of antioxidants on ammonia induced CNS-renal pathobiology in sheep. Free Radic Res 21:35–43
95. Túnez I, Muñoz MC, Medina FJ, Salcedo M, Feijóo M, Montilla P (2007) Comparison of melatonin, vitamin E and L-carnitine in the treatment of neuro- and hepatotoxicity induced by thioacetamide. Cell Biochem Funct 25:119–127
96. Roselló DM, Balestrasse K, Coll C, Coll S, Tallis S, Gurni A et al (2008) Oxidative stress and hippocampus in a low-grade hepatic encephalopathy model: protective effects of curcumin. Hepatol Res 38:1148–1153
97. Subash S, Subramanian P (2009) Morin a flavonoid exerts antioxidant potential in chronic hyperammonemic rats: a biochemical and histopathological study. Mol Cell Biochem 327: 153–161
98. Petronilho F, Périco SR, Vuolo F, Mina F, Constantino L, Comim CM et al (2012) Protective effects of guanosine against sepsis-induced damage in rat brain and cognitive impairment. Brain Behav Immun 26:904–910
99. Schmidt AP, Lara DR, Souza DO (2007) Proposal of a guanine-based purinergic system in the mammalian central nervous system. Pharmacol Ther 116:401–416
100. Paniz LG, Calcagnotto ME, Pandolfo P, Machado DG, Santos GF, Hansel G et al (2014) Neuroprotective effects of guanosine administration on behavioral, brain activity, neurochemical and redox parameters in a rat model of chronic hepatic encephalopathy. Metab Brain Dis 29(3):645–654
101. Bémeur C, Vaquero J, Desjardins P, Butterworth RF (2010) N-acetylcysteine attenuates cerebral complications of non-acetaminophen-induced acute liver failure in mice: antioxidant and anti-inflammatory mechanisms. Metab Brain Dis 25:241–249
102. Harrison PM, Wendon JA, Gimson AE, Alexander GJ, Williams R (1991) Improvement by acetylcysteine of hemodynamics and oxygen transport in fulminant hepatic failure. N Engl J Med 324:1852–1857
103. Jones AL (1998) Mechanism of action and value of N-acetylcysteine in the treatment of early and late acetaminophen poisoning: a critical review. J Toxicol Clin Toxicol 36:277–285
104. Walsh TS, Hopton P, Philips BJ, Mackenzie SJ, Lee A (1998) The effect of N-acetylcysteine on oxygen transport and uptake in patients with fulminant hepatic failure. Hepatology 27: 1332–1340
105. Wendon JA, Harrison PM, Keays R, Williams R (1994) Cerebral blood flow and metabolism in fulminant liver failure. Hepatology 19:1407–1413
106. Globus MY, Alonso O, Dietrich WD, Busto R, Ginsberg MD (1995) Glutamate release and free radical production following brain injury: effects of posttraumatic hypothermia. J Neurochem 65:1704–1711

Chapter 17
Oxidative Stress and Liver Transplantation

Piotr Czubkowski, Piotr Socha, and Joanna Pawłowska

17.1 Introduction

Reactive oxygen species (ROS) are generated during many physiological and pathological processes (Fig. 17.1). Liver transplant recipients are particularly vulnerable due to ischemia–reperfusion (IR) injury, posttransplant complications, immunosuppressive treatment, and imbalanced oxidative status carried from pretransplant conditions [1]. If defense mechanisms are not capable, released ROS cause lipid, protein, and DNA damage. Consequences of prolonged oxidative imbalance are not fully understood, however its importance is evidenced in atherosclerosis, cancers, and neurodegenerative diseases. Especially cardiovascular events are of main concern and were shown to influence survival in adult population but no effective antioxidant prophylaxis exists as yet.

Hepatic ischemia followed by reperfusion is a major problem leading to outburst of ROS and apoptosis of hepatocytes [2]. Moreover, remote organs including the lung, kidney, intestine, pancreas, adrenals, and myocardium among others, may be injured as well [3].

More severe effects of IR should be expected if "marginal graft" is considered. That issue emerged with the extension of criteria for donors as natural response to the shortage of available organs.

So as to alleviate IR injury, a large number of studies, mostly experimental, were carried out in search of optimal organ preservation techniques including modified reperfusion solutions, genetic interventions, and new techniques of mechanical perfusion. These efforts shed new light on our current knowledge and elucidated

P. Czubkowski (✉) • P. Socha • J. Pawłowska
Department of Gastroenterology, Hepatology, Feeding Disorders and Pediatrics,
The Children's Memorial Health Institute, Al. Dzieci Polskich 20, Warsaw 04-730, Poland
e-mail: piotr.czubkowski@mp.pl

E. Albano, M. Parola (eds.), *Studies on Hepatic Disorders*,
Oxidative Stress in Applied Basic Research and Clinical Practice,
DOI 10.1007/978-3-319-15539-5_17

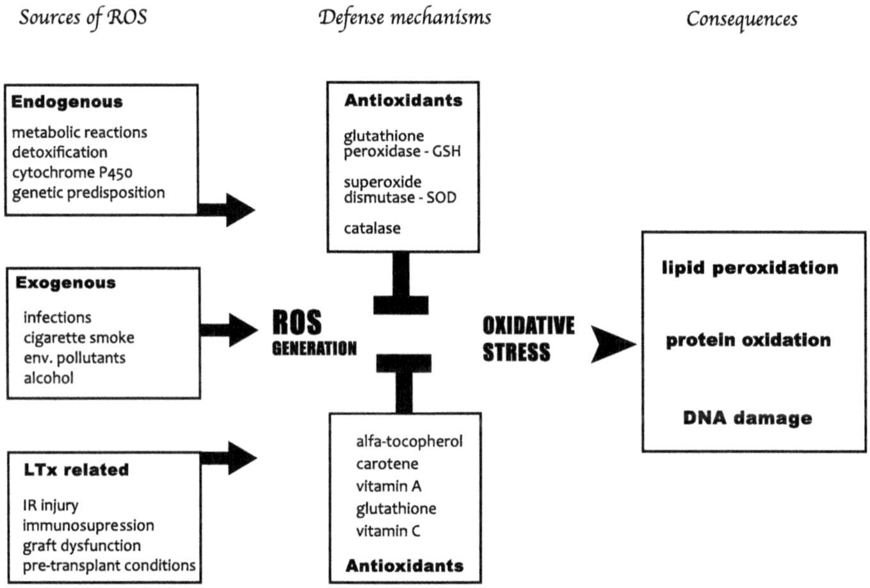

Fig. 17.1 Sources of oxidative stress in liver transplant recipients

potential targets for effective interventions, however still prospective validation of novel approaches is required.

In this chapter we discuss the problem of oxidative stress in liver transplant recipients.

17.2 Pretransplant Conditions

17.2.1 Transplant Candidates

ROS generated before transplantation are not fully eliminated after procedure and may be related to worse graft function. In cirrhosis, increased production of ROS leads to the reduction in the bioavailability of several antioxidants such as carotenoids, α-tocopherol, and glutathione [4]. It was also demonstrated that liver transplantation (LTx) is only partially associated with a long-term improvement of oxidative imbalance observed in patients with end-stage chronic liver disease [5]. Corradini et al. performed multivariate analysis in preoperative recipients which included 7beta-hydroxycholesterol plasma concentration and MELD score. 7beta-hydroxycholesterol, a marker of oxidative stress, was an independent risk factor of poor graft function. Cirrhotic patients presented with upregulated systemic oxidative stress and the markers were still increased 1 month after OLT [6].

Free radical formation is increased in most conditions leading to liver insufficiency, especially in acute liver failure (Chap. 12), alcoholic cirrhosis (Chap. 13), hepatitis C (Chap. 14), and NASH (Chap. 15).

Generally in cholestatic conditions hydrophobic bile acids play a crucial role in cellular pathways involved in toxic injury of hepatocytes. ROS generation subsequently activates MAP kinase pathways, EGFR [epidermal growth factor receptor], and CD95 (Fas receptor). Translocation of Fas to the plasma membrane of hepatocytes results in assembly of the death-inducing signaling complex (DISC), which in turn cleaves and activates caspase 8 and 10 and the Bcl-2 protein [7].

Cholestatic malabsorption of fat-soluble vitamins delivers an extra risk due to lack of basic exogenous scavengers [8]. There are studies reporting decreased concentration of α-tocopherol in patients with primary biliary cirrhosis, primary sclerotic cholangitis, biliary atresia, and alcoholic liver disease [8–10].

The mechanisms leading to increased lipid peroxidation in cirrhotic patients are still not well understood. Probably a decreased production and increased consumption of endogenous antioxidants may play a crucial role. Recently, bacterial translocation was proven to contribute to liver injury. Extrahepatic biliary obstruction promotes bacterial translocation and this process is an important cause of morbidity and mortality in patients with jaundice [11]. In nonalcoholic fatty liver disease (NAFLD) pathogenesis, dietary factors seem to play the role in relation to intestinal translocation of bacterial endotoxin and generation of free radicals [12]. As indicated by the presence of bacterial DNA, a marker of bacterial translocation, it was associated with aggravation of peripheral vasodilation and with worsening of intrahepatic endothelial dysfunction in cirrhotic patients [13]. However, most of the episodes of bacterial infection in cirrhotic patients remain undetected [14].

Early diagnosis of bacterial translocation may be crucial to prevent sepsis [15]. Septicemia results in a rapid free radical release that cannot be managed efficiently by antioxidant mechanisms. The most frequent chronic condition leading to liver transplantation in children is biliary atresia. Recently, it was suggested that early stages of disease are associated with stronger inflammatory reaction, augmented oxidative DNA and mitochondrial damage what was manifested by higher immunoreactive 8-hydroxydeoxyguanosine and decrease in mitochondrial copy number [16]. Interestingly, an activity of protective superoxide dismutase (SOD) was increased significantly in liver tissues of infants with biliary atresia, Alagille syndrome, and portal vein thrombosis but not in alfa1-antytrypsin deficiency or neonatal hepatitis. Simultaneously SOD activity correlated with severity of cholestasis [17]. Interestingly, steatosis itself is well tolerated by hepatocytes [18], however its importance was highlighted with regard to graft state and its function after transplantation. Sokol et al. proposed that hepatocytes sharing steatosis in combination with retention of toxic bile acids may generate the "second hit" through stimulation of ROS generation and signaling cascades that trigger hepatocellular injury [19, 20]. Moreover Children with prolonged cholestasis usually present with malnutrition that results in depletion of antioxidants. The major consideration is lipid-soluble vitamins and vitamin E deficiency seems to be the major problem. The water-soluble vitamin E-TPGS has been used efficiently to normalize vitamin E status in those subjects but it does not improve the increased lipid peroxidation and poor polyunsaturated fatty acid status [10]. As depletion of many antioxidants is expected in chronic cholestatic disease that requires supplementation of many nutrients and thus improvement of one nutrient deficit may not ameliorate free radical injury. Recently, beneficial effect of α-tocopherol and β-carotene supplementation before liver transplantation was showed in the rat model of ischemia–reperfusion. Both supplements were given 21 days before surgery improving oxidative balance and energetic state as measured by ATP [21]. These data suggest that antioxidant therapy before transplantation may ameliorate the effects of reperfusion.

17.2.2 Marginal Donors

The state of the donor organ is regarded the most important indicator of prognosis for graft function and survival. On the other hand the lack of available organs and mortality on waiting lists stimulated introduction of the so-called extended criteria donors. That brought compromising between graft availability and its quality. Although overall posttransplant outcomes are inferior to results with optimal, whole-liver grafts, aggressive utilization of extended criteria donors (ECD) and organs procured via donation after cardiac death (DCD) significantly lowers median wait-times for OLT, MELD scores at OLT and death while awaiting transplantation [22]. Non-heart beating, elderly donors, as well as steatotic and small-for-size grafts are especially vulnerable to IR oxidative injury, early allograft dysfunction (EAD), primary nonfunction (PNF), and rejection [23–25] The incidence of complications is higher if there is more than one risk factor; thus overlapping should be minimized [26]. Liver steatosis, the most frequently encountered issue due to increasing rates of NAFLD, predisposes to increased IR oxidative injury [27]. The factors causing progression of NAFLD to fibrosis and cirrhosis have not been defined in humans as yet. Nevertheless, there is growing evidence supporting the role of proinflammatory cytokines, particularly TNFα and Il-6 [28]. Recent data from USA registry showed that liver steatosis over 30 % were independent risk factors of worse 1-year graft survival. Nevertheless, if other donor risk factors are eliminated (>age 40 years, cold ischemia >5 h, no donation after cardiac death) grafts may be used safely. If CIT is above 11 h mild steatosis worsens survival as well [29]. Recently, significance of hepatic lipid composition rather than size and morphology of lipid droplets was highlighted in IR injury.

Elevated thromboxane A2 (TXA2), vasoactive proinflammatory lipid mediator derived from arachidonic acid (AA), an omega-6 fatty acid (Ω-6FA), impairs sinusoidal perfusion and augments inflammation so as hepatocellular injury after IR. These events may be prevented by preconditioning with dietary Ω-3FA or selective blockage of TXA2 receptors [30]. Thus, restoration of normal Ω-6:Ω-3 FA proportion seems to be a relevant protective condition in steatotic grafts so as donor intervention with pharmacological preconditioning may protect marginal organ [31, 32]. There is an obvious necessity to carry out additional research so as to improve the outcomes in marginal graft usage. Still at the beginning stage, hypothermic machine perfusion (HMP) seems to be a promising preservation option.

Henry et al. demonstrated HMP significantly reduced proinflammatory cytokine expression, relieving the downstream activation of adhesion molecules and migration of leukocytes, including neutrophils and macrophages [33].

17.3 Ischemia-Reperfusion Injury

17.3.1 Pathogenesis

Liver ischemia/reperfusion injury can be divided into two distinct phases. Ischemia as the initial condition, results in the shortage of oxygen supply, alteration in mitochondrial enzymes, and depletion of ATP formation, by failure of aerobic oxidative phosphorylation [34, 35]. State of hypoxia makes the graft vulnerable during reoxygenation. Reduction of ATP causes Na/K/ATPase disturbances, sodium accumulation, and cell death [36, 37].

Generation of ROS during reperfusion is not fully understood. Primarily, xanthine oxidase (XOD)/xanthine dehydrogenase (XDH) was regarded the main ROS generator. This theory was mainly based on antioxidant properties of allopurinol which is XOD inhibitor [38, 39] and currently mitochondria seem to be a key source of ROS [40]. Superoxide massively produced in mitochondria under oxidative stress leads to breakdown of its membrane and cellular death [41].

The main event in the beginning of IR is Kupffer cells activation. It is promoted by ROS, complement and CD4+ lymphocytes [42–44]. Activated Kupffer cells generate ROS and cytokines, mainly TNFα and IL-1, in the initial phase of reperfusion [45, 46].

TNFα is one of the core players in the IR promoting expression of adhesion molecules on vascular endothelial cells and recruitment of neutrophils [47]. Activated neutrophils release ROS and proteases that are responsible for oxidative stress during the late phase of reperfusion injury [48]. The role of oxidative stress in IR injury is multifactorial. ROS may directly regulate apoptotic and necrotic pathways of cell injury. Other effects comprise upregulation of nuclear transcription factors and proinflammatory genes, [49] induction of antioxidant enzymes, [50] and regulation of sinusoidal blood flow [51].

The key events during IR and effects of ROS are summarized in Table 17.1.

17.3.2 Antioxidant Interventions in IR Injury

17.3.2.1 Antioxidant Treatment

Several antioxidant interventions were tested for its potency to reduce IR injury, but very few are candidates for implementation to clinical practice. Most of the current evidence is based on experimental animal models. The main strategies are based on

Table 17.1 Key events in ischemia-reperfusion

• Decrease in oxidative phosphorylation results in ATP depletion
• Intracellular accumulation of sodium and calcium
• ROS formation
• Upregulation of NFkB factor
• TNFα and Il-1 release
• Direct cellular damage through protein, lipid, and DNA oxidation
• Promotion of apoptotic and necrotic death
• Induction of antioxidative mechanisms
• Release of mediators involved in regulation of sinusoidal blood flow and liver regeneration

pharmacologic substances, ischemic preconditioning (IPC) of the graft, gene therapy, and machine perfusion techniques. Antioxidants may be given orally, intravenously, or as additives to preservation solutions.

N-Acetylocysteine (NAC) is a precursor of reduced glutathione and scavenger of hydroxyl radicals which proved to be effective in paracetamol overdose and possibly in other forms of acute liver failure [52, 53]. In pediatric population (12 children in the treatment group and 13 children as controls), NAC was given intravenously in combination with prostaglandin E1 for 6 days after reperfusion. No complications or adverse events of treatment occurred. Treated group presented with improved liver function tests, shortened median postoperative in-hospital stay, and alleviated severity of rejection though the proportion of patients with allograft rejection was not significantly different between the two groups [54]. NAC administered during donor operation did not show effect on IR injury or acute cellular rejection [55]. Trials with variable antioxidants are summarized in Table 17.2 [31, 32, 36, 56–108].

17.3.2.2 Ischemic Preconditioning

IPC is defined as a brief period of liver ischemia followed by reperfusion, and has demonstrated protections against a prolonged IR injury and improved the capacity of regeneration [109]. The concept was originally introduced by Murry et al. in 1986 [110] but exact pathways still remain unclear.

The protective mechanisms of IPC against hepatic IR injury include alterations in energy metabolism, neutrophil accumulation, microcirculatory disturbances, liver regeneration, and formation of proinflammatory mediators [109, 111]. The generation of small amounts of ROS stimulates cell response and promotes anti-apoptotic mechanisms [112, 113]. The clinical evidence is contradictory. There was significant improvement in biochemical markers of liver cell function, reduced need for reoperation in the postoperative period, and lower incidence of primary nonfunction after OLT if IPC was performed [114]. In another study, Koneru et al. [115] randomized 101 deceased donors to 10 min IPC ($n=50$) or no-IPC ($n=51$) groups, and showed increased liver injury after IPC. One-year patient and graft survival in IPC versus no-IPC were 88 % versus 78 % ($p=0.1$)

Table 17.2 Antioxidant agents with protective capacity against liver IR injury

Antioxidant	Category	Species	Injury type	Admin.	Ref.
Albumin	Protein	Rat	CI	*perf.*	[56]
Alfa-tocopherol	Diet	Rat	CI/WI	*im*	[57]
Alfa-tocopherol	Diet	Rat	CI/WI	*iv*	[58]
Alfa-tocopherol/ascorbate	Diet	Human	WI	*Iv*	[59]
Alfa-tocopherol/ascorbate	Diet	Rat	WI	*ip/po*	[60]
Alfa-tocopherol/Beta-carotene	Diet	Rat	CI/WI	*po*	[61]
Allopurinol	XO inhibitor	Rat	WI	*ip*	[62, 63]
Amiodaron	Potassium-CB	Rat	CI	*iv*	[64]
Ascorbate	Diet	Rat	WI	*iv*	[65]
Atrial natriuretic peptide	Hormone	Rat	CI/WI	*iv*	[66, 67]
Bucillamine	Thiol	Rat	CI/WI	*iv*	[68, 69]
Cardiotrophin-1	Cytokine	Pig	CI/WI	*iv*	[70]
Carnitine	QUAT	Rat	WI	*ip*	[71]
Carvedilol	βα-Blocker	Rat	CI	*sol*	[72]
CGRP	Peptide	Rat	CI	*sol*	[73]
Cyclosporin/ibuprofen	CNI/NSAD	Rat	WI	*po*	[74]
CAT derivatives	Int. enzyme	Rat	WI	*iv*	[75]
Coenzyme Q/Penoxyfylline	LMWA	Rat	WI	*iv/ig*	[76]
Desferrioxamine	Iron chelator	Dog	CI/WI	*iv*	[77]
Cyanidin	Plant phenol	Rat	WI	*po*	[78]
Ebselen	Enzyme	Rat	WI	*po*	[79]
GABAR agonist	Receptor	Rat	CI/WI	*iv*	[80]
Green tea extract	Plant extract	Rat	WI	*po*	[81]
GSH	LMWA	Rat	WI	*iv*	[82]
GSH	LMWA	Rat	CI/WI	*iv*	[83]
Idebenone	Coenzyme Q derivative	Rat	CI	*sol*	[84]
Leflunomide	Isoxazole derivative	Rat	CI	*ig*	[85]
Levosimendan	Calcium sensitizer	Rat	WI	*iv*	[86]
Lipoic acid	LMWA	Rat	CI	*iv*	[87]
Lisinopril	ACEI	Rat	CI	*ip*	[88]
Lithium	GSK-3β/NF-κB inh.	Mice	WI	*iv*	[89]
Mangifera indica	Plant extract	Rat	WI	*po*	[90]
MMDP*	Multidrug	Rat	CI	*po/ip/ia*	[32]
Melatonin	Hormone	Rat	WI	*po*	[91, 92]
Methylprednisolone	Steroid	Human	CI/WI	*iv*	[93]
Methylprednisolone	Steroid	Rat	WI	*iv*	[94]
Meloxicam	COX-2 inhibitor	Rat	CI	*sol*	[95]
MK-886	5-Lipoxygenase inh.	Rat	WI	*ip*	[96]
Montelukast	LAR	Rat	WI	*ip*	[96]
N-Acetylocysteine	Thiol	Rabbit	WI	*iv*	[97]

(continued)

Table 17.2 (continued)

Antioxidant	Category	Species	Injury type	Admin.	Ref.
N-Acetylocysteine	Thiol	Human	CI/WI	*iv*	[98]
Nicaraven	CCB	Dog	WI	*iv*	[99]
Picroliv	Plant extract	Rat	WI	*po*	[100]
Propyl gallate	Food preservative	Rat	WI	*iv*	[101]
Relaxin	Peptide hormone	Rat	CI	*sol*	[102]
Quercetin	Plant phenol	Rat	WI	*po*	[103]
Simvastatin	Statin	Rat	WI/CI	*sol*	[31]
Silibinin	Plant extract	Rat	CI	*sol*	[104]
SOD derivatives	Int. enzyme	Rat	WI	*iv*	[105]
Tacrolimus	CNI	Rat	WI	*po*	[36]
Tetrandrine	Plant extract	Mice	WI	*ip*	[106]
Theaflavin	Plant extract	Mice	WI	*ip*	[107]
Triplotide	NF-κB inh.	Mice	WI	*ip*	[108]

WI warm ischemia, *CI* cold ischemia, *ig* intragastrical, *ip* intraperitoneal, *po* oral, *iv* intravenous, *ia* intraarterial, *sol* preservation solution, *CCB* calcium channel blocker, *LAR* leukotriene receptor antagonist, *LMWA* low molecular weight antioxidants, *XO* xanthine oxidase, *NSAD* non-steroidal anti-inflammatory drug, *CGRP* calcitonin gene-related peptide, *GABAR* γ-aminobutyric acid receptor, **MDDP* multidrug donor preconditioning including pentoxyphyline, glycine, deferoxamine, *N*-acetylocysteine, erythropoetine, melatonin, simvastatin, *Potassium-CB* Potassium-channel blockers, *QUAT* quaternary ammonium cation, *ACEI* angiotensin-converting-enzyme inhibitor, *NF-κB* nuclear factor kappa-light-chain-enhancer of activated B cells

and 86 versus 76 % ($p=0.25$), respectively. IPC recipients had fewer moderate and severe rejections within 30 days ($p=0.09$).

IPC may support the regeneration of small-for-size liver grafts by promotion of mitochondrial SOD, heat shock protein HSP70 and heme-oxygenase-1 (HSP32) with concurrent inhibition of IL-1 [116, 117]. Moreover, induction of HSP70 and HO-1 stimulate buffering system which helps the liver to resist ischemic injury [118].

17.3.2.3 Gene Therapy

The experimental use of plasmid DNA shed the new light on possible interventions against IR. In animal models the adenoviral copper/zinc SOD gene transfer decreased hepatic injury-prevented primary nonfunction and promoted liver regeneration after OLT [119–121]. He et al. [122], combined genes of extracellular superoxide dismutase (EC-SOD) and catalase, with polylipid nanoparticles and injected via portal vein into mice. The overexpression of both genes significantly alleviated oxidative IR injury, reduced oxidative stress markers, and improved liver histology. In spite of encouraging animal models, we are far from clinical usage of vectors in humans, mostly due to ethical and safety issues. Moreover, in deceased donors, time limits make the use of vectors unlikely to be successful, and even nonreplicative adenoviral vectors can have side effects and might cause liver injury [123].

17.4 Posttransplant Conditions

17.4.1 Acute Rejection

Endothelialitis, bile duct damage, and portal inflammation with lymphocytes, neutrophils, and eosinophils are classic components of acute rejection. Eosinophil peroxidase (EPO) produced exclusively by eosinophils [124, 125] catalyzes the reaction of H_2O_2 with chloride to generate HOCl. Under oxidative stress HOCl induces apoptosis in the biliary epithelial cells via p53 and p-JNK-related mitochondria-caspase-dependent pathway [126]. Moreover, TNFα mRNA increased simultaneously with 8-hydroxy-2′ deoxyguanosine (8-OHdG), and correlated with the severity of acute graft rejection and oxidative DNA damage [127]. This may indirectly account for increased risk of early graft failure, after severe reperfusion injury during transplant.

17.4.2 Immunosuppression and ROS

Immunosuppressive therapy is regarded as an additional risk factor for increased free radical formation however the evidence is conflicting. The drugs by chemical actions can produce free radicals and recurrent infections, and other complications can also contribute to free radical formation and antioxidant deficiencies. Different drugs were described in respect to oxidative injury. The oxidative status [oxidative stress markers and antioxidants] of 23 children, 1.5–12 years after liver transplantation, receiving cyclosporine CsA ($n = 14$) or tacrolimus ($n = 9$) did not differ significantly from age-matched healthy controls [128]. Christians et al. investigated the effect of combination of cyclosporine or tacrolimus with mTORI on energy production, glucose metabolism, and ROS generation in perfused rat brain slices. Everolimus and sirolimus showed different effects—everolimus ameliorated cyclosporine—induced mitochondrial dysfunction, whereas sirolimus increased CsA's toxicity by inhibition of compensatory anaerobic glycolysis. Authors proposed that everolimus can cross the mitochondrial membrane and may reduce the distribution of CsA into mitochondria [129]. Tacrolimus appears to have antioxidative properties and downregulates free radical tissue levels after severe liver ischemia [130].

The major clinical consideration related to free radical formation is atherosclerotic risk and endothelial dysfunction. That is why immunosuppressive drugs are investigated with regard to cardiovascular risk. In the in vitro study on cultured human endothelial cells it was demonstrated that mycophenolate acid inhibits endothelial O_2 formation by inhibition of NADPH oxidase, whereas CNI increased its activity [131]. Other study showed that methylprednisolone and MMF induced minor changes in endothelial function in comparison with CsA, sirolimus and tacrolimus [132]. As lipid disturbances could significantly contribute to arteriosclerotic risks we investigated lipid metabolism in liver transplant recipients receiving

CNI [133] and sirolimus [134] but neither of immunosuppressive regimens contributed to impaired lipid metabolism nor oxidative stress.

Liver transplantation may increase free radical formation directly by immunosuppressive regimen but indirectly infectious complications and renal problems can be an additional risk factor for free radical injury. There is an interesting case report by Samonte et al. which describes a story of a young girl born with severe liver disease who suffered from liver failure since the first month of her life [135].

Although she underwent successful liver transplantation, she struggled with complications of disease and immunosuppressive regiment, i.e., rejections, infections, and adverse drug effects leading to chronic renal failure and other complications. When awaiting combined liver and kidney transplant she developed scurvy due to vitamin C deficiency. We published an editorial article to comment on this report [136] and discussed the role of antioxidant deficits that may occur after liver transplantation in patients presenting with infectious and/or organ complications. Ascorbic acid spares vitamin E and its deficiency can lead to increased free radical injury. The potential cooperation between vitamin C and E can be of major importance in respect of chronic liver and kidney disease. The sparing effect of glutathione in scurvy was also documented. Thus, single nutrient deficiency can be even more profound if combined with low levels of other antioxidants. Liver transplantation seems to correct most of the deficiencies and antioxidant status after liver transplantation restores its capacities except for glutathione [137]. Nevertheless some patients present with complications which can solely lead to oxidative stress.

Oxidative stress is regarded as important and in some cases, the initiating factor of tissue injury. It plays a role both in graft damage due to inflammatory reaction induced by rejection and in damage of host tissues burned in the fire of oxidative injury. In case of chronically ill patients, who suffer from severe complications of basic disease i.e., infections, major surgery, one may suppose that endogenous system of antioxidants may exhaust. In such situations the real needs for vitamin C, E, and other antioxidants may be much higher as they exceed the normal intake of these vitamins [136].

Nutritional deficiencies can also be related to the poor diet or feeding problems that may occur after liver transplantation. Patients exposed to any chronic or recurrent condition may suffer from multiple nutrition deficits. The case presentation of scurvy brings an important message—some deficiencies may not be suspected based on clinical symptoms and patients at risk should be monitored by measurement of biochemical indicators of nutrient deficiencies.

17.5 Summary

Liver transplant recipients are at higher risk of oxidative stress and related consequences. How far clinical outcomes are influenced remains yet to be established. Cardiovascular complications are one of the most important causes of morbidity and mortality in adult population. Atherosclerosis is the critical condition with strong ROS involvement. These risks should be recognized early, preferably in adolescence and young adults where chances for effective prophylaxis are obviously higher. Continuous progress makes us determined to perform more challenging liver transplantations with higher risk of graft injury.

The state of donor organ is regarded as the most important indicator of prognosis for function and survival. Due to necessity of increase in the donor pool, the future of liver transplantation is inevitably related with increasing number of marginal grafts. Dealing with non-heart beating, elderly donors, and steatotic liver grafts is apparently more challenging because of higher risk of oxidative injury, primary graft dysfunction, and rejection.

In spite of recognizing well, currently we lack effective methods to deal with oxidative stress. Up-to-date research is of limited value in clinical practice and data from animal models can't be extrapolated on human organs. So as to clarify pathways of injury, molecular research should be prefered instead of measurement of injury markers [138].

To know, the biggest scientific and clinical challenges are:

- Better understanding of basic mechanisms of IR injury enabling targeted antioxidative treatment
- Further development of preservation techniques
- The evaluation of antioxidant treatment in prospective long-term clinical trials in which effects and safety can be closely monitored

References

1. Czubkowski P, Socha P, Pawlowska J (2010) Current status of oxidative stress in pediatric liver transplantation. Pediatr Transplant 14:169–177
2. Jaeschke H, Woolbright BL (2012) Current strategies to minimize hepatic ischemia-reperfusion injury by targeting reactive oxygen species. Transplant Rev 26:103–114
3. Nastos C, Kalimeris K, Papoutsidakis N, Tasoulis MK, Lykoudis PM, Theodoraki K, Nastou D, Smyrniotis V, Arkadopoulos N (2014) Global consequences of liver ischemia/reperfusion injury. Oxid Med Cell Longev 2014:1–13
4. Augusto VS, Rodrigues AJ, Reis GS, Silveira APC, de Castro E, Silva O Jr, Mente ED, Jordão AA Jr, Evora PR (2014) Evaluation of oxidative stress in the late postoperative stage of liver transplantation. Transplant Proc 46:1453–1457
5. Trevisani F, Caraceni P, Simoncini M, Micati M, Domenicali M, Dazzani F, Zambruni A, Stefanelli C, Grazi G, Nardo B, Guarnieri C, Bernardi M (2002) Evidence of oxidative imbalance in long-term liver transplant patients. Dig Liver Dis 34:279–284
6. Corradini SG, Micheletta F, Natoli S, Iappelli M, Di Angelantonio E, De Marco R, Elisei W, Siciliano M, Rossi M, Berloco P, Attili AF, Diczfalusy U, Iuliano L (2005) High preoperative recipient plasma 7beta-hydroxycholesterol is associated with initial poor graft function after liver transplantation. Liver Transpl 11:1494–1504
7. Sokol RJ, Devereaux M, Dahl R, Gumpricht E (2006) "Let there be bile" understanding hepatic injury in cholestasis. J Pediatr Gastroenterol Nutr 43:S4–S9
8. Floreani A, Baragiotta A, Martines R, Naccarato R, D'odorico A (2000) Plasma antioxidant levels in chronic cholestatic liver diseases. Aliment Pharmacol Ther 14:353–358

9. Nagita A, Ando M (1997) Assessment of hepatic vitamin E status in adult patients with liver disease. Hepatology 16:392–397
10. Socha P, Koletzko B, Pawlowska J, Proszynska K, Socha J (1997) Treatment of cholestatic children with water-soluble vitamin E [alpha-tocopheryl polyethylene glycol succinate]: effects on serum vitamin E, lipid peroxides, and polyunsaturated fatty acids. J Pediatr Gastroenterol Nutr 24:189–193
11. Sakrak O, Akpinar M, Bedirli A, Akyurek N, Aritas Y (2003) Short and long-term effects of bacterial translocation due to obstructive jaundice on liver damage. Hepatogastroenterology 50:1542–1546
12. Thuy S, Ladurner R, Volynets V, Wagner S, Strahl S, Königsrainer A, Maier KP, Bischoff SC, Bergheim I (2008) Nonalcoholic fatty liver disease in humans is associated with increased plasma endotoxin and plasminogen activator inhibitor 1 concentrations and with fructose intake. J Nutr 138:1452–1455
13. Bellot P, García-Pagán JC, Francés R, Abraldes JG, Navasa M, Pérez-Mateo M, Such J, Bosch J (2010) Bacterial DNA translocation is associated with systemic circulatory abnormalities and intrahepatic endothelial dysfunction in patients with cirrhosis. J Hepatol 52:2044–2052
14. Such J, Francés R, Muñoz C, Zapater P, Casellas JA, Cifuentes A, Rodríguez-Valera F, Pascual S, Sola-Vera J, Carnicer F, Uceda F, Palazón JM, Pérez-Mateo M (2002) Detection and identification of bacterial DNA in patients with cirrhosis and culture-negative, nonneutrocytic ascites. Hepatology 36:135–141
15. Wong F, Bernardi M, Balk R, Christman B, Moreau R, Garcia-Tsao G, Patch D, Soriano G, Hoefs J, Navasa M (2005) Sepsis in cirrhosis: report on the 7th meeting of the International Ascites Club. Gut 54:718–725
16. Tiao MM, Lin TK, Kuo FY, Huang CC, Du YY, Chen CL, Chuang JH (2007) Early stage of biliary atresia is associated with significant changes in 8-hydroxydeoxyguanosine and mitochondrial copy number. J Pediatr Gastroenterol Nutr 45:329–334
17. Broide E, Klinowski E, Koukoulis G, Hadzic N, Portmann B, Baker A, Scapa E, Mieli-Vergani G (2000) Superoxide dismutase activity in children with chronic liver disease. J Hepatol 32:188–192
18. Pessayre D, Mansouri A, Fromenty B (2002) Nonalcoholic steatosis and steatohepatitis. V. Mitochondrial dysfunction in steatohepatitis. Am J Physiol Gastrointest Liver Physiol 282:G193–G199
19. Sokol RJ, Winklhofer-Roob BM, Devereaux MW, McKim JM Jr (1995) Generation of hydroperoxides in isolated rat hepatocytes and hepatic mitochondria exposed to hydrophobic bile acids. Gastroenterology 109:1249–1256
20. Rodrigues CM, Steer CJ (2000) Mitochondrial membrane perturbations in cholestasis. J Hepatol 32(135):141
21. Codoner-Franch P, Muniz P, Gasco E, Domingo JV, Valls-Belles V (2008) Effect of diet supplemented with alpha-tocopherol and β-carotene on ATP and antioxidant levels after hepatic ischemia–reperfusion. J Clin Biochem Nutr 43:13–18
22. Sass DA, Reich DJ (2011) Liver transplantation in 21st century: expanding the donor options. Gastroenterol Clin North Am 40:641–658
23. Chen XB, Xu MQ (2014) Primary graft dysfunction after liver transplantation. Hepatobiliary Pancreat Dis Int 13:125–137
24. Ploeg RJ, D'Alessandro AM, Knechtle SJ, Stegall MD, Pirsch JD, Hoffmann RM, Sasaki T, Sollinger HW, Belzer FO, Kalayoglu M (1993) Risk factors for primary dysfunction after liver transplantation—a multivariate analysis. Transplantation 56:807–813
25. Clavien PA, Harvey PR, Strasberg SM (1992) Preservation and reperfusion injuries in liver allografts. An overview and synthesis of current studies. Transplantation 53:957–978
26. Detry O, Deroover A, Meurisse N, Hans MF, Delwaide J, Lauwick S, Kaba A, Joris J, Meurisse M, Honoré P (2014) Donor age as a risk factor in donation after circulatory death

liver transplantation in a controlled withdrawal protocol programme. Br J Surg 101: 784–792

27. Reiniers MJ, van Golen RF, van Gulik TM, Heger M (2014) Reactive oxygen and nitrogen species in steatotic hepatocytes: a molecular perspective on the pathophysiology of ischemia-reperfusion injury in the fatty liver. Antioxid Redox Signal 21(7):1119–1142
28. Cohen JC, Horton JD, Hobbs HH (2011) Human fatty liver disease: old questions and new insights. Science 332:15190–15123
29. Spitzer AL, Lao OB, Dick AA, Bakthavatsalam R, Halldorson JB, Yeh MM, Upton MP, Reyes JD, Perkins JD (2010) The biopsied donor liver: incorporating macrosteatosis into high-risk donor assessment. Liver Transpl 16:874–884
30. El-Badry AM, Jang JH, Elsherbiny A, Contaldo C, Tian Y, Raptis DA, Laczko E, Moritz W, Graf R, Clavien PA (2011) Chemical composition of hepatic lipids mediates reperfusion injury of the macrosteatotic mouse liver through thromboxane A2. J Hepatol 55:1291–1299
31. Gracia-Sancho J, García-Calderó H, Hide D, Marrone G, Guixé-Muntet S, Peralta C, García-Pagán JC, Abraldes JG, Bosch J (2013) Simvastatin maintains function and viability of steatotic rat livers procured for transplantation. J Hepatol 58:1140–1146
32. Von Heesen M, Seibert K, Husler M, Scheuer C, Wagner M, Menger MD, Schilling MK, Moussavian MR (2012) Multidrug donor preconditioning protects steatotic liver grafts against ischemia-reperfusion injury. Am J Surg 203:168–76
33. Henry SD, Nachber E, Tulipan J, Stone J, Bae C, Reznik L, Kato T, Samstein B, Emond JC, Guarrera JV (2012) Hypothermic machine preservation reduces molecular markers of ischemia/reperfusion injury in human liver transplantation. Am J Transplant 12:2477–86
34. Gonzalez-Flecha B, Cutrin JC, Boveris A (1993) Time course and mechanism of oxidative stress and tissue damage in rat liver subjected to in vivo ischemia–reperfusion. J Clin Invest 91:456–464
35. Nishimura Y, Romer LH, Lemasters JJ (1998) Mitochondrial dysfunction and cytoskeletal disruption during chemical hypoxia to cultured rat hepatic sinusoidal endothelial cells: the pH paradox and cytoprotection by glucose, acidotic pH, and glycine. Hepatology 27: 1039–1049
36. Blum H, Osbakken MD, Johnson RG Jr (1991) Sodium flux and bioenergics in the ischemic rat liver. Magn Reson Med 18:348–357
37. Vollmar B, Glasz J, Leiderer R, Post S, Menger MD (1994) Hepatic microcirculatory perfusion failure is a determinant of liver dysfunction in warm ischemia–reperfusion. Am J Pathol 145:1421–1431
38. Atalla SL, Toledo-Pereyra LH, MacKenzie GH, Cederna JP (1985) Influence of oxygen-derived free radical scavengers on ischemic livers. Transplantation 40:584–590
39. Ishii K, Suita S, Sumimoto H (1990) Effect of verapamil on conversion of xanthine dehydrogenase to oxidase in ischemic rat liver. Res Exp Med 190:389–399
40. Jaeschke H (2002) Xanthine oxidase-induced oxidant stress during hepatic ischemia–reperfusion: are we coming full circle after 20 years? Hepatology 36:761–763
41. Nieminen AL, Byrne AM, Herman B, Lemasters JJ (1997) Mitochondrial permeability transition in hepatocytes induced by t-BuOOH: NAD[P]H and reactive oxygen species. Am J Physiol 272:C1286–C1294
42. Arumugam TV, Magnus T, Woodruff TM, Proctor LM, Shiels IA, Taylor SM (2006) Complement mediators in ischemia-reperfusion injury. Clin Chim Acta 374:33–45
43. Fondevila C, Busuttil RW, Kupiec-Weglinski JW (2003) Hepatic ischemia/reperfusion injury-a fresh look. Exp Mol Pathol 74:86–93
44. Jaeschke H, Farhood A, Bautista AP, Spolarics Z, Spitzer JJ (1993) Complement activates Kupffer cells and neutrophils during reperfusion after hepatic ischemia. Am J Physiol 264:G801–G809
45. Wanner GA, Ertel W, Müller P, Höfer Y, Leiderer R, Menger MD, Messmer K (1996) Liver ischemia and reperfusion induces a systemic inflammatory response through Kupffer cell activation. Shock 5:34–40

46. Lentsch AB, Kato A, Yoshidome H, McMasters KM, Edwards MJ (2000) Inflammatory mechanisms and therapeutic strategies for warm hepatic ischemia/reperfusion injury. Hepatology 32:169–173
47. Rudiger HA, Clavien PA (2002) Tumor necrosis factor-alpha, but not Fas, mediates hepatocellular apoptosis in the murine ischemic liver. Gastroenterology 122:202–210
48. Weiss SJ (1989) Tissue destruction by neutrophils. N Engl J Med 320:365–376
49. Liu TZ, Lee KT, Chern CL, Cheng JT, Stern A, Tsai LY (2001) Free radical-triggered hepatic injury of experimental obstructive jaundice of rats involves overproduction of proinflammatory cytokines and enhanced activation of nuclear factor kappaB. Ann Clin Lab Sci 31:383–390
50. Bauer M, Bauer I (2002) Heme oxygenase-1: Redox regulation and role in the hepatic response to oxidative stress. Antioxid Redox Signal 4:749–758
51. Pannen BH (2002) New insights into the regulation of hepatic blood flow after ischemia and reperfusion. Anesth Analg 94:1448–1457
52. Harrison PM, Keays R, Bray GP, Alexander GJ, Williams R (1990) Improved outcome of paracetamol-induced fulminant hepatic failure by late administration of acetylcysteine. Lancet 335:1572–1573
53. San-Miguel B, Alvarez M, Culebras JM, González-Gallego J, Tuñón MJ (2006) N-acetyl-cysteine protects liver from apoptotic death in an animal model of fulminant hepatic failure. Apoptosis 11:1945–1957
54. Bucuvalas JC, Ryckman FC, Krug S et al (2001) Effect of treatment with prostaglandin E1 and N-acetylcysteine on pediatric liver transplant recipients: a singlecenter study. Pediatr Transplant 5:274–278
55. Khan AW, Fuller BJ, Shah SR, Davidson BR, Rolles K (2005) A prospective randomized trial of N-acetyl cysteine administration during cold preservation of the donor liver for transplantation. Ann Hepatol 4(2):121–126
56. Strubelt O, Younes M, Li Y (1994) Protection by albumin against ischaemia and hypoxia-induced hepatic injury. Pharmacol Toxicol 75:280–284
57. Giakoustidis D, Papageorgiou G, Iliadis S, Kontos N, Kostopoulou E, Papachrestou A, Tsantilas D, Spyridis C, Takoudas D, Botsoglou N, Dimitriadou A, Giakoustidis E (2002) Intramuscular administration, of very high dose of alpha-tocopherol protects liver from severe ischemia/reperfusion injury. World J Surg 26:872–877
58. Gondolesi GE, Lausada N, Schinella G, Semplici AM, Vidal MS, Luna GC, Toledo J, de Buschiazzo PM, Raimondi JC (2002) Reduction of ischemia-reperfusion injury in parenchymal and nonparenchymal liver cells by donor treatment with DL-alpha-tocopherol prior to organ harvest. Transplant Proc 34:1086–1091
59. Cerwenka H, Khoschsorur G, Bacher H, Werkgartner G, El-Shabrawi A, Quehenberger F, Rabl H, Mischinger HJ (1999) Normothermic liver ischemia and antioxidant treatment during hepatic resections. Free Radic Res 30:463–469
60. Soltys K, Dikdan G, Koneru B (2001) Oxidative stress in fatty livers of obese Zucker rats: rapid amelioration and improved tolerance to warm ischemia with tocopherol. Hepatology 34:13–18
61. Codoñer-Franch P, Muñiz P, Gasco E, Domingo JV, Valls-Belles V (2008) Effect of a diet supplemented with alfa-tocopherol and β-carotene on ATP and antioxidant levels after hepatic ischemia-reperfusion injury. J Clin Biochem Nutr 43:13–18
62. Br J, Yeom D, Sm L (2001) Protective effect of allopurinol on hepatic energy metabolism in ischemic and reperfused rat liver. Shock 15:112–117
63. Karwinski W, Soreide O (1997) Allopurinol improves scavenging ability of the liver after ischemia/reperfusion injury. Liver 17:139–143
64. Moussavian MR, Kollmar O, Schmidt M, Scheuer C, Wagner M, Slotta JE, Gronow G, Justinger C, Menger MD, Schilling MK (2009) Amiodarone pretreatment of organ donors exerts anti-oxidative protection but induces excretory dysfunction in liver preservation and reperfusion. Liver Transpl 15:763–75
65. My S, Sm L (2002) Protective effect of low dose of ascorbic acid on hepatobiliary function in hepatic ischemia/reperfusion in rats. J Hepatol 36:72–77

66. Gerbes AL, Vollmar AM, Kiemer AK, Bilzer M (1998) The guanylate cyclase-coupled natriuretic peptide receptor: a new target for prevention of cold ischemia–reperfusion damage of the rat liver. Hepatology 28:1309–1317
67. Bilzer M, Jaeschke H, Vollmar AM, Paumgartner G, Gerbes AL (1999) Prevention of Kupffer cell-induced oxidant injury in rat liver by atrial natriuretic peptide. Am J Physiol 276: G1137–G1144
68. Amersi F, Nelson SK, Shen XD, Kato H, Melinek J, Kupiec-Weglinski JW, Horwitz LD, Busuttil RW, Horwitz MA (2002) Bucillamine, a thiol antioxidant, prevents transplantation-associated reperfusion injury. Proc Natl Acad Sci U S A 99:8915–8920
69. Junnarkar SP, Tapuria N, Mani A, Dijk S, Fuller B, Seifalian AM, Davidson BR (2010) Attenuation of warm ischemia-reperfusion injury in the liver by bucillamine through decreased neutrophil activation and Bax/Bcl-2 modulation. J Gastroenterol Hepatol 25: 1891–1899
70. Aguilar-Melero P, Luque A, Machuca MM, Pérez de Obanos MP, Navarrete R, Rodríguez-García IC, Briceño J, Iñiguez M, Ruiz J, Prieto J, de la Mata M, Gomez-Villamandos RJ, Muntane J, López-Cillero P (2010) Cardiotrophin-1 reduces ischemia/reperfusion injury during liver transplant. J Surg Res 181:e83–91
71. Çekın AH, Gür G, Türkoğlu S, Aldemır D, Yilmaz U, Gürsoy M, Taşkoparan M, Boyacioğlu S (2013) The protective effect of L-carnitine on hepatic ischemia-reperfusion injury in rats. Turk J Gastroenterol 24:51–56
72. Ben Mosbah I, Roselló-Catafau J, Alfany-Fernandez I, Rimola A, Parellada PP, Mitjavila MT, Lojek A, Ben Abdennebi H, Boillot O, Rodés J, Peralta C (2010) Addition of carvedilol to University Wisconsin solution improves rat steatotic and nonsteatotic liver preservation. Liver Transpl 16:163–171
73. Song SW, Guo KI, Shi Y, Chang Y, Liu YF (2009) Pretreatment with calcitonin gene-related peptide attenuates hepatic ischemia/reperfusion injury in rats. Transplant Proc 41: 1493–1498
74. Konukoglu D, Tasci I, Cetinkale O (1998) Effects of cyclosporin A and ibuprofen on liver ischemia-reperfusion injury in the rat. Clin Chim Acta 275:1–8
75. Yabe Y, Koyama Y, Nishikawa M, Takakura Y, Hashida M (1999) Hepatocyte-specific distribution of catalase and its inhibitory effect on hepatic ischemia/reperfusion injury in mice. Free Radic Res 30:265–274
76. Portakal O, Inal-Erden M (1999) Effects of pentoxifylline and coenzyme Q10 in hepatic ischemia/reperfusion injury. Clin Biochem 32:461–466
77. Park K, Chung KY, Sung SH, Kim BR, Kim YS (2003) Protective effect of desferrioxamine during canine liver transplantation: significance of peritransplant liver biopsy. Transplant Proc 35:117–119
78. Tsuda T, Horio F, Kato Y, Osawa T (2002) Cyanidin 3-O-β-D-glucoside attenuates the hepatic ischemia-reperfusion injury through a decrease in the neutrophil chemoattractant production in rats. J Nutr Sci Vitaminol 48:134–141
79. Ozaki M, Nakamura M, Teraoka S, Ota K (1997) Ebselen, a novel anti-oxidant compound, protects the rat liver from ischemia/reperfusion injury. Transpl Int 10:96–102
80. Hori T, Uemoto S, Walden LB, Chen F, Baine AM, Hata T, Nguyen JH (2013) Pretreatment of small-for-size grafts in vivo by γ-aminobutyric acid receptor regulation against oxidative stress-induced injury in rat split orthotopic liver transplantation. Int J Hepatol 2013:149123
81. Zhong Z, Froh M, Connor HD, Li X, Conzelmann LO, Mason RP, Lemasters JJ, Thurman RG (2002) Prevention of hepatic ischemia-reperfusion injury by green tea extract. Am J Physiol Gastrointest Liver Physiol 283:G957–G964
82. Schauer RJ, Gerbes AL, Vonier D, Meissner H, Michl P, Leiderer R, Schildberg FW, Messmer K, Bilzer M (2004) Glutathione protects the rat liver against reperfusion injury after prolonged warm ischemia. Ann Surg 239:220–231
83. Schauer RJ, Kalmuk S, Gerbes AL, Leiderer R, Meissner H, Schildberg FW, Messmer K, Bilzer M (2004) Intravenous administration of glutathione protects parenchymal and non-parenchymal liver cells against reperfusion injury following rat liver transplantation. World J Gastroenterol 10:864–870

84. Wieland E, Schütz E, Armstrong VW, Küthe F, Heller C, Oellerich M (1995) Idebenone protects hepatic microsomes against oxygen radical-mediated damage in organ preservation solutions. Transplantation 60:444–451
85. Karaman A, Fadillioglu E, Turkmen E, Tas E, Yilmaz Z (2006) Protective effects of leflunomide against ischemia-reperfusion injury of the rat liver. Pediatr Surg Int 22:428–434
86. Grossini E, Pollesello P, Bellofatto K, Sigaudo L, Farruggio S, Origlia V, Mombello C, Mary DA, Valente G, Vacca G (2014) Protective effects elicited by levosimendan against liver ischemia/reperfusion injury in anesthetized rats. Liver Transpl 20:361–375
87. Muller C, Dunschede F, Koch E, Vollmar AM, Kiemer AK (2003) Alpha-lipoic acid preconditioning reduces ischemia-reperfusion injury of the rat liver via the PI3-kinase/Akt pathway. Am J Physiol Gastrointest Liver Physiol 285:G769–G778
88. Morsy MA (2011) Protective effect of lisinopril on hepatic ischemia/reperfusion injury in rats. Indian J Pharmacol 43:652–655
89. Xia Y, Rao J, Yao A, Zhang F, Li G, Wang X, Lu L (2012) Lithium exacerbates hepatic ischemia/reperfusion injury by inhibiting GSK-3β/NF-κB-mediated protective signaling in mice. Eur J Pharmacol 697:117–125
90. Sánchez GM, Rodríguez HMA, Giuliani A, Núñez Sellés AJ, Rodríguez NP, León Fernández OS, Re L (2003) Protective effect of Mangifera indica L. extract [Vimang] on the injury associated with hepatic ischaemia reperfusion. Phytother Res 17:197–201
91. Sewerynek E, Reiter RJ, Melchiorri D, Ortiz GG, Lewinski A (1996) Oxidative damage in the liver induced by ischemia–reperfusion: protection by melatonin. Hepatogastroenterology 43:898–905
92. Zaoualí MA, Reiter RJ, Padrissa-Altés S, Boncompagni E, García JJ, Ben Abnennebi H, Freitas I, García-Gil FA, Rosello-Catafau J (2011) Melatonin protects steatotic and nonsteatotic liver grafts against cold ischemia and reperfusion injury. J Pineal Res 50:213–221
93. Kotsch K, Ulrich F, Reutzel-Selke A, Pascher A, Faber W, Warnick P, Hoffman S, Francuski M, Kunert C, Kuecuek O, Schumacher G, Wesslau C, Lun A, Kohler S, Weiss S, Tullius SG, Neuhaus P, Pratschke J (2008) Methylprednisolone therapy in deceased donors reduces inflammation in the donor liver and improves outcome after liver transplantation: a prospective randomized controlled trial. Ann Surg 248:1042–1050
94. Saidi RF, Chang J, Verb S, Brooks S, Nalbantoglu I, Adsay V, Jacobs MJ (2007) The effect of methylprednisolone on warm ischemia-reperfusion injury in the liver. Am J Surg 193: 345–347
95. Stoffels B, Yonezawa K, Yamamoto Y, Schäfer N, Overhaus M, Klinge U, Kalff JC, Minor T, Tolba RH (2011) Meloxicam, COX-2 inhibitor, ameliorates ischemia/reperfusion injury in non-heart-beating donor livers. Eur Surg Res 47:109–117
96. Daglar G, Karaca T, Yuksek YN, Gozalan U, Akbiyik F, Sokmensuer C, Gurel B, Kama NA (2009) Effect of montelukast and MK-886 on hepatic ischemia-reperfusion injury in rats. J Surg Res 153:31–38
97. Glantzounis GK, Yang W, Koti RS, Mikhailidis DP, Seifalian AM, Davidson BR (2004) Continuous infusion of N-acetylcysteine reduces liver warm ischaemia-reperfusion injury. Br J Surg 91:1330–1339
98. Weigand MA, Plachky J, Thies JC, Spies-Martin D, Otto G, Martin E, Bardenheuer HJ (2001) N-Acetylcysteine attenuates the increase in alpha-glutathione S-transferase and circulating ICAM-1 and VCAM-1 after reperfusion in humans undergoing liver transplantation. Transplantation 72:694–698
99. Yokota R, Fukai M, Shimamura T, Suzuki T, Watanabe Y, Nagashima K, Kishida A, Furukawa H, Hayashi T, Todo S (2000) A novel hydroxyl radical scavenger, nicaraven, protects the liver from warm ischemia and reperfusion injury. Surgery 127:661–669
100. Singh AK, Mani H, Seth P, Gaddipati JP, Kumari R, Banuadha KK, Sharma SC, Kulshreshtha DK, Maheshwari RK (2000) Picroliv preconditioning protects the rat liver against ischemia-reperfusion injury. Eur J Pharmacol 395:229–239
101. Wu TW, Fung KP, Wu J, Zeng LH, Nakamura H (1994) Propyl gallate as a hepatoprotector in vitro and in vivo. Biochem Pharmacol 48:419–422

102. Boehnert MU, Armbruster FP, Hilbig H (2009) Relaxin as a protective substance in the preserving solution for liver transplantation. In: Fifth international conference on relaxin and related peptides, vol 1160. Annals of the New York Academy of Sciences. pp 320–321
103. Su JF, Guo CJ, Wei JY, Yang JJ, Jiang YG, Li YF (2003) Protection against hepatic ischemia-reperfusion injury in rats by oral pretreatment with quercetin. Biomed Environ Sci 16:1–8
104. Ligeret H, Brault A, Vallerand D, Haddad Y, Haddad PS (2008) Antioxidant and mitochondrial protective effects of silibinin in cold preservation-warm reperfusion liver injury. J Ethnopharmacol 12:507–514
105. Nguyen WD, Kim DH, Alam HB, Provido HS, Kirkpatrick JR (1999) Polyethylene glycol-superoxide dismutase inhibits lipid peroxidation in hepatic ischemia/reperfusion injury. Crit Care 3:127–130
106. Cheng F, Li Y, Feng L, Li S (2008) Effects of tetrandrine on ischemia/reperfusion injury in mouse liver. Transplant Proc 40:2163–2166
107. Luo XY, Takahara T, Hou J, Kawai K, Sugiyama T, Tsukada K, Takemoto M, Takeuchi M, Zhong L, Li XK (2012) Theaflavin attenuates ischemia-reperfusion injury in a mouse fatty liver model. Biochem Biophys Res Commun 417:287–293
108. Wu C, Wang P, Rao J, Wang Z, Zhang C, Lu L, Zhang F (2011) Triptolide alleviates hepatic ischemia/reperfusion injury by attenuating oxidative stress and inhibiting NF-kB activity in mice. J Surg Res 166:e205–213
109. Yan S, Jin L-M, Liu Y-X, Zhou L, Xie H-Y, Zheng S-S (2010) Outcomes and mechanisms of ischemic preconditioning in liver transplantation. Hepatobiliary Pancreat Dis Int 9:346–354
110. Murry CE, Jennings RB, Reimer KA (1986) Preconditioning with ischemia: a delay of lethal cell injury in ischemic myocardium. Circulation 74:1124–1136
111. Sindram D, Rudiger HA, Upadhya AG, Strasberg SM, Clavien PA (2002) Ischemic preconditioning protects against cold ischemic injury through an oxidative stress dependent mechanism. J Hepatol 36:78–84
112. Rüdiger HA, Graf R, Clavien PA (2003) Sub-lethal oxidative stress triggers the protective effects of ischemic preconditioning in the mouse liver. J Hepatol 39:972–977
113. Terajima H, Enders G, Thiaener A, Hammer C, Kondo T, Thiery J, Yamamoto Y, Yamaoka Y, Messmer K (2000) Impact of hyperthermic preconditioning on postischemic hepatic microcirculatory disturbances in an isolated perfusion model of the rat liver. Hepatology 31:407–415
114. Amador A, Grande L, Martí J, Deulofeu R, Miquel R, Solá A, Rodriguez-Laiz G, Ferrer J, Fondevila C, Charco R, Fuster J, Hotter G, García-Valdecasas JC (2007) Ischemic preconditioning in deceased donor liver transplantation: a prospective randomized clinical trial. Am J Transplant 7:2180–2189
115. Koneru B, Shareef A, Dikdan G, Desai K, Klein KM, Peng B, Wachsberg RH, de la Torre AN, Debroy M, Fisher A, Wilson DJ, Samanta AK (2007) The ischemic preconditioning paradox in deceased donor liver transplantation—evidence from a prospective randomized single blind clinical trial. Am J Transplant 7:2788–2796
116. Franco-Gou R, Roselló-Catafau J, Casillas-Ramirez A, Massip-Salcedo M, Rimola A, Calvo N, Bartrons R, Peralta C (2006) How ischaemic preconditioning protects small liver grafts. J Pathol 208:62–73
117. Rehman H, Connor HD, Ramshesh VK, Theruvath TP, Mason RP, Wright GL, Lemasters JJ, Zhong Z (2008) Ischemic preconditioning prevents free radical production and mitochondrial depolarization in small-for-size rat liver grafts. Transplantation 85:1322–1331
118. Coito AJ, Buelow R, Shen XD, Amersi F, Moore C, Volk HD, Busuttil RW, Kupiec-Weglinski JW (2002) Heme-oxygenase-1 gene transfer inhibits inducible nitric oxide synthase expression and protects genetically fat Zucker rat livers from ischemia reperfusion injury. Transplantation 74:96–102
119. Lehmann TG, Luedde T, Schwabe RF, Bunzendahl H, Samulski RJ, Lemasters JJ, Brenner DA (2006) Minimizing oxidative stress by gene delivery of superoxide dismutase accelerates regeneration after transplantation of reduced-size livers in the rat. Liver Transpl 12:550–559
120. Lehmann TG, Wheeler MD, Schwabe RF, Connor HD, Schoonhoven R, Bunzendahl H, Brenner DA, Jude Samulski R, Zhong Z, Thurman RG (2000) Gene delivery

of Cu/Zn-superoxide dismutase improves graft function after transplantation of fatty livers in the rat. Hepatology 32(6):1255–1264
121. Yabe Y, Kobayashi N, Nishihashi T, Takahashi R, Nishikawa M, Takakura Y, Hashida M (2001) Prevention of neutrophil-mediated hepatic ischemia/reperfusion injury by superoxide dismutase and catalase derivatives. J Pharmacol Exp Ther 298:3520–3526
122. He SQ, Zhang YH, Venugopal SK, Dicus CW, Perez RV, Ramsamooj R, Nantz MH, Zern MA, Wu J (2006) Delivery of antioxidative enzyme genes protects against ischemia/reperfusion-induced live injury in mice. Liver Transpl 12:1869–1879
123. Walther W, Stein U (2000) Viral vectors for gene transfer: a review of their use in the treatment of human diseases. Drugs 60:249–271
124. Nagral A, Ben-Ari Z, Dhillon AP, Burroughs AK (1998) Eosinophils in acute cellular rejection in liver allografts. Liver Transpl Surg 4:355–362
125. Foster PF, Sankary HN, Hart M, Ashmann M, Williams JW (1989) Blood and graft eosinophilia as predictors of rejection in human liver transplantation. Transplantation 47:72–74
126. Salunga TL, Cui ZG, Shimoda S, Zheng HC, Nomoto K, Kondo T, Takano Y, Selmi C, Alpini G, Gershwin ME, Tsuneyama K (2007) Oxidative stress-induced apoptosis of bile duct cells in primary biliary cirrhosis. J Autoimmun 29:78–86
127. Nagakawa Y, Williams GM, Zheng Q, Tsuchida A, Aoki T, Montgomery RA, Klein AS, Sun Z (2005) Oxidative mitochondrial DNA damage and deletion in hepatocytes of rejecting liver allografts in rats: role of TNF-alpha. Hepatology 42:208–215
128. Granot E, Elinav H, Kohen R (2002) Markers of oxidative stress in cyclosporine-treated and tacrolimus-treated children after liver transplantation. Liver Transpl 8:469–475
129. Christians U, Gottschalk S, Miljus J, Hainz C, Benet LZ, Leibfritz D, Serkova N (2004) Alterations in glucose metabolism by cyclosporine in rat brain slices link to oxidative stress: interactions with mTOR inhibitors. Br J Pharmacol 143:388–396
130. Garcia-Criado FJ, Palma-Vargas JM, Valdunciel-Garcia JJ, Toledo AH, Misawa K, Gomez-Alonso A, Toledo-Pereyra LH (1997) Tacrolimus [FK506] down-regulates free radical tissue levels, serum cytokines, and neutrophil infiltration after severe liver ischemia. Transplantation 64:594–598
131. Krötz F, Keller M, Derflinger S, Schmid H, Gloe T, Bassermann F, Duyster J, Cohen CD, Schuhmann C, Klauss V, Pohl U, Stempfle HU, Sohn HY (2007) Mycophenolate acid inhibits endothelial NAD[P]H oxidase activity and superoxide formation by a Rac1-dependent mechanism. Hypertension 49:201–208
132. Trapp A, Weis M (2005) The impact of immunosuppression on endothelial function. J Cardiovasc Pharmacol 45:81–87
133. Wierzbicka A, Pawłowska J, Socha P, Jankowska I, Skorupa E, Teisseyre M, Ismail H, Czubkowski P, Socha J (2007) Lipid, carbohydrate metabolism, and antioxidant status in children after liver transplantation. Transplant Proc 39:1523–1525
134. Jankowska I, Czubkowski P, Socha P, Wierzbicka A, Teisseyre M, Teisseyre J, Pawlowska J (2012) Lipid metabolism and oxidative stress in children after liver transplantation treated with sirolimus. Pediatr Transplant 16:901–906
135. Samonte VA, Sherman PM, Taylor GP, Carricato MN, Fecteau A, Ling SC, Ng VL (2008) Scurvy diagnosed in a paediatric liver transplant patient awaiting combined kidney and liver transplantation. Pediatr Transplant 12:363–367
136. Socha P, Litwin M (2008) Scurvy diagnosed in a pediatric liver transplant awaiting combined kidney and liver transplantation. Pediatr Transplant 12:257–260
137. Hussein MH, Hashimoto T, Daoud GA, Kakita H, Kato S, Goto T, Hibi M, Kato T, Okumura N, Tomishige H, Hara F, Ito T, Fukuda S, Kato I, Suzuki T, Suzuki S, Togari H (2011) Oxidative stress after living related liver transplantation subsides with time in pediatric patients. Pediatr Surg Int 27:17–22
138. Shapiro R, Halloran PF (2008) Organ preservation—can we do it better? Am J Transplant 8:479–480

Chapter 18
Antioxidant Approach to the Therapy of Chronic Liver Diseases

Le Thi Thanh Thuy, Hoang Hai, and Norifumi Kawada

18.1 Introduction

18.1.1 The Antioxidant System

The liver plays a central role in metabolic homeostasis, being responsible for the metabolism, synthesis, storage, and redistribution of nutrients, carbohydrates, fats, and vitamins. The liver is also a significant site of free radical synthesis. Liver enzymes, including diamine oxidase, aldehyde dehydrogenase, tryptophan dual oxidase, liver dehydrogenase, and the cytochrome P450 system, induce oxidation and uncoupling, triggering free radical production [1]. Oxygen-free radicals, more generally termed reactive oxygen species (ROS), and reactive nitrogen species (RNS) are the most important groups of radicals generated in living systems [2].

The body has developed important antioxidant defence mechanisms to protect tissues from free radical-induced damage. An antioxidant is any substance that, when present at low concentrations, as compared to that of an oxidisable substrate, significantly delays or inhibits oxidation of that substrate [3]. Antioxidants can act at several stages of an oxidative sequence; first, by removing oxygen intermediates created during normal oxygen metabolism. Second, antioxidants can remove metal ions required by catalytic proteins (enzymes). Third, antioxidants also remove key ROS including superoxide (O_2^-) and hydrogen peroxide (H_2O_2) that are generated in excess in certain disease states. Fourth, antioxidants scavenge initiator-free radicals including hydroxyl, alkoxyl, and peroxyl species. Fifth, antioxidants can break the chain of an oxidative sequence and quench singlet oxygen [3]. An antioxidant molecule

L.T.T. Thuy • H. Hai • N. Kawada (✉)
Department of Hepatology, Graduate School of Medicine,
Osaka City University, Osaka, Japan
e-mail: thuylt@med.osaka-cu.ac.jp; hhai@med.osaka-cu.ac.jp;
kawadanori@med.osaka-cu.ac.jp

E. Albano, M. Parola (eds.), *Studies on Hepatic Disorders*,
Oxidative Stress in Applied Basic Research and Clinical Practice,
DOI 10.1007/978-3-319-15539-5_18

may have only one of the actions described above. For example, transferrin, lactoferrin, and albumin act as metal-binding proteins. However, many antioxidants have several actions, particularly all of the third, fourth, and the fifth listed above. Thus, the antioxidant enzymes, superoxide dismutase (SOD), catalase, and glutathione peroxidase (GSHPx), for example, exert multiple effects.

Depending on function, antioxidants are located at different sites in cells and can be classified as intracellular, membrane-associated, or extracellular in nature.

18.1.2 Antioxidant Classification

18.1.2.1 Intracellular Antioxidants

Intracellular antioxidant defence systems include SOD, catalase, glutathione (GSH) and an associated family of enzymes, the polypeptide thioredoxin (TRX), and peroxidases of the peroxiredoxin (Prx) family [4].

SOD catalyses the dismutation of superoxide ($O_2^{\bullet-}$) to H_2O_2 and O_2 (18.1):

$$O_2^{\bullet -} + O_2^{\bullet -} + 2H^+ \rightarrow H_2O_2 + O_2 \quad (18.1)$$

A CuZn-SOD is present in the cytosol and in the space between the inner and outer mitochondrial membranes, whereas a manganese-containing SOD is present in the mitochondrial matrix. The actions of both of these enzymes are critical to prevent ROS-induced toxicity [5].

Catalase is found primarily in the peroxisomes of all major organs, especially the liver. The enzyme catalyses a reaction between two H_2O_2 molecules, resulting in formation of water and O_2 [6] (18.2):

$$2H_2O_2 \rightarrow O_2 + 2H_2O \quad (18.2)$$

Catalase action is very important for protecting the cell from oxidative damage caused by ROS; the enzyme can convert millions of molecules of H_2O_2 to water and oxygen each second [7].

GSH and an associated enzyme family (including peroxidases) also play major roles in removing hydrogen peroxide generated by SOD via oxidation of GSH to glutathione disulphide (GSSG) (18.3) [3]:

$$H_2O_2 + 2GSH \rightarrow GSSG + 2H_2O \quad (18.3)$$

GSSH generated, as shown in (18.3), is next re-converted to GSH in a reaction involving nicotinamide adenine dinucleotide phosphate [8]. GSH precursors, including *N*-acetylcysteine (NAC), are widely prescribed commercial antioxidants. The antioxidant effects of NAC are exerted by reducing cystine to cysteine, via synthesis of reduced GSH; by increasing glutathione-*S*-transferase levels; and by capture and neutralisation of ROS [9].

The Prx family of peroxidases is present in organisms of all kingdoms [10]. Prx enzymes exist in multiple isoforms in all eukaryotic cells that reduce hydrogen peroxide and alkyl hydroperoxides to water and alcohol, respectively, with consumption of reducing equivalents. Such equivalents are specifically derived from thiol-containing donor molecules, including TRX. Prx I and II are located in the cytoplasm and Prx III in mitochondria [11].

The TRX cytoplasmic protein occupies a specific intracellular site under pathophysiological conditions [12]. TRX efficiently donates electrons to human GSHPx and members of the Trx superfamily, as mentioned above. TRX per se also acts as an antioxidant or as a scavenger of singlet oxygen, hydroxyl, and hydrogen peroxide radicals [13].

18.1.2.2 Membrane and Extracellular Antioxidants

The extracellular fluids of the human body, including blood plasma, tissue fluid, cerebrospinal fluid, synovial fluid, and seminal plasma, contain little or no catalase activity and only low levels of SOD and selenium-containing GSHPx [14, 15]. Metal-binding proteins are the principal antioxidants of extracellular fluids. Such proteins can sequester iron and copper ions that, in their free forms, catalyse oxidative reactions. Transferrin, lactoferrin, haemoglobin, and myoglobin bind iron, whereas albumin, metallothionein, and ceruloplasmin bind copper [16]. The iron transport protein transferrin is usually one-third loaded with iron and effectively controls plasma iron concentration [3]. Lactoferrin is a glycoprotein exhibiting a high affinity for iron. The capacity of lactoferrin to bind iron is twice that of transferrin; two ferric ions are bound by one lactoferrin molecule [17].

Haemoglobin (and also myoglobin), in their free forms, can mediate oxidative tissue damage by accelerating lipid peroxidation. Haemoglobin-binding proteins such as haptoglobin and heme-binding proteins including haemopexin are serum proteins that function as antioxidants by virtue of their ability to bind to haemoglobin and haem, respectively, thereby preventing oxidative stress that might be triggered by free haemoglobin [18].

The principal copper-binding protein of human plasma is ceruloplasmin, which carries over 95 % of plasma copper [19]. Ceruloplasmin is also an effective chain-breaking antioxidant for a variety of radicals, including those formed via ferroxidase activity [19].

Other extracellular or exogenous antioxidants are derived from food or synthesized in vivo and include vitamin C, vitamin E, bilirubin, melatonin, lipoic acid, coenzyme Q, uric acid, and the melanins. Ascorbic acid (vitamin C) and α-Tocopherol (vitamin E) are food antioxidants that can prevent accumulation of free radicals and lipid peroxidation, in the human body [20, 21]. Ascorbic acid exhibits multiple antioxidant properties including the ability to scavenge $O_2^{\bullet-}$, $HO_2^{\bullet}$, peroxyl ($RO_2^{\bullet}$), thiyl and sulphenyl, and nitricoxide radicals [22]. α-Tocopherol, a fat-soluble vitamin, is a poor antioxidant outside the membrane bilayer but very effective when incorporated into the membrane [23] and combats hydrogen peroxide, lipid peroxides, and other oxidants [20].

Another group of antioxidants are natural phenolic compounds and flavonoids found in vegetables, tea, and red wine [24]. These include quercetin, resveratrol, cathechine, and epigarocathechin. For example, quercetin is a strong antioxidant because the compound chelates metals, scavenges oxygen-free radicals [24, 25], and prevents oxidation of low-density lipoprotein (LDL) in vitro [26].

18.1.2.3 Antioxidants That Modulate Nitric Oxide Metabolism

Nitric oxide (NO), a lipophilic, highly diffusible, and short-lived physiological messenger, regulates many physiological processes including vasodilation, respiration, cell migration, the immune response, and apoptosis [27]. Under pathological conditions, NO production is increased via the action of an inducible isoform of NOS (iNOS). This isoform produces NO in large (potentially damaging) amounts, in a calcium-independent manner. NO can react with O_2^- to produce the harmful molecule peroxynitrite ($ONOO^-$), a form of RNS. Materials including polyphenols and other antioxidants that influence the cell redox balance (including SOD activity) may directly affect the iNOS activity level and the expression thereof [28].

A recently discovered vertebrate globin, cytoglobin (CYGB) [29], the molecular characteristics of which are similar to myoglobin, is an antioxidant by virtue of an ability to scavenge NO. CYGB may facilitate oxygen diffusion through tissues, scavenge NO or other ROS, or exert a protective function during oxidative stress [30]. In vitro studies have shown that *CYGB* overexpression rescues the human neuronal cell line TE671 from pro-oxidant Ro19-8022-induced DNA damage [31]. *CYGB* overexpression also protected human neuroblastoma SH-SY5Y cells from H_2O_2-induced cell death [32, 33]. Furthermore, it has been reported that overexpression of CYGB in rat hepatic stellate cells, either in vitro or in vivo, protected such cells against oxidative stress and inhibited the differentiation thereof to the activated form [34]. Such reports suggest that CYGB may act as a cytoprotective and radical-scavenging molecule in addition to playing a role as gas carrier.

18.2 Antioxidant Therapy for Chronic Viral Hepatitis

As mentioned in Chap. 14, oxidative stress is an important contributor to the development of liver damage during chronic viral hepatitis, a disease in which both the virus per se and the host-mounted immune response can trigger oxidative stress [35], as reflected in increased levels of oxidised proteins and nucleic acids and decreases in antioxidant levels [36, 37].

18.2.1 Vitamins Given Singly or in Combination

Antioxidants have been evaluated over the past 20 years in clinical trials involving chronic hepatitis C (CHC) patients (Table 18.1), commencing with a small open-label study featuring vitamin E supplementation [38]. In the cited work, six CHC patients resistant to interferon (IFN) therapy were given vitamin E 1,200 IU/day for 8 weeks [38]. The vitamin delayed fibrosis progression and reduced oxidative stress, as indicated by reduced activation of hepatic stellate cells and decreased malonaldehyde levels, respectively. However, no change in either serum ALT or HCV-RNA levels was evident, and liver histology was not affected. In another study, 17 CHC patients were given vitamin E (500 mg/day) for 3 months and showed modest reductions in serum ALT levels associated with small decreases in oxidative stress. Thus, TRX levels fell from 59 to 40 ng/mL at the end of treatment [39]. Similar results were found in another study in which 23 CHC patients refractory to IFN therapy were treated; a prospective, randomised, double-blind cross-over design was used. Twelve-week supplementation with vitamin E reduced serum ALT levels. However, serum ALT levels became elevated once more within 1 month of discontinuing vitamin E treatment. Re-treatment of responders (exhibiting ALT decreases of at least 35 %) for 3 months reduced ALT levels once more [40]. Thus, supplementary vitamin E alone did not have any notable effect on significant clinical features. Subsequent studies evaluated antioxidant combinations (Table 18.1) [41–43]. Seven oral antioxidative preparations (glycyrrhizin, schisandra, silymarin, ascorbic acid, lipoic acid, L-glutathione, and alpha-tocopherol) and four intravenous preparations (glycyrrhizin, ascorbic acid, L-glutathione, and B-complex) were given to 50 CHC patients for 20 weeks [41]. Interestingly, at the end of treatment, normalisation of liver enzyme levels was evident in 44 % of patients; ALT levels remained normal throughout the follow-up period in 72.7 % of patients; and a decrease in viral load (of one log or more) was observed in 25 % of patients. However, such encouraging results were not replicated in a larger study of 100 patients treated with antioxidant cocktails for 24 weeks [42]. Another combination therapy, featuring vitamins E, C, and selenium, was tested in 23 CHC patients over 6 months, but no beneficial effect on either HCV-RNA or ALT level, or liver histology, was evident [43].

Mitochondrial damage is a common feature of CHC pathogenesis, and MitoQ was evaluated as a therapeutic agent in the phase II study of Gane et al. [44]. Thirty patients with contraindications to pegylated interferon (PEGIFN) and/or ribavirin (RBV) were randomised to receive MitoQ (40–80 mg/day), or placebo. At the end of 28 days, compared with baseline levels, 40 mg of MitoQ daily decreased both ALT (from 153 to 110 IU/L; $P=0.002$) and AST levels (from 131 to 95 IU/L; $P=0.003$). The figures after administration of 80 mg of MitoQ daily were 131–95 IU/L, $P=0.024$ and 87–75 IU/L, $P=0.017$, respectively. However, no change in HCV-RNA level was noted, suggesting that antioxidants may mildly decrease ALT levels, but that this alone was inadequate to counter CHC.

Table 18.1 Clinical trials of antioxidant therapies for chronic hepatitis C patients

Study no.	Reference [number]	Sample size	Patient details	Antioxidant strategy	Duration	Response
1	Beloqui et al. [48]	24	CHC patients	Oral NAC (600 mg)	5–6 months	Improved ALT levels
2	Houglum et al. [38]	6	CHC patients refractory to IFN therapy	Vitamin E (1,200 UI/day)	8 weeks	Reductions in oxidative stress marker levels; decreased HSC activation; no change in HCV-RNA status
3	Von Herbay et al. [40]	23	CHC patients refractory to IFN therapy	Vitamin E (400 UI/day)	12 weeks	Decreased ALT levels
4	Ideo et al. [46]	120	IFN-non-responders with HCV infections	NAC (1,200 mg/day) + vitamin E (600 mg/day)	6 months	No benefit
5	Look et al. [45]	24	IFN-naïve CHC patients	NAC (1,800 mg/day) + sodium selenite (400 μg/day) ± vitamin E (544 UI/day)	24 weeks	A 2.4-fold greater chance of a complete response and a significantly greater reduction in viral load in test patients
6	Grant et al. [49]	30	CHC patients	Mitoquinone (40 or 80 mg) or placebo	4 weeks	Decreased ALT and AST levels ($P < 0.005$ for both), but no change in HCV-RNA status
7	Neri et al. [50]	77	CHC patients	IFN ± NAC (2.4 g/day)	6 months	Increased time to relapse compared to IFN alone (22 vs. 31 weeks); decreased oxidative stress levels; but the effects were not of long duration
8	Mahmood et al. [39]	17	CHC patients	Vitamin E (500 mg/day)	3 months	Small decreases in ALT levels; reductions in levels of oxidative stress markers
9	Morisco et al. [115]	92	CHC patients	A functional food with a high level of natural antioxidants and high carotenoid bioavailability	3 months	Lower ribavirin levels and higher haemoglobin levels in test patients than controls
10	Saeian et al. [47]	47	CHC patients	Interferon/ribavirin therapy plus vitamin E 800 IU b.d.	24 weeks	No significant difference in ALT or haemoglobin levels, and no sustained antiviral response

11	Melhem et al. [41]	50	CHC patients	Seven antioxidative oral preparations (glycyrrhizin, schisandra, silymarin, ascorbic acid, lipoic acid, L-glutathione, and alpha-tocopherol); and four different intravenous preparations (glycyrrhizin, ascorbic acid, L-glutathione, and B-complex)	20 weeks	Normalisation of liver enzyme levels occurred in 44 % of patients; ALT levels remained normal throughout follow-up in 72.7 % of patients. Decreases in viral loads (of one log or more) were observed in 25 % of patients
12	Groenbaek et al. [43]	23	CHC patients	Ascorbic acid (500 mg/day), D-alpha-tocopherol (945 IU/day) and selenium (200 μg/day); or placebo tablets	6 months	The antioxidant group exhibited significantly higher levels of plasma ascorbic acid and alpha-tocopherol than did the placebo group, and the activity of erythrocyte glutathione peroxidase significantly increased from baseline to month 6 in test patients. No differences were observed in ALT or HCV-RNA levels
13	Gordon et al. [60]	24	CHC patients	Silymarin 600 or 1,200 mg/day	12 weeks	No change
14	Hino et al. [116]	32	CHC patients	IFN/ribavirin + vitamin E 500 mg/day + vitamin C 750 mg/day	26 weeks	Attenuation of the ribavirin-induced decrease in eicosapentaenoic acid levels in erythrocyte membranes
15	Gabbay et al. [42]	100	IFN-failed CHC patients	Combined intravenous and oral antioxidant or placebo; or oral treatment. Oral formulation: glycyrrhiza, 500 mg bid; schizandra, 500 mg tid; ascorbate, 2,000 mg tid; L-glutathione, 150 mg bid; silymarin, 250 mg tid; lipoic acid, 150 mg bid; D- -TOCOPHEROL, 800 IU/day, once daily for 24 weeks. Intravenous formulation: 120 mg glycyrrhiza; 10 g ascorbate; 750 mg L-glutathione; 1 mL B complex, twice weekly	10 weeks	Modest improvement in ALT levels and combined HAI scores upon both oral and iv antioxidant therapy, measured at treatment end. Benefits not sustained after discontinuation of therapy

(continued)

Table 18.1 (continued)

Study no.	Reference [number]	Sample size	Patient details	Antioxidant strategy	Duration	Response
16	Falasca et al. [117]	40	Caucasian CHC patients	SPV complex	3 months	Hepatoprotective efficacy apparent. Reduction in inflammatory cytokine levels and decreased viral load. Decreases in the levels of all of ALT, AST, GGT, alkaline phosphatase, total cholesterol, fasting glucose, insulinemia, HOMA value, and C-reactive protein level, in hepatic steatotic patients
17	Hawke et al. [56]	32	IFN-failed CHC patients	Silymarin (140, 280, 560, or 700 mg every 8 h)	1 week	No change
18	Feld et al. [118]	21	IFN-non-responders among HCV-infected patients	PEG-IFNα2a and ribavirin for 2 weeks followed by SAMe (1,600 mg/day) for 2 weeks and PEGIFN and ribavirin for 48 weeks	2 weeks	Improved early viral kinetics and enhanced induction of ISGs upon SAMe treatment, associated with enhanced Stat1 methylation
19	Filipowicz et al. [119]	29	CHC patients	SAMe, betaine, PEG-IFNα2b, and ribavirin treatment	6 or 12 months	Improved early virological response
20	Yakoot et al. [120]	66	IFN-naïve CHC patients	Spirulina 1,500 mg/day or silymarin 420 mg/day	3 and 6 months	A significantly greater reduction in serum ALT levels in the spirulina- than the silymarin-treated group
21	Biermer et al. [121]	20	IFN-failed CHC patients	Silibinin 1,400 mg/day	2 days	Reduction in HCV-RNA levels; 3/20 patients achieved SVR
22	Fried et al. [122]	154	IFN-failed CHC patients	Silymarin 1,260 or 2,100 mg/day, or placebo	24 weeks	No significant change in either ALT or HCV-RNA levels
23	Adeyemo et al. [123]	32	IFN-non-responders with HCV infections	Silymarin 1,260 or 2,100 mg/day, or placebo	20 weeks	No change in ALT or HCV RNA levels
24	Grant et al. [49]	147	CHC	3MU IFN-alpha three times weekly plus NAC 1,800 mg daily ($n=73$); IFN alone ($n=74$)	6 months	No obvious benefit upon addition of *N*-acetyl cysteine to conventional therapy featuring interferon-alpha

Several trials have evaluated antioxidants as adjuvants to IFN therapy in the era before RBV was introduced. In one such study, 24 treatment-naïve CHC patients were randomised to receive IFN monotherapy alone, or in combination with NAC + sodium selenite or NAC + sodium selenite + vitamin E. After 6 months of treatment, a higher proportion of patients treated with a regimen including vitamin E achieved negative HCV-RNA status, as compared to those receiving IFN monotherapy or IFN + NAC + sodium selenite (6/8 vs. 3/8 vs. 2/8; $P = 0.11$). Comparisons of vitamin E- ($n = 8$) vs. non-treated ($n = 16$) subjects showed that the odds of achieving negative HCV-RNA status at the end of vitamin E treatment was 2.4-fold higher than otherwise ($P = 0.02$). However, no beneficial effect on the sustained viral response (SVR) was evident (2/8 vs. 1/8 vs. 1/8; P=NS). Surprisingly, no effect on oxidative stress marker levels was noted, as assessed by measurement of trolox-equivalent antioxidant capacity and levels of thiobarbituric acid-reactive substances (TBARSs) [45]. The enhanced end-of-treatment response was not confirmed in a placebo-controlled study of 120 CHC patients who were IFN non-responders, randomised to receive IFN with or without NAC (1,200 mg/day) + vitamin E (600 mg/day) for 6 months. ALT normalisation rates were similar at the end of treatment (10.3 % vs. 9.7 %; P=NS) and 6 months after completing treatment (1.3 % vs. 0 %; P=NS). No patient in either group achieved negative HCV-RNA status by the end of treatment [46]. A similar lack of any benefit afforded by vitamin E on HCV-RNA loss or SVR was noted in another randomised controlled trial (RCT) in 47 CHC patients [47].

Thus, vitamins E and/or C alone, or in combination with anti-HCV therapy, tend to reduce serum ALT levels. However, neither HCV titres, nor the histological extent of inflammation, nor the extent of fibrosis, was influenced by the vitamins. Furthermore, in most studies, the decrease in ALT level was marginal and not sustained after treatment cessation, thereby throwing any clinical significance into doubt. Modern treatment regimens feature viral protease inhibitors and other novel antiviral medications, and it remains to be seen if antioxidants can exert beneficial effects as components of such regimens.

18.2.2 N-acetylcysteine

In an open pilot study on 14 CHC patients who were documented IFN non-responders, addition of NAC at 1.8 g/day to IFN improved liver enzyme levels, with a decrease in viral load [48], encouraging testing of the NAC/IFN combination in better-controlled studies [43, 44]. In a placebo-controlled double-blind RCT, addition of 1.8 g/day NAC to IFN did not improve the SVR rate [49]. In another prospective, randomised open-label study, the viral response rates were similar in the test and control groups but the time to relapse after treatment discontinuation was longer after use of NAC (31 vs. 22 weeks; $P < 0.05$). In terms of HCV infection, the goal of treatment (as mentioned earlier) is achievement of SVR, and a 9-week delay in relapse of HCV infection is of minimal clinical significance [50].

18.2.3 *Natural Compounds*

An oral formulation of extracts from the milk thistle, silymarin, is widely used in the USA for treatment of viral hepatitis. A report by [51] showed that, of 1,145 study participants, 33 % had used silymarin either in the past or at the baseline. In terms of function, in vitro studies have shown that silymarin exerts anti-inflammatory and immunomodulatory functions via inhibition of NF-κB (nuclear factor κ-light-chain-enhancer of activated B cells) [52]. Furthermore, silymarin significantly downregulated HCV core mRNA (by 20–36 %) and protein (by 30–60 %) levels in CNS3 cells [53]. Clinical trials showed that silymarin exerted biochemical effects [54], allowing attainment of negative HCV-RNA status [55, 56], and one case report even found an SVR [57]. However, several studies have reported limited effects of, or no consistent benefit afforded by, silymarin in CHC patients [58–61]. Such results may have been compromised by small numbers of patients, short treatment duration, or the low doses used.

In summary, many trials of silymarin have yielded inconsistent results. Silymarin, given as an antioxidant to CHC patients, has been either efficient or inefficient in terms of reducing enzyme levels in vivo and affecting HCV-RNA levels and/or liver histological properties (Table 18.1). Thus, it is difficult to conclude that antioxidants are valuable in CHC patients. Silymarin or NAC may potentially be effective but further longer-term trials with larger numbers of patients are required.

18.3 Antioxidant Therapy for Alcoholic Liver Diseases

Acute or chronic alcohol consumption increases production of both ROS/RNS and other free radical species (e.g., the 1-hydroxyethyl radical), as has been shown in both animal models and patients with alcoholic liver diseases (ALD), as described in Chap. 13. The high-level increase in oxidative stress mediated by alcohol consumption suggests that it is appropriate to use antioxidants to protect against liver damage. Several antioxidants, including vitamin E, *S*-adenosylmethionine (SAMe), polyenylphosphatidylcholine (PPC), and silymarin, have been trialed to this end.

18.3.1 *Vitamins E and C*

A randomised study featuring administration of tocopherol 500 mg daily for 1 year found that no benefit was afforded to ALD patients in terms of either clinical or biochemical function, hospitalisation rate, or mortality [62]. Another double-blind, placebo-controlled randomised trial of vitamin E at 1,000 UI/day was performed in 25 patients with alcoholic hepatitis [63]. Significant decreases in the serum levels of the fibrogenesis marker, hyaluronic acid, were observed in treated patients, suggesting that further investigations should be performed in large patient cohorts.

18.3.2 Polyenylphosphatidylcholine

PPC prevents excess collagen accumulation by enhancing collagenase activity both in cell cultures and animal models [64] and decreases oxidative stress via reduction in, or normalisation of, the levels of 4-hydroxinonenal, isoprostanes, and GSH [65, 66]. PPC (4.5 g/day given orally as 1.5 g tablets three times daily) was tested in 789 veterans (97 % male; mean age 49 years) with biopsy-proven alcoholic cirrhosis in a randomised, double-blind, placebo-controlled multicentre study [67]. The average alcohol intake was comparable in both the control and treatment groups, being about 225 g/day for 19 years before the start of treatment, but was unexpectedly reduced to about 35 g/day during the study period. Liver biopsy was repeated at 24 months, and the main outcome measure was the fibrosis stage, as compared to that at baseline. Two-year biopsies were performed on 412 patients. PPC did not differ from placebo in terms of any effect on the principal outcome. However, a trend toward a reduction in the development of ascites was evident in the PPC group (9 % vs. 14 %; $P<0.057$). Fibrosis progression was more common in those with concomitant HCV infection (32 % vs. 17 %; $P<0.001$). In this latter subgroup, PPC treatment improved liver function, as reflected by the serum levels of liver enzymes and bilirubin [67].

18.3.3 Silymarin

A total of six RCTs have explored the utility of silymarin in the management of ALD patients. Of these, three found that silymarin was effective. In a study of 91 patients with alcoholic cirrhosis, those given silymarin (520 mg/day) on a long-term basis exhibited improvements in 4-year survival, as compared to placebo-treated patients (58 % vs. 39 %; $P=0.03$) [55]. Another study, in which silymarin was given at 420 mg/day, revealed no effect on survival but histological and biochemical parameters improved. However, silymarin was given for only 4 weeks [68]. Lirussi et al. [69] performed an RCT on 60 outpatients; the test group received a combination of silymarin and beta-cyclodextrin (silybin-beta-cyclodextrin) for 6 months. Significant reductions in serum glucose and triglyceride levels were evident, as was a decrease in the levels of an oxidative stress marker, malondialdehyde (MDA), in the test, as compared to the placebo-treated group. Such effects may be attributable to recovery of the levels of energy substrates, consistent with reduced lipid peroxidation and improved insulin activity. No clinically relevant side-effect was observed in either group. Three RCTs ($n=60$–97 subjects; silymarin dose: 280–450 mg/day; duration: 6–24 months) failed to show any biochemical, histological, or survival benefit [70–72], although one study found a significant decrease in the levels of oxidative stress markers [70].

18.3.4 S-adenosylmethionine

SAMe is a promising antioxidant for treatment of ALD patients. SAMe is a precursor of key metabolites including glutathione and polyamines. Oral administration of SAMe (1,200 mg/day for 6 months) significantly increased hepatic glutathione levels in subjects both with and without ALD [73]. More importantly, a long-term, randomised, placebo-controlled, double-blind, multicenter clinical trial of SAMe in 123 patients with alcoholic cirrhosis found that oral SAMe (1,200 mg/day) improved survival compared with placebo [74].

In general, antioxidant therapy for ALD using SAMe or silymarin potentially improves survival if high doses are given long-term. Both agents are well-tolerated; no significant adverse effects were noted [69, 74].

18.4 Antioxidant Therapy for Non-alcoholic Fatty Liver Disease and Non-alcoholic Steatohepatitis

The associations between non-alcoholic fatty liver disease (NAFLD)/non-alcoholic steatohepatitis (NASH) and biomarkers of oxidative stress and lipid oxidation have been assessed in many human and animal studies [75–78]. To date, no established pharmacological treatment for NAFLD/NASH has been reported. Depletion of antioxidants within hepatocytes, thus impairing ROS inactivation, forms the basis of antioxidant supplementation to potentially treat NASH. Several agents have yielded encouraging results; these include vitamin E, vitamin C, betaine, ursodeoxycholic acid (UDCA), pentoxifylline, metformin, and NAC (Table 18.2).

18.4.1 Vitamins E and C

Hasegawa et al. found that the serum alanine transaminase level decreased in NAFLD patients, but not NASH patients, after 6 months of vitamin E supplementation [79]. However, histological findings, including those of steatosis, inflammation, and fibrosis, improved after alpha-tocopherol treatment of NASH patients. Similarly, Kawanaka et al. showed that serum transaminase activities, and the levels of oxidative stress markers including TRX and TBARS, decreased significantly after 6 months of vitamin E therapy [80]. Interestingly, in obese children with NASH, daily oral vitamin E administration for 4–10 months normalised serum aminotransferase and alkaline phosphatase levels [81], although only 11 subjects were involved in the cited study. A large-scale trial involved 247 adults with NASH (but without diabetes) given pioglitazone at 30 mg daily (80 subjects), vitamin E at 800 IU daily (84 subjects), or placebo (83 subjects) over a long duration (96 weeks) [82]. The results clearly showed that, as compared to placebo, vitamin E therapy

Table 18.2 Clinical trials of antioxidant therapies for NALFD/NASH patients

Study no.	Reference [number]	Sample size	Patients	Antioxidant strategy	Duration	Response
1	Hasegawa et al. [79]	22	Non-alcoholic fatty liver ($n=12$); non-alcoholic steatohepatitis ($n=10$)	Dietary instructions for 6 months, following with vitamin E 300 mg/day for 1 year	6 months to 1 year	ALT level decreases in non-alcoholic fatty liver patients, but not non-alcoholic steatohepatitis patients. Improvement of steatosis, inflammation and fibrosis
2	Kawanaka et al. [80]	10	NASH	Vitamin E	6 months	Significant drops in serum transaminase activity, and levels of oxidative stress markers including thioredoxin and thiobarbituric acid-reactive substance (TBARS)
3	Lavine et al. [81]	11	Obese children with NASH	Vitamin E 400–1,200 IU daily	4–10 months	Normalisation of serum aminotransferase and alkaline phosphatase levels
4	Sanyal et al. [82]	247	Non-alcoholic steatohepatitis patients without diabetes	Pioglitazone 30 mg daily ($n=80$); vitamin E 800 IU daily ($n=84$), or placebo ($n=83$)	96 weeks	Compared to placebo, vitamin E therapy was associated with significant improvement in NASH activity scores, with no worsening of fibrosis (43 % vs. 19 %, $P=0.001$). Both drugs reduced liver enzyme levels, local inflammation, and hepatic steatosis
5	Harrison et al. [83]	49	NASH	Combination of alpha-tocopherol (vitamin E) 1,000 IU and vitamin C 1,000 mg	6 months	Significant improvement in fibrosis scores. However, no improvement in necroinflammatory activity or ALT level

(continued)

Table 18.2 (continued)

Study no.	Reference [number]	Sample size	Patients	Antioxidant strategy	Duration	Response
6	Nobili et al. [84]	90	NAFLD children	Combination of vitamins E (600 IU/day) and C (500 mg/day), plus nutritional programming	12 months	Diet and physical exercise afforded significant improvements in both liver function and glucose metabolism
7	Abdelmalek et al. [85]	10	NASH	Betaine (anhydrous) as an oral solution (Cystadane) given in two divided doses daily	12 months	Significant improvements in the serum levels of aspartate aminotransferase ($P = 0.02$); evidently extents of fibrosis and steatosis, and necroinflammatory grade
8	Abdelmalek et al. [86]	55	NASH	Oral betaine (20 g daily) or placebo	12 months	Compared to placebo, neither intra- or inter-group differences nor any changes in either non-alcoholic fatty liver disease activity scores or fibrosis stage were noted. Betaine did not improve hepatic steatosis but may protect against worsening of steatosis
9	Adams et al. [88]	20	NASH	Pentoxifylline (1,600 mg/day)	12 months	Alanine and aspartate aminotransferase levels significantly reduced, as compared to baseline (84 ± 64 vs. 138 ± 76; $P = 0.002$; and 58+/37 vs 102+/62; $P = 0.003$, respectively)
10	Laurin et al. [89]	24	NASH	Ursodeoxycholic acid (UDCA)	12 months	Significant improvement in alkaline phosphatase, ALT, GGT, and hepatic steatosis

11	Lindor et al. [90]	166	NASH	UDCA	2 years	Decreased liver enzyme levels but no change in extent of steatosis, necroinflammation, or fibrosis
12	Dufour et al. [91]	48	NASH	UDCA + vitamin E; UDCA + placebo	2 years	UDCA with vitamin E improved laboratory test values and hepatic steatosis
13	Haukeland et al. [95]	48	NAFLD	Metformin ($n=24$) or placebo ($n=24$)	6 months	Effects of metformin were observed on changes in body-weight ($P<0.001$), serum levels of cholesterol ($P=0.004$), LDL-cholesterol ($P<0.001$), glucose ($P=0.032$) and on HbA1c ($P=0.020$) but not in hepatosteatosis and NAS score
14	Pamuk et al. [99]	35	NASH	NAC 600 mg/day	4 weeks	Improvements in liver enzyme levels
15	de Oliveira et al. [94]	20	NASH	NAC 1.2 g/day and metformin 500 mg/day	12 months	Improvements in liver enzyme levels, insulin resistance, body mass index, and liver histological findings (including those indicating steatosis and fibrosis)

The English in this document has been checked by at least two professional editors, both native speakers of English. For a certificate, please see: http://www.textcheck.com/certificate/l5IZ1G

was associated with significant improvement in NASH, with no worsening of fibrosis (43 % vs. 19 %, $P=0.001$). Both compounds reduced liver enzyme levels, local inflammation, and hepatic steatosis [82].

A combination of daily alpha-tocopherol (vitamin E 1,000 IU) and vitamin C (1,000 mg) given for 6 months in a prospective, double-blind, randomised, placebo-controlled trial significantly improved fibrosis scores in 49 patients; 45 completed the study [83]. However, no improvement in necroinflammatory activity or ALT level was seen using this combination therapy. Not only combinations of vitamin E and C, but also a nutritional programme improved outcomes. Nobili et al. found that both diet and physical exercise in NAFLD children seemed to significantly improve both liver function and glucose metabolism, even in the absence of any antioxidant therapy [84] (Table 18.2).

18.4.2 Betaine

Betaine, the donor of methyl groups for remethylation of homocysteine, may be a therapeutic agent in the context of NASH. Abdelmalek et al. conducted a pilot study on ten patients with NASH who were given an oral solution of anhydrous betaine (Cystadane) in two divided doses daily for 12 months [85]. Significant improvements in the serum levels of aspartate aminotransferase ($P=0.02$), the extent of steatosis, the necroinflammatory grade, and the fibrosis stage were evident during treatment. Eight years later, the cited authors conducted a much larger-scale, randomised placebo-controlled study of 55 patients with biopsy-proven NASH given either oral betaine (20 g daily) or placebo for 12 months [86]. However, as compared to placebo, no intra- or inter-group differences in any of the non-alcoholic fatty liver disease activity scores, or fibrosis stage, were noted. Betaine did not improve hepatic steatosis, but may protect against worsening steatosis [86].

18.4.3 Pentoxifylline

A newer antioxidant drug, pentoxifylline, is a methylxanthine compound inhibiting production of tumor necrosis factor alpha (TNF-α) and has yielded promising results when used to treat NASH patients [87]. Adams et al. performed a 12-month pilot-scale trial to assess the efficacy and safety of pentoxifylline (1,600 mg/day) in 20 patients with NASH. Alanine and aspartate aminotransferase levels were significantly reduced, as compared to those at baseline (84 ± 64 vs. 138 ± 76, $P=0.002$ and 58 ± 37 vs. 102 ± 62, $P=0.003$, respectively) [88]. However, nine patients withdrew from the study, primarily because of nausea, although no serious adverse event occurred [88].

18.4.4 Ursodeoxycholic Acid

UDCA is a naturally occurring bile acid with many hepatoprotective activities and was suggested to benefit 24 NASH patients given the compound for 12 months in the open-label study of Laurin et al. [89]. Later, the same authors conducted a large-scale randomised trial on 166 patients with liver biopsy-proven NASH given UDCA for 2 years. Liver enzyme levels decreased, but no change in the extent of steatosis, necroinflammation, or fibrosis was evident [90]. In contrast, UDCA given in combination with Vitamin E for 2 years yielded results differing from those afforded by UDCA + placebo in that both liver enzyme levels and histology improved in NASH patients [91].

18.4.5 Metformin

As insulin resistance plays roles in the pathogenesis of NAFLD and NASH, insulin-sensitising drugs including metformin have been tested in NASH patients [92–94]. Several open-label, RCTs [95–98] have shown that metformin reduced liver enzyme levels in, and improved histology of, NALFD/NASH patients, as also shown in other studies [92, 94, 96, 98].

18.4.6 N-*acetylcysteine*

NAC has been tested in two studies on NASH patients. Prescription of 600 mg/day NAC given in an open-label, prospective randomised study that ran for 4 weeks improved liver enzyme levels [99]. Another open-label prospective trial included 20 NASH patients given NAC 1.2 g/day and metformin 500 mg/day for 12 months. All of liver enzyme levels, the extent of insulin-resistance, body mass index, and liver histological findings (including steatosis and fibrosis), improved, although NAC had no effect on ballooning or inflammation [94].

In general, oxidative stress and antioxidant defences are both complicated networks of enzymatic and non-enzymatic processes that are not in balance as NALFD/NASH progresses. The effects of antioxidant therapy on such patients may depend on disease stage and severity. As the results of many antioxidant trials are contradictory, the utility of antioxidant therapy in those with NAFLD/NASH requires further evaluation, including the performance of large-scale studies using combinations of two or more antioxidants. Moreover, both lifestyle modifications and pharmaceutical interventions specifically targeting the principal signaling pathways involved not only in oxidative stress but also in inflammation, lipid peroxidation, and fibrosis should be tested.

18.5 Antioxidant Therapy for Liver Fibrosis and Cirrhosis

Cirrhosis, an advanced stage of liver fibrosis, may be reversible not only histologically [100], but also in terms of clinical outcomes [101]. Therefore, therapies that prevent or reverse cirrhosis are in great demand. Oxidative stress is pathogenetically associated with fibrosis development and progression via ROS-induced cellular injury and RNS-induced dysregulation of the hepatic microcirculation. Thus, antioxidants appear to afford great therapeutic potential as treatments of fibrosis/cirrhosis, provided that sufficient levels of antioxidant activity can be delivered to sites of injury within the liver. Experimental models of liver fibrosis/cirrhosis have been used to evaluate antioxidant compounds including food supplements and drugs including polyunsaturated phosphatidylcholine (PPC) [102], peroxisome proliferator-activated receptor (PPAR) α ligand [103], UDCA [104], and resveratrol [105–108].

SAMe, silymarin, and vitamin E have been tested in liver fibrosis/cirrhosis patients. SAMe is required for methylation of many substrates (DNA, proteins, lipids, and many small molecules) and polyamine synthesis. Thus, if the SAMe concentration falls below a certain level, or rises excessively, normal liver function will be compromised [109]. SAMe has exhibited beneficial effects when used to treat alcoholic liver cirrhosis, as discussed in Sect. 18.3.2 [74]. The results of the study cited therein indicated that long-term treatment with SAMe may improve survival or delay the need for liver transplantation in patients with alcoholic liver cirrhosis, especially those with less advanced liver disease. Moreover, treatment with SAMe seemed to be safe and free of side-effects [74]. In a double-blind, placebo-controlled, multicenter clinical trial performed on 220 patients with chronic liver disease (chronic active hepatitis and cirrhosis), SAMe significantly improved serum marker levels (bilirubin and alkaline phosphatase) and subjective symptoms (pruritus and fatigue) associated with cholestasis [110].

Treatment of hepatic cirrhosis with silibinin, the major active constituent of silymarin, improved antioxidant status, enhanced cytoprotection, reversed fibrosis, and stimulated regeneration. Dose-dependent decreases in hepatic enzyme activity after silibinin treatment have been reported [111]. In a placebo-controlled trial, patients with cirrhosis taking silibinin had higher total glutathione concentrations and exhibited concurrent decreases in the level of the N-terminal propeptide of type III collagen, a biomarker of hepatic fibrosis [70]. In RCTs, lower mortality rates have been documented when silibinin was given to patients with cirrhosis [112]. All-cause mortality of cirrhotic patients decreased by 4.4 % and mortality from liver disease by 7.3 % [113]. Although evidence supporting efficacy is, therefore, good, silibinin is not considered suitable for use as a sole treatment but, rather, as an adjunct drug indicated for treatment of a variety of acute and chronic diseases affecting liver function. The available information on silibinin disposition and pharmacodynamics in both small domestic animals and large animal herbivores is limited [114]. Additional pharmacokinetic and pharmacodynamic studies of potentially valuable agents in healthy animals are essential to accumulate evidence-based clinical data.

The effects of vitamin E in alcoholic or NASH patients with fibrosis or cirrhosis have been discussed in Sects. 18.4 and 18.5 above. The most promising results were afforded by the PIVEN trial [82] involving 247 patients treated for 96 months with vitamin E. Histological regression was evident, and fibrosis did not progress.

In summary, although antioxidant treatments may be used as adjunct therapies, prevention or reversion of liver fibrosis/cirrhosis requires elimination of the relevant etiologic agent or disruption of the pathogenic processes causing liver injury. There is as yet but slim evidence that antioxidant treatment alone can achieve these goals.

18.6 Conclusion

The antioxidant defence system is complex, involving the actions of intracellular enzymes, non-enzymatic substances that serve as scavengers, and dietary components. The defence system normally controls production of both ROS and RNS. Oxidative stress occurs when a significant imbalance develops between ROS and RNS production and removal. Thus, antioxidant therapy targets (1) recovery of antioxidant enzyme/compound levels, and (2) reduction in ROS and RNS production. Although antioxidant therapies have yielded valuable results in animal models, the results from human trials remain inconsistent. Today, only silymarin is recommended as a therapy for patients with ALD and vitamin E as an antioxidant for patients with NASH. The discrepancies in treatment outcomes after delivery of antioxidant therapies might be less were several deficiencies of most trials need to be overcome. These include small sample size, short treatment duration, and difficulties in defining clinical endpoints. Moreover, variations in the antioxidants used and the doses thereof fundamentally influence biological outcomes. Although it is difficult to design studies that overcome all of these limitations, especially when funding for such work is sparse, antioxidant therapies remain attractive and promising approaches for the treatment of liver diseases.

References

1. Irshad M (2002) Oxidative stress in liver diseases. Trop Gastroenterol 23(1):6–8
2. Valko M, Izakovic M, Mazur M, Rhodes CJ, Telser J (2004) Role of oxygen radicals in DNA damage and cancer incidence. Mol Cell Biochem 266(1–2):37–56
3. Gutteridge JM (1995) Lipid peroxidation and antioxidants as biomarkers of tissue damage. Clin Chem 41(12 Pt 2):1819–1828
4. Glantzounis GK, Salacinski HJ, Yang W, Davidson BR, Seifalian AM (2005) The contemporary role of antioxidant therapy in attenuating liver ischemia-reperfusion injury: a review. Liver Transpl 11(9):1031–1047
5. Fridovich I (1997) Superoxide anion radical (O2–.), superoxide dismutases, and related matters. J Biol Chem 272(30):18515–18517
6. Chelikani P, Fita I, Loewen PC (2004) Diversity of structures and properties among catalases. Cell Mol Life Sci 61(2):192–208

7. Goodsell DS (2004) Catalase. Molecular of the Month. RCSB Protein Data Bank (Online data base: http://www.rcsb.org/pdb/101/motm.do?momID=57)
8. Meister A (1988) Glutathione metabolism and its selective modification. J Biol Chem 263(33):17205–17208
9. De Vries N, De Flora S (1993) N-acetyl-l-cysteine. J Cell Biochem Suppl 17F:270–277
10. Rhee SG, Kang SW, Chang TS, Jeong W, Kim K (2001) Peroxiredoxin, a novel family of peroxidases. IUBMB Life 52(1–2):35–41
11. Kang SW, Chae HZ, Seo MS, Kim K, Baines IC, Rhee SG (1998) Mammalian peroxiredoxin isoforms can reduce hydrogen peroxide generated in response to growth factors and tumor necrosis factor-alpha. J Biol Chem 273(11):6297–6302
12. Hirota K, Murata M, Sachi Y, Nakamura H, Takeuchi J, Mori K et al (1999) Distinct roles of thioredoxin in the cytoplasm and in the nucleus. A two-step mechanism of redox regulation of transcription factor NF-kappaB. J Biol Chem 274(39):27891–27897
13. Hirota K, Nakamura H, Masutani H, Yodoi J (2002) Thioredoxin superfamily and thioredoxin-inducing agents. Ann N Y Acad Sci 957:189–199
14. Thomson CD (1985) Selenium-dependent and non-selenium-dependent glutathione peroxidase in human tissues of New Zealand residents. Biochem Int 10(4):673–679
15. Blake DR, Hall ND, Treby DA, Halliwell B, Gutteridge JM (1981) Protection against superoxide and hydrogen peroxide in synovial fluid from rheumatoid patients. Clin Sci (Lond) 61(4):483–486
16. Jacob RA (1995) The integrated antioxidant system. Nutr Res 15(6):755–766
17. Legrand D, Pierce A, Elass E, Carpentier M, Mariller C, Mazurier J (2008) Lactoferrin structure and functions. In: Bösze Z (ed) Bioactive components of milk. Springer, New York, pp 163–194
18. Melamed-Frank M, Lache O, Enav BI, Szafranek T, Levy NS, Ricklis RM et al (2001) Structure-function analysis of the antioxidant properties of haptoglobin. Blood 98(13):3693–3698
19. Atanasiu RL, Stea D, Mateescu MA, Vergely C, Dalloz F, Briot F et al (1998) Direct evidence of caeruloplasmin antioxidant properties. Mol Cell Biochem 189(1–2):127–135
20. Burton GW, Ingold KU (1989) Vitamin E as an in vitro and in vivo antioxidant. Ann N Y Acad Sci 570:7–22
21. Kojo S (2004) Vitamin C: basic metabolism and its function as an index of oxidative stress. Curr Med Chem 11(8):1041–1064
22. Halliwell B, Gutteridge JM (1990) The antioxidants of human extracellular fluids. Arch Biochem Biophys 280(1):1–8
23. Burton GW, Joyce A, Ingold KU (1982) First proof that vitamin E is major lipid-soluble, chain-breaking antioxidant in human blood plasma. Lancet 2(8293):327
24. Jang M, Cai L, Udeani GO, Slowing KV, Thomas CF, Beecher CW et al (1997) Cancer chemopreventive activity of resveratrol, a natural product derived from grapes. Science 275(5297):218–220
25. Kandaswami C, Middleton E Jr (1994) Free radical scavenging and antioxidant activity of plant flavonoids. Adv Exp Med Biol 366:351–376
26. de Whalley CV, Rankin SM, Hoult JR, Jessup W, Leake DS (1990) Flavonoids inhibit the oxidative modification of low density lipoproteins by macrophages. Biochem Pharmacol 39(11):1743–1750
27. Muntane J, De la Rosa AJ, Marin LM, Padillo FJ (2013) Nitric oxide and cell death in liver cancer cells. Mitochondrion 13(3):257–262
28. Zaslaver M, Offer S, Kerem Z, Stark AH, Weller JI, Eliraz A et al (2005) Natural compounds derived from foods modulate nitric oxide production and oxidative status in epithelial lung cells. J Agric Food Chem 53(26):9934–9939
29. Kawada N, Kristensen DB, Asahina K, Nakatani K, Minamiyama Y, Seki S et al (2001) Characterization of a stellate cell activation-associated protein (STAP) with peroxidase activity found in rat hepatic stellate cells. J Biol Chem 276(27):25318–25323
30. Sugimoto H, Makino M, Sawai H, Kawada N, Yoshizato K, Shiro Y (2004) Structural basis of human cytoglobin for ligand binding. J Mol Biol 339(4):873–885

31. Hodges NJ, Innocent N, Dhanda S, Graham M (2008) Cellular protection from oxidative DNA damage by over-expression of the novel globin cytoglobin in vitro. Mutagenesis 23(4):293–298
32. Fordel E, Thijs L, Martinet W, Lenjou M, Laufs T, Van Bockstaele D et al (2006) Neuroglobin and cytoglobin overexpression protects human SH-SY5Y neuroblastoma cells against oxidative stress-induced cell death. Neurosci Lett 410(2):146–151
33. Fordel E, Thijs L, Martinet W, Schrijvers D, Moens L, Dewilde S (2007) Anoxia or oxygen and glucose deprivation in SH-SY5Y cells: a step closer to the unraveling of neuroglobin and cytoglobin functions. Gene 398(1–2):114–122
34. Xu R, Harrison PM, Chen M, Li L, Tsui TY, Fung PC et al (2006) Cytoglobin overexpression protects against damage-induced fibrosis. Mol Ther 13(6):1093–1100
35. Medina J, Moreno-Otero R (2005) Pathophysiological basis for antioxidant therapy in chronic liver disease. Drugs 65(17):2445–2461
36. Yadav D, Hertan HI, Schweitzer P, Norkus EP, Pitchumoni CS (2002) Serum and liver micronutrient antioxidants and serum oxidative stress in patients with chronic hepatitis C. Am J Gastroenterol 97(10):2634–2639
37. Ko WS, Guo CH, Yeh MS, Lin LY, Hsu GS, Chen PC et al (2005) Blood micronutrient, oxidative stress, and viral load in patients with chronic hepatitis C. World J Gastroenterol 11(30): 4697–4702
38. Houglum K, Venkataramani A, Lyche K, Chojkier M (1997) A pilot study of the effects of d-alpha-tocopherol on hepatic stellate cell activation in chronic hepatitis C. Gastroenterology 113(4):1069–1073
39. Mahmood S, Yamada G, Niiyama G, Kawanaka M, Togawa K, Sho M et al (2003) Effect of vitamin E on serum aminotransferase and thioredoxin levels in patients with viral hepatitis C. Free Radic Res 37(7):781–785
40. von Herbay A, Stahl W, Niederau C, Sies H (1997) Vitamin E improves the aminotransferase status of patients suffering from viral hepatitis C: a randomized, double-blind, placebo-controlled study. Free Radic Res 27(6):599–605
41. Melhem A, Stern M, Shibolet O, Israeli E, Ackerman Z, Pappo O et al (2005) Treatment of chronic hepatitis C virus infection via antioxidants: results of a phase I clinical trial. J Clin Gastroenterol 39(8):737–742
42. Gabbay E, Zigmond E, Pappo O, Hemed N, Rowe M, Zabrecky G et al (2007) Antioxidant therapy for chronic hepatitis C after failure of interferon: results of phase II randomized, double-blind placebo controlled clinical trial. World J Gastroenterol 13(40):5317–5323
43. Groenbaek K, Friis H, Hansen M, Ring-Larsen H, Krarup HB (2006) The effect of antioxidant supplementation on hepatitis C viral load, transaminases and oxidative status: a randomized trial among chronic hepatitis C virus-infected patients. Eur J Gastroenterol Hepatol 18(9):985–989
44. Gane EJ, Weilert F, Orr DW, Keogh GF, Gibson M, Lockhart MM et al (2010) The mitochondria-targeted anti-oxidant mitoquinone decreases liver damage in a phase II study of hepatitis C patients. Liver Int 30(7):1019–1026
45. Look MP, Gerard A, Rao GS, Sudhop T, Fischer HP, Sauerbruch T et al (1999) Interferon/antioxidant combination therapy for chronic hepatitis C—a controlled pilot trial. Antiviral Res 43(2):113–122
46. Ideo G, Bellobuono A, Tempini S, Mondazzi L, Airoldi A, Benetti G et al (1999) Antioxidant drugs combined with alpha-interferon in chronic hepatitis C not responsive to alpha-interferon alone: a randomized, multicentre study. Eur J Gastroenterol Hepatol 11(11):1203–1207
47. Saeian K, Bajaj JS, Franco J, Knox JF, Daniel J, Peine C et al (2004) High-dose vitamin E supplementation does not diminish ribavirin-associated haemolysis in hepatitis C treatment with combination standard alpha-interferon and ribavirin. Aliment Pharmacol Ther 20(10):1189–1193
48. Beloqui O, Prieto J, Suarez M, Gil B, Qian CH, Garcia N et al (1993) N-acetyl cysteine enhances the response to interferon-alpha in chronic hepatitis C: a pilot study. J Interferon Res 13(4):279–282

49. Grant PR, Black A, Garcia N, Prieto J, Garson JA (2000) Combination therapy with interferon-alpha plus N-acetyl cysteine for chronic hepatitis C: a placebo controlled double-blind multicentre study. J Med Virol 61(4):439–442
50. Neri S, Ierna D, Antoci S, Campanile E, D'Amico RA, Noto R (2000) Association of alpha-interferon and acetyl cysteine in patients with chronic C hepatitis. Panminerva Med 42(3): 187–192
51. Seeff LB, Curto TM, Szabo G, Everson GT, Bonkovsky HL, Dienstag JL et al (2008) Herbal product use by persons enrolled in the hepatitis C Antiviral Long-Term Treatment Against Cirrhosis (HALT-C) Trial. HALT-C Trial Group. Hepatology 47(2):605–612
52. Morishima C, Shuhart MC, Wang CC, Paschal DM, Apodaca MC, Liu Y, et al (2010) Silymarin inhibits in vitro T-cell proliferation and cytokine production in hepatitis C virus infection. Gastroenterology 138(2):671–681, 681.e1–681.e2
53. Bonifaz V, Shan Y, Lambrecht RW, Donohue SE, Moschenross D, Bonkovsky HL (2009) Effects of silymarin on hepatitis C virus and haem oxygenase-1 gene expression in human hepatoma cells. Liver Int 29(3):366–373
54. Buzzelli G, Moscarella S, Giusti A, Duchini A, Marena C, Lampertico M (1993) A pilot study on the liver protective effect of silybin-phosphatidylcholine complex (IdB1016) in chronic active hepatitis. Int J Clin Pharmacol Ther Toxicol 31(9):456–460
55. Ferenci P, Scherzer TM, Kerschner H, Rutter K, Beinhardt S, Hofer H et al (2008) Silibinin is a potent antiviral agent in patients with chronic hepatitis C not responding to pegylated interferon/ribavirin therapy. Gastroenterology 135(5):1561–1567
56. Hawke RL, Schrieber SJ, Soule TA, Wen Z, Smith PC, Reddy KR et al (2010) Silymarin ascending multiple oral dosing phase I study in noncirrhotic patients with chronic hepatitis C. J Clin Pharmacol 50(4):434–449
57. Neumann UP, Biermer M, Eurich D, Neuhaus P, Berg T (2010) Successful prevention of hepatitis C virus (HCV) liver graft reinfection by silibinin mono-therapy. J Hepatol 52(6): 951–952
58. Huber R, Futter I, Ludtke R (2005) Oral silymarin for chronic hepatitis C—a retrospective analysis comparing three dose regimens. Eur J Med Res 10(2):68–70
59. El-Zayadi AR, Attia M, Badran HM, El-Tawil A, Zalata K, Barakat E et al (2005) Non-interferon-based therapy: an option for amelioration of necro-inflammation in hepatitis C patients who cannot afford interferon therapy. Liver Int 25(4):746–751
60. Gordon A, Hobbs DA, Bowden DS, Bailey MJ, Mitchell J, Francis AJ et al (2006) Effects of Silybum marianum on serum hepatitis C virus RNA, alanine aminotransferase levels and well-being in patients with chronic hepatitis C. J Gastroenterol Hepatol 21(1 Pt 2):275–280
61. Polyak SJ, Ferenci P, Pawlotsky JM (2013) Hepatoprotective and antiviral functions of silymarin components in hepatitis C virus infection. Hepatology 57(3):1262–1271
62. de la Maza MP, Petermann M, Bunout D, Hirsch S (1995) Effects of long-term vitamin E supplementation in alcoholic cirrhotics. J Am Coll Nutr 14(2):192–196
63. Mezey E, Potter JJ, Rennie-Tankersley L, Caballeria J, Pares A (2004) A randomized placebo controlled trial of vitamin E for alcoholic hepatitis. J Hepatol 40(1):40–46
64. Li J, Kim CI, Leo MA, Mak KM, Rojkind M, Lieber CS (1992) Polyunsaturated lecithin prevents acetaldehyde-mediated hepatic collagen accumulation by stimulating collagenase activity in cultured lipocytes. Hepatology 15(3):373–381
65. Lieber CS, Leo MA, Aleynik SI, Aleynik MK, DeCarli LM (1997) Polyenylphosphatidylcholine decreases alcohol-induced oxidative stress in the baboon. Alcohol Clin Exp Res 21(2):375–379
66. Aleynik SI, Leo MA, Ma X, Aleynik MK, Lieber CS (1997) Polyenylphosphatidylcholine prevents carbon tetrachloride-induced lipid peroxidation while it attenuates liver fibrosis. J Hepatol 27(3):554–561
67. Lieber CS, Weiss DG, Groszmann R, Paronetto F, Schenker S, Veterans Affairs Cooperative Study 391 Group (2003) II. Veterans Affairs Cooperative Study of polyenylphosphatidylcholine in alcoholic liver disease. Alcohol Clin Exp Res 27(11):1765–1772

68. Salmi HA, Sarna S (1982) Effect of silymarin on chemical, functional, and morphological alterations of the liver. A double-blind controlled study. Scand J Gastroenterol 17(4): 517–521
69. Lirussi F, Beccarello A, Zanette G, De Monte A, Donadon V, Velussi M et al (2002) Silybin-beta-cyclodextrin in the treatment of patients with diabetes mellitus and alcoholic liver disease. Efficacy study of a new preparation of an anti-oxidant agent. Diabetes Nutr Metab 15(4):222–231
70. Lucena MI, Andrade RJ, de la Cruz JP, Rodriguez-Mendizabal M, Blanco E, Sanchez de la Cuesta F (2002) Effects of silymarin MZ-80 on oxidative stress in patients with alcoholic cirrhosis. Results of a randomized, double-blind, placebo-controlled clinical study. Int J Clin Pharmacol Ther 40(1):2–8
71. Bunout D, Hirsch S, Petermann M, de la Maza MP, Silva G, Kelly M et al (1992) [Controlled study of the effect of silymarin on alcoholic liver disease]. Rev Med Chil 120(12): 1370–1375
72. Pares A, Planas R, Torres M, Caballeria J, Viver JM, Acero D et al (1998) Effects of silymarin in alcoholic patients with cirrhosis of the liver: results of a controlled, double-blind, randomized and multicenter trial. J Hepatol 28(4):615–621
73. Vendemiale G, Altomare E, Trizio T, Le Grazie C, Di Padova C, Salerno MT et al (1989) Effects of oral S-adenosyl-L-methionine on hepatic glutathione in patients with liver disease. Scand J Gastroenterol 24(4):407–415
74. Mato JM, Camara J, Fernandez de Paz J, Caballeria L, Coll S, Caballero A et al (1999) S-adenosylmethionine in alcoholic liver cirrhosis: a randomized, placebo-controlled, double-blind, multicenter clinical trial. J Hepatol 30(6):1081–1089
75. Videla LA, Rodrigo R, Orellana M, Fernandez V, Tapia G, Quinones L et al (2004) Oxidative stress-related parameters in the liver of non-alcoholic fatty liver disease patients. Clin Sci (Lond) 106(3):261–268
76. Koek GH, Liedorp PR, Bast A (2011) The role of oxidative stress in non-alcoholic steatohepatitis. Clin Chim Acta 412(15–16):1297–1305
77. Rolo AP, Teodoro JS, Palmeira CM (2012) Role of oxidative stress in the pathogenesis of nonalcoholic steatohepatitis. Free Radic Biol Med 52(1):59–69
78. Zein CO, Yerian LM, Gogate P, Lopez R, Kirwan JP, Feldstein AE et al (2011) Pentoxifylline improves nonalcoholic steatohepatitis: a randomized placebo-controlled trial. Hepatology 54(5):1610–1619
79. Hasegawa T, Yoneda M, Nakamura K, Makino I, Terano A (2001) Plasma transforming growth factor-beta1 level and efficacy of alpha-tocopherol in patients with non-alcoholic steatohepatitis: a pilot study. Aliment Pharmacol Ther 15(10):1667–1672
80. Kawanaka M, Mahmood S, Niiyama G, Izumi A, Kamei A, Ikeda H et al (2004) Control of oxidative stress and reduction in biochemical markers by vitamin E treatment in patients with nonalcoholic steatohepatitis: a pilot study. Hepatol Res 29(1):39–41
81. Lavine JE (2000) Vitamin E treatment of nonalcoholic steatohepatitis in children: a pilot study. J Pediatr 136(6):734–738
82. Sanyal AJ, Chalasani N, Kowdley KV, McCullough A, Diehl AM, Bass NM et al (2010) Pioglitazone, vitamin E, or placebo for nonalcoholic steatohepatitis. N Engl J Med 362(18):1675–1685
83. Harrison SA, Torgerson S, Hayashi P, Ward J, Schenker S (2003) Vitamin E and vitamin C treatment improves fibrosis in patients with nonalcoholic steatohepatitis. Am J Gastroenterol 98(11):2485–2490
84. Nobili V, Manco M, Devito R, Ciampalini P, Piemonte F, Marcellini M (2006) Effect of vitamin E on aminotransferase levels and insulin resistance in children with non-alcoholic fatty liver disease. Aliment Pharmacol Ther 24(11–12):1553–1561
85. Abdelmalek MF, Angulo P, Jorgensen RA, Sylvestre PB, Lindor KD (2001) Betaine, a promising new agent for patients with nonalcoholic steatohepatitis: results of a pilot study. Am J Gastroenterol 96(9):2711–2717

86. Abdelmalek MF, Sanderson SO, Angulo P, Soldevila-Pico C, Liu C, Peter J et al (2009) Betaine for nonalcoholic fatty liver disease: results of a randomized placebo-controlled trial. Hepatology 50(6):1818–1826
87. Satapathy SK, Garg S, Chauhan R, Sakhuja P, Malhotra V, Sharma BC et al (2004) Beneficial effects of tumor necrosis factor-[alpha] inhibition by pentoxifylline on clinical, biochemical, and metabolic parameters of patients with nonalcoholic steatohepatitis. Am J Gastroenterol 99(10):1946–1952
88. Adams LA, Zein CO, Angulo P, Lindor KD (2004) A pilot trial of pentoxifylline in nonalcoholic steatohepatitis. Am J Gastroenterol 99(12):2365–2368
89. Laurin J, Lindor KD, Crippin JS, Gossard A, Gores GJ, Ludwig J et al (1996) Ursodeoxycholic acid or clofibrate in the treatment of non-alcohol-induced steatohepatitis: a pilot study. Hepatology 23(6):1464–1467
90. Lindor KD, Kowdley KV, Heathcote EJ, Harrison ME, Jorgensen R, Angulo P et al (2004) Ursodeoxycholic acid for treatment of nonalcoholic steatohepatitis: results of a randomized trial. Hepatology 39(3):770–778
91. Dufour JF, Oneta CM, Gonvers JJ, Bihl F, Cerny A, Cereda JM et al (2006) Randomized placebo-controlled trial of ursodeoxycholic acid with vitamin e in nonalcoholic steatohepatitis. Clin Gastroenterol Hepatol 4(12):1537–1543
92. Loomba R, Lutchman G, Kleiner DE, Ricks M, Feld JJ, Borg BB et al (2009) Clinical trial: pilot study of metformin for the treatment of non-alcoholic steatohepatitis. Aliment Pharmacol Ther 29(2):172–182
93. Marchesini G, Brizi M, Bianchi G, Tomassetti S, Zoli M, Melchionda N (2001) Metformin in non-alcoholic steatohepatitis. Lancet 358(9285):893–894
94. de Oliveira CP, Stefano JT, de Siqueira ER, Silva LS, de Campos Mazo DF, Lima VM et al (2008) Combination of N-acetylcysteine and metformin improves histological steatosis and fibrosis in patients with non-alcoholic steatohepatitis. Hepatol Res 38(2):159–165
95. Haukeland JW, Konopski Z, Eggesbo HB, von Volkmann HL, Raschpichler G, Bjoro K et al (2009) Metformin in patients with non-alcoholic fatty liver disease: a randomized, controlled trial. Scand J Gastroenterol 44(7):853–860
96. Shields WW, Thompson KE, Grice GA, Harrison SA, Coyle WJ (2009) The effect of metformin and standard therapy versus standard therapy alone in nondiabetic patients with insulin resistance and nonalcoholic steatohepatitis (NASH): a pilot trial. Therap Adv Gastroenterol 2(3):157–163
97. Idilman R, Mizrak D, Corapcioglu D, Bektas M, Doganay B, Sayki M et al (2008) Clinical trial: insulin-sensitizing agents may reduce consequences of insulin resistance in individuals with non-alcoholic steatohepatitis. Aliment Pharmacol Ther 28(2):200–208
98. Uygun A, Kadayifci A, Isik AT, Ozgurtas T, Deveci S, Tuzun A et al (2004) Metformin in the treatment of patients with non-alcoholic steatohepatitis. Aliment Pharmacol Ther 19(5):537–544
99. Pamuk GE, Sonsuz A (2003) N-acetylcysteine in the treatment of non-alcoholic steatohepatitis. J Gastroenterol Hepatol 18(10):1220–1221
100. Friedman SL, Bansal MB (2006) Reversal of hepatic fibrosis—fact or fantasy? Hepatology 43(2 Suppl 1):S82–S88
101. Mallet V, Gilgenkrantz H, Serpaggi J, Verkarre V, Vallet-Pichard A, Fontaine H et al (2008) Brief communication: the relationship of regression of cirrhosis to outcome in chronic hepatitis C. Ann Intern Med 149(6):399–403
102. Lieber CS, Robins SJ, Li J, DeCarli LM, Mak KM, Fasulo JM et al (1994) Phosphatidylcholine protects against fibrosis and cirrhosis in the baboon. Gastroenterology 106(1):152–159
103. Toyama T, Nakamura H, Harano Y, Yamauchi N, Morita A, Kirishima T et al (2004) PPARalpha ligands activate antioxidant enzymes and suppress hepatic fibrosis in rats. Biochem Biophys Res Commun 324(2):697–704
104. Serviddio G, Pereda J, Pallardo FV, Carretero J, Borras C, Cutrin J et al (2004) Ursodeoxycholic acid protects against secondary biliary cirrhosis in rats by preventing mitochondrial oxidative stress. Hepatology 39(3):711–720

105. Kawada N, Seki S, Inoue M, Kuroki T (1998) Effect of antioxidants, resveratrol, quercetin, and N-acetylcysteine, on the functions of cultured rat hepatic stellate cells and Kupffer cells. Hepatology 27(5):1265–1274
106. Di Pascoli M, Divi M, Rodriguez-Vilarrupla A, Rosado E, Gracia-Sancho J, Vilaseca M et al (2013) Resveratrol improves intrahepatic endothelial dysfunction and reduces hepatic fibrosis and portal pressure in cirrhotic rats. J Hepatol 58(5):904–910
107. Lee ES, Shin MO, Yoon S, Moon JO (2010) Resveratrol inhibits dimethylnitrosamine-induced hepatic fibrosis in rats. Arch Pharm Res 33(6):925–932
108. Chan CC, Cheng LY, Lin CL, Huang YH, Lin HC, Lee FY (2011) The protective role of natural phytoalexin resveratrol on inflammation, fibrosis and regeneration in cholestatic liver injury. Mol Nutr Food Res 55(12):1841–1849
109. Mato JM, Martinez-Chantar ML, Lu SC (2013) S-adenosylmethionine metabolism and liver disease. Ann Hepatol 12(2):183–189
110. Frezza M, Surrenti C, Manzillo G, Fiaccadori F, Bortolini M, Di Padova C (1990) Oral S-adenosylmethionine in the symptomatic treatment of intrahepatic cholestasis. A double-blind, placebo-controlled study. Gastroenterology 99(1):211–215
111. Kidd P, Head K (2005) A review of the bioavailability and clinical efficacy of milk thistle phytosome: a silybin-phosphatidylcholine complex (Siliphos). Altern Med Rev 10(3): 193–203
112. Ferenci P, Dragosics B, Dittrich H, Frank H, Benda L, Lochs H et al (1989) Randomized controlled trial of silymarin treatment in patients with cirrhosis of the liver. J Hepatol 9(1):105–113
113. Saller R, Brignoli R, Melzer J, Meier R (2008) An updated systematic review with meta-analysis for the clinical evidence of silymarin. Forsch Komplementmed 15(1):9–20
114. Hackett ES, Twedt DC, Gustafson DL (2013) Milk thistle and its derivative compounds: a review of opportunities for treatment of liver disease. J Vet Intern Med 27(1):10–16
115. Morisco F, Vitaglione P, Carbone A, Stingo S, Scarpati S, Ascione A et al (2004) Tomato-based functional food as interferon adjuvant in HCV eradication therapy. J Clin Gastroenterol 38(6 Suppl):S118–S120
116. Hino K, Murakami Y, Nagai A, Kitase A, Hara Y, Furutani T et al (2006) Alpha-tocopherol [corrected] and ascorbic acid attenuates the ribavirin [corrected] induced decrease of eicosapentaenoic acid in erythrocyte membrane in chronic hepatitis C patients. J Gastroenterol Hepatol 21(8):1269–1275
117. Falasca K, Ucciferri C, Mancino P, Vitacolonna E, De Tullio D, Pizzigallo E et al (2008) Treatment with silybin-vitamin E-phospholipid complex in patients with hepatitis C infection. J Med Virol 80(11):1900–1906
118. Feld JJ, Modi AA, El-Diwany R, Rotman Y, Thomas E, Ahlenstiel G et al (2011) S-adenosyl methionine improves early viral responses and interferon-stimulated gene induction in hepatitis C nonresponders. Gastroenterology 140(3):830–839
119. Filipowicz M, Bernsmeier C, Terracciano L, Duong FH, Heim MH (2010) S-adenosyl-methionine and betaine improve early virological response in chronic hepatitis C patients with previous nonresponse. PLoS One 5(11):e15492
120. Yakoot M, Salem A (2012) Spirulina platensis versus silymarin in the treatment of chronic hepatitis C virus infection. A pilot randomized, comparative clinical trial. BMC Gastroenterol 12:32
121. Biermer M, Schlosser B, Fulop B, van Bommel F, Brodzinski A, Heyne R et al (2012) High-dose silibinin rescue treatment for HCV-infected patients showing suboptimal virologic response to standard combination therapy. J Viral Hepat 19(8):547–553
122. Fried MW, Navarro VJ, Afdhal N, Belle SH, Wahed AS, Hawke RL et al (2012) Effect of silymarin (milk thistle) on liver disease in patients with chronic hepatitis C unsuccessfully treated with interferon therapy: a randomized controlled trial. JAMA 308(3):274–282
123. Adeyemo O, Doi H, Rajender Reddy K, Kaplan DE (2013) Impact of oral silymarin on virus- and non-virus-specific T-cell responses in chronic hepatitis C infection. J Viral Hepat 20(7): 453–462

Part IV
Oxidative Stress and Liver Cancer

Chapter 19
Liver Cancer Classification

Maria Guido, Samantha Sarcognato, Matteo Fassan, and Marco Pizzi

19.1 Introduction

Cancer classification basically aims to provide clinically useful information regarding the outcome as well as the therapeutic approach. A standardized classification is also the prerequisite for designing clinical trials and is necessary for any medical knowledge improvement.

Every pathological classification, ideally, should rely on tumor ontogeny and should integrate histological features with immunohistochemical and molecular data, in order to identify clinically relevant diagnostic categories.

In recent years, this general approach has also been applied to liver neoplasms. Traditional clinico-pathological classification has been thus flanked by molecular studies that thoroughly investigated the molecular biology of liver tumors. These studies have led to significant progresses in our understanding of liver carcinogenesis, despite nowadays a definite integration of molecular, clinical, and pathological features of liver tumors is still lacking.

The current classification of primary liver neoplasms distinguishes each disease entity according to its resemblance to non-neoplastic liver cells and splits them into two main diagnostic categories: (1) hepatocellular carcinoma (HCC), which is composed by cells closely resembling non-neoplastic hepatocytes; and (2) cholangiocarcinoma (CC), which represents the neoplastic counterpart of native bile duct cells.

M. Guido (✉) • S. Sarcognato • M. Pizzi • M. Fassan
Department of Medicine, General Pathology and Cytopathology Unit,
University of Padova, Padova, Italy
e-mail: mguido@unipd.it

E. Albano, M. Parola (eds.), *Studies on Hepatic Disorders*,
Oxidative Stress in Applied Basic Research and Clinical Practice,
DOI 10.1007/978-3-319-15539-5_19

HCC and CC are heterogeneous tumors with several variants that are characterized by peculiar morphological, immunohistochemical, and molecular features. These biological differences correspond to variable clinical behavior and response to therapy. In the following paragraphs, the current definitions and classification of adult primary liver cancers will be outlined.

19.2 Historical Notes

Despite liver masses have been documented since long time, a comprehensive and thorough classification of such lesions has been developed only at the beginning of the twentieth century. G.B. Morgagni himself, in the eighteenth century, could not appropriately discriminate between neoplastic and non-neoplastic liver masses, his knowledge being founded on clinical and macroscopic grounds only. Distinction between primary and secondary (i.e., metastatic) liver neoplasms was even less feasible. It was only at the end of the nineteenth century, with the foundation of histopathology, that reliable and reproducible criteria for the classification of liver tumors were put into place. Primary tumors could be readily distinguished, for the first time, from metastatic disease, and clear-cut concepts regarding liver neoplasms were proposed.

In 1954, the publication of Edmondson and Steiner's seminal paper on "primary carcinoma of the liver" set the basis for any modern classification and grading-system of liver neoplasms [1]. In particular, the histological patterns identified by Edmondson and Steiner still represent the fundament for the modern classification of primary liver tumors. Likewise, the originally proposed grading-system for hepatic malignancies has so deeply influenced the clinical management of patients that it is still in use nowadays in the clinical practice.

Edmondson and Steiner classified primary liver tumors into three main groups, according to their microscopic appearance: (1) liver cell cancers, composed by neoplastic cells very similar to hepatocytes; (2) bile duct cell cancers (with their hilar variant), characterized by an overall glandular architecture; and (3) cancers with features intermediate between liver cell and bile duct cell carcinomas. The latter group had also been previously described by several authors [2], who variously referred to them as "mixed" or "duplex-type tumors," "intermediate-type carcinomas," "carcinoma of dual origin," "cholangiohepatoma," "hepatobiliary carcinoma," and "combined liver cell and bile duct carcinoma."

The extreme morphological heterogeneity of liver carcinomas and the existence of mixed tumors led to propose many hypotheses regarding their cell of origin. Some authors [3] considered hepatocellular and cholangiocellular carcinoma as mere structural variants within the same group of tumors. Such hypothesis was backed up by the common embryological origin of hepatocytes and biliocytes and by the frequent phenotype exchange between hepatocytes and cholangiocytes (i.e., cholangiocellular metaplasia of hepatocytes and hepatocellular metaplasia of cholangiocytes).

This holistic view was criticized by other authors, considering HCC and cholangiocarcinoma as completely unrelated diseases. Such opinion was mainly based on the histopathology of tissue samples ("the origin of hepatocellular and cholangiocellular carcinomas from liver cells and bile ducts respectively is abundantly shown by their microscopic appearances" [2]) and on pathophysiological considerations (the phenotype of neoplastic cells was interpreted based on their mature cell counterpart, rather than their embryological origin).

The origin of hepatocellular and cholangiocellular tumors was surrounded by so many uncertainties that it led several authors to adopt a more prudent positions: among others, Edmondson and Steiner still considered the histogenesis of liver neoplasms an open dilemma and did not propose any interpretation for this specific issue [1]. However, the uncertainty in the pathogenesis did not prevent the medical community to propose international classifications of liver tumors.

In 1957, the World Health Organization (WHO), in particular, took the resolution to standardize the histological definitions, diagnostic criteria, and nomenclature of all neoplastic diseases. This endeavor resulted in the publication of a multivolume series, the International Histological Classification of Tumors (currently known as the "blue books"), that also encompassed gastrointestinal and liver neoplasms. The WHO blue book series (first published between 1967 and 1981) had undergone periodical updates: the fourth (and last) edition of gastrointestinal, pancreatic, and liver tumors has been licensed in 2010 [4] (Table 19.1). Compared to the first edition, many modifications have been introduced regarding both the terminology of premalignant biliary lesions and the definition and subclassification of HCC and mixed tumors.

19.3 Hepatocellular Carcinoma

HCC accounts for 75 % of primary liver malignancies and represents the sixth most common cancer worldwide (estimate incidence: 500,000–1,000,000 cases/year; higher incidence rates in the male gender). Overall, HCC ranks as the third cause of cancer-related death, being the most lethal tumor among cirrhotic patients [4].

The incidence and mortality of HCC feature consistent geographical variability. Such changes are mainly due to differences in the exposure to HCC-carcinogenetic factors. The disease burden is highest in areas with endemic HBV-infection (sub-Saharan Africa and Eastern Asia; incidence rates >20/100,000/year) and lowest in North- and South-America (<5/100,000/year). Mediterranean countries, such as Italy, Spain, and Greece, occupy an intermediate position, with an incidence rate ranging from 10 to 20/100,000/year [5].

Regions with high disease burden are characterized by a mean age at diagnosis significantly lower than that of North America and Europe (peak incidence at 20–40 and 60 years, respectively). Again, this discrepancy can be explained by significant differences in the exposure to HCC etiological agents. In African and Asian countries, the leading cause of HCC is indeed represented by at-birth HBV infection

Table 19.1 Current WHO classification of malignant epithelial liver tumors

Malignant epithelial tumors: hepatocellular
Malignancy-associated and premalignant lesions
– Large cell change (formerly "dysplasia")
– Small cell change (formerly "dysplasia")
– Dysplastic nodules
– Low grade
– High grade
Malignant
– Hepatocellular carcinoma
– Hepatocellular carcinoma, fibrolamellar variant
– Hepatoblastoma, epithelial variants
– Undifferentiated carcinoma
Malignant epithelial tumors: biliary
Premalignant lesions
– Biliary intraepithelial neoplasia, grade 3 (BilIn-3)
– Intraductal papillary neoplasm with low- or intermediate-grade intraepithelial neoplasia
– Intraductal papillary neoplasm with high-grade intraepithelial neoplasia
– Mucinus cystic neoplasm with low- or intermediate-grade intraepithelial neoplasia
– Mucinus cystic neoplasm with high-grade intraepithelial neoplasia
Malignant
– Intrahepatic cholangiocarcinoma
– Intraductal papillary neoplasm with an associated invasive carcinoma
– Mucinus cystic neoplasm with an associated invasive carcinoma
Malignancies of mixed or uncertain origin
– Calcifying nested epithelial stromal tumor
– Carcinosarcoma
– Combined hepatocellular–cholangiocarcinoma
– Hepatoblastoma, mixed epithelial-mesenchymal

and almost 50 % of cases develop in non-cirrhotic livers. Conversely, in Europe and North America, the development of HCC is mainly attributable to late-onset HCV infection, which usually ensues in early adulthood [6]. The vast majority of such cases (>90 %) occurs in the setting of liver cirrhosis.

Ethanol abuse is a well-known risk factor for the development of liver cirrhosis and HCC, despite its potential direct carcinogenic effect is still under debate. Notably, epidemiological evidence indicates that ethanol potentiates the carcinogenic activity of both HCV and HBV, possibly accelerating liver fibrosis and cirrhotic transformation.

In Southern China and sub-Saharan Africa, another important environmental hazard is represented by aflatoxin ingestion. Aflatoxins are a group of structurally related mycotoxins, produced by several *Aspergillus* species and metabolized to carcinogens by liver enzymes. Aflatoxin may contaminate grains, corn, cassava, peanuts, and fermented soybeans, especially when stored at high-moisture conditions. When associated with other HCC-risk factors (e.g., chronic HBV-infection), aflatoxins bear a 60-fold increase in the risk of tumor development [7].

Finally, even systemic metabolic disorders have been linked to liver neoplastic transformation. Non-alcoholic fatty liver disease (NAFLD) and metabolic syndrome (i.e., a systemic metabolic disorder secondary to obesity or diabetes mellitus) have been indeed associated with the development of HCC, especially when cirrhosis has fully developed [8]. Other liver metabolic disorders, such as hereditary hemochromatosis and α-1-antitrypsin deficiency, have been also associated with an increased risk of developing HCC.

19.3.1 Clinical–Pathological Classification

According to the clinical-pathologic profile, two forms of HCC are recognized: (1) early HCC (eHCC); and (2) progressed HCC (pHCC). Such distinction, far from being of only academic importance, is justified by Japanese studies, reporting significant differences in the prognosis of the two entities. eHCC, in fact, is characterized by a significantly longer time-to-recurrence and by a higher overall survival than pHCC [9].

19.3.1.1 Early HCC

The definition of eHCC has been long debated and significant differences did come up between Eastern and Western pathologists regarding the histological criteria for its diagnosis [10]. It was only in recent years that a consensus conference (held under the auspices of the Laennec Liver Pathology Society) solved the discrepancies concerning the diagnosis of such disease [11].

Early HCC is currently defined as a well-differentiated HCC <2 cm, characterized by a vaguely nodular appearance and by an incomplete arterialized circulation. The very low malignant potential of this tumor may lead to consider eHCC as biologically related to other (extra-hepatic) early-stage neoplasms, named as “microinvasive carcinomas,” as such occupying an halfway position between in situ and invasive carcinoma [12].

The macroscopic appearance of eHCC is mainly related to the growth pattern of neoplastic cells, which replace (rather than push aside) the surrounding hepatocyte cords. The incomplete arterialized circulation, with venous vessels still supplying blood, interferes with the sensibility of traditional imaging techniques (CT and NMR), which largely depends on vascular patterns to distinguish benign from malignant lesions. In fact, at CT scans, most eHCCs appear as hypovascular nodular lesions, while at NMR, eHCC is characterized by different patterns that vary according to sequences used: on T1-weighted images, the tumor appears as an isointense

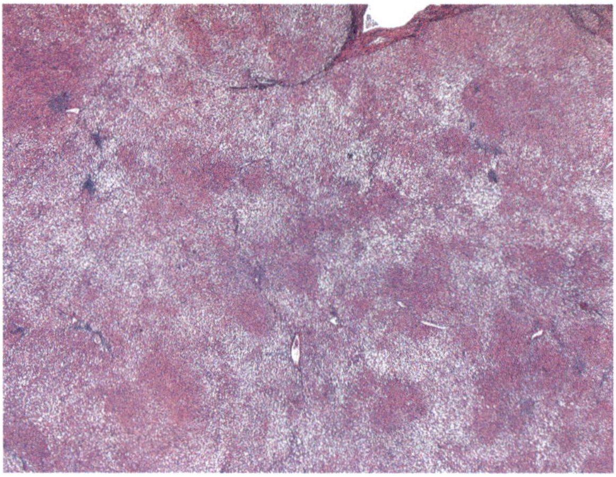

Fig. 19.1 Early-HCC showing diffuse steatosis. Some portal tracts are evident within the tumor. H&E, original magnification 20×

or hyperintense lesion; on T2-weighted images, it is characterized by a mostly isointense pattern [12].

On histology, eHCC is a well-differentiated tumor, with cords of hepatocytes very similar to normal liver, often with diffuse fatty changes, documented in up to 40 % of cases (Fig. 19.1).

Vascular invasion is rarely (if ever) observed, which confirms the extremely low metastatic potential of this tumor [13]. The main differential diagnosis of eHCC is represented by high-grade dysplastic nodule (HGDN). Differentiating HGDN from eHCC can be extremely difficult on morphological grounds only. HGDN and eHCC have indeed several common histological features (i.e., increased cell density compared to the surrounding tissue; presence of intra-tumoral portal tracts; pseudo-glandular growth pattern; diffuse fatty changes; presence of unpaired arteries) [13, 14] and stromal invasion represents the single, most helpful, parameter to differentiate these lesions. Stromal invasion is defined as tumor cell invasion into the portal tracts or fibrous septa within vaguely nodular lesions [10]. In the absence of an evident stromal invasion, additional immunohistochemical stains can be helpful to make the correct diagnosis. Glypican-3 (GPC-3) is a relatively specific marker for HCC in both resected specimens and liver biopsy samples (sensitivity >75 %; specificity >95 %) [15], but, on the other hand, a major limit of GPC-3 immunostaining is represented by its heterogeneous positivity in most HCC cases. This bias can be partially overcome by combining GPC-3 immunostaining with other HCC markers (i.e., heat shock protein 70 and glutamine synthetase) and the positivity for ≥2 of these is indeed highly specific of HCC [16, 17]. In addition, gene expression profiling (GEP) has identified a 3-gene signature [18] encompassing GPC-3, lymphatic vessel endothelial receptor, and survivin, capable to discriminate HGDN and eHCC with a 94 % accuracy, but unfortunately, the application of GEP technology to the clinical practice is still very limited.

19.3.1.2 Progressed HCC

According to the current definitions, pHCC is a well-demarcated tumor which often features a fibrous capsule and is moderately differentiated in most cases. Progressed HCC may develop from a preexisting dysplastic focus or nodule or from an eHCC. The size (<2 cm vs. >2 cm) does not represent a discriminating parameter between eHCC and pHCC, as at the earliest stage, progressed HCC can be as small as less than 2 cm, like eHCC. pHCC lacks portal tracts and its blood supply is entirely provided by newly formed arterial vessels, justifying the characteristic vascular pattern at CT scans (hypervascular appearance with washout in the portal venous phase) [12].

Gross Classification

Despite several macroscopic classifications of pHCC have been proposed, none of them have proven to be of any significant clinical or prognostic relevance. The still in use pHCC gross classification, proposed more than a century ago by Eggel [19], comprises three major cancer subtypes: (1) nodular, (2) massive, and (3) diffuse HCC.

The nodular pattern is the most frequent subtype (usually observed in cirrhotic livers). Nodules may be solitary or multiple, with the latter possibly representing either synchronous-independent tumors or intra-hepatic metastatic diffusion. Multiple nodules should be distinguished from satellite lesions (i.e., nodules <2 cm surrounding a major tumor mass), associated with an adverse prognosis.

The massive subtype consists of large, poorly circumscribed nodules that may occupy an entire hepatic lobe (commonly reported in non-cirrhotic livers).

Finally, the diffuse HCC is the least common subtype characterized by several small nodules, virtually replacing the entire liver and mimicking cirrhosis. Diffuse HCC may represent a true challenge for the clinical/radiological diagnosis [20].

Histologic Classifications

pHCC, in its classical form, is composed of tumor cells recalling normal hepatocytes, separated by sinusoidal-like spaces lined by a single layer of endothelial cells (Fig. 19.2) that stain positive for endothelial markers, such as CD34 and factor VIII. This feature may prove useful in the differential diagnosis of well-differentiated tumors [21].

Sometimes, pHCC displays also diffuse hepatocyte steatosis, even if the prevalence of such changes is significantly lower than those in eHCC. As a general rule, fatty change inversely correlates with the tumor size and is relatively uncommon in HCCs larger than 3 cm [22]. Microvascular invasion (mVI) is another relatively common feature of pHCC. Recent studies indicate that the pattern of mVI variably correlates with prognosis: in particular, invasion of vessels with a muscular

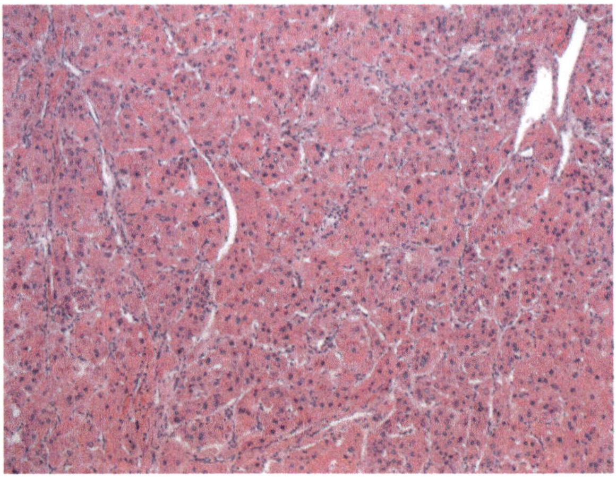

Fig. 19.2 Classical HCC with trabecular pattern. H&E, original magnification 100×

tonaca media and/or more than 1 cm away from the tumor mass has been associated with higher recurrence rates and decreased overall survival [23].

Based on the architectural patterns, HCC is traditionally classified into: (1) trabecular, (2) pseudoglandular, and (3) compact type [22]. These patterns may frequently coexist within the same tumor and correlate with the neoplastic grade of differentiation: the trabecular pattern is usually seen in well-differentiated HCC, while the compact pattern is more frequently reported in moderately to poorly differentiated tumors. Although the architectural classification does not represent an independent prognostic predictor, its knowledge is of pivotal importance to correctly resolve differential diagnosis problems.

Cytological variants of pHCC have been also described [24] and include: (1) clear-cell, (2) pleomorphic, and (3) spindle-cell variant. With the exception of the spindle-cell type (which represents a sarcomatous, poorly differentiated tumor), these cytological variants do not affect the HCC clinical behavior. Again, the rationale for their identification largely relies on possible problems in the differential diagnosis with metastases of tumors of extra-hepatic origin.

As for the immunohistochemical features, about 90 % of HCCs do express Heppar1 antigen (carbamoyl phosphate synthetase-1) [24] with a decreasing positivity in less differentiated tumors. Positive immunostain for cytocheratin 19 (CK19, a marker of stemness) has been correlated with poorer prognosis, higher recurrence rates, and lymph node metastases [25]. A recent international study [26] has indeed demonstrated that CK19-positive HCCs strongly express invasion- and metastasis-related markers, biliary/progenitor cell markers, and members of the miR-200 family (e.g., miR-141 and miR-200c). These data have been confirmed by in vitro analyses, reporting increased invasiveness and resistance to conventional chemotherapy for CK19-positive human HCC tumor cells. These observations may lead to consider CK19-positive HCCs as a separate and more aggressive clinicopathological entity [26].

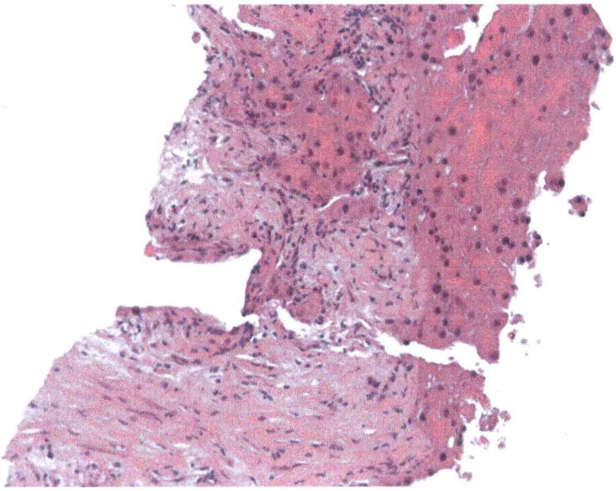

Fig. 19.3 Needle biopsy showing fibrolamellar HCC composed of large eosinophils (oncocytic) cells within fibrous stroma. H&E, original magnification 100×

Beside the classical forms of pHCC, the fibrolamellar variant deserves special consideration. Fibrolamellar HCC (FLC) is a rare entity, representing approximately 5 % of all HCCs (range: 1–8 % of cases, depending on the studies and populations considered) [27]. FLC has peculiar clinical, morphological, and molecular features, as it occurs most frequently in non-cirrhotic livers of children and young adults, with no history of chronic liver disease, and unlike classical HCC, it does not show any sex predominance [27, 28].

On gross examination, the majority of FLCs appears as a yellow to pale nodule with a central scar. Histologically, the tumor consists of nests and cords of large and polygonal hepatocytes surrounded by a collagenous, lamellar stroma (Fig. 19.3). At immunohistochemistry, FLC is positive for HepPar-1, but, unlike classical HCC, it stains for CK7 and the histiocytic marker CD68 [27, 29].

Different studies indicate that the uniqueness of FLC extends to the molecular level. Compared to classical HCC, FLC is genetically more stable, with only few large chromosomal rearrangements and rare mutations/over-expression of *p53* and *β-catenin* genes. Polysomy of chromosome 7 leads to *EGFR* up-regulation, which, in turn, makes FLC a good candidate for anti-EGFR-based therapies [27].

The unique clinico-pathological and molecular characteristics of FLC pose legitimate questions regarding its ontogenesis. A recent study has demonstrated that FLC has a unique transcriptomic signature characterized by the strong expression of specific neuroendocrine genes, suggesting that these tumors may have a cellular origin different from that of HCC [30].

Very recently, a new histological variant of HCC has been described. This variant, designated as "chromophobe HCC with abrupt anaplasia" (CHAA), is composed of cells with smooth contours, chromophobe cytoplasm, and focal nuclear anaplasia. CHAA is often associated, through FISH analysis, with an alternative lengthening of telomere (ALT) phenotype [31].

The differential diagnosis of pHCCs is rarely a problematic one, thanks to their usually straightforward morphological and immunohistochemical features, but poorly differentiated HCC must be distinguished from high-grade cholangiocarcinoma and secondary liver tumors. In such cases, a panel of immunohistochemical markers encompassing Hep-Par1, CD10, low and high molecular-weight cytokeratins is usually helpful [24].

A main diagnostic difficulty is represented by the differential diagnosis between hepatocellular adenoma (HA) and well-differentiated HCC, occurring in non-cirrhotic livers. This differential diagnosis represents a challenge for tumors occurring in unusual clinical settings, such as in a man or an older woman or showing focal atypical morphologic features. Evason and coll. termed these tumors "atypical hepatocellular adenoma-like neoplasms" and demonstrated that most of them show beta-catenin activation and cytogenetic alterations similar to well-differentiated HCC [32]. Bedossa et al. recently suggested the alternative term "well-differentiated hepatocellular neoplasm of uncertain malignant potential" (HUMP), which would imply both a lack of clear-cut diagnostic criteria and the need for a close clinical follow-up [33]. Further studies are required to better characterize this category.

Classification by Histologic Grade

Classical HCC is currently graded according to Edmodson and Steiner's criteria [1], which identified four differentiation grades, according to the severity of nuclear and cytological atypia.

Grade 1 (G1) consists of well-differentiated cells, which are very similar to normal or hyperplastic hepatocytes. Grade 2 (G2) HCC still closely resembles normal hepatic cells, but nuclei are larger and more hyperchromatic than G1. Both G1 and G2 tumors feature a thin trabecular or pseudoglandular growth pattern.

Grade 3 (G3) is characterized by cells with larger and more hyperchromatic nuclei than G2 tumors. Neoplastic cells form broad *trabeculae* (composed of three or more cell layers) and are often arranged in a pseudoglandular growth pattern. In Grade 4 (G4) carcinomas, nuclei are intensely hyperchromatic and occupy the greater part of the cell. G4 HCCs usually display a compact growth pattern, with no evidence of sinusoid-like blood spaces. While occasional spindle cells can be noted, the presence of a diffuse spindle cell component allows to classify the tumor as undifferentiated (or sarcomatous) HCC.

Within the same tumor, different grades of differentiation are often detected, with less-differentiated components being located in the central portion of the lesion (Fig. 19.4). This heterogeneity explains the poor predictive value of needle biopsy in the overall grading of HCC lesions.

From a clinical perspective, HCC grading represents a fundamental parameter to predict prognosis and survival rate after different therapeutic approaches. Some reference centers have also included neoplastic grade in the diagnostic and prognostic work-up for patients suitable to undergo liver transplantation [34–36].

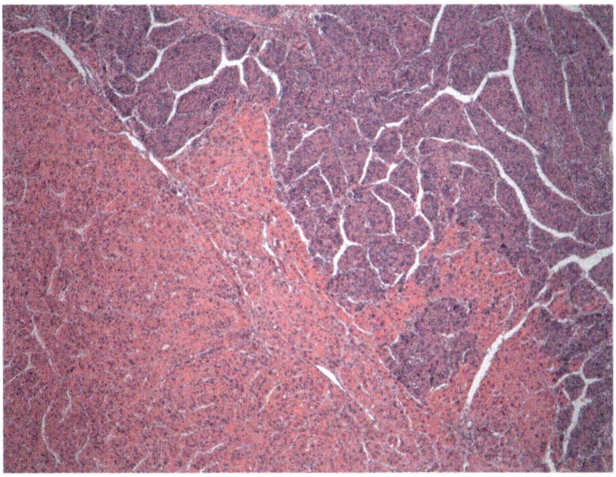

Fig. 19.4 Different grades of differentiation within a same tumor. Edmondson G2 (*left*) and G3 (*right*). H&E, original magnification 40×

19.3.2 Molecular Classification

The emerging of targeted therapies approach strictly requires both an accurate histological and molecular subclassification of HCCs [37]. In this context, the rapid development of next-generation technologies (i.e., next-generation sequencing, microarray analysis) allowed a preliminary extensive characterization of HCC molecular background and helped to face off the increasing complexity of testing several biomarkers on the same tumor sample, which is usually represented by small liver biopsies.

However, despite a huge amount of molecular data being generated, it is not clear yet which of the many genomic, genetic, and epigenetic alterations are the most critical in liver carcinogenesis and/or in defining HCC biological behavior. For this reason, there is currently no molecular subclassification of HCCs that is widely accepted and also routinely applicable in the clinical practice. Nevertheless, a number of molecular genetic markers have been found to correlate with clinical parameters, and in some instances, to have independent prognostic value.

Chromosomal aberrations are common in HCC, but most of them do not probably drive tumor progression [38]. An increased number of chromosomal alterations (mainly losses) are observed in HBV-related tumors. Common alterations include gain of chromosomes 1q, 8q, and 17q, and loss of 4q [39]. Recent data from whole genome analysis showed that the main HCC chromosomal alterations are represented by the gain of 1q, 5, 6p, 7, 8q, 17q, and 20, and the loss of 1p, 4q, 6q, 8p, 13q, 16, 17q, and 21 [40].

Until recent years, only four genes were known to display frequent alterations in HCC: *TP53*, *CTNNB1* (encoding the oncogene β-catenin), *AXIN1*, and *CDKN2A* (encoding the tumor suppressor p16INK4a) [41, 42]. *TP53* mutations are the most frequently observed (20–40 %), with a higher prevalence among Asian patients. Clinically, these alterations have been associated to a poorer patient's prognosis.

Recent whole exome and genome studies underlined the involvement of several other tumor suppressor genes (*TP53*, *PTEN*, *RB*), oncogenes (*c-MYC*, *KRAS*, *MET*, *BRAF*), developmental pathways (Wnt/β-catenin, Hedgehog), or growth factors and their receptors (*IGF2*, *TGFB1*, *FGFR*) [40, 43–47]. As for other tumor types, also in HCC recurrent somatic mutations have been observed in genes annotated as associated with chromatin regulation, such as *ARID1A*, *ARID1B*, *ARID2*, *MLL*, and *MLL3* [48]. Genetic sequencing also confirmed HBV integrations at cancer-related genes such as *TERT*, *MLL4*, and *CCNE1*. Moreover, novel tumor suppressive properties were identified for *IRF2*, whose inactivation led to impaired *TP53* function [40].

Epigenetic mechanisms are key molecular transformation processes involved in the long preneoplastic stages leading to HCC [49–51]. The hypermethylation of CpG islands acts as an alternative, complementary pathway to gene mutation and is an important mechanism involved in carcinogenesis. Both HBV and HCV induce epigenetic changes in specific genes involved in DNA repair, cell cycle control, and apoptosis signaling (*RASSF1A*, *GSTP1*, *CHRNA3*, and *DOK1*) in HCC compared to cirrhotic or normal liver tissues.

An involvement of microRNAs has been recently demonstrated, in HCC, as in other cancers [52], and independent studies described a consistent deregulation of miR-122, miR-199, miR-221, and miR-21 in HCCs. Several miRNA signatures have been found to correlate with important parameters, such as metastatic potential, differentiation, tumor recurrence, and patient survival [53]. Among others, miR-122 is almost exclusively expressed in normal hepatocytes, where it appears to act as a key regulator for their differentiation, via repression of genes not specifically involved in liver biology. Targeting this miRNA for the treatment of HCV infection by LNA-antimiR (miravirsen or SPC3649; Santaris Pharma, Denmark) represents the first example of miRNA-based clinical trial [54, 55].

19.3.3 Classification of Premalignant Hepatocellular Lesions

The vast majority of HCCs (>90 %) develops on a background of chronic hepatitis and/or liver cirrhosis. The hepatocellular carcinogenic cascade is a multistep sequence of both phenotypic and molecular alterations, which has been thoroughly characterized in the cirrhotic setting. Unfortunately, little is still known regarding the neoplastic transformation of hepatocytes in non-cirrhotic livers.

Precancerous hepatocellular lesions comprise: (1) large cell change (LCC); (2) small cell change (SCC); (3) dysplastic foci (DF); and (4) dysplastic nodules (DN) [22]. In the next paragraphs each of these categories will be briefly outlined.

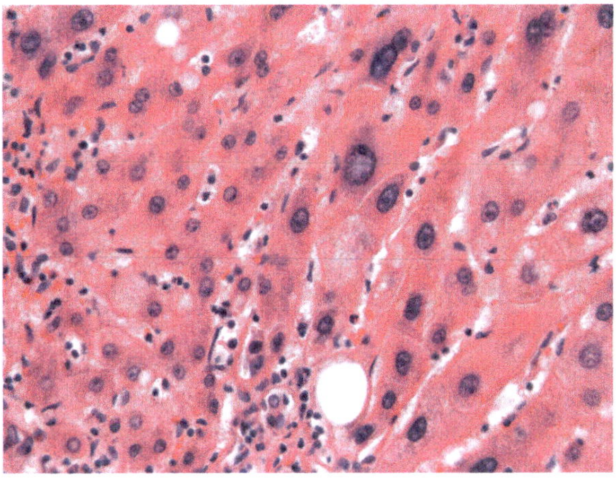

Fig. 19.5 Large cell change in cirrhosis. H&E, original magnification 200×

19.3.3.1 Large Cell Change

LCC is defined as a hepatocellular lesion characterized by increased cell size, nuclear pleomorphism with hyperchromasia, and multinucleation (Fig. 19.5).

LCC was originally described by Anthony and colleagues [56], who named it "liver cell dysplasia" (LCD). This name was deliberately chosen to underline the role of LCC as a preneoplastic lesion, which was inferred from its significant higher frequency in cirrhotic livers and in close proximity to HCC [57].

In recent years, the term "dysplasia" has been substituted by the more noncommittal "LCC," since the premalignant potential of such lesion has not been ultimately proven. In fact, according to the current literature, LCC likely represents an heterogeneous group of lesions, spanning from reactive hepatocyte hyperplasia to true preneoplastic proliferations [58]. This view is sustained by molecular studies, reporting a heterogeneous genetic profile for LCCs. In particular, LCCs associated with HBV-cirrhosis feature a molecular signature quite similar to that commonly observed in HCC (telomere shortening, increased DNA damage, inactivation of the cell cycle checkpoints p16 and p21). Conversely, such genetic changes are never observed in LCCs associated to cholestatic liver disorders [59].

From a practical point of view, however, prospective studies have consistently demonstrated that LCC is associated with an increased neoplastic risk [60–63] and its identification may help to select high-risk cirrhosis.

19.3.3.2 Small Cell Change

SCC, originally referred to as "small cell dysplasia" [64], is a hepatocyte lesion characterized by cells with scant, basophilic cytoplasm, mild nuclear pleomorphism, chromatin hyperchromasia, and increased nuclear-cytoplasmic ratio ("nuclear crowding" at low power-view). The incidence of SCC in cirrhotic livers varies from <1 % (biopsy samples) to up to 50 % (liver explants) [58]. Unlike LCC, SCC is considered a true dysplastic lesion, given its high proliferative activity, frequent chromosomal gains or losses, telomere shortening, and inactivation of the checkpoint protein p21. Despite this, literature studies, investigating the correlation between SCC and the risk of HCC, have reported conflicting results [60, 63, 65, 66].

19.3.3.3 Dysplastic Foci

Dysplastic foci (DF) are small clusters of hepatocytes (major diameter <0.1 cm), characterized by LCC, SCC, or (in hemochromatosis) lacking of cytoplasmic iron [67]. DF may be observed in dysplastic nodules (DNs) and HCCs, but have never been reported in regenerative nodules. Foci of SCC can display immunophenotypic features resembling either progenitor or intermediate hepatocyte-like cells, providing a hint regarding their possible origin [68].

Another clonal-like lesion, first described in 1980, is referred to as "foci of altered hepatocytes" (FAHs). FAHs have been further subclassified in: (1) clear-cell; (2) mixed-cell; and (3) basophilic-cell foci [69, 70]. These lesions have been identified in resections of HCC/hepatic cirrhosis and in numerous animal models of liver carcinogenesis. Clear cell FAH in non-cirrhotic livers feature the activation of signaling pathways similar to those observed in human and rodent hepatocarcinogenesis (AKT/mTOR and Ras/MAPK pathways) [71]. These observations may provide a clue to the still open issue of hepatocarcinogenesis in the non-cirrhotic liver (Fig. 19.6).

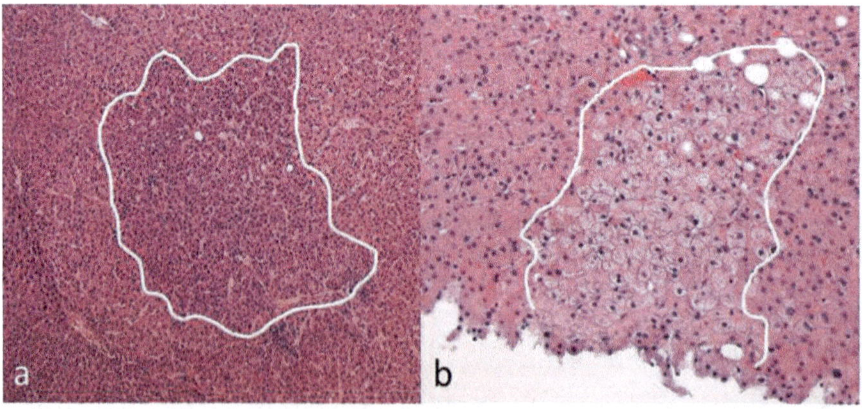

Fig. 19.6 (**a**) dysplastic focus composed of hepatocytes showing small cell change; (**b**) clear cell focus of altered hepatocytes in non-cirrhotic liver. H&E, original magnification 40× (**a**), 100× (**b**)

19.3.3.4 Dysplastic Nodule

Dysplastic nodules (DNs) are real nodular lesions, usually originating in hepatic cirrhosis and only rarely reported in non-cirrhotic chronic liver disorders. DNs are significantly larger than DF (main diameter: 1–2 cm) and can be detected by conventional imaging techniques. On gross examination, DNs consist of clear-cut nodular lesions bulging from the cut surface and sharply demarcated form the surrounding parenchyma [58].

Depending on the degree of cyto-architectural atypia, DNs are classified into (1) low-grade (LGDN) and (2) high-grade (HGDN) lesions.

LGDNs are characterized by mildly increased cell density and scant cytological atypia and may show LCC, but not SCC. Their blood supply is mainly of portal origin, which justifies the hypovascular or isovascular pattern during the arterial-phase of contrast-enhanced imaging. On histological grounds, clonal LGDNs can be barely distinguished from non-clonal regenerative nodules (RNs). Such distinction is, however, of negligible clinical impact, since neither LGDN nor RG need to be surgically treated [72].

In HGDNs, the degree of cytological and/or architectural atypia is more severe than in LGDNs, but falls short for a confident diagnosis of HCC. In HGDNs, SCC is the most frequently documented type of cytological atypia. As previously illustrated, the differential diagnosis between HGDN and eHCC can be extremely difficult (see above).

Several data support the premalignant nature of DNs: clinical–epidemiological studies on cirrhotic patients have reported a strong association between DNs and HCC [73, 74]; histological studies have documented a morphological *continuum* between LGDN, HGDN, and HCC [75]; molecular analyses have demonstrated that DNs and HCC are closely related lesions, characterized by similar epigenetic and genetic changes (e.g., loss of heterozygosis, chromosomal imbalance, telomere shortening, and telomerase reactivation) [76–79].

19.4 Intrahepatic Cholangiocarcinoma

Cholangiocarcinoma (CC) is a heterogeneous malignancy in terms of epidemiology, risk factors, histopathology, molecular pathology, and clinical features. The current WHO classification considers two different pathological entities, i.e. (1) intrahepatic (iCC), and (2) extrahepatic (eCC) CC [22].

iCC is defined as a primary liver neoplasm featuring biliary epithelial differentiation and arising from any portion of the intrahepatic biliary tree. Cancers originating from the hepatic ducts or from their confluence are currently considered within the group of eCC and usually referred to as "hilar CC" (or Klatskin tumor).

Despite its widespread clinical use, this classification has been criticized for two main reasons [80]. Firstly, the diagnosis of CC is frequently made at advanced disease stages, when the intrahepatic or extrahepatic origin of the tumor can be impossible to assess (up to 40 % of CCs are indeed classified as "not otherwise specified"). Secondly, hilar CCs not necessarily originate solely from large biliary structures, but, in fact, perihilar liver parenchyma also contains small bile ducts and ductules, which may theoretically undergo neoplastic transformation. Nonetheless, since the current WHO classification of liver tumors considers these diagnostic categories, it is still advisable to recur to them in the clinical practice. The following paragraphs will be focused on the clinical–pathological features of iCC.

19.4.1 Intrahepatic Cholangiocarcinoma

iCC is the second most common primary hepatic malignancy. The actual incidence and prevalence of iCC are scarcely defined, as few studies addressed this issue and most cancer registries do not discriminate between CC and other hepatobiliary malignancies (i.e., HCC and gallbladder carcinoma). The available literature data report significant geographical differences in the incidence of such tumor, which probably reflect variable exposure to CC-risk factors. The highest incidence rates have been documented in Thailand (113/100,000 in the male population; 50/100,000 in the female population), with Western countries being characterized by a significantly lower number of cases (0.2/100,000 in the male population; 0.1/100,000 in the female population) [81].

Time-trend studies have reported a worldwide increase in the incidence of iCC with little changes in the epidemiology of eCC [82], stirring a debate regarding if these data represent true epidemiological trends or the result of misclassification biases.

iCC usually affects adult patients, with a mean age at diagnosis of about 50 years. A significant exception to this rule is represented by patients affected by primary sclerosing cholangitis (PSC), possibly developing iCC one decade earlier than the general population (mean age at diagnosis: 40 years). iCC has an extremely poor prognosis, with a 5-year overall survival of about 10 % and with no significant improvements in the outcome of this tumor observed in the last decades.

Despite the etiology of iCC is still largely unknown, a common pathogenetic mechanism seems to be represented by chronic inflammation of the biliary tree.

In Eastern countries, *Opisthorchis viverrini* and *Clonorchis sinensis* play a major role in the pathogenesis of iCC, causing chronic cholangiocyte irritation and increased cellular turnover [83]. This, in turn, increases the susceptibility of biliary cells to endogenous and exogenous carcinogens. Other well-known iCC risk factors are represented by hepatolithiasis (2–10 % risk of iCC), bile-duct cysts, and rare congenital disorders of the biliary tree (e.g., cystic dilatation of the extra and/or intrahepatic bile ducts) [81].

Very recent Japanese studies have also reported a significant higher incidence of CC in a cohort of young adults (mean age: 36 years), employed in a printing company and exposed to chlorinated organic solvents. The strength of such association (17 cases out of 111 workers) led the Japanese Government to include this type of CC within the list of occupation-related disorders [84, 85].

In Western countries, PSC is the main recognized risk factor for CC (10 % of overall CC cases) [81]. The lifetime incidence of CC among PSC-patients ranges from 6 to 36 %, with an annual incidence risk of approximately 0.5–1.5 %. Other significant iCC precancerous conditions are represented by liver cirrhosis (irrespective of its etiology), HBV, and HCV chronic viral hepatitis (odds ratios: 22.92, 5.1, and 4.8, respectively) [81].

The clinical challenge lies on the early diagnosis of iCC, since the disease becomes symptomatic only at advanced stages and no specific laboratory markers are available so far. As such, the recognition of iCC requires a high index of suspicion and a multidisciplinary approach, based on clinical, laboratory, endoscopic, radiologic, and histological analyses.

19.4.1.1 Gross Classification

According to its macroscopic appearance, iCC is classified into: (1) mass-forming (MF); (2) periductal-infiltrating (PI); and (3) intraductal-growth (IG) iCC [22].

MF iCC is the most common subtype and presents as a nodular, un-encapsulated, and frequently multi-lobulated mass. On gross examination, it appears as a light-tan lesion with a gray-white central portion, corresponding to extensive fibrous scarring.

PI iCC displays a peculiar growth pattern, with spreading along portal tracts. Neoplastic cells surround and compress native biliary structures. On occasion, the MF and PI subtypes coexist within the same tumor.

IG-iCC appears as a polypoid tumor, growing within the lumen of dilated biliary ducts. Despite the IG subtype has been associated with a better prognosis, the actual clinical impact of such classification is still under discussion.

19.4.1.2 Histological Classification

The histological classification of iCC has been long debated and a general consensus regarding the diagnostic categories to be adopted is still far to come.

Classical examples of iCC are characterized by the proliferation of neoplastic glandular structures within an abundant and dense fibrotic stroma (Fig. 19.7). The latter does not represent a mere desmoplastic reaction to the proliferation of neoplastic glands, but instead it is the result of active fibrous tissue deposition by cancer-associated myofibroblasts. Cancer-associated myofibroblasts are responsible for the production of anti-angiogenic factors, matrix modifying enzymes, and pro-invasive growth factors and cytokines, thus playing a crucial role in promoting iCC aggressiveness and resistance to conventional chemotherapy [86, 87].

Classic iCC have been subdivided according to the architectural growth pattern (papillary vs. tubular iCC) and to the degree of glandular differentiation.

Well-differentiated tumors show evenly neoplastic glands or tubular structures composed by cuboidal epithelial cells, with regular nuclei and slightly acidophilic cytoplasm. Moderately differentiated tumors are characterized by more elongated

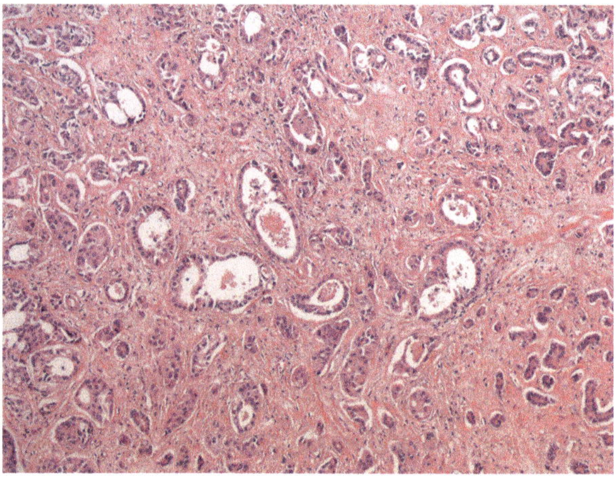

Fig. 19.7 Classical intrahepatic cholangiocarcinoma showing neoplastic glands and fibrous stroma. H&E, original magnification 40×

and tortuous glands, lined by tall, eosinophilic cells with atypical and hyperchromatic nuclei. Poorly differentiated iCC displays nests of solid and occasionally cribriform structures composed by small monomorphic cells, with scant cytoplasm and hyperchromatic nuclei.

Sempoux and colleagues [88] have described a "trabecular" and a "hilar" subtype of iCC, which is frequently associated with HCV-infection and advanced stage fibrosis. Trabecular iCC strongly resembles HCC, while hilar-type iCC displays histological features quite similar to Klatskin tumors. The original Sempoux's series also included a further iCC variant, characterized by very bland ductular structures mimicking ductal plate malformation. Such histotype, previously termed "cholangiocellular carcinoma" and thought to originate from hepatic progenitor cells (HPCs), is now classified as a subtype of combined hepatocellular–cholangiocellular carcinoma (see below).

Komuta and colleagues [89] proposed to classify iCC into "pure mucin producing" (Muc-iCC) and "mixed-iCC." Muc-iCC shows clinical, histological, and molecular features similar to hilar CC and probably originates from mucin-producing cholangiocytes of large bile ducts. Mixed-iCC displays a clinical–pathological profile close to that reported for cholangiocellular carcinoma.

Liau and colleagues [90] subclassified their series of iCC into (1) bile duct-type and (2) cholangiolar-type on the base of morphological features. The two histotypes were different in etiological, clinical, and molecular features. The cholangiolar-type iCC was associated with viral hepatitis, expressed N-cadherin, and was more likely to have IDH1/2 mutations. It was also characterized by higher 5-year survival rate compared to the bile duct-type. The bile duct-type most frequently occurred in patients with a previous history of intrahepatic cholelithiasis, and on histology and immunohistochemistry, closely resembled hilar CC, extrahepatic bile-duct, and pancreatic adenocarcinoma. Liau's bile duct-type clearly corresponds to Komuta's Muc-ICC, while Liau's cholangiolar-type overlaps with Sempoux's cholangiocellular carcinoma and Komuta's mixed-ICC.

The above observations make it clear that much work has still to be done in order to standardize the nomenclature and classification of iCC. A promising approach might be to differentiate intra-hepatic biliary neoplasms according to their specific cell of origin. This would also help to correlate the histogenesis of iCCs with their clinical features and therapeutic responses.

19.4.1.3 Rare Histological Variants of iCC

Rare variants of iCC are characterized by histological features that may closely mimic extra-hepatic tumors [22]. Their knowledge is thus of pivotal importance in order to correctly differentiate primary iCC from secondary liver cancers (i.e., metastases from extra-hepatic carcinomas).

Adenosquamous and squamous carcinomas are characterized by the unequivocal presence of squamous differentiation. Both types are extremely rare and associated with advanced-stage iCCs.

Mucinous iCCs are mainly composed of extracellular collections of mucus ("mucus lakes"), containing occasional neoplastic glands and/or signet ring cells [91]. Pure signet ring carcinomas are extremely rare. Mucinous carcinoma and signet ring cell carcinoma should not be confused with the mucin producing variant of classic iCC.

Sarcomatous iCC is an extremely rare variant composed of spindle-cells, which resembles fibrosarcoma or malignant fibrous histiocytoma. The differential diagnosis with a liver metastasis from a primary soft tissue sarcoma largely relies on the demonstration that spindle cells stain positive for epithelial markers. Sarcomatous iCC bears a significantly worse prognosis than classic iCC.

Lymphoepithelioma-like carcinoma is a peculiar iCC variant composed of aggregates of variably differentiated glands, dispersed within a dense lympho/plasmacytic infiltrate. Neoplastic cells are usually positive for Epstein-Barr virus (EBV)-coded nuclear RNAs (EBER) [92].

Clear-cell iCC is a very rare variant, with only eight cases reported in the literature. Clear-cell iCC closely resembles the more common renal counterpart [93].

Primary liver mucoepidermoid carcinoma (MEC) is an extremely rare and aggressive tumor (only 17 cases described in the literature histologically similar to salivary gland MEC) [94].

19.4.2 A Special Tumor

Mucinous cystic neoplasms (MCNs), previously known as "biliary cysto-adenoma/cysto-adenocarcinoma," are very rare tumors almost exclusively occurring in middle-aged women. MCNs can be either benign (i.e., non-invasive) or malignant (i.e., invasive) tumors. Benign MCNs may eventually progress to invasive neoplasms [95]. The diagnosis of MCN can be extremely difficult, since clinical symptoms, serologic markers, and imaging techniques have extremely high inaccuracy rates (55–100 %). On histology, MCNs consist of multilocular cysts lined by a muco-secreting epithelium. The underlying connective tissue is composed of fascicles of spindle cells similar to ovarian stroma. At immunohistochemistry, ovarian-type stroma is consistently positive for both estrogen and progesterone receptors [96].

Primary liver MCNs seem to be biologically and clinically related to their more frequent pancreatic counterpart.

19.4.3 Molecular Classification

The morphological heterogeneity characterizing biliary tract cancers has limited their comprehensive molecular classification, so far. Previous studies have focused on selected genes, including those altered in pancreatic adenocarcinoma (*KRAS*, *TP53*, *CDKN2A*, and *SMAD4*) [97]. Despite mutations in *PIK3CA*, *PTEN*, *AKT1*, *IDH1*, and *IDH2* have been frequently reported [98], the prevalence of these alterations varies among studies.

iCCs are usually characterized by *IDH1/IDH2* mutations, but (compared to extra-hepatic tumors) less frequently associated to *KRAS* and *TP53* mutations. *IDH1/2*-mutated cancers accumulate 2-hydroxyglutarate in the tumor tissue and release the molecule in the patients' blood [99]. Hence, the measurement of 2-hydroxyglutarate has been proposed as a surrogate biomarker for *IDH1/2* mutational status and as a noninvasive test for the assessment of the iCC tumor burden.

Two recent whole exome-sequencing studies revealed a key role for the chromatin remodeling genes *BAP1*, *ARID1A*, and *PBRM1* in the development of iCC [100]. These data have still a limited clinical impact, but in the next future they may help to identify new diagnostic markers and/or possible targets for tailored therapies.

Several studies have shown that chronic inflammation is critical for the development of iCC, since the massive production of inflammatory cytokines can significantly contribute to the cholangiocytes' malignant transformation. Among others, interleukin 6 (IL-6) is emerging as one of the most important players in this context [101]. IL-6 is a multifunctional cytokine that mediates the response of normal cells to inflammation [102] and serum levels of IL-6 are increased in CC patients, correlating with the tumor burden and decreasing few weeks after surgical resection [103–105]. From a pathogenetic perspective, IL-6 promotes CC-cell survival [105] by up-regulating telomerase activity, Mcl-1, STAT3, p38, and p44/p42 MAPK signaling [106, 107].

19.4.4 Classification of Premalignant Cholangiocellular Lesions

Two major types of precancerous cholangiocellular lesions have been described, namely (1) biliary intraepithelial neoplasia (BilIN, also known as "biliary dysplasia"), and (2) papillary intraductal biliary tumors (i.e., biliary IPN) [22].

While BilIN represents the precancerous lesion of conventional iCC, IPN-B may be associated with either mucinous carcinoma or conventional iCC.

BilIN is currently defined as a multilayered, columnar cell lesion, with variable degree of cytological atypia. BilIN can be observed both in large ducts exposed to chronic inflammation and in small ductular structures, particularly in HCV-related and alcoholic cirrhosis [107, 108].

According to the extent of cytological atypia, BilIN has been graded in a three-tiered scale (from BilIN-1 to BilIN-3) [109].

BilIN-1 corresponds to a low-grade dysplastic lesion and displays a flat or (less frequently) micro-papillary growth pattern. Cytological and nuclear atypia are minimal to mild. BilIN-2 may display a flat, pseudo-papillary or micro-papillary growth pattern. Cytological atypia is more evident than in BilIN-1. BilIN-3 encompasses both high-grade dysplasia and carcinoma in situ [110]. It generally features a pseudo-papillary or micro-papillary growth pattern (flat lesions have been rarely reported). The degree of cytological atypia is close to that observed in invasive carcinoma.

Despite specific criteria have been defined for the diagnosis of BilIN, its recognition and grading display only a moderate inter-observer agreement (60.3 % among expert hepatopathologists; $k=0.45$) [111].

Biliary IPN represents the liver counterpart of pancreatic intraductal papillary mucinous neoplasms (IPMN) [112, 113] and is defined as a prominent papillary proliferation of atypical biliary cells, which fills the bile duct and may lead to its fusiform or cystic enlargement. Biliary IPN, unlike BilIN, is graded in a two-tiered scale (i.e., low- and high-grade lesions) according to architecture complexity, degree of nuclear stratification, and cytological features. At the time of diagnosis, 1/3 to 2/3 of biliary IPNs are associated with invasive adenocarcinoma.

At the molecular level, BilIN shares several key molecular pathways with biliary IPN: in particular *KRAS* mutations are identifiable in low-grade BilIN and slightly increase during progression to BilIN-3 and invasive adenocarcinoma [113, 114].

19.5 Combined Hepatocellular–Cholangiocarcinoma

Combined hepatocellular–cholangiocarcinoma (cHCC-CC) is an uncommon, but increasingly recognized, primary liver tumor featuring both hepatocellular and biliary epithelial differentiation (Fig. 19.8) [115].

Liver tumors with HCC and CC components were first reported by Wells in 1903 [116] and later [117] described accurately by Allen and Lisa (in 1949). Cases of mixed carcinoma (i.e., "hepatobiliary carcinoma") were also observed by Edmondson and Steiner [1].

Allen and Lisa [117] described three forms of mixed HCC/CC tumors: (1) separate tumors (i.e., non-contiguous tumors with different histotypes); (2) contiguous tumors (i.e., histologically unrelated tumors that mingle as they grow); and (3) true mixed tumors (i.e., single lesions featuring both hepatocellular and biliary differentiation).

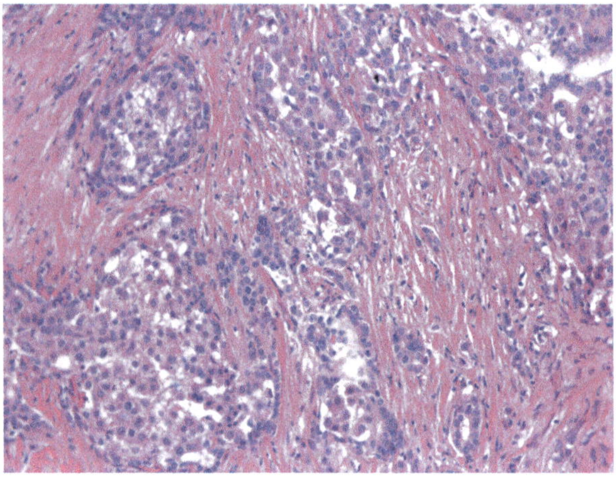

Fig. 19.8 Combined HCC-CC with stem cell features, classical subtype. Nest of neoplastic hepatocytes surrounded by small stem/progenitor like cells. H&E, original magnification 40×

The terminology concerning these tumors has remained unclear for several years, until the third edition of the WHO Classification of Tumors of the Digestive System adopted the term cHCC-CC only for tumors belonging to the "true-mixed" category of Allen and Lisa. As such, cHCC-CC were clearly distinguished from "collision tumors," i.e., independent HCCs and CCs arising simultaneously within the same liver. The accuracy of such nosological approach has been subsequently confirmed by molecular studies reporting that collision tumors originate from independent neoplastic clones [118].

Since the nomenclature and diagnostic criteria of cHCC-CCs have been only recently standardized, data regarding their epidemiology and clinical–pathological features are still scanty. cHCC-CC accounts for about 1.0–14.2 % of all primary liver cancers, with significant geographical and racial differences [119]. Such variability may be, at least partially, attributed to the lack of uniform and standardized diagnostic criteria.

Several authors reported a poor prognosis for cHCC-CC after surgical resection [120]. Despite a very limited experience in liver transplantation, a recent study conducted on a large population has reported similar outcomes for cHCC-CC and HCC in transplanted patients [121].

19.5.1 Gross Classification

No peculiar macroscopic features, differentiate cHCC-CC from pure HCC, but when the cholangiocellular component is largely represented, the tumor may appear firm and sclerotic.

19.5.2 Histological Classification of cHCC-CC

According to the 2010 WHO Classification of Tumors of the Digestive System [22], cHCC-CCs are classified in: (1) classical-type; and (2) cHCC-CC with stem cell features (further subclassified into three variants). Despite this classification provides a standardized terminology and uniform diagnostic criteria, its clinical impact has yet to be studied.

Classical-type cHCC-CC is defined as a tumor with both hepatocellular and cholangiocellular differentiation and with <5 % of cells featuring a stem cell phenotype. The HCC components of cHCC-CC consist of neoplastic hepatocytes with variable degree of differentiation, staining positive for HepPar1, and/or CD10 in the typical canalicular pattern. The CC component is represented by tubular, papillary, or cord-like glandular structures that express biliary cytokeratins (CK7 and CK19) [22].

cHCC-CC with stem cell features includes (1) a "typical subtype" (TS), (2) an "intermediate cell subtype" (ICS), and (3) a "cholangiocellular subtype" (CoS) [22].

TS is composed of nests of hepatocytes surrounded by a population of small cells, with high nucleus-cytoplasma ratio (Fig. 19.8), usually staining positive for CK19. These cells may also express other progenitor cell-markers such as NCAM-1/CD56, c-KIT, and EpCAM [22].

ICS is composed of cells with intermediate features between hepatocytes and cholangiocytes. At immunohistochemistry, they express both HepPar1 and CK19, thus mimicking ductular reaction [22].

CoS consists of neoplastic glands similar to reactive bile ductules or canals of Hering. Such structures are arranged in a cord-like or anastomosing pattern ("antler-like pattern") and embedded in a rich, fibrous stroma (Fig. 19.9). Neoplastic cells are usually positive for CK7 and/or CK19 [22].

A recent clinical–pathological survey conducted on 62 cHCC-CCs revealed that most of these tumors consist of two or more histological subtypes [122]. ICS was most frequently documented and correlated with female gender, tumor size, and the

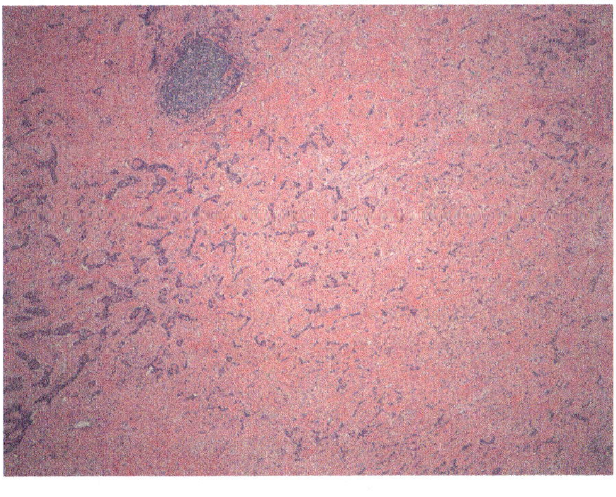

Fig. 19.9 Combined HCC-CC with stem cell features, cholangiocellular type. Tubular structure with "anter-like" appearance and dense fibrous stroma. H&E, original magnification 10×

histological grade of HCC component. The CoS component is directly correlated with the extent of fibrosis and inflammation, but inversely correlated with tumor size and the histological grade of HCC component. Finally, the proportion of TS inversely correlates with the extent of intra-tumor inflammation. Hence, these data suggest that each cHHC-CC subtype bears specific clinical and pathological implications.

Little is still known regarding the putative histogenesis of cHCC-CC. Fujii and colleagues have suggested that such tumor originates from a single cell clone that subsequently undergoes divergent differentiation [118]. This hypothesis has been confirmed by further studies, demonstrating that cHCC-CC cell lines may differentiate to either HCC or CC, under different growth conditions [123]. Theise and colleagues have also reported cHCC-CCs containing subsets of cells morphologically and immunohistochemically resembling HPCs [124]. This observation supports the idea that cHCC-CC may originate from undifferentiated liver stem cells [125]. From a molecular point of view, GEP studies have recently contributed to highlight the relationship between cHCC-CC, pure CC, and pure HCC [126]. In particular, gene expression data indicate that cHCC-CC is more CC alike than HCC (71.4 % of cHCC-CC clusters with CCs by unsupervised analysis). Further studies are required to confirm these results in order to better define the histogenesis of this rare entity.

19.6 Final Notes on the Histogenesis of Liver Cancers

As previously mentioned, the issue of liver cancer histogenesis is still unsettled. Growing evidence suggests that primary liver cancers may arise from adult HPCs or liver cancer stem cells (CSC) [127]. HPCs are bipotent cells capable to differentiate into mature hepatocytes or cholangiocytes [128]. CSCs are neoplastic cells that possess distinct stem-cell properties, which make them crucial for tumor survival and progression. CSCs are thought to originate from tumor progenitor cells, non-neoplastic stem cells, or dedifferentiated cells that have acquired CSC characteristics [129].

Experimental evidence of a direct role for HPCs in HCC histogenesis was provided by Dumble et al. [130]. Thorgeirsson and colleagues have subsequently demonstrated, by GEP analysis, that human HCCs can feature either a hepatocyte-like or a HPC-like signature [131]. At the present day, it is impossible to ascertain, anyway, whether the presence of HPC-like features in a tumor reflects their effective origin from immature hepatocyte precursors or from the de-differentiation of genetically unstable neoplastic cells instead. Studies on HCC animal models have demonstrated that both HPCs, lineage-committed hepatoblasts and differentiated adult hepatocytes, can be targeted by molecular modification triggering neoplastic transformation. Transformed cells may disclose a CSC-phenotype and undergo differentiation toward various cancer histotypes [132]. These results clearly indicate that neoplastic transformation may occur in any phase of hepatocytic differentiation, thus contributing to explain the broad spectrum of clinical and pathological features typically observed in liver tumors. With respect to the relationship between different cancer histotypes, GEP studies have reported the existence of specific molecular signatures for iCC and HCC, but in some cases (e.g., cHCC-CC) GEP patterns may partially overlap, indicating that liver cancers represent, more likely, a biological *continuum* rather than a large group of sharply demarcated clinical entities. These data have been further confirmed by in vitro studies, indicating that simultaneous activation of specific molecular pathways in resting hepatocytes determines their transformation into cholangiocytes and ultimately into CC cells [133].

In conclusion, a more thorough understanding of liver cancer histogenesis and HPCs role will surely help to improve the strategies for liver cancer management.

Acknowledgments The authors are grateful to Dr. Alessio Mussato for his precious technical contribution.

References

1. Edmondson HA, Steiner P (1954) Primary carcinoma of the liver. A study of 100 cases among 48,900 necropsies. Cancer 7:462–503
2. Shanmugaratnam K (1956) Primary carcinomas of the liver and biliary tract. Br J Cancer 2:232–245
3. Berman C (1951) Primary carcinoma of the liver: a study in incidence, clinical manifestations, pathology and aetiology. H. K. Lewis, London
4. Bosman FT, Carneiro F, Hruban RH, Theise ND (2010) WHO classification of tumors of the digestive system. IARC, Lyon
5. Mittal S, El-Serag HB (2013) Epidemiology of hepatocellular carcinoma. Consider the population. J Clin Gastroenterol 47:S2–S6. doi:10.1097/MCG.0b013e3182872f29
6. El-Serag HB (2012) Epidemiology of viral hepatitis and hepatocellular carcinoma. Gastroenterology 142:1264–1273.e1. doi:10.1053/j.gastro.2011.12.061
7. Qian GS, Ross RK, Yu MC, Yuan JM, Gao YT, Henderson BE, Wogan GN, Groopman JD (1994) A follow-up study of urinary markers of aflatoxin exposure and liver cancer risk in Shanghai, People's Republic of China. Cancer Epidemiol Biomarkers Prev 3(1):3–10
8. Ascha MS, Hanouneh IA, Lopez R, Tamimi TA, Feldstein AF, Zein NN (2010) The incidence and risk factors of hepatocellular carcinoma in patients with nonalcoholic steatohepatitis. Hepatology 51:1972–1978. doi:10.1002/hep.23527
9. Takayama T, Makuuchi M, Hirohashi S, Sakamoto M, Yamamoto J, Shimada K, Kosuge T, Okada S, Takayasu K, Yamasaki S (1998) Early hepatocellular carcinoma as an entity with high rate of surgical cure. Hepatology 28:1241–1246
10. Kojiro M (2007) Diagnostic discrepancy of early hepatocellular carcinoma between Japan and West. Hepatol Res 37:S249–S252
11. International Consensus Group for Hepatocellular Neoplasia The International Consensus Group for Hepatocellular Neoplasia (2009) Pathologic diagnosis of early hepatocellular carcinoma: a report of the international consensus group for hepatocellular neoplasia. Hepatology 49:658–664. doi:10.1002/hep.22709
12. Park YN, Kim MJ (2011) Hepatocarcinogenesis: imaging-pathologic correlation. Abdom Imaging 36:232–243. doi:10.1007/s00261-011-9688-y
13. Takayama T, Makuuchi M, Kojiro M, Lauwers GY, Adams RB, Wilson SR, Jang HJ, Charnsangavej C, Taouli B (2008) Early hepatocellular carcinoma: pathology, imaging, and therapy. Ann Surg Oncol 15(4):972–978. doi:10.1245/s10434-007-9685-0
14. Kojiro M, Nakashima O (1999) Histopathologic evaluation of hepatocellular carcinoma with a special reference to small early stage tumor. Semin Liver Dis 19:287–296

15. Libbrecht L, Severi T, Cassiman D, Vander Borght S, Pirenne J, Nevens F, Verslype C, van Pelt J, Roskams T (2006) Glypican-3 expression distinguishes small hepatocellular carcinomas from cirrhosis, dysplastic nodules, and focal nodular hyperplasia-like nodules. Am J Surg Pathol 30(11):1405–1411
16. Di Tommaso L, Franchi G, Park YN, Fiamengo B, Destro A, Morenghi E, Montorsi M, Torzilli G, Tommasini M, Terracciano L, Tornillo L, Vecchione R, Roncalli M (2007) Diagnostic value of HSP70, glypican 3, and glutamine synthetase in hepatocellular nodules in cirrhosis. Hepatology 45(3):725–734
17. Di Tommaso L, Destro A, Seok JY, Balladore E, Terracciano L, Sangiovanni A, Iavarone M, Colombo M, Jang JJ, Yu E, Jin SY, Morenghi E, Park YN, Roncalli M (2009) The application of markers (HSP70 GPC3 and GS) in liver biopsies is useful for detection of hepatocellular carcinoma. J Hepatol 50(4):746–754. doi:10.1016/j.jhep.2008.11.014
18. Llovet JM, Chen Y, Wurmbach E, Roayaie S, Fiel MI, Schwartz M, Thung SN, Khitrov G, Zhang W, Villanueva A, Battiston C, Mazzaferro V, Bruix J, Waxman S, Friedman SL (2006) A molecular signature to discriminate dysplastic nodules from early hepatocellular carcinoma in HCV cirrhosis. Gastroenterology 131(6):1758–1767
19. Eggel H (1901) Ueber das primare carcinom der Leber. Beitr z Path Anat u z allgem Pathol 30:506–604
20. Jakate S, Yabes A, Giusto D, Naini B, Lassman C, Yeh MM, Ferrell LD (2010) Diffuse cirrhosis-like hepatocellular carcinoma: a clinically and radiographically undetected variant mimicking cirrhosis. Am J Surg Pathol 34(7):935–941. doi:10.1097/PAS.0b013e3181ddf52f
21. Tátrai P, Somorácz A, Batmunkh E, Schirmacher P, Kiss A, Schaff Z, Nagy P, Kovalszky I (2009) Agrin and CD34 immunohistochemistry for the discrimination of benign versus malignant hepatocellular lesions. Am J Surg Pathol 33:874–885. doi:10.1097/PAS.0b013e318194b3ea
22. Theise ND, Curado MP, Franceschi S, Hytiroglou P, Kudo M, Park YN, Sakamoto M, Torbenson M, Wee A (2010) Hepatocellular carcinoma. In: Bosman FT, Carneiro F, Hruban RH, Theise ND (eds) WHO classification of tumours of the digestive system. IARC, Lyon
23. Roayaie S, Blume IN, Thung SN, Guido M, Fiel MI, Hiotis S, Labow DM, Llovet JM, Schwartz ME (2009) A system of classifying microvascular invasion to predict outcome after resection in patients with hepatocellular carcinoma. Gastroenterology 137(3):850–855
24. Bedossa P, Paradis V (2011) Hepatocellular carcinoma. In: Saxena R (ed) Practical hepatic pathology. Elsevier, Philadelphia
25. Kim H, Choi GH, Na DC, Ahn EY, Kim GI, Lee JE, Cho JY, Yoo JE, Choi JS, Park YN (2011) Human hepatocellular carcinomas with "Stemness"-related marker expression: keratin 19 expression and a poor prognosis. Hepatology 54(5):1707–1717. doi:10.1002/hep.24559
26. Govaere O, Komuta M, Berkers J, Spee B, Janssen C, de Luca F, Katoonizadeh A, Wouters J, van Kempen LC, Durnez A, Verslype C, De Kock J, Rogiers V, van Grunsven LA, Topal B, Pirenne J, Vankelecom H, Nevens F, van den Oord J, Pinzani M, Roskams T (2014) Keratin 19: a key role player in the invasion of human hepatocellular carcinomas. Gut 63(4):674–685. doi:10.1136/gutjnl-2012-304351
27. Torbenson M (2012) Fibrolamellar carcinoma: 2012 update. Scientifica (Cairo) 2012:743790. doi:10.6064/2012/743790
28. Craig JR, Peters RL, Edmondson HA, Omata M (1980) Fibrolamellar carcinoma of the liver: a tumor of adolescents and young adults with distinctive clinico-pathologic features. Cancer 46(2):372–379
29. Berman MA, Burnham JA, Sheahan DG (1988) Fibrolamellar carcinoma of the liver: an immunohistochemical study of nineteen cases and a review of the literature. Hum Pathol 19(7):784–794
30. Malouf GG, Job S, Paradis V, Fabre M, Brugières L, Saintigny P, Vescovo L, Belghiti J, Branchereau S, Faivre S, de Reyniès A, Raymond E (2014) Transcriptional profiling of pure fibrolamellar hepatocellular carcinoma reveals an endocrine signature. Hepatology 59(6):2228–2237. doi:10.1002/hep.27018
31. Wood LD, Heaphy CM, Daniel HD, Naini BV, Lassman CR, Arroyo MR, Kamel IR, Cosgrove DP, Boitnott JK, Meeker AK, Torbenson MS (2013) Chromophobe hepatocellular

carcinoma with abrupt anaplasia: a proposal for a new subtype of hepatocellular carcinoma with unique morphological and molecular features. Mod Pathol 26(12):1586–1593. doi:10.1038/modpathol.2013.68

32. Evason KJ, Grenert JP, Ferrell LD, Kakar S (2013) Atypical hepatocellular adenoma-like neoplasms with β-catenin activation show cytogenetic alterations similar to well-differentiated hepatocellular carcinomas. Hum Pathol 44(5):750–758. doi:10.1016/j.humpath.2012.07.019
33. Bedossa P, Burt AD, Brunt EM, Callea F, Clouston AD, Dienes HP, Goodman ZD, Gouw AS, Hubscher SG, Roberts EA, Roskams T, Terracciano L, Tiniakos DG, Torbenson MS, Wanless IR (2014) Well-differentiated hepatocellular neoplasm of uncertain malignant potential: proposal for a new diagnostic category. Hum Pathol 45(3):658–660. doi:10.1016/j.humpath.2013.09.020
34. Cillo U, Vitale A, Bassanello M, Boccagni P, Brolese A, Zanus G, Burra P, Fagiuoli S, Farinati F, Rugge M, D'Amico DF (2004) Liver transplantation for the treatment of moderately or well-differentiated hepatocellular carcinoma. Ann Surg 239:150–159
35. Decaens T, Roudot-Thoraval F, Badran H, Wolf P, Durand F, Adam R, Boillot O, Vanlemmens C, Gugenheim J, Dharancy S, Bernard PH, Boudjema K, Calmus Y, Hardwigsen J, Ducerf C, Pageaux GP, Hilleret MN, Chazouillères O, Cherqui D, Mallat A, Duvoux C (2011) Impact of tumour differentiation to select patients before liver transplantation for hepatocellular carcinoma. Liver Int 31:792–801. doi:10.1111/j.1478-3231.2010.02425.x
36. Jonas S, Bechstein WO, Steinmller T, Herrmann M, Radke C, Berg T, Settmacher U, Neuhaus P (2001) Vascular invasion and histopathologic grading determine outcome after liver transplantation for hepatocellular carcinoma in cirrhosis. Hepatology 33:1080–1086
37. Villanueva A (2013) Rethinking future development of molecular therapies in hepatocellular carcinoma: a bottom-up approach. J Hepatol 59(2):392–395. doi:10.1016/j.jhep.2013.03.025
38. Ozen C, Yildiz G, Dagcan AT, Cevik D, Ors A, Keles U, Topel H, Ozturk M (2013) Genetics and epigenetics of liver cancer. N Biotechnol 30(4):381–384. doi:10.1016/j.nbt.2013.01.007
39. Buendia MA (2002) Genetic alterations in hepatoblastoma and hepatocellular carcinoma: common and distinctive aspects. Med Pediatr Oncol 39(5):530–535
40. Guichard C, Amaddeo G, Imbeaud S, Ladeiro Y, Pelletier L, Maad IB, Calderaro J, Bioulac-Sage P, Letexier M, Degos F, Clément B, Balabaud C, Chevet E, Laurent A, Couchy G, Letouzé E, Calvo F, Zucman-Rossi J (2012) Integrated analysis of somatic mutations and focal copy-number changes identifies key genes and pathways in hepatocellular carcinoma. Nat Genet 44(6):694–698. doi:10.1038/ng.2256
41. Ozturk M (1999) Genetic aspects of hepatocellular carcinogenesis. Semin Liver Dis 19(3):235–242
42. Bressac B, Galvin KM, Liang TJ, Isselbacher KJ, Wands JR, Ozturk M (1990) Abnormal structure and expression of p53 gene in human hepatocellular carcinoma. Proc Natl Acad Sci U S A 87(5):1973–1977
43. Fujimoto A, Totoki Y, Abe T, Boroevich KA, Hosoda F, Nguyen HH, Aoki M, Hosono N, Kubo M, Miya F, Arai Y, Takahashi H, Shirakihara T, Nagasaki M, Shibuya T, Nakano K, Watanabe-Makino K, Tanaka H, Nakamura H, Kusuda J, Ojima H, Shimada K, Okusaka T, Ueno M, Shigekawa Y, Kawakami Y, Arihiro K, Ohdan H, Gotoh K, Ishikawa O, Ariizumi S, Yamamoto M, Yamada T, Chayama K, Kosuge T, Yamaue H, Kamatani N, Miyano S, Nakagama H, Nakamura Y, Tsunoda T, Shibata T, Nakagawa H (2012) Whole-genome sequencing of liver cancers identifies etiological influences on mutation patterns and recurrent mutations in chromatin regulators. Nat Genet 44(7):760–764. doi:10.1038/ng.2291
44. Huang J, Deng Q, Wang Q, Li KY, Dai JH, Li N, Zhu ZD, Zhou B, Liu XY, Liu RF, Fei QL, Chen H, Cai B, Zhou B, Xiao HS, Qin LX, Han ZG (2012) Exome sequencing of hepatitis B virus-associated hepatocellular carcinoma. Nat Genet 44(10):1117–1121. doi:10.1038/ng.2391
45. Breuhahn K, Longerich T, Schirmacher P (2006) Dysregulation of growth factor signaling in human hepatocellular carcinoma. Oncogene 25(27):3787–3800
46. Thorgeirsson SS, Grisham JW (2002) Molecular pathogenesis of human hepatocellular carcinoma. Nat Genet 31(4):339–346

47. Yeh MM, Larson AM, Campbell JS, Fausto N, Rulyak SJ, Swanson PE (2007) The expression of transforming growth factor-alpha in cirrhosis, dysplastic nodules, and hepatocellular carcinoma: an immunohistochemical study of 70 cases. Am J Surg Pathol 31(5):681–689
48. Li M, Zhao H, Zhang X, Wood LD, Anders RA, Choti MA, Pawlik TM, Daniel HD, Kannangai R, Offerhaus GJ, Velculescu VE, Wang L, Zhou S, Vogelstein B, Hruban RH, Papadopoulos N, Cai J, Torbenson MS, Kinzler KW (2011) Inactivating mutations of the chromatin remodeling gene ARID2 in hepatocellular carcinoma. Nat Genet 43(9):828–829. doi:10.1038/ng.903
49. Um TH, Kim H, Oh BK, Kim MS, Kim KS, Jung G, Park YN (2011) Aberrant CpG island hypermethylation in dysplastic nodules and early HCC of hepatitis B virus-related human multistep hepatocarcinogenesis. J Hepatol 54(5):939–947. doi:10.1016/j.jhep.2010.08.021
50. Nishida N (2010) Impact of hepatitis virus and aging on DNA methylation in human hepatocarcinogenesis. Histol Histopathol 25(5):647–654
51. Cardin R, Piciocchi M, Bortolami M, Kotsafti A, Barzon L, Lavezzo E, Sinigaglia A, Rodriguez-Castro KI, Rugge M, Farinati F (2014) Oxidative damage in the progression of chronic liver disease to hepatocellular carcinoma: an intricate pathway. World J Gastroenterol 20(12):3078–3086
52. Callegari E, Elamin BK, Sabbioni S, Gramantieri L, Negrini M (2013) Role of microRNAs in hepatocellular carcinoma: a clinical perspective. Onco Targets Ther 6:1167–1178
53. Negrini M, Gramantieri L, Sabbioni S, Croce CM (2011) microRNA involvement in hepatocellular carcinoma. Anticancer Agents Med Chem 11(6):500–521
54. Lanford RE, Hildebrandt-Eriksen ES, Petri A, Persson R, Lindow M, Munk ME, Kauppinen S, Ørum H (2010) Therapeutic silencing of microRNA-122 in primates with chronic hepatitis C virus infection. Science 327(5962):198–201. doi:10.1126/science.1178178
55. Lindow M, Kauppinen S (2012) Discovering the first microRNA-targeted drug. J Cell Biol 199(3):407–412. doi:10.1083/jcb.201208082
56. Anthony PP, Vogel CL, Barker LE (1973) Liver cell dysplasia: a premalignant condition. J Clin Pathol 26(3):217–223
57. Anthony PP, Vogel CL, Sadikali F, Barker LF, Peterson MR (1972) Hepatitis-associated antigen and antibody in Uganda: correlation of serological testing with histopathology. Br Med J 1(5797):403–406
58. Park YN (2011) Update on precursor and early lesions of hepatocellular carcinomas. Arch Pathol Lab Med 135(6):704–715. doi:10.1043/2010-0524-RA
59. Kim H, Park YN (2009) Large liver cell change/dysplasia in hepatitis B virus-related liver cirrhosis. Korean J Hepatol 15(3):375–378. doi:10.3350/kjhep.2009.15.3.375
60. Borzio M, Bruno S, Roncalli M, Mels GC, Ramella G, Borzio F, Leandro G, Servida E, Podda M (1995) Liver cell dysplasia is a major risk factor for hepatocellular carcinoma in cirrhosis: a prospective study. Gastroenterology 108:812–817
61. Ganne-Carrié N, Chastang C, Chapel F, Munz C, Pateron D, Sibony M, Dény P, Trinchet JC, Callard P, Guettier C, Beaugrand M (1996) Predictive score for the development of hepatocellular carcinoma and additional value of liver large cell dysplasia in Western patients with cirrhosis. Hepatology 23:1112–1118
62. Libbrecht L, Craninx M, Nevens F, Desmet V, Roskams T (2001) Predictive value of liver cell dysplasia for development of hepatocellular carcinoma in patients with non-cirrhotic and cirrhotic chronic viral hepatitis. Histopathology 39:66–73
63. Koo JS, Kim H, Park BK, Ahn SH, Han KH, Chon CY, Park C, Park YN (2008) Predictive value of liver cell dysplasia for development of hepatocellular carcinoma in patients with chronic hepatitis B. J Clin Gastroenterol 42(6):738–743. doi:10.1097/MCG.0b013e318038159d
64. Watanabe S, Okita K, Harada T, Kodama T, Numa Y, Takemoto T, Takahashi T (1983) Morphologic studies of the liver cell dysplasia. Cancer 51(12):2197–2205
65. Makino Y, Shiraki K, Sugimoto K, Ito T, Yamanaka T, Fujikawa K, Takase K, Nakano T (2000) Histological features of cirrhosis with hepatitis C virus for prediction of hepatocellular carcinoma development: a prospective study. Anticancer Res 20(5C):3709–3715

66. Hytiroglou P, Park YN, Krinsky G, Theise ND (2007) Hepatic precancerous lesions and small hepatocellular carcinoma. Gastroenterol Clin North Am 36(4):867–887
67. International Working Party (1995) Terminology of nodular hepatocellular lesions. Hepatology 22:983–993
68. Roskams T (2006) Liver stem cells and their implication in hepatocellular and cholangiocarcinoma. Oncogene 25(27):3818–3822
69. Su Q, Benner A, Hofmann WJ, Otto G, Pichlmayr R, Bannasch P (1997) Human hepatic preneoplasia: phenotypes and proliferation kinetics of foci and nodules of altered hepatocytes and their relationship to liver cell dysplasia. Virchows Arch 431:391–406
70. Bannasch P, Mayer D, Hacker HJ (1980) Hepatocellular glycogenosis and hepatocarcinogenesis. Biochim Biophys Acta 605:217–245
71. Ribback S, Calvisi DF, Cigliano A, Sailer V, Peters M, Rausch J, Heidecke CD, Birth B, Dombrowski F (2013) Molecular and metabolic changes in human liver clear cell foci resemble the alterations occurring in rat hepatocarcinogenesis. J Hepatol 58:1147–1156. doi:10.1016/j.jhep.2013.01.013
72. Di Tommaso L, Sangiovanni A, Borzio M, Park YN, Farinati F, Roncalli M (2013) Advanced precancerous lesions in the liver. Best Pract Res Clin Gastroenterol 27(2):269–284. doi:10.1016/j.bpg.2013.03.015
73. Borzio M, Fargion S, Borzio F, Fracanzani AL, Croce AM, Stroffolini T, Oldani S, Cotichini R, Roncalli M (2003) Impact of large regenerative, low grade and high grade dysplastic nodules in hepatocellular carcinoma development. J Hepatol 39:208–214
74. Kobayashi M, Ikeda K, Hosaka T, Sezaki H, Someya T, Akuta N, Suzuki F, Suzuki Y, Saitoh S, Arase Y, Kumada H (2006) Dysplastic nodules frequently develop into hepatocellular carcinoma in patients with chronic viral hepatitis and cirrhosis. Cancer 106(3):636–647
75. Theise ND, Park YN, Kojiro M (2002) Dysplastic nodules and hepatocarcinogenesis. Clin Liver Dis 6(2):497–512
76. Maggioni M, Coggi G, Cassani B, Bianchi P, Romagnoli S, Mandelli A, Borzio M, Colombo P, Roncalli M (2000) Molecular changes in hepatocellular dysplastic nodules on microdissected liver biopsies. Hepatology 32(5):942–946
77. Sun M, Eshleman JR, Ferrell LD, Jacobs G, Sudilovsky EC, Tuthill R, Hussein MR, Sudilovsky O (2001) An early lesion in hepatic carcinogenesis: loss of heterozygosity in human cirrhotic livers and dysplastic nodules at the 1p36-p34 region. Hepatology 33(6):1415–1424
78. Oh BK, Jo Chae K, Park C, Kim K, Jung Lee W, Han KH, Nyun Park Y (2003) Telomere shortening and telomerase reactivation in dysplastic nodules of human hepatocarcinogenesis. J Hepatol 39(5):786–792
79. Lee YH, Oh BK, Yoo JE, Yoon SM, Choi J, Kim KS, Park YN (2009) Chromosomal instability, telomere shortening, and inactivation of p21(WAF1/CIP1) in dysplastic nodules of hepatitis B virus-associated multistep hepatocarcinogenesis. Mod Pathol 22(8):1121–1131. doi:10.1038/modpathol.2009.76
80. Cardinale V, Bragazzi MC, Carpino G, Torrice A, Fraveto A, Gentile R, Pasqualino V, Melandro F, Aliberti C, Bastianelli C, Brunelli R, Berloco PB, Gaudio E, Alvaro D (2013) Cholangiocarcinoma: increasing burden of classifications. Hepatobiliary Surg Nutr 2(5):272–280
81. Shaib Y, El-Serag HB (2004) The epidemiology of cholangiocarcinoma. Semin Liver Dis 24:115–125
82. McGlynn KA, Tarone RE, El-Serag HB (2006) A comparison of trends in the incidence of hepatocellular carcinoma and intrahepatic cholangiocarcinoma in the United States. Cancer Epidemiol Biomarkers Prev 15:1198–1203
83. Kaewpitoon N, Kaewpitoon SJ, Pengsaa P, Sripa B (2008) Opisthorchis viverrini: the carcinogenic human liver fluke. World J Gastroenterol 14:666–674
84. Kubo S, Nakanuma Y, Takemura S, Sakata C, Urata Y, Nozawa A, Nishioka T, Kinoshita M, Hamano G, Terajima H, Tachiyama G, Matsumura Y, Yamada T, Tanaka H, Nakamori S, Arimoto A, Kawada N, Fujikawa M, Fujishima H, Sugawara Y, Tanaka S, Toyokawa H,

Kuwae Y, Ohsawa M, Uehara S, Sato KK, Hayashi T, Endo G (2014) Case series of 17 patients with cholangiocarcinoma among young adult workers of a printing company in Japan. J Hepatobiliary Pancreat Sci 21(7):479–488. doi:10.1002/jhbp.86
85. Fujita T (2012) An unusual risk factor for intrahepatic cholangiocarcinoma. J Hepatol 57(6):1396–1397. doi:10.1016/j.jhep.2012.07.021
86. Sirica AE (2012) The role of cancer-associated myofibroblasts in intrahepatic cholangiocarcinoma. Nat Rev Gastroenterol Hepatol 9:44–54. doi:10.1038/nrgastro.2011.222
87. Sirica AE, Gores GJ (2014) Desmoplastic stroma and cholangiocarcinoma: clinical implications and therapeutic targeting. Hepatology 59(6):2397–2402. doi:10.1002/hep.26762
88. Sempoux C, Jibara G, Ward SC, Fan C, Qin L, Roayaie S, Fiel MI, Schwartz M, Thung SN (2011) Intrahepatic cholangiocarcinoma: new insights in pathology. Semin Liver Dis 31(1):49–60. doi:10.1055/s-0031-1272839
89. Komuta M, Govaere O, Vandecaveye V, Akiba J, Van Steenbergen W, Verslype C, Laleman W, Pirenne J, Aerts R, Yano H, Nevens F, Topal B, Roskams T (2013) Histological diversity in cholangiocellular carcinoma reflects the different cholangiocyte phenotypes. Hepatology 55(6):1876–1888. doi:10.1002/hep.25595
90. Liau JY, Tsai JH, Yuan RH, Chang CN, Lee HJ, Jeng YM (2014) Morphological subclassification of intrahepatic cholangiocarcinoma: etiological, clinicopathological, and molecular features. Mod Pathol 27(8):1163–1173. doi:10.1038/modpathol.2013.241
91. Sasaki M, Nakanuma Y, Shimizu K, Izumi R (1995) Pathological and immunohistochemical findings in a case of mucinous cholangiocarcinoma. Pathol Int 45:781–786
92. Jeng YM, Chen CL, Hsu HC (2001) Lymphoepithelioma-like cholangiocarcinoma: an Epstein-Barr virus-associated tumor. Am J Surg Pathol 25(4):516–520
93. Haas S, Gütgemann I, Wolff M, Fischer HP (2007) Intrahepatic clear cell cholangiocarcinoma: immunohistochemical aspects in a very rare type of cholangiocarcinoma. Am J Surg Pathol 31:902–906
94. Shuangshoti S Jr, Shuangshoti S (2000) Primary mucoepidermoid carcinoma of the intrahepatic bile duct: a case report with review of literature. J Med Assoc Thai 83(2):197–203
95. Simo KA, Mckillop IH, Ahrens WA, Martinie JB, Iannitti DA, Sindram D (2012) Invasive biliary mucinous cystic neoplasm: a review. HPB (Oxford) 14(11):725–740. doi:10.1111/j.1477-2574.2012.00532.x
96. Zen Y, Pedica F, Patcha VR, Capelli P, Zamboni G, Casaril A, Quaglia A, Nakanuma Y, Heaton N, Portmann B (2011) Mucinous cystic neoplasms of the liver: a clinicopathological study and comparison with intraductal papillary neoplasms of the bile duct. Mod Pathol 24:1079–1089. doi:10.1038/modpathol.2011.71
97. Fassan M, Baffa R, Kiss A (2013) Advanced precancerous lesions within the GI tract: the molecular background. Best Pract Res Clin Gastroenterol 27:159–169. doi:10.1016/j.bpg.2013.03.009
98. Borger DR, Tanabe KK, Fan KC, Lopez HU, Fantin VR, Straley KS, Schenkein DP, Hezel AF, Ancukiewicz M, Liebman HM, Kwak EL, Clark JW, Ryan DP, Deshpande V, Dias-Santagata D, Ellisen LW, Zhu AX, Iafrate AJ (2012) Frequent mutation of isocitrate dehydrogenase (IDH)1 and IDH2 in cholangiocarcinoma identified through broad-based tumor genotyping. Oncologist 17:72–79. doi:10.1634/theoncologist. 2011-0386
99. Borger DR, Goyal L, Yau TC, Poon RT, Ancukiewicz M, Deshpande V, Christiani DC, Liebman HM, Yang H, Kim H, Ellwood-Yen K, Faris JE, Iafrate AJ, Kwak EL, Clark JW, Allen JN, Blaszkowsky LS, Murphy JE, Saha SK, Hong TS, Wo JY, Ferrone CR, Tanabe KK, Bardeesy N, Straley KS, Agresta S, Schenkein DP, Ellisen LW, Ryan DP, Zhu AX (2014) Circulating oncometabolite 2-hydroxyglutarate is a potential surrogate biomarker in patients with isocitrate dehydrogenase-mutant intrahepatic cholangiocarcinoma. Clin Cancer Res 20(7):1884–1890. doi:10.1158/1078-0432.CCR-13-2649
100. Jiao Y, Pawlik TM, Anders RA, Selaru FM, Streppel MM, Lucas DJ, Niknafs N, Guthrie VB, Maitra A, Argani P, Offerhaus GJ, Roa JC, Roberts LR, Gores GJ, Popescu I, Alexandrescu ST, Dima S, Fassan M, Simbolo M, Mafficini A, Capelli P, Lawlor RT, Ruzzenente A, Guglielmi A, Tortora G, de Braud F, Scarpa A, Jarnagin W, Klimstra D, Karchin R, Velculescu

VE, Hruban RH, Vogelstein B, Kinzler KW, Papadopoulos N, Wood LD (2013) Exome sequencing identifies frequent inactivating mutations in BAP1, ARID1A and PBRM1 in intrahepatic cholangiocarcinomas. Nat Genet 45:1470–1473. doi:10.1038/ng.2813

101. Johnson C, Han Y, Hughart N, McCarra J, Alpini G, Meng F (2012) Interleukin-6 and its receptor, key players in hepatobiliary inflammation and cancer. Transl Gastrointest Cancer 1(1):58–70
102. Okada K, Shimizu Y, Nambu S, Higuchi K, Watanabe A (1994) Interleukin-6 functions as an autocrine growth factor in a cholangiocarcinoma cell line. J Gastroenterol Hepatol 9:462–467
103. Goydos JS, Brumfield AM, Frezza E, Booth A, Lotze MT, Carty SE (1998) Marked elevation of serum interleukin-6 in patients with cholangiocarcinoma: validation of utility as a clinical marker. Ann Surg 227:398–404
104. Meng F, Yamagiwa Y, Ueno Y, Patel T (2006) Over-expression of interleukin-6 enhances cell survival and transformed cell growth in human malignant cholangiocytes. J Hepatol 44:1055–1065
105. Isomoto H, Kobayashi S, Werneburg NW, Bronk SF, Guicciardi ME, Frank DA, Gores GJ (2005) Interleukin 6 upregulates myeloid cell leukemia-1 expression through a STAT3 pathway in cholangiocarcinoma cells. Hepatology 42(6):1329–1338
106. Yamagiwa Y, Meng F, Patel T (2006) Interleukin-6 decreases senescence and increases telomerase activity in malignant human cholangiocytes. Life Sci 78:2494–2502
107. Rougemont AL, Genevay M, McKee TA, Gremaud M, Mentha G, Rubbia-Brandt L (2010) Extensive biliary intraepithelial neoplasia (BilIN) and multifocal early intrahepatic cholangiocarcinoma in nonbiliary cirrhosis. Virchows Arch 456(6):711–717. doi:10.1007/s00428-010-0899-3
108. Wu TT, Levy M, Correa AM, Rosen CB, Abraham SC (2009) Biliary intraepithelial neoplasia in patients without chronic biliary disease: analysis of liver explants with alcoholic cirrhosis, hepatitis C infection, and noncirrhotic liver diseases. Cancer 115(19):4564–4575. doi:10.1002/cncr.24471
109. Zen Y, Aishima S, Ajioka Y, Haratake J, Kage M, Kondo F, Nimura Y, Sakamoto M, Sasaki M, Shimamatsu K, Wakasa K, Park YN, Chen MF, Atomi Y, Nakanuma Y (2005) Proposal of histological criteria for intraepithelial atypical/proliferative biliary epithelial lesions of the bile duct in hepatolithiasis with respect to cholangiocarcinoma: preliminary report based on interobserver agreement. Pathol Int 55(4):180–188
110. Sato Y, Harada K, Sasaki M, Nakanuma Y (2013) Histological characteristics of biliary intraepithelial neoplasia-3 and intraepithelial spread of cholangiocarcinoma. Virchows Arch 462(4):421–427. doi:10.1007/s00428-013-1384-6
111. Zen Y, Adsay NV, Bardadin K, Colombari R, Ferrell L, Haga H, Hong SM, Hytiroglou P, Klöppel G, Lauwers GY, van Leeuwen DJ, Notohara K, Oshima K, Quaglia A, Sasaki M, Sessa F, Suriawinata A, Tsui W, Atomi Y, Nakanuma Y (2007) Biliary intraepithelial neoplasia: an international interobserver agreement study and proposal for diagnostic criteria. Mod Pathol 20(6):701–709
112. Chen TC, Nakanuma Y, Zen Y, Chen MF, Jan YY, Yeh TS, Chiu CT, Kuo TT, Kamiya J, Oda K, Hamaguchi M, Ohno Y, Hsieh LL, Nimura Y (2001) Intraductal papillary neoplasia of the liver associated with hepatolithiasis. Hepatology 34:651–658
113. Sibulesky L, Nguyen J, Patel T (2012) Preneoplastic conditions underlying bile duct cancer. Langenbecks Arch Surg 397(6):861–867. doi:10.1007/s00423-012-0943-7
114. Schlitter AM, Born D, Bettstetter M, Specht K, Kim-Fuchs C, Riener MO, Jeliazkova P, Sipos B, Siveke JT, Terris B, Zen Y, Schuster T, Höfler H, Perren A, Klöppel G, Esposito I (2013) Intraductal papillary neoplasms of the bile duct: stepwise progression to carcinoma involves common molecular pathways. Mod Pathol 27(1):73–86. doi:10.1038/modpathol.2013.112
115. Singh S, Chakraborty S, Bonthu N, Radio S, Hussain SM, Sasson A (2013) Combined hepatocellular cholangiocarcinoma: a case report and review of literature. Dig Dis Sci 58(7):2114–2123. doi:10.1007/s10620-013-2585-1
116. Wells H (1903) Primary carcinoma of the liver. Am J Sci 126:403–417
117. Allen RA, Lisa J (1949) Combined liver cell and bile duct carcinoma. Am J Pathol 25:647–655

118. Fujii H, Zhu XG, Matsumoto T, Inagaki M, Tokusashi Y, Miyokawa N, Fukusato T, Uekusa T, Takagaki T, Kadowaki N, Shirai T (2000) Genetic classification of combined hepatocellular-cholangiocarcinoma. Hum Pathol 31:1011–1017
119. Jarnagin WR, Weber S, Tickoo SK, Koea JB, Obiekwe S, Fong Y, DeMatteo RP, Blumgart LH, Klimstra D (2002) Combined hepatocellular and cholangiocarcinoma: demographic, clinical, and prognostic factors. Cancer 94:2040–2046
120. Kassahun WT, Hauss J (2008) Management of combined hepatocellular and cholangiocarcinoma. Int J Clin Pract 62:1271–1278. doi:10.1111/j.1742-1241.2007.01694.x
121. Sapisochin G, de Lope CR, Gastaca M, de Urbina JO, López-Andujar R, Palacios F, Ramos E, Fabregat J, Castroagudín JF, Varo E, Pons JA, Parrilla P, González-Diéguez ML, Rodriguez M, Otero A, Vazquez MA, Zozaya G, Herrero JI, Antolin GS, Perez B, Ciria R, Rufian S, Fundora Y, Ferron JA, Guiberteau A, Blanco G, Varona MA, Barrera MA, Suarez MA, Santoyo J, Bruix J, Charco R (2014) Intrahepatic cholangiocarcinoma or mixed hepatocellular-cholangiocarcinoma in patients undergoing liver transplantation: a Spanish matched cohort multicenter study. Ann Surg 259(5):944–952. doi:10.1097/SLA.0000000000000494
122. Sasaki M, Sato H, Kakuda Y, Sato Y, Choi JH, Nakanuma Y (2014) Clinicopathological significance of "subtypes with stem-cell feature" in combined hepatocellular-cholangiocarcinoma. Liv Intern, in press. doi:10.1111/liv.12563
123. Yano H, Iemura A, Haramaki M, Momosaki S, Ogasawara S, Higaki K, Kojiro M (1996) A human combined hepatocellular and cholangiocarcinoma cell line (KMCH-2) that shows the features of hepatocellular carcinoma or cholangiocarcinoma under different growth conditions. J Hepatol 24:413–422
124. Theise ND, Yao JL, Harada K, Hytiroglou P, Portmann B, Thung SN, Tsui W, Ohta H, Nakanuma Y (2003) Hepatic 'stem cell' malignancies in adults: four cases. Histopathology 43(3):263–271
125. Komuta M, Spee B, Vander Borght S, De Vos R, Verslype C, Aerts R, Yano H, Suzuki T, Matsuda M, Fujii H, Desmet VJ, Kojiro M, Roskams T (2008) Clinicopathological study on cholangiocellular carcinoma suggesting hepatic progenitor cell origin. Hepatology 47:1544–1556. doi:10.1002/hep.22238
126. Lo RC, Ng IO (2013) Hepatic progenitor cells: their role and functional significance in the new classification of primary liver cancers. Liver Cancer 2(2):84–92
127. Woo HG, Lee JH, Yoon JH, Kim CY, Lee HS, Jang JJ, Yi NJ, Suh KS, Lee KU, Park ES, Thorgeirsson SS, Kim YJ (2010) Identification of a cholangiocarcinoma-like gene expression trait in hepatocellular carcinoma. Cancer Res 70:3034–3041. doi:10.1158/0008-5472.CAN-09-2823
128. Roskams T, Katoonizadeh A, Komuta M (2010) Hepatic progenitor cells: an update. Clin Liver Dis 14(4):705–718. doi:10.1016/j.cld.2010.08.003
129. Chiba T, Kamiya A, Yokosuka O, Iwama A (2009) Cancer stem cells in hepatocellular carcinoma: recent progress and perspective. Cancer Lett 286:145–153. doi:10.1016/j.canlet.2009.04.027
130. Dumble ML, Croager EJ, Yeoh GC, Quail EA (2002) Generation and characterization of p53 null transformed hepatic progenitor cells: oval cells give rise to hepatocellular carcinoma. Carcinogenesis 23(3):435–445
131. Lee JS, Heo J, Libbrecht L, Chu IS, Kaposi-Novak P, Calvisi DF, Mikaelyan A, Roberts LR, Demetris AJ, Sun Z, Nevens F, Roskams T, Thorgeirsson SS (2006) A novel prognostic subtype of human hepatocellular carcinoma derived from hepatic progenitor cells. Nat Med 12(4):410–416
132. Holczbauer A, Factor VM, Andersen JB, Marquardt JU, Kleiner DE, Raggi C, Kitade M, Seo D, Akita H, Durkin ME, Thorgeirsson SS (2013) Modeling pathogenesis of primary liver cancer in lineage-specific mouse cell types. Gastroenterology 145:221–231. doi:10.1053/j.gastro.2013.03.013
133. Fan B, Malato Y, Calvisi DF, Naqvi S, Razumilava N, Ribback S, Gores GJ, Dombrowski F, Evert M, Chen X, Willenbring H (2012) Cholangiocarcinomas can originate from hepatocytes in mice. J Clin Invest 122:2911–2915. doi:10.1172/JCI63212

Chapter 20
Oxidative Stress Mechanisms in Hepatocarcinogenesis

Raquel Urtasun, Carmen Berasain, and Matías A. Avila

20.1 Introduction

Hepatocellular carcinoma (HCC) is among the five most common causes of cancer worldwide. HCC is an epithelial tumor developing from hepatocytes. HCC arises frequently in patients with chronic liver injury and inflammation, such as chronic viral hepatitis C and B (HCV, HBV), chronic alcoholic injury, and in patients with non-alcoholic steatohepatitis (NASH) [1–3].

Increased evidence indicates that continuous oxidative stress may play a critical role in the development of various human malignancies, including HCC [4]. The majority of HCC cases occur in fibrotic/cirrhotic livers and independently from the etiology the common denominator for hepatocarcinogenesis is chronic inflammation associated with severe oxidative stress [5]. The term oxidative stress describes adverse interactions of molecular oxygen (O_2) or of its reactive derivatives with biomolecules causing a dysbalance between the generation of cell-damaging molecules and the cellular capacity for their detoxification. The molecular players comprise both reactive oxygen (ROS) and reactive nitrogen species (RNS). There is a strong mechanistic link between chronic inflammation and cancer due to the

R. Urtasun
Division of Hepatology, Center for Applied Medical Research, University of Navarra, Pamplona, Spain

C. Berasain (✉) • M.A. Avila (✉)
Division of Hepatology, Center for Applied Medical Research, Universidad de Navarra, Pamplona, Spain

Networked Biomedical Research Center for Hepatic and Digestive Diseases, Clinical University of Navarra, Instituto de Salud Carlos III, Pamplona, Spain
e-mail: cberasain@unav.es; maavila@unav.es

E. Albano, M. Parola (eds.), *Studies on Hepatic Disorders*,
Oxidative Stress in Applied Basic Research and Clinical Practice,
DOI 10.1007/978-3-319-15539-5_20

increased production of free radicals or ROS at the site of inflammation. ROS are potential carcinogens because of their roles in mutagenesis, tumor promotion, and progression [6]. At the molecular level, increased ROS result in molecular changes which include lipid peroxidation (LPO) and oxidative DNA damage [7]. Accumulation of such genetic damage can eventually contribute to increased chromosomal aberrations associated with cell transformation [8] and liver carcinogenesis [4, 9]. Besides, ROS may also activate cellular signalling pathways strongly involved in cell proliferation and survival, such as those mediated by mitogen-activated protein kinase (MAPK), nuclear factor-κB (NF-κB), phosphatidylinositol 3-kinase (PI3K), p53 and β-catenin/Wnt, as well as others associated with angiogenesis [8].

Recent studies have shown that oxidative stress is commonly induced in all forms of chronic liver injury and plays a crucial role in hepatic fibrogenesis [10, 11] and cancer development [12]. In this regard, various groups have shown that HBV infection can induce oxidative stress using HBV transgenic mice or HBV DNA transfection of cells in vitro, while oxidative stress is also common among HBV-infected patients with chronic liver disease [13]. Markers of DNA damage are also commonly elevated in liver of patients with chronic HCV infection and correlate well with the degree of viral infection and inflammation, known risk factors for HCC [14]. Besides, HCV infection is associated with elevated levels of ROS and decreased antioxidant levels in patients [15]. In animal models, oxidative stress has been associated with the development of steatosis and liver tumors in HCV core protein transgenic mice [16]. Moreover, in the literature there are immunohistochemical studies providing evidence that LPO and oxidative DNA damage frequently occur in liver from patients with non-alcoholic fatty liver disease (NAFLD), suggesting the potential role of oxidative stress in some clinical features of NAFLD including liver fibrosis and possibly hepatocarcinogenesis [17].

Several studies have examined the influence of oxidative stress on the onset and development of HCC in humans in terms of the association of oxidative stress with poor histological differentiation and tumor size [18], increased neovascularization [19], and activation of telomerase [20]. All these findings suggest that oxidative stress biomarkers might potentially be useful for predicting the development and recurrence of HCC in patients with chronic liver disease. In this sense, some clinical studies using liver specimens obtained by biopsy or surgery have shown the predictive power of oxidative stress biomarkers on HCC development [12, 21].

In this context, this chapter summarizes the evidence supporting the contribution of oxidative stress to hepatocarcinogenesis in HBV and HCV infection, and NASH. The role of oxidative stress in alcohol-induced liver injury and carcinogenesis will be reviewed elsewhere in this book. However, it is important to bear in mind that although oxidative stress mechanisms are presumably related to carcinogenesis in HCV, HBV, and NASH cases, their actual involvement remains to be directly proven.

20.2 Evidence and Role of Oxidative Stress in HCV- HVB- and NASH -HCC Patients

20.2.1 Oxidative DNA Damage: 8-Hydroxy-2-Deoxyguanosine

ROS have been implicated in a number of pathologies, including aging, inflammatory diseases, and as previously mentioned in the development of cancer, because they can cause oxidative damage to nucleic acids, proteins, and lipids [4]. ROS include oxygen-centred radicals and nonradical compounds. Among the different radicals, the hydroxyl radical ($OH^{\bullet}$) is the most reactive and is also responsible for the formation of 8-hydroxy-2-deoxyguanosine (8-OHdG) [22]. Moreover, the reaction of peroxynitrite ($ONOO^-$) (an oxidant and nitrating agent formed in vivo by the reaction between O_2^- and NO) with guanine residues results as well in the formation of 8-OHdG [6, 23]. 8-OHdG is therefore a DNA base-modified product generated by ROS that is considered an oxidative stress marker and an indicator of oxidative DNA damage [24]. Its presence can induce a G:C to T:A transversion by replacing cytosine with adenine for coupling with guanine via oxidation at the C8 position [24]. There are studies demonstrating that 8-OHdG is implicated in carcinogenesis [24, 25] including hepatocarcinogenesis [26]. These studies in which the 8-OHdG content was measured histologically showed that this oxidation product was more abundant in HCC tissues than in non-cancerous tissue. However, the level of oxidative stress measured as 8-OHdG varies depending on the underlying liver disease (e.g., HBV infection, HCV infection, ALD, NASH, etc.). For instance, one work reported that intrahepatic accumulation of 8-OHdG was related to hepatocarcinogenesis in patients with chronic hepatitis C [27]. The same study demonstrated that a close relationship exists between oxidative DNA damage (measured by 8-OHdG quantitative immunostaining and electrochemical detection), hepatocarcinogenesis, and hepatic iron overload in patients with chronic HCV infection [27]. Later on, a comparative study with chronically HCV- and HBV-infected patients performed by Fujita et al. [28] showed 8-OHdG-positive signals in all the liver tissue, indicating that DNA damage by oxidation occurs frequently in the liver of those patients. They showed that quantitative hepatocytic 8-OHdG immunoreactivity was significantly correlated with serum transaminase levels, both in chronic hepatitis C and B. However, 8-OHdG in HCV-infected livers was significantly higher than in HBV-infected ones, suggesting that HCV infection may cause a more advanced oxidative stress in the liver during chronic infection. Moreover, with the same inflammation score, HCV patients had significantly higher 8-OHdG levels as compared to chronic hepatitis B, indicating that HCV infection itself has a more direct influence on hepatic oxidative stress formation.

A more recent work investigated 8-OHdG levels in HCC tissues from patients with different background diseases, including NASH [29]. They found that the 8-OHdG content measured immunohistochemically in cancerous and noncancerous liver tissues was significantly increased in NASH–HCC patients and comparable to that in patients with HCV-HCC [29]. Interestingly, the results suggest that the accumulation of 8-OHdG in the liver is a risk factor for carcinogenesis in NASH patients, as it is in HCV-infected patients, and that 8-OHdG content significantly increased as liver disease advanced. The degree of hepatic fat deposit seems to be relevant to hepatic oxidative stress formation in NASH because hepatic 8-OHdG levels were positively correlated with the extent of steatosis in NASH. Furthermore, in this study Tanaka et al. showed that hepatic 8-OHdG content was related not only to an increase in the inflammation score, but also to carcinogenesis. They also compared background factors as age and fasting blood sugar in patients with NASH–HCC and NASH without HCC. They concluded that NASH–HCC patients were significantly older and had higher fasting blood sugar, which is consistent with previously reported data. However, multivariate analysis of these factors and 8-OHdG content in the liver tissue revealed that old age and 8-OHdG content were independent risk factors for hepatocarcinogenesis [29].

20.2.2 LPO Markers and Etheno DNA Adducts

ROS can react with polyunsaturated fatty acids derived from membrane phospholipids or from dietary intake, resulting in the generation of LPO products such as *trans*-4-hydroxy-2-nonenal (4-HNE) and malondialdehyde (MDA) [30, 31]. MDA is one of the final products of polyunsaturated fatty acids and is widely used as a marker of oxidative stress in experimental studies. Nevertheless, its determination in plasma as a marker remains controversial as it does not originate exclusively from LPO, it is not metabolically stable, and the colorimetric determination lacks specificity. Frequently, hepatic levels of MDA are assessed by TBARS (thiobarbituric acid-reacting substance) assay [32].

Endogenous LPO products such as MDA and 4-HNE are particularly reactive in forming adducts during periods of oxidative stress. The main MDA-derived DNA adduct is 3-(2-deoxy-β-D-erythro-pentofuranosyl) pyrimido[1,2-α]purin-10(3*H*)-one (M_1dG). M_1dG adducts appear to be promutagenic as they induce mutations in oncogenes and tumor suppressor genes seen in human tumors. M_1dG adduct levels also appear to correlate with altered cell cycle control and gene expression in cultured cells [33, 34]. 4-HNE can react with DNA bases such as deoxyadenosine and deoxycytidine to f*orm the exocyclic etheno-DNA adducts including 1,* N^6-ethenodeoxyadenosine (ε-dA) and 3, N^4-ethenodeoxycytidine (ε-dC) [35]. These adducts are highly mutagenic and for instance can lead to missense mutations in the *TP53* gene which encodes the tumor suppressor p53 [36, 37]. Moreover, 4-HNE is one of the major aldehydic metabolites of LPO and is considered to be one of the most reliable markers of LPO [32].

Increased formation of 4-HNE adducts has been reported in liver tissues from patients with different chronic liver diseases including alcoholic liver injury, NASH, chronic viral hepatitis, hereditary hemochromatosis, and liver cancer [31, 38, 39]. In this sense, Zhou et al. found that ε-dA levels were significantly higher in non-tumor liver tissues with moderate and severe inflammation compared to those with absent and mild inflammation. Moreover, many ε-dA-positive hepatocytes were localized in contact with infiltrating lymphocytes in non-tumor liver tissues. These findings provided evidence of a potential link between ROS-generated ε-dA and inflammation and fibrosis in chronic liver diseases. In this work, the authors hypothesized that chronic inflammation and fibrosis could facilitate continuous accumulation of oxidative ε-DNA damage, which may partly drive chronic inflammatory and fibrotic/cirrhotic liver tissues to HCC. Therefore, measurement of ε-dA in liver tissue from patients could be explored as a potential biomarker for disease progression and cancer risk assessment in patients with cancer-prone liver diseases [40]. Furthermore, different authors carried out an ultrasensitive and specific immunoprecipitation/high performance liquid chromatography (HPLC)-fluorescence detection method for quantifying ε-dA excreted in human urine. Patients with chronic hepatitis, liver cirrhosis, and HCC due to HBV and HCV infection had more than 20 times higher urinary ε-dA levels compared to uninfected individuals without liver disease [31]. These findings show that ε-dA may play a driving role towards hepatocarcinogenesis in chronic liver diseases.

Most interestingly, various studies also reported that hepatic oxidative stress development as assessed by the level of the 4-HNE index (i.e., a semiquantitative 4-HNE immunostaining) was significantly higher in patients with NASH without HCC than in those with simple steatosis, being increased with the progression of histologic fibrosis staging in NASH [17, 29]. Interestingly, steatosis alone could not cause hepatic oxidative stress as suggested by the fact that steatosis and oxidative stress markers did not correlate in simple steatosis patients. Therefore, an additional component is necessary for the progression from simple steatosis to NASH [41].

20.2.3 Iron Overload

It is known from in vitro and in vivo studies that iron overload enhances oxidative stress [42, 43]. Hepatic iron overload is generally moderate and mixed (parenchymal and sinusoidal) in NASH [44] and might be contributory to the second component or "hit" in the "two-hit" theory proposed as an explanation of the wide spectrum of hepatic presentations of this condition [45]. In addition, hepatic iron overload ultimately participate in the progression of liver disease to HCC [46]. Supporting this idea, there are several works showing that iron may indeed be the substrate of oxidative stress and could be responsible for the second "hit" in patients with NASH [47]. In this case, iron overload could worsen oxidative stress in the liver, thus aggravating LPO [45]. Concretely, at the molecular level, the saturation of β-oxidation by excess free fatty acids will lead to the generation of hydrogen peroxide (H_2O_2), which in turn can be converted to highly reactive hydroxyl radicals in the presence of free iron via Fenton reaction [48]. Indeed, there are several findings [41, 44] showing that serum iron, transferrin saturation, and ferritin levels and the grade of hepatic iron staining are significantly higher in NASH compared with simple steatosis. Alike, quantitative analysis demonstrated that hepatocytic 8-OHdG levels were significantly correlated with these iron-related markers in patients with NASH; this correlation strongly indicated that the increase in the body-stored iron is specifically related to increased hepatocytic oxidatively generated damage to DNA in NASH patients [41]. Sorrentino et al. [49] recently reported higher liver iron overload in HCC patients in a cohort of 153 patients with NASH cirrhosis. Liver iron overload was mainly intrasinusoidal in the HCC patients, suggesting that beyond liver iron content, the localization of the deposits might also have an influence on liver cancer development. The link between liver iron overload and the risk of developing HCC secondary to liver cirrhosis is now well established, especially in patients with alcoholic or metabolic liver disease. For example, it has been proven that in patients with NASH, liver iron overload produces necrosis and inflammation, and as a consequence insulin resistance and progression of liver lesions to hepatocarcinoma [45].

Moderate iron overload is sometimes observed in the liver of patients infected with HCV. As previously mentioned, patients with HCV have higher levels of 8-OHdG in their livers than those infected with HBV [28]. A potential mediator of increased hepatic oxidative stress development in HCV is an excessive deposition of iron in the liver parenchyma, which is frequently found in chronic HCV infection [50, 51]. According to this idea, markers of hepatic iron as for example serum ferritin levels, and total iron score, were higher in HCV patients than in HBV. Interestingly, hepatic 8-OHdG levels were significantly correlated with ferritin levels and total iron scores, also indicating a strong relationship between oxidative DNA damage and iron overload status in the liver of patients with chronic hepatitis C. Therefore, it is plausible that ROS production during chronic HCV infection is due to a great extent to the high iron levels found in the hepatic tissues of these patients. This may lead to necrosis and progressive liver inflammation and increased risk for developing liver cancer [45].

However, the mechanisms involved in the modification of iron metabolism associated with chronic HCV infection remain unclear [52]. Iron depletion has demonstrated efficacy in reducing the rate of complications, improving survival, and even bringing about the regression of liver lesions in patients with genetic hemochromatosis [53]. In addition, Kawamura et al. performed a preliminary study using phlebotomy as a means to reduce excessive iron accumulation in the liver in patients with HCV infection. Transaminases levels experienced a 50 % reduction in those patients. However, the authors concluded that even if phlebotomy had beneficial effects, further maintenance therapy was important, including maintenance phlebotomy and restriction of iron intake, as demonstrated in a large number of patients with HCV infection [52].

20.2.4 Nitrosative Stress Markers

Oxidative stress also causes increased activation of transcription factors that favor increased expression of inducible nitric oxide synthase (iNOS), therefore leading to a burst in NO levels and nitrosative stress [54]. Among the key molecules responsible for nitrosative stress, NO and $ONOO^-$ species are the most abundant. NO has a high affinity for thiol (–SH) groups leading to intracellular formation of *S*-nitroso compounds. Moreover, reaction of NO with caspases and calpains, two groups of proteases that carry various cysteine residues crucial for enzymatic activity, may interfere with cellular apoptosis [55]. In addition, high levels of NO may also trigger deamination of nucleotides and thus contribute to the formation of promutagenic lesions in genomic DNA [56]. Accumulating evidence indicates that NO takes part in inflammation-mediated carcinogenesis, tumor angiogenesis, and metastasis [57]. The main cellular targets of $ONOO^-$ are thiol group containing molecules, transition metals, and carbon dioxide. In addition, nitration of protein tyrosine residues generates 3-nitrotyrosine (3-NT) [58], which can be used as a blood biomarker for nitrosative stress. Both iNOS and 3-NT levels are significantly higher in NASH, and a major role for NO during NASH-associated fibrogenesis has been suggested [59]. In addition, the reaction of $ONOO^-$ with guanine residues results in the formation of 8-HO-dG and 8-nitroguanine (DNA oxidative molecules), both markers of oxidative stress [23]. Concretely, Horiike et al. demonstrated that 8-nitroguanine was accumulated in the liver of patients with HCV-HCC patients [60].

Rahman et al. demonstrated that iNOS expression was significantly higher in HCV-positive HCC compared with non-HCV-HCC patients. This study suggested that iNOS may play an important role in prognosis of HCV-positive HCC patients and that this could be partially attributable to modulation of angiogenesis [61]. Experimental tumor models have provided more convincing evidence for a direct role of NO in tumor growth and metastasis. For example, in a rat adenocarcinoma model, in which endothelial cells of the tumor vasculature expressed iNOS, treatment with NG-nitro-L-arginine methyl ester (L-NAME), an iNOS inhibitor, decreased NO production and tumor growth. Similarly, iNOS transduction in a human colonic adenocarcinoma line led to stimulation of tumor growth and vascularity in nude mice [62] was abrogated by treatment with a selective iNOS inhibitor, 1,400 W [63].

20.2.5 Cellular Antioxidant Systems

Cells are protected from oxidative stress by intracellular antioxidant molecules like glutathione (GSH) and thioredoxin (TRX). There are also important antioxidant enzymes such as CuZn superoxide dismutase (CuZnSOD, SOD1) and Mn superoxide dismutase (MnSOD, SOD2), GSH peroxidase (GPX), catalase, and heme oxygenase [64].

20.2.5.1 Superoxide Dismutase

Several clinical studies with patients at different stages of HCC development showed a strong correlation between low levels of SOD and the severity of HCC [65]. Moreover, SOD activity in tumors and surrounding normal tissues have been found to correlate positively with postsurgery survival time in HCC patients [66]. Elevated serum MnSOD levels have been found in patients with HCC and relatively high activity of the enzyme has also been observed in patients with chronic hepatitis and liver cirrhosis [67]. MnSOD is an inducible mitochondrial enzyme that responds to the increase in ROS and acts as an antioxidant enzyme that catalyzes the dismutation of the highly reactive superoxide anion to O_2^- to the less reactive species H_2O_2. Takami et al. previously demonstrated that MnSOD expression was induced in primary cultured hepatocytes that were loaded with H_2O_2 in vitro, and that serum MnSOD levels can be used to distinguish between NASH and simple steatosis in patients with nonalcoholic fatty liver disease [68]. Recently, Tamai et al. showed that serum MnSOD and TRX levels increased as HCV-related chronic liver disease progressed, especially among patients with HCC. Although there was no correlation between serum levels of MnSOD and TRX, higher serum MnSOD levels and lower TRX levels in patients with HCC trended towards an indication of poor patient prognosis. These results suggest that serum MnSOD and TRX levels are not only potential biomarkers for HCV-related liver disease progression, but may also serve as prognostic markers in HCC [69].

On the other hand, overexpression of CuZnSOD (SOD1) in the HCC cell line HepG2 has been shown to suppress cell growth in vitro and to reduce tumor mass in vivo [70]. Therefore, CuZnSOD levels may play an important role in the development and prognosis of HCC. CuZnSOD converts superoxide O_2^- to H_2O_2 [71]. CuZnSOD transgenic and mutant mice have been widely used to study the role of oxygen-free radicals in different experimental systems [72]. Increased levels of CuZnSOD usually confer resistance to acute oxidative insults, whereas lack of expression renders null mice more sensitive to oxidative stress. Elchuri et al. [70] demonstrated the long-term effects of CuZnSOD deficiency in mutant mice. They described histopathological, biochemical, and gene expression alterations that occur during the course of hepatocarcinogenesis in CuZnSOD-deficient mice. Cellular targets known to be sensitive to oxidative damage are affected by CuZnSOD deficiency, and the resulting alterations most likely lead to a disturbance of cellular homeostasis and ultimately to the onset of hepatocarcinogenesis later in life. In humans, decreased protein levels and activity of CuZnSOD have been reported in cirrhosis [73]. In this same regard, a more recent study demonstrated a significant decrease in the levels of total antioxidant capacity and total SOD activity in HCC tissue samples [40]. These data provide evidence that the antioxidant complement of HCC tissue is severely impaired, and that HCC tissues undergo serious oxidative stress. The poorer differentiated HCC had the lower levels of total antioxidant capacity and total SOD activity. On the contrary, the poorer differentiated HCC had the higher prevalence of ε-dA [40].

20.2.5.2 Glutathione (GSH/GSSG)

Glutathione can exist in either a reduced (GSH), or an oxidized (GSSG) form. Within the cell glutathione is present mainly in the reduced form, which can be converted to the oxidized form during oxidative stress. The redox status depends on the relative amount of the reduced and oxidized forms of glutathione (GSH/GSSG), and this ratio is a critical determinant for evaluating the oxidative stress in the cells [74]. Cellular glutathione and related enzymes such as glutathione peroxidase, glutathione *S*-transferase, and glutathione reductase are among the principal protective mechanisms against endogenous and exogenous toxic substances and free radicals-mediated damage in different tissues including liver [75]. Decreased GSH levels have been found in a large number of HCV patients [76], while diminished antioxidant defenses including GSH levels is also a major factor promoting oxidative stress in NASH patients [77].

Hanna et al. demonstrated a decrease in the GSH levels in cirrhotic and cancer tissues compared to healthy liver tissues. GSH homeostasis at the cellular level is maintained by the balance between biosynthesis, uptake, oxidation, and export. Its decrease in cirrhotic and cancer tissues is probably related to a reduced synthesis of the tripeptide by the diseased liver. Indeed, a significant proportion of dietary methionine is converted into cysteine, one of the three GSH amino acids precursors, in the liver. However, methionine metabolism is significantly impaired in the chronically injured and cirrhotic liver, compromising GSH synthesis [78]. This alteration may influence the capability of the liver to provide protection against oxidative damage [79]. Interestingly, another study showed that the most poorly differentiated HCCs had the lowest levels of GSH and total glutathione, as well as the ratios of GSH/GSSG and GSH/total glutathione [80].

20.2.5.3 Nuclear Factor E2-Related Factor 2

Nuclear factor E2-related factor 2 (Nrf2) is a transcription factor protein. It controls the ability of the cell to cope with oxidative stress by elevating the expression of key genes for cellular detoxification. Nrf2 recognizes a conserved antioxidant response element (ARE) within the promoters of the responsive genes [81]. It has been reported that activation of Nrf2 as a result of HCV infection is mediated by the mitogen-activated protein (MAP) kinases p38 MAPK and janus kinase (JAK). In HCV-infected livers, the HCV core protein is found to be a potent regulator of the antioxidant defense Nrf2/ARE pathway [82]. However, in another contradictory study researchers found suppressed Nrf2/ARE pathway in an in vitro cellular HCV infection model [83].

Studies with Nrf2-deficient mice demonstrated the role of this transcription factor in protecting liver from xenobiotic-mediated hepatocarcinogenesis. During long-term treatment with 2-amino-3-methylimidazo[4,5-f]quinoline, a carcinogenic and mutagenic heterocyclic amine derivative, the multiplicity and incidence of liver tumors in male and female mice were significantly higher in Nrf2-knockout than in wild-type animals [84].

Figure 20.1 summarizes the most relevant molecular markers of oxidative stress and cellular antioxidant defense systems described above in the context of HVB, HCV, and NASH-related hepatocarcinogenesis.

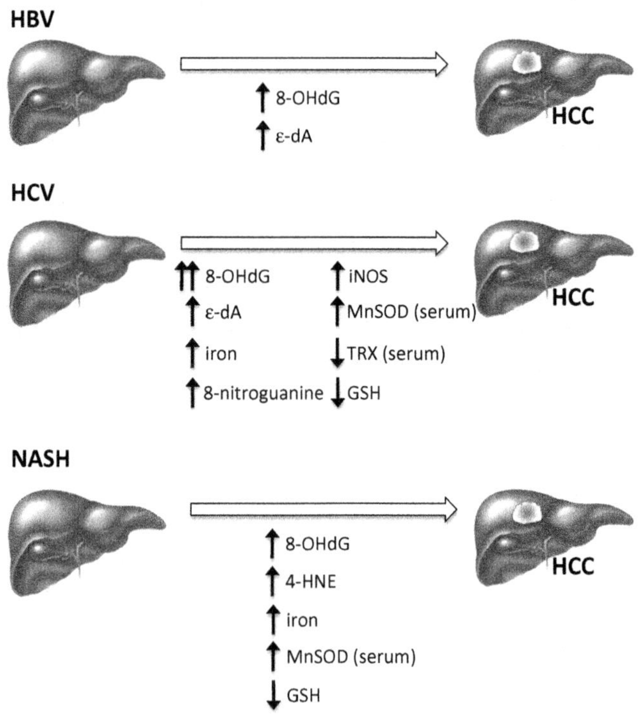

Fig. 20.1 Summary of the most relevant molecular markers of oxidative stress and cellular antioxidant defense systems described the context of HVB, HCV, and NASH-related hepatocarcinogenesis. See text for details

20.3 Oxidative Stress-Activated Cellular Signalling Pathways and HCC

Intracellular signalling pathways are highly sensitive to changes in redox homeostasis through redox sensing iron or thiol groups present in certain proteins. Mechanistic studies have shown that oxidative stress activates signalling cascades, which can seriously influence the regulation of cell growth and transformation processes [85]. ROS are linked with various cellular signalling pathways, including the activation of heat shock response activator protein 1, MAPKs, the nuclear factor kappa B (NF-κB), protein kinase C (PKC), signal transducers and activators of transcription 3 (STAT3), peroxisome proliferator-activated receptor-γ (PPARγ), and as previously described Nrf2, which engage in the regulation of pro-oxidant and antioxidant genes expression, mediating oxidative injury and also antioxidant defense systems regulation [85]. Moreover, ROS and RNS may activate the expression of proto-oncogenes, but also trigger the apoptotic machinery of the cell via mitochondrial or extra-mitochondrial signals [23]. These different cellular outcomes may be dependent on the intensity and duration of the signals, and therefore on the magnitude and persistence of the pro-oxidant situation. Large variations in these responses have been found among different tumor cell types, suggesting that the impact of oxidative stress on cell signaling and cellular responses can be very different depending on the context and the dynamic changes of redox homeostasis [23].

20.3.1 NF-κB and HCC Development

NF-κB is a eukaryotic transcription factor family. It is normally found in the cytoplasm as an inactive dimer, with the most common being the p50–p65 heterodimer, bound to the inhibitory subunit IκB. Upon activation, NF-κB is released from IκB and translocates to the nucleus, where it binds target sequences in the promoter of responsive genes. As mentioned above, ROS are able to trigger both the apoptotic and necrotic cell death depending on the severity of the oxidative stress [86]. One of the possible factors which are responsible for the maintenance of the balance between cell life and death is the activity of the NF-κB pathway. Although the expression of NF-κB target genes in most events promotes cell survival, there are some exceptions when NF-κB activation may lead to cell death [86]. Importantly, NF-κB is a well-known redox-sensitive factor. Through the modulation of different intracellular kinase-mediated signaling pathways, ROS can trigger NF-κB activation in the cytoplasm and its translocation into the nucleus. In contrast, the oxidation (or nitrosylation) of the critical cysteine residue Cys62 in the p50 NF-κB subunit can inhibit its DNA-binding activity [87]. On the other hand, ablation of NF-κB activity by hepatocyte-specific deletion of Nemo, the regulatory subunit of the IκB kinase complex (IKK), resulted in increased oxidative stress and hepatocellular injury upon liver ischemia and reperfusion [88]. Therefore, ROS can modulate NF-κB response, and NF-κB activity can in turn attenuate ROS production and promote survival, in part through the regulation of the expression of antioxidant proteins [89]. In multicellular organisms, activation of the apoptotic program in response to oxidative stress is considered as a mechanism preventing seriously damaged cells from accumulating DNA mutations that could lead to cancer or other diseases. The fragile balance between life and death in that case could depend on a proper activation of the NF-κB pathway. The antiapoptotic activity of these signaling proteins is crucial in numerous human diseases [86]. Specifically, the NF-κB signalling system can play an important part in hepatocarcinogenesis, with different roles depending on the liver cell-type, and also influencing the pathological setting underlying HCC development [90]. NF-κB can act as a tumor-suppressor in hepatocytes, inhibiting liver cancer development by promoting hepatocyte survival during the injury/tumor-initiation phase. When NF-κB is inactivated in hepatocytes, cell death and subsequent compensatory proliferation increase, leading to enhanced carcinogenesis. A critical pro-survival mechanism controlled by NF-κB in hepatocytes involves the oxidative stress response, as demonstrated by experiments utilizing antioxidants in different mouse models [91, 92]. It has been shown that hepatocyte-specific deletion of IKKb, the catalytic subunit of the IKK complex, resulted in markedly increased ROS levels after acute treatment with the hepatic carcinogen diethylnitrosamine (DEN) and in enhanced DEN-induced hepatocarcinogenesis.

This phenotype was attributed in part to the diminished expression of antioxidant enzymes such as SOD2 in the livers of IKKb-deficient mice [93]. However, it should be considered that NF-κB also regulates the expression of inflammatory enzymes, including the inducible form of NO synthase and the inducible form of cyclooxygenase (COX2) in macrophages [94], further stressing the differential roles that NF-κB may play in hepatocarcinogenesis depending on the cell type where it is activated. This is illustrated by the fact that IKKb inactivation in myeloid cells inhibits HCC development in DEN-treated mice [92].

As previously mentioned, in HCV infection there is a decrease in liver antioxidant defenses and an increased generation of ROS. It was directly demonstrated that the HCV core and NS5A nonstructural proteins caused oxidation of mitochondrial glutathione leading to increased ROS in this organelle [95, 96]. Interestingly, NS5A-induced activation of NF-κB can be neutralized by antioxidants [96]. More recently, expression of the full-length open reading frame of an infectious cDNA clone of HCV in non-transformed human hepatocytes resulted in increased ROS levels and concomitantly triggered NF-κB activity, a response that has been suggested to have antiapoptotic effects and favor viral persistence and therefore the risk of developing HCC [97].

20.3.2 p38α MAPK Pathway and HCC

p38α is the major p38/MAPK isoform, being activated in response to inflammation and oxidative stress. p38α negatively regulates cell cycle progression and controls the expression of different inflammatory mediators, as well as antioxidant genes [98]. In an interesting study, Sakurai et al. showed that p38α MAPK pathway activation prevented ROS accumulation in mice acutely exposed to DEN, and that hepatocyte-specific deletion of this MAPK resulted in significantly enhanced DEN-induced hepatocarcinogenesis [93]. Later on, the same authors described that p38α activity is also important for suppression of ROS accumulation upon thioacetamide (TAA) administration, which leads to induction of fibrosis, cirrhosis, and HCC [99]. In both models, p38α activity prevents ROS accumulation by controlling the expression of heat shock protein (HSP) 25, the mouse homolog of human HSP27. Restoration of HSP25 expression in the p38α-deficient liver prevents TAA-induced ROS accumulation and fibrogenesis [99].

20.4 Role of Oxidative Stress in Experimental Hepatocarcinogenesis

Animal models of HCC can help to understand the cellular and molecular mechanisms underlying the pathogenesis of this type of tumor. Within these experimental models, there are genetic models including conditional knock-out or transgenic mice, which are mainly used to study the involvement of specific gene/protein in the carcinogenic process [100, 101]. On the other hand, models of chemically induced HCC can mimic the injury-fibrosis-malignancy cycle if a genotoxic compound is administered together with or followed by a promoting agent. Here we next revise several mouse models useful for HCC investigation in which oxidative stress is involved.

20.4.1 Genetically Modified Mouse Models

20.4.1.1 Transgenic Mice Expressing HBV Proteins

Different studies have shown that overexpression of the HBV large surface antigen (IsAg) and HBV X protein (HBx) in the liver leads to HCC development in transgenic mice [102]. Regarding the potential pro-carcinogenic mechanisms involved, transgenic mice expressing the HBV proteins (including HBx) display elevated hepatic oxidative stress levels compared to nontransgenic controls, with a concurrent increase in oxidative DNA damage [103]. HBx protein directly interacts with mitochondrial membrane proteins, altering mitochondrial membrane potential and mitochondrial electron transport; in consequence, metabolism of hepatocytes is altered and ROS production is increased [104] (Fig. 20.2). These findings were

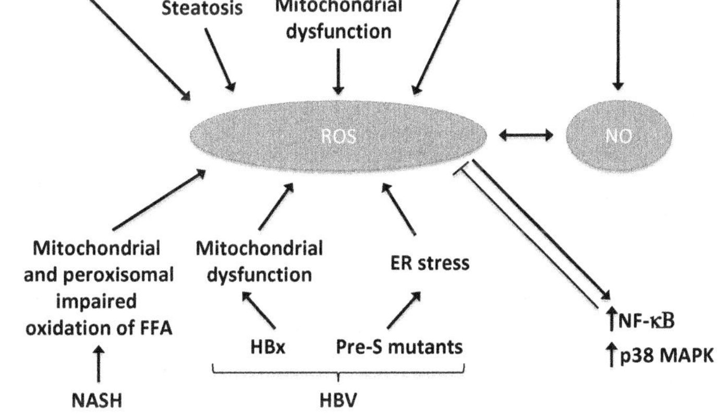

Fig. 20.2 Principal cellular mechanisms leading to the generation of free radicals in the HBV- and HCV-infected liver, as well as in NASH

extended to demonstrate that ROS scavengers were able to inhibit HBx-mediated mitochondrial membrane depolarization and subsequent apoptosis [105]. Later on, silymarin (a natural antioxidant polyphenol) was used to prevent the HCC in a HBx transgenic mouse model [106] demonstrating its beneficial effects on the early stages of liver pathogenesis, preventing and delaying liver carcinogenesis [106].

Wang et al. [107] identified two types of LHBs (large surface proteins of HBV) with deletions at the pre-S1 (ΔS1-LHBs) and pre-S2 (ΔS2-LHBs) in ground glass hepatocytes obtained from patients with advanced diseases of chronic HBV infection. Oxidative DNA damage was observed in livers of transgenic mice carrying pre-S mutant LHBs, as well as in ground glass hepatocytes [108] (Fig. 20.2). Pre-S mutant LHBs induce oxidative stress and lead to oxidative DNA damage of HBV-infected hepatocytes through endoplasmic reticulum stress signalling pathways. The oxidative DNA damage caused by pre-S mutant LHBs may result in genomic instability and mutation of liver cells, and ultimately lead to HCC. The overexpression of LHBs proteins in transgenic mice has been shown to be cytopathic and could lead to liver injury, regenerative hyperplasia, chronic inflammation, oxidative DNA damage, hepatocyte aneuploidy, and eventually progression to HCC [107, 109].

20.4.1.2 HCV Core Protein Transgenic Mice

Transgenic mice models have been widely used to study the biological actions of HCV proteins in the liver, including carcinogenesis. Various transgenic mouse models are available expressing different HCV proteins. Hepatitis C virus gene products have been expressed either alone or in combination in the liver of transgenic mice by using different liver-specific promoters. Early studies engineered transgenic mouse lines harboring genes excised from the cDNA of the HCV genome of genotype 1b [16, 110]. The mouse lines were derived from a C57BL/6 strain, which is known for its rare spontaneous occurrence of HCC. Three different transgenic mouse lines were established, which carried the core gene, envelope genes, or non-structural genes respectively, under the same transcriptional control element. Among these mouse lines, only the transgenic mice carrying the core gene developed HCCs [110, 111]. The expression of HCV core leads to progressive hepatic steatosis in several lines of constitutive transgenic mice, followed by HCC after 80–105 weeks in 32 % of the male mice [111, 112]. These observations indicate that the core protein of HCV per se has an oncogenic potential when expressed in vivo. The envelope genes transgenic mice did not develop hepatocellular injury or neoplastic lesions by 16 months of age, despite high expression levels of both E1 and E2 proteins [113]. Similarly, the transgenic mice carrying the entire non-structural genes did not develop HCC [114]. When the complete viral protein complement is expressed, steatosis and HCC occur in 15 % of the mice after approximately 90–100 weeks [112]. However, other researchers did not find the tumor phenotype, indicating that genetic background of mice and/or oncogene expression level plays an important role in the outcome of these transgenic mice models [52]. There are also studies showing animals which develop only steatosis [115] or different phenotypes, including oxidative injury [95], depending on the promoter used, the context of expression, and the background of the mouse strain used.

In general, it is difficult to clarify the mechanism of carcinogenesis even for a simple model in which only the core protein is expressed in otherwise normal liver tissues. Although the core protein predominantly exists in the cytoplasm associated with lipid droplets, it is also present in the mitochondria and nuclei [110, 111]. As previously mentioned, one consequence of the expression of the core protein is an increased production of oxidative stress in the liver. The localization of the core protein in the nuclei supported the idea that this protein may act as a transcriptional regulator and affect the proliferative ability of the cells, thereby being associated with hepatocarcinogenesis. The core protein alters mitochondrial function and results directly in an increase in the cellular abundance of ROS. The consequences of this state of chronic oxidative stress include a reduction in mitochondrial metabolic processes, which might contribute to the development of steatosis by inhibition of β-oxidation and oxidative damage to both mitochondrial and chromosomal DNA [16, 95] (Fig. 20.2). Hepatic steatosis in hepatitis C may work as fuel for oxidative stress overproduction [116]. Accumulated lipids may be utilized as substrates for ROS production, thereby placing patients with chronic hepatitis C and NASH onto the track to HCC development [116] (Fig. 20.2).

20.4.1.3 Methionine Adenosyltransferase 1a (*Mat1a*) Null Mice

From NASH to cirrhosis and HCC, methionine metabolism is progressively impaired. *S*-adenosylmethionine (SAMe), generated in the first reaction of methionine metabolism catalyzed by methionine adenosyltransferase (MAT), is the principal biological methyl donor and a precursor for GSH and polyamines synthesis [117]. Chronic hepatic SAMe deficiency generates oxidative stress [118, 119] and predisposes to injury and malignant transformation [78, 117]. Different isoforms of MAT exist. MAT II is the product of the *MAT2A* gene and is expressed in fetal hepatocytes and in all adult cell types except the mature hepatocyte. MAT I/III is encoded by the *MAT1A* gene, which is actively transcribed in adult healthy hepatocytes [120]. Importantly, *MAT1A* expression and MATI/III activity are progressively lost in chronic human liver disease, and this gene is ultimately silenced in HCC [78, 121]. Therefore, *MAT1A* expression is considered a marker of the healthy and differentiated liver phenotype. *Mat1A* knockout (KO) mice have increased hepatic oxidative stress, impaired mitochondrial function, and developed steatohepatitis and HCC spontaneously [122]. The oxidative stress displayed by *Mat1A* KO mice is due in part to reduced levels of the antioxidant GSH and to an altered expression of genes involved in lipid oxidation such as *Cyp4a10* and *Cyp4a14* [122, 123].

Concretely, *Mat1A* KO mice show early signs of liver inflammation and spontaneously develop steatohepatitis by 8 mounts of age. Persistent inflammation is accompanied by hepatocellular proliferation, and by 1 year of life, a significant proportion of mice develop liver tumors [122, 123]. Interestingly, *Mat1A* KO mice show increased hepatic expression of Cyp2E1, a cytochrome involved in the metabolism of a variety of xenobiotics, including carcinogens, that favors the generation of ROS. The administration of a Cyp2E1 inhibitor to *Mat1A* KO mice markedly reduced liver injury triggered by carbon tetrachloride (CCl_4), underscoring the importance of oxidative stress in liver injury associated with *Mat1A* deficiency [122].

20.4.1.4 *Acyl-coenzyme A Oxidase* Null Mice

Acyl-coenzyme A oxidase (AOX) is the rate-limiting enzyme of peroxisomal β-oxidation of long chain fatty acids (LCFA). Accordingly, *AOX* KO mice have defective peroxisomal β-oxidation of LCFA, which accumulate in the liver leading to the development of steatohepatitis. *AOX* KO mice begin to exhibit microvesicular fatty change of hepatocytes at 7 days of age [124]. By 30 days of age, the *AOX* KO mouse liver shows a more severe steatosis. Later on, by 4–5 months of age, elevated expression of PPAR-α, cytochrome P450 genes (Cyp4a10 and Cyp4a14), and increased levels of H_2O_2 are observed. However, a compensatory increase of fatty acid oxidation is observed by 6–7 months of age, and hepatic steatosis recovers by regeneration of hepatocytes. Finally, *AOX* KO mice develop hepatocellular adenomas and HCCs by 15 months of age [124]. Interestingly, the increased hepatocellular proliferative activity found in *AOX* KO mice at pre-neoplastic stages (2–8 months of age) was accompanied by a marked intrahepatic production of H_2O_2 [125].

20.4.1.5 *Mdr2* Null Mice

The Mdr2 protein, encoded by *abcb4*, is a phospholipid flippase that promotes biliary secretion of phospholipids. In the bile, phospholipids form mixed micelles with bile acids and cholesterol, thereby protecting the biliary epithelium from the damaging effects of bile acids. *Mdr2*-deficient mice develop cholestasis with features closely resembling critical features of chronic cholestatic liver diseases, including primary sclerosing cholangitis, fibrosis, and with a background-dependent latency of several months, preneoplastic lesions slowly progressing to metastatic liver cancer [126]. *Mdr2*-deficient mice, compared with wild-type mice, included deregulation of genes that control lipid synthesis, storage, and oxidation; decreased serum levels of cholesterol and phospholipids; and reduced hepatic long-chain fatty acyl-CoAs (LCA-CoAs). Low levels of LCA-CoAs could result from enhanced catabolism (e.g., via peroxisomal oxidation), which leads to medium-chain fatty acyl-CoAs (MCA-CoAs) accumulation and oxidative stress via H_2O_2 [127]. Importantly, dietary administration of antioxidant compounds (tannic acid or selenomethionine) had a partial chemopreventive effect on HCC development in *Mdr2* Ko mice [128].

20.4.2 Mouse Models of Chemically Induced Oxidative Stress and HCC

There are few rodent models of HCC arising within a background of regenerative nodules and cirrhosis, and most depend on the administration of hepatotoxic and/or carcinogenic agents. Ideally, these models should reproduce the injury–fibrosis–malignancy cycle seen in chronic human liver diseases. However, most chemotoxic-induced HCC models do not completely recreate the pathophysiology of the human disease. Only those chemicals generating oxidative stress are described in this section.

20.4.2.1 Diethylnitrosamine

The genotoxic drug DEN has been used since the 1960s to induce HCCs in rodents and is the most widely employed chemical to induce liver cancer in mice. DEN is known to induce DNA damage, mutations, oxidative stress, and hepatocyte apoptosis [92, 129]. DEN allows the induction of HCC development in a time- and dose-dependent manner, being also easy to reproduce. The DEN model has been largely used to study the pathophysiology of HCC [92, 130]. However, it is important to bear in mind that in this model the sequence of events leading from chronic inflammation and fibrosis to cirrhosis and tumor development is completely skipped.

DEN is a DNA-alkylating agent leading to the formation of mutagenic DNA adducts. DEN undergoes metabolic activation in hepatocytes by enzymes of the cytochrome P450 family generating ROS. This molecule acts as a complete carcinogen, able to induce damage to DNA, proteins, and lipids, leading to hepatocyte death. Similar to what occurs in patients, HCC development in this model usually follows a slow multistep sequence, where cycles of necrosis and regeneration promote neoplastic transformation. The progression from early dysplastic lesions to fully malignant tumors is associated with an increased occurrence of genomic alterations [25, 131]. DEN is typically administered to mice between 12 and 15 days of age by a single intraperitoneal injection (5–25 μg/g body weight). Using this protocol, originally described by Vesselinovitch and Mihailovich [132], 100 % of B6C3F1 male mice developed HCCs, on average 44 weeks after intraperitoneal injection of DEN [132]. Interestingly, this model also reproduces the increased incidence of HCC in male individuals that is observed in patients [130]. When administered later in life, tumor promotion is required and can be achieved by different ways, for instance by phenobarbital or CCl_4 administration, partial hepatectomy, or as recently demonstrated by high fat diet feeding [133]. These protocols are less efficient in producing HCCs than those previously mentioned, but may resemble better the clinical situation [134]. DEN is metabolized in centrilobular (zone 3) hepatocytes leading to ROS production [93]. Moreover, administration of the chemical antioxidant butylated hydroxyanisole (BHA) to *Ikkβ*Δhep mice, lacking hepatocyte IKKβ (a negatively regulator of ROS formation), prevents DEN-induced ROS accumulation and liver damage, thereby attenuating HCC development [92].

20.4.2.2 Thioacetamide

TAA is a hepatotoxin that can be administered either in drinking water or by intraperitoneal injection. The hepatotoxic action is a result of the pro-oxidant properties of the compound, leading to hepatic oxidative stress and liver damage [135]. TAA causes hepatocellular necrosis after biotransformation to an active metabolite via the flavine adenine dinucleotide monooxygenase pathway resulting in the formation of TAA-S-oxide [136]. Usually administered TAA doses range from 400 to 600 mg/kg and the frequency of administration also varies from once daily for 2 days to once daily for 3 days. More recently, chronic TAA administration has been shown to lead to the development of cirrhosis [137]. In addition, chronic administration of TAA for 6 months to rats that previously received a single injection of DEN (200 mg/kg body weight) resulted in the appearance of dysplastic nodules and tumors [138]. Interestingly, experimental studies in rodents have shown that the administration of the antioxidant polyphenols silymarin and silibinin protects against hepatotoxicity and tumor formation induced by TAA [139].

20.4.2.3 Carbon Tetrachloride

CCl_4 is one of the most potent hepatotoxins [140]. CCl_4 administration is the most commonly employed model of liver disease. CCl_4 can be administered in drinking water, in inhaled gases, or by intraperitoneal injection. The hepatotoxicity of CCl_4 involves two phases [141]. First, CCl_4 is metabolized by Cyp2E1 to form trichloromethyl radicals [142]. These radicals are highly energetic and cause LPO and membrane damage. Second, ROS can activate Kupffer cells (the resident liver macrophages) leading to an inflammatory response, which results in the secretion of cytokines, chemokines, and other pro-inflammatory factors [143]. These factors not only have a direct cytotoxic effect, but also attract and activate monocytes, neutrophils, and lymphocytes, which contribute to the tissue damage.

In general, compounds like phenobarbital, ethanol, and acetone induce microsomal cytochrome p450 and therefore potentiate the hepatotoxicity of CCl_4. For example, weekly injections of CCl_4 accompanied by alcohol administration through drinking water lead to HCC development after 104 weeks [25, 140]. It is believed that the repeated cycle of injury, inflammation, and repair leads to fibrosis and eventually HCC. As previously mentioned, CCl_4 can be combined with DEN and this approach results in faster HCC development (approximately 4 months) on a background of chronic inflammation, fibrosis, and liver injury similar to the natural history of HCC development in humans [144].

20.4.2.4 Choline- and Methionine-Deficient Diets (CMD)

A chronic deficiency of the major dietary methyl group donors, methionine, choline, folic acid, and vitamin B12, can induce the development of liver cancer in rodents. Over the years, a number of varied studies have demonstrated that diets lacking methyl donors may act as carcinogens that can induce liver tumor formation in the absence of any exogenous carcinogens. Dietary methyl deficiency causes the induction of oxidative stress in the livers of exposed rats and mice [145]. Mice subjected to a long-term choline-deficient diet (CDD) develop tumors after 50–52 weeks. A CDD induces steatosis in all mice; nevertheless, fat accumulation can differ considerably between species and is not necessarily strain-dependent. The proposed mechanism of carcinogenicity by CDD is through the formation of oval cells, a result of oxidative DNA damage and chromosomal instability because of the depletion of hepatic antioxidant mechanisms [114]. It has been assumed that oval cells, either directly or indirectly through the generation of hepatocytes, function as tumor progenitors [114]. The CDD can be combined with the administration of a hepatotoxic compound, such as DEN or CCl_4, and serves as a good model for steatohepatitis with further development to HCC [146].

20.4.3 Genetically Modified Models with Enhanced Susceptibility to Chemically Induced Hepatocarcinogenesis

Lin et at. showed that knockout of the toll-like receptor 2 (TLR2) gene enhances DEN-induced hepatocarcinogenesis [147]. They demonstrated that TLR2 signaling plays a defense role against HCC through eliciting intracellular senescence and maintaining autophagy flux in liver cells. In addition, these authors further proved that ROS and endoplasmic reticulum (ER) stress are directly responsible for the aggravation of liver carcinogenesis in TLR2-deficient mice. Interestingly, treatment of these animals with the antioxidant agent *N*-acetyl-cysteine (NAC) can eliminate ROS accumulation and alleviate ER stress in DEN-injured liver tissue [147]. Furthermore, this study showed that pretreatment of TLR2-deficient mice with NAC attenuated the development and progression of HCC.

Maeda et al. demonstrated that mice lacking IKKβ or P38α only in hepatocytes exhibited a marked increase in HCC occurrence following treatment with DEN [92]. P38α and IKKβ negatively regulate ROS accumulation and c-jun kinase (JNK) activity through different mechanisms. All of them maintain hepatocyte viability and suppress liver carcinogenesis as demonstrated by the fact that both IkkβΔhep [92] and p38αΔhep [99] mice accumulate higher levels of ROS in these cells after DEN administration and exhibit increased hepatocellular death than control mice. Yet, neither IkkβΔhep nor p38αΔhep mice show spontaneous liver damage or HCC formation unless challenged with a carcinogen.

20.5 Antioxidants in HCC Chemoprevention

As discussed above, abundant evidences indicate that oxidative stress indeed plays a central role in HBV, HCV, and NASH-associated liver damage and HCC development. This knowledge opens the possibility of devising strategies aimed at the inhibition of oxidative stress, and therefore quell HCC development. Among them, chemoprevention has to be considered as an interesting approach that could lower the current morbidity and mortality associated with HCC [148]. In this regard, there are numerous studies showing that different antioxidants act not only as free radical scavengers, but also as direct modulators of multiple cell signaling pathways through diverse mechanisms [149]. A number of chemopreventive agents have been examined both in animal models of HCC and in humans. Chemoprevention against HCC in mouse models has previously been achieved with antioxidants, including resveratrol [150], vitamin E [151], tannic acid [152], BHA [153], and some selenium compounds [154]. In most cases, the chemopreventive agent was supplemented during several months, but in some models a short prenatal and postnatal exposure resulted in significant decrease of HCC incidence [154]. In particular, from some studies conducted both in vivo and in vitro, resveratrol emerged as a promising molecule that inhibits carcinogenesis [150]. Various effects and mechanisms of action have been proposed, but some investigations indicated that resveratrol can suppress the growth of HCC cells and prevent hepatocarcinogenesis by mitigating oxidative stress [155, 156].

Interestingly, most of the antioxidants used as chemopreventive agents come from diet. The relationship between diet and liver cancer is less studied relative to other gastrointestinal cancers. In addition, there are some contradictory observations regarding the antioxidant effects of dietary antioxidants in preventing HCC. For example, fruits and vegetables are a rich source of antioxidants, such as retinol, carotenoids, vitamin C, vitamin E, and selenium. In general, they are thought to exert protective effects against cancer. However, as shown by different studies, not all antioxidant nutrients might be protective against HCC [157]. For example, in some studies carotenoids were shown to suppress liver cancer, in patients with viral hepatitis and cirrhosis [158, 159]. Moreover, carotenoids appear to play an important role in the prevention of hepatitis virus infection-related liver carcinogenesis. In contrast, vitamin C consumption appeared to be associated with an increased risk of HCC. Although vitamin C has antioxidant potential, it also acts to stimulate the absorption of iron from food [160], and iron overload is considered a risk factor for HCC [161]. Although limited in number, intervention studies using supplemental vitamin E also do not indicate any overall benefit for liver cancer prevention. However, there is some evidence that supplemental vitamin E in individuals with preexisting liver conditions (e.g., hepatitis and HCC) may have some benefits. The most convincing anti-cancer data for vitamin E come from animal studies [162]. Dietary selenium is an essential mineral for both humans and animals. It functions as a component of several proteins including glutathione peroxidases (GPx) and thioredoxin reductases among others. Several studies have examined whether changes in oxidative stress correlated with effects on carcinogenicity by selenium intake. Selenium increases serum and liver GPx activity, but did not decrease the oxidative damage indices. Overall, the role of selenium in human and experimental liver cancer is not clear. Although some studies have shown that feeding higher amounts of selenium inhibits initiation, and promotion of carcinogenesis, others show no effect or even an enhancement of carcinogenesis [162]. There are also controversial data around NAC. NAC is a precursor of glutathione and increases glutathione levels, and it also can scavenge free radicals itself. However, NAC did not prevent DEN-induced HCC [162].

As previously mentioned, reduced hepatic SAMe availability is associated with chronic liver disease progression and hepatocarcinogenesis [78]. Several experimental studies have cogently demonstrated the chemopreventive potential of long-term SAMe administration [163, 164]. The underlying mechanisms to this chemopreventive effect of SAMe are likely to be multifarious, given the high number of different biochemical reactions in which this methyl-donor can participate. Nevertheless, the antioxidant properties of SAMe, due in part to its contribution to GSH synthesis, could participate in the chemopreventive mechanisms of this molecule [163].

Numerous experimental studies have also addressed the chemopreventive activity of polyphenols, flavonoids, and non-flavonoids, against HCC. These compounds found in plants are characterized by the presence of hydroxyl groups bound to aromatic rings and display notorious antitumoral effects both in vitro and in in vivo models of HCC, as previously mentioned for resveratrol [165]. As occurs with SAMe, the mechanisms of the preventive and/or antitumoral effects of polyphenols are likely to be pleiotropic, affecting many cellular signalling pathways and mediators [162, 165]. Nevertheless, and in spite of the chemical heterogeneity of polyphenols, common denominators in their biological effects are the potentiation of antioxidant defense systems, including antioxidant enzymes, and the inhibition of inflammatory pathways and ROS production [162, 165].

In spite of the abundant experimental information on the antitumoral effects of antioxidants in HCC, few clinical trials have been carried out [166]. This paucity of clinical studies is due in part to the lack of standardized biomarkers to monitor efficacy. Toxicity concerns are also present, considering that patients at risk of developing HCC normally have an impaired liver function, and therefore should be carefully selected. Nevertheless, these clinical interventions are worth exploring. The identification of effective chemopreventive strategies able to halt chronic liver disease progression and HCC development is much needed. Moreover, these chemopreventive interventions are also likely to be cost-effective.

Acknowledgments *Work in the authors' laboratory is funded by*: The agreement between FIMA and the "UTE project CIMA"; RTICC-RD06 00200061; CIBEREhd; FIS PI10/02642, PI13/00359, PI10/00038 and PI13/00385 from Instituto de Salud Carlos III. R. Urtasun is supported by a "Torres Quevedo" contract from Ministerio de Educación, Spain.

References

1. Goodgame B, Shaheen NJ, Galanko J, El-Serag HB (2003) The risk of end stage liver disease and hepatocellular carcinoma among persons infected with hepatitis C virus: publication bias? Am J Gastroenterol 98(11):2535–2542
2. Reid AE (2001) Nonalcoholic steatohepatitis. Gastroenterology 121(3):710–723
3. Befeler AS, di Bisceglie AM (2002) Hepatocellular carcinoma: diagnosis and treatment. Gastroenterology 122(6):1609–1619

4. Valko M, Leibfritz D, Moncol J, Cronin MTD, Mazur M, Telser J (2007) Free radicals and antioxidants in normal physiological functions and human disease. Int J Biochem Cell Biol 39(1):44–84
5. Seitz HK, Stickel F (2006) Risk factors and mechanisms of hepatocarcinogenesis with special emphasis on alcohol and oxidative stress. Biol Chem 387(4):349–360
6. Waris G, Ahsan H (2006) Reactive oxygen species: role in the development of cancer and various chronic conditions. J Carcinog 5:14
7. Hussain SP, Hofseth LJ, Harris CC (2003) Radical causes of cancer. Nat Rev Cancer 3(4):276–285
8. Martindale JL, Holbrook NJ (2002) Cellular response to oxidative stress: signaling for suicide and survival. J Cell Physiol 192(1):1–15
9. He G, Yu G-Y, Temkin V, Ogata H, Kuntzen C, Sakurai T et al (2010) Hepatocyte necrosis induced by oxidative stress and IL-1 alpha release mediate carcinogen-induced compensatory proliferation and liver tumorigenesis. Cancer Cell 17(3):286–297
10. Marra F, Gastaldelli A, Svegliati-Baroni G, Tell G, Tiribelli C (2008) Molecular basis and mechanisms of progression of non-alcoholic steatohepatitis. Trends Mol Med 14(2):72–81
11. Brenner DA, Seki E, Taura K, Kisseleva T, Deminicis S, Iwaisako K et al (2011) Non-alcoholic steatohepatitis-induced fibrosis: toll-like receptors, reactive oxygen species and Jun N-terminal kinase. Hepatol Res 41(7):683–686
12. Tanaka S, Mogushi K, Yasen M, Ban D, Noguchi N, Irie T et al (2011) Oxidative stress pathways in noncancerous human liver tissue to predict hepatocellular carcinoma recurrence: a prospective, multicenter study. Hepatology 54(4):1273–1281
13. Hsieh Y-H, Su I-J, Wang H-C, Chang W-W, Lei H-Y, Lai M-D et al (2004) Pre-S mutant surface antigens in chronic hepatitis B virus infection induce oxidative stress and DNA damage. Carcinogenesis 25(10):2023–2032
14. De Maria N, Colantoni A, Fagiuoli S, Liu GJ, Rogers BK, Farinati F et al (1996) Association between reactive oxygen species and disease activity in chronic hepatitis C. Free Radic Biol Med 21(3):291–295
15. Seronello S, Sheikh MY, Choi J (2007) Redox regulation of hepatitis C in nonalcoholic and alcoholic liver. Free Radic Biol Med 43(6):869–882
16. Moriya K, Nakagawa K, Santa T, Shintani Y, Fujie H, Miyoshi H et al (2001) Oxidative stress in the absence of inflammation in a mouse model for hepatitis C virus-associated hepatocarcinogenesis. Cancer Res 61(11):4365–4370
17. Seki S, Kitada T, Sakaguchi H, Nakatani K, Wakasa K (2003) Pathological significance of oxidative cellular damage in human alcoholic liver disease. Histopathology 42(4):365–371
18. Ichiba M, Maeta Y, Mukoyama T, Saeki T, Yasui S, Kanbe T et al (2003) Expression of 8-hydroxy-2′-deoxyguanosine in chronic liver disease and hepatocellular carcinoma. Liver Int 23(5):338–345
19. Jo M, Nishikawa T, Nakajima T, Okada Y, Yamaguchi K, Mitsuyoshi H et al (2011) Oxidative stress is closely associated with tumor angiogenesis of hepatocellular carcinoma. J Gastroenterol 46(6):809–821
20. Nishikawa T, Nakajima T, Katagishi T, Okada Y, Jo M, Kagawa K et al (2009) Oxidative stress may enhance the malignant potential of human hepatocellular carcinoma by telomerase activation. Liver Int 29(6):846–856
21. Chuma M, Hige S, Nakanishi M, Ogawa K, Natsuizaka M, Yamamoto Y et al (2008) 8-Hydroxy-2′-deoxy-guanosine is a risk factor for development of hepatocellular carcinoma in patients with chronic hepatitis C virus infection. J Gastroenterol Hepatol 23(9):1431–1436
22. Shigenaga MK, Gimeno CJ, Ames BN (1989) Urinary 8-hydroxy-2′-deoxyguanosine as a biological marker of in vivo oxidative DNA damage. Proc Natl Acad Sci U S A 86(24):9697–9701
23. Goetz ME, Luch A (2008) Reactive species: a cell damaging rout assisting to chemical carcinogens. Cancer Lett 266(1):73–83
24. Kasai H (1997) Analysis of a form of oxidative DNA damage, 8-hydroxy-2′-deoxyguanosine, as a marker of cellular oxidative stress during carcinogenesis. Mutat Res 387(3):147–163

25. Farazi PA, DePinho RA (2006) Hepatocellular carcinoma pathogenesis: from genes to environment. Nat Rev Cancer 6(9):674–687
26. Nakae D, Kobayashi Y, Akai H, Andoh N, Satoh H, Ohashi K et al (1997) Involvement of 8-hydroxyguanine formation in the initiation of rat liver carcinogenesis by low dose levels of N-nitrosodiethylamine. Cancer Res 57(7):1281–1287
27. Kato J, Kobune M, Nakamura T, Kuroiwa G, Takada K, Takimoto R et al (2001) Normalization of elevated hepatic 8-hydroxy-2′-deoxyguanosine levels in chronic hepatitis C patients by phlebotomy and low iron diet. Cancer Res 61(24):8697–8702
28. Fujita N, Sugimoto R, Ma N, Tanaka H, Iwasa M, Kobayashi Y et al (2008) Comparison of hepatic oxidative DNA damage in patients with chronic hepatitis B and C. J Viral Hepat 15(7):498–507
29. Tanaka S, Miyanishi K, Kobune M, Kawano Y, Hoki T, Kubo T et al (2013) Increased hepatic oxidative DNA damage in patients with nonalcoholic steatohepatitis who develop hepatocellular carcinoma. J Gastroenterol 48(11):1249–1258
30. Bartsch H, Nair J (2006) Chronic inflammation and oxidative stress in the genesis and perpetuation of cancer: role of lipid peroxidation, DNA damage, and repair. Langenbecks Arch Surg 391(5):499–510
31. Nair J, Srivatanakul P, Haas C, Jedpiyawongse A, Khuhaprema T, Seitz HK et al (2010) High urinary excretion of lipid peroxidation-derived DNA damage in patients with cancer-prone liver diseases. Mutat Res 683(1–2):23–28
32. Esterbauer H, Gebicki J, Puhl H, Jürgens G (1992) The role of lipid peroxidation and antioxidants in oxidative modification of LDL. Free Radic Biol Med 13(4):341–390
33. Chaudhary AK, Nokubo M, Reddy GR, Yeola SN, Morrow JD, Blair IA et al (1994) Detection of endogenous malondialdehyde-deoxyguanosine adducts in human liver. Science 265(5178):1580–1582
34. Ji C, Rouzer CA, Marnett LJ, Pietenpol JA (1998) Induction of cell cycle arrest by the endogenous product of lipid peroxidation, malondialdehyde. Carcinogenesis 19(7):1275–1283
35. El Ghissassi F, Barbin A, Nair J, Bartsch H (1995) Formation of 1, N6-ethenoadenine and 3, N4-ethenocytosine by lipid peroxidation products and nucleic acid bases. Chem Res Toxicol 8(2):278–283
36. Horio Y, Chen A, Rice P, Roth JA, Malkinson AM, Schrump DS (1996) Ki-ras and p53 mutations are early and late events, respectively, in urethane-induced pulmonary carcinogenesis in A/J mice. Mol Carcinog 17(4):217–223
37. Barbin A, Froment O, Boivin S, Marion MJ, Belpoggi F, Maltoni C et al (1997) p53 gene mutation pattern in rat liver tumors induced by vinyl chloride. Cancer Res 57(9):1695–1698
38. Paradis V, Kollinger M, Fabre M, Holstege A, Poynard T, Bedossa P (1997) In situ detection of lipid peroxidation by-products in chronic liver diseases. Hepatology 26(1):135–142
39. Weltman MD, Farrell GC, Hall P, Ingelman-Sundberg M, Liddle C (1998) Hepatic cytochrome P450 2E1 is increased in patients with nonalcoholic steatohepatitis. Hepatology 27(1):128–133
40. Zhou L, Yang Y, Tian D, Wang Y (2013) Oxidative stress-induced 1, N6-ethenodeoxyadenosine adduct formation contributes to hepatocarcinogenesis. Oncol Rep 29(3):875–884
41. Fujita N, Miyachi H, Tanaka H, Takeo M, Nakagawa N, Kobayashi Y et al (2009) Iron overload is associated with hepatic oxidative damage to DNA in nonalcoholic steatohepatitis. Cancer Epidemiol Biomarkers Prev 18(2):424–432
42. Kadiiska MB, Burkitt MJ, Xiang QH, Mason RP (1995) Iron supplementation generates hydroxyl radical in vivo. An ESR spin-trapping investigation. J Clin Invest 96(3):1653–1657
43. Cornejo P, Varela P, Videla LA, Fernández V (2005) Chronic iron overload enhances inducible nitric oxide synthase expression in rat liver. Nitric Oxide 13(1):54–61
44. Fargion S, Mattioli M, Fracanzani AL, Sampietro M, Tavazzi D, Fociani P et al (2001) Hyperferritinemia, iron overload, and multiple metabolic alterations identify patients at risk for nonalcoholic steatohepatitis. Am J Gastroenterol 96(8):2448–2455

45. Nahon P, Ganne-Carrié N, Trinchet JC, Beaugrand M (2010) Hepatic iron overload and risk of hepatocellular carcinoma in cirrhosis. Gastroenterol Clin Biol 34(1):1–7
46. Martinelli ALC, Filho ABA, Franco RF, Tavella MH, Ramalho LNZ, Zucoloto S et al (2004) Liver iron deposits in hepatitis B patients: association with severity of liver disease but not with hemochromatosis gene mutations. J Gastroenterol Hepatol 19(9):1036–1041
47. Blendis L, Oren R, Halpern Z (2000) NASH: can we iron out the pathogenesis? Gastroenterology 118(5):981–983
48. Videla LA, Fernández V, Tapia G, Varela P (2003) Oxidative stress-mediated hepatotoxicity of iron and copper: role of Kupffer cells. Biometals 16(1):103–111
49. Sorrentino P, D'Angelo S, Ferbo U, Micheli P, Bracigliano A, Vecchione R (2009) Liver iron excess in patients with hepatocellular carcinoma developed on non-alcoholic steato-hepatitis. J Hepatol 50(2):351–357
50. Sebastiani G, Tempesta D, Alberti A (2012) Hepatic iron overload is common in chronic hepatitis B and is more severe in patients coinfected with hepatitis D virus. J Viral Hepat 19(2):e170–e176
51. Piperno A, Vergani A, Malosio I, Parma L, Fossati L, Ricci A et al (1998) Hepatic iron overload in patients with chronic viral hepatitis: role of HFE gene mutations. Hepatology 28(4):1105–1109
52. Kawamura Y, Akuta N, Sezaki H, Hosaka T, Someya T, Kobayashi M et al (2005) Determinants of serum ALT normalization after phlebotomy in patients with chronic hepatitis C infection. J Gastroenterol 40(9):901–906
53. Falize L, Guillygomarc'h A, Perrin M, Lainé F, Guyader D, Brissot P et al (2006) Reversibility of hepatic fibrosis in treated genetic hemochromatosis: a study of 36 cases. Hepatology 44(2):472–477
54. Marshall HE, Merchant K, Stamler JS (2000) Nitrosation and oxidation in the regulation of gene expression. FASEB J 14(13):1889–1900
55. Li C-Q, Wogan GN (2005) Nitric oxide as a modulator of apoptosis. Cancer Lett 226(1):1–15
56. Cooke MS, Evans MD, Dizdaroglu M, Lunec J (2003) Oxidative DNA damage: mechanisms, mutation, and disease. FASEB J 17(10):1195–1214
57. Lala PK, Chakraborty C (2001) Role of nitric oxide in carcinogenesis and tumour progression. Lancet Oncol 2(3):149–156
58. Urtasun R, Cubero FJ, Vera M, Nieto N (2009) Reactive nitrogen species switch on early extracellular matrix remodeling via induction of MMP1 and TNFalpha. Gastroenterology 136(4):1410–1422, e1–e4
59. Muriel P (2009) Role of free radicals in liver diseases. Hepatol Int 3(4):526–536
60. Horiike S, Kawanishi S, Kaito M, Ma N, Tanaka H, Fujita N et al (2005) Accumulation of 8-nitroguanine in the liver of patients with chronic hepatitis C. J Hepatol 43(3):403–410
61. Rahman MA, Dhar DK, Yamaguchi E, Maruyama S, Sato T, Hayashi H et al (2001) Coexpression of inducible nitric oxide synthase and COX-2 in hepatocellular carcinoma and surrounding liver: possible involvement of COX-2 in the angiogenesis of hepatitis C virus-positive cases. Clin Cancer Res 7(5):1325–1332
62. Roles of nitric oxide in tumor growth (1995). http://www.pnas.org/content/92/10/4392.short
63. Thomsen LL, Scott JM, Topley P, Knowles RG, Keerie AJ, Frend AJ (1997) Selective inhibition of inducible nitric oxide synthase inhibits tumor growth in vivo: studies with 1400W, a novel inhibitor. Cancer Res 57(15):3300–3304
64. Tsan MF, Clark RN, Goyert SM, White JE (2001) Induction of TNF-α and MnSOD by endotoxin: role of membrane CD14 and toll-like receptor-4. Am J Physiol Cell Physiol 280(6):C1422–C1430
65. Casaril M, Corso F, Bassi A, Capra F, Gabrielli GB, Stanzial AM et al (1994) Decreased activity of scavenger enzymes in human hepatocellular carcinoma, but not in liver metastases. Int J Clin Lab Res 24(2):94–97
66. Lin MT, Wang MY, Liaw KY, Lee PH, Chien SF, Tsai JS et al (2001) Superoxide dismutase in hepatocellular carcinoma affects patient prognosis. Hepatogastroenterology 48(40):1102–1105

67. Clemente C, Elba S, Buongiorno G, Guerra V, D'Attoma B, Orlando A et al (2007) Manganese superoxide dismutase activity and incidence of hepatocellular carcinoma in patients with Child-Pugh class A liver cirrhosis: a 7-year follow-up study. Liver Int 27(6):791–797
68. Takami Y, Uto H, Tamai T, Sato Y, Ishida Y-I, Morinaga H et al (2010) Identification of a novel biomarker for oxidative stress induced by hydrogen peroxide in primary human hepatocytes using the 2-nitrobenzenesulfenyl chloride isotope labeling method. Hepatol Res 40(4):438–445
69. Tamai T, Uto H, Takami Y, Oda K, Saishoji A, Hashiguchi M et al (2011) Serum manganese superoxide dismutase and thioredoxin are potential prognostic markers for hepatitis C virus-related hepatocellular carcinoma. World J Gastroenterol 17(44):4890–4898
70. Elchuri S, Oberley TD, Qi W, Eisenstein RS, Jackson Roberts L, Van Remmen H et al (2005) CuZnSOD deficiency leads to persistent and widespread oxidative damage and hepatocarcinogenesis later in life. Oncogene 24(3):367–380
71. Fridovich I (1995) Superoxide radical and superoxide dismutases. Annu Rev Biochem 64:97–112
72. Kruidenier L, van Meeteren ME, Kuiper I, Jaarsma D, Lamers CB, Zijlstra FJ et al (2003) Attenuated mild colonic inflammation and improved survival from severe DSS-colitis of transgenic Cu/Zn-SOD mice. Free Radic Biol Med 34(6):753–765
73. Skrzycki M, Ścibior D, Podsiad M, Czeczot H (2008) Activity and protein level of CuZnSOD and MnSOD in benign and malignant liver tumors. Clin Biochem 41(1–2):91–96
74. Chai YC, Ashraf SS, Rokutan K, Johnston RB, Thomas JA (1994) S-thiolation of individual human neutrophil proteins including actin by stimulation of the respiratory burst: evidence against a role for glutathione disulfide. Arch Biochem Biophys 310(1):273–281
75. Ścibior D, Skrzycki M, Podsiad M, Czeczot H (2008) Glutathione level and glutathione-dependent enzyme activities in blood serum of patients with gastrointestinal tract tumors. Clin Biochem 41(10–11):852–858
76. Ivanov AV, Bartosch B, Smirnova OA, Isaguliants MG, Kochetkov SN (2013) HCV and oxidative stress in the liver. Viruses 5(2):439–469
77. Sumida Y, Niki E, Naito Y, Yoshikawa T (2013) Special issue on "Oxidative stress and redox signaling in the gastrointestinal tract and related organs" for Free Radical Research Involvement of free radicals and oxidative stress in NAFLD/NASH. Free Radic Res 47(11):869–880
78. Lu SC, Mato JM (2012) S-adenosylmethionine in liver health, injury, and cancer. Physiol Rev 92(4):1515–1542
79. Fernandez-Checa JC, Kaplowitz N (2005) Hepatic mitochondrial glutathione: transport and role in disease and toxicity. Toxicol Appl Pharmacol 204(3):263–273
80. Lee K-T, Tsai S-M, Wang S-N, Lin S-K, Wu S-H, Chuang S-C et al (2007) Glutathione status in the blood and tissues of patients with virus-originated hepatocellular carcinoma. Clin Biochem 40(15):1157–1162
81. Motohashi H, Yamamoto M (2004) Nrf2-Keap1 defines a physiologically important stress response mechanism. Trends Mol Med 10(11):549–557
82. Ivanov AV, Smirnova OA, Ivanova ON, Masalova OV, Kochetkov SN, Isaguliants MG (2011) Hepatitis C virus proteins activate NRF2/ARE pathway by distinct ROS-dependent and independent mechanisms in HUH7 cells. PLoS One 6(9):e24957
83. Carvajal-Yepes M, Himmelsbach K, Schaedler S, Ploen D, Krause J, Ludwig L et al (2011) Hepatitis C virus impairs the induction of cytoprotective Nrf2 target genes by delocalization of small Maf proteins. J Biol Chem 286(11):8941–8951
84. Kitamura Y, Umemura T, Kanki K, Kodama Y, Kitamoto S, Saito K et al (2007) Increased susceptibility to hepatocarcinogenicity of Nrf2-deficient mice exposed to 2-amino-3-methylimidazo[4,5-f]quinoline. Cancer Sci 98(1):19–24
85. Anisimov VN (2007) Biology of aging and cancer. Cancer Control 14(1):23–31
86. Siomek A (2012) NF-κB signaling pathway and free radical impact. Acta Biochim Pol 59(3):323–331

87. Kabe Y, Ando K, Hirao S, Yoshida M, Handa H (2005) Redox regulation of NF-kappaB activation: distinct redox regulation between the cytoplasm and the nucleus. Antioxid Redox Signal 7(3–4):395–403
88. Beraza N, Lüdde T, Assmus U, Roskams T, Vander Borght S, Trautwein C (2007) Hepatocyte-specific IKK gamma/NEMO expression determines the degree of liver injury. Gastroenterology 132(7):2504–2517
89. Kamata H, Honda S-I, Maeda S, Chang L, Hirata H, Karin M (2005) Reactive oxygen species promote TNFalpha-induced death and sustained JNK activation by inhibiting MAP kinase phosphatases. Cell 120(5):649–661
90. DiDonato JA, Mercurio F, Karin M (2012) NF-κB and the link between inflammation and cancer. Immunol Rev 246(1):379–400
91. Luedde T, Beraza N, Kotsikoris V, van Loo G, Nenci A, De Vos R et al (2007) Deletion of NEMO/IKKgamma in liver parenchymal cells causes steatohepatitis and hepatocellular carcinoma. Cancer Cell 11(2):119–132
92. Maeda S, Kamata H, Luo J-L, Leffert H, Karin M (2005) IKKbeta couples hepatocyte death to cytokine-driven compensatory proliferation that promotes chemical hepatocarcinogenesis. Cell 121(7):977–990
93. Sakurai T, He G, Matsuzawa A, Yu G-Y, Maeda S, Hardiman G et al (2008) Hepatocyte necrosis induced by oxidative stress and IL-1 alpha release mediate carcinogen-induced compensatory proliferation and liver tumorigenesis. Cancer Cell 14(2):156–165
94. Glauert HP, Calfee-Mason K, Li Y, Nilakantan V, Twaroski ML, Tharappel J et al (2009) The role of NF-κB in PPAR α-mediated hepatocarcinogenesis. PPAR Res 2008:286249
95. Okuda M, Li K, Beard MR, Showalter LA, Scholle F, Lemon SM et al (2002) Mitochondrial injury, oxidative stress, and antioxidant gene expression are induced by hepatitis C virus core protein. Gastroenterology 122(2):366–375
96. Gong G, Waris G, Tanveer R, Siddiqui A (2001) Human hepatitis C virus NS5A protein alters intracellular calcium levels, induces oxidative stress, and activates STAT-3 and NF-kappa B. Proc Natl Acad Sci U S A 98(17):9599–9604
97. Tang W, Lázaro CA, Campbell JS, Parks WT, Katze MG, Fausto N (2007) Responses of nontransformed human hepatocytes to conditional expression of full-length hepatitis C virus open reading frame. Am J Pathol 171(6):1831–1846
98. Nakagawa H, Maeda S (2012) Molecular mechanisms of liver injury and hepatocarcinogenesis: focusing on the role of stress-activated MAPK. Patholog Res Int 2012:172894
99. Sakurai T, Kudo M, Umemura A, He G, Elsharkawy AM, Seki E et al (2013) p38α inhibits liver fibrogenesis and consequent hepatocarcinogenesis by curtailing accumulation of reactive oxygen species. Cancer Res 73(1):215–224
100. Feng H, Hu B, Liu K-W, Li Y, Lu X, Cheng T et al (2011) Activation of Rac1 by Src-dependent phosphorylation of Dock180(Y1811) mediates PDGFRα-stimulated glioma tumorigenesis in mice and humans. J Clin Invest 121(12):4670–4684
101. Koch KS, Maeda S, He G, Karin M, Leffert HL (2009) Targeted deletion of hepatocyte Ikkbeta confers growth advantages. Biochem Biophys Res Commun 380(2):349–354
102. Chisari FV, Klopchin K, Moriyama T, Pasquinelli C, Dunsford HA, Sell S et al (1989) Molecular pathogenesis of hepatocellular carcinoma in hepatitis B virus transgenic mice. Cell 59(6):1145–1156
103. Hagen TM, Huang S, Curnutte J, Fowler P, Martinez V, Wehr CM et al (1994) Extensive oxidative DNA damage in hepatocytes of transgenic mice with chronic active hepatitis destined to develop hepatocellular carcinoma. Proc Natl Acad Sci U S A 91(26):12808–12812
104. Lee YI, Hwang JM, Im JH, Lee YI, Kim NS, Kim DG et al (2004) Human hepatitis B virus-X protein alters mitochondrial function and physiology in human liver cells. J Biol Chem 279(15):15460–15471
105. Shirakata Y, Koike K (2003) Hepatitis B virus X protein induces cell death by causing loss of mitochondrial membrane potential. J Biol Chem 278(24):22071–22078

106. Wu Y-F, Fu S-L, Kao C-H, Yang C-W, Lin C-H, Hsu M-T et al (2008) Chemopreventive effect of silymarin on liver pathology in HBV X protein transgenic mice. Cancer Res 68(6):2033–2042
107. Wang H-C, Huang W, Lai M-D, Su I-J (2006) Hepatitis B virus pre-S mutants, endoplasmic reticulum stress and hepatocarcinogenesis. Cancer Sci 97(8):683–688
108. Su I-J, Wang H-C, Wu H-C, Huang W-Y (2008) Ground glass hepatocytes contain pre-S mutants and represent preneoplastic lesions in chronic hepatitis B virus infection. J Gastroenterol Hepatol 23(8 Pt 1):1169–1174
109. Huang SN, Chisari FV (1995) Strong, sustained hepatocellular proliferation precedes hepatocarcinogenesis in hepatitis B surface antigen transgenic mice. Hepatology 21(3):620–626
110. Moriya K, Yotsuyanagi H, Shintani Y, Fujie H, Ishibashi K, Matsuura Y et al (1997) Hepatitis C virus core protein induces hepatic steatosis in transgenic mice. J Gen Virol 78(Pt 7):1527–1531
111. Moriya K, Fujie H, Shintani Y, Yotsuyanagi H, Tsutsumi T, Ishibashi K et al (1998) The core protein of hepatitis C virus induces hepatocellular carcinoma in transgenic mice. Nat Med 4(9):1065–1067
112. Lerat H, Honda M, Beard MR, Loesch K, Sun J, Yang Y et al (2002) Steatosis and liver cancer in transgenic mice expressing the structural and nonstructural proteins of hepatitis C virus. Gastroenterology 122(2):352–365
113. Koike K, Moriya K, Ishibashi K, Matsuura Y, Suzuki T, Saito I et al (1995) Expression of hepatitis C virus envelope proteins in transgenic mice. J Gen Virol 76(Pt 12):3031–3038
114. Heindryckx F, Colle I, Van Vlierberghe H (2009) Experimental mouse models for hepatocellular carcinoma research. Int J Exp Pathol 90(4):367–386
115. Perlemuter G, Sabile A, Letteron P, Vona G, Topilco A, Chrétien Y et al (2002) Hepatitis C virus core protein inhibits microsomal triglyceride transfer protein activity and very low density lipoprotein secretion: a model of viral-related steatosis. FASEB J 16(2):185–194
116. Koike K (2005) Steatosis in chronic hepatitis C: fuel for overproduction of oxidative stress? J Gastroenterol 40(6):664–665
117. Baric I, Cuk M, Fumić K, Vugrek O, Allen RH, Glenn B et al (2005) S-Adenosylhomocysteine hydrolase deficiency: a second patient, the younger brother of the index patient, and outcomes during therapy. J Inherit Metab Dis 28(6):885–902
118. Avila MA, Mingorance J, Martínez-Chantar M-L, Casado M, Martín-Sanz P, Boscá L et al (1997) Regulation of rat liver S-adenosylmethionine synthetase during septic shock: role of nitric oxide. Hepatology 25(2):391–396
119. Avila MA, García-Trevijano ER, Lu SC, Corrales FJ, Mato JM (2004) Methylthioadenosine. Int J Biochem Cell Biol 36(11):2125–2130
120. Latasa MU, Boukaba A, García-Trevijano ER, Torres L, Rodríguez JL, Caballeria J et al (2001) Hepatocyte growth factor induces MAT2A expression and histone acetylation in rat hepatocytes: role in liver regeneration. FASEB J 15(7):1248–1250
121. Avila MA, Berasain C, Torres L, Martin-Duce A, Corrales FJ, Yang H et al (2000) Reduced mRNA abundance of the main enzymes involved in methionine metabolism in human liver cirrhosis and hepatocellular carcinoma. J Hepatol 33(6):907–914
122. Martínez-Chantar ML, Corrales FJ, Martínez-Cruz LA, García-Trevijano ER, Huang Z-Z, Chen L et al (2002) Spontaneous oxidative stress and liver tumors in mice lacking methionine adenosyltransferase 1A. FASEB J 16(10):1292–1294
123. Lu SC, Alvarez L, Huang Z-Z, Chen L, An W, Corrales FJ et al (2001) Methionine adenosyltransferase 1A knockout mice are predisposed to liver injury and exhibit increased expression of genes involved in proliferation. Proc Natl Acad Sci U S A 98(10):5560–5565
124. Cook WS, Jain S, Jia Y, Cao WQ, Yeldandi AV, Reddy JK et al (2001) Peroxisome proliferator-activated receptor alpha-responsive genes induced in the newborn but not prenatal liver of peroxisomal fatty acyl-CoA oxidase null mice. Exp Cell Res 268(1):70–76
125. Fan C-Y, Pan J, Usuda N, Yeldandi AV, Rao MS, Reddy JK (1998) Steatohepatitis, spontaneous peroxisome proliferation and liver tumors in mice lacking peroxisomal fatty acyl-CoA

oxidase: implications for peroxisome proliferator-activated receptor α natural ligand metabolism. J Biol Chem 273(25):15639–15645
126. Mauad TH, van Nieuwkerk CM, Dingemans KP, Smit JJ, Schinkel AH, Notenboom RG et al (1994) Mice with homozygous disruption of the mdr2 P-glycoprotein gene. A novel animal model for studies of nonsuppurative inflammatory cholangitis and hepatocarcinogenesis. Am J Pathol 145(5):1237–1245
127. Moustafa T, Fickert P, Magnes C, Guelly C, Thueringer A, Frank S, et al (2012) Alterations in lipid metabolism mediate inflammation, fibrosis, and proliferation in a mouse model of chronic cholestatic liver injury. Gastroenterology 142(1):140–151.e12
128. Katzenellenbogen M, Mizrahi L, Pappo O, Klopstock N, Olam D, Barash H et al (2007) Molecular mechanisms of the chemopreventive effect on hepatocellular carcinoma development in Mdr2 knockout mice. Mol Cancer Ther 6(4):1283–1291
129. Kakehashi A, Wei M, Fukushima S, Wanibuchi H (2013) Oxidative stress in the carcinogenicity of chemical carcinogens. Cancers (Basel) 5(4):1332–1354
130. Naugler WE, Sakurai T, Kim S, Maeda S, Kim K, Elsharkawy AM et al (2007) Gender disparity in liver cancer due to sex differences in MyD88-dependent IL-6 production. Science 317(5834):121–124
131. Thorgeirsson SS, Grisham JW (2002) Molecular pathogenesis of human hepatocellular carcinoma. Nat Genet 31(4):339–346
132. Vesselinovitch SD, Mihailovich N (1983) Kinetics of diethylnitrosamine hepatocarcinogenesis in the infant mouse. Cancer Res 43(9):4253–4259
133. Park S, Choi Y, Um S-J, Yoon SK, Park T (2011) Oleuropein attenuates hepatic steatosis induced by high-fat diet in mice. J Hepatol 54(5):984–993
134. Lee G-H (2000) Review article: paradoxical effects of phenobarbital on mouse hepatocarcinogenesis. Toxicol Pathol 28(2):215–225
135. Ingawale DK, Mandlik SK, Naik SR (2014) Models of hepatotoxicity and the underlying cellular, biochemical and immunological mechanism(s): a critical discussion. Environ Toxicol Pharmacol 37(1):118–133
136. Chieli E, Malvaldi G (1984) Role of the microsomal FAD-containing monooxygenase in the liver toxicity of thioacetamide S-oxide. Toxicology 31(1):41–52
137. Petermann H, Heymann S, Vogl S, Dargel R (1996) Phagocytic function and metabolite production in thioacetamide-induced liver cirrhosis: a comparative study in perfused livers and cultured Kupffer cells. J Hepatol 24(4):468–477
138. Gervasi PG, Longo V, Marzano M, Saviozzi M, Malvaldi G (1989) Chronic liver injury by thioacetamide and promotion of hepatic carcinogenesis. J Cancer Res Clin Oncol 115(1):29–35
139. Ahangar-Darabi M, Setorki M, Rafieian-kopaei M (2013) The effect of silymarin on liver injury induced by thioacetamide in rats. J HerbMed Pharmacol 2(2):29–33
140. Weisburger EK (1977) Carcinogenicity studies on halogenated hydrocarbons. Environ Health Perspect 21:7–16
141. Avasarala S, Yang L, Sun Y, Leung AW-C, Chan W-Y, Cheung W-T et al (2006) A temporal study on the histopathological, biochemical and molecular responses of CCl(4)-induced hepatotoxicity in Cyp2e1-null mice. Toxicology 228(2–3):310–322
142. Sheweita SA, El-Gabar MA, Bastawy M (2001) Carbon tetrachloride changes the activity of cytochrome P450 system in the liver of male rats: role of antioxidants. Toxicology 169(2):83–92
143. Luckey SW, Petersen DR (2001) Activation of Kupffer cells during the course of carbon tetrachloride-induced liver injury and fibrosis in rats. Exp Mol Pathol 71(3):226–240
144. Dapito DH, Mencin A, Gwak G-Y, Pradere J-P, Jang M-K, Mederacke I et al (2012) Promotion of hepatocellular carcinoma by the intestinal microbiota and TLR4. Cancer Cell 21(4):504–516
145. Pogribny IP, James SJ, Beland FA (2012) Molecular alterations in hepatocarcinogenesis induced by dietary methyl deficiency. Mol Nutr Food Res 56(1):116–125

146. de Lima VMR, Oliveira CPMS, Alves VAF, Chammas MC, Oliveira EP, Stefano JT et al (2008) A rodent model of NASH with cirrhosis, oval cell proliferation and hepatocellular carcinoma. J Hepatol 49(6):1055–1061
147. Lin H, Liu X-B, Yu J-J, Hua F, Hu Z-W (2013) Antioxidant N-acetylcysteine attenuates hepatocarcinogenesis by inhibiting ROS/ER stress in TLR2 deficient mouse. PLoS One 8(10):e74130
148. Ueno Y, Sollano JD, Farrell GC (2009) Prevention of hepatocellular carcinoma complicating chronic hepatitis C. J Gastroenterol Hepatol 24(4):531–536
149. Hayes JD, McMahon M (2001) Molecular basis for the contribution of the antioxidant responsive element to cancer chemoprevention. Cancer Lett 174(2):103–113
150. Bishayee A, Politis T, Darvesh AS (2010) Resveratrol in the chemoprevention and treatment of hepatocellular carcinoma. Cancer Treat Rev 36(1):43–53
151. Calvisi DF, Ladu S, Hironaka K, Factor VM, Thorgeirsson SS (2004) Vitamin E down-modulates iNOS and NADPH oxidase in c-Myc/TGF-alpha transgenic mouse model of liver cancer. J Hepatol 41(5):815–822
152. Nepka C, Sivridis E, Antonoglou O, Kortsaris A, Georgellis A, Taitzoglou I et al (1999) Chemopreventive activity of very low dose dietary tannic acid administration in hepatoma bearing C3H male mice. Cancer Lett 141(1–2):57–62
153. Sun B, Karin M (2008) NF-kappaB signaling, liver disease and hepatoprotective agents. Oncogene 27(48):6228–6244
154. Popova NV (2002) Perinatal selenium exposure decreases spontaneous liver tumorogenesis in CBA mice. Cancer Lett 179(1):39–42
155. Pirola L, Fröjdö S (2008) Resveratrol: one molecule, many targets. IUBMB Life 60(5):323–332
156. Athar M, Back JH, Kopelovich L, Bickers DR, Kim AL (2009) Multiple molecular targets of resveratrol: anti-carcinogenic mechanisms. Arch Biochem Biophys 486(2):95–102
157. Bjelakovic G, Nikolova D, Simonetti RG, Gluud C (2004) Antioxidant supplements for prevention of gastrointestinal cancers: a systematic review and meta-analysis. Lancet 364(9441):1219–1228
158. Murakoshi M, Nishino H, Satomi Y, Takayasu J, Hasegawa T, Tokuda H et al (1992) Potent preventive action of alpha-carotene against carcinogenesis: spontaneous liver carcinogenesis and promoting stage of lung and skin carcinogenesis in mice are suppressed more effectively by alpha-carotene than by beta-carotene. Cancer Res 52(23):6583–6587
159. Krinsky NI (1989) Carotenoids as chemopreventive agents. Prev Med 18(5):592–602
160. Lynch SR (1997) Interaction of iron with other nutrients. Nutr Rev 55(4):102–110
161. Kowdley KV (2004) Iron, hemochromatosis, and hepatocellular carcinoma. Gastroenterology 127(5):S79–S86
162. Glauert HP, Calfee-Mason K, Stemm DN, Tharappel JC, Spear BT (2010) Dietary antioxidants in the prevention of hepatocarcinogenesis: a review. Mol Nutr Food Res 54(7): 875–896
163. Pascale RM, Simile MM, De Miglio MR, Feo F (2002) Chemoprevention of hepatocarcinogenesis: S-adenosyl-L-methionine. Alcohol 27(3):193–198
164. Lu SC, Ramani K, Ou X, Lin M, Yu V, Ko K et al (2009) S-adenosylmethionine in the chemoprevention and treatment of hepatocellular carcinoma in a rat model. Hepatology 50(2):462–471
165. Stagos D, Amoutzias GD, Matakos A, Spyrou A, Tsatsakis AM, Kouretas D (2012) Chemoprevention of liver cancer by plant polyphenols. Food Chem Toxicol 50(6):2155–2170
166. Kwon KH, Barve A, Yu S, Huang M-T, Kong A-NT (2007) Cancer chemoprevention by phytochemicals: potential molecular targets, biomarkers and animal models. Acta Pharmacol Sin 28(9):1409–1421

Chapter 21
The Role of Oxidative Stress in Hepatocarcinogenesis

Helmut K. Seitz and Sebastian Mueller

21.1 Introduction

The incidence of hepatocellular cancer (HCC) is rising worldwide. HCC is the most frequent complication of hepatic cirrhosis and its increase may also be explained by the fact that the therapy of liver cirrhosis is improved and therefore, patients with cirrhosis live longer [1]. Viral hepatitis B and C are still the major problem worldwide with high incidence of HCC. In addition, in the Western world alcoholic liver disease (ALD) as well as non-alcoholic fatty liver disease (NAFLD) are predominant, and both liver disease lead to HCC.

Chronic alcohol misuse is a major health problem worldwide. It was concluded that alcohol accounts for approximately 1.8 million deaths per year (3.2 % of all deaths) [2, 3]. One of the most significant diseases caused by chronic alcohol consumption is cancer. In February 2007, an international group of experts met at the International Agency for Research on Cancer (IARC) in Lyon, France, to evaluate the role of alcohol and its first metabolite acetaldehyde as potential carcinogens. This working group has concluded that the occurrence of various tumors including HCC is related to the consumption of alcoholic beverages [4].

The major focus of this review is on mechanisms in alcohol-mediated hepatocarcinogenesis with special emphasis on oxidative stress in the liver which includes the action of acetaldehyde as well as the generation of reactive oxidative species (ROS) during inflammation as well as mediated by Cytochrome P450-2E1 (CYP2E1). Finally, the role of iron as a catalytic factor in the production of oxidative stress and the role of alcohol in iron regulation will be discussed. With respect to other mechanisms, it is referred to recent review articles [5, 6].

H.K. Seitz, M.D. (✉) • S. Mueller
Department of Medicine, Salem Medical Centre and Alcohol Research Centre, University of Heidelberg, Heidelberg, Germany
e-mail: helmut_karl.seitz@urz.uni-heidelberg.de

E. Albano, M. Parola (eds.), *Studies on Hepatic Disorders*, Oxidative Stress in Applied Basic Research and Clinical Practice,
DOI 10.1007/978-3-319-15539-5_21

21.2 Animal Experiments

It has been believed that alcohol per se is not carcinogenic rather than a tumor promoter or a co-carcinogen, since administration of alcohol alone did not induce tumors. However, Beland and coworkers [7] succeeded to produce HCC in B6C3F1 mice of both sexes within 104 weeks following continuous alcohol consumption of 2.5 and 5 % in their drinking water without any additional carcinogen. There was a significant dose-related trend in the development of hepatocellular adenomas and HCC. Before, a great number of studies have investigated the modifying effect of alcohol on chemically induced carcinogenesis using mice and rats. In most studies, the administration of alcohol increased chemically induced carcinogenesis [8]. With respect to hepatocarcinogenesis, most of the studies have been performed with nitrosamine as inducing agent. Almost all these studies have shown an inhibition of carcinogenesis with alcohol, but on the other hand an enhancement in incidence of extrahepatic tumors. In addition, when manipulations were added such as the administration of a methyl-deficient low carbohydrate diet or partial hepatectomy, hepatic carcinogenesis was stimulated by alcohol [8]. Striking enhancement of hepatic carcinogenesis was also observed when alcohol and the procarcinogen were given strictly alternatively to avoid an interaction between alcohol and carcinogen metabolism [8].

In two recent animal experiments, it has been shown that CYP2E1 is primarily responsible for the effect of alcohol on the development of HCC. When ethanol was given to rats after they have received a single dose of a low amount of nitrosamine, these animals developed hepatic precarcinogenic lesions [9] as well as hepatic adenomas after 10 months of feeding [10]. However, when CYP2E1 was blocked by the specific CYP2E1 inhibitor chlormethiazole, no tumors could be detected [10]. A similar observation was made by Tsuchishima et al. [11], who showed the development of hepatic tumors by chronic alcohol feeding without the application of an additional carcinogen. Upregulation of CYP2E1 with consecutive ROS generation and the expression of c-Myc seems to be of major importance in this animal model.

In various animal models, attempts have been made to correlate the stages of initiation, promotion, and progression in hepatocarcinogenesis with specific precancerous histological features. Thus, sites of focal growth have been observed which show a number of metabolic alterations, for example, enzyme-altered foci and preneoplastic nodules [10–12]. Interestingly, MDB formation is high in HCC and the incidence of HCC is significantly higher in cirrhosis with MDBs than without [13]. It was therefore hypothesized that MDBs may represent an initial phenotypical alteration in the carcinogenic transformation of hepatocytes. Another histological abnormality observed in experimental hepatocarinogenesis is the occurrence of oval cells which originate from the portal triads after long-term alcohol exposure [14]. These cells do also appear after administration of a choline-deficient ethionine supplemented diet which is known to stimulate hepatocarcinogenesis. Recently, the occurrence of oval cells has also been observed in patients with chronic ALD. Oval cell proliferation is enhanced by TNFα, by TGFß-1, and by various cytokines, all of them increased in ALD [15].

21.3 Epidemiology

HCC is among those cancers that present with a rising incidence worldwide, particularly in Western industrialized countries. For example in the U.S., HCC is the fastest growing cause of cancer-related death in men with incidence rates increasing more than twofold between 1985 and 2002 [16, 17]. Overall, HCC is the fifth most common cancer and the third most frequent cause of cancer mortality, only surpassed by cancers of the lungs and the stomach [18]. Parallel to the increased incidence of HCC is the worldwide increase in alcohol consumption [19].

The relative risk for HCC is between 4.5 and 7.3 when more than 80 g alcohol per day is consumed, compared with abstinence or consumption of less than 40 g per day [20–22]. A dose–response relationship for the amount of alcohol consumed and the risk of HCC has been shown. The major risk factor is cirrhosis of the liver per se [20, 23, 24]. The occurrence of HCC without cirrhosis has been reported; however, it is extremely rare [25, 26]. One to two percent of alcoholic cirrhotics develop HCC per year.

Although alcohol itself leads to liver cirrhosis and promotes HCC, it is also a co-factor for the development of HCC in other chronic liver diseases. Thus, chronic alcohol misuse may enhance and/or accelerate hepatocarcinogenesis in patients with HBV [27] and HCV [28] infection, with hereditary hemochromatosis [29], and with NAFLD [30]. With respect to viral hepatitis, alcohol may stimulate oxidative stress and may, therefore, contribute to inflammation. It has been shown that chronic alcohol consumption of more than 25 g per day leads to a 10-year earlier occurrence of HCC in a Japanese population, indicating an accelerating effect of alcohol in HBV-driven hepatocarcinogenesis [27]. Chronic alcohol misuse also increases the risk for HCV infection [28]. Whether this is due to an impaired function of the immune system following alcohol ingestion or relates to a risky lifestyle of alcoholics is still unknown. In addition, alcohol may increase viral replication possibly by immunosuppression. Finally, alcohol may stimulate inflammation and, thus, oxidative stress [28].

In hereditary hemochromatosis, hepatic iron overload is a major factor in hepatocarcinogenesis and alcohol enhances iron deposition in the liver resulting in increased oxidative stress (see below) [29].

With respect to NAFLD, it became clear that type 2 diabetics have an increased risk for HCC [31]. The pathogenesis of NAFLD includes the accumulation of fat in the liver which may be predominantly induced by hyperinsulinemia due to peripheral insulin resistance. Free fatty acids induce cytochrome P4502E1 (CYP2E1) and lead to the generation of ROS. It has been shown that the relative risk for HCC in type 2 diabetics is approximately 4, and it increases to almost 10 when more than 80 g per day are consumed [20, 21]. Patients with NASH who consume daily alcohol even at a social level have significant increased risk to develop HCC as compared to NASH patients who are abstinent [30]. This is one reason why the intake of alcohol of any type and dose is not recommended in NASH patients [32].

21.4 Pathogenesis

21.4.1 General Aspects

Without any doubt, one major prerequisite in the pathogenesis of HCC is the presence of cirrhosis [20, 24]. The fact that chronic alcohol consumption results in HCC via cirrhosis can be either explained by [1] cirrhosis-associated mechanisms, [2] a long time process leading parallel to cirrhosis and HCC, or [3] an early initiation of HCC followed by tumor promotion due to cirrhosis.

It is interesting that former drinker who had been abstinent for less than 10 years carries a higher risk of developing HCC than those who continue to drink [33]. Explanations could be that cessation of drinking rather reflects advanced liver cirrhosis which is per se associated with HCC occurrence, or stimulated liver cell regeneration following alcohol abstinence enhancing cell turnover, expansion of dysplastic cell clones, and the likelihood of tumor initiation.

Carcinogenesis by alcohol is complex and still poorly understood for many reasons [5, 6, 34–38]. One major reason is due to the fact that it is rather the metabolites and side products of alcohol turnover rather than direct actions of alcohol itself. Indeed, alcohol is rapidly converted into more reactive, toxic, or even carcinogenic compounds such as acetaldehyde or reactive oxygen species (ROS) via non-enzymatic or enzymatic reactions. The production of these metabolites during alcohol consumption is complicated since it is dependent on many additional factors that include the genetic background or the tissue-specific induction of detoxifying enzymes [36–38]. The complete understanding of alcohol-mediated carcinogenesis is also hampered by the fact that side products such as ROS are very difficult to measure and appropriate models to study ROS and ROS-mediated carcinogenesis are still lacking [39]. Finally, commonly used animal models such as rodents significantly differ from humans with regard to alcohol sensitivity, metabolism, and consequently with regard to cancer development. Although iron and its accumulation in the liver during alcohol consumption has been appreciated for a long time, a better understanding of the underlying molecular mechanisms has been only gained in the last decade with the discoveries of key regulatory molecules such as hepcidin and the HFE gene.

Except for cirrhosis of the liver which will be discussed elsewhere in this book, various mechanisms may act in concert in hepatocarcinogenesis including (1) oxidative stress (2) the action of acetaldehyde, the first oxidative metabolite of ethanol oxidation with its effect on the antioxidative defense system, DNA repair systems, and epigenetics (reduced availability of methyl groups associated with altered methylation of DNA and histones), (3) additional epigenic alterations due to effects of ethanol on DNA/histone acetylation and deacetylation, (4) the reduction of retinol and retinoic acid associated with hyperproliferation and the loss of cell differentiation, (5) changes in intercellular signal transduction pathways, and (6) immunosuppression. Recent review articles have covered these aspects in detail [5, 6].

We will discuss here only mechanisms due to oxidative stress either directly (e.g., generation of ROS) or indirectly (e.g., inhibition of the antioxidative defense system or the DNA repair system).

21.4.2 The Role Oxidative Stress

The formation of ROS such as superoxide anion and hydrogen peroxide causes oxidative injury. Several enzyme systems are capable to produce ROS, including the cytochrome P-450 2E1 (CYP2E1)-dependent microsomal mono-oxygenase system, the mitochondrial respiratory chain, and the cytosolic enzymes xanthine oxidase and aldehyde oxidase [40]. Ethanol-mediated ROS formation may be due to an increased electron leakage from the mitochondrial reparatory chain associated with the stimulation of reduced nicotinamide adenine dinucleotide (NADH) shuttling into mitochondria [41] and to the interaction between *N*-acetylsphingosine (from tumor necrosis factor alpha) and mitochondria [42]. The induction of sphingomyelinase by TNF-α increases the levels of ceramide, an inhibitor of the activity of the mitochondrial electron transport chain, leading to increased mitochondrial production of ROS. ROS can also be generated in alcoholic hepatitis with activated hepatic phagocytes [43]. Hepatic iron accumulation as observed in ALD increases ROS and finally nitric oxide production because ethanol-mediated stimulation of inducible nitric oxide synthase results in the formation of peroxynitrite which is highly reactive [44].

21.4.2.1 Acetaldehyde, a Possible Pathogenic Factor in Hepatocarcinogenesis

According to the IARC, there is sufficient evidence to classify acetaldehyde as a carcinogen in experimental animals and humans [4].

Acetaldehyde is highly toxic, mutagenic, and carcinogenic [38]. It interferes with DNA synthesis and DNA repair. In vivo and in vitro experiments in prokaryotic and eukaryotic cell cultures as well as in animal models have shown that acetaldehyde has direct mutagenic and carcinogenic effects, causes point mutations in the hypoxanthine-guanine-phosphoribosyl transferase locus in human lymphocytes, and induces sister chromatid exchanges and gross chromosomal aberrations [45–48]. Acetaldehyde induces inflammation and metaplasia of tracheal epithelium, delays cell cycle progression, and enhances cell injury associated with cellular hyperregeneration in the mucosa of the esophagus and colon [38]. Acetaldehyde also binds to protein and DNA [49–55]. Thus, structure and function of various proteins are altered including the antioxidative defense system with glutathione, DNA repair enzymes, and cell organelles such as mitochondria and microtubules [38]. Decreased mitochondrial function results in inhibition of fatty acid oxidation and ATP formation. Decreased microtubular function leads to inhibition of the secretion of macromolecules such as very low-density lipoproteins from the liver. Both factors favor the generation of fatty liver. In addition, apoptosis as well as survival factors such as NFκB are induced [37]; acetaldehyde directly inhibits O6-methylguanosyl transferase, an enzyme that repairs DNA adducts [56].

Most importantly, however, acetaldehyde binds to DNA and forms stable adducts [49–55]. Binding to DNA represents one mechanism by which acetaldehyde could trigger replication errors and/or mutations in oncogenes and tumor suppressor genes. It has been shown that the major stable DNA adduct N2-ethyl-desoxyguanosine (N2-Et-dG) serves as a substrate of eukaryotic DNA polymerase. However, N2-Et-dG seems rather a marker for chronic ethanol consumption than a major risk lesion for cancer. In addition, another DNA adduct of acetaldehyde, 1,N2-propano-desoxyguanosine (PdG), has been identified, especially in the presence of basic amino acids, histones, and polyamines. While N2-Et-dG is non-mutagenic and may represent a marker of chronic alcohol ingestion, PdG has mutagenic properties. These acetaldehyde-associated effects occurred at AA concentrations from 40 to 1,000 μM.

Apart from genetic changes along with chronic alcoholism, i.e., mutations, DNA cross-links, or impaired DNA repair, chronic and acute alcohol intake may affect epigenetic mechanisms of gene expression such as methylation of DNA. DNA methylation is an important determinant in controlling gene expression whereby hypermethylation has a silencing effect on genes and hypomethylation may lead to increased gene expression. And indeed, alcohol intercepts with these epigenetic mechanisms [57, 58].

Alcohol interacts with absorption, storage, biologic transformation, and excretion of compounds which are essential for methyl group transfer including folate, vitamin B6, and certain lipotropes. Especially, the production of *S*-adenosyl-L-METHIONINE (SAMe), the universal methyl group donor in methylation reactions, is impaired. Acetaldehyde interacts with SAMe synthesis through inhibition of crucial enzymes involved in SAMe generation. This can lead to compromised formation of endogenous antioxidants such as glutathione and also lead to impaired cellular membrane stability.

In addition, alcohol interacts with methylation of certain genes and thereby contributes to liver damage and tumor development. Accordingly, alcohol-induced depletion of lipotropes may cause hypomethylation of oncogenes leading to their activation. The decrease in methylation capacity caused by chronic alcohol consumption can therefore contribute to epigenetic alterations of genes involved in hepatocarcinogenesis [58].

Genetic linkage studies with alcoholics have provided strong support for the assumption that AA plays a central role in alcohol-associated carcinogenesis. These studies found that individuals who accumulate AA because they carry certain alleles of the genes encoding alcohol dehydrogenase (ADH) or ALDH have an increased cancer risk [59]. Briefly, class I isozymes account for most of the alcohol metabolism. For both the ADH1B and the ADH1C genes, several alleles exist that result in differences in the activity of the ADH molecules they encode. While polymorphisms at the ADH1B gene affect drinking behavior, polymorphism at the ADH1C gene may affect disease risk. The effects of the different ADH1C alleles on alcohol metabolism and, consequently, on drinking levels and alcohol-related carcinogenesis, are subtle. They can best be studied in Caucasian populations in which the highly active ADH1B*2 allele is rare. Studies on the relationship between ADH1C alleles and cancer occurrence in Caucasians have led to contradictory results with respect to cancer of the upper alimentary tract [38] and the female breast [60].

We determined ADH1C polymorphisms in more than 400 heavy drinkers with daily alcohol intake of more than 60 g and various cancers of the upper aerodigestive tract, liver, and breast. Cases of cancer patients were compared with carefully matched control patients with alcohol-related diseases (e.g., cirrhosis of the liver, pancreatitis, and alcohol dependence) but no cancer [61]. Cancer patients and control subjects were of similar age and had similar histories of alcohol consumption and cigarette smoking. In this study, significantly more patients with alcohol-related cancers either had at least one ADH1C*1 allele, or were homozygous for ADH1C*1, than did patients with other alcohol-related diseases. Statistical analyses determined a significant association between ADH1C*1 allele frequency and rate of homozygosity and an increased risk for alcohol-related cancer ($p = 0.001$) including HCC. Individuals homozygous for ADH1C*1 had a relative risk of developing HCC of 3.6 compared with people homozygous for ADH1C*2 [61]. At lower levels of alcohol consumption, the difference in cancer risk between the various gene carriers was less striking. This is not surprising, however, because higher levels of alcohol consumption also result in production of more AA which then can act on DNA. Other studies did not confirm the effect of AA accumulation on HCC development [62].

Acetaldehyde also accumulates when its degradation is inadequate. The main enzyme that breaks down acetaldehyde is ALDH2. It is encoded by the ALDH2 gene, for which there are two main alleles, ALDH2*1 and ALDH2*2. The ALDH2*2 allele differs from the normal ALDH2*1 allele by an nucleotide substitution (G(A) in exon 12 of the ALDH2 gene, resulting in an almost inactive ALDH enzyme. People who are homozygous for ALDH2*2 have an extremely low ALDH activity; when these people drink alcohol, acetaldehyde accumulates and the "flushing syndrome" develops. These subjects do not tolerate alcohol at all and are, therefore, generally protected against developing alcoholism. Those who are heterozygous also have greatly reduced (i.e., less than 10 %) ALDH2 activity. Nevertheless, they tolerate alcohol ingestion and may even become heavy drinkers and alcoholics.

Several epidemiological studies have demonstrated that the risk of alcohol-associated cancer of the aerodigestive tract is significantly elevated in people with low ALDH2 activity, with a relative risk of >10 for oropharyngeal, laryngeal, and esophageal cancer [62]. Individuals with low ALDH activity also exhibited an increase in acetaldehyde DNA adducts [63]. However, this was not reported for the liver, possibly due to the fact that cirrhosis of the liver is more important in ethanol-mediated hepatocarcinogenesis than acetaldehyde, and that acetaldehyde is possibly not a major pathogenetic factor in the development of cirrhosis.

21.4.2.2 The Role of Cytochrome P4502E1

Most important, however, is the production of ROS via CYP2E1. It has been shown that alcohol induces CYP2E1 in the liver. This induction is an adaptive process and is associated with an increased metabolism of ethanol to acetaldehyde and also to ROS. The induction differs individually and is most likely due to the fact that the degradation of CYP2E1 by the ubiquitine proteasome pathway is inadequate since alcohol has an effect on this pathway [64]. A significant increase in hepatic CYP2E1 activity occurs already following the ingestion of 40 g of ethanol daily for 1 week, which is further enhanced after 4 weeks [65]. However, this occurs not in all individuals.

The induction of CYP2E1 occurs not only following chronic ethanol ingestion, but also in NAFLD (see below). The consequences of CYP2E1 induction are [35] (1) increased ethanol metabolism to AA, (2) Generation of ROS as a byproduct of ethanol metabolism, (3) interaction with the metabolism of drugs leading sometimes to enhanced drug toxicity, (4) enhanced procarcinogen activation to ultimative carcinogens, and (5) enhanced degradation of retinol and retinoic acid. We will discuss here the importance of CYP2E1 in ROS production and retinoid metabolism.

CYP2E1 and Generation of ROS and Carcinogenic DNA Adducts

As pointed out, ethanol metabolism through CYP2E1 not only produces acetaldehyde, but also, especially when induced, generates various ROS including H_2O_2, OH–, and carbon-centered OH– [40]. In animal experiments, the induction of CYP2E1 correlates with NAD phosphate (NADPH) oxidase activity, the generation of hydroxyethyl radicals, lipid peroxidation, and the severity of hepatic damage, all of which could be prevented by the CYP2E1 inhibitor chlormethiazole [66]. In addition, DNA lesions have been found to be lower in CYP2E1 knock-out mice as compared to wild-type mice [67] and hepatic injury was significantly increased in transgenic mice that overexpressed CYP2E1 [68].

These ROS can be neutralized by a potent antioxidative defense system. However, chronic alcohol consumption injures this system mostly due to acetaldehyde and leads among others to a significant decrease of glutathione [37, 38]. ROS not only activates c-Jun N-terminal kinase (JNK) with consecutive expression of the AP-1 gene leading to cellular hyperregeneration, a procarcinogenic state, but also causes lipid peroxidation [69]. Various lipid peroxidation products including 4-hydroxynonenal may bind to various purine and pyrimidin bases forming exocyclic DNA adducts. It has been shown that these adducts are highly mutagenic and carcinogenic [70–72]. These adducts have been identified in the liver of patients with ALD and other types of liver disease associated with inflammation and oxidative stress [73] as well as in the urine of patients with viral hepatitis [74]. Using HPLC for determination of these adducts, we found increased concentrations not only in patients with viral hepatitis such as hepatitis B and C, but also in patients with ALD [74]. Thus, measurement of exocyclic etheno-DNA adducts in the urine of patients with ALD could be a predictive marker for risk assessment of HCC in the alcoholic.

In a series of experiments using CYP2E1 overexpressing HepG2 cells, we could show that these etheno-DNA adducts correlate significantly with CYP2E1 and 4HNE [75]. This is a time- and concentration-dependent process. Adduct formation could be blocked by the use of the specific CYP2E1 inhibitor CMZ [70]. Most exocyclic etheno-DNA adducts have been observed in cells with a high expression of CYP2E1 and a low concentration of mitochondrial glutathione. Thus, both factors may play an important role in the production of this important mutagenic DNA adduct. Furthermore, CYP2E1, 4HNE, and etheno-DNA adducts also correlated significantly in liver biopsy samples from patients with ALD [75].

Most recently, we have investigated 97 biopsies from patients with non-cirrhotic ALD of various severities including pure fatty liver, steatohepatitis, and fibrosis. We found a significant correlation between CYP2E1 and etheno DNA adducts ($p=0.0001$) and also between CYP2E1 and fibrosis ($p=0.03$) (Seitz and Mueller, personal communication). These data show clearly that CYP2E1 is at least one driving force in the progression of ALD possibly by increasing oxidative stress.

In an animal model using Lieber–DeCarli alcohol-containing and control diets, hepatocarcinogenesis was induced by a single small dose of diethylnitrosamine given prior to the alcohol administration. One month of ethanol feeding resulted in a significant increase of preneoplastic lesions in the liver associated with a significant increase in hepatic CYP2E1, nuclear accumulation of NFκB protein, and cellular regeneration which was not observed in control animals [9]. However, when chlormethiazole, a strong selective CYP2E1 inhibitor, was given to those animals who received alcohol, the nuclear accumulation of NFκB protein and the occurrence of hepatic PGST foci were inhibited. In addition, some of the animals developed hepatic adenoma under alcohol which were completely inhibited by the concomitant administration of chlormethiazole, and thus, by the inhibition of CYP2E1 [10]. These animal experiments contributed to the understanding of the underlining mechanisms of the co-carcinogenic effect of alcohol. Both the induction of CYP2E1 by chronic alcohol consumption results in oxidative stress as well as the depletion of retinoic acid and may be responsible for hepatocarcinogenesis.

CYP2E1 and Its Possible Role in NAFLD

CYP2E1 is also induced in the livers of patients with NAFLD possibly due to the inducing effect of free fatty acids and/or acetone when diabetes is present [76]. Both CYP2E1 and etheno-DNA adducts were also measured in leptin-deficient, insulin-resistant Zucker rats which develop obesity and NASH and represent a genetic model for NASH. Etheno-DNA adducts were found to correlate significantly with hepatic fat and with CYP2E1 [75]. However, more interesting was the observation that when alcohol was administered to these animals, CYP2E1 expression as well as DNA-adduct formation was further enhanced emphasizing that additional alcohol consumption even at moderate levels in NASH may be deleterious [31].

The negative role of ethanol was further underlined in a dietary animal model resembling NASH, where the addition of daily alcohol even at a level of 17 % of daily calories (much less than in the classic Lieber DeCarli diet with 36 % of daily calories) resulted in significant morphological injuries associated with molecular changes [77].

Most recently, we found in 3 out of 21 liver biopsies from children with NASH a massive etheno DNA adduct formation, which correlated with inflammatory activity. Similar DNA lesions were also noted in liver biopsies from adult patients with NASH (Seitz and Mueller, personal communication).

Since it has been reported that even social alcohol consumption increases the risk for HCC in NASH significantly [30], alcohol consumption of any type and amount was not recommended in NASH patients [31].

CYP2E1 and Retinoids

Retinoic acid (RA), one of the most active forms of retinoids, is an important factor in the regulation of cell growth, apoptosis, and cell differentiation. Thus, reduction of RA may lead to uncontrolled cellular proliferation, loss of cell differentiation, and dysregulated apoptosis, which can act to promote the process of carcinogenesis. It has been reported that chronic ethanol consumption results in a significant depletion of retinol in the liver of patients with ALD [78].

The mechanism of alcohol-associated decrease in retinol and retinoic acid has multiple causes. Since ADH and ALDH share the common substrates ethanol and retinol as well as AA and retinal to form retinoic acid, an interaction at these enzyme sites is not surprising. It has been demonstrated that ethanol acts as a competitive inhibitor of retinol oxidation [69, 79]. Beside the fact that ethanol competes with retinol for the binding side of ADH, there are other mechanisms explaining the decrease in retinoic acid. Since chronic ethanol consumption increases CYP2E1 activity, an enhanced catabolism of vitamin A and retinoic acid into polar metabolites due to an induction of cytochrome P-450 2E1 occurs [80]. Although a variety of cytochrome isoenzymes such as CYP1A1, CYP2B4, CYP2C3, CYP2C7, CYP2E1, and CYP26 are involved in retinoic metabolism, CYP2E1 seems of major importance [80].

It was demonstrated that rat hepatic microsomes from ethanol-fed rats when incubated with RA showed an enhanced degradation of RA, resulting in the generation of polar RA metabolites such as 18-OH-RA and 4-oxo-RA as compared to microsomes from control animals [81]. It has also been shown that this in vitro metabolism of RA can be inhibited by CMZ and by CYP2E1 antibodies demonstrating the involvement of CYP2E1 in the metabolism of RA. In addition, rats which received ethanol chronically showed decreased levels of hepatic RA associated with an increase in CYP2E1, which were almost completely restored when CMZ was added to the diet [82]. The data suggest that CMZ can restore both hepatic retinol and retinyl ester concentrations to normal levels in ethanol-fed rats through blocking both enhanced degradation of vitamin A and mobilization of vitamin A from the liver into the circulation, indicating that CYP2E1 is the major enzyme responsible for the alcohol-enhanced catabolism of retinoids in hepatic tissue after exposure to alcohol. Furthermore, not only hepatic concentrations of RA normalized with CMZ, but also changes seen in cell proliferation and cell cycle behavior observed after ethanol administration [83].

Moreover, the impaired retinoid homeostasis leads to aberrant retinoid receptor signaling through upregulation of the JNK signaling pathways. Since ethanol-mediated CYP2E1 induction not only results in reduced RA levels but also in increased oxidative stress, both increased oxidative stress and low RA (via decreased mitogen-activated protein kinase phosphatase-1 (MKP-1)) result in an activation of the JNK pathway. Since there is cross talk between JNK pathway and RXR/RAR receptors, the activated JNK pathway activates AP-1 gene resulting in an increase in c-fos and c-jun (which are 14-fold increased in the livers of ethanol-fed rats as compared to controls) [69]. The overall effect of excessive alcohol ingestion is dysregulated apoptosis, cellular proliferation, immune function, and inflammation, which can act to promote the process of carcinogenesis.

Unidentified polar metabolites of CYP2E1-mediated RA metabolism observed when microsomes from ethanol-fed rats were incubated with RA had clear apoptotic properties, since these metabolites resulted in a change of the hepatic mitochondrial membrane potential, a release of cytochrome C, and an activation of the caspase cascade [81]. Taking this observation into consideration, the intake of vitamin A or β-carotene, a precursor of vitamin A together with ethanol, may lead to increased hepatic apoptosis and cellular injury which has been described in the baboon [84].

In summary, CYP2E1 is primarily involved in the degradation of retinol and RA resulting in loss of RA and in the generation of apoptotic unidentified polar metabolites. Chronic ethanol consumption induces CYP2E1 with a consecutive enhancement of these consequences. The loss of RA may be at least one factor in ethanol-mediated hepatocarcinogenesis.

21.4.2.3 Hepatic Iron as an Important Factor in Oxidative Stress

Iron Toxicity and Carcinogenesis

Iron toxicity has been well known for many decades and usually attributed to Fenton chemistry, first described in 1896 by Fenton and referring to the reaction of reduced ferrous iron with H_2O_2 to ferric iron and hydroxyl radicals [85]. These radicals are highly and universally reactive to all cellular compounds, including DNA, proteins, lipids, and carbohydrates. They cause oxidative damage to DNA, including base modifications and DNA strand breaks, and may change the structure of proteins and lipids, eventually leading to mutagenesis.

Although the direct chemical nature of the iron toxicity in Fenton's chemistry has been recently questioned by Saran et al. [86], it is generally assumed that at least Fenton-like reactions contribute to the toxicity of iron. For this reason, we will not further differentiate both reactions in this review. In vitro studies have shown that liver viability decreases in the presence of high amounts of iron and that the depletion of iron by iron chelators can protect cells from peroxide toxicity. Hepatic iron overload and the risk for liver cancer have also been extensively studied using in vivo experiments in the past. In patients with hereditary hemochromatosis due to HFE mutations, the hepatic iron overload ultimately determines the overall survival [87]. In a large transplant registry study, Ko et al. showed that iron alone increased the risk of HCC by a factor of 2.2 [88]. Turlin et al. showed in 1995 that the iron-associated risk of HCC was not dependent on whether the patient had cirrhosis or not [89]. In an interesting study from the NIH, Zacharski et al. could demonstrate that iron depletion by phlebotomy may have a protective effect on general carcinogenesis [90]. He studied prospectively 641 controlled subjects without iron depletion and 636 subjects in an iron reduction group. Iron reduction was achieved by monthly phlebotomy. Patients were followed up for an average for 4.5 years. Interestingly, the risk of new visceral malignancies was lower in the iron reduction group than controlled group (38 vs. 60), and among patients with new cancers, those in the iron reduction group had lower cancer-specific and all-cause mortality. Mean ferritin levels across all 6-monthly visits were lower among all patients who did not develop cancer. Notably, in a 1988 animal study, Hann et al. showed reduced tumor growth in iron-deficient mice [91] and Tsukamoto et al. in 1995 showed in rodents that a mixture of high fat diet and alcohol together with supplemented iron led to pronounced cirrhosis [92]. In a meta analysis by D'Amico et al., it was concluded that liver iron is one among several independent prognostic factors of survival in liver cirrhosis [93]. Ganne-Carrie showed in 229 patients with ALD or hepatitis C virus-related cirrhosis that iron is the best parameter for predicting mortality in ALD patients ($p=0.007$) followed over 57 months [94]. This study found no prognostic significance of hepatic iron for HCV.

Taken together, there seems to be enough evidence from human and animal studies to suggest that hepatic iron overload leads to hepatocellular damage. This in turn leads to increased fibrosis progression, risk for hepatocarcinogenesis, and ultimately iron seems to be an important prognostic factor for overall survival in patients with ALD. The most evident and generally accepted mechanisms of iron carcinogenesis include Fenton-like reactions that lead to the highly aggressive hydroxyl radicals.

Prevalence of Iron Overload in ALD

A significant iron deposition in patients with chronic alcohol consumption has been appreciated for a long time [95]. Although the introduction of percutaneous needle biopsy of the liver in the fifties added a new tool for exact quantitation of hepatic iron deposition, it further led to considerable confusion since histological stainable iron deposits are common in cirrhotic livers [96, 97]. This is especially the case in alcoholic subjects and some investigators have even equated mild to moderate alcoholic cirrhosis with hemochromatosis. When analyzing the Finch data for differences between alcoholic and non-alcoholic patients, Powell noted that laboratory signs of hepatocellular damage showed a significant higher prevalence in the alcoholic group (serum transaminase elevation 35.5 % vs. 15.4 %). Interestingly, most large studies seem to indicate that mutations in the HFE gene [98] are not responsible for the iron accumulation [99–101] in ALD patients. Moreover, since hereditary hemochromatosis shows a very weak clinical penetration of about 10 %, it still remains open what exact role alcohol plays in patients with hemochromatosis and whether it is an important but underestimated disease modifier.

A significantly higher mean liver iron concentration was found in alcoholics as compared to controls (156.4 vs. 53 μg per 100 mg dry weight) [102]. However, liver iron was much higher in the hemochromatosis group (2,094.5 μg). No relationship between liver iron concentration and the amount of alcohol consumed was noted. Serum ferritin concentrations seem to reflect the iron overload in patients with hemochromatosis and mild ALD, while no association was found with serum ferritin in patients with severe ALD. Serum ferritin is generally more frequently elevated in ALD than in other liver diseases [103] and decreases after 1.5–6 weeks of abstaining from alcohol. In a recent study from Japan, excess iron accumulation was found in 22 hepatic tissues with ALDs, but not in any normal hepatic tissues [104].

We have recently characterized a large cohort at the Heidelberg Salem Medical Centre using biopsy, but also various newer non-invasive tests including transient elastography [105]. From a cohort of 235 patients with ALD, of whom 86 had underwent liver biopsy, with detailed iron characterization regarding to intensity and location (macrophages vs. hepatocytes), it can be summarized that ca. 50 % of patients with ALD have significant pathological iron deposition. Interestingly, iron deposits are found both at the same rate in macrophages and hepatocytes. About 2/3 of these patients with advanced fibrosis stage have elevated ferritin levels higher 400 μg/mL and an increased transferrin saturation higher in 45 %. This is especially important to consider, since ferritin levels and transferrin saturation are commonly used to screen patients with hereditary hemochromatosis. Taken together, patients with ALD show a significantly increased histological iron deposition in about half of the population that increases with disease progression towards cirrhosis. This iron deposition is reflected by serum ferritin levels, but unfortunately over-layered by additional inflammatory conditions. In fact, according to our analysis, serum ferritin levels reflect hepatocyte iron deposition more closely than macrophage iron deposition in ALD. Moreover and often overlooked, a significant proportion of patients show elevated levels of transferrin saturation and ferritin levels that can be easily mistaken for hemochromatosis.

Systemic Iron Homeostasis

Enormous progress has been made in the last two decades to better understand the molecular mechanisms of iron regulation and homeostasis both at the cellular and systemic level [106–108]. These milestones include the discovery of iron regulatory proteins (IRPs) and their posttranscriptional regulation of cellular iron homeostasis in the late 1980s and early 1990s as well as identification of the systemic master switch protein hepcidin about 10 years ago [109]. Despite this progress, numerous aspects of iron regulation still remain unexplained in humans. The discovery of the HFE gene that causes the majority of hereditary hemochromatosis forms has clearly helped to resolve the role of hemochromatosis in alcoholics [98]. Interestingly, most large studies seem to indicate that mutations in the HFE gene [98] are not responsible for the iron accumulation in patients with ALD [99–101]. These studies thus seem to rule out HFE as important genetic disease modifier. In contrast, as briefly discussed above, alcohol could be an important disease modifier in patients with hemochromatosis, but typically with weak clinical penetration. In addition, at the systemic level, ALD appears rather complex since alcohol affects not only the liver but also many other organ systems including the bone marrow and the immune system so that iron regulation obviously is altered at many different levels and locations throughout the body.

The majority of body iron, roughly 2 g in humans, is distributed in the oxygen carrier hemoglobin of red blood cells and developing erythroid cells (Fig. 21.1). Excess iron is usually stored in the liver, which normally contains about 1 g of iron, predominantly in the form of ferritin [107]. It is generally believed that mammals lose iron from regular sloughing of the mucosa and skin cells or during bleeding and do not possess any mechanisms for iron excretion from the body [106]. Therefore, the balance is maintained by the tight control of dietary iron absorption at the brush border of enterocytes in the proximal duodenum (Fig. 21.1). Dietary iron uptake involves the reduction of ferric iron to ferrous iron in the intestinal lumen by ferric reductases, such as duodenal cytochrome B (Dcytb), and subsequent transport of iron via the apical membrane of enterocytes by divalent metal transporters (such as DMT1 encoded by SCL11A2 (solute carrier family 11, member 2) gene). Cytosolic iron is exported by the basolateral iron transporter ferroportin (FP-1 or SLC40A1) from the enterocytes to the blood compartment and then bound to transferrin, the major abundant iron carrying protein within serum. Before binding, ferrous iron is oxidized by ferroxidases (e.g., hephaestin/ceruloplasmin) to the ferric form. Iron is then distributed within the body and used in various pathways, but mainly utilized in the bone marrow for the synthesis of new heme.

About 1–2 mg per day of iron is absorbed to keep a stable iron balance. Of note, iron undergoes an efficient recycling. Senescent erythrocytes that have a life span of 120 days are sequestered by macrophages and the iron is re-used for new heme synthesis. This recycling machinery accounts for about 90 % of newly synthesized hemoglobin. It is important to note that the iron export pump, ferroportin, is not only found on enterocytes, but also on macrophages and hepatocytes. The ferroportin-mediated efflux of ferrous iron from enterocytes and macrophages into the serum is critical for systemic iron homeostasis and mainly controlled via the liver-secreted 25 AS peptide hormone, hepcidin. Mechanistically, hepcidin binds to ferroportin and promotes its phosphorylation, internalization, and lysosomal degradation [110, 111]. Hepcidin is primarily expressed in hepatocytes as a precursor pro-peptide, although other locations of secretion have been described, such as macrophages [112] and to a lesser extent cardiomyocytes. Hepcidin efficiently blocks ferroportin, which leads to accumulation of iron within macrophages and blocks the iron absorption via enterocytes (Fig. 21.1).

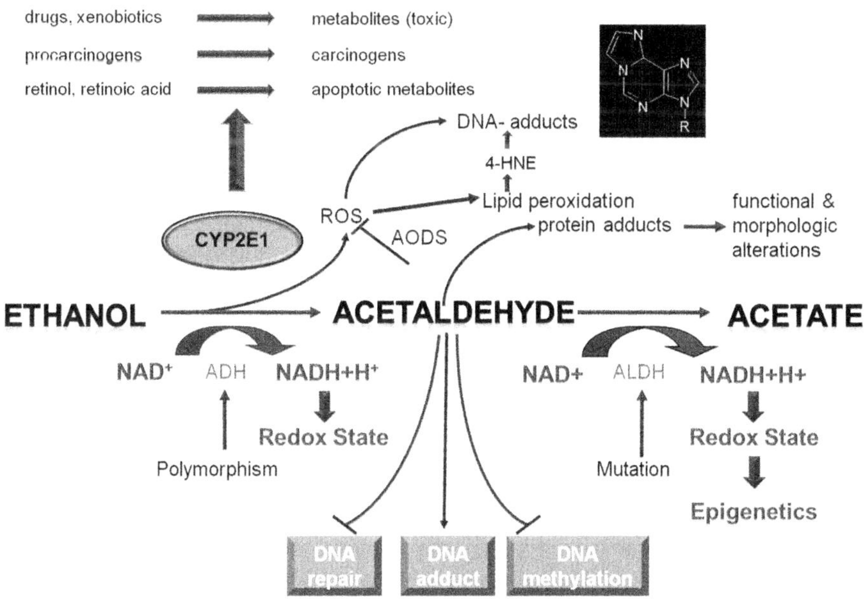

Fig. 21.1 Ethanol metabolism via alcohol dehydrogenase (ADH), acetaldehyde dehydrogenase (ALDH), and cytochrome P-4502E1 and the effect of acetaldehyde, and reactive oxygen species (ROS) on carcinogenesis. Ethanol is metabolized to acetaldehyde via ADH and further metabolized to acetate via ALDH. Acetaldehyde has toxic, mutagenic, and carcinogenic properties, inhibits DNA repair, decreases DNA methylation, binds to proteins with functional and morphologic alterations, and forms DNA adducts. Thus, acetaldehyde accumulation is associated with increased carcinogenesis as with ADH isoenzymes that reveal increased alcohol metabolizing activity (ADH 1B*2 and ADH 1C*1). Acetaldehyde may also accumulate when ALDH activity is low as observed in 50 % of Asians. Both the ADH and the ALDH reactions result in the generation of NADH leading to an altered redox state of the hepatocyte associated with epigenetic changes. Ethanol is also metabolized via cytochrome P2E1 (CYP2E1) to ROS which leads to lipid peroxidation and lipid peroxidation products such as 4-hydroxynonenal (4-HNE) bind to DNA resulting in highly carcinogenic DNA adducts. There is an interaction with acetaldehyde since acetaldehyde inhibits the antioxidant defense system (AODS) resulting in a further increase of ROS. Finally, alcohol-induced CYP2E1 may also activate procarcinogens to carcinogens such as nitrosamines and can transform retinoic acid to inactive metabolites, and/or to reactive polar metabolites which cause liver cell apoptosis. Retinoic acid depletion can promote cell proliferation and, thus, carcinogenesis. For more details see text

An important evolutionary-conserved mechanism to induce hepcidin is an infection or inflammatory state. Via cytokines, namely IL-6 but also microbial molecules (such as lipopolysaccharide), hepcidin is strongly induced leading to a rapid decrease of serum iron, which is thought to function as anti-bacterial defense mechanism. More recently, another important inflammatory cofactor, H_2O_2 has been also identified as potent inducer of hepcidin. In contrast, hepcidin levels seem to be suppressed in patients with genetic hemochromatosis, leading to increased uptake of iron via the duodenum and increased release of iron through macrophages. Cytokine-mediated induction of hepcidin is thought to be the reason for anemia of chronic disease [113], while the disruption of hepcidin is generally associated with the systemic iron overload (e.g., genetic hemochromatosis). Despite the progress and the discovery of various upstream regulators of hepcidin (such as C/EBPα, BMP6, SMAD 1, 5, 8 and 4, TMPRSS6, IL-6, CREBH, CHOP, and TLR4), an overall and conclusive understanding of the regulatory network with respect to the control of iron is not yet completely understood. In addition, conflicting findings of the response of hepcidin towards ROS have been presented for which we recently identified a partial explanation [114]. Thus, the central ROS, H_2O_2 induces hepcidin in hepatocytes independent of IL-6 when exposed in continuous manner. Bolus treatments, however, which reflect an artificially high H_2O_2 exposure, blocked hepcidin expression [115]. They are even toxic at H_2O_2 concentrations higher than 50 μmol and hepcidin suppression at such conditions was due to unspecific inhibition of the mRNA transcription machinery [115].

Cellular Iron Regulation

Iron is also tightly regulated at the cellular level. Developing erythrocytes, as well as most other cell types, require iron from plasma transferrin. Iron is loaded onto transferrin with a capacity of two atoms of ferric iron per molecule. Transferrin binds with a very high affinity to cell surface transferrin receptor 1 (TfR1) ubiquitously expressed. The TF-TfR1 complex is endocytosed via clathrin-coated pits. The release of ferrous iron into the intracellular compartment requires acidification of the endosome using a proton pump and the reduction of ferric iron to the ferrous form via Steap3. DMT1 transports the ferrous iron across the endosomal membrane into the cytosol where it can be stored in ferritin complexes in non-erythroid cells or incorporated into hemoglobin in erythroid cells [116]. Iron is then either re-used for various synthesis pathways such as intracellular heme synthesis and iron cluster protein synthesis or transported into the major iron storage protein ferritin.

The expressions of DMT1 and ferritin are coordinately post-transcriptionally regulated by binding of trans-acting iron responsive proteins (IRP1 and IRP2) to iron responsive elements (IREs) in the untranslated regions (UTRs) of their respective mRNAs [117–119]. In iron-starved cells, IRPs bind with high affinity to cognate. IREs are evolutionary conserved hairpin structures of about 30 nucleotides with characteristic sequences. The effect of IRP binding to IREs is dependent on their position. TfR1 mRNA contains five IREs within its long 3′ UTR that stabilizes and protects the transcript from degradation leading to protein upregulation, but other mRNAs, for example, mRNAs encoding H and L ferritin, contain a single IRE in their 5′ UTRs where binding results in decreased protein translation by steric blockade. As a result, increased TfR1 levels stimulated acquisition of iron from plasma transferrin to counteract iron deficiency. The inhibition of ferritin synthesis leads to decreased abundance of this protein as iron storage becomes obsolete under these conditions. Conversely, in cells with high iron content, both IRP1 and IRP2 become unavailable for IRE-binding lowering TfR1mRNA degradation and ferritin mRNA translation. Thus, when iron supply exceeds cellular means, the IRE-IRP switch minimizes further iron uptake via TfR1 and favors the storage of excess iron in newly synthesized ferritin. Other IREs have been discovered in the genes of ALAS2, mitochondrial aconitase, ferroportin, HIF2α, βAPP, and α-synuclein, which in turn control iron storage and erythroid iron utilization, energy homeostasis, hypoxia responses, and neurological pathways, respectively.

Both IRP1 and IRP2 are sensitive to ROS and RNS (reactive nitrogen species) (reviewed in [120, 121]). This is evident through the redox regulation of IRP1 through its iron sulfur cluster (ISC). Exposure of cells to micromolar concentrations of ROS and RNS (especially NO) also leads to the destabilization of the IRP1-ISC complex with subsequent induction of IRE-binding activity via an incompletely characterized mechanism. This response can be antagonized by MPO-derived hypochlorite [122]. In vitro studies showed that ROS and RNS remove iron of the ISC and convert it to a non-functional cluster. IRP also responds to NO, which likewise induces IRE-binding at the expense of its aconitase activity [123]. However, ROS and RNS may also modify potential cluster destabilizing factors rather than the IRP1-ISC complex itself. Furthermore, some discrepancies are found in the literature about the regulation of IRP2 and NO. NO may either positively or negatively regulate IRP2, which could be explained by differential effects of the numerous NO species. Taken together, the redox regulation of IRE-IRP directly links the iron metabolism to inflammatory processes, hypoxia, and oxidative stress.

Hepatic Causes of Iron Overload in ALD

The complete picture of the molecular changes in iron metabolism during chronic alcohol consumption is not yet completely understood (Fig. 21.2). There are various studies that focused on the central systemic master switch hepcidin. Both acute and chronic alcohol exposure seem to suppress hepcidin expression in the liver and in sera from patients with ALD and also pro-hepcidin levels are reduced. Ohtake and coworkers demonstrated that serum hepcidin is decreased in patients with ALD [124]. Most notably, Harrison-Findik et al. showed in their elegant study using an alcohol mouse model that the expression of hepatic hepcidin is rapidly suppressed upon exposure to alcohol [125]. 4-methylpyrazole, a competitive inhibitor of alcohol-metabolizing enzymes (e.g., ADHs), abolished this effect. However, ethanol did not alter the expression of TfR1 and ferritin or the activation of iron regulatory mRNA-binding proteins IRP1 and IRP2. Mice maintained on 10–20 % ethanol for 7 days displayed a downregulation of liver hepcidin without changes in liver triglycerides or histology. This was accompanied by elevated DMT1 and ferroportin expression in the duodenum [125]. Ethanol downregulated hepcidin promotor activity and the DNA binding activity of the CCAT/NH-binding protein α (C/EBPα) but not β [126]. The same author showed later that the effect of alcohol and hepcidin was independent of Kupffer cell activation and TNFα signaling [127]. Conversely, there are observational studies showing that the hepatic expression of TfR1 is increased under conditions of alcohol consumption. In 2002, Suzuki et al. showed that TfR1 is increased in patients with ALD and that abstinence from drinking decreased TfR1 [104]. Furthermore, ethanol increased transferrin bound iron uptake

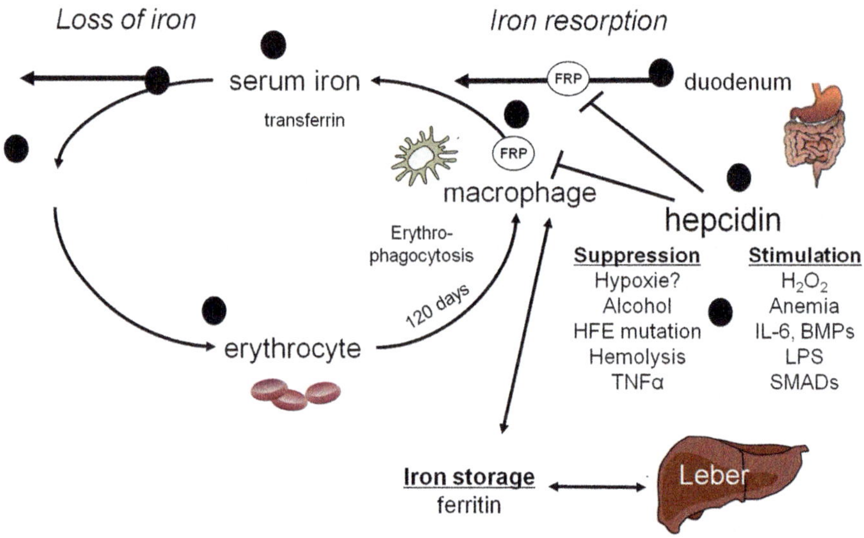

Fig. 21.2 Systemic iron homeostasis and utilization in the body. *Black circles* indicate potential target sites of iron regulation affected by chronic or acute alcohol consumption. Dietary iron is absorbed in the duodenum and bound to transferrin. Iron is then mainly delivered to the bone marrow for erythropoiesis, while senescent erythrocytes are phagocytosed by macrophages. This efficient mechanism recycles >90 % of all iron for new heme synthesis. Excess iron is stored in hepatic ferritin or temporarily in macrophage ferritin. Regulation of iron metabolism by hepcidin and upstream factors are also shown (modified after [132]). *FRP* ferroportin, *HFE* gene that is mutated in hereditary hemochromatosis, *LPS* lipopolysaccharide, *IL-6* Interleukine 6, *BMP* bone morphogenetic protein, *TNF-α* tumor necrosis factor α

into hepatocytes, possibly due to ethanol-induced TfR1 expression, and partly mediated by activation of IRP1 [128, 129]. TfR1 upregulation may be directly related to translational effects of ROS and not due to IRP1 [130].

Notably, we could not find differences in the expression of various iron-related mRNAs in a preliminary microarray study comparing iron overload and normal ALD patients (Mueller et al., unpublished). We also noted a high variability in the expression of these mRNAs, implicating some caution in the interpretation of liver samples from ALD patients. It seems that ALD lesions are quite heterogeneously scattered throughout the liver. Therefore, it remains an open question whether the effects observed in an acute alcohol-exposure model in rodents are directly related to alcohol and hepcidin. Additional indirect conditions on the perfusion of the splanchnic system could also be involved. Two recent publications have tried to further elucidate the mechanisms of alcohol-mediated hepcidin expression [115, 131]. Both studies showed that bolus H_2O_2 treatments, a major ROS insult, could drastically suppress hepcidin expression. However, we recently found that this effect is rather due to the artificial exposure to high H_2O_2 levels (>50 μM) than peroxide itself [114]. If H_2O_2 is released continuously at low concentrations (between 0.3 and 6 μM) to hepatoma and primary liver cells, as they are typically released by

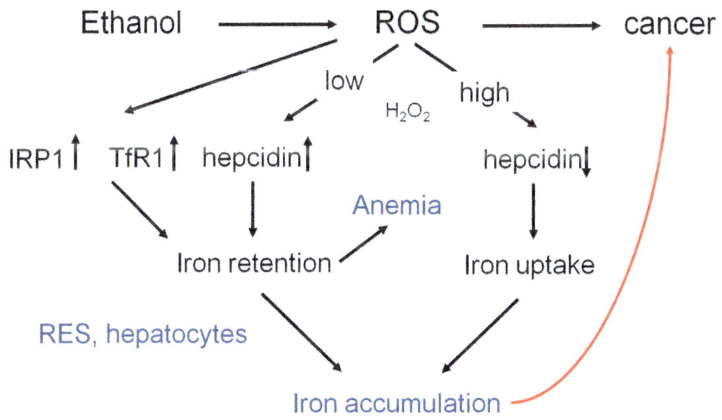

Fig. 21.3 Present understanding of systemic and cellular changes of iron homeostasis in ALD ultimately leading to cancer (modified after [132]). *IRP1* iron regulatory protein 1, *TfR1* transferrin receptor 1, *RES* reticulo endothelial system

inflammatory cells or during cellular metabolism, a strong hepcidin upregulation rather than suppression was observed. H_2O_2 synergistically stimulates hepcidin promoter activity in combination with recombinant IL-6 or BMP-6 in a manner that requires the functional STAT3 responsive element. The H_2O_2-mediated hepcidin induction requires STAT3 phosphorylation and is effectively blocked by siRNA-mediated STAT3 silencing, overexpression of SOCS3 (suppressor of cytokine signaling 3), and antioxidants such as *N*-acetylcysteine [114].

In summary, numerous rodent and human studies suggest the involvement of key iron molecules in the iron overload of ALD patients. These molecules may include hepcidin, TfR1, IRP1, and ROS. However, no definite conclusions can be made due to methodological challenges and complex interactions among these molecules (Fig. 21.3).

Therapy of ALD in the Context of Iron Overload

Based on our preliminary knowledge of underlying mechanisms of iron disturbances in patients with ALD, no treatment options are readily available. The idea of iron depletion by phlebotomy has been controversially discussed in patients with HCV that equally develop iron overload; however, phlebotomy is highly propagated in certain countries, such as Japan. There has been one trial initiated in France in 2012 and titled, "phlebotomy in risk of hepatocellular carcinoma in patients with compensated alcoholic cirrhosis" (Tirrox). Unfortunately, this study has stopped recruiting patients due to delayed recruitment compared to that anticipated (Ref: ClinicalTrials.gov NCT01342705). For the moment, it can be regularly perceived that iron overload is an important factor in the progression of ALD, also in combination with other liver diseases, such as HCV. It is an important challenge for the future to clearly identify the underlying molecular mechanisms and to develop novel targeted therapeutic strategies and to improve the early detection of hepatic iron overload in ALD (e.g., measurement of liver iron by susceptometry or hepcidin-ELISA for diagnostic purposes). Future studies will continue with the goal of preventing fibrosis progression and to prevent HCC development, thereby improving overall survival by targeted therapeutic approaches that may include a pharmacological intervention leading to stimulation of hepcidin.

References

1. Bosch FX, Ribes J, Diaz M, Cleries R (2004) Primary liver cancer: worldwide incidence and trends. Gastroenterology 127:S5–S16
2. Rehm J, Room R, Monteiro R et al (2004) Global and regional burden of disease attributable to selected major risk factors. In: Ezatti M, Murray C, Lopez AD, Rodgers A, Murray C (eds) Comparative quantification of health risks. World Health Organization, Geneva
3. Rehm J, Klotsche J, Patra J (2007) Comparative quantification of alcohol exposure as risk factor for global burden of disease. Int J Methods Psychiatr Res 16:66–76
4. Baan R, Straif K, Grosse Y et al (2007) Carcinogenicity of alcoholic beverages. Lancet Oncol 8:292–293
5. Seitz HK, Stickel F (2007) Molecular mechanisms of alcohol-mediated carcinogenesis. Nat Rev Cancer 7:599–612
6. Zakhari S, Vasiliou V, Guo M (2011) Alcohol and cancer. Springer, New York
7. Beland FA, Benson RW, Mellick PW et al (2005) Effect of ethanol on the tumorigenicity of urethane (ethyl carbamate) in B6C3F1 mice. Food Chem Toxicol 43:1–19
8. IARC (2007) IARC monographs on the evaluation of carcinogenic risks to humans. In: Alcohol consumption and ethyl carbamate, vol 96. IARC, Lyon
9. Chavez PR, Lian F, Chung J et al (2011) Long-term ethanol consumption promotes hepatic tumorigenesis but impairs normal hepatocyte proliferation in rats. J Nutr 141:1049–1055
10. Ye Q, Lian F, Chavez PR et al (2012) Cytochrome P450 2E1 inhibition prevents hepatic carcinogenesis induced by diethylnitrosamine in alcohol-fed rats. Hepatobiliary Surg Nutr 1:5–18
11. Tsuchishima M, George J, Shiroeda H et al (2013) Chronic ingestion of ethanol induces hepatocellular carcinoma in mice without additional hepatic insult. Dig Dis Sci 58:1923–1933
12. De Lima VM, Oliveira CP, Alves VA et al (2008) A rodent model of NASH with cirrhosis, oval cell proliferation and hepatocellular carcinoma. J Hepatol 49:1055–1061
13. Nakanuma Y, Ohta G (1985) Is mallory body formation a preneoplastic change? A study of 181 cases of liver bearing hepatocellular carcinoma and 82 cases of cirrhosis. Cancer 55:2400–2405
14. Smith P, Tee LBG, Yeoh GCT (1996) Appearance of oval cells in the liver of rats after long-term exposure to ethanol. Hepatology 23:145–154
15. Seitz HK, Mueller S (2010) Alcoholic liver disease. In: Dancygier H (ed) Clinical hepatology. Principle and practice of hepatobiliary diseases, vol 2. Springer, Heidelberg, pp 1111–1151
16. El-Serag HB (2004) Hepatocellular carcinoma: recent trends in the United States. Gastroenterology 127:S27–S34
17. El-Serag HB, Rudolph KL (2011) Hepatocellular carcinoma: epidemiology and molecular carcinogenesis. Gastroenterology 132:2557–2576
18. WHO. Mortality database. http://www.who.int/whosis/en

19. Rehm J, Samokhvalov AV, Shield KD (2013) Global burden of alcoholic liver disease. J Hepatol 59:160–168
20. Morgan TR, Mandayam S, Jamal MM (2004) Alcohol and hepatocellular cancer. Gastroenterology 127:S87–S97
21. Hassan MM, Hwang LY, Hatten CJ et al (2002) Risk factors for hepatocellular carcinoma: synergism of alcohol with viral hepatitis and diabetes mellitus. Hepatology 36:1206–1213
22. Tagger A, Donato F, Ribero ML et al (1999) Case control study on hepatitis C virus (HCV) as a risk factor for hepatocellular carcinoma: the role of HCV genotypes and the synergism with hepatitis B virus and alcohol. Int J Cancer 81:695–699
23. DeBac C, Stroffolini T, Gaeta GB et al (1994) Pathogenetic factors in cirrhosis with and without hepatocellular carcinoma: a multicentre Italian study. Hepatology 20:1225–1230
24. Fattovich G, Stroffolini T, Zagni I, Donato F (2004) Hepatocellular carcinoma in cirrhosis: incidence and risk factors. Gastroenterology 127(5 Suppl 1):S35–S50
25. Nzeako UC, Goodman ZD, Ishak KG (1996) Hepatocellular carcinoma in cirrhotic and non-cirrhotic livers. A clinico-histopathologic study of 804 North American patients. Am J Clin Pathol 105:65–75
26. Chiesa R, Donato F, Tagger A et al (2000) Etiology of hepatocellular carcinoma in Italian patients with and without cirrhosis. Cancer Epidemiol Biomarkers Prev 9:213–216
27. Ohnishi K, Iida S, Iwama S et al (1982) The effect of chronic habitual alcohol intake on the development of liver cirrhosis and hepatocellular carcinoma: relation to hepatitis B surface antigen carriage. Cancer 49:672–677
28. Mueller S, Millonig G, Seitz HK (2009) Alcoholic liver disease and hepatitis C: a frequently underestimated combination. World J Gastroenterol 15:3462–3471
29. Kowdley KV (2004) Iron, hemochromatosis, and hepatocellular carcinoma. Gastroenterology 127:S79–S86
30. Ascha MS, Hanouneh IA, Lopez R et al (2010) The incidence and risk factors of hepatocellular carcinoma in patients with nonalcoholic steatohepatitis. Hepatology 51:1972–1978
31. ElSerag HB, Tran T, Everhart JE (2004) Diabetes increases the risk of chronic liver disease and hepatocellular cancer. Gastroenterology 126:460–468
32. Liangpusakul S, Chalasani N (2012) What should we recommend to our patients with NAFLD regarding alcohol use? Am J Gastroenterol 107:976–978
33. Donato F, Tagger A, Gelatti U et al (2002) Alcohol and hepatocellular carcinoma: the effect of lifetime intake and hepatitis virus infections in men and women. Am J Epidemiol 155:323–331
34. Seitz HK, Stickel F (2013) Ethanol and hepatocarcinogenesis. In: Watson RR, Preedy VR, Zibadi S (eds) Alcohol, nutrition, and health consequences. Springer, New York, pp 411–427
35. Seitz HK, Wang XD (2013) The role of cytochrome P-4502E1 in ethanol-mediated carcinogenesis. In: Dey A (ed) Cytochrome P4502E1: its role in disease and drug metabolism, vol 67, Subcellular biochemistry. Springer, New York, pp 131–144
36. Seitz HK, Mueller S (2011) Ethanol metabolism and its consequences. In: Anzenbacher P, Zanger U (eds) Metabolism of drugs and xenobiotics. Wiley, Weinheim
37. Seitz HK, Stickel F (2006) Risk factors and mechanisms of hepatocarcinogenesis with special emphasis on alcohol and oxidative stress. Biol Chem 387:349–360
38. Seitz HK, Stickel F (2009) Acetaldehyde as an underestimated risk factor for cancer development: role of genetics in ethanol metabolism. Genes Nutr 5:121–128
39. Mueller S, Millonig G, Seitz HK, Waite GN (2012) Chemiluminescence detection of hydrogen peroxide. In: Pantopoulos K, Schipper H (eds) Principles of free radical biomedicine. Nova Publishers, New York, pp 283–302
40. Albano E (2002) Free radical mechanisms in immune reactions associated with alcoholic liver disease. Free Radic Biol Med 32:110–114
41. Bailey SM, Cunningham CC (2002) Contribution of mitochondria to oxidative stress associated with alcoholic liver disease. Free Radic Biol Med 32:11–16

42. Garcia-Ruiz C, Colell A, Paris R, Fernandez-Checa JC (2000) Direct interaction of GD3 ganglioside with mitochondria generates reactive oxygen species followed by mitochondrial permeability transition, cytochrome c release, and caspase activation. FASEB J 14:847–858
43. Bautista AP (2002) Neutrophilic infiltration in alcoholic hepatitis. Alcohol 27:17–21
44. Chamulitrat W, Spitzer JJ (1996) Nitric oxide and liver injury in alcohol-fed rats after lipopolysaccharide administration. Alcohol Clin Exp Res 20:1065–1070
45. Obe G, Jonas R, Schmidt S (1986) Metabolism of ethanol in vitro produces a compound which induces sister-chromatid exchanges in human peripheral lymphocytes in vitro: acetaldehyde not ethanol is mutagenic. Mutat Res 174:47–51
46. Helander A, Lindahl-Kiessling K (1991) Increased frequency of acetaldehyde-induced sister-chromatid exchanges in human lymphocytes treated with an aldehyde dehydrogenase inhibitor. Mutat Res 264:103–107
47. Maffei F, Fimognari C, Castelli E et al (2000) Increased cytogenetic damage detected by FISH analysis on micronuclei in peripheral lymphocytes from alcoholics. Mutagenesis 15:517–523
48. Matsuda T, Kawanishi M, Yagi T et al (1998) Specific tandem GG to TT base substitutions induced by acetaldehyde are due to intra-strand crosslinks between adjacent guanine bases. Nucleic Acids Res 26:1769–1774
49. Theruvathu JA, Jaruga P, Nath RG et al (2005) Polyamines stimulate the formation of mutagenic 1, N2-propanodeoxyguanosine adducts from acetaldehyde. Nucleic Acids Res 33:3513–3520
50. Fang JL, Vaca CE (1995) Development of a 32P-postlabelling method for the analysis of adducts arising through the reaction of acetaldehyde with 2′-deoxyguanosine-3′-monophosphate and DNA. Carcinogenesis 16:2177–2185
51. Fang JL, Vaca CE (1997) Detection of DNA adducts of acetaldehyde in peripheral white blood cells of alcohol abusers. Carcinogenesis 18:627–632
52. Matsuda T, Matsumoto A, Uchida M et al (2007) Increased formation of hepatic N2-ethylidene-2'-deoxyguanosine DNA adducts in aldehyde dehydrogenase 2-knockout mice treated with ethanol. Carcinogenesis 28:2363–2366
53. Wang M, Yu N, Chen L et al (2006) Identification of an acetaldehyde adduct in human liver DNA and quantitation as N2-ethyldeoxyguanosine. Chem Res Toxicol 19:319–324
54. Wang M, McIntee EJ, Cheng G et al (2000) Identification of DNA adducts of acetaldehyde. Chem Res Toxicol 13:1149–1157
55. Stein S, Lao Y, Yang IY et al (2006) Genotoxicity of acetaldehyde- and crotonaldehyde-induced 1, N2-propanodeoxyguanosine DNA adducts in human cells. Mutat Res 608:1–7
56. Garro AJ, Espina N, Farinati F, Salvagnini M (1986) The effects of chronic ethanol consumption on carcinogen metabolism and on O6-methylguanine transferase-mediated repair of alkylated DNA. Alcohol Clin Exp Res 10:73S–77S
57. French SW (2013) Epigenetic events in liver cancer resulting from alcoholic liver disease. Alcohol Res 35:57–68
58. Stickel F, Herold C, Seitz HK, Schuppan D (2006) Alcohol and methyl transfer: implication for alcohol related hepatocarcinogenesis. In: Ali S, Friedman SL, Mann DA (eds) Liver disease: biochemical mechanisms and new therapeutic insights. Science Publisher, Enfield, pp 45–58
59. Seitz HK, Becker P (2007) Alcohol metabolism and cancer risk. Alcohol Res Health 30:38–47
60. Seitz HK (2012) Alcohol and breast cancer. Breast 21:426–427
61. Homann N, Stickel F, Konig IR et al (2006) Alcohol dehydrogenase 1C*1 allele is a genetic marker for alcohol-associated cancer in heavy drinkers. Int J Cancer 118:1998–2002
62. Yokoyama A, Muramatsu T, Ohmori T et al (1998) Alcohol-related cancers and aldehyde dehydrogenase-2 in Japanese alcoholics. Carcinogenesis 19:1383–1387
63. Matsuda T, Yabushita H, Kanaly RA et al (2006) Increased DNA damage in ALDH2-deficient alcoholics. Chem Res Toxicol 19:1374–1378

64. French SM (2013) The importance of CYP2E1 in the pathogenesis of alcoholic liver disease and drug toxicity and the role of the proteasome. In: Dey A (ed) Cytochrome P4502E1: its role in disease and drug metabolism. Springer, New York, pp 131–144
65. Oneta CM, Lieber CS, Li J et al (2002) Dynamics of cytochrome P4502E1 activity in man: induction by ethanol and disappearance during withdrawal phase. J Hepatol 36:47–52
66. Gouillon Z, Lucas D, Li J et al (2000) Inhibition of ethanol-induced liver disease in the intragastric feeding rat model by chlormethiazole. Proc Soc Exp Biol Med 224:302–308
67. Bradford BU, Kono H, Isayama F et al (2005) Cytochrome P450 CYP2E1, but not nicotinamide adenine dinucleotide phosphate oxidase, is required for ethanol-induced oxidative DNA damage in rodent liver. Hepatology 41:336–344
68. Morgan K, French SW, Morgan TR (2002) Production of a cytochrome P450 2E1 transgenic mouse and initial evaluation of alcoholic liver damage. Hepatology 36:122–134
69. Wang XD, Seitz HK (2004) Alcohol and retinoid interaction. In: Watson RR, Preedy VR (eds) Nutrition and alcohol: linking nutrient interactions and dietary intake. CRC, Boca Raton, pp 313–321
70. El Ghissassi F, Barbin A, Bartsch H (1998) Metabolic activation of vinyl chloride by rat liver microsomes: low-dose kinetics and involvement of cytochrome P450 2E1. Biochem Pharmacol 55:1445–1452
71. Pandya GA, Moriya M (1996) 1, N6-ethenodeoxyadenosine, a DNA adduct highly mutagenic in mammalian cells. Biochemistry 35:11487–11492
72. Levine RL, Yang IY, Hossain M et al (2000) Mutagenesis induced by a single 1, N6-ethenodeoxyadenosine adduct in human cells. Cancer Res 60:4098–4104
73. Frank A, Seitz HK, Bartsch H et al (2004) Immunohistochemical detection of 1, N6-ethenodeoxyadenosine in nuclei of human liver affected by diseases predisposing to hepato-carcinogenesis. Carcinogenesis 25:1027–1031
74. Nair J, Srivatanakul P, Haas C et al (2010) High urinary excretion of lipid peroxidation-derived DNA damage in patients with cancer-prone liver diseases. Mutat Res 683:23–28
75. Wang Y, Millonig G, Nair J et al (2009) Ethanol-induced cytochrome P4502E1 causes carcinogenic etheno-DNA lesions in alcoholic liver disease. Hepatology 50:453–461
76. Lieber CS (2004) CYP2E1: from ASH to NASH. Hepatol Res 28:1–11
77. Wang Y, Seitz H, Wang X (2010) Moderate alcohol consumption aggravates high-fat diet induced steatohepatitis in rats. Alcohol Clin Exp Res 34:567–573
78. Leo MA, Lieber CS (1982) Hepatic vitamin A depletion in alcoholic liver injury. N Engl J Med 307:597–601
79. Wang XD, Liu C, Chung J et al (1998) Chronic alcohol intake reduces retinoic acid concentration and enhances AP-1 (c-Jun and c-Fos) expression in rat liver. Hepatology 28:744–750
80. Liu C, Russell RM, Seitz HK, Wang XD (2001) Ethanol enhances retinoic acid metabolism into polar metabolites in rat liver via induction of cytochrome P4502E1. Gastroenterology 120:179–189
81. Dan Z, Popov Y, Patsenker E et al (2005) Hepatotoxicity of alcohol-induced polar retinol metabolites involves apoptosis via loss of mitochondrial membrane potential. FASEB J 19:845–847
82. Liu C, Chung J, Seitz HK et al (2002) Chlormethiazole treatment prevents reduced hepatic vitamin A levels in ethanol-fed rats. Alcohol Clin Exp Res 26:1703–1709
83. Chung J, Liu C, Smith DE et al (2001) Restoration of retinoic acid concentration suppresses ethanol-enhanced c-Jun expression and hepatocyte proliferation in rat liver. Carcinogenesis 22:1213–1219
84. Leo MA, Kim C, Lowe N, Lieber CS (1992) Interaction of ethanol with beta-carotene: delayed blood clearance and enhanced hepatotoxicity. Hepatology 15:883–891
85. Fenton HJ (1894) Oxidation of tartaric acid in presence of iron. J Chem Soc 65:899–910
86. Saran M, Michel C, Stettmaier K, Bors W (2000) Arguments against the significance of the Fenton reaction contributing to signal pathways under in vivo conditions. Free Radic Res 33:567–579

87. Niederau C, Fischer R, Purschel A et al (1996) Long-term survival in patients with hereditary hemochromatosis. Gastroenterology 110:1107–1119
88. Ko C, Siddaiah N, Berger J et al (2007) Prevalence of hepatic iron overload and association with hepatocellular cancer in end-stage liver disease: results from the National Hemochromatosis Transplant Registry. Liver Int 27:1394–1401
89. Turlin B, Juguet F, Moirand R et al (1995) Increased liver iron stores in patients with hepatocellular carcinoma developed on a noncirrhotic liver. Hepatology 22:446–450
90. Zacharski LR, Chow BK, Howes PS et al (2008) Decreased cancer risk after iron reduction in patients with peripheral arterial disease: results from a randomized trial. J Natl Cancer Inst 100:996–1002
91. Hann HW, Stahlhut MW, Blumberg BS (1988) Iron nutrition and tumor growth: decreased tumor growth in iron-deficient mice. Cancer Res 48:4168–4170
92. Tsukamoto H, Horne W, Kamimura S et al (1995) Experimental liver cirrhosis induced by alcohol and iron. J Clin Invest 96:620–630
93. D'Amico G, Garcia-Tsao G, Pagliaro L (2006) Natural history and prognostic indicators of survival in cirrhosis: a systematic review of 118 studies. J Hepatol 44:217–231
94. Ganne-Carrie N, Christidis C, Chastang C et al (2000) Liver iron is predictive of death in alcoholic cirrhosis: a multivariate study of 229 consecutive patients with alcoholic and/or hepatitis C virus cirrhosis: a prospective follow up study. Gut 46:277–282
95. Powell LW (1975) The role of alcoholism in hepatic storage disease. In: Seixas FA, Williams K, Seggleston S (eds) Medical consequences of alcoholism. New York Academy of Sciences, New York, pp 124–134
96. Bell ET (1955) The relation of portal cirrhosis to hemochromatosis and to diabetes mellitus. Diabetes 4:435–446
97. Conrad ME Jr, Berman A, Crosby WH (1962) Iron kinetics in Laennec's cirrhosis. Gastroenterology 43:385–390
98. Feder JN, Gnirke A, Thomas W et al (1996) A novel MHC class I-like gene is mutated in patients with hereditary haemochromatosis. Nat Genet 13:399–408
99. Pascoe A, Kerlin P, Steadman C et al (1999) Spur cell anaemia and hepatic iron stores in patients with alcoholic liver disease undergoing orthotopic liver transplantation. Gut 45:301–305
100. Sohda T, Takeyama Y, Irie M et al (1999) Putative hemochromatosis gene mutations and alcoholic liver disease with iron overload in Japan. Alcohol Clin Exp Res 23:21S–23S
101. Dostalikova-Cimburova M, Kratka K, Stransky J et al (2012) Iron overload and HFE gene mutations in Czech patients with chronic liver diseases. Dis Markers 32:65–72
102. Chapman RW, Morgan MY, Laulicht M et al (1982) Hepatic iron stores and markers of iron overload in alcoholics and patients with idiopathic hemochromatosis. Dig Dis Sci 27:909–916
103. Bell H, Skinningsrud A, Raknerud N, Try K (1994) Serum ferritin and transferrin saturation in patients with chronic alcoholic and non-alcoholic liver diseases. J Intern Med 236:315–322
104. Suzuki Y, Saito H, Suzuki M, Hosoki Y, Sakurai S, Fujimoto Y, Kohgo Y (2002) Up-regulation of transferrin receptor expression in hepatocytes by habitual alcohol drinking is implicated in hepatic iron overload in alcoholic liver disease. Alcohol Clin Exp Res 26:26S–31S
105. Mueller S (2013) Non-invasive assessment of patients with alcoholic liver disease. Clin Liver Dis 2:68–71
106. Wang J, Pantopoulos K (2011) Regulation of cellular iron metabolism. Biochem J 434:365–381
107. Andrews NC (1999) Disorders of iron metabolism. N Engl J Med 341:1986–1995
108. Ganz T (2007) Molecular control of iron transport. J Am Soc Nephrol 18:394–400
109. Nicolas G, Bennoun M, Devaux I et al (2001) Lack of hepcidin gene expression and severe tissue iron overload in upstream stimulatory factor 2 (USF2) knockout mice. Proc Natl Acad Sci U S A 98:8780–8785

110. Nemeth E, Ganz T (2009) The role of hepcidin in iron metabolism. Acta Haematol 122:78–86
111. Hentze MW, Muckenthaler MU, Galy B, Camaschella C (2010) Two to tango: regulation of Mammalian iron metabolism. Cell 142:24–38
112. Sow FB, Florence WC, Satoskar AR et al (2007) Expression and localization of hepcidin in macrophages: a role in host defense against tuberculosis. J Leukoc Biol 82:934–945
113. Weiss G, Goodnough LT (2005) Anemia of chronic disease. N Engl J Med 352:1011–1023
114. Millonig G, Ganzleben I, Peccerella T et al (2012) Sustained submicromolar H2O2 levels induce hepcidin via signal transducer and activator of transcription 3 (STAT3). J Biol Chem 287:37472–37482
115. Miura K, Taura K, Kodama Y, Schnabl B, Brenner DA (2008) Hepatitis C virus-induced oxidative stress suppresses hepcidin expression through increased histone deacetylase activity. Hepatology 48:1420–1429
116. Richardson DR, Lane DJ, Becker EM et al (2010) Mitochondrial iron trafficking and the integration of iron metabolism between the mitochondrion and cytosol. Proc Natl Acad Sci U S A 107:10775–10782
117. Rouault TA (2006) The role of iron regulatory proteins in mammalian iron homeostasis and disease. Nat Chem Biol 2:406–414
118. Recalcati S, Minotti G, Cairo G (2010) Iron regulatory proteins: from molecular mechanisms to drug development. Antioxid Redox Signal 13:1593–1616
119. Wallander ML, Leibold EA, Eisenstein RS (2006) Molecular control of vertebrate iron homeostasis by iron regulatory proteins. Biochim Biophys Acta 1763:668–689
120. Fillebeen C, Pantopoulos K (2002) Redox control of iron regulatory proteins. Redox Rep 7:15–22
121. Mueller S (2005) Iron regulatory protein 1 as a sensor of reactive oxygen species. Biofactors 24:171–181
122. Mutze S, Hebling U, Stremmel W et al (2003) Myeloperoxidase-derived hypochlorous acid antagonizes the oxidative stress-mediated activation of iron regulatory protein 1. J Biol Chem 278:40542–40549
123. Watts RN, Hawkins C, Ponka P, Richardson DR (2006) Nitrogen monoxide (NO)-mediated iron release from cells is linked to NO-induced glutathione efflux via multidrug resistance-associated protein 1. Proc Natl Acad Sci U S A 103:7670–7675
124. Ohtake T, Saito H, Hosoki Y et al (2007) Hepcidin is down-regulated in alcohol loading. Alcohol Clin Exp Res 31:S2–S8
125. Harrison-Findik DD, Schafer D, Klein E et al (2006) Alcohol metabolism-mediated oxidative stress down-regulates hepcidin transcription and leads to increased duodenal iron transporter expression. J Biol Chem 281:22974–22982
126. Harrison-Findik DD, Klein E, Crist C et al (2007) Iron-mediated regulation of liver hepcidin expression in rats and mice is abolished by alcohol. Hepatology 46:1979–1985
127. Harrison-Findik DD, Klein E, Evans J, Gollan J (2009) Regulation of liver hepcidin expression by alcohol in vivo does not involve Kupffer cell activation or TNF-alpha signaling. Am J Physiol Gastrointest Liver Physiol 296:G112–G118
128. Shindo M, Torimoto Y, Saito H et al (2006) Functional role of DMT1 in transferrin-independent iron uptake by human hepatocyte and hepatocellular carcinoma cell, HLF. Hepatol Res 35:152–162
129. Suzuki M, Fujimoto Y, Suzuki Y et al (2004) Induction of transferrin receptor by ethanol in rat primary hepatocyte culture. Alcohol Clin Exp Res 28:98S–105S
130. Andriopoulos B, Hegedusch S, Mangin J et al (2007) Sustained hydrogen peroxide induces iron uptake by transferrin receptor-1 independent of the iron regulatory protein/iron-responsive element network. J Biol Chem 282:20301–20308
131. Nishina S, Hino K, Korenaga M et al (2008) Hepatitis C virus-induced reactive oxygen species raise hepatic iron level in mice by reducing hepcidin transcription. Gastroenterology 134:226–238
132. Mueller S, Rausch V (2015) The role of iron in alcohol-mediated hepatocarcinogenesis. Adv Exp Med Biol 815:89–112

Index

A
Acetaldehyde, 483–485
Acetaminophen (APAP)
  hepatotoxicity, 200–203
    P450-mediated APAP metabolism, 201
    ROS production and oxidative stress, 200, 201
    xanthine oxidoreductase, 202
  liver diseases induced by, 102
  toxin injury, 136–137
Activating transcription factor-6 (ATF6), 114–115
Activator protein-1 (AP-1), 100
Acute liver failure (ALF)
  autoimmune hepatitis, 206
  causes, 200–207
  drug-induced liver injury
    acetaminophen hepatotoxicity, 200–203
    non-acetaminophen, 203–204
  oxidative stress in liver regeneration, 207
  viral hepatitis, 204–205
  Wilson disease, 206
Acyl-coenzyme A oxidase (AOX), 464
Adiponectin, 286
Alarmins, 315
Alcohol dehydrogenase (ADH), 216
Alcoholic cirrhosis, 215, 230–231
Alcoholic fibrosis, 230–231
Alcoholic liver disease (ALD)
  alcoholic steatosis, 224–225
  alcohol-induced oxidative stress
    hydroxyethyl-free radicals, 220–221
    iron overload, 221–222
    liver antioxidant defenses, 222–224
    nitric oxide, 221
    reactive oxygen species, 219–220
  antioxidant therapy
    polyenylphosphatidylcholine, 399
    S-adenosylmethionine, 400
    silymarin, 399
    vitamins E and C, 398
  CYP2E1, 131
  fibrosis and cirrhosis, 230–231
  hepatocellular cancer (HCC), 479
  hepatocellular injury, 228–229
  HSF-1, 132
  lobular inflammation, 225–228
  NF-κB activation, 131
  Nox and XO, 132
  pathogenesis, 217–218
  secondary iron overload, 352
  SOD, 132
Alcoholic steatosis, 224–225
Alcohol-induced oxidative stress
  hydroxyethyl-free radicals, 220–221
  iron overload, 221–222
  liver antioxidant defenses, 222–224
  nitric oxide, 221
  reactive oxygen species, 219–220
Aldehyde oxidase (AO), 202
Allopurinol, 332
α-tocopherol, chemical structure, 52
Ammonia-induced oxidative stress, hepatic encephalopathy
  human studies, 362–363
  in vitro evidence, 360–361
  in vivo evidence, 361–362
AMP-kinase (AMPK), 24

E. Albano, M. Parola (eds.), *Studies on Hepatic Disorders*,
Oxidative Stress in Applied Basic Research and Clinical Practice,
DOI 10.1007/978-3-319-15539-5

Angiogenesis, fibrogenesis and, 187–188
Antagonistic pleiotropy theory, 67
Antioxidants
  chemoprevention, hepatocellular carcinoma, 467–469
  defense mechanism
    hepatocellular injury, 99–101
    unfolded protein response, 116–117
  enzymatic, 12–14
  non-enzymatic, 10–11
Antioxidant therapy
  alcoholic liver diseases, 398–400
  autoimmune hepatitis, 332
  chronic viral hepatitis, 392–397
  classification, 390–392
  IgG4-related sclerosing cholangitis (IgG4-SC), 335
  liver fibrosis and cirrhosis, 406–407
  NAFLD/NASH, 400–405
  natural and pharmacological antioxidant defense mechanisms, 331–332
  nitric oxide metabolism, 392
  overlap syndromes in autoimmune liver disease, 335
  primary biliary cirrhosis, 332–334
  primary sclerosing cholangitis, 334–335
  system, 389–390
Apoptosis, 105
Apoptosis-inducing factor (AIF), 203
Apoptosis signal-regulating kinase 1 (ASK1), 25, 203
Ataxia-telangiectasia mutated (ATM), 24–25
ATP7B, 206
Autoimmune hepatitis (AIH), 328–329, 332
Autoimmune liver disease, oxidative stress
  antioxidant therapy, 331–335
  betaine, 337
  calcineurin inhibitors, 338
  curcumin, 338
  docosahexaenoic acid (DHA), 336
  D-penicillamine, 337
  fibric acid derivates, 339
  immunopathogenesis, 327–330
  melatonin, 336
  mitoquinone (Mite), 337
  N-acetylcysteine (NAC), 337
  pentoxifylline, 337
  PPC, 337
  selenium (selenocysteine), 338
  silimaryn, 336
  sirolimus, 339
  vitamin C/ascorbic acid, 339
  vitamin D, 339
  vitamin E (alpha-tocopherol), 338
  zinc/polaprezinc (combination of zinc and l-carnosine), 338
Autophagy, 106, 164

B

Bax inhibitor 1, 152
β-actin, 20
β-carotene, chemical structure, 52
Betaine, 337
Bezafibrate and fenofibrate, 333
Bile duct ligation (BDL), 102, 137
Bile secretion, 315–316
Biliary fibrosis, 177
Biliary intraepithelial neoplasia (BilIN), 436–437
Biosphere evolution and oxygen, 65–66

C

Calcineurin inhibitors, 338
cAMP response element-binding protein (CREB), 318
Carbamazepine, 165
Carbon tetrachloride, 466
Catalase, 12
Cell cycle progression, 69
Cell death. *See* Apoptosis; Autophagy; Necrosis
Cell enlargement, 68
Cell senescence
  antagonistic pleiotropy theory, 67
  biomarkers
    cell cycle progression, 69
    cell enlargement, 68
    DNA damage foci, 69
    SA-β-gal activity, 68
  DNA damage
    DDR, 73, 74
    by genotoxic agents, 70
    by oncogenic/hyper-proliferative signals, 71
    by oxidative stress, 70–71
  during embryonic development, 68
  gene expression, 72
  in liver
    cytokines and growth factors, 75
    disease prognosis, 75
    ductular epithelium, 77
    histological staining, 76
    pre-malignant senescent hepatocytes, 76
    stellate cells, 77

ROS production and phenotype stabilization, 73
secretory phenotype, 70–71
telomere uncapping, 69
Cellular antioxidant systems, 455–458
Cellular redox state, 43
Cellular signalling pathways and
NF-κB and HCC development, 459–460
p38α MAPK pathway and HCC, 460
Central nervous system complications, chronic liver failure and oxidative stress in. *See* Hepatic encephalopathy
Centrilobular fibrosis, 177
Ceramide, 290, 293
Chaperones, 114, 115
Cholangiocarcinoma (CC)
gross classification, 433
hilar CC, 431
intrahepatic, 431–435
molecular classification, 436
mucinous cystic neoplasms (MCNs), 435–436
NO, cholangiopathies, 317–320
premalignant cholangiocellular lesions, 436–437
Cholangiocytes
anatomy and functions, 310–311
and bile production, 311
and liver repair, 312–313
Cholangiopathies
classification, 309, 310
hepatic progenitor cell (HPC) compartment, 310
reactive oxidative species in, 313–320
Cholestasis, 102
Cholestatic liver disease, 137
Cholesterol
and ischemia-reperfusion injury
CD diet, 160, 161
ER, 161
hypercholesterolemia, 160
synthesis, 162
Choline-and methionine-deficient diets (CMD), 466–467
Chromatin remodelling, 29
Chronic hepatitis C (CHC)
antioxidant therapies, clinical trials, 393–396
milk thistle, 398
N-acetylcysteine, 397
secondary iron overload, 351
silymarin, 398
Chronic liver diseases (CLD). *See also* Fibrogenesis
antioxidant therapy
alcoholic liver diseases, 398–400
antioxidant classification, 390–392
antioxidant system, 389–390
chronic viral hepatitis, 392–397
liver fibrosis and cirrhosis, 406–407
NAFLD/NASH, 400–405
causes, 171
fibrogenic progression
aetiology, 173
epidemiological data, 172
hepatic cell populations, 174
hepatocytes, 173
Kupffer cells and macrophages, 173
oxidative stress and ROS, 171–172
redox mechanisms, 171–172
Chronic liver failure
etiologic factors, 358
oxidative stress and central nervous system complications, 357 (*see also* Hepatic encephalopathy)
Chronic viral hepatitis (CVH)
Cirrhosis, 280. *See also* Non-alcoholic steatohepatitis (NASH)
antioxidant therapy, 406–407
Closed circular DNA (cccDNA), 261
Colchicine, 333, 334
Combined hepatocellular-cholangiocarcinoma (cHCC-CC)
gross classification, 438
hepatobiliary carcinoma, 437
histological classification, 439–440
nomenclature and diagnostic criteria, 438
Curcumin, 338
Cyclic-adenosine monophosphate (cAMP), 311
Cyclin-dependent kinase inhibitor (CDKI), 69
Cyclin dependent kinases (CDKs), 79
Cyclooxygenase (COX), 7–8
CYP2E1, 9
alcoholic liver disease, 131
hepatitis C virus, 253
Cysteine oxidation, 17
Cytochrome P450, 8–9
xenobiotic-induced liver diseases, 102
Cytochrome P4502E1 (CYP2E1), 479
alkylation by HER, 227
animal experiments, 480
ethanol metabolism, 216–217
HepG2 cells, 228
with NADPH-oxidase, 219
NAFLD, 487–488

Cytochrome P4502E1 (CYP2E1) (*cont.*)
  and retinoids, 488–489
  ROS generation and carcinogenic DNA adducts, 486–487
Cytoglobin (CYGB), 392
Cytokine-induced neutrophil chemoattractant (CINC), 128

D
Damage-associated molecular patterns (DAMPs), 124–125, 315
De novo lipogenesis (DNL) process, 284
Diethylnitrosamine, 465
Direct acting antiviral (DAA) therapies, 244
DNA damage
  DDR, 73, 74
  foci, 69
  by genotoxic agents, 70
  by oncogenic/hyper-proliferative signals, 71
  by oxidative stress, 70–71
DNA polymerase, 69
Docosahexaenoic acid (DHA), 336
Drug-induced liver injury (DILI)
  acetaminophen hepatotoxicity, 200–203
  non-acetaminophen, 203–204
Ductular reaction, 310, 312
DUOX1 and DUOX2, 5
Dysplastic nodules (DNs), 431

E
EGF-R, 84
Eicosanoids, 108
Encephalomyocarditis virus (EMCV), 248
Encephalopathy, hepatic. *See* Hepatic encephalopathy
Endoplasmic reticulum (ER). *See also* Unfolded protein response (UPR)
  alcoholic steatosis, 225
  ER-mitochondrial interactions in UPR, 118–119
  in eukaryotic cells, 113–114
  oxidative damage on, 104
  smooth ER, 113–114
Endothelial cells, 107–108
Endothelial nitric oxide synthase (eNOS), 108
Endothelin-1 (ET-1), 108
Endotoxin-associated liver inflammation, 133–134
Enzymatic antioxidant, 12–14
  defenses
    CAT, 48
    GSTs, 51
    GTPx, 48
    PRXs, 49
    SOD, 47–48
    thiol modifications, 50
    TRXs, 49
Epidermal growth factor receptor (EGFR), 82, 83, 318
ER oxidase 1 (Ero1), 117
Ethanol, 118
  metabolism, 216–217

F
Farnesoid X receptor (FXR), 333, 335
Fatty acid uptake, 282–283
Fibric acid derivates, 339
Fibrogenesis
  aetiology, 173
  angiogenesis and, 187–188
  causes, 171
  epidemiological data, 172
  hepatic cell populations, 174
  hepatocytes, 173
  hypoxia, 187–188
  Kupffer cells and macrophages, 173
  myofibroblasts
    activated HSCs and, 188–189
    activation, 175–176
    origin, 175
    phenotypic responses, 175, 176
    pro-fibrogenic functional target cells, 181–182
  NADPH-oxidase isoforms
    intracellular ROS generation, 184–187
    and pro-fibrogenic responses, 182–183
  oxidative stress
    experimental and clinical studies, 178
    genesis, 179
    intracellular ROS levels, 178, 179
    ROS elevation, 179–180
  oxidative stress and ROS, 171–172
  redox mechanisms, 171–172
  therapeutic strategies, 189–191
Fibrosis, 134–136
  aetiology and pro-fibrogenic mechanisms, 176–178
  alcoholic, 230–231
Forkhead box O (FOXO), 26–27, 80–81
Free cholesterol (FC) deposition, mitochondria, 289
Free fatty acids (FFA), 118
  oxidation, 283
Free radicals, 43
FVIII (coagulation factor VIII), 117

G
GADD34 expression, 114, 115
Gene expression profiling (GEP), 422
Genetically modified models, susceptibility, 467
Glutaredoxin (GRX), 13–14
Glutathione (GSH), 11
  chemical structure, 52
Glutathione peroxidases (GPx), 12, 100
Glutathionylation, 16
Glyceraldehyde-3-phosphate dehydrogenase (GAPDH), 23–24
Glypican-3 (GPC-3), 422
GSSG, 202
Gut-derived endotoxin, 285–286

H
Hayflick's limit, 67
HCV core protein transgenic mice, 462–463
Heme oxygenase-1 (Ho-1), 56
Hepatic encephalopathy
  antioxidant strategies, 364–365
  oxidative stress
    ammonia-induced, 360–363
    manganese-induced, 363–364
    neuropathology, 359–360
  pathogenesis, 359
Hepatic fibrosis, HCV, 254–256
Hepatic iron
  ALD, iron overload, 490–491
    hepatic causes, 495–497
    therapy, 497–498
  cellular iron regulation, 494–495
  iron toxicity and carcinogenesis, 489–490
  overload, oxidative stress and
    biological functions, 347
    chemical properties, 347–349
    hepatocellular carcinoma and, 353
    iron metabolism, 345–347
    iron toxicity, 347–349
    primary iron overload, 349–350
    secondary iron overload, 350–352
  systemic iron homeostasis, 491–494
Hepatic progenitor cells (HPCs), 229, 310
Hepatic steatosis
  HCV, 254–256
  metabolic pathways
    fatty acid uptake, 282–283
    FFA oxidation, 283
    very low-density lipoprotein-triglyceride secretion, 285
Hepatitis A virus (HAV), 204–205
Hepatitis B virus (HBV), 204–205
  genome, 261
  oxidative stress, 262
  pathogenesis, 260–261
  rcplication, 261
Hepatitis C virus (HCV), 134, 205
  acute and chronic infection, 243–244
  direct acting antiviral (DAA) therapies, 244
  genome, 244, 245
  hepatic fibrosis and cirrhosis, 255–256
  hepatic steatosis, 254–255
  hepatocellular carcinoma, 256–259
  life cycle
    attachment, 244
    viral assembly and release, 244–246
    viral RNA translation, 246
  metabolic syndrome, 259
  mixed cryoglobulinemia, 260
  oxidative stress
    CYP2E1, 253
    decreased antioxidant defense, 253–254
    effects, 259–260
    iNOS, 253
    iron overload, 250–251
    mitochondrial ROS, 251
    Nox enzymes, 251–252
    signaling pathways of, 250
  prevalence, 242
  structural and nonstructural proteins, 246
  transmission, 242–243
  in vitro and in vivo models, 247–249
Hepatocellular carcinoma (HCC), 136
  animal experiments, 480
  antioxidants, chemoprevention, 467–469
  cellular signalling pathways and
    NF-κB and HCC development, 459–460
    p38α MAPK pathway and HCC, 460
  clinical-pathological classification, 419–427
    early HCC (eHCC), 421–422
    progressed HCC (pHCC), 423–427
  early stage
    definition, 421
    gene expression profiling (GEP), 422
    glypican-3 (GPC-3), 422
    high-grade dysplastic nodule, 422
    macroscopic appearance, 421
    stromal invasion, 422
  epidemiology, 481
  experimental hepatocarcinogenesis
    acyl-coenzyme A oxidase (AOX), 464
    carbon tetrachloride, 466
    chemically induced oxidative stress, 465–467
    choline-and methionine-deficient diets (CMD), 466–467

Hepatocellular carcinoma (HCC) (*cont.*)
- diethylnitrosamine, 465
- genetically modified models, susceptibility, 467
- HCV core protein transgenic mice, 462–463
- *Mdr2* null mice, 464
- methionine adenosyltransferase 1a (Mat1a) null mice, 463–464
- thioacetamide, 466
- transgenic mice expressing HBV proteins, 461–462

HCV, 256–259
hepatic iron overload, oxidative stress and, 353
incidence, 419, 479
molecular classification, 427–428
mortality, 419
oxidative stress, HCV-HVB-and NASH-HCC Patients
- cellular antioxidant systems, 455–458
- 8-hydroxy-2-deoxyguanosine, 451–452
- iron overload, 453–454
- LPO markers and etheno DNA adducts, 452–453
- nitrosative stress markers, 454–455

pathogenesis
- aspects of, 482–483
- role oxidative stress, 483–498

premalignant hepatocellular lesions, 428–431
- dysplastic foci, 430
- dysplastic nodules (DNs), 431
- large cell change, 429
- small cell change, 429–430

progressed stage
- chromophobe HCC with abrupt anaplasia (CHAA), 425
- classical forms, 425
- differential diagnosis, 426
- grade 1 (G1), 426
- grade 2 (G2), 426
- grade 3 (G3), 426
- grade 4 (G4), 426
- gross classification, 423
- histologic classifications, 423–426

Hepatocellular injury, 228–229
- endothelial cells, 107–108
- hepatic stellate cells, 108–109
- hepatocytes, 105–107
- Kupffer cells, 107
- oxidative stress and, 99
- signaling pathways
  - in endoplasmic reticulum, 104
  - in lysosomes, 104–105
  - in mitochondria, 103
  - in plasma membrane, 103

Hepatocyte
- fibrogenesis, 173
- injury, 105–107
- transplantation, 75

Hepatocyte growth factor (HGF), 82
Hepatotoxicity, acetaminophen, 200–203
HepG2 cells, 228
HIFs, 27–28
High mobility group box 1 (HMGB1), 124–125
HMG-CoA reductase (HMGCR), 161
HSF-1, alcoholic liver disease, 132
Hydrogen peroxide (H_2O_2), 4, 154
8-Hydroxy-2-deoxyguanosine, 451–452
Hydroxyl radical (•OH), 4
Hypercholesterolemia, 160
Hypoxia, fibrogenesis, 187–188
Hypoxia-inducible factors (HIFs), 187

I

Idiosyncratic DILI (iDILI), 203
IgG4-related sclerosing cholangitis (IgG4-SC), 335
IL-6 and IL-8, 71–72
Immunopathogenesis
- autoimmune hepatitis, 328–329
- overlap syndromes, 327–328
- primary biliary cirrhosis, 329
- primary sclerosing cholangitis, 330

Immunosuppression and ROS, 379–380
Inducible nitric oxide synthase (iNOS), 294
Inducible NO synthetase (iNOS), 221
Inflammation, liver
- alcoholic liver disease
  - CYP2E1, 131
  - HSF-1, 132
  - NF-κB activation, 131
  - Nox and XO, 132
  - SOD, 132
- cholestatic liver disease, 137
- endotoxin-associated, 133–134
- fibrosis, 134–136
- hepatocellular carcinoma, 136
- immune system, 137–138
- ischemia-reperfusion injury
  - cytokine-induced neutrophil chemoattractant (CINC), 128
  - KC inactivation, 126–127
  - macrophage function inhibition, 126
  - neutrophil recruitment, 127
  - xanthine oxidase, 127–128

leukocytes, ROS production in, 124–126
non-alcoholic fatty liver disease, 129–131
reactive oxygen species and, 123–124
toxin injury, 136–137
viral hepatitis, 134
Inflammatory cytokine/adipokines, 286–287
Inositol-requiring 1α (IRE1α), 114–116
Insulin receptor kinase (IRK), 21
Interleukin-6 (IL-6), 286
Internal ribosomal entry site (IRES), 246
International Agency for Research on Cancer (IARC), 479
Intracellular antioxidants, 390–391
Iron
metabolism, oxidative stress and
distribution and absorption, 345–346
hepcidin and systemic iron homeostasis, 346–347
storage and recycling, 346
overload, 453–454
primary, 349–350
secondary, 350–352
toxicity, 347–349
Ischemia-reperfusion (IR) injury
autophagy, 164
cholesterol and
CD diet, 160, 161
ER, 161
hypercholesterolemia, 160
cytokine-induced neutrophil chemoattractant (CINC), 128
fatty liver sensitizes to, 159–160
inflammatory cytokines generation, 158–159
KC inactivation, 126–127
liver transplantation
antioxidant treatment, 375–376
gene therapy, 378
ischemic preconditioning, 376–378
pathogenesis, 375
macrophage function inhibition, 126
neutrophil recruitment, 127
NF-κB activation, 158–159
ROS generation and defense
ATP depletion, 151
hydrogen peroxide, 154
MnSOD, 163–164
MPT, 151
peroxynitrite, 154–155
superoxide formation, 152–154
ROS sources
extramitochondrial sources, 157–158
mitochondrial metabolism, 155–157
xanthine oxidase, 127–128
ISHEN classification, 358

J
JNK1/2, 185
JNK (c-Jun N-terminal kinase), 103, 202

K
Keap1-Nrf2 signalling, 15
Kruppel-like factor 2 (KLF2), 161
Kupffer cells
damages, 107
inactivation, 126–127
and macrophages, 173
ROS production, 125

L
Leukocytes, ROS production in, 124–126
Lipid peroxidation, 46, 47, 103
Lipogenesis process, de novo, 284
Lipotoxicity, 287–290
Lipoxygenase (LOX), 7–8
Liver cancer classification
cholangiocarcinoma
gross classification, 433
hilar CC, 431
intrahepatic, 431–435
molecular classification, 436
mucinous cystic neoplasms (MCNs), 435–436
premalignant cholangiocellular lesions, 436–437
combined hepatocellular-cholangiocarcinoma (cHCC-CC)
gross classification, 438
hepatobiliary carcinoma, 437
histological classification, 439–440
nomenclature and diagnostic criteria, 438
hepatocellular carcinoma
clinical-pathological classification, 419–427
incidence and mortality, 419
molecular classification, 427–428
premalignant hepatocellular lesions, 428–431
histogenesis, liver cancers, 440–441
historical notes, 418–419
Liver damages, by ROS
base degradation, 44
carbonylation, 45
hydroxylation, 45
lipid peroxidation, 46, 47
mutations, 44
nitration, 45
protein modification, 46–47
sulfoxidation, 45

Liver diseases
  cholestasis, 102
  non-alcoholic fatty liver disease, 101–102
  xenobiotic-induced, 102–103
Liver failure. *See* Acute liver failure (ALF)
Liver fibrosis, 406–407
Liver regeneration, 207
  cell cycle, 79–80
  epidermal growth factor receptor (EGF-R), 82, 83
  hepatocyte growth factor (HGF), 82
  partial hepatectomy (PH), 78
  redox balance and sensors, 80–81
  ROS
    and cell cycle, 82
    EGF-R, 84
    Nrf2, 84–85
    p38-MAPK, 85
Liver repair, 316
Liver transplantation, oxidative stress and ischemia-reperfusion injury
    antioxidant interventions, 375–378
    pathogenesis, 375
  marginal donors, 374–375
  posttransplant conditions
    acute rejection, 379
    immunosuppression and ROS, 379–380
  pretransplant conditions
    marginal donors, 374–375
    transplant candidates, 372–374
Lobular inflammation, 225–228
LPO markers and etheno DNA adducts, 452–453
Lyn, 20
Lysosomes, 104–105, 298

M
Macrophage, function inhibition, 126
Malondialdehyde (MDA), 130–131
Manganese-induced oxidative stress, hepatic encephalopathy, 363–364
Manganese superoxide dismutase (MnSOD), 163, 204. *See also* SOD2
MAPK pathway, 25
MATI and MATIII, 222–223
Mdr2 null mice, 464
Melatonin, 336
Membrane and extracellular antioxidants, 391–392
Metabolic syndrome, 259
Metalloproteinase-2 (MMP2), 318
Methionine adenosyltransferase 1a (Mat1a) null mice, 463–464
Methionine oxidation, 16
Mevalonate pathway, 162
Microsomes, 295–296
Mitochondria, 292–295
  oxidative damage on, 103
  ROS generation in, 6
Mitochondrial permeability transition (MPT), 151
Mitofusin-1 and-2, 119
Mitogen-activated protein kinase (MAPK), 202–203
Mitoquinone (Mite), 337
Mixed cryoglobulinemia, 259
Mixed lineage kinase 3 (MLK3), 203
MnSOD, 163–164
Modification of UDCA (norUDCA), 333–334
Monocyte chemotactic protein (MCP)-1, 130
MPT, 151
Mucinous cystic neoplasms (MCNs), 435–436
Multiple parallel hit theory, NAFLD, 281
Myeloperoxidase (MPO), 125
Myofibroblasts, fibrogenesis
  activated HSCs and, 188–189
  activation, 175–176
  origin, 175
  phenotypic responses, 175, 176
  pro-fibrogenic functional target cells, 181–182

N
N-acetylcysteine (NAC), 164, 337
  antioxidant therapies, 397
NAD(P)H:quinone oxidoreductase 1 (Nqo1), 56
NADPH oxidase (NOX), 6–7, 107, 109
  alcoholic liver disease, 132
  enzymes, 6–7
  fibrogenesis
    intracellular ROS generation, 184–187
    and pro-fibrogenic responses, 182–183
Necrosis, 105–106
Neutrophils, 130
Nitric oxide (NO), 5, 108
  cholangiopathies
    and bile secretion, 315–316
    and cholangiocarcinoma, 317–320
    and liver repair, 316
Nitric oxide synthase (NOS), 44
Nitrosative stress markers, 454–455
N-nitrosodiethylamine (DEN), 136
Non-acetaminophen drug-induced liver injury, 203–204
Non-alcoholic fatty liver disease (NAFLD), 101–102
  antioxidant therapy
    betaine, 404
    metformin, 405
    N-acetylcysteine, 405

pentoxifylline, 404
ursodeoxycholic acid, 405
vitamins E and C, 400–404
clinical importance, 280
epidemiology, 280
hepatic steatosis, metabolic pathways
fatty acid uptake, 282–283
FFA oxidation, 283
very low-density lipoprotein-triglyceride secretion, 285
hepatocellular cancer (HCC), 479
multiple parallel hit theory, 281
NASH, 129–131
oxidative stress and mitochondrial dysfunction
lysosomes, 298
microsomes, 295–296
mitochondria, 292–295
peroxisomes, 296–297
prevalence, 280
secondary iron overload, 351–352
steatosis to steatohepatitis
gut-derived endotoxin and toll-like receptors, 285–286
inflammatory cytokine/adipokines, 286–287
lipotoxicity, 287–290
symptoms, 280
Non-alcoholic steatohepatitis (NASH), 101
adaptive immune system, 130–131
adiponectin, 129
antioxidant therapy
leptin, 130
neutrophils, 130
prevalence, 280
ROS generation by KCs, 130
Non-autoimmune biliary diseases, oxidative stress. *See* Cholangiopathies
Non-enzymatic antioxidant, 10–11
defenses
β-carotene, 53, 54
glutathione (GSH), 53
vitamin C, 53
vitamin E/α-tocopherol, 52, 53
Non-proteins substrates, 29
NOX. *See* NADPH oxidase (NOX)
NOX1, 109
NOX2, 109
Nrf2, 15, 28–29
antioxidant defense of, 55–57
liver regeneration, 84–85
Nuclear erythroid-related factor 2 (Nrf-2), 100
Nuclear factor-kB (NF-kB), 26, 100–101
activation
alcoholic liver disease, 131
ischemia-reperfusion injury, 158–159

O

Overlap syndromes, 327–328
AIH-PBC, 335
AIH-PSC, 335
in autoimmune liver disease, 335
Oxidative stress. *See also* Reactive oxygen species (ROS)
antioxidant defenses, 99
autoimmune liver disease
antioxidant therapy, 331–335
betaine, 337
calcineurin inhibitors, 338
curcumin, 338
docosahexaenoic acid (DHA), 336
D-Penicillamine, 337
fibric acid derivates, 339
immunopathogenesis, 327–330
melatonin, 336
mitoquinone (Mite), 337
N-acetylcysteine (NAC), 337
pentoxifylline, 337
PPC, 337
selenium (selenocysteine), 338
silimaryn, 336
sirolimus, 339
vitamin C/ascorbic acid, 339
vitamin D, 339
vitamin E (alpha-tocopherol), 338
zinc/polaprezinc (combination of zinc and l-carnosine), 338
chemically induced, 465–467
chronic liver failure, central nervous system complications (*see* Hepatic encephalopathy)
fibrogenesis
experimental and clinical studies, 178
genesis, 179
intracellular ROS levels, 178, 179
and ROS, 171–172
ROS elevation, 179–180
HCV
CYP2E1, 253
decreased antioxidant defense, 253–254
effects, 259–260
iNOS, 253
iron overload, 250–251
mitochondrial ROS, 251
Nox enzymes, 251–252
signaling pathways of, 250
HCV-HVB-and NASH-HCC Patients
cellular antioxidant systems, 455–458
8-hydroxy-2-deoxyguanosine, 451–452
iron overload, 453–454

Oxidative stress. *See also* Reactive oxygen species (ROS) (*cont.*)
LPO markers and etheno DNA adducts, 452–453
nitrosative stress markers, 454–455
and hepatic iron overload
biological functions, 347
chemical properties, 347–349
hepatocellular carcinoma and, 353
iron metabolism, 345–347
iron toxicity, 347–349
primary iron overload, 349–350
secondary iron overload, 350–352
hepatocarcinogenesis
antioxidants, chemoprevention, 467–469
cellular signalling pathways and, 458–460
experimental hepatocarcinogenesis, 460–467
role in HCV-HVB-and NASH-HCC Patients, 451–458
hepatocellular injury (*See* Hepatocellular injury)
in liver diseases (*See* Liver diseases)
liver inflammation (*See* Inflammation, liver)
and liver transplantation
ischemia-reperfusion injury, 375–378
marginal donors, 374–375
posttransplant conditions, 379–380
pretransplant conditions, 372–374
and mitochondrial dysfunction, NAFLD
lysosomes, 298
microsomes, 295–296
mitochondria, 292–295
peroxisomes, 296–297
nonautoimmune biliary diseases (*see* Cholangiopathies)
pathogenesis role, HCC
acetaldehyde, 483–485
cytochrome P4502E1, 486–489
hepatic iron, 489–498
unfolded protein response, 118
2-Oxoglutarate (2OG) dioxygenase, 17
Oxygen-centered non-radicals, 4, 5
Oxygen-centered radicals, 3–5

P

Paracetamol, liver diseases induction, 102
Partial hepatectomy (PH), 78
D-Penicillamine, 337
Pentoxifylline, 337
Pericellular/perisinusoidal fibrosis, 177
Peroxiredoxins (PRX), 12–13
Peroxisomal enzymes, 9
Peroxisome, 296–297
Peroxisome proliferator-activated receptor family (PPAR), 319
Peroxyl and alkoxyl radicals, 4–5
Peroxynitrite ($OONO^-$), 5, 154–155
3-Phosphatase and tensin homolog (PTEN), 19
Phospholipases, 313
PKM2, 23
Plasma membrane, oxidative damage on, 103
Platelet-derived growth factor (PDGF), 109
p38-MAPK
liver regeneration, 85
Post-necrotic/bridging fibrosis, 177
PPC, 337
Primary biliary cirrhosis (PBS), 329, 332–334
Primary sclerosing cholangitis (PSC), 330, 334–335
Prostaglandins, 7, 174, 313
Protein folding, 114
Protein kinase-like ER kinase (PERK), 114–116
Protein tyrosine phosphatases, 18–20
Pyruvate kinase (PK), 23

R

Reactive ductular cells (RDC), 312
Reactive nitrogen species (RNS), 43–44
Reactive oxidative species, 479
cholangiopathies, 313–314
inflammation effect, 314–315
nitric oxide (NO) effects, 315–320
generation and carcinogenic DNA adducts, CYP2E1, 486–487
Reactive oxygen species (ROS), 123–124, 371
and cell senescence (*see* Cell senescence)
cellular sources, 43–44
damages generated by
base degradation, 44
carbonylation, 45
hydroxylation, 45
lipid peroxidation, 46, 47
mutations, 44
nitration, 45
protein modification, 46–47
sulfoxidation, 45
generation, mechanism
cytochrome P450, 8–9
lipoxygenase and cyclooxygenase, 7–8
in mitochondria, 6
NOX enzymes, 6–7
peroxisomal enzymes, 9
xanthine oxidase, 8
hydrogen peroxide (H_2O_2), 4
hydroxyl radical ($^{\bullet}OH$), 4

ischemia-reperfusion injury
ATP depletion, 151
extramitochondrial sources, 157–158
hydrogen peroxide, 154
mitochondrial metabolism, 155–157
MnSOD, 163–164
MPT, 151
peroxynitrite, 154–155
superoxide formation, 152–154
nitric oxide (NO•), 5
peroxyl and alkoxyl radicals, 4–5
peroxynitrite ($OONO^-$), 5
scavenging
enzymatic antioxidants, 12–14
keap1-Nrf2 signalling, 15
non-enzymatic antioxidants, 10–11
singlet oxygen ($^1O_2$), 4
superoxide anion (•O2-), 3–4
Receptor-interacting protein kinase 3 (RIP3), 203
Receptor tyrosine kinases (RTKs), 19, 21–22
Redox mechanisms, fibrogenesis, 171–172
Redox signaling. *See also* Reactive oxygen species (ROS)
cysteine oxidation, 17
glutathionylation, 16
methionine oxidation, 16
non-proteins substrates, 29
proteins, redox-sensitive targets in
AMP-kinase, 24
ASK1, 25
ATM, 24–25
β-actin, 20
GAPDH, 23–24
MAPK pathway, 25
PKM2, 23
protein tyrosine phosphatases, 18–20
receptor tyrosine kinases, 21–22
Src family kinases, 20–21
spatial regulation, 17–18
transcription factors
chromatin remodelling, 29
FoxOs, 26–27
HIFs, 27–28
NF-kB, 26
Nrf2, 28–29
Renin-angiotensin system (RAS), 186

S
SA-β-gal activity, 68
S-adenosylmethionine (SAMe), 222–223
Selenium (selenocysteine), 338
Signal transduction pathways
in endoplasmic reticulum, 104
in lysosomes, 104–105
in mitochondria, 103
in plasma membrane, 103
Silimaryn, 336
Singlet oxygen ($^1O_2$), 4
Sirolimus, 339
Sod2, 204
Src family kinases, 20–21
SREBP-1, 225
Steatosis
hepatic, metabolic pathways
fatty acid uptake, 282–283
FFA oxidation, 283
very low-density lipoprotein-triglyceride secretion, 285
to steatohepatitis, NAFLD
gut-derived endotoxin and toll-like receptors, 285–286
inflammatory cytokine/adipokines, 286–287
lipotoxicity, 287–290
Stellate cells injury
autocrine effects, 109
paracrine effects, 108
Sulfiredoxin 1 (Srxn1), 56
Sulfiredoxin (SRX), 13
Superoxide anion (•O2-), 3–4
hepatocyte injury, 106
Superoxide dismutase (SOD), 12, 100
alcoholic liver disease, 132
Superoxide formation, 152–154

T
Telomere uncapping, 69
Thioacetamide, 466
Thioredoxin (TRX), 11, 13–14
Toll-like receptor 4 (TLR4), 258
Toll-like receptor (TLR), 186, 285–286
Toxin injury, 136–137
Transgenic mice expressing HBV proteins, 461–462
TRX reductase (TRXR), 11
Tumor necrosis factor (TNF), 151
Tumor necrosis factor-alpha (TNF-α), 286

U
UDCA, 332
Uncoupling protein-2 (UCP2), 294

Unfolded protein response (UPR), 104
  alcoholic steatosis, 225
  antioxidant defense, 116–117
  eIF2α dephosphorylation, 115
  ER lumen, 117–118
  ER-mitochondrial interactions, 118–119
  GADD34 expression, 114, 115
  IRE1α, 114–116
  oxidative stress as activator, 118
  PERK-mediated phosphorylation, 114, 115

V

Vasoconstriction, 108
Very low-density lipoprotein-triglyceride (VLDL) secretion, 285
Viral hepatitis, 134
  acute liver failure, 204–205
Vitamin C (Ascorbic acid), 10, 339
  antioxidant defenses, 53
  chemical structure, 52
Vitamin D, 339
Vitamin E, 52
  alpha-tocopherol), 338
  beta-tocopherol), 10

W

Wilson disease, 206

X

Xanthine dehydrogenase, 202
Xanthine oxidase (XO), 8, 202
  alcoholic liver disease, 127–128, 132
Xanthine oxidoreductase (XOR), 202
X-box-binding protein-1 (XBP1s), 115, 117
Xenobiotic-induced, 102–103

Z

Zinc/polaprezinc (combination of zinc and l-carnosine), 338

CPSIA information can be obtained at www.ICGtesting.com
Printed in the USA
LVOW02*1144300615

444407LV00002B/78/P

9 783319 155388